Springer Series on

Wave Phenomena 16

Guest Editor: T. Tamir

Springer
Berlin
Heidelberg
New York
Barcelona
Budapest
Hong Kong
London
Milan
Paris
Santa Clara
Singapore
Tokyo

Springer Series on

Wave Phenomena

Editors: L. M. Brekhovskikh L. B. Felsen H. A. Haus
Managing Editor: H. K. V. Lotsch

Volume 19
Laser Ultrasonics
By L. M. Lyamshev

Volume 20
Surface Acoustic Waves in Inhomogeneous Media
By S. V. Biryukov, Yu. V. Gulyaev, V. V. Krylov and V. P. Plessky

Volume 1–18 are listed at the end of the book

K.E. Oughstun G.C. Sherman

Electromagnetic Pulse Propagation in Causal Dielectrics

With 121 Figures

Dr. K.E. Oughstun
Professor of Electrical Engineering
and Mathematics
College of Engineering and Mathematics
University of Vermont
Burlington, VT 05405–0156, USA

Dr. G.C. Sherman
Department of Physics
Fullerton College
321 East Chapman Avenue
Fullerton, CA 92832–1351, USA

Guest Editor:
Professor T. Tamir, Ph.D.
Polytechnic University
Department of Electrical Engineering
333 Jay Street
Brooklyn, NY 11202, USA

First Edition 1994 - Corrected Printing 1997

ISSN 0931-7252

ISBN-13: 978-3-642-64753-6

Library of Congress Cataloging-in-Publication Data applied for.

Die Deutsche Bibliothek - CIP-Einheitsaufnahme
Oughstun, Kurt E.: Electromagnetic pulse propagation in causal dielectrics/ K.E. Oughstun ; G. C. Sherman. - Corr. printing. - Berlin ; Heidelberg ; New York ; Barcelona ; Budapest ; Hong Kong ; London ; Milan ; Paris ; Santa Clara ; Singapore ; Tokyo : Springer, 1997
(Springer series on wave phenomena ; 16)
ISBN-13: 978-3-642-64753-6

ISBN-13: 978-3-642-64753-6 e-ISBN-13: 978-3-642-61227-5
DOI: 10.1007/978-3-642-61227-5

Typesetting: Asco Trade Typesetting Ltd., Hong Kong
SPIN. 10575586 54/3144 - 5 4 3 2 1 0 - Printed on acid-free paper

Dedicated to
Professor Emil Wolf,
our teacher and mentor.

In addition, Kurt Oughstun dedicates this book to his late parents Edmund and Ruth (Kinat) Oughstun, his closest friend and wife Joyce (Maguder) Oughstun, his children Marcianna Oughstun and Kristen Oughstun, extended family members Arienne Marie-Therese Smith and Darwin Sagastume Guzman, and his close friend Richard Albanese. George Sherman dedicates this book to his parents Robert and Augusta Sherman, his children Rebecca Harman, Laura Sherman, Wayne Sherman, Wendy Middleton and Alison Sherman, and their mother Barbara Sherman, and his closest friends Dee Johnson, Lauretta Kelty, and Anthony Devaney.

Preface

This research monograph presents a systematic treatment of the theory of the propagation of transient electromagnetic fields (such as optical pulses) through dielectric media which exhibit both dispersion and absorption. The work divides naturally into two parts. Part I presents a summary of the fundamental theory of the radiation and propagation of rather general electromagnetic waves in causal, linear media which are homogeneous and isotropic but which otherwise have rather general dispersive and absorbing properties. In Part II, we specialize on the propagation of a plane, transient electromagnetic field in a homogeneous dielectric. Although we have made some contributions to the fundamental theory given in Part I, most of the results of our own research appear in Part II.

The purpose of the theory presented in Part II is to predict and to explain in explicit detail the dynamics of the field after it has propagated far enough through the medium to be in the mature-dispersion regime. It is the subject of a classic theory, based on the research conducted by A. Sommerfeld and L. Brillouin in the early 1900s and described in modern textbooks on advanced electrodynamics. Over the past seventeen years, we have improved the classic theory in several ways. In particular, we have increased the accuracy of many of the approximations, we have applied modern mathematical methods to clarify some of the confusion present in the older theory, we have extended the theory to include the field dynamics in space-time regions that were excluded by the classic theory, we have applied numerical techniques with computers to guide our analysis and to check the accuracy of our analytic results, we have studied the dynamics of a number of special pulses, and we have developed a physical model which provides a simplified quantitative algorithm for obtaining all of the dynamics of the field and explains all of those dynamics in physical terms. All of these advances are included in this book.

Since our primary interest is in the effects of dispersion and absorption on the propagation of the field, we exclude other effects by studying the propagation of a plane wave in a linear, homogeneous, isotropic dielectric. We assume that the plane wave is incident normally on the dielectric which fills a half-space. The electromagnetic field is known for all time on the plane boundary. As in the classic theory, we employ the Lorentz model to describe the interaction between the field and material medium. Our goal is to determine the details of the field behavior within the medium.

It is straightforward to obtain an exact integral expression for the field within the dielectric in terms of the known field on the plane boundary. The integral cannot be evaluated analytically, however, so it is necessary to either evaluate the integral numerically for specific cases or to develop analytic approximations of the integral in regimes of interest. The bulk of the theory presented herein is concerned with obtaining asymptotic approximations of the integral valid after the field has propagated sufficiently far into the material, a space-time region known as the mature-dispersion regime. To do so, we apply a version of the saddle-point method, due to F.W.J. Olver, which provides improved insight into the meaning of the approximations. Since the results break down in certain space-time regions, we apply uniform asymptotic methods to describe the pulse dynamics in those regions. A summary of these advanced asymptotic techniques is included in Part I.

In the final chapter, we apply our mathematical results to present a physical model of pulse dynamics in the mature-dispersion regime. The model provides a simple mathematical algorithm for determining field dynamics and a simple physical explanation for all of its features. Such a model has not been available in the classic theory. Indeed, to our knowledge, the only dispersive systems which have been described by such a physical model are those which do not include any absorption or gain at any frequency. Such systems are necessarily noncausal. Thus, we believe that the physical model presented herein is the first such model available for a causal dispersive system.

Our research in this area began in the early 1970s while Oughstun was a graduate student and Sherman was a faculty member at the Institute of Optics of the University of Rochester in Rochester, New York. We would like to acknowledge financial support during that period by The Institute of Optics, the National Science Foundation, the Center for Naval Research and the Corning Glass Works Foundation. This phase of our work culminated with the completion of Oughstun's Ph.D. thesis in 1978. Since that time, we have continued to work on the subject part time while Oughstun was working with United Technologies Corporation, the University of Wisconsin-Madison and the University of Vermont and Sherman was working with Itek Corporation, Schlumberger Doll Research, Mission Research Corporation, and the Rocketdyne Division of Rockwell International Corporation. Oughstun acknowledges the support of his recent research on dispersive pulse propagation by the Applied Mathematics Group of the U.S. Air Force Office of Scientific Research. We have published the results of our research in several papers that are referred to in the list of references. Some of the material presented herein is excerpted directly from those papers. Assistance with updated calculations using IMSL software of the uniform asymptotic description of the pole contribution and the rectangular pulse modulated signal by Paul D. Smith and Judith Elizabeth Keil Laurens at the University of Vermont are gratefully acknowledged.

Burlington — Kurt E. Oughstun
Santa Barbara — George C. Sherman
June 1994

Contents

1 Introduction

1.1 Motivation

The dynamic evolution of an electromagnetic pulse as it propagates through a linear, temporally dispersive medium is a classical problem of electromagnetism. If the medium was nondispersive, an arbitrary pulse would propagate unaltered at the phase velocity of the wave field in the medium. In a dispersive medium, however, the pulse is modified as it propagates due to two fundamentally interconnected effects. First of all, each monochromatic spectral component of the initial pulse propagates through the dispersive medium with its own phase velocity so that the phasal relationship between the various spectral components of the pulse changes with the propagation distance. Secondly, each monochromatic spectral component is absorbed at its own rate so that the relative amplitudes between the spectral components of the pulse change with the propagation distance. These two simple effects then result in a complicated change in the dynamical structure of the propagated field. The rigorous analysis of dispersive pulse propagation phenomena is complicated by the simple fact that the dispersive and absorptive parts of the medium response are connected through the physical requirement of causality [1.1]. For an initial pulse with a sufficiently rapid rise-time these effects manifest themselves through the formation of well-defined precursor fields [1.2–4] whose evolution has been shown [1.5–11] to be completely determined by the dispersive and absorptive properties of the medium. The precursor fields are readily distinguished in the dynamical evolution of the propagated field by the fact that the range of their frequency of oscillation is typically quite different from that of the input field and their attenuation is typically much less than that at the carrier frequency of the input field.

The precursor fields (or forerunners) were first described by *Sommerfeld* [1.2] and *Brillouin* [1.3, 4] in their seminal analysis of optical signal propagation in a locally linear, isotropic, causally dispersive medium that is described by a single resonance Lorentz model. Unfortunately, their analysis errantly concluded that the amplitudes of these precursor fields were, for the most part, negligible in comparison to the main signal evolution and that the main signal arrival occurred with a sudden rise in amplitude of the field. Because of this it was further concluded that the signal arrival could not be given an unambiguous physical definition. These misconceptions have, unfortunately, settled into the standard literature on electromagnetic wave theory [1.12, 13]. The recent analysis [1.5–11] of linear dispersive pulse propagation that is based upon modern asymptotic techniques [1.14–19] has provided a complete, rigorous description of the dynamical field evolution in both a single and double resonance Lorentz medium. In particular, this analysis has shown that the precursor fields that result from an input unit step-function modulated signal are a dominant feature of the field evolution in the mature dispersion regime. The mature dispersion regime has been found [1.20] to include all propagation distances that are greater than one absorption depth in the medium at the signal frequency of the input field. In addition, this modern asymptotic description [1.5–7] has also provided both a precise definition and physical interpretation of the signal velocity in the dispersive medium. This proper description of the signal velocity is critically dependent upon the correct description and interpretation of the precursor fields.

The central importance that the precursor fields hold in both the analysis and interpretation of linear dispersive pulse-propagation phenomena is also realized in the study of ultrashort pulse dynamics. The asymptotic theory clearly shows that the resultant pulse distortion due to an input rectangular modulated pulse is primarily due to the precursor fields that are associated with the leading and trailing edges of the input pulse envelope of arbitrary duration [1.8]. The interference between these two sets of precursor fields naturally leads to asymmetric pulse distortion. In addition, recent results for ultrashort Gaussian pulse propagation [1.21] reinforce the fundamental role that the precursor fields play in dispersive pulse dynamics.

The subject of electromagnetic pulse propagation in dispersive media has been and continues to be of considerable practical importance in several areas of contemporary optics and engineering electromagnetics. For example, the effects of dispersion are prevalent in all fiber-optic communication and integrated-optics systems. Furthermore, with the current experimental capabilities of producing ultrashort (near femtosecond) optical pulses [1.22] and ultrawideband microwave pulses, a new technology is rapidly developing in which the importance of the effects of dispersion is greatly magnified.

This research monograph develops a unified, detailed analysis of electromagnetic pulse propagation in linear, causally dispersive media. The fundamental theory of the electromagnetic field in a general temporally dispersive medium is presented in Part I of the monograph. The theoretical development is performed in the general situation in which both the dielectric permittivity and electric conductivity of a homogeneous, isotropic medium may be frequency dependent. The rigorous angular spectrum of plane waves representation of the pulsed radiation field (in which the spatio-temporal behavior of the current source is specified) as well as the pulsed electromagnetic beam field (in which the spatio-temporal behavior of either the electric or magnetic field vector is specified over a plane) are developed here without approximation. The first form of the angular spectrum representation is most appropriate for the analysis of antenna systems with a prescribed current distribution while the second form is most appropriate for optical field type problems in which the input field is known across some specified plane. With the problem of temporal pulse dispersion in mind, Part I concludes with an overview of the modern saddle point methods that may be used for the asymptotic evaluation of the temporal frequency contour integral that is contained in the angular spectrum representation.

The asymptotic theory is applied in Part II of the monograph which provides a detailed development of the asymptotic description of plane wave pulse propagation in a causally dispersive medium. A plane wave field is considered in order that attention may be focused on the temporal pulse evolution alone. Since the asymptotic theory depends upon specific saddle point dynamics in the complex frequency plane, and since the saddle point dynamics depend upon the specific form of the medium dispersion relation, a particular model of the material dispersion must be chosen. Because it is a causal model, and because of its applicability to the description of dielectric dispersion in either the microwave or optical frequency domains, as well as its historical connection to the classic research of *Sommerfeld* and *Brillouin*, a single resonance Lorentz model of a dielectric medium is considered here. Although the detailed results of the asymptotic description developed here are then specific to this form of the medium dispersion, the same methodology is applicable to other forms of the medium dispersion. The asymptotic description developed here is then used to develop a physical interpretation of dispersive pulse dynamics in a Lorentz model medium that may be extendable to other dispersive systems. This physical description of dispersive pulse propagation phenomena supplants the previous group velocity description and reduces to it in the limit as the medium absorption vanishes.

1.2 History of Previous Research

The historical record of research conducted in the area of dispersive wave propagation is extensive and quite varied in both content and depth of mathematical rigor. This section provides a brief survey of only the most important published literature on the subject. This includes papers that are concerned with such closely related topics as electromagnetic pulse propagation in waveguiding systems and plasma media, as well as research on nonelectromagnetic pulse propagation phenomena that may only be peripherally related to the central topic of this research monograph. A brief description of the shortcomings of much of this previous research is presented throughout the section.

The earliest introductory research effort into the problem of dispersive wave propagation was conducted by *Hamilton* [1.23] in 1839 when the concept of group velocity seems to have been first introduced. Subsequently in 1877 and 1881, *Rayleigh* [1.24, 25] distinguished between the phase and group velocities. However, these early consideration are primarily conceptual and oversimplified.

At the turn of the century, the development of the theory of the dispersive properties of dielectric media was begun in terms of a classical atomistic model that culminated in Lorentz's classical work [1.26–29]. *Drude* [1.30] indicates in a footnote that Maxwell (1869) was the first to base the theory of anomalous dispersion upon such an atomistic model. In research independent of Maxwell's, Sellmeier, v. Helmholtz, and Ketteler also used this model for the basis of a theory for material dispersion. A modern, rigorous theory of dielectric dispersion may be found in the research monograph of *Böttcher* and *Bordewijk* [1.31]. A brief review of the classical Lorentz model of dielectric dispersion is presented in Sect. 1.4.

With the Lorentz model in place as a causal model of the dielectric response, *Sommerfeld* [1.2] then considered the propagation of a unit-step-function modulated scalar-wave signal in a single resonance Lorentz model medium. His classic analysis was the first to demonstrate that the signal arrival does not always propagate with the group velocity in the dispersive medium and that even though the group velocity may exceed the vacuum speed of light c in the region of anomalous dispersion, the wave field arrives with a velocity that is always less than or equal to c. In an important extension of *Sommerfeld*'s results, *Brillouin* [1.3] also employed a Fourier-integral representation of a unit-step-function modulated plane-wave signal that is propagating through a semi-infinite, single resonance Lorentz medium. Due to the presence of a complex-valued phase function $\phi(\omega)$ appearing in the exponential propagation factor $\exp[\phi(\omega)z/c]$ in the integrand, *Brillouin* applied the then recently developed method of steepest descent [1.32] in order to obtain the asymptotic behavior of the propagated signal for large propagation distances z. In using this method, the original contour of integration, which extended from negative to positive infinity in the upper-half of the complex ω-plane, is deformed through the saddle points (or stationary points) of $\phi(\omega)$ along the path of steepest descent; the dominant contribution to the asymptotic approximation of the integral representation of the field then is due to the immediate neighborhood about the dominant saddle point as the point of observation moves off to infinity with the field. Under a first-order approximation of the behavior of the phase function in the complex ω-plane, the time behavior of the approximate locations of the saddle points along with their relative importance was deduced, giving rise to the following succession of events. First of all, in complete agreement with the principle of causality, for $\theta = ct/z < 1$, no field is present due to the behavior of $\phi(\omega)$ at $|\omega| = \infty$ and the fact that the integrand appearing in the integral representation of the propagated field is analytic in the upper-half of the complex ω-plane. Between $\theta = ct/z \geqslant 1$ and larger values of this parameter, two sets of precursors or forerunners are present for a single resonance Lorentz medium. From a physical point of view, these precursor fields arise from those Fourier components comprising the initial pulse shape whose velocities of propagation through the dispersive medium are greater than the velocity of propagation of the Fourier component at the applied signal frequency ω_c. The main signal at the applied signal frequency ω_c will then arrive sometime during the evolution of these precursor fields. From a purely mathematical point of view, this main signal arrival is dependent upon the crossing of the deformed contour of integration with the singularity at $\omega = \omega_c$ that appears in the integrand of the integral representation of the propagated field. At such a crossing, the integral may then be evaluated by use of the residue theorem. As a consequence, the main signal is found to arrive with a mathematically well-defined (but physically incorrect) signal velocity which may, in some special cases, coincide with the group velocity. Thus, in an involved analysis, *Brillouin* has presented a basic asymptotic description of pulse propagation in a classical dispersive medum that is now a classic of electromagnetic theory. The role this research played in *Brillouin*'s scientific career may be found in the biographical article by *Mosseri* [1.33].

Because of the unnecessary constraint imposed on the deformed contour of integration by the method of steepest descent in *Brillouin*'s analysis, the signal arrival was defined to occur when the path of steepest descent moved across the simple pole singularity at $\omega = \omega_c$. This misconception resulted in a frequency dependence of the signal velocity that erroneously peaks to the vacuum speed of light c near to the medium resonance frequency ω_0 and that is incomplete in its description when $\omega > \omega_0$. This was corrected in part by *Baerwald* [1.34] in 1930 who was the first to show that the signal velocity is at a minimum near the medium resonance frequency. In a more recent consideration of the signal velocity [1.35], it was concluded that the measurable signal velocity curve should lie somewhere between the curves predicted by *Brillouin* and *Baerwald*. However, no clear physical interpretation of the signal velocity was provided by any of these descriptions. In addition, the velocity of propagation of each spectral component in the dispersive medium was left undefined.

Subsequent to this foundational work by *Sommerfeld*, *Brillouin*, and *Baerwald*, several other methods were developed to describe dispersive wave and pulse propagation phenomena in a variety of media and waveguiding structures. An excellent review of the methods of approximate solution of one-dimensional wave equations may be found in the article by *Eckart* [1.36]. The principal subject of this paper is an exposition of the mathematical foundation for the Hamilton-Jacobi ray theory as based upon the method of stationary phase. Of equal importance are the papers by *Whitham* [1.37] and *Lighthill* [1.38]] on the general mathematical properties of three-dimensional wave propagation and the group velocity for ship-wave patterns and magnetohydrodynamic waves. Both papers solve their appropriate boundary value problem by utilizing a plane-wave expansion representation to which the method of stationary phase is applied. Their approach, however, is useful only for nonabsorbing media, thereby limiting the types of dispersion relations that may be considered.

This analysis naturally led to the description of propagation in a dispersive medium by means of ray techniques [1.39–43] which provides several alternate approaches to the asymptotic approximation of exact integral representations. The first is the direct-ray method, which is applicable to solving partial differential equations with the appropriately specified boundary or initial conditions. This is accomplished by assuming an asymptotic series for the solution which is then substituted into the partial differential equation. Families of rays are introduced along which the functional terms of the assumed series then satisfy ordinary differential equations, which can then be solved. Alternatively, there is the space-time ray theory which employs the plotting of rays and dispersion surfaces along with the initial or boundary values of the field to demonstrate the propagation phenomena and develop the asymptotic representation.

Both of these ray techniques are useful in certain applications; however, they are heuristic in origin since they require additional unnecessary hypotheses about the nature of the solution. Moreover, the applicability of the results is limited because the theory does not generate error terms for the resultant asymptotic representations. Further evidence of such limitations of validity is pointed out by *Felsen* and coworkers [1.44–46], who have applied space-time ray and dispersion surface techniques to the problems of propagation in dispersive media with applications to isotropic (cold) plasma media. *Felsen* noted the existence of certain transition regions wherein the ray-optic technique fails and one must resort to an exact integral representation and subsequent rigorous asymptotic analysis.

Consider next the integral-representation techniques [1.47–53] that have been employed in connection with the one-dimensional boundary-value problem for pulse propagation in hollow waveguiding structures. The general method utilized in this type of problem, along with its limitations, is presented in Sect. 7.9 of *Jackson* [1.13]. *Elliott* [1.49], *Forrer* [1.50], and *Wanselow* [1.51] considered the problem of one-dimensional pulse distortion in a waveguide. Applying Fourier-transform techniques, an integral representation for the field is obtained, which is then solved by expanding the propagation factor by Taylor series into a quadratic approximation. *Knop* and *Cohn* [1.52] pointed out that this quadratic approximation applies only in the quasimonochromatic case, and that if the initial pulse rise time is short, this quasimonochromatic assumption no longer applies and the resulting transfer function for the system yields an output before an input is applied, violating the principle of causality. However, approximate numerical solutions for this problem with the exact form of the phase function indicate that for times $t > \frac{L}{c}$ (L being the propagation length of the guide), the approximately degraded waveforms obtained by *Elliott* (and corrected by *Knop* and *Cohn*) are good approximations to the actual output pulse shapes of the waveguide. In order to overcome these difficulties, *Knop* [1.47] derived an exact solution to the problem of the propagation of a rectangular-pulse-modulated carrier input in a simple plasma medium. By employing transfer function techniques and the Laplace transform, a closed form analytic solution for the output was obtained in the form of a series summation over odd-order Bessel functions. *Case* and *Haskell* [1.48] simplified the result somewhat by indicating that this series summation could be written as the sum of two Lommel functions. Finally, the paper by *Vogler* [1.53] generalized these results to arbitrary waveforms in ideal waveguides. However, these results apply only to idealized situations.

A problem of related interest is the transient response as treated in the papers by *Wait* [1.54], and *Haskell* and *Case* [1.55]. *Wait* dealt with approximate methods for determining the distortion of a pulse propagating through a dispersive channel such as a hollow waveguide. Through use of the transfer function approach, the one-dimensional output pulse is obtained as a superposition over all frequencies of monochromatic plane waves. The stationary phase method is then applied to evaluate this integral representation for a general phase function $\phi(\omega)$, the propagation of a pulse in an ideal waveguide being given as an introductory example. The stationary phase method in mode theory is then developed in an application to a more realistic model wherein the group velocity has a maximum or minimum (the existence of a maximum corresponds to the buildup of the field just prior to the arrival of the main signal). The quasimonochromatic pulse is considered next, followed by a discussion of pulse distortion where the leading edge of the envelope of a step-modulated carrier is analyzed. The Taylor series expansion of $\phi(\omega)$ in the asymptotic expansion clearly shows that the even-order derivatives of $\phi(\omega)$ result in symmetrical pulse distortion while any asymmetrical pulse distortion is due to the odd-order derivatives of $\phi(\omega)$. Finally, a more generalized discussion of the leading edge of the signal was undertaken (i.e., the transient solution) for wave propagation in an idealized waveguide. In a reexamination of Sommerfeld's problem, *Haskell* and *Case* [1.55] presented the complete analysis of the arrival of the main signal in an isotropic plasma medium. By employing the saddle-point method of integration, uniform asymptotic solutions for the transient response have been obtained that agree with the principle of causality. This solution can be divided into three successive regions: a region before the saddle point crosses the pole (the anterior transient), a region when the saddle point is in the neighborhood of the pole (the main signal buildup), and a region after the saddle point has crossed the pole (the posterior transient). Consequently, the one-dimensional boundary-value problem of a step-function modulated carrier propagating in an isotropic plasma medium has been completely solved for the transient response. A brief review of the applications of such uniform asymptotic expansion techniques for radiation and diffraction problems may be found in the paper by *Ludwig* [1.56].

A somewhat different application of the problem of pulse modification in a dispersive medium is presented in the paper by *Wait* [1.57] which employs the Laplace transform to obtain an integral representation of a one-dimensional electromagnetic pulse propagating in a geological medium described by a frequency-dependent conductivity. Although no closed-form solutions are obtained and only numerical evaluations of the integral representation are presented, this paper does indicate the applicability of this theory to geological measurements by the analysis of resulting pulse deformations. *Wait* [1.58] also has considered the exact solution for the electromagnetic field radiated by a unit step-function pulsed dipole source in a cold collisionless electron plasma by employing Laplace transform techniques.

A generalization of these integral representation techniques for the problem of pulse modification in a dispersive medium may be found in the analysis due to *Jones* [1.59] of the one-dimensional boundary-value problem with a general phase function (i.e., for propagation in a general dispersive medium or system). By employing the Fourier integral representation of the pulse and assuming that the medium has a quadratic dispersion relation (which is equivalent to the quasimonochromatic approximation, which yields noncausal results if the initial pulse rise time is short), it is then shown that the resultant integral can be evaluated using the properties of Fresnel integrals. Furthermore, after a sufficiently long time, the pulse shape is shown to be proportional to the Fourier transform of the initial pulse shape, in addition to the result that the pulse propagates at the group velocity at the applied carrier frequency. Consequently, an analogy can be drawn between the degraded pulse envelope shape and the diffraction produced when light passes through a small aperture in an opaque screen whose aperture function is identical to the initial pulse envelope shape.

An equally important problem in dispersive wave propagation, which has received considerable attention, concerns the propagation velocities of light in dispersive and absorptive media [1.61–68]. The review paper by *Smith* [1.63] discusses the definitions, physical significances, interrelationships, and observabilities of seven different velocities of light, with an outlook to determine the proper velocity to be used in describing dispersive pulse propagation. It is shown here that "there is no observable physical quantity associated with the phase of a light wave", and that "a phase velocity cannot be attributed to a wave packet or to any wave except a monochromatic wave". As a consequence, the concept of phase velocity is useful only in determining the phase of a monochromatic wave in space and time given the phase at some other position and time. Similarly, it is shown that in a region of anomalous dispersion the standard definition of the group velocity fails to describe the motion of the peak in the envelope of an arbitrary pulse. As a consequence, it has been proposed to replace this standard definition with the velocity of motion of the temporal center of gravity of the amplitude of the wave packet as the definition of a generalized group velocity. The velocity of energy transport, defined as the Poynting vector divided by the total energy density, is incorrectly criticized as not corresponding to the propagation of any real observable physical quantity, as is the signal velocity introduced by *Sommerfeld* and *Brillouin*. For completeness, the author also considers briefly the relativistic velocity constant and the ratio of units velocity appearing in Maxwell's equations. Finally, the author introduces a new definition for the velocity of light, called the centrovelocity, which is the velocity of motion of the temporal center of gravity of the intensity of the radiation. It is important to note here that both the centrovelocity and the generalized definition of the group velocity are undefined for a step-function modulated signal.

Of further interest in this regard are the experimental results of *Faxvog* et al. [1.69], and *Basov* et al. [1.70] concerning the apparent measured pulse velocities in dispersive media which exceed the vacuum speed of light c, in apparent violation of the special theory of relativity. The latter paper points out that, on the basis of the rate equations for resonant radiative energy transfer in a two-level active medium, as soon as the energy density in the leading portion of the pulse reaches a sufficient level, all of the active particles will produce stimulated emission at the leading edge of the pulse, while its trailing edge is propagated through the medium with either a much lower amplification or even attenuation. As a result, the peak of the pulse will undergo an additional shift forward, resulting in an effective peak velocity exceeding c. However, this does not contradict the relativistic principle of causality since such a motion of the peak of the pulse occurs due to the deformation of the initially weak leading edge of nonzero intensity. That is, the energy associated with the pulse has not propagated with a velocity exceeding c, but only the pulse shape. Note also in this case that the centrovelocity and generalized group velocity lead to results describing the motion of the peak, and hence, exceed the vacuum speed of light c. More recent experimental results [1.71] are similarly flawed [1.72].

The concept of the energy transport velocity [1.61–66] has received considerable attention recently, the most important paper of which is that by *Loudon* [1.64]. By calculating the total energy density associated with the electromagnetic disturbance in the dispersive medium as the sum of the energy density of the wave plus the energy stored in the medium, Loudon arrived at a closed form solution for the energy velocity for propagation of monochromatic waves in a classical Lorentz medium which is causal (i.e., is less than or equal to the vacuum speed of light c for all frequencies ω of radiation). This important result is shown in this monograph to yield results which are entirely consistent with the concept of the signal velocity of non-monochromatic waves in such causal media.

More recently published research concerned with the propagation of wave packets in dispersive and absorptive media [1.73–77] have employed the technique of expanding the phase function appearing in the integral representation of the field by a Taylor series. This may also be coupled with a recursive technique to obtain correction terms of arbitrary dispersive and absorptive orders for the resultant envelope function and pulse velocity. Such an analysis again assumes the quasimonochromatic approximation, and hence, can only be applied to study the evolution of pulses with slowly varying envelope functions. In addition, ray and dispersion surface techniques have been extended [1.78–81] to include more general media as well as shorter initial pulse widths but still require that the medium be weakly dispersive.

It is clear that the most promising approach to obtain a complete, accurate description of dispersive pulse propagation is through well-defined asymptotic-expansion techniques and related approaches. Based upon the seminal analyses of *Sommerfeld* [1.2], *Brillouin* [1.3, 4], and *Baerwald* [1.34], this approach has been followed by a number of researchers [1.5–11, 18–21, 82–94] in a variety of applications with great success. This analysis has also been accompained by a few well-defined experiments [1.95–97] which validate the accuracy of this more complicated mathematical description.

1.3 Organization of the Book

The subject of classical electromagnetic pulse propagation in a linear, temporally dispersive medium is considered here in two major movements: the fundamental electromagnetic theory and the ensuing analytical description of the propagated field through the use of mathematically well-defined asymptotic expansion techniques. Other major topics, such as inverse methods for the determination of the dispersive medium properties from the propagated field characteristics, are not considered. Nevertheless, the proper solution of the forward problem in a homogeneous, isotropic, locally linear, causal, temporally dispersive medium contains enough material to fill many volumes, each dedicated to a different physical model of the dispersive medium so as to treat purely dielectric, purely conductive, and semiconducting medium. The fundamental electromagnetic theory presented here is developed for a general dispersive medium with both a frequency dependent dielectric permittivity $\varepsilon(\omega)$ and a frequency dependent conductivity $\sigma(\omega)$. This then provides the underlying theory for pulse propagation in purely dielectric, purely conductive, or semiconducting media. The asymptotic approximation of the propagated field is developed here for a single resonance Lorentz model dielectric medium. This description provides the framework for the description of pulse propagation phenomena in other types of dispersive media.

The fundamental theory developed in Part I of the monograph begins with the defining electromagnetic field equations in a general temporally dispersive medium with complex-valued dielectric permittivity $\varepsilon(\omega)$ and electrical conductivity $\sigma(\omega)$. For convenience, both MKS and Gaussian (cgs) units are used throughout the book through the use of a conversion factor that appears in each equation (if required). This basic theory is followed in Chap. 3 by a completely rigorous development of the angular spectrum of plane waves representation of the pulsed radiation field in a general, temporally dispersive medium. This analysis includes a rigorous exposition of Weyl's proof and subsequent derivation of Weyl's integral representation, as well as both Sommerfeld's and Ott's integral representations, of a spherical wave in terms of a superposition of plane waves. The angular spectrum representation of pulsed electromagnetic beam fields is then considered in Chap. 4 along with its relationship to Huygen's principle via the Rayleigh-Sommerfeld diffraction integrals. The general polarization characteristics of the field vectors are also considered in detail. The chapter concludes with a derivation of the specific integral representation of the pulsed plane wave electromagnetic field whose asymptotic description forms the subject of Part II of the monograph. For comparison, the quasimonochromatic approximation of this integral representation that leads to the heuristic theory of plane wave pulse propagation is presented at the end of this chapter. Finally, in preparation for the asymptotic analysis of Part II, Part I concludes with a complete review of those advanced saddle point methods that are appropriate for the asymptotic evaluation of the type of single contour integral that appears in the integral representation of the propagated plane wave pulse field.[1]

Because the asymptotic analysis requires that the behavior of the complex phase propagation function that appears in the integral representation of the propagated field be known throughout the complex ω-plane, this asymptotic analysis must be carried out with a specific model of the complex dielectric permittivity. Because it is a causal model, as well being of general interest from the optical to the microwave domain, not to mention its historical place as the medium used by both *Sommerfeld* and *Brillouin* in their classic analysis, the single resonance Lorentz model of a dielectric medium is considered here. Although specific to this model, the asymptotic analysis presented here serves as a general guide for other types of material dispersion.

This analysis begins in Chap. 6 with a thorough study of the behavior of the complex phase propagation function that appears in the integral representation in order to obtain the information required for the asymptotic analysis of this contour integral. With the aid of computer graphics, illustrations of the dynamic behavior of the phase function as a function of time at any fixed propagation distance are obtained which are much more complete than that which has previously been available. New approximate expressions for the locations of the relevant saddle points and the associated complex phase behavior at them are derived and shown to be much more accurate than those used by *Brillouin* [1.3, 4].

With these new and more accurate results, the uniform asymptotic behavior of the basic integral arising in the description of dispersive pulse propagation is next considered. This analysis is begun in Chap. 7, wherein the uniform asymptotic behavior of the Sommerfeld and Brillouin precursor fields is considered with detailed examples given for the various types of initial pulse shapes considered in this research. The first forerunner (Sommerfeld's precursor), which is due to the distant pair of saddle points, is analyzed first according to a direct application of *Olver*'s saddle-point method [1.14], which is the alternative approach to the method of steepest descents. However, since the distant saddle points are initially located at infinity in the complex ω-plane, *Olver*'s method is not applicable to obtaining the asymptotic behavior of the front of the first precursor field. For that purpose, the uniform asymptotic technique due to *Handelsman* and *Bleistein* [1.18] is employed. The result is an asymptotic approximation of the first precursor field that is valid uniformly over the entire duration of the field. The second forerunner (Brillouin's precursor), which is due to the pair of saddle points located in the region of the complex ω-plane about the origin, is analyzed next according to a direct application of *Olver*'s saddle-point method. In this case, the two near first-order saddle points, which are

[1] Much of the material presented here in Part I has been used as the basis of an advanced graduate level course on physical optics at the Institute of Optics of the University of Rochester and is currently used as part of an advanced graduate level course in electromagnetic wave theory in the Electrical Engineering Department of the University of Vermont.

initially located along the imaginary axis in the complex ω-plane, approach one another along this axis as θ increases, and coalesce into a single second-order saddle point at $\theta = \theta_1$, after which they separate symmetrically with respect to the imaginary axis. As a consequence, Olver's method yields discontinuous results at $\theta = \theta_1$ and is not applicable to obtaining the asymptotic behavior of the second precursor field for small values of $|\theta - \theta_1|$ with $\theta \neq \theta_1$. For that purpose one must apply the appropriate uniform asymptotic expansion [1.15] for values of θ about the critical value θ_1. The result is an asymptotic approximation of the second precursor field that is uniformly valid and continuous over the entire duration of the field. This then completes the rigorous asymptotic evaluation of the separate precursor fields that are characteristic of a single resonance Lorentz medium.

The asymptotic behavior of the main signal is considered next in Chap. 8. This final contribution to the asymptotic approximation of the total propagated field is due to the interaction of the deformed contour of integration through the relevant saddle points with any pole singularities appearing in the integrand of the integral representation of the propagated field. The analysis is begun with the nonuniform asymptotic approximation of the pole contribution, which is based upon a simple application of Cauchy's residue theorem. This nonuniform approximation is valid only if the interacting saddle point and pole remain isolated from each other. If the saddle point and pole come within close proximity of each other, however, the nonuniform approximation becomes invalid and must be replaced by the uniform asymptotic approximation of the pole contribution. That analysis of the main signal evolution is considered next and is compared to the results of the nonuniform approximation.

With these new and more accurate results, the complete, continuous dynamic evolution of the total propagated field is considered in Chap. 9. The continuous evolution of the total precursor field is considered first, followed by the interaction of that field with the main signal contribution. The signal arrival and its associated signal velocity is then obtained from a consideration of the interaction of these two sets of component fields. In an important modification of Brillouin's approach, the signal arrival is shown to occur at the instant when the real exponential behavior associated with the simple pole singularity appearing in the integrand becomes less (in magnitude) than the real exponential behavior associated with the interacting saddle point. With this new interpretation of the signal arrival, an accurate definition of the signal velocity is obtained. It is important to point out here that Brillouin's slight misinterpretation concerning the signal arrival yields incorrect results for the signal velocity in the absorption band of the medium. As a consequence, it is shown here that the signal velocity does not peak to the vacuum speed of light c near resonance, but rather attains a minimum there. Finally, employing approximate analytic expressions for the signal velocity in the dispersive medium, it is shown that the signal velocity of an arbitrary pulse as defined here is related to the velocity of energy propagation for monochromatic plane waves, as derived by *Loudon* [1.64]. This extremely important result makes possible a better physical understanding of the basic dynamics of dispersive pulse propagation. With these results, the complete dynamic evolution of the various initial pulse shapes considered in this research (e.g., the unit step-function modulated sine wave, the rectangular pulse modulated sine wave, the delta-function pulse, and the hyperbolic-tangent modulated sine wave) in propagating through a dispersive Lorentz medium are given. This detailed discussion provides an overview of all of the important aspects of dispersive pulse propagation obtained in this research. The monograph concludes with the new physical interpretation and subsequent physical model of dispersive pulse dynamics. This model supplants the previous group velocity model that is restricted to nonabsorptive media and reduces to it in the limit as the medium absorption is allowed to vanish.

2 Fundamental Field Equations in a Temporally Dispersive Medium

In this chapter the fundamental medium and plane-wave propagation properties that are necessary for the general analysis of electromagnetic radiation and wave propagation in a homogeneous, isotropic, locally linear, temporally dispersive medium that occupies all space is presented. The general frequency dependence of both the dielectric permittivity and the electric conductivity is included in the analysis so that the resultant field equations apply to both perfect and imperfect dielectrics, conductors and semiconductors over the entire frequency domain. The analysis here focuses on the source-free field equations and plane-wave solutions in such simple, temporally dispersive media. Both Gaussian (cgs) and MKS units are employed throughout this book; this is accomplished through the use of a conversion factor that appears in double brackets $[\![\cdot]\!]$ in each equation. If this factor is included in the equation it is then in cgs units provided that one also lets $\varepsilon_0 = \mu_0 = 1$, while if this factor is omitted the equation is then in MKS units. Finally, if no such factor appears then the equation is correct in both systems of units.

2.1 Fundamental Field Equations in a Temporally Dispersive Medium

In a homogeneous, isotropic, locally linear, temporally dispersive medium occupying all of three-dimensional space with no externally supplied charge or current sources, the Maxwell field equations are

$$\nabla\cdot\mathcal{D}(\mathbf{r},t) = [\![4\pi]\!]\varrho_c(\mathbf{r},t)\ , \tag{2.1.1a}$$

$$\nabla\times\mathcal{E}(\mathbf{r},t) = -\left[\!\!\left[\frac{1}{c}\right]\!\!\right]\frac{\partial\mathcal{B}(\mathbf{r},t)}{\partial t}\ , \tag{2.1.1b}$$

$$\nabla\cdot\mathcal{B}(\mathbf{r},t) = 0\ , \tag{2.1.1c}$$

$$\nabla\times\mathcal{B}(\mathbf{r},t) = \left[\!\!\left[\frac{1}{c}\right]\!\!\right]\mu\frac{\partial\mathcal{D}(\mathbf{r},t)}{\partial t} + \left[\!\!\left[\frac{4\pi}{c}\right]\!\!\right]\mu\mathcal{J}_c(\mathbf{r},t)\ , \tag{2.1.1d}$$

which are appropriate for nonhysteretic materials for which $\mathcal{B}(\mathbf{r},t) = \mu\mathcal{H}(\mathbf{r},t)$

with the magnetic permeability μ a constant. The remaining constitutive relations are

$$\mathcal{J}_c(\mathbf{r},t) = \int_{-\infty}^{t} \hat{\sigma}(t-t')\mathcal{E}(\mathbf{r},t')\,dt' \ , \tag{2.1.2a}$$

$$\mathcal{D}(\mathbf{r},t) = \int_{-\infty}^{t} \hat{\varepsilon}(t-t')\mathcal{E}(\mathbf{r},t')\,dt' \ , \tag{2.1.2b}$$

which are appropriate for homogeneous, isotropic, locally linear, temporally dispersive media. Here $\hat{\sigma}(t)$ denotes the (real-valued) electric conductivity and $\hat{\varepsilon}(t)$ denotes the (real-valued) dielectric permittivity of the general dispersive medium. By causality, $\hat{\sigma}(t-t')=0$ and $\hat{\varepsilon}(t-t')=0$ for $t'>t$, as exhibited in the upper limit of integration in the above two relations. The dependence of the conduction current density $\mathcal{J}_c$ on the magnetic field $\mathcal{B}$ (known as the Hall effect) is, in most cases, small and is not considered here. The conduction current density $\mathcal{J}_c(\mathbf{r},t)$ and charge density $\varrho_c(\mathbf{r},t)$ are related by the equation of continuity

$$\nabla\cdot\mathcal{J}_c(\mathbf{r},t) + \frac{\partial\varrho_c(\mathbf{r},t)}{\partial t} = 0 \ , \tag{2.1.3}$$

which follows from the divergence of (2.1.1d) and substitution from (2.1.1a).

The temporal Fourier transform of a function $\mathcal{F}(\mathbf{r},t)$ of both position $\mathbf{r}$ and time t is defined here by the pair of equations

$$\tilde{\mathbf{F}}(\mathbf{r},\omega) = \int_{-\infty}^{\infty} \mathcal{F}(\mathbf{r},t)\mathrm{e}^{\mathrm{i}\omega t}\,dt \ , \tag{2.1.4a}$$

$$\mathcal{F}(\mathbf{r},t) = \frac{1}{2\pi}\int_{-\infty}^{\infty} \tilde{\mathbf{F}}(\mathbf{r},\omega)\mathrm{e}^{-\mathrm{i}\omega t}\,d\omega \ . \tag{2.1.4b}$$

By the Fourier integral theorem, the sufficient conditions for the existence of the Fourier transform pair relationship (2.1.4) are that the function $\mathcal{F}(\mathbf{r},t)$ be piecewise smooth in any finite time interval and that $\mathcal{F}(\mathbf{r},t)$ be absolutely integrable over the entire time domain. However, these conditions are not necessary and it is often convenient to represent simplified physical fields by idealized mathematical functions for which either of the existence conditions may be violated. As *Bracewell* [2.1] has pointed out, "physical possibility is a valid sufficient condition for the existence of a transform." In light of this, it is assumed throughout this analysis that all such transforms exist, at least in the generalized function sense [2.2].

The temporal Fourier transform of the equation of continuity (2.1.3) yields

$$\nabla\cdot\tilde{\mathbf{J}}_c(\mathbf{r},\omega) = \mathrm{i}\omega\tilde{\rho}_c(\mathbf{r},\omega) \ ,$$

while the temporal Fourier transform of the constitutive relation (2.1.2a) yields, with the application of the convolution theorem and the requirement of causality,

$$\tilde{\mathbf{J}}_c(\mathbf{r},\omega) = \sigma(\omega)\tilde{\mathbf{E}}(\mathbf{r},\omega) \ , \tag{2.1.5a}$$

so that

$$\tilde{\rho}_c(\mathbf{r},\omega) = \frac{\sigma(\omega)}{i\omega}\nabla\cdot\tilde{\mathbf{E}}(\mathbf{r},\omega) \ .$$

Furthermore, the temporal Fourier transform of the constitutive relation (2.1.2b) yields, with the application of the convolution theorem and the requirement of causality,

$$\tilde{\mathbf{D}}(\mathbf{r},\omega) = \varepsilon(\omega)\tilde{\mathbf{E}}(\mathbf{r},\omega) \ , \tag{2.1.5b}$$

so that the temporal Fourier transform of the divergence relation (2.1.1a) gives, with substitution of the above expression for $\tilde{\rho}_c(\mathbf{r},\omega)$,

$$\begin{aligned}\varepsilon(\omega)\nabla\cdot\tilde{\mathbf{E}}(\mathbf{r},\omega) &= [\![4\pi]\!]\tilde{\rho}_c(\mathbf{r},\omega)\\ &= [\![4\pi]\!]\frac{\sigma(\omega)}{i\omega}\nabla\cdot\tilde{\mathbf{E}}(\mathbf{r},\omega) \ .\end{aligned}$$

In order that this final expression be satisfied, either $\nabla\cdot\tilde{\mathbf{E}}(\mathbf{r},\omega) = 0$ or, if not, then $[\![4\pi]\!]\sigma(\omega)/(\omega\varepsilon(\omega)) = i$, which will not be satisfied for general expressions of $\sigma(\omega)$ and $\varepsilon(\omega)$. One must then have that $\nabla\cdot\tilde{\mathbf{E}}(\mathbf{r},\omega) = 0$ so that $\tilde{\rho}_c(\mathbf{r},\omega) = 0$, which in turn implies that $\rho_c(\mathbf{r},t) = 0$. This result is a direct consequence of the continuity equation (2.1.3), the divergence relation (2.1.1a) for the electric displacement vector, and, most importantly, the two constitutive relations (2.1.2a and 2b) which introduce the conductivity and dielectric permittivity of the homogeneous, isotropic, locally linear, temporally dispersive medium.

The final space-time domain form of the set of Maxwell's field equations to be considered here is then given by

$$\nabla\cdot\mathcal{D}(\mathbf{r},t) = 0 \ , \tag{2.1.6a}$$

$$\nabla\times\mathcal{E}(\mathbf{r},t) = -\left[\!\!\left[\frac{1}{c}\right]\!\!\right]\frac{\partial\mathcal{B}(\mathbf{r},t)}{\partial t} \ , \tag{2.1.6b}$$

$$\nabla\cdot\mathcal{B}(\mathbf{r},t) = 0 \ , \tag{2.1.6c}$$

$$\nabla\times\mathcal{B}(\mathbf{r},t) = \left[\!\!\left[\frac{1}{c}\right]\!\!\right]\mu\frac{\partial\mathcal{D}(\mathbf{r},t)}{\partial t} + \left[\!\!\left[\frac{4\pi}{c}\right]\!\!\right]\mu\mathcal{J}_c(\mathbf{r},t) \ , \tag{2.1.6d}$$

with no external charge or current sources, with the constitutive relations (2.1.2) and the equation of continuity

$$\nabla \cdot \mathcal{J}_c(\mathbf{r}, t) = 0 \ . \tag{2.1.7}$$

Hence, the electric displacement vector $\mathcal{D}(\mathbf{r}, t)$, the electric field intensity vector $\mathcal{E}(\mathbf{r}, t)$, the magnetic induction vector $\mathcal{B}(\mathbf{r}, t)$, the magnetic field intensity vector $\mathcal{H}(\mathbf{r}, t)$, and the conduction current density vector $\mathcal{J}_c(\mathbf{r}, t)$ are all solenoidal vector fields in the source free, infinitely extended dispersive medium. Since the medium properties are spatially continuous everywhere, it then follows, for example, that

$$\oiint_S \mathcal{J}_c(\mathbf{r}, t) \cdot \hat{n} \, da = 0$$

for any closed surface S.

2.1.1 Temporal Frequency Domain Representation of the Field and Medium Properties

The temporal frequency spectra of the electric field intensity and magnetic induction field vectors are defined by the Fourier transformations

$$\tilde{\mathbf{E}}(\mathbf{r}, \omega) = \int_{-\infty}^{\infty} \mathcal{E}(\mathbf{r}, t) e^{i\omega t} \, dt \ , \tag{2.1.8a}$$

$$\tilde{\mathbf{B}}(\mathbf{r}, \omega) = \int_{-\infty}^{\infty} \mathcal{B}(\mathbf{r}, t) e^{i\omega t} \, dt \ , \tag{2.1.8b}$$

with the corresponding inverse transformations

$$\mathcal{E}(\mathbf{r}, t) = \frac{1}{2\pi} \int_{-\infty}^{\infty} \tilde{\mathbf{E}}(\mathbf{r}, \omega) e^{-i\omega t} \, d\omega \ , \tag{2.1.9a}$$

$$\mathcal{B}(\mathbf{r}, t) = \frac{1}{2\pi} \int_{-\infty}^{\infty} \tilde{\mathbf{B}}(\mathbf{r}, \omega) e^{-i\omega t} \, d\omega \ . \tag{2.1.9b}$$

Since both of the field vectors $\mathcal{E}$ and $\mathcal{B}$ are real-valued vector functions of both position and time, then their temporal frequency spectra satisfy the symmetry property

$$\begin{aligned} \tilde{\mathbf{E}}^*(\mathbf{r}, \omega) &= \tilde{\mathbf{E}}(\mathbf{r}, -\omega) \ , \\ \tilde{\mathbf{B}}^*(\mathbf{r}, \omega) &= \tilde{\mathbf{B}}(\mathbf{r}, -\omega) \ , \end{aligned} \tag{2.1.10}$$

for real ω. As a consequence, the real part of each spectral function is an even function of ω while the imaginary part is an odd function of ω.

The temporal frequency spectra of the electric displacement vector and the conduction current density are, with the constitutive relations (1.1.2),

$$\tilde{\mathbf{D}}(\mathbf{r},\omega) = \int_{-\infty}^{\infty} \mathscr{D}(\mathbf{r},t)e^{i\omega t}\,dt = \varepsilon(\omega)\tilde{\mathbf{E}}(\mathbf{r},\omega) \ , \tag{2.1.11a}$$

$$\tilde{\mathbf{J}}_c(\mathbf{r},\omega) = \int_{-\infty}^{\infty} \mathscr{J}_c(\mathbf{r},t)e^{i\omega t}\,dt = \sigma(\omega)\tilde{\mathbf{E}}(\mathbf{r},\omega) \ , \tag{2.1.11b}$$

where

$$\varepsilon(\omega) = \int_{-\infty}^{\infty} \hat{\varepsilon}(t)e^{i\omega t}\,dt \ , \tag{2.1.12a}$$

$$\sigma(\omega) = \int_{-\infty}^{\infty} \hat{\sigma}(t)e^{i\omega t}\,dt \ , \tag{2.1.12b}$$

are the temporal frequency spectra of the dielectric permittivity and electric conductivity of the dispersive medium. Since both the field vectors $\mathscr{D}$ and $\mathscr{J}_c$ are real-valued vector functions of both position and time, then their temporal frequency spectra satisfy the same symmetry property stated in (2.1.10) so that

$$\varepsilon^*(\omega)\tilde{\mathbf{E}}^*(\mathbf{r},\omega) = \varepsilon(-\omega)\tilde{\mathbf{E}}(\mathbf{r},-\omega) \ , \tag{2.1.13a}$$

$$\sigma^*(\omega)\tilde{\mathbf{E}}^*(\mathbf{r},\omega) = \sigma(-\omega)\tilde{\mathbf{E}}(\mathbf{r},-\omega) \ , \tag{2.1.13b}$$

and hence

$$\varepsilon^*(\omega) = \varepsilon(-\omega) \ , \tag{2.1.14a}$$

$$\sigma^*(\omega) = \sigma(-\omega) \ , \tag{2.1.14b}$$

for real ω. The last set of symmetry relations also follows from (2.1.12) since $\hat{\varepsilon}(t)$ and $\hat{\sigma}(t)$ are both real valued. As a consequence, the real parts of both $\varepsilon(\omega)$ and $\sigma(\omega)$ are even functions of ω, while their imaginary parts are both odd functions ω.

Upon returning to the fundamental constitutive relations (2.1.2) it is seen that at any given point in space the conduction current and displacement vector both depend upon the past history of the electric field intensity at that point through the conductivity function $\hat{\sigma}(t-t')$ and the dielectric permittivity function $\hat{\varepsilon}(t-t')$, respectively. From a physical standpoint it seems reasonable to expect that the medium response is significant only in some neighborhood preceeding the instant of time t at which the field vectors $\mathscr{J}_c(\mathbf{r},t)$ and $\mathscr{D}(\mathbf{r},t)$ are calculated; however, this is not necessary for the ensuing mathmatical development. It is then seen appropriate to expand the electric field intensity $\mathscr{E}(\mathbf{r},t')$ in a Taylor series about the instant $t'=t$ as

$$\mathscr{E}(\mathbf{r},t') = \sum_{n=0}^{\infty} \frac{1}{n!}\frac{\partial^n \mathscr{E}(\mathbf{r},t)}{\partial t^n}(t'-t)^n \ ,$$

provided that $\mathscr{E}(\mathbf{r},t')$ and all of its time derivatives exist at each instant $t' \leqslant t$.

This Taylor series expansion represents $\mathscr{E}(\mathbf{r},t')$ for all $t' \leqslant t$ if and only if the remainder $R_N(t')$ after N terms approaches zero as $N \to \infty$. Substitution of this expansion into the constitutive relation (2.1.2a) then gives

$$\mathscr{J}_c(\mathbf{r},t) = \sum_{n=0}^{\infty} \sigma^{(n)} \frac{\partial^n \mathscr{E}(\mathbf{r},t)}{\partial t^n} , \tag{2.1.15a}$$

where

$$\sigma^{(n)} = \frac{(-1)^n}{n!} \int_0^{\infty} \hat{\sigma}(\tau)\tau^n \, d\tau . \tag{2.1.15b}$$

Similarly, with this substitution the constitutive relation (2.1.2b) becomes

$$\mathscr{D}(\mathbf{r},t) = \sum_{n=0}^{\infty} \varepsilon^{(n)} \frac{\partial^n \mathscr{E}(\mathbf{r},t)}{\partial t^n} , \tag{2.1.16a}$$

with

$$\varepsilon^{(n)} = \frac{(-1)^n}{n!} \int_0^{\infty} \hat{\varepsilon}(\tau)\tau^n \, d\tau . \tag{2.1.16b}$$

Notice that the coefficients $\sigma^{(n)}$ and $\varepsilon^{(n)}$ are proportional to the n'th. order moments of the conductivity and dielectric permittivity functions, respectively. These forms (2.1.15a and 16a) of the constitutive relations explicitly exhibit the local, linear dependence of the derived field vectors $\mathscr{J}_c$ and $\mathscr{D}$ on the derivatives of the electric field intensity vector. For a quasi-static field these relations may be approximated as

$$\mathscr{J}_c(\mathbf{r},t) \simeq \sigma^{(0)} \mathscr{E}(\mathbf{r},t) + \sigma^{(1)} \frac{\partial \mathscr{E}(\mathbf{r},t)}{\partial t} ,$$

$$\mathscr{D}(\mathbf{r},t) \simeq \varepsilon^{(0)} \mathscr{E}(\mathbf{r},t) + \varepsilon^{(1)} \frac{\partial \mathscr{E}(\mathbf{r},t)}{\partial t} ,$$

from which it is seen that $\sigma^{(0)}$ and $\varepsilon^{(0)}$ are just the static conductivity and dielectric permittivity of the medium.

The temporal Fourier transform of (2.1.15a and 2.1.16a) yield, with (2.1.11),

$$\tilde{\mathbf{D}}(\mathbf{r},\omega) = \varepsilon(\omega)\tilde{\mathbf{E}}(\mathbf{r},\omega) = \left(\sum_{n=0}^{\infty} (-\mathrm{i}\omega)^n \varepsilon^{(n)} \right) \tilde{\mathbf{E}}(\mathbf{r},\omega) ,$$

$$\tilde{\mathbf{J}}_c(\mathbf{r},\omega) = \sigma(\omega)\tilde{\mathbf{E}}(\mathbf{r},\omega) = \left(\sum_{n=0}^{\infty} (-\mathrm{i}\omega)^n \sigma^{(n)} \right) \tilde{\mathbf{E}}(\mathbf{r},\omega) ,$$

so that the temporal frequency spectra of the dielectric permittivity and electric conductivity of the dispersive medium are given by

$$\varepsilon(\omega) = \sum_{n=0}^{\infty} \varepsilon^{(n)}(-i\omega)^n \ , \tag{2.1.17a}$$

$$\sigma(\omega) = \sum_{n=0}^{\infty} \sigma^{(n)}(-i\omega)^n \ . \tag{2.1.17b}$$

The real and imaginary parts of these two expressions then give

$$\varepsilon_r(\omega) = \mathbb{R}\mathrm{e}\{\varepsilon(\omega)\} = \varepsilon^{(0)} - \varepsilon^{(2)}\omega^2 + \varepsilon^{(4)}\omega^4 - \varepsilon^{(6)}\omega^6 + \cdots \ , \tag{2.1.18a}$$

$$\varepsilon_i(\omega) = \mathbb{I}\mathrm{m}\{\varepsilon(\omega)\} = -\varepsilon^{(1)}\omega + \varepsilon^{(3)}\omega^3 - \varepsilon^{(5)}\omega^5 + \cdots \ , \tag{2.1.18b}$$

$$\sigma_r(\omega) = \mathbb{R}\mathrm{e}\{\sigma(\omega)\} = \sigma^{(0)} - \sigma^{(2)}\omega^2 + \sigma^{(4)}\omega^4 - \sigma^{(6)}\omega^6 + \cdots \ , \tag{2.1.19a}$$

$$\sigma_i(\omega) = \mathbb{I}\mathrm{m}\{\sigma(\omega)\} = -\sigma^{(1)}\omega + \sigma^{(3)}\omega^3 - \sigma^{(5)}\omega^5 + \cdots \ , \tag{2.1.19b}$$

Once again it is seen that the real parts of both $\varepsilon(\omega)$ and $\sigma(\omega)$ are even functions of ω while the imaginary parts are odd functions of ω. Furthermore, it is seen that

$$\varepsilon^{(n)} = \frac{i^n}{n!} \frac{\partial^n \varepsilon(\omega)}{\partial \omega^n}\bigg|_{\omega=0} \ , \tag{2.1.20a}$$

$$\sigma^{(n)} = \frac{i^n}{n!} \frac{\partial^n \sigma(\omega)}{\partial \omega^n}\bigg|_{\omega=0} \ . \tag{2.1.20b}$$

Finally, the temporal frequency spectrum form of Maxwell's equations (2.1.6) are

$$\nabla \cdot \tilde{\mathbf{E}}(\mathbf{r},\omega) = 0 \ , \tag{2.1.21a}$$

$$\nabla \times \tilde{\mathbf{E}}(\mathbf{r},\omega) = \left[\!\left[\frac{1}{c}\right]\!\right] i\omega \tilde{\mathbf{B}}(\mathbf{r},\omega) \ , \tag{2.1.21b}$$

$$\nabla \cdot \tilde{\mathbf{B}}(\mathbf{r},\omega) = 0 \ , \tag{2.1.21c}$$

$$\nabla \times \tilde{\mathbf{B}}(\mathbf{r},\omega) = -\left[\!\left[\frac{1}{c}\right]\!\right] i\omega\mu\varepsilon_c(\omega) \tilde{\mathbf{E}}(\mathbf{r},\omega) \ , \tag{2.1.21d}$$

where

$$\varepsilon_c(\omega) = \varepsilon(\omega) + i[\![4\pi]\!]\frac{\sigma(\omega)}{\omega} \tag{2.1.22}$$

is called the *complex permittivity* of the dispersive medium.

2.1.2 Complex Time-Harmonic Form of the Field Quantities

The complex form of the electric and magnetic field vectors for a completely time-harmonic field of fixed angular frequency ω is given by

$$\mathbf{E}_\omega(\mathbf{r}, t) = \mathbf{E}(\mathbf{r})e^{-i\omega t} , \tag{2.1.23a}$$

$$\mathbf{B}_\omega(\mathbf{r}, t) = \mathbf{B}(\mathbf{r})e^{-i\omega t} , \tag{2.1.23b}$$

where each complex field vector is related to its real-valued counterpart by

$$\mathscr{E}(\mathbf{r}, t) = \mathbb{R}e\{\mathbf{E}_\omega(\mathbf{r}, t)\} = \mathbb{R}e\{\mathbf{E}(\mathbf{r})e^{-i\omega t}\} , \tag{2.1.24a}$$

$$\mathscr{B}(\mathbf{r}, t) = \mathbb{R}e\{\mathbf{B}_\omega(\mathbf{r}, t)\} = \mathbb{R}e\{\mathbf{B}(\mathbf{r})e^{-i\omega t}\} . \tag{2.1.24b}$$

With this substitution, the constitutive relations given in (2.1.2) become

$$\begin{aligned}\mathscr{J}_c(\mathbf{r}, t) &= \mathbb{R}e\left\{\mathbf{E}(\mathbf{r}) \int_{-\infty}^{\infty} \hat{\sigma}(t - t')e^{-i\omega t'} dt'\right\} \\ &= \mathbb{R}e\{\sigma(\omega)\mathbf{E}(\mathbf{r})e^{-i\omega t}\} ,\end{aligned} \tag{2.1.25a}$$

and similarly

$$\mathscr{D}(\mathbf{r}, t) = \mathbb{R}e\{\varepsilon(\omega)\mathbf{E}(\mathbf{r})e^{-i\omega t}\} . \tag{2.1.25b}$$

The appropriate complex form of these two field vectors is the given by

$$\mathbf{J}_{c\omega}(\mathbf{r}, t) = \mathbf{J}_c(\mathbf{r})e^{-i\omega t} = \sigma(\omega)\mathbf{E}(\mathbf{r})e^{-i\omega t} , \tag{2.1.26a}$$

$$\mathbf{D}_\omega(\mathbf{r}, t) = \mathbf{D}(\mathbf{r})e^{-i\omega t} = \varepsilon(\omega)\mathbf{E}(\mathbf{r})e^{-i\omega t} . \tag{2.1.26b}$$

Consider now determining the relationship between these complex field vectors for a time-harmonic field and the temporal frequency spectra of the real-valued field vectors. With substitution of (2.1.24a) into (2.1.8a) for the temporal frequency spectrum of the electric field intensity vector one obtains

$$\begin{aligned}\tilde{\mathbf{E}}(\mathbf{r}, \omega') &= \int_{-\infty}^{\infty} \mathbb{R}e\{\mathbf{E}(\mathbf{r})e^{-i\omega t}\}e^{i\omega' t} dt \\ &= \tfrac{1}{2}\left\{\mathbf{E}(\mathbf{r}) \int_{-\infty}^{\infty} e^{i(\omega' - \omega)t} dt + \mathbf{E}^*(\mathbf{r}) \int_{-\infty}^{\infty} e^{i(\omega' + \omega)t} dt\right\} \\ &= \pi\{\mathbf{E}(\mathbf{r})\delta(\omega' - \omega) + \mathbf{E}^*(\mathbf{r})\delta(\omega' + \omega)\} ,\end{aligned} \tag{2.1.27}$$

with similar expressions for the other field vectors. Hence, the frequency spectrum of the complex monochromatic field given in (2.1.23) with real field vectors given in (2.1.24) is comprised of a pair of spectral lines at $\omega' = \pm\omega$.

Substitution of (2.1.24 and 25) into Maxwell's field equations (2.1.6) results in the set of equations

$$\mathrm{Re}\{\varepsilon(\omega)\nabla\cdot\mathbf{E}(\mathbf{r})e^{-i\omega t}\} = 0 \ ,$$

$$\mathrm{Re}\{\nabla\times\mathbf{E}(\mathbf{r})e^{-i\omega t}\} = \left[\frac{1}{c}\right]\mathrm{Re}\{i\omega\mathbf{B}(\mathbf{r})e^{-i\omega t}\} \ ,$$

$$\mathrm{Re}\{\nabla\cdot\mathbf{B}(\mathbf{r})e^{-i\omega t}\} = 0 \ ,$$

$$\mathrm{Re}\{\nabla\times\mathbf{B}(\mathbf{r})e^{-i\omega t}\} = -\left[\frac{1}{c}\right]\mathrm{Re}\{i\omega\mu\varepsilon_c(\omega)\mathbf{E}(\mathbf{r})e^{-i\omega t}\} \ ,$$

where $\varepsilon_c(\omega)$ is the complex permittivity, as defined in (2.1.22). Since it is only the real parts of the above quantities that enter into the determination of the real-valued field vectors through (2.1.24 and 25), one is free to impose the additional condition that the imaginary parts of these quantities also satisfy the same relations. One then obtains the complex time-harmonic differential form of Maxwell's equations

$$\nabla\cdot\mathbf{E}(\mathbf{r}) = 0 \ , \qquad (2.1.28a)$$

$$\nabla\times\mathbf{E}(\mathbf{r}) = \left[\frac{1}{c}\right]i\omega\mathbf{B}(\mathbf{r}) \ , \qquad (2.1.28b)$$

$$\nabla\cdot\mathbf{B}(\mathbf{r}) = 0 \ , \qquad (2.1.28c)$$

$$\nabla\times\mathbf{B}(\mathbf{r}) = -\left[\frac{1}{c}\right]i\omega\mu\varepsilon_c(\omega)\mathbf{E}(\mathbf{r}) \ , \qquad (2.1.28d)$$

which are the same as the temporal-frequency domain form given in (2.1.21); the difference between these two sets of equations lies entirely in the interpretation of the field vectors.

Upon taking the curl of (2.1.28b) and substituting from (2.1.28d) one obtains

$$\nabla\times\nabla\times\mathbf{E}(\mathbf{r}) = \left[\frac{1}{c^2}\right]\omega^2\mu\varepsilon_c(\omega)\mathbf{E}(\mathbf{r}) \ ,$$

which, with the vector identity for the curlcurl operator and (2.1.28a), becomes

$$\nabla^2\mathbf{E}(\mathbf{r}) + \left[\frac{1}{c^2}\right]\omega^2\mu\varepsilon_c(\omega)\mathbf{E}(\mathbf{r}) = \mathbf{0} \ . \qquad (2.1.29)$$

In a similar fashion, upon taking the curl of (2.1.28d) and substituting from (2.1.28b) and using the fact that the magnetic-field vector is solenoidal, one obtains

$$\nabla^2\mathbf{B}(\mathbf{r}) + \left[\frac{1}{c^2}\right]\omega^2\mu\varepsilon_c(\omega)\mathbf{B}(\mathbf{r}) = \mathbf{0} \ . \qquad (2.1.30)$$

In Gaussian (or cgs) units the factor in the double-line brackets $[\![\cdot]\!]$ remains intact in these (and all) equations. In addition, $\varepsilon_0 = \mu_0 = 1$ so that μ and ε are the relative magnetic permeability and the relative dielectric permittivity of the medium, respectively. With the *vacuum wave number* of the field defined as

$$k_0 \equiv \frac{\omega}{c} , \tag{2.1.31}$$

the harmonic wave equations (2.1.29 and 30) take the form

$$\nabla^2 \mathbf{E}(\mathbf{r}) + k_0^2 \mu \varepsilon_c(\omega) \mathbf{E}(\mathbf{r}) = \mathbf{0} , \tag{2.1.32a}$$

$$\nabla^2 \mathbf{B}(\mathbf{r}) + k_0^2 \mu \varepsilon_c(\omega) \mathbf{B}(\mathbf{r}) = \mathbf{0} , \tag{2.1.32b}$$

when cgs units are employed.

In MKS units the factor in the double-line brackets $[\![\cdot]\!]$ is replaced by unity in these (and all) equations. In addition, μ and ε are the absolute magnetic permeability and absolute dielectric permittivity of the dispersive medium, respectively, the relative values being given by

$$\mu^{(r)} = \mu/\mu_0 , \qquad \varepsilon^{(r)}(\omega) = \varepsilon(\omega)/\varepsilon_0 .$$

Since $c = (\mu_0 \varepsilon_0)^{-1/2}$, the harmonic wave equations in MKS units may be written as

$$\nabla^2 \mathbf{E}(\mathbf{r}) + k_0^2 \frac{\mu \varepsilon_c(\omega)}{\mu_0 \varepsilon_0} \mathbf{E}(\mathbf{r}) = \mathbf{0} , \tag{2.1.33a}$$

$$\nabla^2 \mathbf{B}(\mathbf{r}) + k_0^2 \frac{\mu \varepsilon_c(\omega)}{\mu_0 \varepsilon_0} \mathbf{B}(\mathbf{r}) = \mathbf{0} , \tag{2.1.33b}$$

where k_0 is again the vacuum wave number, as given in (2.1.31).

The *complex index of refraction* of the dispersive medium is defined as

$$n(\omega) = \left(\frac{\mu \varepsilon_c(\omega)}{\mu_0 \varepsilon_0}\right)^{1/2} = [\mu^{(r)} \varepsilon_c^{(r)}(\omega)]^{1/2} , \tag{2.1.34}$$

where $\varepsilon_c^{(r)}(\omega) = \varepsilon_c(\omega)/\varepsilon_0$ is the relative complex permittivity of the medium. Hence, in either system of units the harmonic wave equations are given by

$$\nabla^2 \mathbf{E}(\mathbf{r}) + k_0^2 n^2(\omega) \mathbf{E}(\mathbf{r}) = \mathbf{0} , \tag{2.1.35a}$$

$$\nabla^2 \mathbf{B}(\mathbf{r}) + k_0^2 n^2(\omega) \mathbf{B}(\mathbf{r}) = \mathbf{0} . \tag{2.1.35b}$$

These are the *vector Helmholtz equations* for homogeneous, isotropic, linear, temporally dispersive media.

2.2 Electromagnetic Energy and Energy Flow in a Temporally Dispersive Medium

Of considerable importance to the physical interpretation of electromagnetic-wave propagation is the concept of electromagnetic energy and energy flow in the dispersive host medium. The statement of the conservation of energy in the coupled-field medium system, as embodied in Poynting's theorem, is readily obtained as a direct mathematical consequence of the real space-time form of Maxwell's field equations. Although subject to interpretation, this relation provides a mathematically consistent formulation of energy flow in the electromagnetic field that is extremely useful in the analysis and physical interpretation of electromagnetic information transmission in temporally dispersive media and systems [2.3]. However, a detailed description of the distribution of energy in the coupled field-medium system requires a detailed understanding of the field-medium interaction. For the present, only the general formulation and physical interpretation are considered.

2.2.1 Poynting's Theorem and the Conservation of Energy

The analysis begins with the pair of curl relations (2.1.6b and 6d) of the real space-time domain form of Maxwell's equations with no external charge or current sources, which may be written as

$$\nabla \times \mathscr{E}(\mathbf{r},t) = -\left[\frac{1}{c}\right]\frac{\partial \mathscr{B}(\mathbf{r},t)}{\partial t} ,$$

$$\nabla \times \mathscr{H}(\mathbf{r},t) = \left[\frac{1}{c}\right]\frac{\partial \mathscr{D}(\mathbf{r},t)}{\partial t} + \left[\frac{4\pi}{c}\right]\mathscr{J}_c(\mathbf{r},t) ,$$

where $\mathscr{B}(\mathbf{r},t) = \mu\mathscr{H}(\mathbf{r},t)$. Upon taking the scalar product of the first equation with $\mathscr{H}(\mathbf{r},t)$ and the second equation with $\mathscr{E}(\mathbf{r},t)$ and taking their difference, one obtains

$$\mathscr{E}\cdot\nabla\times\mathscr{H} - \mathscr{H}\cdot\nabla\times\mathscr{E} = \left[\frac{1}{c}\right]\left(\mathscr{E}\cdot\frac{\partial\mathscr{D}}{\partial t} + \mathscr{H}\cdot\frac{\partial\mathscr{B}}{\partial t}\right) + \left[\frac{4\pi}{c}\right]\mathscr{J}_c\cdot\mathscr{E} ,$$

which may be rewritten as

$$\mathscr{J}_c\cdot\mathscr{E} = -\left[\frac{c}{4\pi}\right]\nabla\cdot(\mathscr{E}\times\mathscr{H}) - \left[\frac{1}{4\pi}\right]\left(\mathscr{E}\cdot\frac{\partial\mathscr{D}}{\partial t} + \mathscr{H}\cdot\frac{\partial\mathscr{B}}{\partial t}\right) . \tag{2.2.1}$$

The *Poynting vector* $\mathscr{S}(\mathbf{r},t)$ of the electromagnetic field is defined at each space-time point as

$$\mathscr{S}(\mathbf{r},t) \equiv \left[\frac{c}{4\pi}\right]\mathscr{E}(\mathbf{r},t)\times\mathscr{H}(\mathbf{r},t) \tag{2.2.2}$$

which has the dimensional units of power per unit area. In addition, one may define the scalar quantities $\mathscr{U}_e'(\mathbf{r},t)$ and $\mathscr{U}_m'(\mathbf{r},t)$ that are associated with the electric and magnetic fields, respectively, by the general relations

$$\frac{\partial\mathscr{U}_e'(\mathbf{r},t)}{\partial t} \equiv \left[\frac{1}{4\pi}\right] \mathscr{E}(\mathbf{r},t)\cdot\frac{\partial\mathscr{D}(\mathbf{r},t)}{\partial t} , \tag{2.2.3a}$$

$$\frac{\partial\mathscr{U}_m'(\mathbf{r},t)}{\partial t} \equiv \left[\frac{1}{4\pi}\right] \mathscr{H}(\mathbf{r},t)\cdot\frac{\partial\mathscr{B}(\mathbf{r},t)}{\partial t} , \tag{2.2.3b}$$

which have the dimensions of energy per unit volume, with sum

$$\mathscr{U}'(\mathbf{r},t) = \mathscr{U}_e'(\mathbf{r},t) + \mathscr{U}_m'(\mathbf{r},t) . \tag{2.2.4}$$

With these definitions (2.2.1) becomes

$$\mathscr{J}_c(\mathbf{r},t)\cdot\mathscr{E}(\mathbf{r},t) = -\nabla\cdot\mathscr{S}(\mathbf{r},t) - \frac{\partial\mathscr{U}'(\mathbf{r},t)}{\partial t} \tag{2.2.5}$$

which is the *differential form of Poynting's theorem.* Integration of this expression over an arbitrary volume V bounded by a closed surface S and application of the divergence theorem yields

$$\iiint_V \mathscr{J}_c(\mathbf{r},t)\cdot\mathscr{E}(\mathbf{r},t)\,d^3r = -\oint\!\!\oint_S \mathscr{S}(\mathbf{r},t)\cdot\hat{n}\,da - \frac{d}{dt}\iiint_V \mathscr{U}'(\mathbf{r},t)\,d^3r , \tag{2.2.6}$$

where $\hat{n}$ is the unit outward normal vector to S. This is the *integral form of Poynting's theorem* and may be taken as a mathematical expression of the conservation of energy of the combined system of the medium and electromagnetic field. Because of this interpretation, the volume integral on the left-hand side of this equation is interpreted as the time rate of work being done on the conduction current in V by the electromagnetic field, the volume integral on the right-hand side of this equation is interpreted as the time rate of energy loss by the electromagnetic field in V, and the surface integral is interpreted as the time rate of field energy flow out of V across the surface S.

The usual (i.e., accepted) physical interpretation of the scalar quantity $\mathscr{U}'(\mathbf{r},t)$ is that it represents the total electromagnetic energy density of the coupled field-medium system at the space-time point $(\mathbf{r},t)$, measured in ergs per cm^3 in cgs units and joules per m^3 in MKS units. The separate scalar quantities $\mathscr{U}_e'(\mathbf{r},t)$ and $\mathscr{U}_m'(\mathbf{r},t)$ then represent the electric and magnetic energy densities of the coupled field-medium system. If one assumes that as $t\to-\infty$, $\mathscr{E}(-\infty) = \mathscr{H}(-\infty) = \mathbf{0}$ and that $\mathscr{U}'(-\infty) = 0$ everywhere in space, then (2.2.3) yields (this condition is clearly not satisfied by a monochromatic field and thus, the following pair of equations do not have unrestricted applicability)

$$\mathcal{U}'_e(\mathbf{r},t) = \left[\!\left[\frac{1}{4\pi}\right]\!\right] \int \mathcal{E}(\mathbf{r},t)\cdot\frac{\partial\mathcal{D}(\mathbf{r},t)}{\partial t}\,dt\ , \tag{2.2.7a}$$

$$\mathcal{U}'_m(\mathbf{r},t) = \left[\!\left[\frac{1}{4\pi}\right]\!\right] \int \mathcal{H}(\mathbf{r},t)\cdot\frac{\partial\mathcal{B}(\mathbf{r},t)}{\partial t}\,dt\ . \tag{2.2.7b}$$

The general interpretation of the Poynting vector $\mathcal{S}(\mathbf{r},t)$ is that at any point of observation of an electromagnetic field at which $|\mathcal{S}|$ is different from zero one can assert that electromagnetic energy is flowing in the direction of $\mathcal{S}$ such that across an elemental plane surface perpendicular to the direction of $\mathcal{S}$ at that point the rate of flow of energy in the field is $|\mathcal{S}|$ ergs per second per cm^2 (in cgs units) or joules per second per m^2 (in MKS units). However, it must be emphasized that the local quantities $\mathcal{U}'(\mathbf{r},t)$ and $\mathcal{S}(\mathbf{r},t)$ merely provide a physical interpretation (albeit a useful one) of the density and transfer of energy in the coupled electromagnetic field and medium system. The physically proper statement of the transfer of electromagnetic energy is embodied in Poynting's theorem given in either of (2.2.5 or 6). Even though the total flow of electromagnetic energy through a closed surface S may be correctly represented by $-\oint\!\!\oint_S \mathcal{S}\cdot\hat{n}\,da$, it cannot be definitely concluded that the time rate of energy flow at a point is uniquely given by the value of the Poynting vector $\mathcal{S}(\mathbf{r},t)$ at that point, for one may add to the Poynting vector any solenoidal vector field without affecting the statement of conservation of energy expressed by the integral form (2.2.6) of Poynting's theorem. Hence, there is no strictly valid justification for the accepted interpretation of the Poynting vector except that it is useful and seldom leads to erroneous results provided that proper care is taken in its application and interpretation. When using this interpretation, however, one should keep in mind that the flow of electromagnetic energy implied by the value of the Poynting vector $\mathcal{S}$ need not coincide with any intuitive notion of energy flow.

The classical interpretation of Poynting's theorem as a statement of the conservation of energy is then seen to depend to a considerable degree on hypothesis. As stated by *Stratton* [2.4], "A theory is not an absolute truth but a self-consistent analytical formulation of the relations governing a group of natural phenomena." With the acceptance of the validity of the macroscopic Maxwell theory, Poynting's theorem is clearly a valid, self-consistent relationship. In the absence of the conductive current density $\mathcal{J}_c$ (i.e. in a nonconducting medium) the differential form of Poynting's theorem becomes

$$\nabla\cdot\mathcal{S}(\mathbf{r},t) + \frac{\partial\mathcal{U}'(\mathbf{r},t)}{\partial t} = 0\ , \tag{2.2.8}$$

which has exactly the same form as the equation of continuity (2.1.3) for the conservation of charge. By analogy with this equation the current density is now the Poynting vector $\mathcal{S}(\mathbf{r},t)$ and the volume density of the conserved quantity is the total energy density $\mathcal{U}'(\mathbf{r},t)$ of the coupled electromagnetic field-medium system.

This final point is clearly illustrated by considering the explicit form of the energy density in a homogeneous, isotropic, locally linear, temporally dispersive medium. Since the magnetic field is taken here to satisfy the simple constitutive relation $\mathscr{B}(\mathbf{r},t) = \mu\mathscr{H}(\mathbf{r},t)$, one immediately obtains for the magnetic energy density that

$$\begin{aligned}\mathscr{U}'_m(\mathbf{r},t) &= \left[\!\!\left[\frac{1}{4\pi}\right]\!\!\right]\mu\int\mathscr{H}\cdot\frac{\partial\mathscr{H}}{\partial t}dt = \left[\!\!\left[\frac{1}{4\pi}\right]\!\!\right]\frac{1}{2}\mu\int\frac{\partial}{\partial t}[\mathscr{H}\cdot\mathscr{H}]\,dt \\ &= \left[\!\!\left[\frac{1}{4\pi}\right]\!\!\right]\frac{1}{2}\mu\mathscr{H}(\mathbf{r},t)\cdot\mathscr{H}(\mathbf{r},t) \\ &= \left[\!\!\left[\frac{1}{4\pi}\right]\!\!\right]\frac{1}{2}\mathscr{B}(\mathbf{r},t)\cdot\mathscr{H}(\mathbf{r},t)\ .\end{aligned} \tag{2.2.9}$$

For a general homogeneous, isotropic, locally linear, temporally dispersive medium satisfying the constitutive relation (2.1.2b) one has the multipole expansion [2.5, 6]

$$\mathscr{D}(\mathbf{r},t) = \varepsilon_0\mathscr{E}(\mathbf{r},t) + [\![4\pi]\!]\mathscr{P}(\mathbf{r},t) - [\![4\pi]\!]\nabla\cdot\mathscr{Q}(\mathbf{r},t) + \cdots, \tag{2.2.10}$$

where $\mathscr{P}(\mathbf{r},t)$ is the macroscopic polarization density vector and $\mathscr{Q}(\mathbf{r},t)$ is the macroscopic quadrupole moment density tensor of the temporally dispersive medium. If the medium response is dominated by the medium polarizability, then (2.1.2b) and (2.2.10) yield

$$\varepsilon_0\mathscr{E}(\mathbf{r},t) + [\![4\pi]\!]\mathscr{P}(\mathbf{r},t) = \int_{-\infty}^{\infty}\hat{\varepsilon}(t-t')\mathscr{E}(\mathbf{r},t')\,dt'$$

to a good degree of approximation, so that

$$\begin{aligned}\mathscr{P}(\mathbf{r},t) &= \left[\!\!\left[\frac{1}{4\pi}\right]\!\!\right]\int_{-\infty}^{\infty}[\hat{\varepsilon}(t-t') - \varepsilon_0\delta(t-t')]\mathscr{E}(\mathbf{r},t')\,dt' \\ &= \int_{-\infty}^{\infty}\hat{\chi}_e(t-t')\mathscr{E}(\mathbf{r},t')\,dt'\ .\end{aligned} \tag{2.2.11}$$

Here

$$[\![4\pi]\!]\hat{\chi}_e(t) = \hat{\varepsilon}(t) - \varepsilon_0\delta(t) \tag{2.2.12}$$

is the electric susceptibility of the medium. The electric energy density (2.2.7a) of the coupled field-medium system is then given by

$$\mathscr{U}'_e(\mathbf{r},t) = \left[\!\!\left[\frac{1}{4\pi}\right]\!\!\right]\frac{1}{2}\varepsilon_0\mathscr{E}(\mathbf{r},t)\cdot\mathscr{E}(\mathbf{r},t) + \int\mathscr{E}(\mathbf{r},t)\cdot\frac{\partial\mathscr{P}(\mathbf{r},t)}{\partial t}dt\ . \tag{2.2.13}$$

This expression explicitly states that a portion of the electric field energy resides in the dispersive medium. As for (2.2.7a), this equation does not have unrestricted applicability; it is valid if the field vanishes at $t \to -\infty$. Substitution of (2.2.9 and 13) combined with (2.2.4) into the differential form (2.2.5) of Poynting's theorem then gives

$$\nabla \cdot \mathcal{S}(\mathbf{r},t) = -\left[\!\left[\frac{1}{4\pi}\right]\!\right]\left\{\mu\mathcal{H}(\mathbf{r},t)\cdot\frac{\partial\mathcal{H}(\mathbf{r},t)}{\partial t} + \varepsilon_0\mathcal{E}(\mathbf{r},t)\cdot\frac{\partial\mathcal{E}(\mathbf{r},t)}{\partial t}\right.$$
$$\left. + [\![4\pi]\!]\mathcal{E}(\mathbf{r},t)\cdot\left(\mathcal{J}_c(\mathbf{r},t) + \frac{\partial\mathcal{P}(\mathbf{r},t)}{\partial t}\right)\right\} . \qquad (2.2.14)$$

This is the appropriate differential form of Poynting's theorem for a temporally dispersive polarizable medium with dispersive conductivity in the absence of external current sources. Since its derivation does not rely on any assumption of the field behavior as $t \to -\infty$, this expression then has general applicability. Furthermore, it is readily found that the same equation holds with an external current source if $\mathcal{J}_c(\mathbf{r},t)$ is replaced by the total current density $\mathcal{J}(\mathbf{r},t) = \mathcal{J}_c(\mathbf{r},t) + \mathcal{J}_{ext}(\mathbf{r},t)$.

It is evident that one may associate the first two terms on the right-hand side of (2.2.14) with the energy density (or, more specifically, the rate of change of the energy density) in the field alone and the last term with the electric energy density that is interacting with the dispersive medium. A certain portion of this interaction energy is "reactively" stored in the medium and the remainder is absorbed, giving the evolved heat in the absorptive medium. The total energy density in the field is then given by the sum of the first two terms plus the reactively stored portion of the last term. Unfortunately, this separation is found in the general case to be intractable analytically.

2.2.2 The Energy Density and Evolved Heat in a Dispersive and Absorptive Medium

Attention is now given to the precise physical definition of the expressions for the electric and magnetic energy densities $\mathcal{U}_e$ and $\mathcal{U}_m$ as well as to the evolved heat (or dissipation) $\mathcal{Q}$ in the electrodynamics of a dispersive and absorptive medium. Without the formal definitions given in (2.2.3a and 3b), the differential form of Poynting's theorem may be written as

$$-\nabla \cdot \mathcal{S}(\mathbf{r},t) = \left[\!\left[\frac{1}{4\pi}\right]\!\right]\left(\mathcal{E}(\mathbf{r},t)\cdot\frac{\partial\mathcal{D}(\mathbf{r},t)}{\partial t} + \mathcal{H}(\mathbf{r},t)\cdot\frac{\partial\mathcal{B}(\mathbf{r},t)}{\partial t}\right) + \mathcal{J}(\mathbf{r},t)\cdot\mathcal{E}(\mathbf{r},t) , \qquad (2.2.15)$$

where $\mathcal{J}(\mathbf{r},t) = \mathcal{J}_c(\mathbf{r},t) + \mathcal{J}_{ext}(\mathbf{r},t)$ is here the *total current density*.

In the simplest case of a nonmagnetic and nondispersive medium the constitutive relations give $\mathcal{B}(\mathbf{r},t) = \mu\mathcal{H}(\mathbf{r},t)$ and $\mathcal{D}(\mathbf{r},t) = \varepsilon\mathcal{E}(\mathbf{r},t)$, where μ and ε are simple scalar constants. In that case the quantity

$$
\begin{aligned}
\mathscr{U}_m(\mathbf{r},t) &= \left[\!\left[\frac{1}{4\pi}\right]\!\right] \int \mathscr{H}(\mathbf{r},t)\cdot\frac{\partial\mathscr{B}(\mathbf{r},t)}{\partial t}\,dt \\
&= \left[\!\left[\frac{1}{4\pi}\right]\!\right] \frac{\mu}{2} \int \frac{\partial}{\partial t}(\mathscr{H}(\mathbf{r},t)\cdot\mathscr{H}(\mathbf{r},t))\,dt \\
&= \left[\!\left[\frac{1}{4\pi}\right]\!\right] \frac{\mu}{2} |\mathscr{H}(\mathbf{r},t)|^2 \\
&= \left[\!\left[\frac{1}{4\pi}\right]\!\right] \frac{1}{2} \mathscr{H}(\mathbf{r},t)\cdot\mathscr{B}(\mathbf{r},t)
\end{aligned}
\tag{2.2.16}
$$

is immediately identified with the energy density of the magnetic field and is the same as that given by (2.2.7b), while the quantity

$$
\begin{aligned}
\mathscr{U}_e(\mathbf{r},t) &= \left[\!\left[\frac{1}{4\pi}\right]\!\right] \int \mathscr{E}(\mathbf{r},t)\cdot\frac{\partial\mathscr{D}(\mathbf{r},t)}{\partial t}\,dt \\
&= \left[\!\left[\frac{1}{4\pi}\right]\!\right] \frac{\varepsilon}{2} \int \frac{\partial}{\partial t}(\mathscr{E}(\mathbf{r},t)\cdot\mathscr{E}(\mathbf{r},t))\,dt \\
&= \left[\!\left[\frac{1}{4\pi}\right]\!\right] \frac{\varepsilon}{2} |\mathscr{E}(\mathbf{r},t)|^2 \\
&= \left[\!\left[\frac{1}{4\pi}\right]\!\right] \frac{1}{2} \mathscr{E}(\mathbf{r},t)\cdot\mathscr{D}(\mathbf{r},t)
\end{aligned}
\tag{2.2.17}
$$

is immediately identified with the energy density of the electric field and is the same as that given by (2.2.7a). The total electromagnetic energy density is then written as

$$
\begin{aligned}
\mathscr{U}(\mathbf{r},t) &= \mathscr{U}_e(\mathbf{r},t) + \mathscr{U}_m(\mathbf{r},t) \\
&= \left[\!\left[\frac{1}{4\pi}\right]\!\right] \frac{1}{2} [\varepsilon|\mathscr{E}(\mathbf{r},t)|^2 + \mu|\mathscr{H}(\mathbf{r},t)|^2] \ ,
\end{aligned}
\tag{2.2.18}
$$

and the Poynting vector $\mathscr{S}(\mathbf{r},t) = [\![c/4\pi]\!]\mathscr{E}(\mathbf{r},t) \times \mathscr{H}(\mathbf{r},t)$ has the physical meaning of the total electromagnetic energy flux across a unit area normal to the direction of $\mathscr{S}(\mathbf{r},t)$.

For a nonmagnetic medium with frequency-dependent dielectric permittivity $\varepsilon(\omega) = \varepsilon_r(\omega) + i\varepsilon_i(\omega)$ and a frequency-dependent electrical conductivity $\sigma(\omega) = \sigma_r(\omega) + i\sigma_i(\omega)$ the magnetic permeability is a real constant so that (2.2.16) remains valid for the energy density of the magnetic field and $\mathscr{U}'_m(\mathbf{r},t) = \mathscr{U}_m(\mathbf{r},t)$. In addition, the interpretation of the Poynting vector $\mathscr{S}(\mathbf{r},t)$ given in (2.2.2) as the total electromagnetic energy flux density remains intact. However, the expression given in (2.2.17) for $\mathscr{U}_e(\mathbf{r}, t)$ no longer represents the energy density of the electric field. Furthermore, the quantity $\mathscr{U}'_e(\mathbf{r}, t)$ defined in (2.2.7a) is now interpreted as a generalized electric energy density of the coupled field-medium system, as is clearly evident in (2.2.13) for a dispersive medium that is dominated by the medium polarizability. Nevertheless, a formal separation of this coupled field-medium quantity can be made by rewriting (2.2.3a) as [2.7]

$$\frac{\partial \mathscr{U}_e(\mathbf{r}, t)}{\partial t} + \mathscr{Q}(\mathbf{r}, t) = \left[\!\left[\frac{1}{4\pi}\right]\!\right] \mathscr{E}(\mathbf{r}, t) \cdot \frac{\partial \mathscr{D}(\mathbf{r}, t)}{\partial t} + \mathscr{J}_c(\mathbf{r}, t) \cdot \mathscr{E}(\mathbf{r}, t) \tag{2.2.19}$$

where $\mathscr{U}_e(\mathbf{r}, t)$ now represents just the electric energy density both in the field and reactively stored in the dispersive medium, and where $\mathscr{Q}(\mathbf{r}, t)$ represents the *evolved heat* or *dissipation* in the medium. Comparison with (2.2.3a) shows that

$$\frac{\partial \mathscr{U}_e}{\partial t} + \mathscr{Q} = \frac{\partial \mathscr{U}'_e}{\partial t} + \mathscr{J}_c \cdot \mathscr{E}$$

from which it is seen that $\mathscr{U}_e = \mathscr{U}'_e$ only in the absence of all loss mechanisms. With (2.2.19) Poynting's theorem (2.2.15) becomes

$$\frac{\partial}{\partial t}[\mathscr{U}_e(\mathbf{r}, t) + \mathscr{U}_m(\mathbf{r}, t)] + \mathscr{Q}(\mathbf{r}, t) = -\nabla \cdot \mathscr{S}(\mathbf{r}, t) - \mathscr{J}_{\text{ext}}(\mathbf{r}, t) \cdot \mathscr{E}(\mathbf{r}, t) \ . \tag{2.2.20}$$

All that remains now is to obtain separate expressions for $\mathscr{U}_e(\mathbf{r}, t)$ and $\mathscr{Q}(\mathbf{r}, t)$. To that end, substitution of (2.1.15 and 16) into (2.2.19) yields the general expression

$$\begin{aligned} \frac{\partial \mathscr{U}_e(\mathbf{r}, t)}{\partial t} + \mathscr{Q}(\mathbf{r}, t) &= \left[\!\left[\frac{1}{4\pi}\right]\!\right] \mathscr{E}(\mathbf{r}, t) \cdot \frac{\partial}{\partial t} \sum_{n=0}^{\infty} \varepsilon^{(n)} \frac{\partial^n \mathscr{E}(\mathbf{r}, t)}{\partial t^n} + \mathscr{E}(\mathbf{r}, t) \cdot \sum_{n=0}^{\infty} \sigma^{(n)} \frac{\partial^n \mathscr{E}(\mathbf{r}, t)}{\partial t^n} \\ &= \sigma^{(0)} \mathscr{E}^2(\mathbf{r}, t) + \frac{1}{2}\left(\left[\!\left[\frac{1}{4\pi}\right]\!\right] \varepsilon^{(0)} + \sigma^{(1)}\right) \frac{\partial \mathscr{E}^2(\mathbf{r}, t)}{\partial t} \\ &\quad + \frac{1}{2}\left(\left[\!\left[\frac{1}{4\pi}\right]\!\right] \varepsilon^{(1)} + \sigma^{(2)}\right) \frac{\partial^2 \mathscr{E}^2(\mathbf{r}, t)}{\partial t^2} \\ &\quad - \left(\left[\!\left[\frac{1}{4\pi}\right]\!\right] \varepsilon^{(1)} + \sigma^{(2)}\right) \left(\frac{\partial \mathscr{E}(\mathbf{r}, t)}{\partial t}\right)^2 + \cdots , \end{aligned} \tag{2.2.21}$$

where the (real-valued) coefficients $\sigma^{(n)}$ and $\varepsilon^{(n)}$ are proportional to the n'th order moments of the conductivity and dielectric permittivity functions, respectively, see (2.1.15b and 16b). In general, one then has that

$$\mathscr{U}_e(\mathbf{r}, t) = \frac{1}{2}\left[\!\left[\frac{1}{4\pi}\right]\!\right] \alpha_1 \mathscr{E}^2(\mathbf{r}, t) + \cdots , \tag{2.2.22}$$

$$\mathscr{Q}(\mathbf{r}, t) = \sigma^{(0)} \mathscr{E}^2(\mathbf{r}, t) + \frac{1}{2}\left[\!\left[\frac{1}{4\pi}\right]\!\right] \beta_1 \frac{\partial \mathscr{E}^2(\mathbf{r}, t)}{\partial t} + \cdots , \tag{2.2.23}$$

with

$$\alpha_1 + \beta_1 = \varepsilon^{(0)} + [\![4\pi]\!]\sigma^{(1)} , \tag{2.2.24}$$

and so-on for higher-order coefficients. However, in general, the coefficients α_1 and β_1 cannot be expressed separately in terms of the dielectric permittivity and conductivity, even for the simple case of a nondispersive medium for which $\varepsilon = \varepsilon^{(0)}$ and $\sigma = \sigma^{(0)}$; the single exception is that of a strict electrostatic field. If the electric field vector $\mathscr{E}(\mathbf{r}, t)$ varies sufficiently slow with time such that $|\beta_1|/T \ll \sigma^{(0)}$, where T is the characteristic time for the variation of the field, then (2.2.23) for the evolved heat takes the form

$$\mathscr{Q}(\mathbf{r}, t) \cong \sigma^{(0)}\mathscr{E}^2(\mathbf{r}, t) \ ; \qquad \frac{|\beta_1|}{T} \ll \sigma^{(0)} \ .$$

In addition, if $|\alpha_1| \gg |\beta_1|$, then (2.2.22) for the electric energy density becomes

$$\mathscr{U}_e(\mathbf{r}, t) \cong \frac{1}{2}\left[\!\!\left[\frac{1}{4\pi}\right]\!\!\right]\varepsilon^{(0)}\mathscr{E}^2(\mathbf{r}, t) \ ; \qquad \frac{|\beta_1|}{T} \ll \sigma^{(0)} \ , \qquad |\alpha_1| \gg |\beta_1| \ .$$

However, it is by no means certain that the condition $|\alpha_1| \gg |\beta_1|$ will be satisfied and it is entirely possible that $|\alpha_1| \lesssim |\beta_1|$.

It is therefore clear that *one cannot, in general, express the electric energy density* $\mathscr{U}_e(\mathbf{r}, t)$ *and the dissipation* $\mathscr{Q}(\mathbf{r}, t)$ *separately in terms of the dielectric permittivity and electrical conductivity of a dispersive medium.* To unambiguously determine these quantities it is necessary to employ a specific physical model of the dispersive medium, [2.8, 9] e.g., the equation of motion at the microscopic level.

2.2.3 Complex Time-Harmonic Form of Poynting's Theorem

For a completely time-harmonic field of fixed angular frequency ω, each real-valued field quantity may be expressed in the form, cf. (2.1.23–26),

$$\mathscr{A}(\mathbf{r}, t) = \mathrm{Re}\{\mathbf{A}(\mathbf{r})e^{-i\omega t}\} = \tfrac{1}{2}[\mathbf{A}(\mathbf{r})e^{-i\omega t} + \mathbf{A}^*(\mathbf{r})e^{i\omega t}] , \tag{2.2.25}$$

where $\mathbf{A}(\mathbf{r})$ is in general complex-valued and is related to the positive frequency component of the temporal frequency spectrum of $\mathscr{A}(\mathbf{r}, t)$, cf. (2.1.27). The scalar product of two such real-valued vector fields with the same oscillation frequency ω is then given by

$$\mathscr{A}(\mathbf{r}, t)\cdot\mathscr{B}(\mathbf{r}, t) = \tfrac{1}{4}[\mathbf{A}(\mathbf{r})e^{-i\omega t} + \mathbf{A}^*(\mathbf{r})e^{i\omega t}]\cdot[\mathbf{B}(\mathbf{r})e^{-i\omega t} + \mathbf{B}^*(\mathbf{r})e^{i\omega t}]$$

$$= \tfrac{1}{2}\mathrm{Re}\{\mathbf{A}(\mathbf{r})\cdot\mathbf{B}^*(\mathbf{r}) + \mathbf{A}(\mathbf{r})\cdot\mathbf{B}(\mathbf{r})e^{-i2\omega t}\} \tag{2.2.26a}$$

$$= \tfrac{1}{2}\mathrm{Re}\{\mathbf{A}^*(\mathbf{r})\cdot\mathbf{B}(\mathbf{r}) + \mathbf{A}(\mathbf{r})\cdot\mathbf{B}(\mathbf{r})e^{-i2\omega t}\} , \tag{2.2.26b}$$

and their vector product is given by

$$\mathscr{A}(\mathbf{r},t) \times \mathscr{B}(\mathbf{r},t) = \tfrac{1}{4}[\mathbf{A}(\mathbf{r})e^{-i\omega t} + \mathbf{A}^*(\mathbf{r})e^{i\omega t}] \times [\mathbf{B}(\mathbf{r})e^{-i\omega t} + \mathbf{B}^*(\mathbf{r})e^{i\omega t}]$$

$$= \tfrac{1}{2}\mathbb{R}\mathrm{e}\{\mathbf{A}(\mathbf{r}) \times \mathbf{B}^*(\mathbf{r}) + \mathbf{A}(\mathbf{r}) \times \mathbf{B}(\mathbf{r})e^{-i2\omega t}\} \tag{2.2.27a}$$

$$= \tfrac{1}{2}\mathbb{R}\mathrm{e}\{\mathbf{A}^*(\mathbf{r}) \times \mathbf{B}(\mathbf{r}) + \mathbf{A}(\mathbf{r}) \times \mathbf{B}(\mathbf{r})e^{-i2\omega t}\} . \tag{2.2.27b}$$

The *time-average of a periodic function* $\mathscr{A}(\mathbf{r},t)$ is simply defined by the integral

$$\langle \mathscr{A}(\mathbf{r},t) \rangle \equiv \frac{1}{T} \int_{-T/2}^{T/2} \mathscr{A}(\mathbf{r},t)\, dt , \tag{2.2.28}$$

where T is the fundamental period of oscillation of the function; for an aperiodic function one takes the limit as $T \to \infty$ in the above expression. If $\mathscr{A}(\mathbf{r},t)$ is a simple harmonic function of time with fixed angular frequency ω, then $T = 2\pi/\omega$ and its time-average is zero. The time-average of the scalar and vector products of two such harmonic vector fields with the same oscillation frequency do not vanish, however, as seen from (2.2.26 and 27), and are given by

$$\langle \mathscr{A}(\mathbf{r},t) \cdot \mathscr{B}(\mathbf{r},t) \rangle = \tfrac{1}{2}\mathbb{R}\mathrm{e}\{\mathbf{A}(\mathbf{r}) \cdot \mathbf{B}^*(\mathbf{r})\}$$

$$= \tfrac{1}{2}\mathbb{R}\mathrm{e}\{\mathbf{A}^*(\mathbf{r}) \cdot \mathbf{B}(\mathbf{r})\} , \tag{2.2.29a}$$

$$\langle \mathscr{A}(\mathbf{r},t) \times \mathscr{B}(\mathbf{r},t) \rangle = \tfrac{1}{2}\mathbb{R}\mathrm{e}\{\mathbf{A}(\mathbf{r}) \times \mathbf{B}^*(\mathbf{r})\}$$

$$= \tfrac{1}{2}\mathbb{R}\mathrm{e}\{\mathbf{A}^*(\mathbf{r}) \times \mathbf{B}(\mathbf{r})\} . \tag{2.2.29b}$$

The time-average of the Poynting vector for a time-harmonic field is then given by

$$\langle \mathscr{S}(\mathbf{r},t) \rangle = \left[\!\left[\frac{c}{4\pi}\right]\!\right] \langle \mathscr{E}(\mathbf{r},t) \times \mathscr{H}(\mathbf{r},t) \rangle$$

$$= \frac{1}{2}\left[\!\left[\frac{c}{4\pi}\right]\!\right] \mathbb{R}\mathrm{e}\{\mathbf{E}(\mathbf{r}) \times \mathbf{H}^*(\mathbf{r})\}$$

$$= \mathbb{R}\mathrm{e}\{S(\mathbf{r})\} , \tag{2.2.30}$$

where

$$\mathbf{S}(\mathbf{r}) \equiv \frac{1}{2}\left[\!\left[\frac{c}{4\pi}\right]\!\right] \mathbf{E}(\mathbf{r}) \times \mathbf{H}^*(\mathbf{r}) \tag{2.2.31}$$

is the *complex Poynting vector* of the time-harmonic field. Notice that some authors choose to leave the factor $\frac{1}{2}$ in (2.2.30) when MKS units are employed, in which case $\mathbf{S} = \mathbf{E} \times \mathbf{H}^*$ and $\langle \mathscr{S} \rangle = \frac{1}{2}\mathbb{R}\mathrm{e}\{\mathbf{S}\}$.

Consider now the complex, time-harmonic differential form of Maxwell's equations, as given in (2.1.28), whose curl relations may be written as

$$\nabla \times \mathbf{E}(\mathbf{r}) = \left[\left[\frac{1}{c}\right]\right] \mathrm{i}\omega \mathbf{B}(\mathbf{r}) \ , \tag{2.2.32a}$$

$$\nabla \times \mathbf{H}(\mathbf{r}) = -\left[\left[\frac{1}{c}\right]\right] \mathrm{i}\omega\varepsilon(\omega)\mathbf{E}(\mathbf{r}) + \left[\left[\frac{4\pi}{c}\right]\right] \mathbf{J}_c(\mathbf{r}) \ , \tag{2.2.32b}$$

in the absence of any external current source. The complex conjugate of the second curl relation gives

$$\mathbf{J}_c^*(\mathbf{r}) = \left[\left[\frac{c}{4\pi}\right]\right] \nabla \times \mathbf{H}^*(\mathbf{r}) - \left[\left[\frac{1}{4\pi}\right]\right] \mathrm{i}\omega\varepsilon^*(\omega)\mathbf{E}^*(\mathbf{r}) \ ,$$

where ω is assumed here to be real-valued. The scalar product of this expression with the complex electric field vector $\mathbf{E}(\mathbf{r})$ yields, with the first curl relation (2.2.32a),

$$\begin{aligned} \mathbf{J}_c^* \cdot \mathbf{E} &= \left[\left[\frac{c}{4\pi}\right]\right] \mathbf{E} \cdot (\nabla \times \mathbf{H}^*) - \left[\left[\frac{1}{4\pi}\right]\right] \mathrm{i}\omega \mathbf{E} \cdot \mathbf{D}^* \\ &= -\left[\left[\frac{c}{4\pi}\right]\right] \nabla \cdot (\mathbf{E} \times \mathbf{H}^*) + \left[\left[\frac{c}{4\pi}\right]\right] \mathbf{H}^* \cdot (\nabla \times \mathbf{E}) - \left[\left[\frac{1}{4\pi}\right]\right] \mathrm{i}\omega \mathbf{E} \cdot \mathbf{D}^* \\ &= -\left[\left[\frac{c}{4\pi}\right]\right] \nabla \cdot (\mathbf{E} \times \mathbf{H}^*) + \left[\left[\frac{1}{4\pi}\right]\right] \mathrm{i}\omega(\mathbf{B} \cdot \mathbf{H}^* - \mathbf{E} \cdot \mathbf{D}^*) \ . \end{aligned}$$

With the definition (2.2.31) of the complex Poynting vector one finally obtains

$$\tfrac{1}{2}\mathbf{J}_c^*(\mathbf{r}) \cdot \mathbf{E}(\mathbf{r}) = -\nabla \cdot \mathbf{S}(\mathbf{r}) - 2\mathrm{i}\omega[u_e(\mathbf{r}) - u_m(\mathbf{r})] \tag{2.2.33}$$

which is the analog of the differential form (2.2.6) of Poynting's theorem for time-harmonic fields. The quantity $u_m(\mathbf{r})$, defined as

$$u_m(\mathbf{r}) \equiv \frac{1}{4}\left[\left[\frac{1}{4\pi}\right]\right] \mathbf{B}(\mathbf{r}) \cdot \mathbf{H}^*(\mathbf{r}) \tag{2.2.34}$$

is commonly referred to as the "*harmonic magnetic energy density*", and the quantity $u_e(\mathbf{r})$, defined as

$$u_e(\mathbf{r}) \equiv \frac{1}{4}\left[\left[\frac{1}{4\pi}\right]\right] \mathbf{E}(\mathbf{r}) \cdot \mathbf{D}^*(\mathbf{r}) \tag{2.2.35}$$

is commonly referred to as the "*harmonic electric energy density*". However, neither of these two defined quantities $u_m(\mathbf{r})$ and $u_e(\mathbf{r})$ can be rigorously related to the respective magnetic and electric energy densities [i.e., $\mathscr{U}_m'(\mathbf{r}, t) = \mathscr{U}_m(\mathbf{r}, t)$

and $\mathscr{U}_e(\mathbf{r}, t)$] of the coupled field-medium system since (2.2.7a and 7b) are not strictly applicable for a time-harmonic field. If one ignores this fact (as is commonly done in many treatments) then (2.2.9) immediately yields the relation $\langle \mathscr{U}_m(\mathbf{r}, t)\rangle = \mathbb{R}\mathrm{e}\{u_m(\mathbf{r})\}$ which, although physically appealing, is simply not rigorously valid. Further difficulties arise if one continues to push this argument for the electric energy density through (2.2.7a). Nevertheless, (2.2.3a and 3b) are applicable for the time-harmonic case and provide a partial connection of the complex densities (2.2.34 and 35) with physical reality. The time-average of (2.2.3b) gives

$$\left\langle \frac{\partial \mathscr{U}_m(\mathbf{r}, t)}{\partial t} \right\rangle = \left[\!\left[\frac{1}{4\pi}\right]\!\right] \left\langle \mathscr{H}(\mathbf{r}, t) \cdot \frac{\partial \mathscr{B}(\mathbf{r}, t)}{\partial t} \right\rangle$$

$$= \left[\!\left[\frac{1}{4\pi}\right]\!\right] \frac{\mu}{4} \langle (\mathbf{H}e^{-i\omega t} + \mathbf{H}^* e^{i\omega t}) \cdot (-i\omega \mathbf{H} e^{-i\omega t} + i\omega \mathbf{H}^* e^{i\omega t}) \rangle$$

so that

$$\left\langle \frac{\partial \mathscr{U}_m(\mathbf{r}, t)}{\partial t} \right\rangle = \left[\!\left[\frac{1}{4\pi}\right]\!\right] \frac{\mu}{4} (i\omega \mathbf{H} \cdot \mathbf{H}^* - i\omega \mathbf{H}^* \cdot \mathbf{H}) = 0 \ , \tag{2.2.36}$$

which simply states that the time-average magnetic energy density is constant for a time-harmonic field when the magnetic permeability μ is real-valued. Unfortunately, this does not provide any connection between $u_m(\mathbf{r})$ and $\langle \partial \mathscr{U}_m(\mathbf{r}, t)/\partial t \rangle$; however, it is interesting to note that the intermediate expression appearing just prior to (2.2.36) is equal to $2\,\mathbb{R}\mathrm{e}\{i\omega u_m(\mathbf{r})\}$. The time-average of (2.2.3a) gives

$$\left\langle \frac{\partial \mathscr{U}'_e(\mathbf{r}, t)}{\partial t} \right\rangle = \left[\!\left[\frac{1}{4\pi}\right]\!\right] \left\langle \mathscr{E}(\mathbf{r}, t) \cdot \frac{\partial \mathscr{D}(\mathbf{r}, t)}{\partial t} \right\rangle$$

$$= \left[\!\left[\frac{1}{4\pi}\right]\!\right] \frac{1}{4} \langle (\mathbf{E}e^{-i\omega t} + \mathbf{E}^* e^{i\omega t}) \cdot (-i\omega \varepsilon(\omega) \mathbf{E} e^{-i\omega t} + i\omega \varepsilon^*(\omega) \mathbf{E}^* e^{i\omega t}) \rangle$$

$$= \left[\!\left[\frac{1}{4\pi}\right]\!\right] \frac{1}{4} (i\omega \mathbf{E} \cdot \mathbf{D}^* - i\omega \mathbf{E}^* \cdot \mathbf{D})$$

$$= \left[\!\left[\frac{1}{4\pi}\right]\!\right] \frac{1}{2} \mathbb{R}\mathrm{e}\{i\omega \mathbf{E}(\mathbf{r}) \cdot \mathbf{D}^*(\mathbf{r})\} = 2\,\mathbb{R}\mathrm{e}\{i\omega u_e(\mathbf{r})\} \ , \tag{2.2.37}$$

so that

$$\mathbb{I}\mathrm{m}\{u_e(\mathbf{r})\} = -\frac{1}{2\omega} \left\langle \frac{\partial \mathscr{U}'_e(\mathbf{r}, t)}{\partial t} \right\rangle \ , \tag{2.2.38}$$

and the imaginary part of the harmonic electric energy density is related to the time-average of the time-rate of change of the electric energy density of the coupled field-medium system. Unfortunately, no relationship of this type results for the real part of $u_e(\mathbf{r})$.

Integration of (2.2.33) over an arbitrary volume V bounded by a closed surface S and application of the divergence theorem yields

$$\frac{1}{2}\iiint_V \mathbf{J}_c^*(\mathbf{r})\cdot\mathbf{E}(\mathbf{r})\,d^3r = -\oiint_S \mathbf{S}(\mathbf{r})\cdot\hat{n}\,da - 2i\omega\iiint_V [u_e(\mathbf{r}) - u_m(\mathbf{r})]\,d^3r\ , \tag{2.2.39}$$

which is the analog of the integral form (2.2.6) of Poynting's theorem for time-harmonic fields. The time-harmonic forms (2.2.33 and 39) of Poynting's theorem are, just as are their general counterparts (2.2.5 and 6), a rigorous consequence of Maxwell's equations and are therefore self-consistent relationships within the framework of classical electrodynamics. As before, it is the physical interpretation of the quantities appearing in these relations that depend to a certain degree on hypothesis.

The real part of (2.2.39) is, with (2.2.30),

$$\begin{aligned}\frac{1}{2}\iiint_V &\mathrm{Re}\{\mathbf{J}_c^*(\mathbf{r})\cdot\mathbf{E}(\mathbf{r})\}\,d^3r\\ &= -\oiint_S \langle\mathscr{S}(\mathbf{r},t)\rangle\cdot\hat{n}\,da + 2\omega\iiint_V \mathrm{Im}\{u_e(\mathbf{r}) - u_m(\mathbf{r})\}\,d^3r\ ,\end{aligned} \tag{2.2.40}$$

which is just a statement of the conservation of energy in the coupled field-medium system. The imaginary part of (2.2.39) is given by

$$\begin{aligned}\frac{1}{2}\iiint_V &\mathrm{Im}\{\mathbf{J}_c^*(\mathbf{r})\cdot\mathbf{E}(\mathbf{r})\}\,d^3r\\ &= -\oiint_S \mathrm{Im}\{\mathbf{S}(\mathbf{r})\}\cdot\hat{n}\,da + 2\omega\iiint_V \mathrm{Re}\{u_e(\mathbf{r}) - u_m(\mathbf{r})\}\,d^3r\ .\end{aligned} \tag{2.2.41}$$

From (2.1.26a)

$$\mathbf{J}_c^*(\mathbf{r})\cdot\mathbf{E}(\mathbf{r}) = \sigma^*(\omega)\mathbf{E}^*(\mathbf{r})\cdot\mathbf{E}(\mathbf{r})\ ,$$

so that the real and imaginary parts are expressible as

$$\mathrm{Re}\{\mathbf{J}_c^*(\mathbf{r})\cdot\mathbf{E}(\mathbf{r})\} = \sigma_r(\omega)|\mathbf{E}(\mathbf{r})|^2\ ,$$

$$\mathrm{Im}\{\mathbf{J}_c^*(\mathbf{r})\cdot\mathbf{E}(\mathbf{r})\} = -\sigma_i(\omega)|\mathbf{E}(\mathbf{r})|^2\ .$$

Again, since the magnetic permeability μ is real-valued, then

$$\mathrm{Re}\{u_m(\mathbf{r})\} = u_m(\mathbf{r})\ ,$$

$$\mathrm{Im}\{u_m(\mathbf{r})\} = 0\ .$$

Finally, from (2.2.35)

$$u_e(\mathbf{r}) = \frac{1}{4}\left[\frac{1}{4\pi}\right]\varepsilon^*(\omega)\mathbf{E}^*(\mathbf{r})\cdot\mathbf{E}(\mathbf{r})\ ,$$

so that the real and imaginary parts are given by

$$\mathbb{R}\mathrm{e}\{u_e(\mathbf{r})\} = \frac{1}{4}\left[\!\left[\frac{1}{4\pi}\right]\!\right]\varepsilon_r(\omega)|\mathbf{E}(\mathbf{r})|^2 ,$$

$$\mathbb{I}\mathrm{m}\{u_e(\mathbf{r})\} = -\frac{1}{4}\left[\!\left[\frac{1}{4\pi}\right]\!\right]\varepsilon_i(\omega)|\mathbf{E}(\mathbf{r})|^2 .$$

Substitution of these expressions into (2.2.40) yields

$$\begin{aligned} -\oint\!\!\oint_S \langle\mathscr{S}(\mathbf{r},t)\rangle\cdot\hat{n}\,da &= \frac{\omega}{2}\left[\!\left[\frac{1}{4\pi}\right]\!\right]\iiint_V \left[\varepsilon_i(\omega) + [\![4\pi]\!]\frac{\sigma_r(\omega)}{\omega}\right]|\mathbf{E}(\mathbf{r})|^2\,d^3r \\ &= \frac{\omega}{2}\left[\!\left[\frac{1}{4\pi}\right]\!\right]\iiint_V \mathbb{I}\mathrm{m}\{\varepsilon_c(\omega)\}|\mathbf{E}(\mathbf{r})|^2\,d^3r . \end{aligned} \tag{2.2.42}$$

The left-hand side of this equation is interpreted as the time-average electromagnetic-power flow into the region V across the boundary surface S, so that the right-hand side may be interpreted as the time-average rate of energy that is dissipated in the medium contained within the region V. In addition, substitution of the above expressions into (2.2.41) yields

$$\begin{aligned} &\oint\!\!\oint_S \mathbb{I}\mathrm{m}\{\mathbf{S}(\mathbf{r})\}\cdot\hat{n}\,da \\ &\quad = 2\omega\iiint_V\left[u_m(\mathbf{r}) - \frac{1}{4}\left[\!\left[\frac{1}{4\pi}\right]\!\right]\left(\varepsilon_r(\omega) - [\![4\pi]\!]\frac{\sigma_i(\omega)}{\omega}\right)|\mathbf{E}(\mathbf{r})|^2\right]d^3r \\ &\quad = 2\omega\iiint_V\left[u_m(\mathbf{r}) - \frac{1}{4}\left[\!\left[\frac{1}{4\pi}\right]\!\right]\mathbb{R}\mathrm{e}\{\varepsilon_c(\omega)\}|\mathbf{E}(\mathbf{r})|^2\right]d^3r . \end{aligned} \tag{2.2.43}$$

The right-hand side of this expression may be interpreted as the time-average of the reactive or stored energy within the medium enclosed by the surface S. It would seem that these interpretations provide separate identifications of both the time-average electric energy density $\mathscr{U}_e(\mathbf{r},t)$ and dissipation $\mathscr{Q}(\mathbf{r},t)$ for a time-harmonic field in terms of the dielectric permittivity and conductivity of the dispersive medium. However, there is no rigorous connection between the integrands appearing on the right-hand sides of (2.2.42 and 43) and the time-average quantities $\langle\mathscr{Q}(\mathbf{r},t)\rangle$, $\langle\mathscr{U}_e(\mathbf{r},t)\rangle$ and $\langle\mathscr{U}_m(\mathbf{r},t)\rangle$; as a consequence, these identifications cannot be made, in agreement with the conclusion of Sect. 2.2.2.

For an attenuative medium, the time-average electromagnetic power flow into any given region V in the medium across the boundary surface S of V must be positive for any nonzero field so that

$$-\oint\!\!\oint_S \langle\mathscr{S}(\mathbf{r},t)\rangle\cdot\hat{n}\,da > 0 ,$$

and (2.2.42) then gives

$$\omega \iiint_V \mathbb{I}\mathrm{m}\{\varepsilon_c(\omega)\} |\mathbf{E}(\mathbf{r})|^2 d^3r > 0 \ ,$$

which is valid for all nonstatic fields ($\omega \neq 0$)[1]. Since this inequality holds for an arbitrary region V in the medium, then $\mathbb{I}\mathrm{m}\{\varepsilon_c(\omega)\} > 0$ and consequently

$$\varepsilon_i(\omega) + [\![4\pi]\!]\frac{\sigma_r(\omega)}{\omega} > 0 \ , \tag{2.2.44}$$

where ω is the real-valued angular frequency of the (nonstatic) time-harmonic field so that $\omega > 0$. This general inequality allows for amplification in one medium parameter provided that it is exceeded by the loss associated with the other medium parameter. If the dielectric permittivity and the medium conductivity are separately attenuative, then

$$\begin{aligned} &\varepsilon_i(\omega) > 0 \ , \\ &\sigma_r(\omega) > 0 \ , \end{aligned} \tag{2.2.45}$$

for all finite, positive $\omega > 0$. However, the signs for the real part $\varepsilon_r(\omega)$ of the dielectric permittivity and the imaginary part $\sigma_i(\omega)$ of the medium conductivity are not subject to any physical restriction. They must, nevertheless, satisfy the appropriate dispersion relations that are required by causality.

2.3 The Harmonic Electromagnetic Plane Wave Field

The simplest electromagnetic field is that of a harmonic electromagnetic plane wave for which the complex field vectors appearing in (2.1.24) are given by

$$\mathbf{E}(\mathbf{r}) = \mathbf{e}e^{i\tilde{\mathbf{k}}\cdot\mathbf{r}} \ , \tag{2.3.1a}$$

$$\mathbf{B}(\mathbf{r}) = \mathbf{b}e^{i\tilde{\mathbf{k}}\cdot\mathbf{r}} \ , \tag{2.3.1b}$$

where $\mathbf{e}$ and $\mathbf{b}$ are fixed complex-valued vectors (i.e., are both independent of the coordinate position $\mathbf{r}$). The *surfaces of constant phase* for this field are then given by $\mathbb{R}\mathrm{e}\{\tilde{\mathbf{k}}\cdot\mathbf{r}) = \text{constant}$ which yields a family of parallel plane surfaces whose normal vector is parallel to $\mathbb{R}\mathrm{e}\{\tilde{\mathbf{k}}\}$. Substitution of either of these expressions for the complex field vectors into the appropriate vector Helmholtz equation (2.1.35) then results in the identification

$$\tilde{\mathbf{k}}\cdot\tilde{\mathbf{k}} = k_0^2 n^2(\omega) \ ,$$

[1] There is no loss for a strictly static field, as is clearly evident from (2.2.42).

where $k_0 = \omega/c$ is the vacuum wave number of the field. Let $\tilde{\mathbf{k}} = \tilde{k}\hat{s}$ where $\hat{s}$ is a (real-valued) unit vector in the direction of the wavevector $\tilde{\mathbf{k}}$, and where

$$\tilde{k}(\omega) = k_0 n(\omega) \tag{2.3.2}$$

is the *complex wave number* of the field in the homogeneous, isotropic, linear, temporally dispersive medium.

Substitution of the field vectors (2.3.1) into their appropriate divergence relation of the complex time-harmonic differential form of Maxwell's equations (2.1.28) then yields the *transversality relation*

$$\hat{s}\cdot\mathbf{e} = \hat{s}\cdot\mathbf{b} = 0 \ . \tag{2.3.3}$$

Furthermore, the curl relation (2.1.28b) gives

$$\tilde{\mathbf{k}} \times \mathbf{e} = \left[\!\left[\frac{1}{c}\right]\!\right]\omega\mathbf{b}$$

so that

$$\mathbf{b} = [\![c]\!]\frac{n(\omega)}{c}\hat{s} \times \mathbf{e} \ , \tag{2.3.4}$$

while the curl relation (2.1.28d) yields

$$\tilde{\mathbf{k}} \times \mathbf{b} = -\left[\!\left[\frac{1}{c}\right]\!\right]\omega\mu\varepsilon_c(\omega)\mathbf{e}$$

so that

$$\mathbf{e} = -\left[\!\left[\frac{1}{c}\right]\!\right]\frac{c}{n(\omega)}\hat{s} \times \mathbf{b} \ . \tag{2.3.5}$$

These three equations (2.3.3–5) show that the set of vectors $\{\mathbf{e}, \mathbf{b}, \hat{s}\}$ form a right-handed orthogonal triad at any point of space. The field vectors **e** and **b** therefore lie in the surfaces of constant phase and their complex-amplitudes are related by

$$b = [\![c]\!]\frac{n(\omega)}{c}\mathrm{e} \tag{2.3.6}$$

at all points of space. The *complex intrinsic impedance* of the dispersive medium is defined as the ratio of the complex electric field amplitude to the complex magnetic field intensity amplitude $h = b/\mu$, so that with (2.3.6) and (2.1.34),

$$\eta(\omega) \equiv \frac{\mathrm{e}}{h} = \mu\frac{\mathrm{e}}{b} = \left(\frac{\mu}{\varepsilon_c(\omega)}\right)^{1/2} \ . \tag{2.3.7}$$

For free space the intrinsic impedance is given by $\eta_0 = (\mu_0/\varepsilon_0)^{1/2} = 376.73\Omega$ in MKS units while it is unity in cgs units.

From (2.1.22, 34) and (2.3.2), the complex wave number of the harmonic electromagnetic plane wave field is found to be given by

$$\tilde{k}(\omega) = \left[\!\left[\frac{1}{c}\right]\!\right]\omega(\mu\varepsilon_c(\omega))^{1/2}$$

$$= \left[\!\left[\frac{1}{c}\right]\!\right]\omega\left\{\mu\left[\varepsilon(\omega) + \mathrm{i}[\![4\pi]\!]\frac{\sigma(\omega)}{\omega}\right]\right\}^{1/2} . \tag{2.3.8}$$

Let

$$\tilde{k}(\omega) = \beta(\omega) + \mathrm{i}\alpha(\omega) , \tag{2.3.9}$$

where $\beta(\omega) = \mathbb{R}\mathrm{e}\{\tilde{k}(\omega)\}$ is the *propagation factor* and $\alpha(\omega) = \mathbb{I}\mathrm{m}\{\tilde{k}(\omega)\}$ is the *attenuation factor* for the harmonic plane wave of frequency ω in the dispersive medium. Since $\varepsilon(\omega) = \varepsilon_r(\omega) + \mathrm{i}\varepsilon_i(\omega)$ and $\sigma(\omega) = \sigma_r(\omega) + \mathrm{i}\sigma_i(\omega)$, then the real and imaginary parts of the complex permittivity are respectively given by

$$\varepsilon_c'(\omega) = \mathbb{R}\mathrm{e}\{\varepsilon_c(\omega)\} = \varepsilon_r(\omega) - [\![4\pi]\!]\frac{\sigma_i(\omega)}{\omega} , \tag{2.3.10a}$$

$$\varepsilon_c''(\omega) = \mathbb{I}\mathrm{m}\{\varepsilon_c(\omega)\} = \varepsilon_i(\omega) + [\![4\pi]\!]\frac{\sigma_r(\omega)}{\omega} . \tag{2.3.10b}$$

The real and imaginary parts of the complex wave number (2.3.8) are then determined from the relation

$$\beta^2 - \alpha^2 + 2\mathrm{i}\alpha\beta = \left[\!\left[\frac{1}{c^2}\right]\!\right]\omega^2\mu(\varepsilon_c' + \mathrm{i}\varepsilon_c'') ,$$

which, upon equating real and imaginary parts, yields the pair of relations

$$\beta^2 - \alpha^2 = \left[\!\left[\frac{1}{c^2}\right]\!\right]\omega^2\mu\varepsilon_c' ,$$

$$2\alpha\beta = \left[\!\left[\frac{1}{c^2}\right]\!\right]\omega^2\mu\varepsilon_c'' .$$

Upon solving the last equation for β and substituting in the preceding equation, are obtains the fourth-order polynomial equation

$$\alpha^4 + \left[\!\left[\frac{1}{c^2}\right]\!\right]\omega^2\mu\varepsilon_c'\alpha^2 - \left[\!\left[\frac{1}{c^4}\right]\!\right]\frac{1}{4}\omega^4\mu^2\varepsilon_c''^2 = 0$$

with the solution (assuming that $\varepsilon_c' > 0$)

$$\alpha^2 = \left[\frac{1}{c^2}\right] \frac{1}{2} \omega^2 \mu \varepsilon_c' \left\{ \left[1 + \left(\frac{\varepsilon_c''}{\varepsilon_c'} \right)^2 \right]^{1/2} - 1 \right\},$$

and consequently

$$\beta^2 = \left[\frac{1}{c^2}\right] \omega^2 \mu \varepsilon_c' + \alpha^2$$

$$= \left[\frac{1}{c^2}\right] \frac{1}{2} \omega^2 \mu \varepsilon_c' \left\{ \left[1 + \left(\frac{\varepsilon_c''}{\varepsilon_c'} \right)^2 \right]^{1/2} + 1 \right\}.$$

The plane wave propagation and attenuation factors in the homogeneous, isotropic, locally linear, temporally dispersive medium are then given by

$$\beta(\omega) = \left[\frac{1}{c}\right] \omega \left[\frac{1}{2} \mu \varepsilon_c'(\omega) \right]^{1/2} \left\{ \left[1 + \left(\frac{\varepsilon_c''(\omega)}{\varepsilon_c'(\omega)} \right)^2 \right]^{1/2} + 1 \right\}^{1/2}, \qquad (2.3.11a)$$

$$\alpha(\omega) = \left[\frac{1}{c}\right] \omega \left[\frac{1}{2} \mu \varepsilon_c'(\omega) \right]^{1/2} \left\{ \left[1 + \left(\frac{\varepsilon_c''(\omega)}{\varepsilon_c'(\omega)} \right)^2 \right]^{1/2} - 1 \right\}^{1/2}. \qquad (2.3.11b)$$

These relations may then be used to classify dispersive materials according to the origin of the loss over the frequency domain of interest. If $\varepsilon_c' < 0$ then ε_c' is replaced by $|\varepsilon_c'|$ and the above expressions for α and β are interchanged.

For low-loss materials in a given spectral domain, which may be referred to as *near-ideal dielectrics*[2], the inequality

$$\frac{\varepsilon_c''(\omega)}{\varepsilon_c'(\omega)} \ll 1 \qquad (2.3.12)$$

is satisfied over the frequency domain of interest. The plane wave propagation and attenuation factors may then be approximated as

$$\beta(\omega) \approx k_n(\omega) \left[1 + \frac{1}{8} \left(\frac{\varepsilon_c''(\omega)}{\varepsilon_c'(\omega)} \right)^2 \right], \qquad (2.3.13a)$$

$$\alpha(\omega) \approx \frac{1}{2} k_n(\omega) \frac{\varepsilon_c''(\omega)}{\varepsilon_c'(\omega)}, \qquad (2.3.13b)$$

where

$$k_n(\omega) = \left[\frac{1}{c}\right] c [\mu \varepsilon_c'(\omega)]^{1/2} k_0 = n_r(\omega) k_0 \qquad (2.3.14)$$

---

[2] Some texts refer to such low-loss materials as being imperfect dielectrics. This terminology is not used here as all dispersive dielectric media have loss over some frequency domain and consequently are imperfect.

is the wave number in a fictitious lossless medium (i.e., an ideal dielectric) with real-valued refractive index $n_r(\omega)$. The complex intrinsic impedance of the harmonic plane wave field is, from (2.3.7), found as

$$\begin{aligned}\eta(\omega) &= \left(\frac{\mu}{\varepsilon_c'(\omega)[1 + i\varepsilon_c''(\omega)/\varepsilon_c'(\omega)]}\right)^{1/2} \\ &= \eta_n(\omega)\left(1 + i\frac{\varepsilon_c''(\omega)}{\varepsilon_c'(\omega)}\right)^{-1/2} \\ &\approx \eta_n(\omega)\left\{\left[1 - \frac{3}{8}\left(\frac{\varepsilon_c''(\omega)}{\varepsilon_c'(\omega)}\right)^2\right] - i\frac{\varepsilon_c''(\omega)}{2\varepsilon_c'(\omega)}\right\},\end{aligned} \tag{2.3.15}$$

where

$$\eta_n(\omega) = \left(\frac{\mu}{\varepsilon_c'(\omega)}\right)^{1/2} \tag{2.3.16}$$

is the intrinsic impedance in the fictitious lossless medium with refractive index $n_r(\omega)$. The phase difference δ between the complex electric and magnetic field amplitudes is then given by

$$\tan\delta \approx \frac{\varepsilon_c''(\omega)}{2\varepsilon_c'(\omega)} \tag{2.3.17}$$

where $E = He^{-i\delta}$. Finally, the phase velocity of a harmonic plane wave field in a near-ideal dielectric is given by

$$v_p(\omega) = \frac{\omega}{\beta(\omega)} \approx \frac{c/n_r(\omega)}{1 + \frac{1}{8}\left(\frac{\varepsilon_c''(\omega)}{\varepsilon_c'(\omega)}\right)^2}, \tag{2.3.18}$$

where $c/n_r(\omega)$ is the phase velocity in the ideal lossless medium. Once again, notice that the above expressions (2.3.13)–(2.3.18) are valid approximations only within those frequency domains where the inequality (2.3.12) is satisfied and the medium behaves as a near-ideal dielectric.

For a material such as a *semiconductor* where the losses are primarily conductive, the complex wave number (2.3.8) of a harmonic electromagnetic plane wave field may be approximated as

$$\tilde{k}(\omega) \approx \left[\!\left[\frac{1}{c}\right]\!\right]\omega\left\{\mu\left[\varepsilon_c'(\omega) + i[\![4\pi]\!]\frac{\sigma_r(\omega)}{\omega}\right]\right\}^{1/2}, \tag{2.3.19}$$

where $\varepsilon_c'(\omega) = \varepsilon_r(\omega) - [\![4\pi]\!]\sigma_i(\omega)/\omega$. For $\omega \gg \sigma_r(\omega)/\varepsilon_c'(\omega)$ the loss term in the above expression is small so that the previous approximations apply with $\varepsilon_c'' = [\![4\pi]\!]\sigma_r(\omega)/\omega$ and the semiconductor behaves as a near-ideal dielectric. On the other hand, for $\omega \ll \sigma_r(\omega)/\varepsilon_c'(\omega)$ the semiconductor material behavior is dominated by the real-part $\sigma_r(\omega)$ of the conductivity and may then be considered as a good conductor. In that low-frequency domain the complex wave number (2.3.19) may be further approximated as

$$\tilde{k}(\omega) \approx \left[\frac{2\pi^{1/2}}{c}\right] e^{(i\pi/4)} [\mu\omega\sigma_r(\omega)]^{1/2} = \frac{1+i}{d_p}, \tag{2.3.20}$$

where

$$d_p \equiv \left[\frac{c}{2\pi^{1/2}}\right] \frac{1}{[\frac{1}{2}\omega\mu\sigma_r(\omega)]^{1/2}} \tag{2.3.21}$$

is referred to as the *penetration depth* of the conductor. From (2.3.20) it is seen that the electric and magnetic fields are $\pi/4$ out of time phase in a good conductor. The complex intrinsic impedance of a time-harmonic plane-wave field in the low-frequency domain of a semiconductor material may be approximated as

$$\eta(\omega) \approx \left[\frac{1}{2\pi^{1/2}}\right] e^{-(i\pi/4)} \left(\frac{\mu\omega}{\sigma_r(\omega)}\right)^{1/2}. \tag{2.3.22}$$

Since $\sigma_r(\omega) \approx \sigma^{(0)} - \sigma^{(2)}\omega^2$ as $\omega \to 0$, then $\eta(\omega) \to 0$ as $\omega \to 0$. Finally, the phase velocity is approximately given by

$$\begin{aligned} v_p(\omega) &= \frac{\omega}{\beta(\omega)} \\ &\approx \omega d_p(\omega) = \left[\frac{c}{2\pi^{1/2}}\right]\left(\frac{2\omega}{\mu\sigma_r(\omega)}\right)^{1/2}, \end{aligned} \tag{2.3.23}$$

which is very small in a good conductor.

Consider finally the harmonic plane wave energy densities and Poynting vector in a general dispersive medium. From (2.2.34 and 35) with substitution from (2.3.1a and b) the harmonic electric and magnetic energy densities are found as

$$\begin{aligned} u_e(\mathbf{r}) &\equiv \frac{1}{4}\left[\frac{1}{4\pi}\right] \mathbf{E}(\mathbf{r}) \cdot \mathbf{D}^*(\mathbf{r}) \\ &= \frac{1}{4}\left[\frac{1}{4\pi}\right] \varepsilon^*(\omega) |\mathrm{e}|^2 e^{-2\alpha(\omega)\hat{\mathbf{s}}\cdot\mathbf{r}}, \end{aligned} \tag{2.3.24}$$

$$u_m(\mathbf{r}) \equiv \frac{1}{4}\left[\frac{1}{4\pi}\right] \mathbf{B}(\mathbf{r}) \cdot \mathbf{H}^*(\mathbf{r})$$

$$= \frac{1}{4}\left[\frac{1}{4\pi}\right] \frac{1}{\mu} |b|^2 e^{-2\alpha(\omega)\hat{\mathbf{s}}\cdot\mathbf{r}} \tag{2.3.25a}$$

$$= \frac{1}{4}\left[\frac{1}{4\pi}\right] |\varepsilon_c^{1/2}(\omega)|^2 |\mathrm{e}|^2 e^{-2\alpha(\omega)\hat{\mathbf{s}}\cdot\mathbf{r}}, \tag{2.3.25b}$$

where the second form of the above equation follows by substitution from (2.3.7). Notice that, from (2.1.34)

$$|\varepsilon_c^{1/2}(\omega)|^2 = [\varepsilon_c^{1/2}(\omega)][\varepsilon_c^{1/2}(\omega)]^*$$
$$= \frac{\varepsilon_0\mu_0}{\mu}|n(\omega)|^2 = \frac{[\![c^2]\!]}{\mu c^2}|n(\omega)|^2 , \tag{2.3.26}$$

where $n(\omega)$ is the complex index of refraction of the dispersive, conducting medium. With this substitution Eq. (2.3.25b) for the harmonic plane wave magnetic energy density becomes

$$u_m(\mathbf{r}) = \frac{1}{4}\left[\!\!\left[\frac{c^2}{4\pi}\right]\!\!\right]\frac{1}{\mu c^2}|n(\omega)|^2|\mathrm{e}|^2 e^{-2\alpha(\omega)\hat{s}\cdot\mathbf{r}} . \tag{2.3.27}$$

From (2.2.31), (2.3.1), and (2.3.4) with (2.3.7) the complex Poynting vector of the harmonic plane wave field is given by

$$\mathbf{S}(\mathbf{r}) = \frac{1}{2}\left[\!\!\left[\frac{c}{4\pi}\right]\!\!\right]\mathbf{E}(\mathbf{r}) \times \mathbf{H}^*(\mathbf{r})$$
$$= \left[\!\!\left[\frac{c}{4\pi}\right]\!\!\right]\frac{1}{2\mu^{1/2}}[\varepsilon_c^{1/2}(\omega)]^*\mathbf{e} \times (\hat{s} \times \mathbf{e}^*)e^{-2\alpha(\omega)\hat{s}\cdot\mathbf{r}}$$
$$= \left[\!\!\left[\frac{c}{4\pi}\right]\!\!\right]\frac{1}{2\mu^{1/2}}[\varepsilon_c^{1/2}(\omega)]^*|\mathrm{e}|^2 e^{-2\alpha(\omega)\hat{s}\cdot\mathbf{r}}\hat{s} , \tag{2.3.28}$$

and the time-average Poynting vector in the dispersive medium is then

$$\langle\mathscr{S}(\mathbf{r},t)\rangle = \left[\!\!\left[\frac{c}{4\pi}\right]\!\!\right]\frac{1}{2\mu^{1/2}}\mathrm{Re}\{\varepsilon_c^{1/2}(\omega)\}^*|\mathrm{e}|^2 e^{-2\alpha(\omega)\hat{s}\cdot\mathbf{r}}\hat{s} . \tag{2.3.29}$$

Since

$$\varepsilon_c^{1/2}(\omega) = \frac{(\varepsilon_0\mu_0)^{1/2}}{\mu^{1/2}}n(\omega) = \frac{[\![c]\!]}{\mu^{1/2}c}n(\omega) , \tag{2.3.30}$$

then (2.3.29) may be rewritten as

$$\langle\mathscr{S}(\mathbf{r},t)\rangle = \left[\!\!\left[\frac{c^2}{4\pi}\right]\!\!\right]\frac{1}{2\mu c}n_r(\omega)|\mathrm{e}|^2 e^{-2\alpha(\omega)\hat{s}\cdot\mathbf{r}}\hat{s} , \tag{2.3.31}$$

where $n_r(\omega)$ is the real part of the complex index of refraction of the dispersive, conducting medium.

2.4 The Lorentz Model of the Material Dispersion

The classical theory of resonance absorption and temporal frequency dispersion in solid, liquid, and gaseous media is due mainly to *Lorentz* [2.10, 11] and *Drude* [2.12], the Lorentz model being applicable to dielectric media while the Drude model applies to free-electron metals. Although both of these models are phenomenological in their origin, they do provide accurate expressions for describing the dispersive properties of such media from the infrared to the optical region of the electromagnetic spectrum. Furthermore, they both provide a classical basis for their quantum-mechanical counterparts. Most importantly, they yield causal response functions [2.13] and so are appropriate for this rigorous description of electromagnetic-wave propagation in dispersive media. For these reasons and the fact that the Lorentz theory is of considerable importance in optics, it has been chosen in this research as the model for which the asymptotic theory of dispersive-pulse propagation is developed in Part II. This phenomenological model is now briefly reviewed and the velocity of energy flow of a monochromatic field in such a multiple-resonance medium is derived.

2.4.1 The Classical Lorentz Model of Dielectric Resonance

The Lorentz model of linear dispersive media describes dielectric-type media as a collection of neutral atoms with elastically bound electrons to the nucleus (i.e., as a collection of Lorentz oscillators), where each electron is bound to the nucleus by a Hooke's law restoring force. Under the action of an applied electromagnetic field, the equation of motion of a typical bound electron is given by

$$m\left(\frac{d^2\mathbf{r}_j}{dt^2} + 2\delta_j\frac{d\mathbf{r}_j}{dt} + \omega_j^2\mathbf{r}_j\right) = -\mathrm{e}\mathcal{E}_{\mathrm{loc}}\ , \tag{2.4.1}$$

where m is the mass of the electron and e is the magnitude of the electronic charge. The quantity ω_j is the *undamped resonance frequency* of this oscillator type (denoted by the index j), and δ_j is the associated *phenomenological damping constant* of the oscillator. The applied field $\mathcal{E}_{\mathrm{loc}}$ is the local microscopic electric field intensity which acts on the electron as a driving force; the additional force $-[\![1/c]\!]\mathrm{e}\mathbf{v}_j \times \mathcal{B}_{\mathrm{loc}}$ arising from the interaction of the electron with the local microscopic magnetic field is negligible in comparison to the electric field interaction (due to the smallness of the velocity of the electron in comparison with the vacuum speed of light c) and is consequently neglected. Finally, the term $2m\delta_j\, d\mathbf{r}_j/dt$ represents a phenomenological damping term for the motion of the electron of the jth oscillator type. The actual loss mechanism for a free atom is radiation damping, but it arises from a variety of scattering mechanisms in a solid. By Titchmarsh's theorem [2.13], the system response is causal for all $\delta_j > 0$.

Under the action of an applied monochromatic field of fixed regular frequency ω with complex form $\mathbf{E}_\omega = \mathbf{E}_{\text{loc}} \exp(-i\omega t)$ one obtains the phasor solution to (2.4.1) as

$$\mathbf{r}_j = \frac{e/m}{\omega^2 - \omega_j^2 + 2i\delta_j\omega} \mathbf{E}_{\text{loc}} , \tag{2.4.2}$$

and the local induced dipole moment is then given by

$$\mathbf{p}_j = -e\mathbf{r}_j = \frac{-e^2/m}{\omega^2 - \omega_j^2 + 2i\delta_j\omega} \mathbf{E}_{\text{loc}} . \tag{2.4.3}$$

If there are N_j Lorentz oscillators per unit volume ($j = 0, 2, 4, \ldots$) characterized by the undamped resonance frequencies ω_j and damping constants δ_j, then the macroscopic polarization of the medium is given by the summation over all oscillator types as

$$\mathbf{P} = \sum_j N_j \langle\langle \mathbf{p}_j \rangle\rangle = \langle\langle \mathbf{E}_{\text{loc}} \rangle\rangle \sum_j N_j \alpha_j(\omega) , \tag{2.4.4}$$

where the double angle brackets $\langle\langle\ \rangle\rangle$ indicate a spatial average over atomic sites. Here

$$\alpha_j(\omega) = \frac{-e^2/m}{\omega^2 - \omega_j^2 + 2i\delta_j\omega} \tag{2.4.5}$$

is the *atomic polarizability* of the Lorentz oscillator that is characterized by ω_j and δ_j with number density N_j. Furthermore,

$$N = \sum_j N_j \tag{2.4.6}$$

is the total number of electrons per unit volume that interact with the applied local electric field.

The *electric susceptibility* $\chi_e(\omega)$ of the linear isotropic dielectric medium is defined by the relation

$$\mathbf{P}(\omega) \equiv \varepsilon_0 \chi_e(\omega) \mathbf{E}(\omega) \tag{2.4.7}$$

where $\mathbf{E}$ is the macroscopic electric field vector. In order to relate the microscopic atomic polarizability to the macroscopic electric susceptibility, it is necessary to know the relationship between the microscopic field $\mathbf{E}_{\text{loc}}$ and the macroscopic field $\mathbf{E}$. In general, $\langle\langle \mathbf{E}_{\text{loc}} \rangle\rangle$ is a spatial average over atomic sites and not over the region between atomic sites, so that usually $\langle\langle \mathbf{E}_{\text{loc}} \rangle\rangle \neq \mathbf{E}$. If the number density of the Lorentz oscillators is not too large, however, it is a good approximation to take $\langle\langle \mathbf{E}_{\text{loc}} \rangle\rangle = \mathbf{E}$; the model that results from this assumption is found to contain all of the essential features needed to describe the optical properties of the dispersive dielectric medium [2.5]. Under this assumption there then results

$$\chi_e(\omega) = \frac{1}{\varepsilon_0}\sum_j N_j \alpha_j(\omega) = -\frac{1}{\varepsilon_0}\sum_j \frac{N_j e^2/m}{\omega^2 - \omega_j^2 + 2i\delta_j\omega} \; . \tag{2.4.8}$$

The complex-valued *dielectric permittivity* $\varepsilon(\omega)$ of the medium is then given as

$$\begin{aligned}\varepsilon(\omega) &= \varepsilon_0[1 + [\![4\pi]\!]\chi_e(\omega)] \\ &= \varepsilon_0\left(1 - \sum_j \frac{b_j^2}{\omega^2 - \omega_j^2 + 2i\delta_j\omega}\right)\end{aligned} \tag{2.4.9}$$

where

$$b_j = \left(\frac{[\![4\pi]\!]}{\varepsilon_0} N_j \frac{e^2}{m}\right)^{1/2} \tag{2.4.10}$$

is the *plasma frequency* with number density N_j. Finally, from (2.1.34), the *complex index of refraction* $n(\omega) = n_r(\omega) + in_i(\omega)$ of the dispersive dielectric medium with $\mu = \mu_0$ is given by

$$n(\omega) = \left(\frac{\varepsilon(\omega)}{\varepsilon_0}\right)^{1/2} = \left(1 - \sum_j \frac{b_j^2}{\omega^2 - \omega_j^2 + 2i\delta_j\omega}\right)^{1/2} \; . \tag{2.4.11}$$

The real part of this expression gives the real refractive index $n_r(\omega)$ of the dielectric medium, whereas the imaginary part $n_i(\omega)$ is related to the attenuation factor, cf. (2.3.8–9), by

$$\alpha(\omega) = \frac{\omega}{c} n_i(\omega) \; . \tag{2.4.12}$$

This quantity should not be confused with the atomic polarizability $\alpha_j(\omega)$ given in (2.4.5).

From a quantum-mechanical approach [2.14] it is found that the dielectric response function of the medium can be written as

$$\varepsilon(\omega) = \varepsilon_0\left(1 - \frac{[\![4\pi]\!]}{\varepsilon_0}\frac{e^2}{m}\sum_j \frac{Nf_j}{\omega^2 - \omega_j^2 + 2i\delta_j\omega}\right) \; . \tag{2.4.13}$$

Although there is a formal similarity between (2.4.9 and 13), the physical interpretation of some of the terms is quite different. Classically, ω_j is the resonance frequency of a bound electron, whereas quantum mechanically it is the transition frequency of a bound electron between two atomic states that are separated in energy by the amount $\hbar\omega_j$. The parameter f_j, called the oscillator strength, is a measure of the relative probability of a quantum mechanical transition taking place with energy change $\hbar\omega_j$, which satisfies the sum rule (compare with (2.4.6))

$$\sum_j f_j = 1. \tag{2.4.14}$$

Consequently, although the physical theory upon which the Lorentz model is based is simplistic, the proper form of the dielectric response of the medium is obtained. The model parameters b_j, ω_j, and δ_j can not be obtained from the classical theory, but can be evaluated by fitting them to experimental measurements of the dispersive properties of the medium under consideration. As a consequence, (2.4.13) provides an accurate model of the optical properties of dielectric media.

The behavior of the real and imaginary parts of the complex index of refraction along the positive real frequency axis is illustrated in Fig. 2.1a for a single-resonance Lorentz medium. The frequency regions wherein the real index of refraction $n_r(\omega)$ increases with ω (i.e., where $n_r(\omega)$ has a positive slope) are termed *normally dispersive*, while the region where $n_r(\omega)$ decreases with increasing ω

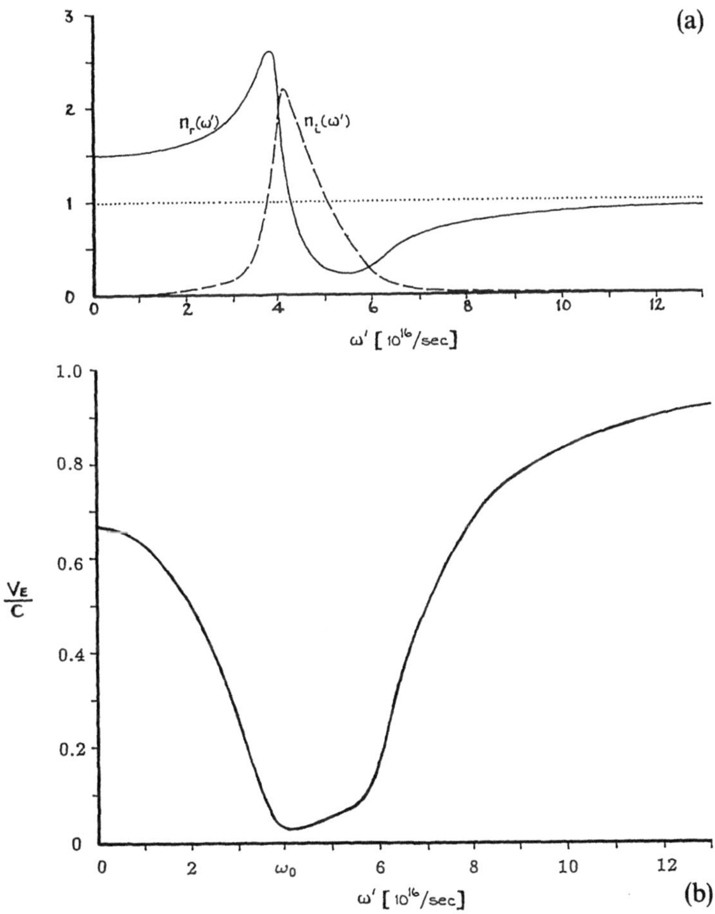

Fig. 2.1a, b. Frequency dispersion of **(a)** the real and imaginary parts of the complex index of refraction and **(b)** the relative time-average energy velocity v_E/c, for a single resonance Lorentz medium with medium parameters $\omega_0 = 4 \times 10^{16}$/s, $\delta_0 = 0.28 \times 10^{16}$/s, $b_0^2 = 20 \times 10^{32}$/s^2.

(i.e., where $n_r(\omega)$ has a negative slope) is said to exhibit *anomalous dispersion*. Notice that the real index of refraction $n_r(\omega)$ varies rapidly with ω within the region of anomalous dispersion, and that this region essentially coincides with the region of strong absorption of the medium.

2.4.2 The Velocity of Energy Flow of a Monochromatic Field in a Multiple-Resonance Lorentz Medium

A phenomenological point of view that is often useful is to consider the electromagnetic field to have associated with it, electromagnetic energy which flows through space with a velocity $\mathbf{v}_E$. The flow direction and the rate of flow of energy associated with the electromagnetic field per unit area through a surface normal to the direction of flow is taken to be given by the Poynting vector, cf. (2.2.2),

$$\mathcal{S}(\mathbf{r},t) = \left[\!\left[\frac{c}{4\pi}\right]\!\right] \mathcal{E}(\mathbf{r},t) \times \mathcal{H}(\mathbf{r},t) \ , \tag{2.4.15}$$

where $\mathcal{E}$ and $\mathcal{H}$ denote the real-valued electric and magnetic field vectors, respectively, of the electromagnetic field. Consequently, the velocity of propagation of the energy is given by

$$\mathbf{v}_E = \frac{\mathcal{S}}{\mathcal{U}_{\text{tot}}} \ , \tag{2.4.16}$$

where $\mathcal{U}_{\text{tot}}$ is the volume density of the total energy associated with the electromagnetic field. When the electromagnetic field propagates either in vacuum or in a nondispersive (and nonmagnetic) medium, it is straightforward to show that the velocity of energy propagation $v_E = |\mathbf{v}_E|$ is given by c and $c(\varepsilon^{(r)}\mu^{(r)})^{-1/2}$ in the respective cases [for example, consider the nondispersive forms of (2.3.24, 25 and 29)]. The calculation is much more difficult, however, when the field propagates in a dispersive medium because it is impossible to express the electric energy density and the dissipation separately in terms of the general dielectric permittivity (and electrical conductivity) of a dispersive medium, cf. Sect. 2.2.2. Nevertheless, the calculation is possible if a specific dynamical model for the medium response is given, such as that provided by the Lorentz model. The generalization of the theory given by *Loudon* [2.8–9] for a single resonance Lorentz medium to a multiple resonance medium is now presented for a monochromatic field. The connection between the resulting velocity of energy flow and the propagation velocity of a pulse in a Lorentz medium is not apparent from the theory. The fact that there is a close connection is demonstrated in Part II for a single-resonance Lorentz medium and in [2.15] for a double-resonance medium. One of the most spectacular results of this theory [2.15] is that the frequency dependence of the velocity of energy flow of a monochromatic wave of angular frequency ω_c is shown to be an upper envelope to the frequency dependence of the signal velocity of a pulse with signal frequency ω_c.

The analysis begins with the differential form of Poynting's theorem which may be written as, cf. (2.2.14),

$$\nabla \cdot \mathscr{S} = -\left[\!\left[\frac{1}{4\pi}\right]\!\right]\left(\mu_0 \mathscr{H} \cdot \frac{\partial \mathscr{H}}{\partial t} + \varepsilon_0 \mathscr{E} \cdot \frac{\partial \mathscr{E}}{\partial t} + [\![4\pi]\!]\mathscr{E} \cdot \frac{\partial \mathscr{P}}{\partial t}\right) \tag{2.4.17}$$

for a nonconducting, dispersive dielectric with $\mu = \mu_0$. First, consider obtaining an expression for the term $\mathscr{E} \cdot \partial\mathscr{P}/\partial t$ for a multiple resonance Lorentz medium that is interacting with a monochromatic plane wave field. From (2.4.1) one obtains

$$\mathscr{E} = -\frac{m}{\mathrm{e}}\left(\frac{d^2\mathbf{r}_j}{dt^2} + 2\delta_j\frac{d\mathbf{r}_j}{dt} + \omega_j^2\mathbf{r}_j\right) , \tag{2.4.18}$$

and from (2.4.3, 4) the macroscopic polarization vector is seen to be given by

$$\mathscr{P} = -\sum_j N_j \mathrm{e}\mathbf{r}_j \ . \tag{2.4.19}$$

These two expressions then yield

$$\mathscr{E} \cdot \frac{\partial \mathscr{P}}{\partial t} = \sum_j N_j m\left[\frac{1}{2}\frac{d}{dt}\left(\frac{d\mathbf{r}_j}{dt}\right)^2 + 2\delta_j\left(\frac{d\mathbf{r}_j}{dt}\right)^2 + \frac{1}{2}\omega_j^2\frac{d}{dt}(\mathbf{r}_j)^2\right] . \tag{2.4.20}$$

The term on the right-hand side of Poynting's theorem (2.4.17) is then seen to be the sum of a perfect time-differential and the term in δ_j which corresponds to the dissipation mechanism. If one then defines a *total energy density*

$$\mathscr{U} = \mathscr{U}_f + \mathscr{U}_{\mathrm{osc}} \tag{2.4.21}$$

where

$$\mathscr{U}_f \equiv \left[\!\left[\frac{1}{4\pi}\right]\!\right]\frac{1}{2}(\varepsilon_0|\mathscr{E}|^2 + \mu_0|\mathscr{H}|^2) \tag{2.4.22}$$

is the *energy density stored in the field alone*, and where

$$\mathscr{U}_{\mathrm{osc}} \equiv \frac{1}{2}\sum_j N_j m\left[\left(\frac{d\mathbf{r}_j}{dt}\right)^2 + \omega_j^2(\mathbf{r}_j)^2\right] \tag{2.4.23}$$

is the *energy density stored in the multiple-resonance Lorentz medium*, then the differential form of Poynting's theorem (2.4.17) becomes

$$\nabla \cdot \mathscr{S} + 2\sum_j N_j m\delta_j\left(\frac{d\mathbf{r}_j}{dt}\right)^2 = -\frac{d\mathscr{U}}{dt} \ . \tag{2.4.24a}$$

Integration of this expression over an arbitrary volume V bounded by a closed surface S and application of the divergence theorem yields the expression

$$\oiint_S \mathcal{S}\cdot\hat{n}\,da + 2\sum_j \iiint_V N_j m\delta_j\left(\frac{d\mathbf{r}_j}{dt}\right)^2 d^3r = -\frac{d}{dt}\iiint_V \mathcal{U}\,d^3r\ , \tag{2.4.24b}$$

where $\hat{n}$ is the outward unit normal vector to the surface S. This equation expresses the conservation of energy in the dispersive dielectric, where the two terms of the left-hand side represent the rate of energy loss in the region V by leakage across its surface S and by dissipation in the medium contained in V, respectively, whereas the integral on the right-hand side represents the rate of change of the total electromagnetic energy stored within the region V.

For a time-harmonic field with angular frequency ω, the time-average energy density stored in the multiple-resonance Lorentz medium is found from (2.4.23) with (2.4.2) as

$$\langle\mathcal{U}_{\text{osc}}\rangle = \frac{1}{4}|\mathbf{E}|^2\sum_j \frac{N_j e^2}{m}\frac{\omega^2+\omega_j^2}{(\omega^2-\omega_j^2)^2+4\delta_j^2\omega^2}\ ,$$

which, with (2.4.10), becomes

$$\langle\mathcal{U}_{\text{osc}}\rangle = \left[\frac{1}{4\pi}\right]\frac{\varepsilon_0}{4}|\mathbf{E}|^2\sum_j \frac{b_j^2(\omega^2+\omega_j^2)}{(\omega^2-\omega_j^2)^2+4\delta_j^2\omega^2}\ . \tag{2.4.25}$$

The time-average value of the energy density stored in the monochromatic plane-wave electromagnetic field is obtained from (2.4.22) with (2.3.7) as

$$\langle\mathcal{U}_f\rangle = \left[\frac{1}{4\pi}\right]\frac{1}{4}(\varepsilon_0+|\varepsilon(\omega)|)|\mathbf{E}|^2\ ,$$

which may be written in terms of the complex index of refraction $n(\omega) = [\varepsilon(\omega)/\varepsilon_0]^{1/2}$ as

$$\begin{aligned}\langle\mathcal{U}_f\rangle &= \left[\frac{1}{4\pi}\right]\frac{\varepsilon_0}{4}(1+|n^2(\omega)|)|\mathbf{E}|^2\\ &= \left[\frac{1}{4\pi}\right]\frac{\varepsilon_0}{4}[n_r^2(\omega)+n_i^2(\omega)+1]|\mathbf{E}|^2\ .\end{aligned} \tag{2.4.26}$$

With use of the relation, cf. (2.4.11),

$$n_r^2 - n_i^2 = \frac{\varepsilon_r(\omega)}{\varepsilon_0} = 1 - \sum_j \frac{b_j^2(\omega^2-\omega_j^2)}{(\omega^2-\omega_j^2)^2+4\delta_j^2\omega^2}\ , \tag{2.4.27}$$

the total time-average electromagnetic energy density stored in both the field

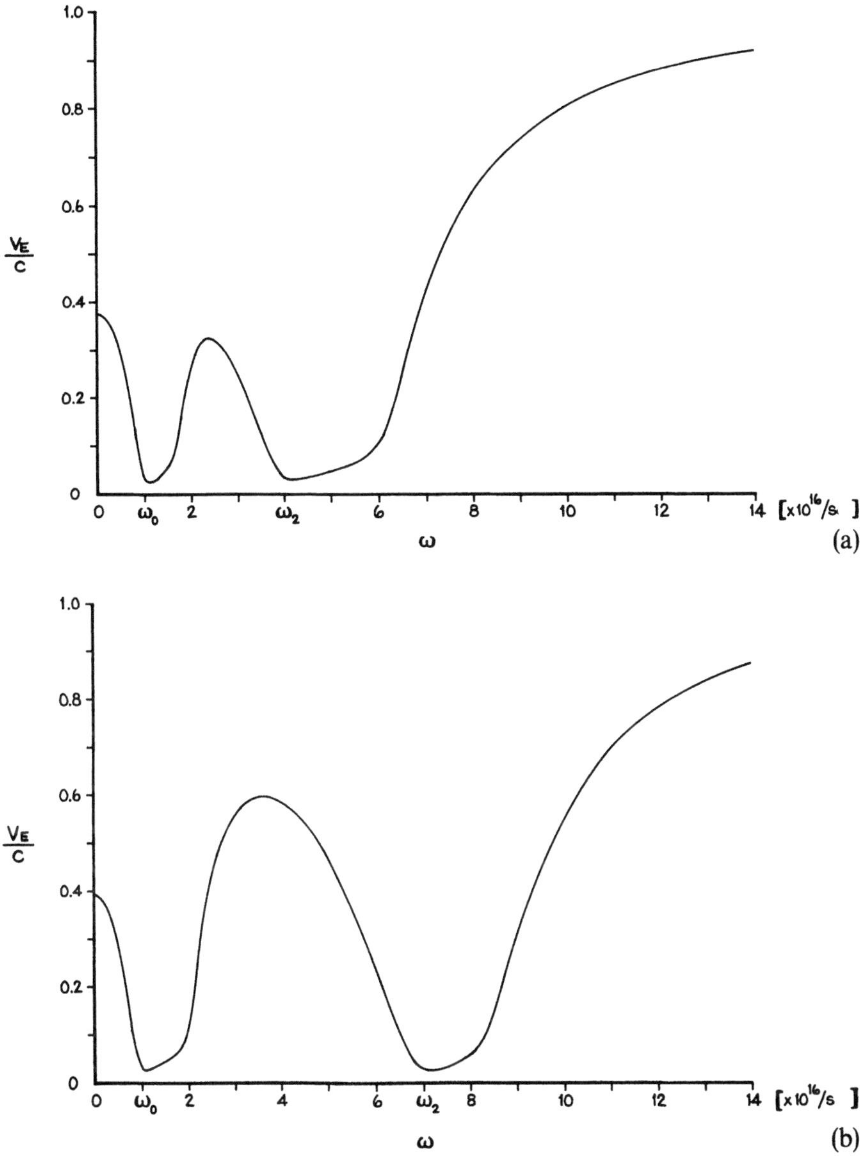

Fig. 2.2a, b. Frequency dispersion of the relative time-average energy-transport velocity v_E/c in a double-resonance Lorentz medium. The medium parameters in part **(a)** are: $\omega_0 = 1 \times 10^{16}$/s, $b_0^2 = 5 \times 10^{32}/\mathrm{s}^2$, $\delta_0 = 0.1 \times 10^{16}$/s, $\omega_2 = 4 \times 10^{16}$/s, $b_2^2 = 20 \times 10^{32}/\mathrm{s}^2$, $\delta_2 = 0.28 \times 10^{16}$/s. The medium parameters in part **(b)** are the same except that $\omega_2 = 7 \times 10^{16}$/s.

and the medium is found to be given by

$$\begin{aligned}\langle \mathcal{U} \rangle &= \langle \mathcal{U}_f \rangle + \langle \mathcal{U}_{\text{osc}} \rangle \\ &= \left[\frac{1}{4\pi}\right] \frac{\varepsilon_0}{2} |\mathbf{E}|^2 \left(n_r^2(\omega) + \sum_j \frac{b_j^2 \omega^2}{(\omega^2 - \omega_j^2)^2 + 4\delta_j^2 \omega^2} \right),\end{aligned} \tag{2.4.28}$$

where the summation extends over all the medium resonance frequencies.

The time-average value of the magnitude of the Poynting vector for the monochromatic plane-wave field is readily found to be given by, cf. (2.3.31),

$$\langle |\mathcal{S}| \rangle = \left[\frac{c^2}{4\pi}\right] \frac{1}{2\mu_0 c} n_r(\omega) |\mathcal{E}|^2 . \tag{2.4.29}$$

The *time-averaged velocity of energy transport* in a multiple-resonance Lorentz medium is then given by the ratio of these two time-averaged quantities, so that

$$v_E \equiv \frac{\langle |\mathcal{S}| \rangle}{\langle \mathcal{U} \rangle} = \frac{c}{n_r(\omega) + \dfrac{1}{n_r(\omega)} \displaystyle\sum_j \frac{b_j^2 \omega^2}{(\omega^2 - \omega_j^2)^2 + 4\delta_j^2 \omega^2}} . \tag{2.4.30}$$

This important result is a generalization of the well-known expression for the energy velocity in a single-resonance Lorentz medium due to *Loudon* [2.8] and reduces to his expression in that special case. Since the denominator appearing in the final expression of (2.4.30) is greater than or equal to unity, then for all values of the angular frequency of oscillation of the monochromatic field, the energy transport velocity given by (2.4.30) yields results that are in agreement with the relativistic principle of causality. Brillouin's derivation of the energy velocity in [2.16] is in error due to the neglect of the electromagnetic energy stored in the Lorentz oscillators.

The frequency dependence of the time-average energy velocity in a single-resonance Lorentz medium is illustrated in Fig. 2.1b for the same medium parameters used in part (a) of the figure for the frequency dispersion of the real and imaginary parts of the complex index of refraction. Notice that the energy velocity attains a minimum value just above the resonance frequency ω_0 near where $n_i(\omega)$ attains its maximum value and that v_E remains small through the region of anomalous dispersion. For comparison, the frequency dependence of the time-average energy velocity in two different double-resonance Lorentz media is illustrated in Fig. 2.2. The medium parameters in part (b) are the same as those in part (a) with the single exception that the second resonance frequency ω_2 has been increased from $\omega_2 = 4 \times 10^{16}$/s to $\omega_2 = 7 \times 10^{16}$/s. This increase not only shifts the location of the second minimum in the energy velocity to a higher frequency but also shifts the location of the local maximum in the energy velocity that occurs between the two resonance frequencies ω_0 and ω_2 to a higher frequency as well as increases the value of the energy velocity of this relative maximum.

3 The Angular Spectrum Representation of the Pulsed Radiation Field

3.1 Introduction and Mathematical Preliminaries

Attention is now directed to the rigorous solution of the electromagnetic field that is radiated by a general current source in a homogeneous, isotropic, locally linear, temporally dispersive medium characterized by the time-dependent dielectric permittivity $\hat{\varepsilon}(t)$, electrical conductivity $\hat{\sigma}(t)$, and constant magnetic permeability μ that occupies all of space. The applied current source $\mathscr{J}_0(\mathbf{r}, t)$ is assumed here to be a well-behaved known function of both position $\mathbf{r}$ and time t that vanishes for $|z| \geqslant Z$, where Z is a positive constant, as illustrated in Fig. 3.1. Furthermore, it is assumed that the current source is turned on at time $t = 0$, so that

$$\mathscr{J}_0(\mathbf{r}, t) = \mathbf{0} \ ; \qquad t \leqslant 0 \ . \tag{3.1.1}$$

Since the present analysis is concerned only with the electromagnetic field that is radiated by this current source, it is then required that both field vectors $\mathscr{E}(\mathbf{r}, t)$ and $\mathscr{B}(\mathbf{r}, t)$ also vanish for $t \leqslant 0$, viz.

$$\left.\begin{array}{l}\mathscr{E}(\mathbf{r}, t) = \mathbf{0} \\ \mathscr{B}(\mathbf{r}, t) = \mathbf{0}\end{array}\right\} \qquad t \leqslant 0 \ . \tag{3.1.2}$$

These requirements on the current source are not restrictive in any physical sense since all radiation problems may be cast so as to satisfy them.

The complex temporal frequency spectrum of a function $\mathbf{f}(\mathbf{r}, t)$ of both position $\mathbf{r}$ and time t that vanishes for $t < 0$ is of critical importance to the solution of the above posed radiation problem and is now considered in some detail following, in part, the treatment by *Stratton* [3.1]. The Laplace transform of $\mathbf{f}(\mathbf{r}, t)$ with respect to the time t is defined here as

$$\mathscr{L}\{\mathbf{f}(\mathbf{r}, t)\} \equiv \int_0^\infty \mathbf{f}(\mathbf{r}, t) e^{i\omega t} dt \ , \tag{3.1.3}$$

which is simply a Fourier transform with complex ω that is taken over only the positive time interval. Let $\mathbf{f}'(\mathbf{r}, t)$ be another function of both position and time such that

$$\mathbf{f}'(\mathbf{r}, t) = \mathbf{f}(\mathbf{r}, t) \ ; \qquad t > 0 \ , \tag{3.1.4}$$

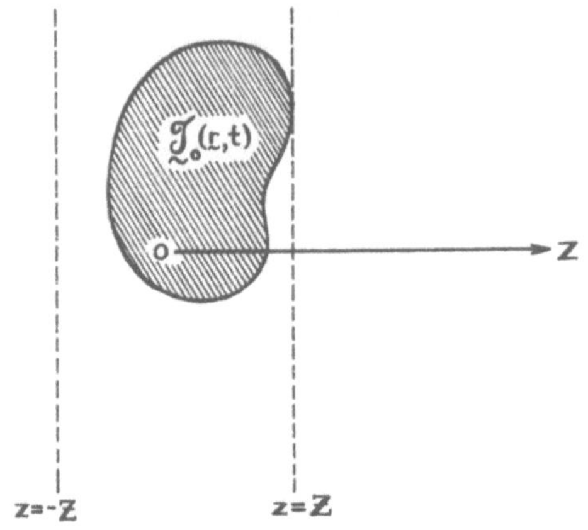

Fig. 3.1. General current source imbedded in a dispersive medium.

but which may not vanish for $t \leqslant 0$. The Laplace transform (3.1.3) may then be written as

$$\mathscr{L}\{\mathbf{f}(\mathbf{r},t)\} = \int_{-\infty}^{\infty} u(t)\mathbf{f}'(\mathbf{r},t)\mathrm{e}^{\mathrm{i}\omega t}\,dt \ , \tag{3.1.5}$$

where $u(t) = 0$ for $t < 0$ and $u(t) = 1$ for $t > 0$ is the *Heaviside unit-step function.* It is then seen that, for real ω, the Laplace transform of $\mathbf{f}(\mathbf{r},t)$ is equal to the Fourier transform of $u(t)\mathbf{f}'(\mathbf{r},t)$, viz.

$$\mathscr{L}\{\mathbf{f}(\mathbf{r},t)\} = \mathscr{F}_{\omega}\{u(t)\mathbf{f}'(\mathbf{r},t)\} \ ; \qquad \text{for real } \omega \ , \tag{3.1.6}$$

where the subscript ω indicates that it is the Fourier transform variable. The inverse Fourier transform of (3.1.6) then yields

$$u(t)\mathbf{f}'(\mathbf{r},t) = \mathscr{F}^{-1}\{\mathscr{L}[\mathbf{f}(\mathbf{r},t)]\} \tag{3.1.7}$$

for real ω.

For complex ω, let $\omega = \omega' + \mathrm{i}\omega''$ where $\omega' = \mathbb{R}\mathrm{e}(\omega)$ and $\omega'' = \mathbb{I}\mathrm{m}(\omega)$. The Laplace transform (3.1.5) then becomes

$$\begin{aligned}\mathscr{L}\{\mathbf{f}(\mathbf{r},t)\} &= \int_{-\infty}^{\infty} [u(t)\mathbf{f}'(\mathbf{r},t)\mathrm{e}^{-\omega'' t}]\mathrm{e}^{\mathrm{i}\omega' t}\,dt \\ &= \mathscr{F}_{\omega'}\{u(t)\mathbf{f}'(\mathbf{r},t)\mathrm{e}^{-\omega'' t}\} \ .\end{aligned} \tag{3.1.8}$$

The inverse Fourier transform of (3.1.8) then yields

$$\begin{aligned}u(t)\mathbf{f}'(\mathbf{r},t)\mathrm{e}^{-\omega'' t} &= \mathscr{F}_{\omega'}^{-1}\{\mathscr{L}[\mathbf{f}(\mathbf{r},t)]\} \\ &= \frac{1}{2\pi}\int_{\infty}^{\infty} \mathscr{L}[\mathbf{f}(\mathbf{r},t)]\mathrm{e}^{-\mathrm{i}\omega' t}\,d\omega' \ ,\end{aligned} \tag{3.1.9}$$

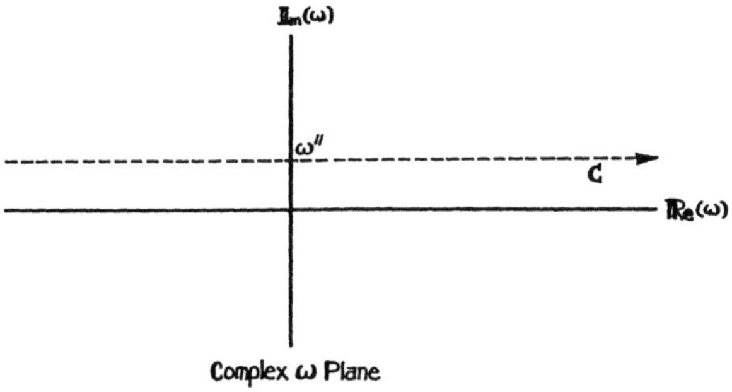

Fig. 3.2. Straight line contour C in the complex ω-plane.

which may be rewritten as

$$u(t)\mathbf{f}'(\mathbf{r},t) = \frac{1}{2\pi}\int_{-\infty}^{\infty} \mathscr{L}[\mathbf{f}(\mathbf{r},t)]\mathrm{e}^{-\mathrm{i}(\omega'+\mathrm{i}\omega'')t}\,d\omega'$$

$$= \frac{1}{2\pi}\int_{-\infty+\mathrm{i}\omega''}^{\infty+\mathrm{i}\omega''} \mathscr{L}[\mathbf{f}(\mathbf{r},t)]\mathrm{e}^{-\mathrm{i}\omega t}\,d\omega$$

$$= \frac{1}{2\pi}\int_{C} \mathscr{L}[\mathbf{f}(\mathbf{r},t)]\mathrm{e}^{-\mathrm{i}\omega t}\,d\omega \; . \tag{3.1.10}$$

Here C is the straight line contour $\omega = \omega' + \mathrm{i}\omega''$ with ω'' fixed and ω' ranging from $-\infty$ to $+\infty$, as illustrated in Fig. 3.2. Since $\mathbf{f}(\mathbf{r},t) = u(t)\mathbf{f}'(\mathbf{r},t)$, Eqs. (3.1.3 and 10) then define the Laplace transform pair relationship

$$\tilde{\mathbf{f}}(\mathbf{r},\omega) \equiv \mathscr{L}\{\mathbf{f}(\mathbf{r},t)\} = \int_{0}^{\infty} \mathbf{f}(\mathbf{r},t)\mathrm{e}^{\mathrm{i}\omega t}\,dt \; , \tag{3.1.11a}$$

$$\mathbf{f}(\mathbf{r},t) \equiv \mathscr{L}^{-1}\{\tilde{\mathbf{f}}(\mathbf{r},\omega)\} = \frac{1}{2\pi}\int_{C} \tilde{\mathbf{f}}(\mathbf{r},\omega)\mathrm{e}^{-\mathrm{i}\omega t}\,d\omega \; , \tag{3.1.11b}$$

where $\tilde{\mathbf{f}}(\mathbf{r},\omega)$ is the complex temporal frequency spectrum of $\mathbf{f}(\mathbf{r},t)$ with $\omega = \omega' + \mathrm{i}\omega''$. Notice that $\omega'' = \mathrm{Im}\{\omega\}$ plays a passive role in the Laplace transform operation since it remains constant in both the forward and inverse transformations. Nevertheless, its presence can be important because the factor $\mathrm{e}^{-\omega'' t}$ appearing in the integrand of the transformation (3.1.11a) may serve as a convergence factor when $\omega'' > 0$. In particular,

$$\tilde{\mathbf{f}}(\mathbf{r},\omega) = \int_{0}^{\infty} \mathbf{f}(\mathbf{r},t)\mathrm{e}^{-\omega'' t}\mathrm{e}^{\mathrm{i}\omega' t}\,dt \tag{3.1.12}$$

is just the Fourier transform $\mathscr{F}_{\omega'}\{\mathbf{f}(\mathbf{r},t)\mathrm{e}^{-\omega'' t}\}$. The Fourier transform of $\mathbf{f}(\mathbf{r},t)$

alone is

$$\mathscr{F}_{\omega'}\{\mathbf{f}(\mathbf{r},t)\} = \int_0^\infty \mathbf{f}(\mathbf{r},t)e^{i\omega' t}\,dt \ ,$$

which exists provided that $\mathbf{f}(\mathbf{r},t)$ is absolutely integrable, viz.

$$\lim_{T\to\infty} \int_0^T |\mathbf{f}(\mathbf{r},t)|\,dt < \infty \ .$$

If $\mathbf{f}(\mathbf{r},t)$ does not vanish properly at infinity then the above integral fails to converge and the existence of the Fourier transform $\mathscr{F}_{\omega'}\{\mathbf{f}(\mathbf{r},t)\}$ is not guaranteed. However, if there exists a real number γ such that

$$\lim_{T\to\infty} \int_0^T |\mathbf{f}(\mathbf{r},t)e^{-\gamma t}|\,dt < \infty \ , \tag{3.1.13}$$

then $\mathbf{f}(\mathbf{r},t)$ is transformable for all $\omega'' \geqslant \gamma$ and its temporal frequency spectrum is given by the Laplace trasnsform (3.1.11a). The lower bound γ_a of all of the values of γ which satisfy the inequality in Eq. (3.1.13) is called the *abscissa of absolute convergence* for the function $\mathbf{f}(\mathbf{r},t)$.

The Laplace transform of the time derivative $\partial\mathbf{f}(\mathbf{r},t)/\partial t$ can be related to the Laplace transform of $\mathbf{f}(\mathbf{r},t)$ through integration by parts as follows:

$$\begin{aligned}\mathscr{L}\left\{\frac{\partial\mathbf{f}(\mathbf{r},t)}{\partial t}\right\} &= \int_0^\infty \frac{\partial\mathbf{f}(\mathbf{r},t)}{\partial t}e^{i\omega t}\,dt \\ &= [\mathbf{f}(\mathbf{r},t)e^{i\omega t}]_0^\infty - i\omega\int_0^\infty \mathbf{f}(\mathbf{r},t)e^{i\omega t}\,dt \\ &= -\mathbf{f}(\mathbf{r},0) - i\omega\mathscr{L}\{\mathbf{f}(\mathbf{r},t)\} \ ,\end{aligned} \tag{3.1.14}$$

where the fact that $|\mathbf{f}(\mathbf{r},t)e^{i\omega t}| = |\mathbf{f}(\mathbf{r},t)e^{-\omega'' t}|$ must vanish as $t\to\infty$ for all $\omega'' \geqslant \gamma_a$ has been used.

For the appropriate form of the convolution theorem for the Laplace transform, consider determining the function whose Laplace transform is equal to the product $\tilde{f}_1(\omega)\tilde{f}_2(\omega) = \mathscr{L}\{f_1(t)\}\mathscr{L}\{f_2(t)\}$ so that

$$\begin{aligned}\mathscr{L}^{-1}\{\tilde{f}_1(\omega)\tilde{f}_2(\omega)\} &= \frac{1}{2\pi}\int_C \tilde{f}_1(\omega)\tilde{f}_2(\omega)e^{-i\omega t}\,d\omega \\ &= \frac{1}{2\pi}\int_C \tilde{f}_1(\omega)\left[\int_0^\infty f_2(\tau)e^{i\omega\tau}\,d\tau\right]e^{-i\omega t}\,d\omega \\ &= \int_0^\infty d\tau\cdot f_2(\tau)\left[\frac{1}{2\pi}\int_C \tilde{f}_1(\omega)e^{-i\omega(t-\tau)}\,d\omega\right] \ ,\end{aligned}$$

and consequently

$$\mathscr{L}^{-1}\{\tilde{f}_1(\omega)\tilde{f}_2(\omega)\} = \int_0^\infty f_1(t-\tau)f_2(\tau)u(t-\tau)\,d\tau \ ,$$

where the unit step function $u(t-\tau)$ is explicitly included to emphasize the fact that $f_1(t)$ vanishes for $t < 0$. Because $u(t-\tau)$ vanishes for negative values of its argument, the upper limit of integration in τ must be t and the above equation becomes

$$\mathscr{L}^{-1}\{\tilde{f}_1(\omega)\tilde{f}_2(\omega)\} = \int_0^t f_1(t-\tau)f_2(\tau)\,d\tau \ , \tag{3.1.15}$$

which may be rewritten as

$$\mathscr{L}\left\{\int_0^t f_1(t-\tau)f_2(\tau)\,d\tau\right\} = \mathscr{L}\{f_1(t)\}\mathscr{L}\{f_2(t)\} \ , \tag{3.1.16}$$

which is the convolution theorem for the Laplace transform.

The spatio-temporal Fourier-Laplace transform of a vector function $\mathscr{F}(\mathbf{r},t)$ of both position and time that vanishes for $t < 0$ is defined here by the pair of equations

$$\mathbf{F}(\mathbf{k},\omega) = \mathscr{F}\mathscr{L}\{\mathscr{F}(\mathbf{r},t)\} = \iiint_{-\infty}^{\infty} d^3r \int_0^\infty dt \cdot \mathscr{F}(\mathbf{r},t)\mathrm{e}^{-\mathrm{i}(\mathbf{k}\cdot\mathbf{r}-\omega t)} \ , \tag{3.1.17a}$$

$$\mathscr{F}(\mathbf{r},t) = \mathscr{F}^{-1}\mathscr{L}^{-1}\{\mathbf{F}(\mathbf{k},\omega)\} = \frac{1}{(2\pi)^4}\iiint_{-\infty}^{\infty} d^3k \int_C d\omega \cdot \mathbf{F}(\mathbf{k},\omega)\mathrm{e}^{\mathrm{i}(\mathbf{k}\cdot\mathbf{r}-\omega t)} \ , \tag{3.1.17b}$$

where $\mathbf{k} = \hat{1}_x k_x + \hat{1}_y k_y + \hat{1}_z k_z$ and $\mathbf{r} = \hat{1}_x x + \hat{1}_y y + \hat{1}_z z$. Since, from (3.1.17b),

$$\begin{aligned}\frac{\partial\mathscr{F}(\mathbf{r},t)}{\partial x_j} &= \frac{1}{(2\pi)^4}\frac{\partial}{\partial x_j}\iiint_{-\infty}^{\infty} d^3k \int_C d\omega \cdot \mathbf{F}(\mathbf{k},\omega)\mathrm{e}^{\mathrm{i}(\mathbf{k}\cdot\mathbf{r}-\omega t)} \\ &= \frac{1}{(2\pi)^4}\iiint_{-\infty}^{\infty} d^3k \int_C d\omega \cdot \mathrm{i}k_j\mathbf{F}(\mathbf{k},\omega)\mathrm{e}^{\mathrm{i}(\mathbf{k}\cdot\mathbf{r}-\omega t)} \ ,\end{aligned}$$

then the transforms of the first spatial derivatives of $\mathscr{F}(\mathbf{r},t)$ are given by

$$\mathscr{F}\mathscr{L}\left\{\frac{\partial\mathscr{F}(\mathbf{r},t)}{\partial x_j}\right\} = \mathrm{i}k_j\mathbf{F}(\mathbf{k},\omega) \ , \tag{3.1.18}$$

where $x_1 = x$, $x_2 = y$, $x_3 = z$ and $k_1 = k_x$, $k_2 = k_y$, $k_3 = k_z$. With this result the spatio-temporal transform of the divergence of the vector field $\mathscr{F}(\mathbf{r},t)$ is found to be

$$\mathscr{FL}\{\nabla\cdot\mathscr{F}(\mathbf{r},t)\} = \mathscr{FL}\left\{\hat{1}_x\cdot\frac{\partial\mathscr{F}(\mathbf{r},t)}{\partial x} + \hat{1}_y\cdot\frac{\partial\mathscr{F}(\mathbf{r},t)}{\partial y} + \hat{1}_z\cdot\frac{\partial\mathscr{F}(\mathbf{r},t)}{\partial z}\right\}$$

$$= i\mathbf{k}\cdot\mathbf{F}(\mathbf{k},\omega)\ , \tag{3.1.19}$$

and the spatio-temporal transform of the curl of $\mathscr{F}(\mathbf{r},t)$ is given by

$$\mathscr{FL}\{\nabla\times\mathscr{F}(\mathbf{r},t)\} = \mathscr{FL}\left\{\hat{1}_x\times\frac{\partial\mathscr{F}(\mathbf{r},t)}{\partial x} + \hat{1}_y\times\frac{\partial\mathscr{F}(\mathbf{r},t)}{\partial y} + \hat{1}_z\times\frac{\partial\mathscr{F}(\mathbf{r},t)}{\partial z}\right\}$$

$$= i\mathbf{k}\times\mathbf{F}(\mathbf{k},\omega)\ . \tag{3.1.20}$$

The spatio-temporal transforms of higher-order spatial derivatives of $\mathscr{F}(\mathbf{r},t)$ may then be obtained through repeated application of the above relations.

3.2 The Fourier-Laplace Representation of the Radiation Field

From the constitutive relations (2.1.2) and the conditions (3.1.1 and 2), after the current source has been turned on (i.e. for $t > 0$) the total current density in the homogeneous, isotropic, locally linear, temporally dispersive medium is given by

$$\mathscr{J}(\mathbf{r},t) = \mathscr{J}_0(\mathbf{r},t) + \int_0^t \hat{\sigma}(t-t')\mathscr{E}(\mathbf{r},t')\,dt'\ , \tag{3.2.1}$$

and the electric displacement vector is

$$\mathscr{D}(\mathbf{r},t) = \int_0^t \hat{\varepsilon}(t-t')\mathscr{E}(\mathbf{r},t')\,dt'\ . \tag{3.2.2}$$

The Fourier-Laplace transform of these relations then gives, with application of the convolution theorem (3.1.16) for the Laplace transform,

$$\mathbf{J}(\mathbf{k},\omega) = \mathbf{J}_0(\mathbf{k},\omega) + \sigma(\omega)\mathbf{E}(\mathbf{k},\omega)\ , \tag{3.2.3}$$

$$\mathbf{D}(\mathbf{k},\omega) = \varepsilon(\omega)\mathbf{E}(\mathbf{k},\omega)\ , \tag{3.2.4}$$

where $\mathbf{J}_0(\mathbf{k},\omega)$ is the spatio-temporal frequency spectrum of the applied current source.

The Maxwell's equations for the electric and magnetic field vectors in a homogeneous, isotropic, locally linear, temporally dispersive medium occupying all of three-dimensional space with charge and current sources are

$$\nabla \cdot \mathscr{D}(\mathbf{r},t) = [\![4\pi]\!]\varrho(\mathbf{r},t) \ , \tag{3.2.5a}$$

$$\nabla \times \mathscr{E}(\mathbf{r},t) = -\left[\!\!\left[\frac{1}{c}\right]\!\!\right]\frac{\partial \mathscr{B}(\mathbf{r},t)}{\partial t} \ , \tag{3.2.5b}$$

$$\nabla \cdot \mathscr{B}(\mathbf{r},t) = 0 \ , \tag{3.2.5c}$$

$$\nabla \times \mathscr{B}(\mathbf{r},t) = \left[\!\!\left[\frac{1}{c}\right]\!\!\right]\mu\frac{\partial \mathscr{D}(\mathbf{r},t)}{\partial t} + \left[\!\!\left[\frac{4\pi}{c}\right]\!\!\right]\mu\mathscr{J}(\mathbf{r},t) \ , \tag{3.2.5d}$$

where the magnetic permeability μ is a constant. The Fourier-Laplace transform of these relations then gives, with application of (3.1.14) and the initial conditions (3.1.2), the set of algebraic vector equations, using (3.1.19 and 20),

$$\mathrm{i}\mathbf{k} \cdot \mathbf{D}(\mathbf{k},\omega) = [\![4\pi]\!]\rho(\mathbf{k},\omega) \ , \tag{3.2.6a}$$

$$\mathrm{i}\mathbf{k} \times \mathbf{E}(\mathbf{k},\omega) = \left[\!\!\left[\frac{1}{c}\right]\!\!\right]\mathrm{i}\omega\mathbf{B}(\mathbf{k},\omega) \ , \tag{3.2.6b}$$

$$\mathrm{i}\mathbf{k} \cdot \mathbf{B}(\mathbf{k},\omega) = 0 \ , \tag{3.2.6c}$$

$$\mathrm{i}\mathbf{k} \times \mathbf{B}(\mathbf{k},\omega) = -\left[\!\!\left[\frac{1}{c}\right]\!\!\right]\mathrm{i}\omega\mu\mathbf{D}(\mathbf{k},\omega) + \left[\!\!\left[\frac{4\pi}{c}\right]\!\!\right]\mu\mathbf{J}(\mathbf{k},\omega) \ . \tag{3.2.6d}$$

With (3.2.4), the first of the above set of equations may be expressed as

$$\mathbf{k} \cdot \mathbf{E}(\mathbf{k},\omega) = -[\![4\pi]\!]\mathrm{i}\frac{\rho(\mathbf{k},\omega)}{\varepsilon(\omega)} \ . \tag{3.2.7}$$

In addition, subsitution of (3.2.3 and 4) into (3.2.6d) yields

$$\mathbf{k} \times \mathbf{B}(\mathbf{k},\omega) = -\left[\!\!\left[\frac{1}{c}\right]\!\!\right]\omega\mu\varepsilon_c(\omega)\mathbf{E}(\mathbf{k},\omega) - \mathrm{i}\left[\!\!\left[\frac{4\pi}{c}\right]\!\!\right]\mu\mathbf{J}_0(\mathbf{k},\omega) \ , \tag{3.2.8}$$

where

$$\varepsilon_c(\omega) = \varepsilon(\omega) + \mathrm{i}[\![4\pi]\!]\frac{\sigma(\omega)}{\omega} \tag{3.2.9}$$

is the *complex permittivity* of the dispersive, conducting medium, as defined in (2.1.22).

The spectrum of the charge density may be expressed in terms of the spectrum of the current density through the use of the equation of continuity

$$\nabla \cdot \mathscr{J}(\mathbf{r},t) = -\frac{\partial \varrho(\mathbf{r},t)}{\partial t} \ . \tag{3.2.10}$$

The Fourier-Laplace transform of this equation then gives, with the fact that the initial value of ϱ vanishes,

$$\mathrm{i}\mathbf{k}\cdot\mathbf{J}(\mathbf{k},\omega)=\mathrm{i}\omega\rho(\mathbf{k},\omega)\ .$$

Substitution of (3.2.3) into this equation and solving for the spectrum of the charge density then yields

$$\rho(\mathbf{k},\omega)=\frac{\mathbf{k}\cdot\mathbf{J}_0(\mathbf{k},\omega)+\sigma(\omega)\mathbf{k}\cdot\mathbf{E}(\mathbf{k},\omega)}{\omega}\ . \tag{3.2.11}$$

With this result (3.2.7) becomes

$$\mathbf{k}\cdot\mathbf{E}(\mathbf{k},\omega)=-[\![4\pi]\!]\mathrm{i}\frac{\mathbf{k}\cdot\mathbf{J}_0(\mathbf{k},\omega)+\sigma(\omega)\mathbf{k}\cdot\mathbf{E}(\mathbf{k},\omega)}{\omega\varepsilon(\omega)}\ ,$$

so that

$$\mathbf{k}\cdot\mathbf{E}(\mathbf{k},\omega)=-[\![4\pi]\!]\mathrm{i}\frac{1}{\omega\varepsilon_c(\omega)}\mathbf{k}\cdot\mathbf{J}_0(\mathbf{k},\omega)\ , \tag{3.2.12}$$

where $\varepsilon_c(\omega)$ is again the complex permittivity of the dispersive, conducting medium.

With (3.2.8 and 12), the spatio-temporal frequency form of the set of Maxwell's equations (3.2.6) take the final form

$$\mathbf{k}\cdot\mathbf{E}(\mathbf{k},\omega)=-[\![4\pi]\!]\mathrm{i}\frac{1}{\omega\varepsilon_c(\omega)}\mathbf{k}\cdot\mathbf{J}_0(\mathbf{k},\omega)\ , \tag{3.2.13a}$$

$$\mathbf{k}\times\mathbf{E}(\mathbf{k},\omega)=\left[\!\!\left[\frac{1}{c}\right]\!\!\right]\omega\mathbf{B}(\mathbf{k},\omega)\ , \tag{3.2.13b}$$

$$\mathbf{k}\cdot\mathbf{B}(\mathbf{k},\omega)=0\ , \tag{3.2.13c}$$

$$\mathbf{k}\times\mathbf{B}(\mathbf{k},\omega)=-\left[\!\!\left[\frac{1}{c}\right]\!\!\right]\omega\mu\varepsilon_c(\omega)\mathbf{E}(\mathbf{k},\omega)-\left[\!\!\left[\frac{4\pi}{c}\right]\!\!\right]\mathrm{i}\mu\mathbf{J}_0(\mathbf{k},\omega)\ . \tag{3.2.13d}$$

The original set of differential relations for the electric and magnetic field vectors have thus been replaced by a set of algebraic equations in which the spatio-temporal spectra of the electric and magnetic field vectors are the only remaining unknown quantities.

Consider first the solution for the spatio-temporal frequency spectrum of the electric field vector. To that end, the vector product of **k** with (3.2.13b) gives

$$\mathbf{k}\times(\mathbf{k}\times\mathbf{E})=\left[\!\!\left[\frac{1}{c}\right]\!\!\right]\omega\mathbf{k}\times\mathbf{B}\ . \tag{3.2.14}$$

With the vector identity for the triple cross product the left-hand side of this equation may be written as

$$\mathbf{k} \times (\mathbf{k} \times \mathbf{E}) = (\mathbf{k} \cdot \mathbf{E})\mathbf{k} - k^2\mathbf{E}$$

$$= -[\![4\pi]\!]\mathrm{i}\frac{1}{\omega\varepsilon_c}(\mathbf{k} \cdot \mathbf{J}_0)\mathbf{k} - k^2\mathbf{E} \ ,$$

with substitution from (3.2.13a). Furthermore, the right-hand side of (3.2.14) can be expressed in terms of $\mathbf{E}(\mathbf{k}, \omega)$ with (3.2.13d). With these substitutions, Eq. (3.2.14) becomes

$$\left(k^2 - \frac{\mu\varepsilon_c}{[\![c^2]\!]}\omega^2\right)\mathbf{E} = \left[\!\!\left[\frac{4\pi}{c^2}\right]\!\!\right]\mathrm{i}\omega\mu\mathbf{J}_0 - [\![4\pi]\!]\mathrm{i}\frac{\mathbf{k} \cdot \mathbf{J}_0}{\omega\varepsilon_c}\mathbf{k} \ .$$

If one then defines the *complex velocity* in the dispersive, conducting medium with complex permittivity $\varepsilon_c(\omega)$ as

$$v^2(\omega) \equiv \frac{[\![c^2]\!]}{\mu\varepsilon_c(\omega)} \tag{3.2.15}$$

one finally obtains

$$\mathbf{E}(\mathbf{k}, \omega) = [\![4\pi]\!]\mathrm{i}\frac{\left[\!\!\left[\frac{1}{c^2}\right]\!\!\right]\omega\mu\mathbf{J}_0(\mathbf{k}, \omega) - \dfrac{\mathbf{k} \cdot \mathbf{J}_0(\mathbf{k}, \omega)}{\omega\varepsilon_c(\omega)}\mathbf{k}}{k^2 - \dfrac{\omega^2}{v^2(\omega)}} \ . \tag{3.2.16}$$

This equation can be somewhat simplified through the use of the vector identity

$$(\mathbf{k} \cdot \mathbf{J}_0)\mathbf{k} = \mathbf{k} \times (\mathbf{k} \times \mathbf{J}_0) + k^2\mathbf{J}_0 \ ,$$

with which the numerator in (3.2.16) becomes

$$\left[\!\!\left[\frac{1}{c^2}\right]\!\!\right]\omega\mu\mathbf{J}_0 - \frac{\mathbf{k} \cdot \mathbf{J}_0}{\omega\varepsilon_c}\mathbf{k} = \left[\!\!\left[\frac{1}{c^2}\right]\!\!\right]\omega\mu\mathbf{J}_0 - \frac{\mathbf{k} \times (\mathbf{k} \times \mathbf{J}_0)}{\omega\varepsilon_c} - \frac{k^2}{\omega\varepsilon_c}\mathbf{J}_0$$

$$= \frac{1}{\omega\varepsilon_c}\left[\left(\frac{\omega^2}{v^2} - k^2\right)\mathbf{J}_0 - \mathbf{k} \times (\mathbf{k} \times \mathbf{J}_0)\right] .$$

With this substitution, Eq. (3.2.16) may be rewritten as

$$\mathbf{E}(\mathbf{k}, \omega) = [\![4\pi]\!]\frac{\mathrm{i}}{\omega\varepsilon_c(\omega)}\left\{\frac{\mathbf{k} \times (\mathbf{k} \times \mathbf{J}_0(\mathbf{k}, \omega))}{\dfrac{\omega^2}{v^2(\omega)} - k^2} - \mathbf{J}_0(\mathbf{k}, \omega)\right\} . \tag{3.2.17}$$

The solution for the spatio-temporal frequency spectrum of the magnetic field vector is obtained by taking the vector product of $\mathbf{k}$ with (3.2.13d) and applying (3.2.13b and c) as follows: Beginning with the vector identity

$$\mathbf{k} \times (\mathbf{k} \times \mathbf{B}) = (\mathbf{k} \cdot \mathbf{B})\mathbf{k} - k^2 \mathbf{B} = -k^2 B \;,$$

substitution from (3.2.13d) then gives

$$\begin{aligned} -k^2 \mathbf{B} &= -\left[\!\left[\frac{1}{c}\right]\!\right] \omega\mu\varepsilon_c \mathbf{k} \times \mathbf{E} - \left[\!\left[\frac{4\pi}{c}\right]\!\right] \mathrm{i}\mu \mathbf{k} \times \mathbf{J}_0 \\ &= -\frac{\omega^2}{v^2}\mathbf{B} - \left[\!\left[\frac{4\pi}{c}\right]\!\right] \mathrm{i}\mu \mathbf{k} \times \mathbf{J}_0 \;, \end{aligned}$$

and hence

$$\mathbf{B}(\mathbf{k},\omega) = \left[\!\left[\frac{4\pi}{c}\right]\!\right] \mathrm{i}\mu \frac{\mathbf{k} \times \mathbf{J}_0(\mathbf{k},\omega)}{k^2 - \dfrac{\omega^2}{v^2(\omega)}} \;. \tag{3.2.18}$$

The inverse Fourier-Laplace transform of (3.2.16–18) then yields the following *Fourier-Laplace representation of the radiation field* in a homogeneous, isotropic, locally linear, temporally dispersive, conducting medium:

$$\mathscr{E}(\mathbf{r},t) = \mathrm{i}\frac{[\![4\pi]\!]}{(2\pi)^4} \int_C d\omega \iiint_{-\infty}^{\infty} d^3k \frac{\left[\!\left[\dfrac{1}{c^2}\right]\!\right] \omega\mu \mathbf{J}_0(\mathbf{k},\omega) - \dfrac{\mathbf{k}\cdot \mathbf{J}_0(\mathbf{k},\omega)}{\omega\varepsilon_c(\omega)}\mathbf{k}}{k^2 - \dfrac{\omega^2}{v^2(\omega)}} e^{\mathrm{i}(\mathbf{k}\cdot\mathbf{r}-\omega t)} \tag{3.2.19a}$$

$$= -\mathrm{i}\frac{[\![4\pi]\!]}{(2\pi)^4} \int_C \frac{d\omega}{\omega\varepsilon_c(\omega)} \iiint_{-\infty}^{\infty} d^3k \left[\mathbf{J}_0(\mathbf{k},\omega) + \frac{\mathbf{k} \times (\mathbf{k} \times \mathbf{J}_0(\mathbf{k},\omega))}{k^2 - \dfrac{\omega^2}{v^2(\omega)}} \right] e^{\mathrm{i}(\mathbf{k}\cdot\mathbf{r}-\omega t)}, \tag{3.2.19b}$$

$$\mathscr{B}(\mathbf{r},t) = \mathrm{i}\left[\!\left[\frac{4\pi}{c}\right]\!\right] \frac{\mu}{(2\pi)^4} \int_C d\omega \iiint_{-\infty}^{\infty} d^3k \frac{\mathbf{k} \times \mathbf{J}_0(\mathbf{k},\omega)}{k^2 - \dfrac{\omega^2}{v^2(\omega)}} e^{\mathrm{i}(\mathbf{k}\cdot\mathbf{r}-\omega t)} \;. \tag{3.2.20}$$

In order to apply contour integration techniques to simplify (3.2.19 and 20), it is necessary to express the ω-integral in a form that involves only positive real values of ω. To that end, consider the sum

$$\mathscr{I} - \mathscr{I}_+ + \mathscr{I}_- \;, \tag{3.2.21}$$

where

$$\mathscr{I}_\pm = \int_{C_\pm} d\omega \iiint_{-\infty}^{\infty} d^3k \cdot \mathbf{A}(\mathbf{k},\omega) e^{i(\mathbf{k}\cdot\mathbf{r}-\omega t)} , \tag{3.2.22}$$

where C_+ is the contour $\omega = \omega' + ia$ where ω' varies from 0 to ∞, and where C_- is the contour $\omega = \omega' + ia$ where ω' varies from $-\infty$ to 0. If the spectral function $\mathbf{A}(\mathbf{k},\omega)$ has the property that

$$\mathbf{A}(-\mathbf{k}, -\omega' + ia) = \mathbf{A}^*(\mathbf{k}, \omega' + ia) , \tag{3.2.23}$$

then

$$\begin{aligned}\mathscr{I}_+^* &= \int_0^\infty d(\omega' + ia)^* \iiint_{-\infty}^{\infty} d^3k \cdot \mathbf{A}^*(\mathbf{k}, \omega' + ia) e^{-i[\mathbf{k}\cdot\mathbf{r}-(\omega'+ia)^* t]} \\ &= \int_0^\infty d(\omega' - ia) \iiint_{-\infty}^{\infty} d^3k \cdot \mathbf{A}^*(\mathbf{k}, \omega' + ia) e^{-i[\mathbf{k}\cdot\mathbf{r}-(\omega'-ia)t]} .\end{aligned}$$

Under the change of variable $\mathbf{k} \to -\mathbf{k}$, $\omega' \to -\omega'$ this integral becomes

$$\begin{aligned}\mathscr{I}_+^* &= \int_0^{-\infty} d(-\omega' - ia) \iiint_{-\infty}^{\infty} d^3k \cdot \mathbf{A}^*(-\mathbf{k}, -\omega' + ia) e^{-i[-\mathbf{k}\cdot\mathbf{r}+(\omega'+ia)t]} \\ &= \int_{-\infty}^{0} d(\omega' + ia) \iiint_{-\infty}^{\infty} d^3k \cdot \mathbf{A}(\mathbf{k}, \omega' + ia) e^{i[\mathbf{k}\cdot\mathbf{r}-(\omega'+ia)t]} \\ &= \int_{ia-\infty}^{ia} d\omega \iiint_{-\infty}^{\infty} d^3k \cdot \mathbf{A}(\mathbf{k},\omega) e^{i(\mathbf{k}\cdot\mathbf{r}-\omega t)} = \mathscr{I}_- .\end{aligned} \tag{3.2.24}$$

Hence, if the property specified in (3.2.23) is satisfied, then

$$\begin{aligned}\mathscr{I} &= \mathscr{I}_+ + \mathscr{I}_- = \mathscr{I}_+ + \mathscr{I}_+^* \\ &= 2\,\mathrm{Re}\{\mathscr{I}_+\} ,\end{aligned} \tag{3.2.25}$$

where the ω-integral in $\mathscr{I}_+$ involves only positive real values of ω.

In order to determine if the spectral functions appearing in (3.2.19 and 20) satisfy the property (3.2.23), consider first the Fourier-Laplace transform $\mathbf{b}(\mathbf{k},\omega)$ of a real-valued function $\mathscr{b}(\mathbf{r},t)$, viz.

$$\mathbf{b}(\mathbf{k},\omega) = \int_0^\infty dt \iiint_{-\infty}^{\infty} d^3r \cdot \mathscr{b}(\mathbf{r},t) e^{-i(\mathbf{k}\cdot\mathbf{r}-\omega t)} ,$$

in which case

$$\begin{aligned}\mathbf{b}(-\mathbf{k}, -\omega' + ia) &= \int_0^\infty dt \iiint_{-\infty}^{\infty} d^3r \cdot \mathscr{b}(\mathbf{r},t) e^{+i[\mathbf{k}\cdot\mathbf{r}-(\omega'-ia)t]} \\ &= \mathbf{b}^*(\mathbf{k}, \omega' + ia) = \mathbf{b}^*(\mathbf{k},\omega)\end{aligned} \tag{3.2.26}$$

with $\omega = \omega' + ia$. Since $\mathscr{J}_0(\mathbf{r}, t)$, $\hat{\varepsilon}(t)$, and $\hat{\sigma}(t)$ are all real-valued functions, their spectra satisfy the symmetry relations

$$\mathbf{J}_0(-\mathbf{k}, -\omega' + ia) = \mathbf{J}_0^*(\mathbf{k}, \omega) \ , \tag{3.2.27}$$

$$\varepsilon(-\omega' + ia) = \varepsilon^*(\omega) \ , \tag{3.2.28}$$

$$\sigma(-\omega' + ia) = \sigma^*(\omega) \ . \tag{3.2.29}$$

One then has for the complex permittivity

$$\begin{aligned}
\varepsilon_c(-\omega' + ia) &= \varepsilon(-\omega' + ia) + i[\![4\pi]\!]\frac{\sigma(-\omega' + ia)}{-\omega' + ia} \\
&= \varepsilon^*(\omega) - i[\![4\pi]\!]\frac{\sigma^*(\omega)}{\omega' - ia} \\
&= \left[\varepsilon(\omega) + i[\![4\pi]\!]\frac{\sigma(\omega)}{\omega}\right]^* = \varepsilon_c^*(\omega) \ .
\end{aligned} \tag{3.2.30}$$

As a consequence of this result one finds that

$$\begin{aligned}
i(-\omega' + ia)\varepsilon_c(-\omega' + ia) &= [i(\omega' + ia)\varepsilon_c(\omega)]^* \\
&= [i\omega\varepsilon_c(\omega)]^* \ ,
\end{aligned} \tag{3.2.31}$$

and hence

$$\begin{aligned}
\frac{(-\omega' + ia)^2}{v^2(-\omega' + ia)} &= \left[\!\left[\frac{1}{c^2}\right]\!\right]\mu\varepsilon_c(-\omega' + ia)(-\omega' + ia)^2 \\
&= \left[\!\left[\frac{1}{c^2}\right]\!\right]\mu\varepsilon_c^*(\omega)(\omega' - ia)^2 \\
&= \left[\!\left[\frac{1}{c^2}\right]\!\right]\mu[\varepsilon_c(\omega)(\omega' + ia)^2]^* \\
&= \left[\!\left[\frac{1}{c^2}\right]\!\right]\mu[\omega^2\varepsilon_c(\omega)]^* = \left(\frac{\omega^2}{v^2(\omega)}\right)^* \ .
\end{aligned} \tag{3.2.32}$$

It follows immediately from the relations expressed in (3.2.27–32) that the spectral functions appearing in (3.2.19 and 20) satisfy the condition specified in (3.2.23). The Fourier-Laplace representation of the radiation field in a homogeneous, isotropic, locally linear, temporally dispersive, conducting medium may then be expressed as

$$\mathscr{E}(\mathbf{r},t) = \frac{[\![4\pi]\!]}{8\pi^4}\mathbb{R}\mathrm{e}\left\{ \mathrm{i}\int_{C_+} d\omega \iiint_{-\infty}^{\infty} d^3k \frac{\left[\!\!\left[\frac{1}{c^2}\right]\!\!\right]\omega\mu\mathbf{J}_0(\mathbf{k},\omega) - \frac{\mathbf{k}\cdot\mathbf{J}_0(\mathbf{k},\omega)}{\omega\varepsilon_c(\omega)}\mathbf{k}}{k^2 - \frac{\omega^2}{v^2(\omega)}} \mathrm{e}^{\mathrm{i}(\mathbf{k}\cdot\mathbf{r}-\omega t)} \right\} \tag{3.2.33a}$$

$$= -\frac{[\![4\pi]\!]}{8\pi^4}\mathbb{R}\mathrm{e}\left\{ \mathrm{i}\int_{C_+} \frac{d\omega}{\omega\varepsilon_c(\omega)} \iiint_{-\infty}^{\infty} d^3k \left[\mathbf{J}_0(\mathbf{k},\omega) + \frac{\mathbf{k}\times[\mathbf{k}\times\mathbf{J}_0(\mathbf{k},\omega)]}{k^2 - \frac{\omega^2}{v^2(\omega)}} \right] \mathrm{e}^{\mathrm{i}(\mathbf{k}\cdot\mathbf{r}-\omega t)} \right\} \tag{3.2.33b}$$

$$\mathscr{B}(\mathbf{r},t) = \left[\!\!\left[\frac{4\pi}{c}\right]\!\!\right]\frac{\mu}{8\pi^4}\mathbb{R}\mathrm{e}\left\{ \mathrm{i}\int_{C_+} d\omega \iiint_{-\infty}^{\infty} d^3k \frac{\mathbf{k}\times\mathbf{J}_0(\mathbf{k},\omega)}{k^2 - \frac{\omega^2}{v^2(\omega)}} \mathrm{e}^{\mathrm{i}(\mathbf{k}\cdot\mathbf{r}-\omega t)} \right\} \tag{3.2.34}$$

for $t > 0$, where C_+ is the contour $\omega = \omega' + \mathrm{i}a$, where ω' varies from 0 to ∞ and where a is fixed at a value greater than the abscissa of absolute convergence for the radiation problem. This final form of the Fourier-Laplace representations of the electric and magnetic field vectors explicitly exhibits their real-valued character.

3.3 The Scalar and Vector Potentials of the Radiation Field

Since the magnetic induction field $\mathscr{B}(\mathbf{r},t)$ is a transverse (or solenoidal) vector field, it can be expressed as the curl of another vector field $\mathscr{A}(\mathbf{r},t)$ as

$$\mathscr{B}(\mathbf{r},t) = \nabla\times\mathscr{A}(\mathbf{r},t)\ , \tag{3.3.1}$$

where $\mathscr{A}(\mathbf{r},t)$ is the *vector potential*. It is then seen from (3.2.20 and 34) that such a vector potential for the radiation field is given by

$$\mathscr{A}(\mathbf{r},t) = \left[\!\!\left[\frac{4\pi}{c}\right]\!\!\right]\frac{\mu}{(2\pi)^4}\int_{C} d\omega \iiint_{-\infty}^{\infty} d^3k \frac{\mathbf{J}_0(\mathbf{k},\omega)}{k^2 - \frac{\omega^2}{v^2(\omega)}} \mathrm{e}^{\mathrm{i}(\mathbf{k}\cdot\mathbf{r}-\omega t)} \tag{3.3.2a}$$

$$= \left[\!\!\left[\frac{4\pi}{c}\right]\!\!\right]\frac{\mu}{8\pi^4}\mathbb{R}\mathrm{e}\left\{ \int_{C_+} d\omega \iiint_{-\infty}^{\infty} d^3k \frac{\mathbf{J}_0(\mathbf{k},\omega)}{k^2 - \frac{\omega^2}{v^2(\omega)}} \mathrm{e}^{\mathrm{i}(\mathbf{k}\cdot\mathbf{r}-\omega t)} \right\} \tag{3.3.2b}$$

Comparison of (3.3.2a) with (3.2.19a), and (3.3.2b) with (3.2.33a) shows that the electric field vector $\mathscr{E}(\mathbf{r}, t)$ can be expressed as

$$\mathscr{E}(\mathbf{r}, t) = -i\frac{[\![4\pi]\!]}{(2\pi)^4}\int_C \frac{d\omega}{\omega\varepsilon_c(\omega)}\iiint_{-\infty}^{\infty} d^3k \frac{(\mathbf{k}\cdot\mathbf{J}_0(\mathbf{k},\omega))\mathbf{k}}{k^2 - \dfrac{\omega^2}{v^2(\omega)}} e^{i(\mathbf{k}\cdot\mathbf{r}-\omega t)} - \left[\!\!\left[\frac{1}{c}\right]\!\!\right]\frac{\partial \mathscr{A}(\mathbf{r}, t)}{\partial t}$$

$$= -\frac{[\![4\pi]\!]}{8\pi^4}\mathrm{Re}\left\{ i\int_{C_+} \frac{d\omega}{\omega\varepsilon_c(\omega)}\iiint_{-\infty}^{\infty} d^3k \frac{(\mathbf{k}\cdot\mathbf{J}_0(\mathbf{k},\omega))\mathbf{k}}{k^2 - \dfrac{\omega^2}{v^2(\omega)}} e^{i(\mathbf{k}\cdot\mathbf{r}-\omega t)} \right.$$

$$\left. - \left[\!\!\left[\frac{1}{c}\right]\!\!\right]\frac{\partial \mathscr{A}(\mathbf{r}, t)}{\partial t} \right\}, \tag{3.3.3}$$

which may be written in the form

$$\mathscr{E}(\mathbf{r}, t) = -\nabla\varphi(\mathbf{r}, t) - \left[\!\!\left[\frac{1}{c}\right]\!\!\right]\frac{\partial \mathscr{A}(\mathbf{r}, t)}{\partial t}, \tag{3.3.4}$$

where

$$\varphi(\mathbf{r}, t) = \frac{[\![4\pi]\!]}{(2\pi)^4}\int_C \frac{d\omega}{\omega\varepsilon_c(\omega)}\iiint_{-\infty}^{\infty} d^3k \frac{\mathbf{k}\cdot\mathbf{J}_0(\mathbf{k},\omega)}{k^2 - \dfrac{\omega^2}{v^2(\omega)}} e^{i(\mathbf{k}\cdot\mathbf{r}-\omega t)} \tag{3.3.5a}$$

$$= \frac{[\![4\pi]\!]}{8\pi^4}\mathrm{Re}\left\{ \int_{C_+} \frac{d\omega}{\omega\varepsilon_c(\omega)}\iiint_{-\infty}^{\infty} d^3k \frac{\mathbf{k}\cdot\mathbf{J}_0(\mathbf{k},\omega)}{k^2 - \dfrac{\omega^2}{v^2(\omega)}} e^{i(\mathbf{k}\cdot\mathbf{r}-\omega t)} \right\} \tag{3.3.5b}$$

is a *scalar potential* for the radiation field. Hence, by determining the vector and scalar potentials $\mathscr{A}(\mathbf{r}, t)$ and $\varphi(\mathbf{r}, t)$, respectively, the electric and magnetic field vectors of the radiation field are directly obtained from (3.3.4 and 1), respectively. The problem of determining $\mathscr{E}(\mathbf{r}, t)$ and $\mathscr{B}(\mathbf{r}, t)$ for a given current source $\mathscr{J}_0(\mathbf{r}, t)$ has thus been reduced to determining $\mathscr{A}(\mathbf{r}, t)$ and $\varphi(\mathbf{r}, t)$, which are specified by somewhat simpler expressions.

Since the magnetic induction field $\mathscr{B}(\mathbf{r}, t)$ is defined through (3.3.1) in terms of the curl of the vector potential field $\mathscr{A}(\mathbf{r}, t)$, it is then clear that the vector potential is arbitrary to the extent that the gradient of some scalar function $\Lambda(\mathbf{r}, t)$ can be added to $\mathscr{A}(\mathbf{r}, t)$ without affecting $\mathscr{B}(\mathbf{r}, t)$. That is, $\mathscr{B}(\mathbf{r}, t)$ is left unchanged by the transformation

$$\mathscr{A}(\mathbf{r}, t) \to \mathscr{A}'(\mathbf{r}, t) = \mathscr{A}(\mathbf{r}, t) + \nabla\Lambda(\mathbf{r}, t). \tag{3.3.6}$$

Under this transformation the electric field vector (3.3.4) becomes

$$\mathscr{E}(\mathbf{r},t) = -\nabla\varphi(\mathbf{r},t) - \left[\!\left[\frac{1}{c}\right]\!\right]\frac{\partial\mathscr{A}'(\mathbf{r},t)}{\partial t}$$

$$= -\nabla\varphi(\mathbf{r},t) - \left[\!\left[\frac{1}{c}\right]\!\right]\frac{\partial\mathscr{A}(\mathbf{r},t)}{\partial t} - \left[\!\left[\frac{1}{c}\right]\!\right]\nabla\frac{\partial\Lambda(\mathbf{r},t)}{\partial t} \ .$$

Hence, in order that the electric field remain unchanged under the transformation (3.3.6), the scalar potential must be simutaneously transformed as

$$\varphi(\mathbf{r},t) \to \varphi'(\mathbf{r},t) = \varphi(\mathbf{r},t) - \left[\!\left[\frac{1}{c}\right]\!\right]\frac{\partial\Lambda(\mathbf{r},t)}{\partial t} \ . \tag{3.3.7}$$

The complete transformation specified by the pair of equations (3.3.6 and 7) is called a *gauge transformation* and the invariance of the field vectors $\mathscr{E}$ and $\mathscr{B}$ under such a transformation is referred to as *gauge invariance*. The scalar function $\Lambda(\mathbf{r},t)$ is called the *gauge function* of the transformation.

The freedom of choice that is implied by (3.3.6, 7) means that one can always choose a set of potentials $\{\mathscr{A}', \varphi'\}$ such that [3.2]

$$\nabla\cdot\mathscr{A}'(\mathbf{r},t) + \frac{[\![c]\!]}{c^2}\frac{\partial\varphi'(\mathbf{r},t)}{\partial t} = 0 \ , \tag{3.3.8}$$

which is called the *Lorentz condition.* In order to prove that a pair of potentials $\{\mathscr{A}', \varphi'\}$ can always be found to satisfy the Lorentz condition, suppose that the original pair of potentials $\{\mathscr{A}, \varphi\}$ do not satisfy (3.3.8). One may then undertake a gauge transformation to the new pair of potentials $\{\mathscr{A}', \varphi'\}$ and demand that $\mathscr{A}'$ and φ' satisfy the Lorentz condition, viz.

$$\nabla\cdot\mathscr{A}'(\mathbf{r},t) + \frac{[\![c]\!]}{c^2}\frac{\partial\varphi'(\mathbf{r},t)}{\partial t} = 0$$

so that

$$\nabla^2\Lambda(\mathbf{r},t) - \frac{1}{c^2}\frac{\partial^2\Lambda(\mathbf{r},t)}{\partial t^2} = -\left(\nabla\cdot\mathscr{A}(\mathbf{r},t) + \frac{[\![c]\!]}{c^2}\frac{\partial\varphi(\mathbf{r},t)}{\partial t}\right) . \tag{3.3.9}$$

Thus, provided that a gauge function $\Lambda(\mathbf{r},t)$ can be found that satisfies (3.3.9), the new pair of potentials $\{\mathscr{A}', \varphi'\}$ will satisfy the Lorentz condition.

Even for a pair of potentials that satisfy the Lorentz condition (3.3.8) there still remains a certain degree of arbitrariness. This is because the *restricted gauge transformation*

$$\begin{aligned} \mathscr{A}(\mathbf{r},t) &\to \mathscr{A}'(\mathbf{r},t) = \mathscr{A}(\mathbf{r},t) + \nabla\psi(\mathbf{r},t) \ , \\ \varphi(\mathbf{r},t) &\to \varphi'(\mathbf{r},t) = \varphi(\mathbf{r},t) - \left[\!\left[\frac{1}{c}\right]\!\right]\frac{\partial\psi(\mathbf{r},t)}{\partial t} \ , \end{aligned} \tag{3.3.10}$$

with

$$\nabla^2\psi(\mathbf{r},t) - \frac{1}{c^2}\frac{\partial^2\psi(\mathbf{r},t)}{\partial t^2} = 0 \tag{3.3.11}$$

preserves the Lorentz condition provided that $\{\mathscr{A},\varphi\}$ satisfy it initially, as is readily evident from (3.3.9). All pairs of potentials $\{\mathscr{A},\varphi\}$ in this restricted class belong to the so-called *Lorentz gauge.*

The divergence of the vector potential (3.3.2) for the radiation field is found as

$$\nabla\cdot\mathscr{A}(\mathbf{r},t) = \left[\!\left[\frac{4\pi}{c}\right]\!\right]\frac{\mu}{(2\pi)^4}\mathrm{i}\int_C d\omega \iiint_{-\infty}^{\infty} d^3k \frac{\mathbf{k}\cdot\mathbf{J}_0(\mathbf{k},\omega)}{k^2 - \dfrac{\omega^2}{v^2(\omega)}}\mathrm{e}^{\mathrm{i}(\mathbf{k}\cdot\mathbf{r}-\omega t)}\,, \tag{3.3.12}$$

and the partial time derivative of the scalar potential (3.3.5) for the radiation field is

$$\begin{aligned}\frac{\partial\varphi(\mathbf{r},t)}{\partial t} &= -\frac{[\![4\pi]\!]}{(2\pi)^4}\mathrm{i}\int_C \frac{d\omega}{\varepsilon_c(\omega)} \iiint_{-\infty}^{\infty} d^3k \frac{\mathbf{k}\cdot\mathbf{J}_0(\mathbf{k},\omega)}{k^2 - \dfrac{\omega^2}{v^2(\omega)}}\mathrm{e}^{\mathrm{i}(\mathbf{k}\cdot\mathbf{r}-\omega t)}\\ &= -\left[\!\left[\frac{4\pi}{c^2}\right]\!\right]\frac{\mu}{(2\pi)^4}\mathrm{i}\int_C d\omega\cdot v^2(\omega) \iiint_{-\infty}^{\infty} d^3k \frac{\mathbf{k}\cdot\mathbf{J}_0(\mathbf{k},\omega)}{k^2 - \dfrac{\omega^2}{v^2(\omega)}}\mathrm{e}^{\mathrm{i}(\mathbf{k}\cdot\mathbf{r}-\omega t)}\,,\end{aligned} \tag{3.3.13}$$

where $v^2(\omega)$ is the complex velocity in the dispersive, conducting medium, defined in (3.2.15). With these two relations the left-hand side of the Lorentz condition (3.3.8) is found to be

$$\begin{aligned}&\nabla\cdot\mathscr{A}(\mathbf{r},t) + \frac{[\![c]\!]}{c^2}\frac{\partial\varphi(\mathbf{r},t)}{\partial t}\\ &= \left[\!\left[\frac{4\pi}{c}\right]\!\right]\frac{\mu}{(2\pi)^4}\mathrm{i}\int_C d\omega\cdot\left(1-\frac{v^2(\omega)}{c^2}\right) \iiint_{-\infty}^{\infty} d^3k \frac{\mathbf{k}\cdot\mathbf{J}_0(\mathbf{k},\omega)}{k^2 - \dfrac{\omega^2}{v^2(\omega)}}\mathrm{e}^{\mathrm{i}(\mathbf{k}\cdot\mathbf{r}-\omega t)}\\ &= \left[\!\left[\frac{4\pi}{c}\right]\!\right]\frac{\mu}{8\pi^4}\mathbb{R}\mathrm{e}\left\{\mathrm{i}\int_{C_+} d\omega\cdot\left(1-\frac{v^2(\omega)}{c^2}\right) \iiint_{-\infty}^{\infty} d^3k \frac{\mathbf{k}\cdot\mathbf{J}_0(\mathbf{k},\omega)}{k^2 - \dfrac{\omega^2}{v^2(\omega)}}\mathrm{e}^{\mathrm{i}(\mathbf{k}\cdot\mathbf{r}-\omega t)}\right\}\,,\end{aligned} \tag{3.3.14}$$

which does not, in general, vanish. Hence, the vector and scalar potentials (3.3.2, 5) for the radiation field do not, in general, satisfy the Lorentz condition

(3.3.8). Nevertheless, a gauge function $\Lambda(\mathbf{r},t)$ can always be determined from (3.3.9) so as to define a gauge transformation to a new pair of potentials $\{\mathscr{A}', \varphi'\}$ that will satisfy the Lorentz condition. However, the vector and scalar potentials defined in (3.3.2 and 5) are clearly the simplest and most natural forms of the potentials for the radiation problem.

3.3.1 The Special Case of a Nonconducting, Nondispersive Medium

It is because of the frequency dependence of the complex velocity appearing in (3.3.13) that $\nabla \cdot \mathscr{A}$ and $\partial\varphi/\partial t$ for the radiation problem cannot be directly related (without a proper gauge transformation) except in certain special cases. To that end, consider an idealized nondispersive, nonconducting ($\sigma = 0$) dielectric medium with constant, real-valued dielectric permittivity ε, so that $\varepsilon_c(\omega) = \varepsilon$ and $v(\omega) = v$, where

$$v^2 = \frac{[\![c^2]\!]}{\mu\varepsilon} . \tag{3.3.15}$$

The vector and scalar potentials (3.3.2 and 5) for the radiation field are then seen to satisfy the generalized Lorentz condition

$$\nabla \cdot \mathscr{A}(\mathbf{r},t) + \frac{[\![c]\!]}{v^2}\frac{\partial\varphi(\mathbf{r},t)}{\partial t} = 0 , \tag{3.3.16}$$

so that $\varphi(\mathbf{r},t)$ can be determined directly from $\mathscr{A}(\mathbf{r},t)$.

From (3.3.2a) one has that

$$\begin{aligned}
\nabla^2\mathscr{A}(\mathbf{r},t) - \frac{1}{v^2}\frac{\partial^2\mathscr{A}(\mathbf{r},t)}{\partial t^2}
&= \left[\!\left[\frac{4\pi}{c}\right]\!\right]\frac{\mu}{(2\pi)^4}\int_C d\omega \iiint_{-\infty}^{\infty} d^3k \frac{\mathbf{J}_0(\mathbf{k},\omega)}{k^2 - \omega^2/v^2}\left(-k^2 + \frac{\omega^2}{v^2}\right)e^{i(\mathbf{k}\cdot\mathbf{r}-\omega t)} \\
&= -\left[\!\left[\frac{4\pi}{c}\right]\!\right]\frac{\mu}{(2\pi)^4}\int_C d\omega \iiint_{-\infty}^{\infty} d^3k \cdot \mathbf{J}_0(\mathbf{k},\omega)e^{i(\mathbf{k}\cdot\mathbf{r}-\omega t)} \\
&= -\left[\!\left[\frac{4\pi}{c}\right]\!\right]\mu\mathscr{F}^{-1}\mathscr{L}^{-1}\{\mathbf{J}_0(\mathbf{k},\omega)\} ,
\end{aligned}$$

and the vector potential is then seen to satisfy the inhomogeneous wave equation

$$\nabla^2\mathscr{A}(\mathbf{r},t) - \frac{1}{v^2}\frac{\partial^2\mathscr{A}(\mathbf{r},t)}{\partial t^2} = -\left[\!\left[\frac{4\pi}{c}\right]\!\right]\mu\mathscr{J}_0(\mathbf{r},t) \tag{3.3.17}$$

when the medium is nondispersive and nonconducting. Similarly, from (3.3.5a) one has that

$$\nabla^2\varphi(\mathbf{r},t)-\frac{1}{v^2}\frac{\partial^2\varphi(\mathbf{r},t)}{\partial t^2}$$

$$=\frac{[\![4\pi]\!]}{(2\pi)^4}\frac{1}{\varepsilon}\int_C\frac{d\omega}{\omega}\iiint_{-\infty}^{\infty}d^3k\frac{\mathbf{k}\cdot\mathbf{J}_0(\mathbf{k},\omega)}{k^2-\omega^2/v^2}\left(-k^2+\frac{\omega^2}{v^2}\right)e^{i(\mathbf{k}\cdot\mathbf{r}-\omega t)}$$

$$=-\frac{[\![4\pi]\!]}{(2\pi)^4}\frac{1}{\varepsilon}\int_C d\omega\iiint_{-\infty}^{\infty}d^3k\frac{\mathbf{k}\cdot\mathbf{J}_0(\mathbf{k},\omega)}{\omega}e^{i(\mathbf{k}\cdot\mathbf{r}-\omega t)}\ .$$

From (3.2.11) with $\sigma=0$ one finds

$$\rho_0(\mathbf{k},\omega)=\frac{\mathbf{k}\cdot\mathbf{J}_0(\mathbf{k},\omega)}{\omega}\ , \tag{3.3.18}$$

and the scalar potential is then seen to satisfy the inhomogeneous wave equation

$$\nabla^2\varphi(\mathbf{r},t)-\frac{1}{v^2}\frac{\partial^2\varphi(\mathbf{r},t)}{\partial t^2}=-\frac{[\![4\pi]\!]}{\varepsilon}\varrho_0(\mathbf{r},t)\ , \tag{3.3.19}$$

where

$$\varrho_0(\mathbf{r},t)=\mathscr{F}^{-1}\mathscr{L}^{-1}\left\{\frac{\mathbf{k}\cdot\mathbf{J}_0(\mathbf{k},\omega)}{\omega}\right\} \tag{3.3.20}$$

is the charge density associated with the external current source $\mathscr{J}_0(\mathbf{r},t)$ as required by the equation of continuity in the nondispersive, nonconducting medium. The relations (3.3.16, 17 and 20) clearly show that $\mathscr{A}(\mathbf{r},t)$ and $\varphi(\mathbf{r},t)$ are the standard vector and scalar potentials, respectively, in the Lorentz gauge.

3.3.2 The Spectral Lorentz Condition for Dispersive, Conducting Media

The temporal frequency transforms of the vector and scalar potentials given in (3.3.2 and 5), respectively, for the radiation field in a homogeneous, isotropic, locally linear, dispersive and conducting medium are found to be

$$\tilde{\mathscr{A}}(\mathbf{r},\omega)=\left[\!\!\left[\frac{4\pi}{c}\right]\!\!\right]\frac{\mu}{(2\pi)^3}\iiint_{-\infty}^{\infty}d^3k\frac{\mathbf{J}_0(\mathbf{k},\omega)}{k^2-\dfrac{\omega^2}{v^2(\omega)}}e^{i\mathbf{k}\cdot\mathbf{r}}\ , \tag{3.3.21}$$

$$\tilde{\varphi}(\mathbf{r},\omega)=\left[\!\!\left[\frac{4\pi}{c^2}\right]\!\!\right]\frac{\mu v^2(\omega)}{(2\pi)^3\omega}\iiint_{-\infty}^{\infty}d^3k\frac{\mathbf{k}\cdot\mathbf{J}_0(\mathbf{k},\omega)}{k^2-\dfrac{\omega^2}{v^2(\omega)}}e^{i\mathbf{k}\cdot\mathbf{r}}\ . \tag{3.3.22}$$

Consider then the temporal spectral domain form of (3.3.16) with a frequency-dependent complex velocity as given by (3.2.15). With (3.3.21, 22) one has

$$\nabla\cdot\tilde{\mathscr{A}}(\mathbf{r},\omega) - \mathrm{i}\omega\frac{[\![c]\!]}{v^2(\omega)}\tilde{\varphi}(\mathbf{r},\omega) = 0 \ , \tag{3.3.23a}$$

which may be rewritten as

$$\nabla\cdot\tilde{\mathscr{A}}(\mathbf{r},\omega) - \left[\!\!\left[\frac{1}{c}\right]\!\!\right]\mathrm{i}\omega\mu\varepsilon_c(\omega)\tilde{\varphi}(\mathbf{r},\omega) = 0 \ . \tag{3.3.23b}$$

It is seen that the temporal frequency spectra of the vector and scalar potentials for the radiation field in a conducting, dispersive medium satisfy a *generalized spectral Lorentz condition.* This condition reduces to the temporal frequency domain form of (3.3.16) when the medium is nonconducting and nondispersive, as well as the temporal frequency domain form of the Lorentz condition (3.3.8) in free space.

Consider now obtaining the individual wave equations satisfied by the radiation potentials. From (3.3.21) one has that

$$\begin{aligned}
&\nabla^2\tilde{\mathscr{A}}(\mathbf{r},\omega) + \frac{\omega^2}{v^2(\omega)}\tilde{\mathscr{A}}(\mathbf{r},\omega) \\
&= \left[\!\!\left[\frac{4\pi}{c}\right]\!\!\right]\frac{\mu}{(2\pi)^3}\iiint_{-\infty}^{\infty} d^3k\frac{\mathbf{J}_0(\mathbf{k},\omega)}{k^2 - \dfrac{\omega^2}{v^2(\omega)}}\left(-k^2 + \frac{\omega^2}{v^2(\omega)}\right)\mathrm{e}^{\mathrm{i}\mathbf{k}\cdot\mathbf{r}} \\
&= -\left[\!\!\left[\frac{4\pi}{c}\right]\!\!\right]\frac{\mu}{(2\pi)^3}\iiint_{-\infty}^{\infty} d^3k\,\mathbf{J}_0(\mathbf{k},\omega)\mathrm{e}^{\mathrm{i}\mathbf{k}\cdot\mathbf{r}} \\
&= -\left[\!\!\left[\frac{4\pi}{c}\right]\!\!\right]\mu\mathscr{F}^{-1}\{\mathbf{J}_0(\mathbf{k},\omega)\} \ ,
\end{aligned}$$

and the temporal frequency spectrum of the vector potential satisfies the inhomogeneous, dispersive wave equation

$$\nabla^2\tilde{\mathscr{A}}(\mathbf{r},\omega) + \frac{\omega^2}{v^2(\omega)}\tilde{\mathscr{A}}(\mathbf{r},\omega) = -\left[\!\!\left[\frac{4\pi}{c}\right]\!\!\right]\mu\tilde{\mathscr{J}}_0(\mathbf{r},\omega) \ . \tag{3.3.24}$$

Similarly, from (3.3.22) one finds that

$$\nabla^2\tilde{\varphi}(\mathbf{r},\omega) + \frac{\omega^2}{v^2(\omega)}\tilde{\varphi}(\mathbf{r},\omega) = -\left[\!\!\left[\frac{4\pi}{c^2}\right]\!\!\right]\frac{\mu v^2(\omega)}{(2\pi)^3\omega}\iiint_{-\infty}^{\infty} d^3k(\mathbf{k}\cdot\mathbf{J}_0(\mathbf{k},\omega))\mathrm{e}^{\mathrm{i}\mathbf{k}\cdot\mathbf{r}} \ .$$

From (3.2.11) it is seen that the spectrum of the charge density may be written as

$$\rho(\mathbf{k},\omega) = \rho_0(\mathbf{k},\omega) + \rho_c(\mathbf{k},\omega) \tag{3.3.25}$$

where

$$\rho_0(\mathbf{k}, \omega) = \frac{\mathbf{k} \cdot \mathbf{J}_0(\mathbf{k}, \omega)}{\omega} \tag{3.3.26}$$

is the *nonconductive charge density*, and where

$$\rho_c(\mathbf{k}, \omega) = \frac{\sigma(\omega)}{\omega}\mathbf{k} \cdot \mathbf{E}(\mathbf{k}, \omega) \tag{3.3.27}$$

is the *conductive charge density*. With substitution of (3.3.26) there results

$$\begin{aligned} \nabla^2 \tilde{\varphi}(\mathbf{r}, \omega) + \frac{\omega^2}{v^2(\omega)}\tilde{\varphi}(\mathbf{r}, \omega) &= -\left[\!\left[\frac{4\pi}{c^2}\right]\!\right]\frac{\mu v^2(\omega)}{(2\pi)^3}\iiint\limits_{-\infty}^{\infty} d^3k \rho_0(\mathbf{k}, \omega) e^{i\mathbf{k} \cdot \mathbf{r}} \\ &= -\left[\!\left[\frac{4\pi}{c^2}\right]\!\right]\mu v^2(\omega)\tilde{\varrho}_0(\mathbf{r}, \omega) , \end{aligned}$$

so that the temporal frequency spectrum of the scalar potential satisfies the inhomogeneous, dispersive wave equation

$$\nabla^2 \tilde{\varphi}(\mathbf{r}, \omega) + \frac{\omega^2}{v^2(\omega)}\tilde{\varphi}(\mathbf{r}, \omega) = -\frac{[\![4\pi]\!]}{\varepsilon_c(\omega)}\tilde{\varrho}_0(\mathbf{r}, \omega) . \tag{3.3.28}$$

Here

$$\tilde{\varrho}_0(\mathbf{r}, \omega) = \mathscr{F}^{-1}\left\{\frac{\mathbf{k} \cdot \mathbf{J}_0(\mathbf{k}, \omega)}{\omega}\right\} \tag{3.3.29}$$

is the temporal frequency spectrum of the nonconductive charge density.

3.4 The Angular Spectrum of Plane Waves Representation of the Radiation Field

The Fourier-Laplace representations (3.2.33, 34) for the electromagnetic field vectors and (3.3.2–5) for the vector and scalar potentials for the radiation field in a homogeneous, isotropic, locally linear, dispersive and conducting medium occupying all of space are each of the general form

$$\mathscr{G}(\mathbf{r}, t) = \frac{i}{(2\pi)^4}\int\limits_C d\omega \iiint\limits_{-\infty}^{\infty} d^3k \frac{\mathbf{F}(\mathbf{k}, \omega)}{k^2 - \dfrac{\omega^2}{v^2(\omega)}} e^{i(\mathbf{k} \cdot \mathbf{r} - \omega t)} \tag{3.4.1a}$$

$$= \frac{1}{8\pi^4}\mathrm{Re}\left\{ i \int\limits_{C_+} d\omega \iiint\limits_{-\infty}^{\infty} d^3k \frac{\mathbf{F}(\mathbf{k}, \omega)}{k^2 - \dfrac{\omega^2}{v^2(\omega)}} e^{i(\mathbf{k} \cdot \mathbf{r} - \omega t)} \right\} , \tag{3.4.1b}$$

where C is the contour $\omega = \omega' + ia$ with ω' varying from $-\infty$ to $+\infty$, while C_+ is the contour $\omega = \omega' + ia$ with ω' varying from 0 to $+\infty$, and where the spectral function $\mathbf{F}(\mathbf{k}, \omega)$ satisfies the property, cf. (3.2.23),

$$\mathbf{F}(-\mathbf{k}, -\omega' + ia) = \mathbf{F}^*(\mathbf{k}, \omega' + ia) \ . \tag{3.4.2}$$

For each equation represented by the generic form (3.4.1), the spectral function $\mathbf{F}(\mathbf{k}, \omega)$ is given by either the scalar or vector product of $\mathbf{k}$ with $\mathbf{J}_0(\mathbf{k}, \omega) = \mathscr{F}\mathscr{L}\{\mathscr{J}_0(\mathbf{r}, t)\}$. Since $\mathscr{J}_0(\mathbf{r}, t)$ and its first-order spatial derivatives vanish for $|z| \geqslant Z$, where Z is a positive constant, it then follows that $\mathscr{F}(\mathbf{r}, t) = \mathscr{F}^{-1}\mathscr{L}^{-1}\{\mathbf{F}(\mathbf{k}, \omega)\}$ also vanishes for $|z| \geqslant Z$, where

$$\mathbf{F}(\mathbf{k}, \omega) = \int_0^\infty dt \iiint_{-\infty}^{\infty} d^3r\, \mathscr{F}(\mathbf{r}, t) e^{-i(\mathbf{k}\cdot\mathbf{r} - \omega t)} \ . \tag{3.4.3}$$

Let $k_z = k_z' + ik_z''$ and $\omega = \omega' + ia$ where k_z', k_z'', ω' and a are all real-valued. With these substitutions (3.4.3) becomes

$$\begin{aligned}\mathbf{F}(\mathbf{k}, \omega) &= \int_{-Z}^{Z} dz \cdot e^{-ik_z'z} e^{k_z''z} \int_0^\infty dt \iint_{-\infty}^{\infty} dx\, dy \cdot \mathscr{F}(\mathbf{r}, t) e^{-i(k_x x + k_y y - \omega' t)} e^{-at} \\ &= \int_{-Z}^{Z} \mathbf{G}(k_x, k_y, \omega, z) e^{-ik_z'z} e^{k_z''z}\, dz \ .\end{aligned} \tag{3.4.4}$$

Since the only integral involving $k_z = k_z' + ik_z''$ is taken over a bounded region, and since the integrand is an analytic function of k_z for all k_z and is also a continuous function of the variable of integration (provided that $\mathscr{F}(\mathbf{r}, t)$ is continuous), then $\mathbf{F}(\mathbf{k}, \omega)$ is an entire function of the complex variable k_z (i.e., it is analytic for all k_z) [3.3]. As a consequence, each component of $\mathbf{F}(\mathbf{k}, \omega)$ is bounded in magnitude as

$$\begin{aligned}|F_i(\mathbf{k}, \omega)| &\leqslant \int_{-Z}^{Z} dz \cdot e^{k_z''z} \int_0^\infty dt \iint_{-\infty}^{\infty} dx\, dy \cdot |\mathscr{F}_i(\mathbf{r}, t)| \\ &\leqslant e^{|k_z''|Z} \int_0^\infty dt \iint_{-\infty}^{\infty} dx\, dy \cdot |\mathscr{F}_i(\mathbf{r}, t)| \\ &\leqslant M_i e^{|k_z''|Z} \ ,\end{aligned} \tag{3.4.5}$$

where M_i is independent of k_z.

Because $\mathbf{F}(\mathbf{k}, \omega)$ is an entire function of k_z, the only singularities of the integrand in (3.4.1) are located where

$$k^2 - \frac{\omega^2}{v^2(\omega)} = 0 \ ,$$

and hence, the integrand has two simple poles that are located at

$$k_z = \pm\left(\frac{\omega^2}{v^2(\omega)} - k_T^2\right)^{1/2} \ , \tag{3.4.6}$$

where $k_T^2 \equiv k_x^2 + k_y^2$. Since $|\mathbf{F}(\mathbf{k}, \omega)|$ is bounded, the k_z-integration in (3.4.1) may then be evaluated by the use of Cauchy's residue theorem once the locations of the simple poles given in (3.4.6) are determined.

In order to determine the location of the simple poles in (3.4.6) in the complex k_z-plane it is necessary to determine the sign of the imaginary part of $\omega^2/v^2(\omega)$. From (3.2.9 and 15) one has that

$$\frac{\omega^2}{v^2(\omega)} = \omega^2 \frac{\mu\varepsilon_c(\omega)}{[\![c^2]\!]} = \frac{\mu}{[\![c^2]\!]}\omega^2\left[\varepsilon(\omega) + \mathrm{i}[\![4\pi]\!]\frac{\sigma(\omega)}{\omega}\right]$$

$$= \frac{\mu}{[\![c^2]\!]}[\omega^2\varepsilon(\omega) + \mathrm{i}[\![4\pi]\!]\omega\sigma(\omega)] \ ,$$

so that with $\omega = \omega' + ia$, where ω' and a are both real-valued,

$$\frac{\omega^2}{v^2(\omega)} = \frac{\mu}{[\![c^2]\!]}\{(\omega'^2 - a^2)\varepsilon_r(\omega) - 2a\omega'\varepsilon_i(\omega) - [\![4\pi]\!]a\sigma_r(\omega) - [\![4\pi]\!]\omega'\sigma_i(\omega)$$

$$+ \mathrm{i}[(\omega'^2 - a^2)\varepsilon_i(\omega) + 2a\omega'\varepsilon_r(\omega) + [\![4\pi]\!]\omega'\sigma_r(\omega) - [\![4\pi]\!]a\sigma_i(\omega)]\} \ , \tag{3.4.7}$$

where $\varepsilon_r(\omega) = \mathrm{Re}\{\varepsilon(\omega)\}$, $\varepsilon_i(\omega) = \mathrm{Im}\{\varepsilon(\omega)\}$, $\sigma_r(\omega) = \mathrm{Re}\{\sigma(\omega)\}$, and $\sigma_i(\omega) = \mathrm{Im}\{\sigma(\omega)\}$. The only situation to be considered here is the one for which the inequalities

$$\varepsilon_i(\omega) \geqslant 0 \ , \tag{3.4.8a}$$

$$\sigma_r(\omega) \geqslant 0 \ , \tag{3.4.8b}$$

are satisfied so that there is no amplification of the field in either the pure dielectric or pure conductor limit (that is, the medium is absorptive in both its dielectric and conductive parts). Furthermore, $\omega' \geqslant 0$ in the representation (3.4.1b) and $a \geqslant 0$ where a is a fixed positive constant that is greater than the abscissa of absolute convergence for the radiation problem. The inequality

$$\mathrm{Im}\left\{\frac{\omega^2}{v^2(\omega)}\right\} > 0 \ , \tag{3.4.9a}$$

and hence (since k_x and k_y are both real-valued)

$$\mathrm{Im}\left\{\frac{\omega^2}{v^2(\omega)} - k_T^2\right\} > 0 \ , \tag{3.4.9b}$$

will then be satisfied in any of the following three cases:

(i) If $a = 0$, then

$$\omega' \varepsilon_i(\omega) + [\![4\pi]\!] \sigma_r(\omega) > 0 \ . \tag{3.4.10a}$$

Because of the inequalities (3.4.8a and b) one must have that $\varepsilon_i(\omega) > 0$ at every frequency value where $\sigma_r(\omega) = 0$, and $\sigma_r(\omega) > 0$ at every frequency value where $\varepsilon_i(\omega) = 0$.

(ii) If $\sigma_r(\omega) = \sigma_i(\omega) = 0$ so that one has a nonconducting medium, then

$$\varepsilon_r(\omega) > -\frac{(\omega'^2 - a^2)\varepsilon_i(\omega)}{2a\omega'} \ . \tag{3.4.10b}$$

(iii) If $\varepsilon_i(\omega) = 0$ so that the dielectric permittivity is real-valued, then

$$2a\omega' \varepsilon_r(\omega) + [\![4\pi]\!] \omega' \sigma_r(\omega) - [\![4\pi]\!] a \sigma_i(\omega) > 0 \ .$$

If $a = 0$ this inequality requires that

$$\sigma_r(\omega) > 0 \ , \tag{3.4.10c}$$

while if $a > 0$ there results

$$\sigma_i(\omega) < \frac{2\omega'}{[\![4\pi]\!]} \varepsilon_r(\omega) + \frac{\omega'}{a} \sigma_r(\omega) \ . \tag{3.4.10d}$$

For a nondispersive medium ε_r is a positive constant and $\varepsilon_i = 0$. If, in addition, $\sigma_r = \sigma_i = 0$ so that the medium is also nonconducting, then case (ii) applies if one takes $a > 0$. On the other hand, if the medium is conducting so that $\sigma_r > 0$ then one can take $a = 0$ and (3.4.10c) of case (iii) applies. Hence, for a nondispersive medium the inequality (3.4.9) can always be satisfied.

For an absorptive, temporally dispersive medium $\varepsilon_r(\omega)$ can be negative but $\varepsilon_i(\omega)$ is always nonnegative. In that case $\mathbb{I}\mathrm{m}\{\omega^2/v^2(\omega)\}$ will be negative for sufficiently small $\omega' < a$, as seen from (3.4.7). One can then ensure that the inequality (3.4.9) is satisfied by taking $a = 0$, as in case (i). If this is not permissable because of a positive abscissa of absolute convergence for the particular radiation problem, one can always take the lower limit of the ω-integration in (3.4.1b) to be $\omega' = a$ and proceed to the limit $a \to 0$ after completion of the integration. In any event, ω' will always be considered to be positive when the k_z-integral in (3.4.1b) is evaluated and then the ω-integral possesses a positive lower limit which one then lets pass to zero in a limiting process.

To proceed with the k_z-integration in (3.4.1b) it is assumed that the inequality (3.4.9) is satisfied. The quantity

$$\gamma \equiv \left(\frac{\omega^2}{v^2(\omega)} - k_T^2 \right)^{1/2} , \tag{3.4.11}$$

where $k_T^2 = k_x^2 + k_y^2$, is taken as the principal branch of the square root that is defined in the following manner: let

$$\frac{\omega^2}{v^2(\omega)} - k_T^2 = \Gamma e^{i\theta} \; ; \qquad 0 \leqslant \theta < 2\pi \; , \tag{3.4.12a}$$

and define

$$\gamma = \sqrt{\Gamma} e^{i\theta/2} \; ; \qquad 0 \leqslant \theta < 2\pi \; . \tag{3.4.12b}$$

Then γ always has a positive imaginary part since $0 \leqslant \theta/2 < \pi$. The pole at $k_z = \gamma$ is then always in the upper-half of the complex k_z-plane and the pole at $k_z = -\gamma$ is always in the lower-half of the complex k_z-plane.

Because of the inequality (3.4.5), the i'th component of the integrand in (3.4.1b) is bounded in the following manner:

$$\left| \frac{F_i(\mathbf{k}, \omega)}{k^2 - \dfrac{\omega^2}{v^2(\omega)}} e^{i(\mathbf{k}\cdot\mathbf{r} - \omega t)} \right| = \frac{|F_i(\mathbf{k}, \omega)|}{\left| k^2 - \dfrac{\omega^2}{v^2(\omega)} \right|} e^{-k_z'' z} e^{at}$$

$$\leqslant \frac{M_i}{\left| k^2 - \dfrac{\omega^2}{v^2(\omega)} \right|} e^{|k_z''| Z - k_z'' z} \; , \tag{3.4.13}$$

with $a \to 0$. Hence, in the upper-half of the complex k_z-plane the integrand of (3.4.1b) decays exponentially and goes to zero as $|k_z''| \to \infty$ if $z > Z > 0$. The same exponential decay is present in the lower-half of the complex k_z-plane if $z < -Z$. It then follows from Jordan's lemma [3.4] that the k_z-integral in (3.4.1b) may be evaluated by application of Cauchy's residue theorem either by completing the contour into a closed circuit using the semicircular contour C_1 in the upper-half plane if $z > Z$, as illustrated in Fig. 3.3, or by using the semicircular contour C_2 in the lower-half plane if $z < -Z$, where the radius of each is allowed to go to infinity. If $|z| < Z$, the integral cannot be evaluated by completing the contour.

For the radiation field in the positive half-space $z > Z$ one completes the contour in the upper-half of the complex k_z-plane and the only contribution to the k_z-integral is due to the simple pole at $k_z = \gamma$. For the radiation field in the negative half-space $z < -Z$ one completes the contour in the lower-half plane and the only contribution to the k_z-integral is due to the simple pole at $k_z = -\gamma$. Since

$$k^2 - \frac{\omega^2}{v^2(\omega)} = (k_z - \gamma)(k_z + \gamma) \; , \tag{3.4.14}$$

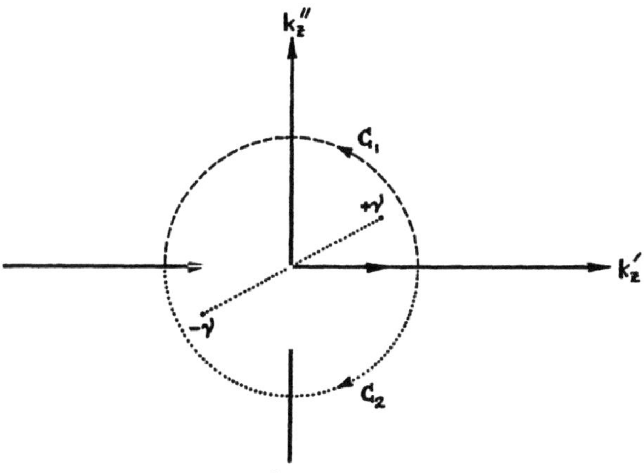

Fig. 3.3. The semicircular contours C_1 and C_2 where C_1 is contained entirely in the upper-half plane while C_2 is contained entirely in the lower-half plane.

then the residue of the simple pole at $k_z = \gamma$ is given by

$$\text{Residue}(k_z = \gamma)$$

$$= \frac{1}{8\pi^4} \lim_{k_z \to \gamma} (k_z - \gamma) \int_{C_+} d\omega \iint_{-\infty}^{\infty} dk_x dk_y \frac{\mathbf{F}(\mathbf{k}, \omega)}{(k_z - \gamma)(k_z + \gamma)} e^{i(\mathbf{k} \cdot \mathbf{r} - \omega t)}$$

$$= \frac{1}{8\pi^4} \int_{C_+} d\omega \iint_{-\infty}^{\infty} dk_x dk_y \frac{\mathbf{F}(\tilde{\mathbf{k}}^+, \omega)}{2\gamma} e^{i(\tilde{\mathbf{k}}^+ \cdot \mathbf{r} - \omega t)} , \tag{3.4.15a}$$

while the residue of the simple pole at $k_z = -\gamma$ is given by

$$\text{Residue}(k_z = -\gamma) = \frac{1}{8\pi^4} \int_{C_+} d\omega \iint_{-\infty}^{\infty} dk_x dk_y \frac{\mathbf{F}(\tilde{\mathbf{k}}^-, \omega)}{-2\gamma} e^{i(\tilde{\mathbf{k}}^- \cdot \mathbf{r} - \omega t)} . \tag{3.4.15b}$$

Here

$$\tilde{\mathbf{k}}^{\pm} \equiv \hat{1}_x k_x + \hat{1}_y k_y \pm \gamma \hat{1}_z \tag{3.4.16}$$

is the *complex wavevector*, where

$$\tilde{\mathbf{k}}^{\pm} \cdot \tilde{\mathbf{k}}^{\pm} = k_x^2 + k_y^2 + \gamma^2 = k_x^2 + k_y^2 + \frac{\omega^2}{v^2(\omega)} - k_x^2 - k_y^2$$

$$= \frac{\omega^2}{v^2(\omega)} = \tilde{k}^2(\omega) , \tag{3.4.17}$$

so that

$$\tilde{k}(\omega) = (\tilde{\mathbf{k}}^{\pm} \cdot \tilde{\mathbf{k}}^{\pm})^{1/2} = \frac{\omega}{v(\omega)}$$

$$= \left[\frac{1}{c}\right] \omega [\mu \varepsilon_c(\omega)]^{1/2} = k_0 n(\omega) \tag{3.4.18}$$

is the *complex wavenumber*, as given in (2.3.8) for the time-harmonic plane-wave field. Here $k_0 = \omega/c$ is the vacuum wavenumber and $n(\omega)$ is the complex index of refraction of the dispersive medium, as given in (2.1.34). Therefore, by the residue theorem [3.5], (3.4.15a) must be multiplied by $2\pi i$ and (3.4.15b) by $-2\pi i$ to obtain the final expressions for the k_z-integration, with the final result

$$\mathscr{G}(\mathbf{r}, t) = \frac{1}{8\pi^4} \mathbb{R}\mathrm{e}\left\{ i(\pm 2\pi i) \int_{C_+} d\omega \iint_{-\infty}^{\infty} dk_x\, dk_y \frac{\mathbf{F}(\tilde{\mathbf{k}}^{\pm}, \omega)}{\pm 2\gamma} e^{i(\tilde{\mathbf{k}}^{\pm} \cdot \mathbf{r} - \omega t)} \right\}$$

$$= -\frac{1}{(2\pi)^3} \mathbb{R}\mathrm{e}\left\{ \int_{C_+} d\omega \iint_{-\infty}^{\infty} dk_x\, dk_y \frac{\mathbf{F}(\tilde{\mathbf{k}}^{\pm}, \omega)}{\gamma} e^{i(\tilde{\mathbf{k}}^{\pm} \cdot \mathbf{r} - \omega t)} \right\} \tag{3.4.19}$$

for the generic integral (3.4.1). This final expression is referred to as the angular spectrum of plane wave representation of the field $\mathscr{G}(\mathbf{r}, t)$; the origin of this terminology is made clear in the following section when (3.4.19) is transformed to polar coordinates in $k_x k_y$-space.

For the electric and magnetic field vectors given in (3.2.33 and 34), respectively, for the radiation field, the k_z-integration of each Fourier-Laplace representation then yields the pair of expressions

$$\mathscr{E}(\mathbf{r}, t) = \frac{[4\pi]}{(2\pi)^3} \mathbb{R}\mathrm{e}\left\{ \int_{C_+} \frac{d\omega}{\omega \varepsilon_c(\omega)} \iint_{-\infty}^{\infty} dk_x\, dk_y \frac{1}{\gamma} \tilde{\mathbf{k}}^{\pm} \times [\tilde{\mathbf{k}}^{\pm} \times \mathbf{J}_0(\tilde{\mathbf{k}}^{\pm}, \omega)] e^{i(\tilde{\mathbf{k}}^{\pm} \cdot \mathbf{r} - \omega t)} \right\} \tag{3.4.20}$$

$$\mathscr{B}(\mathbf{r}, t) = -\left[\frac{4\pi}{c}\right] \frac{\mu}{(2\pi)^3} \mathbb{R}\mathrm{e}\left\{ \int_{C_+} d\omega \iint_{-\infty}^{\infty} dk_x\, dk_y \frac{1}{\gamma} [\tilde{\mathbf{k}}^{\pm} \times \mathbf{J}_0(\tilde{\mathbf{k}}^{\pm}, \omega)] e^{i(\tilde{\mathbf{k}}^{\pm} \cdot \mathbf{r} - \omega t)} \right\} \tag{3.4.21}$$

for $t > 0$, where one uses $\tilde{\mathbf{k}}^{+}$ for $z > Z > 0$ and $\tilde{\mathbf{k}}^{-}$ for $z < -Z$. This pair of equations is the *angular spectrum of plane waves representation* of the electromagnetic field vectors for the radiation field in a homogeneous, isotropic, locally linear, temporally dispersive and conducting medium occupying all of space. For the vector and scalar potentials given in (3.3.2 and 5), respectively, for the radiation field, the k_z-integration of each Fourier-Laplace representation then yields the angular spectrum of plane waves representations

$$\mathscr{A}(\mathbf{r},t)=\left[\!\left[\frac{4\pi}{c}\right]\!\right]\frac{\mu}{(2\pi)^3}\mathbb{R}\mathrm{e}\left\{\mathrm{i}\int\limits_{C_+}d\omega\iint\limits_{-\infty}^{\infty}dk_x\,dk_y\frac{1}{\gamma}\mathbf{J}_0(\tilde{\mathbf{k}}^{\pm},\omega)\mathrm{e}^{\mathrm{i}(\tilde{\mathbf{k}}^{\pm}\cdot\mathbf{r}-\omega t)}\right\} \tag{3.4.22}$$

$$\varphi(\mathbf{r},t)=\frac{[\![4\pi]\!]}{(2\pi)^3}\mathbb{R}\mathrm{e}\left\{\mathrm{i}\int\limits_{C_+}\frac{d\omega}{\omega\varepsilon_c(\omega)}\iint\limits_{-\infty}^{\infty}dk_x\,dk_y\frac{1}{\gamma}\tilde{\mathbf{k}}^{\pm}\cdot\mathbf{J}_0(\tilde{\mathbf{k}}^{\pm},\omega)\mathrm{e}^{\mathrm{i}(\tilde{\mathbf{k}}^{\pm}\cdot\mathbf{r}-\omega t)}\right\} \tag{3.4.23}$$

for $t > 0$. The parameter a for the contour C_+ ($\omega = \omega' + ia$) can be taken to be zero in (3.4.21, 22) since $\gamma = 0$ is an integrable singularity [3.6], and can also be taken to be zero in (3.4.20, 23) provided that $\omega\varepsilon_c(\omega) = 0$ is also an integrable singularity.

The exponential factor appearing in the angular spectrum of plane waves representations (3.4.20–23) is, with (3.4.16), given by

$$\mathrm{e}^{\mathrm{i}(\tilde{\mathbf{k}}^{\pm}\cdot\mathbf{r}-\omega t)}=\mathrm{e}^{\pm\mathrm{i}\gamma z}\mathrm{e}^{\mathrm{i}(k_x x+k_y y-\omega t)}\ , \tag{3.4.24}$$

which are either homogeneous or inhomogeneous plane waves depending upon whether γ is real, imaginary, or complex-valued. From (3.2.9, 15) one has that

$$\frac{\omega^2}{v^2(\omega)}=\left[\!\left[\frac{1}{c^2}\right]\!\right]\omega^2\mu\varepsilon(\omega)+\left[\!\left[\frac{4\pi}{c^2}\right]\!\right]\mathrm{i}\omega\mu\sigma(\omega)\ . \tag{3.4.25}$$

Consider the simplest case where ω is real-valued (taking $a = 0$) and where ε is real-valued and $\sigma = 0$, so that

$$\frac{\omega^2}{v^2}=\left[\!\left[\frac{1}{c^2}\right]\!\right]\omega^2\mu\varepsilon\ .$$

In this case

$$\gamma=\left(\frac{\omega^2}{v^2}-k_T^2\right)^{1/2}$$

is either pure real or pure imaginary. If $k_T^2 = k_x^2 + k_y^2 \leqslant \omega^2/v^2$, then γ is real and positive and (3.4.24) represents pure oscillatory plane waves. On the other hand, if $k_T^2 > \omega^2/v^2$, then γ is pure imaginary with $\gamma = \mathrm{i}|\gamma|$ so that $\mathrm{i}\gamma$ is real and negative and the expression in (3.4.24) possesses exponential decay in the z-direction for $|z| > Z$, since then

$$\mathrm{e}^{\pm\mathrm{i}\gamma z}=\mathrm{e}^{-|\gamma||z|}\ .$$

Thus, for $k_T^2 \leqslant \omega^2/v^2$, (3.4.24) represents homogeneous plane waves for which the surfaces of constant phase coincide with the surfaces of constant amplitude, while for $k_T^2 > \omega^2/v^2$ one has inhomogeneous plane waves where the phase is constant in planes parallel to the z-axis ($k_x x + k_y y = \mathscr{k}_1$) while the amplitude is constant in planes perpendicular to the z-axis ($\gamma z = \mathscr{k}_2$). That the plane waves given in (3.4.24) only decay along the z-direction has no real physical interpretation and is simply a consequence of an imposed dependency due to the performance of the k_z-integral in the Fourier-Laplace representations.

It is then clear that, in general, the angular spectrum of plane waves representations for both the electromagnetic field vectors, given in (3.4.20, 21), and the vector and scalar potentials, given in (3.4.22, 23), combine both homogeneous and inhomogeneous plane waves in the 2π-solid angle about the positive z-axis for $z > Z$, as well as in the 2π-solid angle about the negative z-axis for $z < -Z$. These specific forms of the plane wave representation are known as *Weyl-type expansions* [3.7]; each is valid only in its particular half-space with plane waves propagating only within the 2π solid angle into the respective half-space. When the source of the field ceases to radiate one then has an initial value problem and can use a representation that possesses only homogeneous plane waves in the entire 4π solid angle about the chosen origin. This specific form of the plane wave representation is known as a *Whittaker-type expansion* [3.8]. A comparison of these two expansions for the optical wave field in a nondispersive, nonconducting medium may be found in the published papers by *Sherman* et al. [3.9], *Sherman* [3.10], and *Devaney* and *Sherman* [3.11]. Further discussion of the angular spectrum representation may be found in the research monograph by *Clemmow* [3.12]; however, the field behavior in temporally dispersive media is only cursorally treated.

3.5 Polar Coordinate Form of the Angular Spectrum Representation

Attention is now given to the spatial angular spectrum integral

$$u(\mathbf{r}) = \iint_{-\infty}^{\infty} dk_x\, dk_y \frac{U(k_x, k_y)}{\gamma} e^{i\tilde{\mathbf{k}}^{\pm}\cdot\mathbf{r}} \tag{3.5.1}$$

which appears in the angular spectrum representation (3.4.20) for the electric field intensity vector with $U(k_x, k_y) = \tilde{\mathbf{k}}^{\pm} \times (\tilde{\mathbf{k}}^{\pm} \times \mathbf{J}_0(\tilde{\mathbf{k}}^{\pm}, \omega))$, in the angular spectrum representation (3.4.21) for the magnetic induction field vector with $U(k_x, k_y) = \tilde{\mathbf{k}}^{\pm} \times \mathbf{J}_0(\tilde{\mathbf{k}}^{\pm}, \omega)$, in the angular spectrum representation (3.4.22) for the vector potential field with $U(k_x, k_y) = \mathbf{J}(\tilde{\mathbf{k}}^{\pm}, \omega)$, and in the angular spectrum representation (3.4.23) for the scalar potential field with $U(k_x, k_y) = \tilde{\mathbf{k}}^{\pm} \cdot \mathbf{J}_0(\tilde{\mathbf{k}}^{\pm}, \omega)$. Here $\mathbf{r} = \hat{1}_x x + \hat{1}_y y + \hat{1}_z z$ is the position vector for the field, and

$$\tilde{\mathbf{k}}^{\pm} = \hat{1}_x k_x + \hat{1}_y k_y \pm \hat{1}_z \gamma \tag{3.5.2}$$

is the complex wavevector with

$$\gamma = [\tilde{k}^2(\omega) - k_x^2 - k_y^2]^{1/2} \,. \tag{3.5.3}$$

and

$$\tilde{k}(\omega) = \frac{\omega}{v(\omega)} \tag{3.5.4}$$

is the complex wave number. One employs $\tilde{\mathbf{k}}^+$ for $z > Z > 0$ and $\tilde{\mathbf{k}}^-$ for $z < -Z$. Finally,

$$U(k_x, k_y) = f(k_x, k_y, \gamma) \tag{3.5.5}$$

where $f(k_x, k_y, k_z)$ is an entire function of complex k_x, k_y, and k_z [see the discussion in connection with (3.4.4)]; however, $f(k_x, k_y, \gamma)$ is not an entire function of complex k_x and k_y.

Consider now the change of variable to polar coordinates

$$\begin{aligned} k_x &= \tilde{k}(\omega) \sin\alpha \cos\beta \ , \\ k_y &= \tilde{k}(\omega) \sin\alpha \sin\beta \ , \end{aligned} \tag{3.5.6}$$

where k_x and k_y are both real and ranging over the domain from $-\infty$ to $+\infty$, $\tilde{k}(\omega)$ is the fixed complex number given by (3.5.4), and where α and β are, in general, complex-valued angles whose domains must be determined so as to yield the proper ranges for both k_x and k_y. The Jacobian of this coordinate transformation can be expressed as

$$\frac{\partial(k_x, k_y)}{\partial(\alpha, \beta)} = \begin{vmatrix} \dfrac{\partial k_x}{\partial \alpha} & \dfrac{\partial k_y}{\partial \alpha} \\ \dfrac{\partial k_x}{\partial \beta} & \dfrac{\partial k_y}{\partial \beta} \end{vmatrix} = \tilde{k}^2(\omega) \sin\alpha \cos\alpha \ . \tag{3.5.7}$$

Furthermore, under this coordinate transformation

$$\gamma = \pm\tilde{k}(\omega) \cos\alpha \ , \tag{3.5.8}$$

where the correct sign choice will be determined by the domain of integration.

The derivation now proceeds to the determination of the contours of integration along which the complex angles α and β vary in such a manner that both k_x and k_y are kept real-valued and varying from $-\infty$ to $+\infty$. To begin, the simplest case where $\tilde{k}$ is real-valued is considered, followed by the general case where $\tilde{k}(\omega)$ is complex-valued. Clearly, the first case is a limiting situation of the latter and so lends important guidance to the latter, general derivation.

Consider then the case where $\tilde{k} = k$ is real-valued. It is then clear from (3.5.6) that β must be real-valued in order that both k_x and k_y are real-valued. The angle α must then be complex-valued such that $\sin\alpha$ is real-valued and ranges from 0 to ∞. The real angle β ranging over 0 to 2π will then give k_x and k_y a range over $-\infty$ to $+\infty$. As a consequence, let $\alpha = \alpha' + i\alpha''$ where both α'

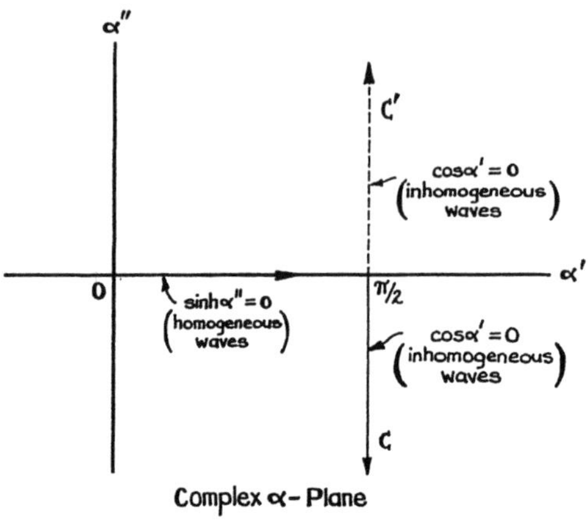

Fig. 3.4. The contours C and C' in the complex α-plane.

and α'' are real-valued, so that

$$\sin\alpha = \sin(\alpha' + \mathrm{i}\alpha'') = \sin\alpha' \cosh\alpha'' + \mathrm{i}\cos\alpha' \sinh\alpha'' \ .$$

Since $\sin\alpha$ must be real-valued, then one obtains the condition

$$\cos\alpha' \sinh\alpha'' = 0 \ ,$$

which is satisfied along either of the following two contours in the complex α-plane: the contour C where α varies from 0 to $\pi/2$ along the α'-axis, and then α varies from $\alpha = \pi/2$ to $\pi/2 - \mathrm{i}\infty$ along the line $\alpha' = \pi/2$; the contour C' where α varies from 0 to $\pi/2$ along the α'-axis, and then α varies from $\alpha = \pi/2$ to $\pi/2 + \mathrm{i}\infty$ along the line $\alpha' = \pi/2$. These two contours are illustrated in Fig. 3.4. Along the $\alpha' = \mathrm{Re}\{\alpha\}$ axis, $\sin\alpha = \sin\alpha'$ so that

$$\gamma = k(1 - \sin^2\alpha')^{1/2} = \pm k\cos\alpha'$$

and the factor $e^{\mathrm{i}\mathbf{k}^{\pm}\cdot\mathbf{r}}$ appearing in the integrand of (3.5.1) represents homogeneous plane waves along this portion of the contour. Along the line $\alpha' = \pi/2$, $\sin\alpha = \cosh\alpha''$ so that

$$\gamma = k(1 - \cosh^2\alpha'')^{1/2} = k(-\sinh^2\alpha'')^{1/2}$$
$$= \pm ik\sinh\alpha''$$

and the factor $e^{\mathrm{i}\mathbf{k}^{\pm}\cdot\mathbf{r}}$ represents inhomogeneous plane waves along this portion of the contour (for either C or C'). As α' varies from 0 to $\pi/2$ along $\alpha'' = 0$, $\sin\alpha = \sin\alpha'$ varies from 0 to 1, and as α varies from $\pi/2$ to $\pi/2 - \mathrm{i}\infty$ along $\alpha' = \pi/2$, $\sin\alpha = \cosh\alpha''$ varies from 1 to $+\infty$, while as α varies from $\pi/2$ to $\pi/2 + \mathrm{i}\infty$ along $\alpha' = \pi/2$, $\sin\alpha = \cosh\alpha''$ also varies from 1 to $+\infty$. Along the contour C, $\cos\alpha$ varies as follows:

$$\alpha'' = 0\ , \qquad \alpha' = 0 \to \frac{\pi}{2}: \qquad \cos\alpha = \cos\alpha' = 1 \to 0\ ,$$

$$\alpha' = \frac{\pi}{2}\ , \qquad \alpha'' = 0 \to -\infty\ : \qquad \cos\alpha = -\mathrm{i}\sinh\alpha'' = 0 \to +\mathrm{i}\infty\ ,$$

and one must then choose the positive branch for γ when α varies along C, viz.

$$\gamma = +k\cos\alpha\ ; \qquad \alpha \in C\ . \tag{3.5.9}$$

On the other hand, when α'' varies from 0 to $+\infty$ with $\alpha' = \pi/2$ along the contour C', $\cos\alpha = -\mathrm{i}\sinh\alpha''$ varies from 0 to $-\mathrm{i}\infty$ and one must then choose the negative branch for γ (viz. $\gamma = -k\cos\alpha$; $\alpha \in C'$). For simplicity, the positive branch as specified in (3.5.9) is chosen here.

Consider now the case where $\tilde{k}(\omega)$ is complex-valued, and let

$$\tilde{k}(\omega) = k(\omega)\mathrm{e}^{\mathrm{i}\kappa(\omega)} \tag{3.5.10}$$

where $k(\omega) \equiv |\tilde{k}(\omega)|$ is the magnitude and $\kappa(\omega)$ is the argument (or phase) of the complex wave number with $0 \leqslant \kappa < \pi/2$, which will always be satisfied for an attenuative medium. Let $\alpha = \alpha' + \mathrm{i}\alpha''$ and $\beta = \beta' + \mathrm{i}\beta''$ with $\alpha' = \mathbb{R}\mathrm{e}\{\alpha\}$, $\alpha'' = \mathbb{I}\mathrm{m}\{\alpha\}$, $\beta' = \mathbb{R}\mathrm{e}\{\beta\}$, and $\beta'' = \mathbb{I}\mathrm{m}\{\beta\}$. With these substitutions (3.5.6) becomes

$$\begin{aligned}\frac{k_x(\omega)}{k(\omega)} &= \mathrm{e}^{\mathrm{i}\kappa(\omega)}\sin(\alpha' + \mathrm{i}\alpha'')\cos(\beta' + \mathrm{i}\beta'')\\
&= (\cos\kappa + \mathrm{i}\sin\kappa)(\sin\alpha'\cosh\alpha'' + \mathrm{i}\cos\alpha'\sinh\alpha'')(\cos\beta'\cosh\beta''\\
&\quad - \mathrm{i}\sin\beta'\sinh\beta'')\\
&= [(\sin\alpha'\cosh\alpha''\cos\beta'\cosh\beta'' + \cos\alpha'\sinh\alpha''\sin\beta'\sinh\beta'')\cos\kappa\\
&\quad - (\cos\alpha'\sinh\alpha''\cos\beta'\cosh\beta'' - \sin\alpha'\cosh\alpha''\sin\beta'\sinh\beta'')\sin\kappa]\\
&\quad + \mathrm{i}[(\sin\alpha'\cosh\alpha''\cos\beta'\cosh\beta'' + \cos\alpha'\sinh\alpha''\sin\beta'\sinh\beta'')\sin\kappa\\
&\quad + (\cos\alpha'\sinh\alpha''\cos\beta'\cosh\beta'' - \sin\alpha'\cosh\alpha''\sin\beta'\sinh\beta'')\cos\kappa]\ ,\end{aligned} \tag{3.5.11a}$$

and

$$\begin{aligned}\frac{k_y(\omega)}{k(\omega)} &= \mathrm{e}^{\mathrm{i}\kappa(\omega)}\sin(\alpha' + \mathrm{i}\alpha'')\sin(\beta' + \mathrm{i}\beta'')\\
&= (\cos\kappa + \mathrm{i}\sin\kappa)(\sin\alpha'\cosh\alpha'' + \mathrm{i}\cos\alpha'\sinh\alpha'')(\sin\beta'\cosh\beta''\\
&\quad + \mathrm{i}\cos\beta'\sinh\beta'')\end{aligned}$$

$$\begin{aligned}&= [(\sin\alpha'\cosh\alpha''\sin\beta'\cosh\beta'' - \cos\alpha'\sinh\alpha''\cos\beta'\sinh\beta'')\cos\kappa\\&\quad - (\sin\alpha'\cosh\alpha''\cos\beta'\sinh\beta'' + \cos\alpha'\sinh\alpha''\sin\beta'\cosh\beta'')\sin\kappa]\\&\quad + \mathrm{i}[(\sin\alpha'\cosh\alpha''\sin\beta'\cosh\beta'' - \cos\alpha'\sinh\alpha''\cos\beta'\sinh\beta'')\sin\kappa\\&\quad + (\sin\alpha'\cosh\alpha''\cos\beta'\sinh\beta'' + \cos\alpha'\sinh\alpha''\sin\beta'\cosh\beta'')\cos\kappa] \ .\end{aligned} \tag{3.5.11b}$$

Since the left-hand sides of both of these expressions must be real-valued, then the imaginary parts appearing on the right-hand sides must both vanish, and consequently one obtains the pair of expressions

$$\begin{aligned}&(\sin\alpha'\cosh\alpha''\cos\beta'\cosh\beta'' + \cos\alpha'\sinh\alpha''\sin\beta'\sinh\beta'')\sin\kappa\\&\quad + (\cos\alpha'\sinh\alpha''\cos\beta'\cosh\beta'' - \sin\alpha'\cosh\alpha''\sin\beta'\sinh\beta'')\cos\kappa = 0 \ ,\end{aligned} \tag{3.5.12a}$$

$$\begin{aligned}&(\sin\alpha'\cosh\alpha''\sin\beta'\cosh\beta'' - \cos\alpha'\sinh\alpha''\cos\beta'\sinh\beta'')\sin\kappa\\&\quad + (\sin\alpha'\cosh\alpha''\cos\beta'\sinh\beta'' + \cos\alpha'\sinh\alpha''\sin\beta'\cosh\beta'')\cos\kappa = 0 \ .\end{aligned} \tag{3.5.12b}$$

Since $\sin\kappa$ and $\cos\kappa$ are known to exist, this pair of simultaneous linear equations in $\sin\kappa$ and $\cos\kappa$ must have a solution, which in turn requires that the determinant of its coefficients must vanish, so that

$$\begin{vmatrix}\begin{pmatrix}\sin\alpha'\cosh\alpha''\cos\beta'\cosh\beta''\\ \quad + \cos\alpha'\sinh\alpha''\sin\beta'\sinh\beta''\end{pmatrix} & \begin{pmatrix}\cos\alpha'\sinh\alpha''\cos\beta'\cosh\beta''\\ \quad - \sin\alpha'\cosh\alpha''\sin\beta'\sinh\beta''\end{pmatrix}\\ \begin{pmatrix}\sin\alpha'\cosh\alpha''\sin\beta'\cosh\beta''\\ \quad - \cos\alpha'\sinh\alpha''\cos\beta'\sinh\beta''\end{pmatrix} & \begin{pmatrix}\sin\alpha'\cosh\alpha''\cos\beta'\sinh\beta''\\ \quad + \cos\alpha'\sinh\alpha''\sin\beta'\cosh\beta''\end{pmatrix}\end{vmatrix}$$

$$= 0 \ ,$$

which leads to the simple expression

$$\sinh\beta''\cosh\beta''(\sin^2\alpha'\cosh^2\alpha'' + \cos^2\alpha'\sinh^2\alpha'') = 0 \ .$$

Since $\cosh\beta'' \neq 0$ for real β'', and since $\sin^2\alpha'\cosh^2\alpha'' + \cos^2\alpha'\sinh^2\alpha'' \neq 0$ for all values of α' and α'', it then follows that

$$\sinh\beta'' = 0 \Rightarrow \beta'' = 0 \tag{3.5.13}$$

and $\beta = \beta'$ is real-valued. With this substitution (3.5.12a, b) become

$$\cos\beta(\sin\alpha'\cosh\alpha''\sin\kappa + \cos\alpha'\sinh\alpha''\cos\kappa) = 0 \ , \tag{3.5.14a}$$

$$\sin\beta(\sin\alpha'\cosh\alpha''\sin\kappa + \cos\alpha'\sinh\alpha''\cos\kappa) = 0 \ , \tag{3.5.14b}$$

and since $\cos\beta = \cos\beta'$ and $\sin\beta = \sin\beta'$ cannot both be equal to zero, one must then have that

$$\sin\alpha'\tan\kappa + \cos\alpha'\tanh\alpha'' = 0\ ; \qquad 0 \leqslant \kappa < \frac{\pi}{2} \tag{3.5.15}$$

Furthermore, (3.5.11a and b) become (with $\beta = \beta'$)

$$\frac{k_x(\omega)}{k(\omega)} = \cos\beta(\sin\alpha'\cosh\alpha''\cos\kappa - \cos\alpha'\sinh\alpha''\sin\kappa) \tag{3.5.16a}$$

$$\frac{k_y(\omega)}{k(\omega)} = \sin\beta(\sin\alpha'\cosh\alpha''\cos\kappa - \cos\alpha'\sinh\alpha''\sin\kappa) \tag{3.5.16b}$$

where α', α'', and $\beta = \beta'$ must vary in such a manner that k_x and k_y both vary over the domain from $-\infty$ to $+\infty$.

Consider now the family of contours in the complex α-plane that are described by (3.5.15). In the limit as $\alpha'' \to +\infty$, $\tanh\alpha'' \to 1$ so that (3.5.15) becomes

$$\sin\alpha'\tan\kappa + \cos\alpha' = 0\ ,$$

so that

$$\tan\alpha' = -\cot\kappa = -\tan\left(\frac{\pi}{2} - \kappa\right) = \tan\left(\kappa - \frac{\pi}{2}\right)$$

and consequently

$$\alpha' \to \kappa - \frac{\pi}{2}\,(\mathrm{mod}\,2\pi) \quad \text{as} \quad \alpha'' \to +\infty\ . \tag{3.5.17}$$

On the other hand, in the limit as $\alpha'' \to -\infty$, $\tanh\alpha'' \to -1$ so that (3.5.15) becomes

$$\sin\alpha'\tan\kappa - \cos\alpha' = 0$$

so that

$$\tan\alpha' = \cot\kappa = \tan\left(\frac{\pi}{2} - \kappa\right)$$

and consequently

$$\alpha' \to \frac{\pi}{2} - \kappa\,(\mathrm{mod}\,2\pi) \quad \text{as} \quad \alpha'' \to -\infty\ . \tag{3.5.18}$$

Furthermore, when $\alpha'' = 0$, $\tanh\alpha'' = 0$ and (3.5.15) gives

$$\sin\alpha'\tan\kappa = 0\ ,$$

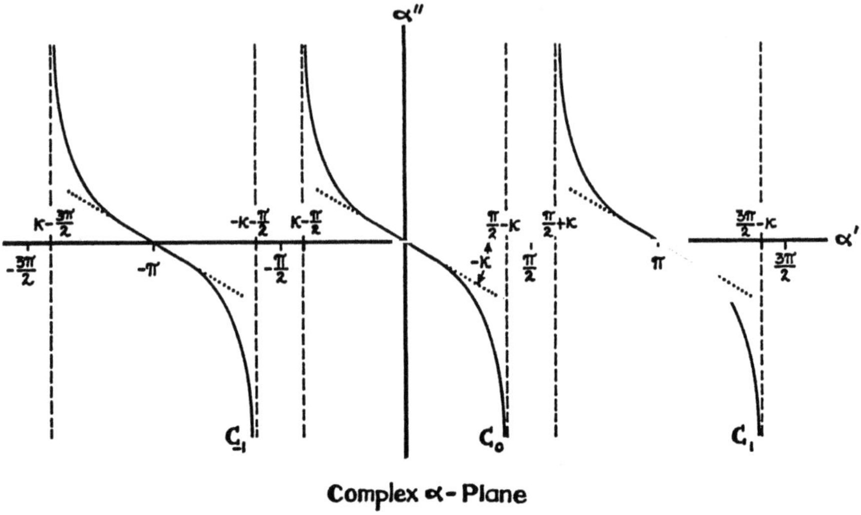

Fig. 3.5. Family of contours C_m in the complex α-plane.

and since $0 \leqslant \kappa < \pi/2$ for complex $\tilde{k}(\omega)$, then $\sin\alpha' = 0$ and

$$\alpha' = \pm m\pi \; ; \qquad m = 0, 1, 2, 3, \ldots \qquad \text{when } \alpha'' = 0 \; . \tag{3.5.19}$$

Hence, each member of the family of contours described by (3.5.15) begins at $\alpha' = m\pi - \pi/2 + \kappa$, $\alpha'' = +\infty$, passses across the α'-axis at $\alpha' = m\pi$, and ends at $\alpha' = m\pi + \pi/2 - \kappa$ for each positive and negative integer value of m, as illustrated in Fig. 3.5. The angle of slope of each member C_m of this family of contours as the curve crosses the α'-axis (i.e., at $\alpha'' = 0$) is obtained by differentiating (3.5.15) to obtain

$$\frac{d\alpha''}{d\alpha'} = \frac{\sin\alpha' \tanh\alpha'' - \tan\kappa\cos\alpha'}{\cos\alpha' \operatorname{sech}^2\alpha''} \, ,$$

so that

$$\operatorname{Tan}^{-1}\left\{\left(\frac{d\alpha''}{d\alpha'}\right)_{\alpha''=0}\right\} = \operatorname{Tan}^{-1}\{-\tan\kappa\} = -\kappa \; , \tag{3.5.20}$$

and the angle of slope is given by minus the phase of the complex wave number.

In the special case when $\kappa = 0$ and $\tilde{k}(\omega)$ is real-valued, (3.5.15) reduces to the equation $\cos\alpha' \sinh\alpha'' = 0$ that was obtained when $\tilde{k}(\omega) = k$ was assumed to be real-valued, and the lower-half of the contour C_0 reduces to the contour C that was obtained there. Therefore, the general contour C to be employed here is the lower-half of C_0 and is directed from $\alpha = 0$ to $\alpha = \pi/2 - \kappa - i\infty$, as illustrated in Fig. 3.6. Along this contour the real and imaginary parts of $\alpha = \alpha' + i\alpha''$

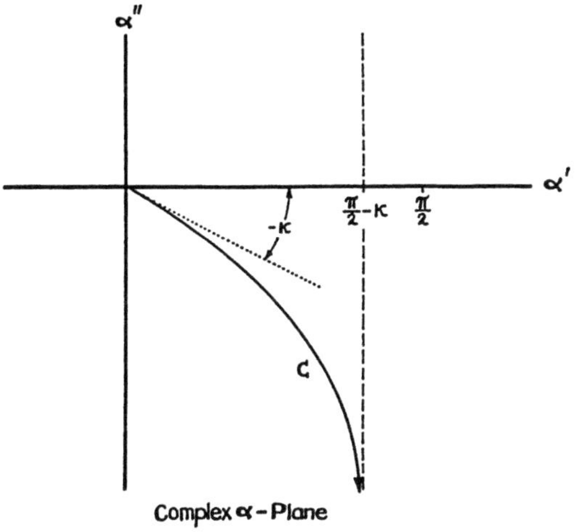

Fig. 3.6. The contour C in the complex α-plane.

vary as

$$\alpha' = 0 \to \frac{\pi}{2} - \kappa \Rightarrow \sin\alpha' = 0 \to \sin\left(\frac{\pi}{2} - \kappa\right) = \cos\kappa \geqslant 0 \ ,$$

$$\cos\alpha' = 1 \to \cos\left(\frac{\pi}{2} - \kappa\right) = \sin\kappa \geqslant 0 \ ,$$

$$\alpha'' = 0 \to -\infty \Rightarrow \sinh\alpha'' = 0 \to -\infty \ ,$$

$$\cosh\alpha'' = 1 \to +\infty \ ,$$

so that k_x and k_y vary along C as

$$\frac{k_x(\omega)}{k(\omega)} = A\cos\beta \ ,$$

$$\frac{k_y(\omega)}{k(\omega)} = A\sin\beta \ ,$$

where A varies from 0 to $+\infty$. Thus, with $\alpha = \alpha' + \mathrm{i}\alpha''$ varying along the contour C and with $\beta = \beta'$ varying between 0 and 2π, both k_x and k_y vary between $-\infty$ to $+\infty$ and are real-valued, as desired. Furthermore, along the contour C, $\cos\alpha' \geqslant 0$, $\sin\alpha' \geqslant 0$, $\cosh\alpha'' \geqslant 0$, and $\sinh\alpha'' \leqslant 0$ so that

$$\cos\alpha = \cos\alpha' \cosh\alpha'' + \mathrm{i}|\sin\alpha' \sinh\alpha''| \qquad (3.5.21)$$

and the phase of $\cos\alpha$ is always between 0 and π. Once again, the positive branch of γ must be taken when α varies along C, see (3.5.9), so that

$$\gamma = +\tilde{k}(\omega)\cos\alpha \; ; \qquad \alpha \in C \; . \tag{3.5.22}$$

Hence, under the change of variable (3.5.6) to polar coordinates the angular spectrum integral (3.5.1) becomes

$$u(\mathbf{r}) = \tilde{k}(\omega) \int_0^{2\pi} d\beta \int_C \sin\alpha \, d\alpha \cdot U(k_x, k_y) e^{i\tilde{\mathbf{k}}^{\pm}\cdot\mathbf{r}} \tag{3.5.23}$$

where

$$\tilde{\mathbf{k}}^{\pm}(\omega) = \tilde{k}(\omega)[\hat{1}_x \sin\alpha\cos\beta + \hat{1}_y \sin\alpha\sin\beta \pm \hat{1}_z \cos\alpha] \tag{3.5.24}$$

and where the polar axis has been chosen to be along the k_z-axis.

3.5.1 Transformation to an Arbitrary Polar Axis

From a physical point of view it is evident that the choice of the polar axis to be along the k_z-axis is simply a matter of convenience that is unnecessary and that the expression (3.5.23) for the angular spectrum integral should be independent of this choice. The rigorous, mathematical proof of this simple observation is, however, another matter. To that end, consider choosing some other polar axis that is specified relative to the k_x, k_y, k_z-coordinate axes by the (arbitrary) angles λ and δ, where λ is the angle of declination from the k_z-axis and δ is the azimuthal angle of the normal projection of the polar axis onto the $k_x k_y$-plane, measured from the k_x-axis, as illustrated in Fig. 3.7 (the wavevector $\mathbf{k}$ depicted

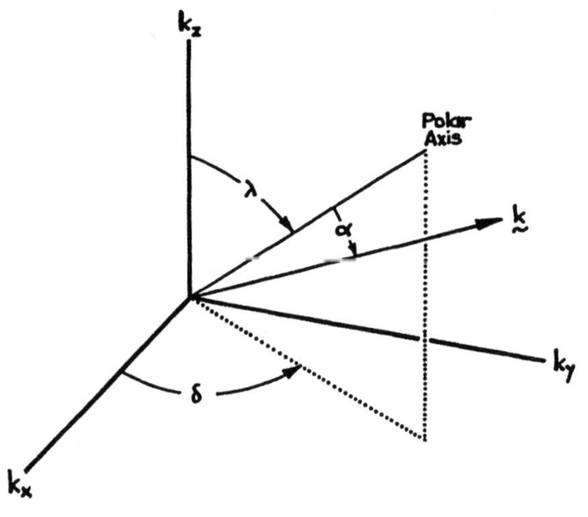

Fig. 3.7. Relation of the polar axis to the k_x, k_y, k_z coordinate axes when $\mathbf{k}$ is real-valued.

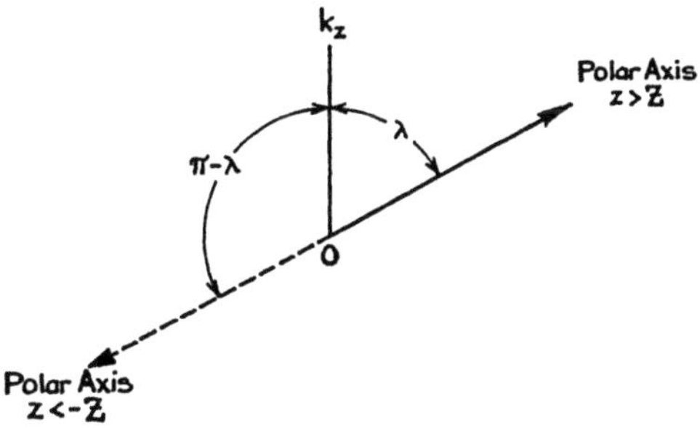

Fig. 3.8. Polar axis choices when either $z > Z$ or $z < -Z$.

here is real-valued for the purpose of illustration). Under this transformation to a new polar axis the expressions in (3.5.6) become (retaining the complex variables α and β as the new angular coordinates of the complex wavevector $\tilde{\mathbf{k}}(\omega)$)

$$k_x(\omega) = \tilde{k}(\omega)(\cos\alpha\sin\lambda\cos\delta \pm \sin\alpha\cos\beta\cos\lambda\cos\delta - \sin\alpha\sin\beta\sin\delta) \ , \tag{3.5.25}$$

$$k_y(\omega) = \tilde{k}(\omega)(\cos\alpha\sin\lambda\sin\delta \pm \sin\alpha\cos\beta\cos\lambda\sin\delta + \sin\alpha\sin\beta\cos\delta) \ ,$$

where the $\pm$ sign appearing in these expressions corresponds to the $\pm$ sign appearing in $\tilde{\mathbf{k}}^{\pm}$ and takes into account that for $z < -Z$, the polar axis is chosen in the opposite direction from that for $z > Z$ so that $\lambda \to \pi - \lambda$, as illustrated in Fig. 3.8 (remember that the polar axis is no longer fixed as it was in the preceeding analysis).

For the Jacobian $\frac{\partial(k_x, k_y)}{\partial(\alpha, \beta)}$ of this transformation, since

$$\frac{1}{\tilde{k}(\omega)}\frac{\partial k_x(\omega)}{\partial\alpha} = -\sin\alpha\sin\lambda\cos\delta \pm \cos\alpha\cos\beta\cos\lambda\cos\delta - \cos\alpha\sin\beta\sin\delta \ ,$$

$$\frac{1}{\tilde{k}(\omega)}\frac{\partial k_y(\omega)}{\partial\beta} = \mp\sin\alpha\sin\beta\cos\lambda\sin\delta + \sin\alpha\cos\beta\cos\delta \ ,$$

$$\frac{1}{\tilde{k}(\omega)}\frac{\partial k_y(\omega)}{\partial\alpha} = -\sin\alpha\sin\lambda\sin\delta \pm \cos\alpha\cos\beta\cos\lambda\sin\delta + \cos\alpha\sin\beta\cos\delta \ ,$$

$$\frac{1}{\tilde{k}(\omega)}\frac{\partial k_x(\omega)}{\partial\beta} = \mp\sin\alpha\sin\beta\cos\lambda\cos\delta - \sin\alpha\cos\beta\sin\delta \ ,$$

one then has that

$$\frac{1}{\tilde{k}^2(\omega)}\frac{\partial(k_x, k_y)}{\partial(\alpha, \beta)} = \begin{vmatrix} \tilde{k}^{-1}\partial k_x/\partial\alpha & \tilde{k}^{-1}\partial k_y/\partial\alpha \\ \tilde{k}^{-1}\partial k_x/\partial\beta & \tilde{k}^{-1}\partial k_y/\partial\beta \end{vmatrix}$$

$$= -\sin^2\alpha\cos\beta\sin\lambda \pm \sin\alpha\cos\alpha\cos\lambda \ . \tag{3.5.26}$$

Furthermore, under the transformation (3.5.25) one has that

$$\begin{aligned}1-\frac{\gamma^2}{\tilde{k}^2}=\frac{k_x^2}{\tilde{k}^2}+\frac{k_y^2}{\tilde{k}^2}\\
&=(\cos\alpha\sin\lambda\cos\delta\pm\sin\alpha\cos\beta\cos\lambda\cos\delta-\sin\alpha\sin\beta\sin\delta)^2\\
&\quad+(\cos\alpha\sin\lambda\sin\delta\pm\sin\alpha\cos\beta\cos\lambda\sin\delta+\sin\alpha\sin\beta\cos\delta)^2\\
&=\cos^2\alpha\sin^2\lambda+\sin^2\alpha\cos^2\beta\cos^2\lambda+\sin^2\alpha\sin^2\beta\\
&\quad\pm 2\sin\alpha\cos\alpha\cos\beta\sin\lambda\cos\lambda\\
&=\cos^2\alpha-\cos^2\alpha\cos^2\lambda+\sin^2\alpha\cos^2\beta-\sin^2\alpha\cos^2\beta\sin^2\lambda\\
&\quad+\sin^2\alpha\sin^2\beta\pm 2\sin\alpha\cos\alpha\cos\beta\sin\lambda\cos\lambda\end{aligned}$$

so that

$$1-\frac{\gamma^2}{\tilde{k}^2}=1-(\cos\alpha\cos\lambda\mp\sin\alpha\cos\beta\sin\lambda)^2\ ,$$

and hence

$$\gamma=\pm\tilde{k}(\omega)(\cos\alpha\cos\lambda\mp\sin\alpha\cos\beta\sin\lambda)\ .$$

When $\lambda=0$ one must have $\gamma=+\tilde{k}(\omega)\cos\alpha$ in accordance with (3.5.22), so that one finally obtains

$$\gamma=\tilde{k}(\omega)(\cos\alpha\cos\lambda\mp\sin\alpha\cos\beta\sin\lambda) \tag{3.5.27}$$

and (3.5.26) becomes

$$\frac{\partial(k_x,k_y)}{\partial(\alpha,\beta)}=\pm\tilde{k}(\omega)\gamma\sin\alpha\ , \tag{3.5.28}$$

where the sign choice appearing in these expressions corresponds to the sign choice appearing in $\tilde{\mathbf{k}}^{\pm}$.

Hence, under the change of variable (3.5.25) to polar coordinates with an arbitrary polar axis direction the angular spectrum integral (3.5.1) becomes

$$u(\mathbf{r})=\pm\tilde{k}(\omega)\iint_{D(\alpha,\beta)}U(k_x,k_y)e^{i\tilde{\mathbf{k}}^{\pm}\cdot\mathbf{r}}\sin\alpha\,d\alpha\,d\beta \tag{3.5.29}$$

where the polar coordinate variables α and β must now both be complex-valued, in general, viz.

$$\alpha=\alpha'+i\alpha''\ ,$$

$$\beta=\beta'+i\beta''\ ,$$

and where the domain of integration $D(\alpha, \beta)$ is some two-dimensional surface in the four-dimensional $\alpha', \alpha'', \beta', \beta''$-space. However, the determination of this integration domain in the general case is a very difficult problem. Nevertheless, it is shown in Sect. 3.5.2 that, given certain restrictions on the analyticity of the spectral function $U(k_x, k_y)$, the angular domain of integration $D(\alpha, \beta)$ can be taken to be the same domain of integration as appearing in (3.5.23). Hence, subject to these restrictions on the analyticity of $U(k_x, k_y)$,

$$u(\mathbf{r}) = \pm\tilde{k}(\omega) \int_0^{2\pi} d\beta \int_C \sin\alpha\, d\alpha \cdot U(k_x, k_y) e^{i\tilde{\mathbf{k}}^{\pm}\cdot\mathbf{r}} \tag{3.5.30}$$

where

$$\begin{aligned}\tilde{\mathbf{k}}^{\pm}(\omega) = \tilde{k}(\omega)[\hat{1}_x(&\cos\alpha\sin\lambda\cos\delta \pm \sin\alpha\cos\beta\cos\lambda\cos\delta - \sin\alpha\sin\beta\sin\delta)\\ &+ \hat{1}_y(\cos\alpha\sin\lambda\sin\delta \pm \sin\alpha\cos\beta\cos\lambda\sin\delta + \sin\alpha\sin\beta\cos\delta)\\ &\pm \hat{1}_z(\cos\alpha\cos\lambda \mp \sin\alpha\cos\beta\sin\lambda)] \ ,\end{aligned} \tag{3.5.31}$$

and where the contour C is that branch of the family of contours in the complex $\alpha = \alpha' + i\alpha''$ plane described by (3.5.15) that begins at $\alpha = 0$ and extends to $\alpha = \pi/2 - \kappa - i\infty$, where $\kappa = \arg(\tilde{k}(\omega))$, $0 \leqslant \kappa < \pi/2$.

3.5.2 Weyl's Proof

The proof that the angular spectrum integral appearing in (3.5.29) is independent of the transformation angles λ and δ, and, subject to certain conditions on the analyticity of the spectral function $U(k_x, k_y)$, reduces to (3.5.30) is now given [3.7]. This important proof begins with the spatial angular spectrum integral (3.5.1) under the general coordinate transformation (3.5.25), which may be written in the general form

$$I(\mathbf{r}, \lambda, \delta) = \int_0^{2\pi} d\beta \int_C d\alpha \frac{1}{\gamma} U(k_x, k_y) e^{i\tilde{\mathbf{k}}^{\pm}\cdot\mathbf{r}} \frac{\partial(k_x, k_y)}{\partial(\alpha, \beta)} \ , \tag{3.5.32}$$

where C is the contour from $\alpha = 0$ to $\alpha = \pi/2 - \kappa - i\infty$ that is described by (3.5.15). This particular form of the integral takes into account the polar coordinate form of the domain of integration for k_x and k_y as expressed in (3.5.23) when the polar axis is along k_z, which is then modified by the transformation to an arbitrary polar axis direction specified by the transformation angles λ and δ through the Jacobian

$$J(k_x, k_y; \alpha, \beta) = \frac{\partial(k_x, k_y)}{\partial(\alpha, \beta)} \ . \tag{3.5.33}$$

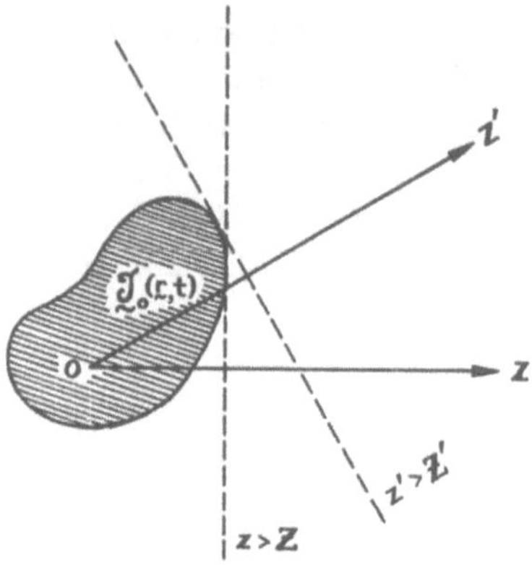

Fig. 3.9. Two possible half-spaces and their associated polar axis directions.

The proof then entails showing that $I(\mathbf{r}, \lambda, \delta)$ is independent of the parameters λ and δ by determining the partial derivatives $\partial I/\partial\lambda$ and $\partial I/\partial\delta$ and showing that they both vanish.

From (3.5.5), the spectral amplitude function appearing (3.5.32) is given by

$$U(k_x, k_y) = f(k_x, k_y, \gamma) \ , \tag{3.5.34}$$

where

$$f(k_x, k_y, k_z) = \iiint\limits_{-\infty}^{\infty} f(x, y, z) e^{-i(k_x x + k_y y + k_z z)} \, dx \, dy \, dz \tag{3.5.35}$$

is an entire function of complex k_x, k_y, k_z. From the discussion associated with (3.4.4) it is seen that the parameters λ and δ, through their specification of the polar axis direction, determine the half-space $|z| > Z$ wherein the integral appearing in (3.5.35) converges, since the polar axis direction specifies the orientation of the z-axis. Two such possible half-spaces and their associated z-axis directions are illustrated in Fig. 3.9 for a current source with compact spatial support.

Define the quantity

$$\psi = \psi(k_x, k_y, k_z) \equiv \frac{1}{k_z} f(k_x, k_y, k_z) e^{i\mathbf{k}\cdot\mathbf{r}} \ , \tag{3.5.36}$$

and let

$$\tilde{\psi} \equiv \psi(k_x, k_y, \gamma) \ . \tag{3.5.37}$$

Then $\tilde{\psi}$ has an integrable singularity at $\gamma = 0$ and $\tilde{\psi}$ vanishes at $\alpha = \pi/2 - \kappa - i\infty$. With this substitution the integral $I(\mathbf{r}, \lambda, \delta)$ becomes

$$I(\mathbf{r}, \lambda, \delta) = \int_0^{2\pi} d\beta \int_C d\alpha \cdot \tilde{\psi} J(k_x, k_y; \alpha, \beta) \ . \tag{3.5.38}$$

Consider first the dependence of $I(\mathbf{r}, \lambda, \delta)$ on the parameter λ. Upon differentiating (3.5.38) with respect to λ and assuming that the orders of differentiation and integration can be interchanged, one obtains

$$\frac{\partial I}{\partial \lambda} = \int_0^{2\pi} d\beta \int_C d\alpha \left(\frac{\partial \tilde{\psi}}{\partial \lambda} J + \tilde{\psi} \frac{\partial J}{\partial \lambda} \right) \equiv I_1 + I_2 , \tag{3.5.39}$$

where

$$I_1 \equiv \int_0^{2\pi} d\beta \int_C d\alpha \cdot \tilde{\psi} \frac{\partial J}{\partial \lambda} , \tag{3.5.40}$$

$$I_2 \equiv \int_0^{2\pi} d\beta \int_C d\alpha \cdot \frac{\partial \tilde{\psi}}{\partial \lambda} J . \tag{3.5.41}$$

Attention is given first to the integral I_1. The Jacobian appearing here is given by

$$J(k_x, k_y; \alpha, \beta) = \frac{\partial k_x}{\partial \alpha} \frac{\partial k_y}{\partial \beta} - \frac{\partial k_y}{\partial \alpha} \frac{\partial k_x}{\partial \beta} ,$$

so that

$$\frac{\partial J}{\partial \lambda} = \left(\frac{\partial}{\partial \alpha} \frac{\partial k_x}{\partial \lambda} \right) \frac{\partial k_y}{\partial \beta} + \left(\frac{\partial}{\partial \beta} \frac{\partial k_y}{\partial \lambda} \right) \frac{\partial k_x}{\partial \alpha} - \left(\frac{\partial}{\partial \alpha} \frac{\partial k_y}{\partial \lambda} \right) \frac{\partial k_x}{\partial \beta} - \left(\frac{\partial}{\partial \beta} \frac{\partial k_x}{\partial \lambda} \right) \frac{\partial k_y}{\partial \alpha} ,$$

where it has been assumed that the orders of differentiation can be interchanged. With this substitution the integral I_1 becomes

$$\begin{aligned} I_1 = & \int_0^{2\pi} d\beta \int_C d\alpha \cdot \tilde{\psi} \left[\left(\frac{\partial}{\partial \alpha} \frac{\partial k_x}{\partial \lambda} \right) \frac{\partial k_y}{\partial \beta} - \left(\frac{\partial}{\partial \alpha} \frac{\partial k_y}{\partial \lambda} \right) \frac{\partial k_x}{\partial \beta} + \left(\frac{\partial}{\partial \beta} \frac{\partial k_y}{\partial \lambda} \right) \frac{\partial k_x}{\partial \alpha} \right. \\ & \left. - \left(\frac{\partial}{\partial \beta} \frac{\partial k_x}{\partial \lambda} \right) \frac{\partial k_y}{\partial \alpha} \right] \\ = & \int_0^{2\pi} d\beta \left\{ \left[\tilde{\psi} \frac{\partial k_x}{\partial \lambda} \frac{\partial k_y}{\partial \beta} - \tilde{\psi} \frac{\partial k_y}{\partial \lambda} \frac{\partial k_x}{\partial \beta} \right]_0^{\pi/2 - k - i\infty} \right. \\ & \left. - \int_C d\alpha \left[\frac{\partial k_x}{\partial \lambda} \frac{\partial}{\partial \alpha} \left(\tilde{\psi} \frac{\partial k_y}{\partial \beta} \right) - \frac{\partial k_y}{\partial \lambda} \frac{\partial}{\partial \alpha} \left(\tilde{\psi} \frac{\partial k_x}{\partial \beta} \right) \right] \right\} \\ & + \int_C d\alpha \left\{ \left[\tilde{\psi} \frac{\partial k_y}{\partial \lambda} \frac{\partial k_x}{\partial \alpha} - \tilde{\psi} \frac{\partial k_x}{\partial \lambda} \frac{\partial k_y}{\partial \alpha} \right]_0^{2\pi} \right. \\ & \left. - \int_0^{2\pi} d\beta \left[\frac{\partial k_y}{\partial \lambda} \frac{\partial}{\partial \beta} \left(\tilde{\psi} \frac{\partial k_x}{\partial \alpha} \right) - \frac{\partial k_x}{\partial \lambda} \frac{\partial}{\partial \beta} \left(\tilde{\psi} \frac{\partial k_y}{\partial \alpha} \right) \right] \right\} \end{aligned} \tag{3.5.42}$$

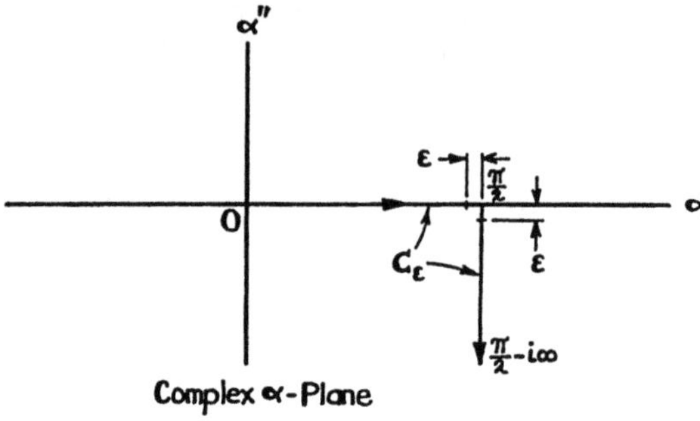

Fig. 3.10. The contour C_ε in the complex α-plane in the special case when $\tilde{k}(\omega)$ is real.

upon integration by parts. To proceed, special consideration must be allowed for the point $\alpha = \pi/2$ on the contour C in the special case when $\tilde{k}(\omega)$ is real (i.e., when $\kappa = 0$). As illustrated in Fig. 3.10, in this special case the contour C must be separated into the contour C_ε which excludes the point $\alpha = \pi/2$ by an infinitesimally small ε-neighborhood about this point where the slope of the contour is discontinuous. Of course, when $\tilde{k}(\omega)$ is complex-valued ($0 < \kappa < \pi/2$) the slope of the contour C is continuous along its entirety and this separation is unnecessary. To include this special case, (3.5.42) must be expanded as

$$
\begin{aligned}
I_1 = \lim_{\varepsilon \to 0} \int_0^{2\pi} d\beta \Bigg\{ & \left[\tilde{\psi} \frac{\partial k_x}{\partial \lambda} \frac{\partial k_y}{\partial \beta} - \tilde{\psi} \frac{\partial k_y}{\partial \lambda} \frac{\partial k_x}{\partial \beta} \right]_{\alpha=0}^{\alpha=\pi/2-\varepsilon} \\
& + \left[\tilde{\psi} \frac{\partial k_x}{\partial \lambda} \frac{\partial k_y}{\partial \beta} - \tilde{\psi} \frac{\partial k_y}{\partial \lambda} \frac{\partial k_x}{\partial \beta} \right]_{\alpha=\pi/2-i\varepsilon}^{\alpha=\pi/2-i\infty} \\
& - \int_{C_\varepsilon} d\alpha \left[\frac{\partial k_x}{\partial \lambda} \frac{\partial}{\partial \alpha} \left(\tilde{\psi} \frac{\partial k_y}{\partial \beta} \right) - \frac{\partial k_y}{\partial \lambda} \frac{\partial}{\partial \alpha} \left(\tilde{\psi} \frac{\partial k_x}{\partial \beta} \right) \right] \Bigg\} \\
& + \lim_{\varepsilon \to 0} \int_{C_\varepsilon} d\alpha \Bigg\{ \left[\tilde{\psi} \frac{\partial k_y}{\partial \lambda} \frac{\partial k_x}{\partial \alpha} - \tilde{\psi} \frac{\partial k_x}{\partial \lambda} \frac{\partial k_y}{\partial \alpha} \right]_0^{2\pi} - \int_0^{2\pi} d\beta \left[\frac{\partial k_y}{\partial \lambda} \frac{\partial}{\partial \beta} \left(\tilde{\psi} \frac{\partial k_x}{\partial \alpha} \right) \right. \\
& \left. - \frac{\partial k_x}{\partial \lambda} \frac{\partial}{\partial \beta} \left(\tilde{\psi} \frac{\partial k_y}{\partial \alpha} \right) \right] \Bigg\}
\end{aligned}
\tag{3.5.43}
$$

when $\kappa = 0$.

Consider first the "surface terms" appearing in (3.5.42, 43). First of all, since the functions $\tilde{\psi}$, k_x, and k_y are all periodic in β with period 2π, then

$$
\left[\tilde{\psi} \frac{\partial k_y}{\partial \lambda} \frac{\partial k_x}{\partial \alpha} - \tilde{\psi} \frac{\partial k_x}{\partial \lambda} \frac{\partial k_y}{\partial \alpha} \right]_0^{2\pi} = 0 \ , \tag{3.5.44}
$$

independent of the contour C. Furthermore, from (3.5.31) there results.

$$\frac{\partial k_x}{\partial \beta} = \tilde{k}(\omega)\sin\alpha(\mp\sin\beta\cos\lambda\cos\delta - \cos\beta\sin\delta) ,$$

$$\frac{\partial k_y}{\partial \beta} = \tilde{k}(\omega)\sin\alpha(\mp\sin\beta\cos\lambda\sin\delta + \cos\beta\cos\delta) ,$$

so that at $\alpha = 0$,

$$\left.\frac{\partial k_x}{\partial \beta}\right|_{\alpha=0} = \left.\frac{\partial k_y}{\partial \beta}\right|_{\alpha=0} = 0 ,$$

and since $\tilde{\psi}$ vanishes at $\alpha = \pi/2 - \kappa - \mathrm{i}\infty$, $0 \leqslant \kappa < \pi/2$, then when $\kappa \neq 0$ one has that

$$\left[\tilde{\psi}\frac{\partial k_x}{\partial \lambda}\frac{\partial k_y}{\partial \beta} - \tilde{\psi}\frac{\partial k_y}{\partial \lambda}\frac{\partial k_x}{\partial \beta}\right]_{\alpha=0}^{\alpha=\pi/2-\kappa-\mathrm{i}\infty} = 0 , \tag{3.5.45}$$

while for $\kappa = 0$ one has that

$$\lim_{\varepsilon\to 0}\left\{\left[\tilde{\psi}\frac{\partial k_x}{\partial \lambda}\frac{\partial k_y}{\partial \beta} - \tilde{\psi}\frac{\partial k_y}{\partial \lambda}\frac{\partial k_x}{\partial \beta}\right]_{\alpha=0}^{\alpha=\pi/2-\varepsilon} + \left[\tilde{\psi}\frac{\partial k_x}{\partial \lambda}\frac{\partial k_y}{\partial \beta} - \tilde{\psi}\frac{\partial k_y}{\partial \lambda}\frac{\partial k_x}{\partial \beta}\right]_{\alpha=\pi/2-\mathrm{i}\varepsilon}^{\alpha=\pi/2-\mathrm{i}\infty}\right\}$$

$$= \lim_{\varepsilon\to 0}\left[\tilde{\psi}\frac{\partial k_x}{\partial \lambda}\frac{\partial k_y}{\partial \beta} - \tilde{\psi}\frac{\partial k_y}{\partial \lambda}\frac{\partial k_x}{\partial \beta}\right]_{\alpha=\pi/2-\mathrm{i}\varepsilon}^{\alpha=\pi/2-\varepsilon} . \tag{3.5.46}$$

This "surface term" will also vanish provided that the quantity to be evaluated is continuous at $\varepsilon = 0$, since then the limit as $\varepsilon \to 0$ can be taken by setting $\varepsilon = 0$. From (3.5.27, 31)

$$\frac{\partial k_x}{\partial \lambda} = \tilde{k}(\omega)(\cos\alpha\cos\lambda \mp \sin\alpha\cos\beta\sin\lambda)\cos\delta = \gamma\cos\delta ,$$

$$\frac{\partial k_y}{\partial \lambda} = \tilde{k}(\omega)(\cos\alpha\cos\lambda \mp \sin\alpha\cos\beta\sin\lambda)\sin\delta = \gamma\sin\delta ,$$

and since $\tilde{\psi} = (1/\gamma)f(k_x, k_y, \gamma)\mathrm{e}^{\mathrm{i}\mathbf{k}^{\pm}\cdot\mathbf{r}}$, then the final quantity to be evaluated in (3.5.46) is indeed continuous at $\varepsilon = 0$ and this "surface term" vanishes.

Thus, all of the "surface terms" in both (3.5.42 and 43) vanish, and these equations can both be written as

$$I_1 = -\lim_{\varepsilon\to 0}\int_0^{2\pi} d\beta\int_C d\alpha\left\{\frac{\partial\tilde{\psi}}{\partial\alpha}\frac{\partial k_x}{\partial \lambda}\frac{\partial k_y}{\partial \beta} + \tilde{\psi}\frac{\partial k_x}{\partial \lambda}\frac{\partial^2 k_y}{\partial\alpha\partial\beta} - \frac{\partial\tilde{\psi}}{\partial\alpha}\frac{\partial k_x}{\partial \beta}\frac{\partial k_y}{\partial \lambda}\right.$$

$$\left. - \tilde{\psi}\frac{\partial^2 k_x}{\partial\alpha\partial\beta}\frac{\partial k_y}{\partial \lambda} + \frac{\partial\tilde{\psi}}{\partial\beta}\frac{\partial k_x}{\partial \alpha}\frac{\partial k_y}{\partial \lambda} + \tilde{\psi}\frac{\partial^2 k_x}{\partial\beta\partial\alpha}\frac{\partial k_y}{\partial \lambda} - \frac{\partial\tilde{\psi}}{\partial\beta}\frac{\partial k_x}{\partial \lambda}\frac{\partial k_y}{\partial \alpha} - \tilde{\psi}\frac{\partial k_x}{\partial \lambda}\frac{\partial^2 k_y}{\partial\beta\partial\alpha}\right\}$$

$$= -\lim_{\varepsilon\to 0} \int_0^{2\pi} d\beta \int_C d\alpha \left\{ \frac{\partial\tilde{\psi}}{\partial\alpha}\left(\frac{\partial k_x}{\partial\lambda}\frac{\partial k_y}{\partial\beta} - \frac{\partial k_x}{\partial\beta}\frac{\partial k_y}{\partial\lambda}\right) + \frac{\partial\tilde{\psi}}{\partial\beta}\left(\frac{\partial k_x}{\partial\alpha}\frac{\partial k_y}{\partial\lambda} - \frac{\partial k_x}{\partial\lambda}\frac{\partial k_y}{\partial\alpha}\right)\right\}, \tag{3.5.47}$$

where $C = C_\varepsilon$ when $\kappa = 0$, and where the limit is ignored when $0 < \kappa < \pi/2$.

Since $\psi(k_x, k_y, k_z)$ has an integrable singularity at $k_z = 0$, but is otherwise an analytic function of complex k_z, and is an entire function of complex k_x, k_y so that $\tilde{\psi} = \psi(k_x, k_y, \gamma)$ is an analytic function on the contour C with an integrable singularity at $\gamma = 0$ (where it also has a branch point), then

$$\begin{aligned}
\frac{\partial\tilde{\psi}}{\partial\alpha} &= \left[\frac{\partial\psi}{\partial k_x}\frac{\partial k_x}{\partial\alpha} + \frac{\partial\psi}{\partial k_y}\frac{\partial k_y}{\partial\alpha} + \frac{\partial\psi}{\partial k_z}\frac{\partial k_z}{\partial\alpha}\right]_{k_z=\gamma} \\
&= \left[\frac{\partial\psi}{\partial k_x}\frac{\partial k_x}{\partial\alpha} + \frac{\partial\psi}{\partial k_y}\frac{\partial k_y}{\partial\alpha} + \frac{\partial\psi}{\partial k_z}\frac{\partial\gamma}{\partial\alpha}\right]_{k_z=\gamma}, \\
\frac{\partial\tilde{\psi}}{\partial\beta} &= \left[\frac{\partial\psi}{\partial k_x}\frac{\partial k_x}{\partial\beta} + \frac{\partial\psi}{\partial k_y}\frac{\partial k_y}{\partial\beta} + \frac{\partial\psi}{\partial k_z}\frac{\partial k_z}{\partial\beta}\right]_{k_z=\gamma} \\
&= \left[\frac{\partial\psi}{\partial k_x}\frac{\partial k_x}{\partial\beta} + \frac{\partial\psi}{\partial k_y}\frac{\partial k_y}{\partial\beta} + \frac{\partial\psi}{\partial k_z}\frac{\partial\gamma}{\partial\beta}\right]_{k_z=\gamma},
\end{aligned} \tag{3.5.48}$$

since γ depends upon the parameters α and β in precisely the same manner that k_z does along the contour C. With these substitutions (3.5.47) becomes

$$\begin{aligned}
I_1 = -\lim_{\varepsilon\to 0} \int_0^{2\pi} d\beta \int_C d\alpha \Bigg\{ & \frac{\partial\psi}{\partial k_x}\frac{\partial k_x}{\partial\alpha}\frac{\partial k_x}{\partial\lambda}\frac{\partial k_y}{\partial\beta} + \frac{\partial\psi}{\partial k_y}\frac{\partial k_x}{\partial\lambda}\frac{\partial k_y}{\partial\alpha}\frac{\partial k_y}{\partial\beta} + \frac{\partial\psi}{\partial k_z}\frac{\partial k_x}{\partial\lambda}\frac{\partial k_y}{\partial\beta}\frac{\partial\gamma}{\partial\alpha} \\
& - \frac{\partial\psi}{\partial k_x}\frac{\partial k_x}{\partial\alpha}\frac{\partial k_x}{\partial\beta}\frac{\partial k_y}{\partial\lambda} - \frac{\partial\psi}{\partial k_y}\frac{\partial k_x}{\partial\beta}\frac{\partial k_y}{\partial\alpha}\frac{\partial k_y}{\partial\lambda} - \frac{\partial\psi}{\partial k_z}\frac{\partial k_x}{\partial\beta}\frac{\partial k_y}{\partial\lambda}\frac{\partial\gamma}{\partial\alpha} \\
& + \frac{\partial\psi}{\partial k_x}\frac{\partial k_x}{\partial\beta}\frac{\partial k_x}{\partial\alpha}\frac{\partial k_y}{\partial\lambda} + \frac{\partial\psi}{\partial k_y}\frac{\partial k_x}{\partial\alpha}\frac{\partial k_y}{\partial\beta}\frac{\partial k_y}{\partial\lambda} + \frac{\partial\psi}{\partial k_z}\frac{\partial k_x}{\partial\alpha}\frac{\partial k_y}{\partial\lambda}\frac{\partial\gamma}{\partial\beta} \\
& - \frac{\partial\psi}{\partial k_x}\frac{\partial k_x}{\partial\beta}\frac{\partial k_x}{\partial\lambda}\frac{\partial k_y}{\partial\alpha} - \frac{\partial\psi}{\partial k_y}\frac{\partial k_x}{\partial\lambda}\frac{\partial k_y}{\partial\beta}\frac{\partial k_y}{\partial\alpha} - \frac{\partial\psi}{\partial k_z}\frac{\partial k_x}{\partial\lambda}\frac{\partial k_y}{\partial\alpha}\frac{\partial\gamma}{\partial\beta}\Bigg\}_{k_z=\gamma} \\
= -\lim_{\varepsilon\to 0} \int_0^{2\pi} d\beta \int_C d\alpha \Bigg\{ & \frac{\partial\psi}{\partial k_x}\frac{\partial k_x}{\partial\lambda}\left(\frac{\partial k_x}{\partial\alpha}\frac{\partial k_y}{\partial\beta} - \frac{\partial k_x}{\partial\beta}\frac{\partial k_y}{\partial\alpha}\right) \\
& + \frac{\partial\psi}{\partial k_y}\frac{\partial k_y}{\partial\lambda}\left(\frac{\partial k_x}{\partial\alpha}\frac{\partial k_y}{\partial\beta} - \frac{\partial k_x}{\partial\beta}\frac{\partial k_y}{\partial\alpha}\right) \\
& + \frac{\partial\psi}{\partial k_z}\left[\frac{\partial k_x}{\partial\lambda}\left(\frac{\partial k_y}{\partial\beta}\frac{\partial\gamma}{\partial\alpha} - \frac{\partial k_y}{\partial\alpha}\frac{\partial\gamma}{\partial\beta}\right) + \frac{\partial k_y}{\partial\lambda}\left(\frac{\partial k_x}{\partial\alpha}\frac{\partial\gamma}{\partial\beta} - \frac{\partial k_x}{\partial\beta}\frac{\partial\gamma}{\partial\alpha}\right)\right]\Bigg\}_{k_z=\gamma}
\end{aligned}$$

$$= -\lim_{\varepsilon \to 0} \int_0^{2\pi} d\beta \int_C d\alpha \left\{ \left(\frac{\partial \psi}{\partial k_x} \frac{\partial k_x}{\partial \lambda} + \frac{\partial \psi}{\partial k_y} \frac{\partial k_y}{\partial \lambda} \right) J(k_x, k_y; \alpha, \beta) \right.$$

$$\left. + \frac{\partial \psi}{\partial k_z} \left[\frac{\partial k_x}{\partial \lambda} \left(\frac{\partial k_y}{\partial \beta} \frac{\partial \gamma}{\partial \alpha} - \frac{\partial k_y}{\partial \alpha} \frac{\partial \gamma}{\partial \beta} \right) + \frac{\partial k_y}{\partial \lambda} \left(\frac{\partial k_x}{\partial \alpha} \frac{\partial \gamma}{\partial \beta} - \frac{\partial k_x}{\partial \beta} \frac{\partial \gamma}{\partial \alpha} \right) \right] \right\}_{k_z = \gamma} . \tag{3.5.49}$$

To evaluate the last two terms appearing in the integrand of the above expression one returns to the definition $\gamma^2 = \tilde{k}^2 - k_x^2 - k_y^2$ which yields, upon differentiation with ω held fixed,

$$\gamma \, d\gamma = -k_x \, dk_x - k_y \, dk_y \ ,$$

so that

$$\frac{\partial \gamma}{\partial \alpha} = -\frac{1}{\gamma} \left(k_x \frac{\partial k_x}{\partial \alpha} + k_y \frac{\partial k_y}{\partial \alpha} \right) ,$$

$$\frac{\partial \gamma}{\partial \beta} = -\frac{1}{\gamma} \left(k_x \frac{\partial k_x}{\partial \beta} + k_y \frac{\partial k_y}{\partial \beta} \right) .$$

One then has that

$$\frac{\partial k_y}{\partial \beta} \frac{\partial \gamma}{\partial \alpha} - \frac{\partial k_y}{\partial \alpha} \frac{\partial \gamma}{\partial \beta} = -\frac{1}{\gamma} \frac{\partial k_y}{\partial \beta} \left(k_x \frac{\partial k_x}{\partial \alpha} + k_y \frac{\partial k_y}{\partial \alpha} \right) + \frac{1}{\gamma} \frac{\partial k_y}{\partial \alpha} \left(k_x \frac{\partial k_x}{\partial \beta} + k_y \frac{\partial k_y}{\partial \beta} \right)$$

$$= -\frac{k_x}{\gamma} \left(\frac{\partial k_x}{\partial \alpha} \frac{\partial k_y}{\partial \beta} - \frac{\partial k_x}{\partial \beta} \frac{\partial k_y}{\partial \alpha} \right) = -\frac{k_x}{\gamma} J(k_x, k_y; \alpha, \beta) , \tag{3.5.50a}$$

and

$$\frac{\partial k_x}{\partial \alpha} \frac{\partial \gamma}{\partial \beta} - \frac{\partial k_x}{\partial \beta} \frac{\partial \gamma}{\partial \alpha} = -\frac{1}{\gamma} \frac{\partial k_x}{\partial \alpha} \left(k_x \frac{\partial k_x}{\partial \beta} + k_y \frac{\partial k_y}{\partial \beta} \right) + \frac{1}{\gamma} \frac{\partial k_x}{\partial \beta} \left(k_x \frac{\partial k_x}{\partial \alpha} + k_y \frac{\partial k_y}{\partial \alpha} \right)$$

$$= -\frac{k_y}{\gamma} \left(\frac{\partial k_x}{\partial \alpha} \frac{\partial k_y}{\partial \beta} - \frac{\partial k_x}{\partial \beta} \frac{\partial k_y}{\partial \alpha} \right) = -\frac{k_y}{\gamma} J(k_x, k_y; \alpha, \beta) . \tag{3.5.50b}$$

With these final substitutions (3.5.49) becomes

$$I_1 = -\lim_{\varepsilon \to 0} \int_0^{2\pi} d\beta \int_C d\alpha \left\{ \left(\frac{\partial \psi}{\partial k_x} \frac{\partial k_x}{\partial \lambda} + \frac{\partial \psi}{\partial k_y} \frac{\partial k_y}{\partial \lambda} \right) \right.$$

$$\left. - \frac{1}{\gamma} \frac{\partial \psi}{\partial k_z} \left(k_x \frac{\partial k_x}{\partial \lambda} + k_y \frac{\partial k_y}{\partial \lambda} \right) \right\}_{k_z = \gamma} J(k_x, k_y; \alpha, \beta) ,$$

and since

$$-\gamma\frac{\partial\gamma}{\partial\lambda} = k_x\frac{\partial k_x}{\partial\lambda} + k_y\frac{\partial k_y}{\partial\lambda}$$

there finally results

$$\begin{aligned} I_1 &= -\lim_{\varepsilon\to 0}\int_0^{2\pi} d\beta \int_C d\alpha \left\{\frac{\partial\psi}{\partial k_x}\frac{\partial k_x}{\partial\lambda} + \frac{\partial\psi}{\partial k_y}\frac{\partial k_y}{\partial\lambda} + \frac{\partial\psi}{\partial k_z}\frac{\partial\gamma}{\partial\lambda}\right\}_{k_z=\gamma} J(k_x,k_y;\alpha,\beta) \\ &= -\lim_{\varepsilon\to 0}\int_0^{2\pi} d\beta \int_C d\alpha \frac{\partial\tilde{\psi}}{\partial\lambda} J(k_x,k_y;\alpha,\beta) = -I_2 \ , \end{aligned} \tag{3.5.51}$$

upon comparison with (3.5.41). Hence, from (3.5.39) one finds the desired result that $I(\mathbf{r},\lambda,\delta)$ is independent of the angular parameter λ. Similarly, by changing λ to δ everywhere in this proof one also obtains that $I(\mathbf{r},\lambda,\delta)$ is independent of the angular parameter δ. Hence

$$\frac{\partial I}{\partial\lambda} = \frac{\partial I}{\partial\delta} = 0 \tag{3.5.52}$$

and this completes Weyl's proof for a dispersive and absorptive medium. As a consequence, *(3.5.30) is valid for any choice of the angles λ and δ such that the point of observation of the field is within the region of convergence of the integral.*

3.5.3 Weyl's Integral Representation

If one chooses the polar axis to be along the observation direction that is specified by the field point position vector $\mathbf{r} = \mathbf{r}(r,\phi,\theta)$, then the transformation angles δ and λ are equal to the azimuthal angle $\delta = \phi$ and the angle of declination $\lambda = \theta$ of the observation direction. With this choice the positive sign is used in front of the integral appearing in (3.5.30) and

$$\tilde{\mathbf{k}}^{\pm}(\omega)\cdot\mathbf{r} = \tilde{k}(\omega) r\cos\alpha \ , \tag{3.5.53}$$

where α (when it is real-valued) is the angle between the polar axis and the wave vector (when it is also real-valued), as illustrated in Fig. 3.11. With these substitutions the angular spectrum integral (3.5.30) becomes

$$u(\mathbf{r}) = \tilde{k}(\omega)\int_0^{2\pi} d\beta \int_C \sin\alpha\, d\alpha\cdot U(k_x,k_y) e^{i\tilde{k}(\omega) r\cos\alpha} \ . \tag{3.5.54}$$

If one now takes the spectral amplitude function $U(k_x,k_y)$ to be unity, viz.

$$U(k_x,k_y) - 1 \ ,$$

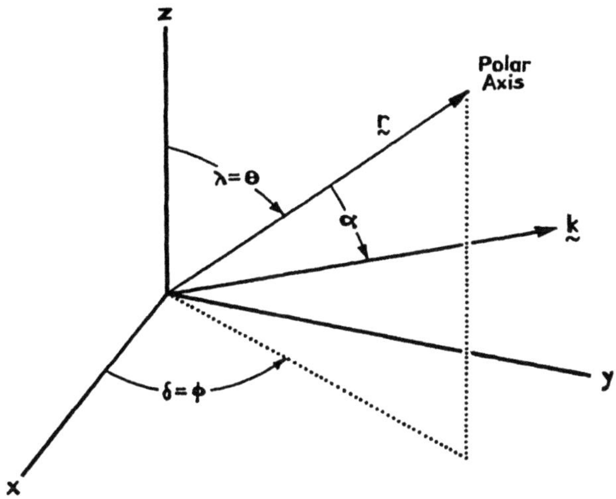

Fig. 3.11. Choice of the polar axis in Weyl's integral.

then (3.5.54) becomes [with the change of variable $\xi = \cos\alpha$ so that $\alpha = 0$ corresponds to $\xi = 1$ and $\alpha = \pi/2 - \kappa - \mathrm{i}\infty$ corresponds to $\xi = (\infty)\mathrm{e}^{i(\pi/2-\kappa)}$]

$$\begin{aligned} u(\mathbf{r}) &= 2\pi\tilde{k}(\omega)\int\limits_C \sin\alpha\cdot \mathrm{e}^{\mathrm{i}\tilde{k}(\omega)r\cos\alpha}\,d\alpha \\ &= -2\pi\tilde{k}(\omega)\int\limits_1^{\infty\,\mathrm{e}^{\mathrm{i}(\pi/2-\kappa)}} \mathrm{e}^{\mathrm{i}\tilde{k}(\omega)r\xi}\,d\xi \\ &= -2\pi\tilde{k}(\omega)\frac{\mathrm{e}^{\mathrm{i}\tilde{k}(\omega)r\xi}}{\mathrm{i}\tilde{k}(\omega)r}\Bigg|_{\xi=1}^{\xi=\infty\,\mathrm{e}^{\mathrm{i}(\pi/2-\kappa)}} \\ &= -2\pi\mathrm{i}\frac{\mathrm{e}^{\mathrm{i}\tilde{k}(\omega)r}}{r}\ , \end{aligned}$$

where the function vanishes at the upper limit of integration independently of $\kappa(\omega) = \arg\{\tilde{k}(\omega)\}$. Substitution of these two expressions in (3.5.1) then yields the general result

$$\frac{\mathrm{e}^{\mathrm{i}\tilde{k}(\omega)r}}{r} = \frac{\mathrm{i}}{2\pi}\iint\limits_{-\infty}^{\infty}\frac{1}{\gamma}\mathrm{e}^{\mathrm{i}\tilde{\mathbf{k}}^{\pm}(\omega)\cdot\mathbf{r}}\,dk_x\,dk_y \tag{3.5.55}$$

known as *Weyl's integral* [3.7], which expresses a spherical wave in terms of a superposition of plane waves in the dispersive medium. Since

$$\tilde{\mathbf{k}}^{\pm}\cdot\mathbf{r} = k_x x + k_y y \pm \gamma z\ ,$$

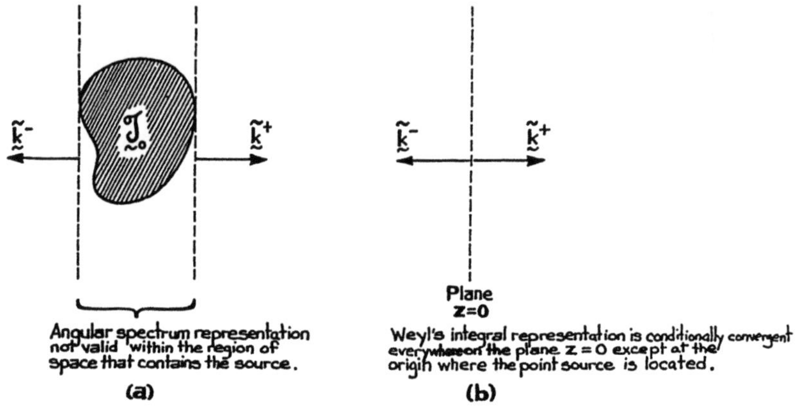

Fig. 3.12. Regions of validity of the angular spectrum and Weyl's integral representations.

then on the plane $z = 0$ there is no distinction between $\tilde{\mathbf{k}}^+$ and $\tilde{\mathbf{k}}^-$. However, the term γz provides exponential decay for the convergence of Weyl's integral. On the plane $z = 0$ the only decay factor appearing in the integrand of (3.5.55) is due to the term γ^{-1}, which goes to zero as γ goes off to infinity, but not with sufficient rapidity for absolute convergence requirements to be satisfied. Consequently, Weyl's integral (3.5.55) converges everywhere except on the plane $z = 0$, it is conditionally convergent everywhere on the plane $z = 0$ except at the origin, and it is divergent at the origin, as illustrated in Fig. 3.12 for (a) the general angular spectrum representation (3.4.20–23) and (b) Weyl's integral (3.5.55).

An alternate representation of Weyl's integral may be directly obtained from (3.5.30) with $U(k_x, k_y) = 1$, so that in place of (3.5.55) one obtains

$$\frac{e^{i\tilde{k}(\omega)r}}{r} = \frac{i\tilde{k}(\omega)}{2\pi} \int_{-\pi}^{\pi} d\beta \int_C \sin\alpha\, d\alpha \cdot e^{i[k_x(\omega)x + k_y(\omega)y \pm \gamma(\omega)z]} ,$$

where the change in the integration domain from $\beta = 0 \to 2\pi$ to $\beta = -\pi \to \pi$ is justified since it results in the same range of values for k_x and k_y. From (3.5.6, 22),

$$k_x = \tilde{k}(\omega) \sin\alpha \cos\beta ,$$

$$k_y = \tilde{k}(\omega) \sin\alpha \sin\beta ,$$

$$\gamma = \tilde{k}(\omega) \cos\alpha ,$$

and the above expression becomes

$$\frac{e^{i\tilde{k}(\omega)r}}{r} = \frac{i\tilde{k}(\omega)}{2\pi} \int_{-\pi}^{\pi} d\beta \int_C \sin\alpha\, d\alpha \cdot e^{i\tilde{k}(\omega)(x \sin\alpha \cos\beta + y \sin\alpha \sin\beta \pm z \cos\alpha)} \tag{3.5.56}$$

which is the polar coordinate form of *Weyl's integral representation*, where the positive sign appearing in the exponential of the integrand is employed when

$z > 0$ and the negative sign when $z < 0$. This representation clearly shows that Weyl's integral expresses a spherical wave as a superposition of inhomogeneous plane waves, where the ordered triple $(\sin\alpha\cos\beta, \sin\alpha\sin\beta, \pm\cos\alpha)$ are the complex direction cosines of the elementary plane wave normals.

3.5.4 Sommerfeld's Integral Representation

Consider the change of variable

$$\zeta = \tilde{k}(\omega)\sin\alpha \tag{3.5.57}$$

in Weyl's integral representation (3.5.56), so that as $\alpha = \alpha' + i\alpha''$ varies from 0 to $\pi/2 - \kappa - i\infty$ over the contour C, ζ varies from 0 to $\tilde{k}(\omega)e^{-i\kappa(\omega)}(\infty) = k\cdot\infty$. Define the quantity

$$\Gamma(\omega) = [\zeta^2 - \tilde{k}^2(\omega)]^{1/2} = -i\tilde{k}(\omega)\cos\alpha = -i\gamma(\omega)\ , \tag{3.5.58}$$

where the branch of the square root is determined such that $\mathbb{R}\mathrm{e}\{\Gamma\} \geqslant 0$ as α varies over the contour C as follows:

$$\begin{aligned}\mathbb{R}\mathrm{e}\{\Gamma\} &= \mathbb{R}\mathrm{e}\{-i\tilde{k}\cos(\alpha' + i\alpha'')\}\\ &= -k(\cos\kappa\sin\alpha'\sinh\alpha'' - \sin\kappa\cos\alpha'\cosh\alpha'')\ .\end{aligned}$$

Since $0 \leqslant \kappa < \pi/2$ so that both $\cos\kappa$ and $\sin\kappa$ are nonnegative, and since, as α varies over the contour C, $\sin\alpha'$ varies from 0 to $\cos\kappa$, $\sinh\alpha''$ varies from 0 to $-\infty$, $\cos\alpha'$ varies from 1 to $\sin\kappa$, and $\cosh\alpha''$ varies from 1 to $+\infty$, then

$$\mathbb{R}\mathrm{e}\{\Gamma\} = Ak\ ,$$

where A varies from 0 to $+\infty$ as α varies over C. In addition, along the contour C

$$\begin{aligned}\mathbb{I}\mathrm{m}\{\zeta\} &= \mathbb{I}\mathrm{m}\{\tilde{k}\sin\alpha\}\\ &= k\,\mathbb{I}\mathrm{m}\{(\cos\kappa + i\sin\kappa)(\sin\alpha'\cosh\alpha'' + i\cos\alpha'\sinh\alpha'')\}\\ &= k\cos\kappa\cosh\alpha''(\cos\alpha'\tanh\alpha'' + \tan\kappa\sin\alpha') = 0\end{aligned}$$

from (3.5.15), so that ζ is real-valued as α varies over C.

With the substitutions (3.5.57 and 58), Weyl's integral representation (3.5.56) becomes

$$\begin{aligned}\frac{e^{i\tilde{k}(\omega)r}}{r} &= \frac{1}{2\pi}\int_0^\infty i\frac{\zeta\,d\zeta}{\tilde{k}\cos\alpha}\int_{-\pi}^{\pi} d\beta\cdot e^{i\zeta(x\cos\beta + y\sin\beta)}e^{\mp\Gamma z}\\ &= \frac{1}{2\pi}\int_0^\infty \frac{1}{\Gamma}e^{\mp\Gamma z}\zeta\,d\zeta\int_{-\pi}^{\pi} d\beta\cdot e^{i\zeta(x\cos\beta + y\sin\beta)}\ ,\end{aligned}$$

where the upper sign choice is employed for $z > 0$ and the lower sign choice for $z < 0$. If one now transforms the spatial coordinates (x, y, z) in the cartesian representation to the cylindrical coordinates (r, φ, z), where $x = r\cos\varphi$, $y = r\sin\varphi$, and $r = (x^2 + y^2)^{1/2}$ is the cylindrical coordinate representation of the spherical radius r, then the above expression becomes

$$\frac{e^{i\tilde{k}(\omega)r}}{r} = \frac{1}{2\pi}\int_0^\infty \frac{1}{\Gamma} e^{\mp\Gamma z}\zeta\, d\zeta \int_{-\pi}^{\pi} e^{i\zeta r\cos(\beta-\varphi)}\, d\beta\ ,$$

and consequently,

$$\frac{e^{i\tilde{k}(\omega)r}}{r} = \int_0^\infty \frac{1}{\Gamma} e^{\mp\Gamma z} J_0(\zeta r)\zeta\, d\zeta \tag{3.5.59}$$

where $J_0(\xi)$ denotes the Bessel function of the first kind of order zero. This result may be written in a more explicit form as

$$\frac{e^{i\tilde{k}(\omega)r}}{r} = \int_0^\infty \frac{1}{[\zeta^2 - \tilde{k}^2(\omega)]^{1/2}} e^{\mp[\zeta^2-\tilde{k}^2(\omega)]^{1/2}z} J_0(\zeta r)\zeta\, d\zeta\ . \tag{3.5.60}$$

It is then seen that the path of integration extends from the origin along the positive, real ζ-axis with a suitable indentation into the complex ζ-plane about the branch point at $\zeta = k$ when $\tilde{k}$ assumes a real value k.

If one now uses the relation [3.13]

$$J_0(\xi) = \frac{1}{2}(H_0^{(1)}(\xi) - H_0^{(1)}(-\xi))$$

which expresses the Bessel function of the first kind of order zero in terms of the Hankel function of the first kind (or Bessel function of the third kind) of order zero, the above representation (3.5.59) becomes

$$\begin{aligned}\frac{e^{i\tilde{k}(\omega)r}}{r} &= \frac{1}{2}\left\{\int_0^\infty \frac{1}{\Gamma} e^{\mp\Gamma z} H_0^{(1)}(\zeta r)\zeta\, d\zeta - \int_0^\infty \frac{1}{\Gamma} e^{\mp\Gamma z} H_0^{(1)}(-\zeta r)\zeta\, d\zeta\right\}\\ &= \frac{1}{2}\left\{\int_0^\infty \frac{1}{\Gamma} e^{\mp\Gamma z} H_0^{(1)}(\zeta r)\zeta\, d\zeta + \int_{-\infty}^0 \frac{1}{\Gamma} e^{\mp\Gamma z} H_0^{(1)}(\zeta r)\zeta\, d\zeta\right\}\end{aligned}$$

under the transformation $\zeta \to -\zeta$ in the second integral. One then finally has the classic result [3.14–15]

$$\begin{aligned}\frac{e^{i\tilde{k}(\omega)r}}{r} &= \frac{1}{2}\int_{-\infty}^\infty \frac{1}{\Gamma} e^{\mp\Gamma z} H_0^{(1)}(\zeta r)\zeta\, d\zeta\\ &= \frac{1}{2}\int_{-\infty}^\infty \frac{1}{[\zeta^2 - \tilde{k}^2(\omega)]^{1/2}} e^{\mp(\zeta^2-\tilde{k}^2(\omega))^{1/2}z} H_0^{(1)}(\zeta r)\zeta\, d\zeta\end{aligned} \tag{3.5.61}$$

which is known as *Sommerfeld's integral representation*, where the upper-sign choice appearing in the exponential is selected for $z > 0$ and the lower-sign choice for $z < 0$. The contour of integration now extends along the real axis from $-\infty$ to $+\infty$ in the complex ζ-plane, except for a suitable indentation into the upper half-plane at the origin in order to avoid the logarithmic branch point that the Hankel function exhibits at that point, in addition to any suitable identation about the branch point $\zeta = k$ when $\tilde{k}$ assumes a real value k. Finally, notice that in this representation the polar axis is fixed along the z-axis.

It is not necessary that the contour of integration appearing in either of (3.5.59 or 61) lie along the real ζ-axis, but rather that the contour for each representation must only lie within a specific domain of analyticity defined by each integrand. For the representation (3.5.59 or 60) it is readily seen that the path of integration must lie within the strip of analyticity

$$-\mathbb{Im}\{\tilde{k}(\omega)\} < \mathbb{Im}\{\zeta\} < \mathbb{Im}\{\tilde{k}(\omega)\}$$

defined by the branch points at complex $\zeta = \pm\tilde{k}(\omega)$. Because of the logarithmic branch point that the Hankel function $H_0^{(1)}(\xi)$ exhibits at the origin, the path of integration for Sommerfeld's integral representation (3.5.61) must lie within the strip of analyticity

$$0 < \mathbb{Im}\{\zeta\} < \mathbb{Im}\{\tilde{k}(\omega)\} .$$

Of course, there are allowable deformations of the contour of integration outside of the strip of analyticity through the use of analytic continuation. Indeed, the path along the real ζ-axis that is described in the discussion following (3.5.61) is a trivial example of one such path.

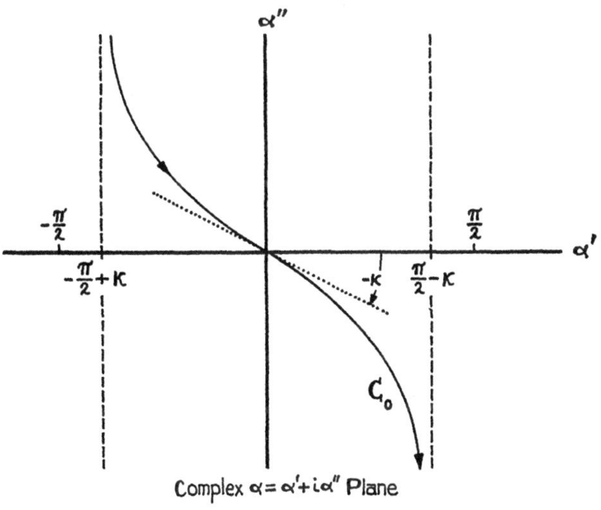

Fig. 3.13. Contour of integration for Ott's integral representation.

3.5.5 Ott's Integral Representation

The major difficulty in both the application and extension of Weyl's integral representation (3.5.56) is the simple fact that there are two integrations to perform. This is somewhat offset by the generality offerred by the arbitrary choice of direction for the polar axis that is not provided by Sommerfeld's integral representation (3.5.61) which entails only a single integration. In order to partially overcome these difficulties *Ott* [3.16] applied the transformation (3.5.57) with (3.5.58) to Sommerfeld's integral representation (3.5.61). Under this transformation, viz.

$$\zeta = \tilde{k}(\omega) \sin \alpha \ ,$$

$$d\zeta = \tilde{k}(\omega) \cos \alpha \, d\alpha \ ,$$

the path of integration along the real ζ-axis is transformed to the contour C_0 that is described by (3.5.15) and extends from $-\pi/2 + \kappa + \mathrm{i}\infty$ through the origin to $\pi/2 - \kappa - \mathrm{i}\infty$ in the complex α-plane, as illustrated in Fig. 3.13, where $\kappa(\omega) = \arg\{\tilde{k}(\omega)\}$. Under this transformation the representation (3.5.61) becomes

$$\frac{e^{i\tilde{k}(\omega)r}}{r} = \frac{1}{2} \int_{C_0} \frac{1}{-i\tilde{k}(\omega) \cos \alpha} e^{\pm i\tilde{k}(\omega)z \cos \alpha} H_0^{(1)}(\tilde{k}(\omega) r \sin \alpha) \tilde{k}(\omega) \sin \alpha \tilde{k}(\omega) \cos \alpha \, d\alpha$$

and consequently

$$\frac{e^{i\tilde{k}(\omega)r}}{r} = i \frac{\tilde{k}(\omega)}{2} \int_{C_0} H_0^{(1)}(\tilde{k}(\omega) r \sin \alpha) e^{\pm i\tilde{k}(\omega)z \cos \alpha} \sin \alpha \, d\alpha \tag{7.4.62}$$

which is known as *Ott's integral representation.* This representation forms the basis of *Baños'* classical research on dipole radiation [2.15].

4 The Angular Spectrum Representation of Pulsed Electromagnetic Beam Fields

A completely general representation of the propagation of a freely-propagating electromagnetic field into the half-space $z \geqslant z_0 > Z > 0$ of a homogeneous, isotropic, locally linear, temporally dispersive medium is now considered. The term "free" is used here[1] to indicate that there are no externally supplied charge or current sources for the field present in this half-space, the field source residing somewhere in the region $|z| \leqslant Z$. It is unnecessary to know what this source is provided that the pair of electromagnetic field vectors are known functions of time and the transverse position vector $\mathbf{r}_T = \hat{1}_x x + \hat{1}_y y$ in the plane $z = z_0$, as illustrated in Fig. 4.1. The rigorous, formal solution of this planar boundary value problem for the electromagnetic field in the half-space $z \geqslant z_0$ forms the basis of investigation for a wide class of pulsed electromagnetic beam field problems.

4.1 The Angular Spectrum Representation of the Freely Propagating Electromagnetic Field

Consider an electromagnetic field that is propagating into the half-space $z \geqslant z_0 > Z > 0$ and let the field vectors on the plane $z = z_0$, denoted by

$$\mathscr{E}(\mathbf{r}_T, z_0, t) = \mathscr{E}_0(\mathbf{r}_T, t) \ , \tag{4.1.1a}$$

$$\mathscr{B}(\mathbf{r}_T, z_0, t) = \mathscr{B}_0(\mathbf{r}_T, t) \ , \tag{4.1.1b}$$

be known functions of time and the transverse position vector $\mathbf{r}_T = \hat{1}_x x + \hat{1}_y y$ in the plane $z = z_0$, as indicated by the 0 subscript. It is assumed here that the two-dimensional spatial Fourier transform in the transverse coordinates as well as the temporal Fourier-Laplace transform of each field vector on the plane $z = z_0$ exists, where

[1] This should not be confused with a free-field whose externally supplied sources have all been turned off.

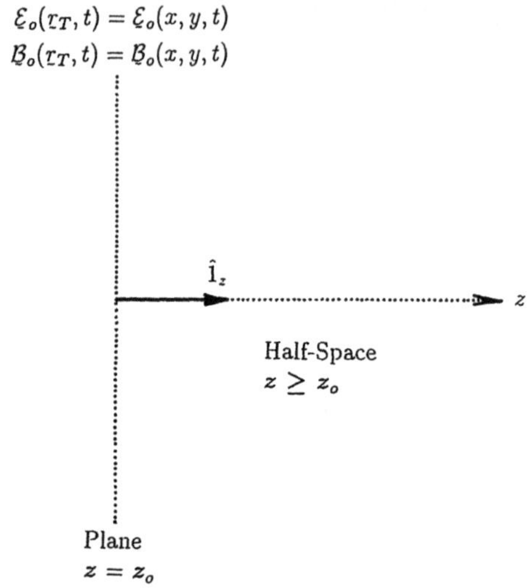

Fig. 4.1. Geometry of the planar boundary value problem.

$$\mathbf{E}_0(\mathbf{k}_T, \omega) = \int_{-\infty}^{\infty} dt \iint_{-\infty}^{\infty} dx\, dy\, \mathscr{E}_0(\mathbf{r}_T, t) e^{-i(\mathbf{k}_T \cdot \mathbf{r}_T - \omega t)} , \tag{4.1.2a}$$

$$\mathscr{E}_0(\mathbf{r}_T, t) = \frac{2}{(2\pi)^3} \mathrm{Re} \left\{ \int_{C_+} d\omega \iint_{-\infty}^{\infty} dk_x\, dk_y\, \mathbf{E}_0(\mathbf{k}_T, \omega) e^{i(\mathbf{k}_T \cdot \mathbf{r}_T - \omega t)} \right\} , \tag{4.1.2b}$$

and

$$\mathbf{B}_0(\mathbf{k}_T, \omega) = \int_{-\infty}^{\infty} dt \iint_{-\infty}^{\infty} dx\, dy\, \mathscr{B}_0(\mathbf{r}_T, t) e^{i(\mathbf{k}_T \cdot \mathbf{r}_T - \omega t)} , \tag{4.1.3a}$$

$$\mathscr{B}_0(\mathbf{r}_T, t) = \frac{2}{(2\pi)^3} \mathrm{Re} \left\{ \int_{C_+} d\omega \iint_{-\infty}^{\infty} dk_x\, dk_y\, \mathbf{B}_0(\mathbf{k}_T, \omega) e^{i(\mathbf{k}_T \cdot \mathbf{r}_T - \omega t)} \right\} . \tag{4.1.3b}$$

Here

$$\mathbf{r}_T = \hat{1}_x x + \hat{1}_y y \tag{4.1.4}$$

is the transverse position vector and

$$\mathbf{k}_T = \hat{1}_x k_x + \hat{1}_y k_y \tag{4.1.5}$$

is the transverse wavevector. If the initial time dependence of the field vectors on the plane $z = z_0$ is such that both field vectors $\mathscr{E}_0(\mathbf{r}_T, t)$ and $\mathscr{B}_0(\mathbf{r}_T, t)$ vanish for all $t < t_0$ for some finite value of t_0, then the time-frequency transform pairs appearing in (4.1.2) and (4.1.3) are both Laplace transformations and the contour C_+ is the straight-line path $\omega = \omega' + ia$ with a being greater than the abscissa of absolute convergence for the initial time evolution of the field; if not, then they are both Fourier transformations.

From (3.3.20, 21) the electromagnetic field vectors for the radiation field throughout the half-space $z > Z > 0$ are given by

$$\mathscr{E}(\mathbf{r}, t) = \frac{[\![4\pi]\!]}{(2\pi)^3} \mathbb{R}\mathrm{e} \int_{C_+} \frac{d\omega}{\omega\varepsilon_c(\omega)} \iint_{-\infty}^{\infty} dk_x \, dk_y \frac{1}{\gamma} \tilde{\mathbf{k}}^+ \times [\tilde{\mathbf{k}}^+ \times \mathbf{J}_0(\tilde{\mathbf{k}}^+, \omega)] e^{i(\tilde{\mathbf{k}}^+ \cdot \mathbf{r} - \omega t)} , \tag{4.1.6}$$

$$\mathscr{B}(\mathbf{r}, t) = -\left[\!\!\left[\frac{4\pi}{c}\right]\!\!\right] \frac{\mu}{(2\pi)^3} \mathbb{R}\mathrm{e} \left\{ \int_{C_+} d\omega \iint_{-\infty}^{\infty} dk_x \, dk_y \frac{1}{\gamma} [\tilde{\mathbf{k}}^+ \times \mathbf{J}_0(\tilde{\mathbf{k}}^+, \omega)] e^{i(\tilde{\mathbf{k}}^+ \cdot \mathbf{r} - \omega t)} \right\} , \tag{4.1.7}$$

where

$$\tilde{\mathbf{k}}^+ = \hat{1}_x k_x + \hat{1}_y k_y + \hat{1}_z \gamma \tag{4.1.8}$$

is the complex wavevector for propagation into the positive half-space $z > Z > 0$, with the associated complex wavenumber

$$\tilde{k}(\omega) = (\tilde{\mathbf{k}}^+ \cdot \tilde{\mathbf{k}}^+)^{1/2}$$

$$= k_0 n(\omega) = \left[\!\!\left[\frac{1}{c}\right]\!\!\right] \omega [\mu \varepsilon_c(\omega)]^{1/2} , \tag{4.1.9}$$

and where $\gamma = \gamma(\omega)$ is defined as the principal branch of the expression, cf. (3.3.11, 12),

$$\gamma = [\tilde{k}^2(\omega) - k_T^2]^{1/2} , \tag{4.1.10}$$

with $k_T^2 = k_x^2 + k_y^2$. On the plane $z = z_0 > Z$, $\mathbf{r} = \mathbf{r}_T + \hat{1}_z z_0$ and the electric field vector is given by, with (4.1.1a),

$$\mathscr{E}_0(\mathbf{r}_T, t) = \frac{[\![4\pi]\!]}{(2\pi)^3} \mathbb{R}\mathrm{e} \left\{ \int_{C_+} \frac{d\omega}{\omega\varepsilon_c(\omega)} \iint_{-\infty}^{\infty} dk_x \, dk_y \frac{1}{\gamma} \tilde{\mathbf{k}}^+ \right.$$
$$\left. \times (\tilde{\mathbf{k}}^+ \times \mathbf{J}_0(\tilde{\mathbf{k}}^+, \omega)) e^{i\gamma z_0} e^{i(\mathbf{k}_T \cdot \mathbf{r}_T - \omega t)} \right\} .$$

Comparison of this expression with (4.1.2b) shows that the spatio-temporal spectrum of the electric field vector on the plane $z = z_0$ is given by

$$\mathbf{E}_0(\mathbf{k}_T, \omega) = \frac{[\![4\pi]\!]}{2\omega\varepsilon_c(\omega)} \frac{1}{\gamma} \tilde{\mathbf{k}}^+ \times [\tilde{\mathbf{k}}^+ \times \mathbf{J}_0(\tilde{\mathbf{k}}^+, \omega)] e^{i\gamma z_0} . \tag{4.1.11}$$

Consequently, on any plane $z \geqslant z_0$, $\mathbf{r} = \mathbf{r}_T + \hat{1}_z z$ and the electric field vector is given by

$$\mathscr{E}(\mathbf{r}_T, z, t) = \frac{[\![4\pi]\!]}{(2\pi)^3} \mathrm{Re}\left\{ \int_{C_+} \frac{d\omega}{\omega\varepsilon_c(\omega)} \iint_{-\infty}^{\infty} dk_x dk_y \frac{1}{\gamma} \tilde{\mathbf{k}}^+ \right.$$

$$\left. \times [\tilde{\mathbf{k}}^+ \times \mathbf{J}_0(\tilde{\mathbf{k}}^+, \omega)] e^{i\gamma z} e^{i(\mathbf{k}_T \cdot \mathbf{r}_T - \omega t)} \right\}$$

$$= \frac{2}{(2\pi)^3} \mathrm{Re}\left\{ \int_{C_+} d\omega \iint_{-\infty}^{\infty} dk_x dk_y \mathbf{E}_0(\mathbf{k}_T, \omega) e^{i\gamma(z - z_0)} e^{i(\mathbf{k}_T \cdot \mathbf{r}_T - \omega t)} \right\}, \tag{4.1.12}$$

which is the desired angular-spectrum representation of the propagated electric field vector in terms of the spectrum of its planar boundary value. Similarly, on the plane $z = z_0 > Z$ the magnetic induction vector is given by, with (4.1.1b),

$$\mathscr{B}_0(\mathbf{r}_T, t)$$

$$= -\left[\!\!\left[\frac{4\pi}{c}\right]\!\!\right] \frac{\mu}{(2\pi)^3} \mathrm{Re}\left\{ \int_{C_+} d\omega \iint_{-\infty}^{\infty} dk_x dk_y \frac{1}{\gamma} [\tilde{\mathbf{k}}^+ \times \mathbf{J}_0(\tilde{\mathbf{k}}^+, \omega)] e^{i\gamma z_0} e^{i(\mathbf{k}_T \cdot \mathbf{r}_T - \omega t)} \right\}.$$

Comparison of this expression with (4.1.3b) shows that the spatio-temporal spectrum of the magnetic induction field vector on the plane $z = z_0$ is given by

$$\mathbf{B}_0(\mathbf{k}_T, \omega) = -\frac{1}{2}\left[\!\!\left[\frac{4\pi}{c}\right]\!\!\right] \mu \frac{1}{\gamma} [\tilde{\mathbf{k}}^+ \times \mathbf{J}_0(\tilde{\mathbf{k}}^+, \omega)] e^{i\gamma z_0}. \tag{4.1.13}$$

Consequently, on any plane $z \geqslant z_0$, $\mathbf{r} = \mathbf{r}_T + \hat{1}_z z$ and the magnetic induction field vector is given by

$$\mathscr{B}(\mathbf{r}_T, z, t)$$

$$= -\left[\!\!\left[\frac{4\pi}{c}\right]\!\!\right] \frac{\mu}{(2\pi)^3} \mathrm{Re}\left\{ \int_{C_+} d\omega \iint_{-\infty}^{\infty} dk_x dk_y \frac{1}{\gamma} [\tilde{\mathbf{k}}^+ \times \mathbf{J}_0(\tilde{\mathbf{k}}^+, \omega)] e^{i\gamma z} e^{i(\mathbf{k}_T \cdot \mathbf{r}_T - \omega t)} \right\}$$

$$= \frac{2}{(2\pi)^3} \mathrm{Re}\left\{ \int_{C_+} d\omega \iint_{-\infty}^{\infty} dk_x dk_y \mathbf{B}_0(\mathbf{k}_T, \omega) e^{i\gamma(z - z_0)} e^{i(\mathbf{k}_T \cdot \mathbf{r}_T - \omega t)} \right\}, \tag{4.1.14}$$

which is the desired angular spectrum representation of the propagated magnetic induction field vector in terms of the spectrum of its planar boundary value. One therefore has the *angular spectrum representation of the freely-propagating electromagnetic field*

$$\mathscr{E}(\mathbf{r}_T, z, t) = \frac{2}{(2\pi)^3}\,\mathrm{Re}\left\{\int_{C_+} d\omega \iint_{-\infty}^{\infty} dk_x\, dk_y\, \mathbf{E}_0(\mathbf{k}_T, \omega)\, e^{i[\mathbf{k}_T\cdot\mathbf{r}_T + \gamma(z-z_0) - \omega t]}\right\} \tag{4.1.15a}$$

$$\mathscr{B}(\mathbf{r}_T, z, t) = \frac{2}{(2\pi)^3}\,\mathrm{Re}\left\{\int_{C_+} d\omega \iint_{-\infty}^{\infty} dk_x\, dk_y\, \mathbf{B}_0(\mathbf{k}_T, \omega)\, e^{i[\mathbf{k}_T\cdot\mathbf{r}_T + \gamma(z-z_0) - \omega t]}\right\} \tag{4.1.15b}$$

for all $z \geqslant z_0 > Z > 0$.

The spatio-temporal spectra of the electromagnetic field vectors at the plane $z = z_0$ cannot be chosen independently of each other since both are ultimately determined from the same radiation source. Indeed, substitution of (4.1.13) into (4.1.11) immediately yields

$$\mathbf{E}_0(\mathbf{k}_T, \omega) = -\frac{[\![c]\!]}{\omega\mu\varepsilon_c(\omega)}\tilde{\mathbf{k}}^+ \times \mathbf{B}_0(\mathbf{k}_T, \omega) \ , \tag{4.1.16}$$

and consequently

$$\mathbf{B}_0(\mathbf{k}_T, \omega) = \frac{[\![c]\!]}{\omega}\tilde{\mathbf{k}}^+ \times \mathbf{E}_0(\mathbf{k}_T, \omega) \ . \tag{4.1.17}$$

In addition, the transversality condition

$$\tilde{\mathbf{k}}^+ \cdot \mathbf{E}_0(\mathbf{k}_T, \omega) = \tilde{\mathbf{k}}^+ \cdot \mathbf{B}_0(\mathbf{k}_T, \omega) = 0 \tag{4.1.18}$$

is satisfied. These equations (4.1.16–18) are precisely the relations that hold between both field vectors and the associated wave vector for a time-harmonic electromagnetic plane wave field in a homogeneous, isotropic, locally linear, temporally dispersive semiconducting medium, cf. (4.3.3–6). The pair of integrands appearing in the angular spectrum representation (4.1.15), given by

$$\mathbf{E}_0(\mathbf{k}_T, \omega)e^{i[\mathbf{k}_T\cdot\mathbf{r}_T + \gamma(z-z_0) - \omega t]} = \mathbf{E}_0(\mathbf{k}_T, \omega)e^{i[\tilde{\mathbf{k}}^+\cdot(\hat{\mathbf{1}}_x x + \hat{\mathbf{1}}_y y + \hat{\mathbf{1}}_z(z-z_0)) - \omega t]} \ ,$$

$$\mathbf{B}_0(\mathbf{k}_T, \omega)e^{i[\mathbf{k}_T\cdot\mathbf{r}_T + \gamma(z-z_0) - \omega t]} = \mathbf{B}_0(\mathbf{k}_T, \omega)e^{i[\tilde{\mathbf{k}}^+\cdot(\hat{\mathbf{1}}_x x + \hat{\mathbf{1}}_y y + \hat{\mathbf{1}}_z(z-z_0)) - \omega t]} \ ,$$

are thus seen to correspond to a time-harmonic electromagnetic plane wave field that is propagating away from the plane $z = z_0$ at each angular frequency ω and transverse wave vector $\mathbf{k}_T = \hat{\mathbf{1}}_x k_x + \hat{\mathbf{1}}_y k_y$ that is present in the initial spectral amplitude vectors $\{\mathbf{E}_0(\mathbf{k}_T, \omega), \mathbf{B}_0(\mathbf{k}_T, \omega)\}$ at that plane, with but one significant difference: the wavevector components k_x and k_y are always real-valued and independent of ω while $\gamma(\omega) = [\tilde{k}^2(\omega) - k_x^2 - k_y^2]^{1/2}$ is, in general, complex-valued. Hence, *each spectral plane wave component is attenuated in the z-direction alone*; this is just a mathematical consequence of the evaluation of the k_z-integral when the angular spectrum representation was derived in Sect.

3.3. The change in the amount of attenuation with propagation distance in different directions between any two parallel planes z = constant is accounted for solely by the dependence of γ upon k_x and k_y.

4.1.1 Geometric Form of the Angular Spectrum Representation

The plane wave spectral components appearing in the angular spectrum representation (4.1.15) of the propagated electromagnetic field into the positive half-space $z \geqslant z_0$ may be cast into a more geometric form by setting

$$
\begin{aligned}
k_x &= \tilde{k}(\omega)p \ , \\
k_y &= \tilde{k}(\omega)q \ , \\
\gamma(\omega) &= \tilde{k}(\omega)m \ ,
\end{aligned}
\tag{4.1.19}
$$

where

$$
\tilde{k}(\omega) = k_0 n(\omega) \tag{4.1.20}
$$

is the complex wavenumber given in (4.1.9), $k_0 = \omega/c$ is the vacuum wavenumber, and

$$
n(\omega) = \frac{c}{[\![c]\!]}[\mu\varepsilon_c(\omega)]^{1/2} \tag{4.1.21}
$$

is the complex index of refraction of the dispersive medium, cf. (2.1.34). Equation (4.1.10) then requires that

$$
m = (1 - p^2 - q^2)^{1/2} \ , \tag{4.1.22}
$$

where the principal branch of the square root expression is to be taken, as defined in (3.3.11, 12). The spatial phase term appearing in the exponential factor of (4.1.15) is then given by

$$
\mathbf{k}_T \cdot \mathbf{r}_T + \gamma(z - z_0) = \tilde{k}(\omega)[px + qy + m(z - z_0)] \ .
$$

The ordered triple of complex numbers (p, q, m) is then seen to be the set of *complex direction cosines* of the complex wavevector $\tilde{\mathbf{k}}^+ = \hat{\mathbf{1}}_x k_x + \hat{\mathbf{1}}_y k_y + \hat{\mathbf{1}}_z \gamma$, since

$$
\begin{aligned}
\tilde{\mathbf{k}}^+ \cdot \mathbf{r} = k_x x + k_y y + \gamma z &= \tilde{k}(\omega)\left(\frac{k_x}{\tilde{k}(\omega)}x + \frac{k_y}{\tilde{k}(\omega)}y + \frac{\gamma(\omega)}{\tilde{k}(\omega)}z\right) \\
&= \tilde{k}(\omega)(px + qy + mz) \ .
\end{aligned}
$$

That these generalized direction cosines are complex-valued follows directly from the fact that, for a causally dispersive medium the wavenumber $\tilde{k}(\omega)$ is complex-valued and may be written as, cf. (2.3.9),

$$\tilde{k}(\omega) = \beta(\omega) + i\alpha(\omega) \ , \tag{4.1.23}$$

where $\beta(\omega)$ is the (real-valued) *plane wave propagation factor* and $\alpha(\omega)$ is the (real-valued) *plane wave attenuation factor*, given in (2.3.11a and b), respectively. Since k_x and k_y must both be real-valued quantities, then the generalized direction cosines p and q must, in general, be complex-valued, so that, with $p = p' + ip''$ and $q = q' + iq''$,

$$\begin{aligned} k_x &= \tilde{k}(\omega)p = [\beta(\omega) + i\alpha(\omega)](p' + ip'') \\ &= [\beta(\omega)p' - \alpha(\omega)p''] + i[\beta(\omega)p'' + \alpha(\omega)p'] \ , \end{aligned} \tag{4.1.24a}$$

$$\begin{aligned} k_y &= \tilde{k}(\omega)q = [\beta(\omega) + i\alpha(\omega)](q' + iq'') \\ &= [\beta(\omega)q' - \alpha(\omega)q''] + i[\beta(\omega)q'' + \alpha(\omega)q'] \ . \end{aligned} \tag{4.1.24b}$$

Hence, in order that both k_x and k_y are real-valued for all values of ω, it is required that

$$p'' = -\frac{\alpha(\omega)}{\beta(\omega)}p' \ , \tag{4.1.25a}$$

$$q'' = -\frac{\alpha(\omega)}{\beta(\omega)}q' \ . \tag{4.1.25b}$$

With these two substitutions, Eqs. (4.1.24a and b) become, respectively

$$k_x = \beta(\omega)\left(1 + \frac{\alpha^2(\omega)}{\beta^2(\omega)}\right)p' \ , \tag{4.1.26a}$$

$$k_y = \beta(\omega)\left(1 + \frac{\alpha^2(\omega)}{\beta^2(\omega)}\right)q' \ . \tag{4.1.26b}$$

Hence, in order to determine the manner in which p' and q' must be allowed to vary so that k_x and k_y both vary from $-\infty$ to $+\infty$ for all $\omega = \omega' + ia$, $\omega' \geqslant 0$, it is necessary to know the behavior of $\beta(\omega)$ along the contour C_+ in the complex ω-plane. Since $\beta(\omega)$ is proportional to the real part of the complex index of refraction which (at least for the medium models considered in this research) is positive for all ω on C_+, then p' and q' must both vary from $-\infty$ to $+\infty$. Furthermore, since $\alpha(\omega) \geqslant 0$ for all ω on C_+ (again, at least for the medium models considered in this research), then p'' varies from $+\infty$ to $-\infty$ as p' varies from $-\infty$ to $+\infty$, while q'' varies from $+\infty$ to $-\infty$ as q' varies from $-\infty$ to $+\infty$. Hence, the contour $C_p(C_q)$ that $p = p' + ip''$ ($q = q' + iq''$) varies over in the complex $p(q)$-plane is a straight line path through the origin at the angle $-\tan^{-1}\{\alpha(\omega)/\beta(\omega)\}$ to the real axis, as illustrated in Fig. 4.2. Notice that C_p and

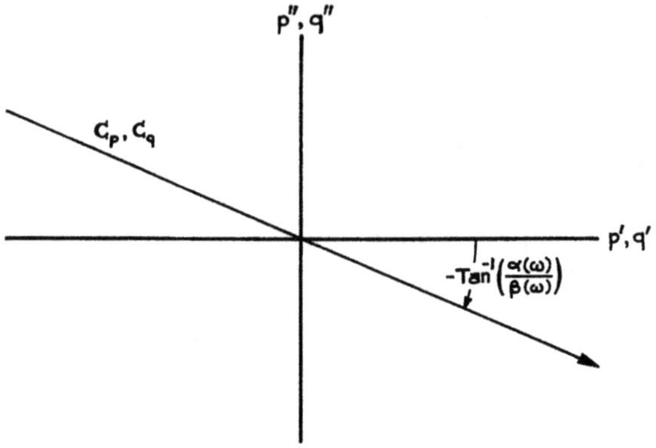

Fig. 4.2. Complex p, q-plane.

C_q both depend upon the angular frequency ω. When the medium is lossless, α vanishes and the contour lies along the real axis; this may occur, for example, at certain frequencies of a dispersive medium or at all frequencies of a non-dispersive medium.

With p'' and q'' as given in (4.1.25), the expression (4.1.22) for the complex direction cosine m becomes

$$m = (1 - p^2 - q^2)^{1/2} = [1 - (p' + ip'')^2 - (q' + iq'')^2]^{1/2}$$

$$= \left[1 - \left(1 - \frac{\alpha^2(\omega)}{\beta^2(\omega)}\right)(p'^2 + q'^2) + 2i\frac{\alpha(\omega)}{\beta(\omega)}(p'^2 + q'^2)\right]^{1/2} . \tag{4.1.27}$$

In order to evaluate this expression with the appropriate branch choice as set forth in (3.3.11, 12), let

$$m = \zeta^{1/2} \tag{4.1.28}$$

with $\zeta = \zeta' + i\zeta''$, where

$$\zeta' = \mathbb{R}\mathrm{e}\{\zeta\} = 1 - \left(1 - \frac{\alpha^2(\omega)}{\beta^2(\omega)}\right)(p'^2 + q'^2) , \tag{4.1.29a}$$

$$\zeta'' = \mathbb{I}\mathrm{m}\{\zeta\} = 2\frac{\alpha(\omega)}{\beta(\omega)}(p'^2 + q'^2) . \tag{4.1.29b}$$

Since $\alpha(\omega)/\beta(\omega) \geqslant 0$ for all ω on C_+, then

$$\zeta''(\omega) \geqslant 0 , \qquad \forall \omega \in C_+ . \tag{4.1.30}$$

Furthermore,

$$\zeta'(\omega) > 0 \ , \qquad \text{when } p'^2 + q'^2 < \frac{1}{1 - \alpha^2(\omega)/\beta^2(\omega)} \ , \tag{4.1.31a}$$

$$\zeta'(\omega) = 0 \ , \qquad \text{when } p'^2 + q'^2 = \frac{1}{1 - \alpha^2(\omega)/\beta^2(\omega)} \ , \tag{4.1.31b}$$

$$\zeta'(\omega) < 0 \ , \qquad \text{when } p'^2 + q'^2 > \frac{1}{1 - \alpha^2(\omega)/\beta^2(\omega)} \ . \tag{4.1.31c}$$

It is then clear that the argument of $\zeta(\omega)$ satisfies the inequality

$$0 \leqslant \arg\{\zeta(\omega)\} \leqslant \pi \ , \tag{4.1.32}$$

and hence, that the appropriate branch of the argument of $m = \zeta^{1/2}$ satisfies the inequality

$$0 \leqslant \arg\{m(\omega)\} \leqslant \frac{\pi}{2} \ , \tag{4.1.33}$$

for all ω on C_+, as illustrated in the sequence of figures depicted in Fig. 4.3. As a consequence, the real and imaginary parts of the complex direction cosine m satisfy the inequalities

$$m' = \mathbb{R}\mathrm{e}\{m(\omega)\} \geqslant 0 \ , \tag{4.1.34a}$$

$$m'' = \mathbb{I}\mathrm{m}\{m(\omega)\} \geqslant 0 \ , \tag{4.1.34b}$$

for all ω on C_+.

Explicit expressions for both m' and m'' may be obtained in the following manner. First of all, since

$$p^2 + q^2 + m^2 = 1 \ ,$$

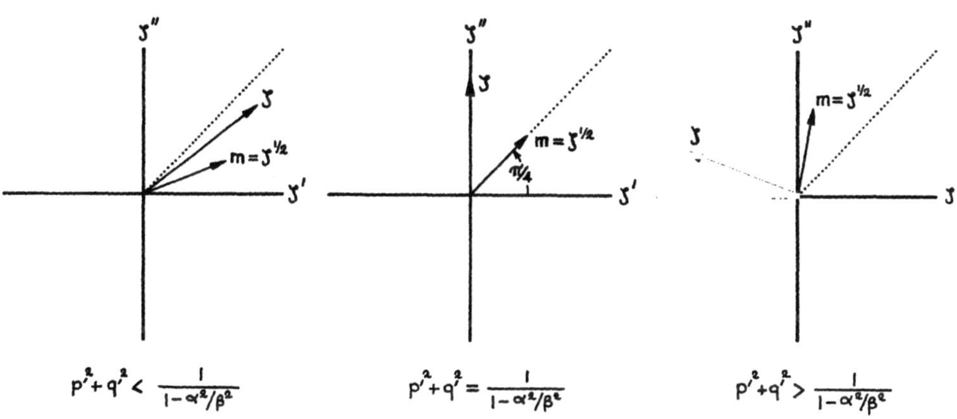

Fig. 4.3. Allowed behavior of the complex direction cosine $m(\omega)$.

with $p = p' + ip''$, $q = q' + iq''$, $m = m' + im'$, and substitution from (4.1.25a, b) for p'' and q'', there results

$$\left(1 - i\frac{\alpha}{\beta}\right)(p'^2 + q'^2) + (m' + im'')^2 = 1 \ ,$$

and consequently

$$\left(1 - \frac{\alpha^2}{\beta^2}\right)(p'^2 + q'^2) + m'^2 - m''^2 - 2i\left[\frac{\alpha}{\beta}(p'^2 + q'^2) - m'm''\right] = 1.$$

Upon equating real and imaginary parts in the above expression there results the pair of simultaneous equations

$$\left(1 - \frac{\alpha^2}{\beta^2}\right)(p'^2 + q'^2) + m'^2 - m''^2 = 1 \ , \tag{4.1.35a}$$

$$m'm'' = \frac{\alpha}{\beta}(p'^2 + q'^2) \ . \tag{4.1.35b}$$

Substitution of m'' from (4.1.35b into a) then yields the simple quartic equation

$$m'^4 + \left[\left(1 - \frac{\alpha^2}{\beta^2}\right)(p'^2 + q'^2) - 1\right]m'^2 - \frac{\alpha^2}{\beta^2}(p'^2 + q'^2)^2 = 0 \ .$$

The general solution for m'^2 is then given by

$$\begin{aligned}
m'^2 &= \frac{1}{2}\left[1 - \left(1 - \frac{\alpha^2}{\beta^2}\right)(p'^2 + q'^2)\right] \\
&\quad \pm \frac{1}{2}\left\{\left[1 - \left(1 - \frac{\alpha^2}{\beta^2}\right)(p'^2 + q'^2)\right]^2 + 4\frac{\alpha^2}{\beta^2}(p'^2 + q'^2)^2\right\}^{1/2} \\
&= \frac{1}{2}\left[1 - \left(1 - \frac{\alpha^2}{\beta^2}\right)(p'^2 + q'^2)\right] \\
&\quad \pm \frac{1}{2}\left[1 - 2\left(1 - \frac{\alpha^2}{\beta^2}\right)(p'^2 + q'^2) + \left(1 + \frac{\alpha^2}{\beta^2}\right)^2 (p'^2 + q'^2)^2\right]^{1/2} .
\end{aligned}$$

Since m' is real-valued, then $m'^2 \geqslant 0$ and only the positive sign choice in the above expression is appropriate. Hence,

$$\begin{aligned}
m' = &\left\{\frac{1}{2}\left[1 - \left(1 - \frac{\alpha^2}{\beta^2}\right)(p'^2 + q'^2)\right]\right. \\
&\left. + \frac{1}{2}\left[1 - 2\left(1 - \frac{\alpha^2}{\beta^2}\right)(p'^2 + q'^2) + \left(1 + \frac{\alpha^2}{\beta^2}\right)^2 (p'^2 + q'^2)^2\right]^{1/2}\right\}^{1/2} ,
\end{aligned} \tag{4.1.36a}$$

and m'' is then given by

$$m'' = \frac{\alpha}{\beta} \cdot \frac{p'^2 + q'^2}{m'} , \tag{4.1.36b}$$

provided that $\alpha \neq 0$. When $\alpha = 0$ (4.1.35a) becomes

$$m'^2 - m''^2 = 1 - (p'^2 + q'^2)$$

and (4.1.35b) states that either $m' = 0$ or $m'' = 0$. Hence, $m'' = 0$ and

$$m' = [1 - (p'^2 + q'^2)]^{1/2} \tag{4.1.37a}$$

when $p'^2 + q'^2 \leqslant 1$, while $m' = 0$ and

$$m'' = [(p'^2 + q'^2) - 1]^{1/2} \tag{4.1.37b}$$

when $p'^2 + q'^2 > 1$.

With these results the complex phase term appearing in the plane wave propagation factor $e^{i\tilde{\mathbf{k}}^+ \cdot \mathbf{r}}$ may be expressed as

$$\begin{aligned} \tilde{\mathbf{k}}^+ \cdot \mathbf{r} &= k_x x + k_y y + \gamma(\omega)\Delta z = k_x x + k_y y + \tilde{k}(\omega) m \Delta z \\ &= \beta(\omega)\left(1 + \frac{\alpha^2(\omega)}{\beta^2(\omega)}\right)(p'x + q'y) + [\beta(\omega)m' - \alpha(\omega)m'']\Delta z \\ &\quad + i[\alpha(\omega)m' + \beta(\omega)m'']\Delta z , \end{aligned}$$

so that

$$e^{i\tilde{\mathbf{k}}^+ \cdot \mathbf{r}} = e^{-[\alpha(\omega)m' + \beta(\omega)m'']\Delta z} \cdot e^{i\beta(\omega)[(1+\alpha^2(\omega)/\beta^2(\omega))(p'x+q'y)+(m'-(\alpha(\omega)/\beta(\omega))m'')\Delta z]} . \tag{4.1.38}$$

If $p' = q' = 0$, then $e^{i\tilde{\mathbf{k}}^+ \cdot \mathbf{r}}$ represents the spatial part of a *homogeneous plane wave* of angular frequency ω since the surfaces of constant amplitude coincide with the surfaces of constant phase, given by Δz = constant. If $\alpha(\omega) = 0$, then (4.1.38) represents the spatial part of a homogeneous plane wave when $p'^2 + q'^2 \leqslant 1$ (in which case $m'' = 0$), while it represents an *evanescent wave* when $p'^2 + q'^2 > 1$ (in which case $m' = 0$) since the surface of constant amplitude are then orthogonal to the equiphase surfaces. In general, $\alpha(\omega) \neq 0$ and either $p' \neq 0$ or $q' \neq 0$; (4.1.38) then represents the spatial part of an *inhomogeneous plane wave* of angular frequency ω since the surfaces of constant amplitude, given by

$$\Delta z = z - z_0 = \text{constant} \tag{4.1.39}$$

are different from the surfaces of constant phase, given by

$$\left(1 + \frac{\alpha^2(\omega)}{\beta^2(\omega)}\right)(p'x + q'y) + \left(m' - \frac{\alpha(\omega)}{\beta(\omega)} m''\right)\Delta z = \text{constant} . \tag{4.1.40}$$

These inhomogeneous plane wave phase fronts propagate in the direction that is specified by the real-valued vector

$$\mathbf{s} = \left(1 + \frac{\alpha^2(\omega)}{\beta^2(\omega)}\right)(p'\hat{1}_x + q'\hat{1}_y) + \left(m' - \frac{\alpha(\omega)}{\beta(\omega)}m''\right)\hat{1}_z \tag{4.1.41}$$

with the magnitude

$$s = \left[\left(1 + \frac{\alpha^2}{\beta^2}\right)^2 (p'^2 + q'^2) + \left(m' - \frac{\alpha}{\beta}m''\right)^2\right]^{1/2} \tag{4.1.42}$$

which is, in general, not equal to unity. The plane phase fronts of the inhomogeneous plane wave given in (4.1.38) are then seen to propagate in the direction that is specified by the set of real-valued direction cosines

$$\left[\frac{1}{s}\left(1 + \frac{\alpha^2}{\beta^2}\right)p', \frac{1}{s}\left(1 + \frac{\alpha^2}{\beta^2}\right)q', \frac{1}{s}\left(m' - \frac{\alpha}{\beta}m''\right)\right], \tag{4.1.43}$$

as illustrated in Fig. 4.4. These inhomogeneous plane wave phase fronts advance into the positive half-space $\Delta z > 0$ when the inequality

$$m' - \frac{\alpha(\omega)}{\beta(\omega)}m'' > 0 \tag{4.1.44}$$

is satisfied. Substitution of (4.1.36b) into this expression then yields the equivalent inequality

$$m'\left(1 - \frac{\alpha^2(\omega)}{\beta^2(\omega)} \cdot \frac{(p'^2 + q'^2)}{m'^2}\right) > 0 .$$

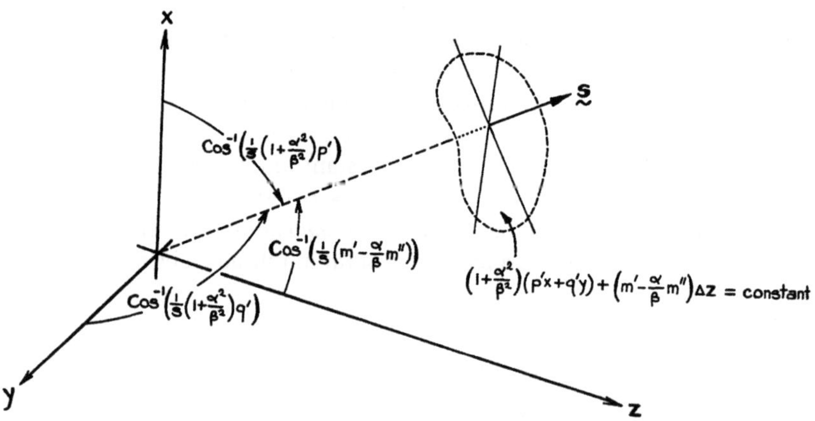

Fig. 4.4. Inhomogeneous plane wave phase front propagating in the direction s.

Since $m' > 0$ when $\alpha(\omega) \neq 0$, then the inequality in (4.1.44) is satisfied if

$$\frac{p'^2 + q'^2}{m'^2} < \frac{\beta^2(\omega)}{\alpha^2(\omega)}, \tag{4.1.45}$$

where the right-hand side of this inequality depends solely upon the dispersive properties of the medium. For a nondispersive medium, $\alpha(\omega) = 0$ for all ω and the inequality (4.1.44) simply becomes $m' > 0$, which is satisfied by the homogeneous plane waves in a loss-free medium.

Consider now obtaining the conditions (if indeed any do exist in the general case) under which the inequality given in (4.1.45) is satisfied. In order to accomplish this (at least in part), return to the quartic equation in m' following (4.1.35), viz.

$$m'^4 + \left[\left(1 - \frac{\alpha^2}{\beta^2}\right)(p'^2 + q'^2) - 1\right]m'^2 = \frac{\alpha^2}{\beta^2}(p'^2 + q'^2)^2 .$$

Two inequalities may then be obtained from this equation dependent upon the sign of the coefficient of m'^2, as follows:

(i) If this coefficient is negative, i.e., if

$$p'^2 + q'^2 < \frac{1}{1 - \alpha^2/\beta^2},$$

which occurs when $0 \leqslant \arg\{m(\omega)\} < \pi/4$, then the above quartic equation implies that

$$\frac{p'^2 + q'^2}{m'^2} < \frac{\beta(\omega)}{\alpha(\omega)} . \tag{4.1.46a}$$

Hence, if $\beta(\omega) \geqslant \alpha(\omega)$, then the inequality appearing in (4.1.46a) implies that the inequality appearing in (4.1.45) is satisfied. On the other hand, if $\alpha(\omega) > \beta(\omega)$, then the inequality appearing in (4.1.45) may still be satisfied, but there is no guarantee from this method of argument.

(ii) If this coefficient is positive, i.e., if

$$p'^2 + q'^2 > \frac{1}{1 - \alpha^2/\beta^2},$$

which occurs when $\pi/4 < \arg\{m(\omega)\} \leqslant \pi/2$, then the above quartic equation implies that

$$\frac{p'^2 + q'^2}{m'^2} > \frac{\beta(\omega)}{\alpha(\omega)} . \tag{4.1.46b}$$

If $\beta(\omega) > \alpha(\omega)$ then the inequality appearing (4.1.45) may be satisfied, but there is no guarantee from this method of argument. However, it is definitely not satisfied if $\alpha(\omega) \leqslant \beta(\omega)$.

Notice that when $\alpha(\omega) = 0$, the inequality specified in case (i) reduces to the inequality $p'^2 + q'^2 < 1$ which specifies the homogeneous plane wave components in a loss-free medium. On the other hand, the inequality specified in case (ii) becomes, when $\alpha(\omega) = 0$, the inequality $p'^2 + q'^2 > 1$ which specifies the evanescent plane wave components in a loss-free medium.

The geometric form of the angular spectrum representation of the freely-propagating electromagnetic field is given by

$$\mathscr{E}(x, y, z, t)$$
$$= \frac{1}{4\pi^3} \mathrm{Re}\left\{ \int_{C_+} d\omega \cdot \mathrm{e}^{-\mathrm{i}\omega t} \iint_{C_p C_q} \mathbf{E}_0(p, q, \omega) \mathrm{e}^{\mathrm{i}\tilde{k}(\omega)(px+qy+m\Delta z)} \tilde{k}^2(\omega)\, dp\, dq \right\}, \tag{4.1.47a}$$

$$\mathscr{B}(x, y, z, t)$$
$$= \frac{1}{4\pi^3} \mathrm{Re}\left\{ \int_{C_+} d\omega \cdot \mathrm{e}^{-\mathrm{i}\omega t} \iint_{C_p C_q} \mathbf{B}_0(p, q, \omega) \mathrm{e}^{\mathrm{i}\tilde{k}(\omega)(px+qy+m\Delta z)} \tilde{k}^2(\omega)\, dp\, dq \right\}, \tag{4.1.47b}$$

where $p = p' + \mathrm{i}p''$ varies over the contour C_p and $q = q' + \mathrm{i}q''$ varies over the contour C_q, as illustrated in Fig. 4.2. Because the spatio-temporal frequency spectra of the boundary values of the field vectors are related by (4.1.16, 17), then the pair of relations given in (4.1.47) for the propagated field vectors in the half-space $z \geqslant z_0$ may be expressed entirely in terms of either $\mathbf{E}_0(p, q, \omega)$ or $\mathbf{B}_0(p, q, \omega)$. Similar results for the formal solution of such boundary value problems for monochromatic wave propagation may be found in the published work of *Goodman* [4.2] and *Bouwkamp* [4.3] for the scalar optical field, and *Carter* [4.4] and *Stamnes* [4.5] for the diffraction theory of the electromagnetic field.

As a final point, notice that the generalized direction cosines p and q may be made to be real-valued if the defining set of relations given in (4.1.19) are replaced

$$k_x = \tilde{k}(\omega)\mathrm{e}^{-\mathrm{i}\psi(\omega)} p ,$$

$$k_y = \tilde{k}(\omega)\mathrm{e}^{-\mathrm{i}\psi(\omega)} q ,$$

$$\gamma(\omega) = \tilde{k}(\omega)\mathrm{e}^{-\mathrm{i}\psi(\omega)} m ,$$

where

$$\psi(\omega) = \tan^{-1}\left(\frac{\alpha(\omega)}{\beta(\omega)}\right) ,$$

and where $\tilde{k}(\omega) = k_0 n(\omega)$, as before. In terms of this new set of generalized direction cosines, the angular spectrum representation given in (4.1.49) becomes

$$\mathscr{E}(x,y,z,t)$$

$$= \frac{1}{4\pi^3}\,\mathrm{Re}\left\{\int_{C_+} d\omega \cdot \mathrm{e}^{-\mathrm{i}\omega t} \iint_{-\infty}^{\infty} \mathbf{E}_0(p,q,\omega)\mathrm{e}^{\mathrm{i}K(\omega)(px+qy+m\Delta z)} K^2(\omega)\,dp\,dq\right\},$$

$$\mathscr{B}(x,y,z,t)$$

$$= \frac{1}{4\pi^3}\,\mathrm{Re}\left\{\int_{C_+} d\omega \cdot \mathrm{e}^{-\mathrm{i}\omega t} \iint_{-\infty}^{\infty} \mathbf{B}_0(p,q,\omega)\mathrm{e}^{\mathrm{i}K(\omega)(px+qy+m\Delta z)} K^2(\omega)\,dp\,dq\right\},$$

where

$$K(\omega) = \tilde{k}(\omega)\mathrm{e}^{-\mathrm{i}\psi(\omega)}$$

is a real-valued quantity.

4.1.2 The Angular-Spectrum Representation and Huygen's Principle

It is well known that the solution of the plane-boundary value problem that is considered in this chapter can also be obtained through a superposition of spherical waves. The solution in terms of spherical waves has its physical origin in Huygen's principle [4.6] and has been (and continues to be) a central theme in classical diffraction theory. It is therefore of some importance to demonstrate the connection between the angular spectrum of plane waves representation and the mathematical embodiment of Huygen's principle as found in classical diffraction theory. It is clear that Weyl's integral (3.5.55), which expresses a spherical wave in terms of a superposition of plane waves, provides this connection, which is now derived following the treatment by *Sherman* [4.7, 8].

The derivation begins with the angular-spectrum representation of the freely-propagating electromagnetic field which, in its simplest form, may be written as

$$\mathscr{E}(x,y,z,t)$$

$$= \frac{1}{4\pi^3}\,\mathrm{Re}\left\{\int_{C_+} d\omega \cdot \mathrm{e}^{-\mathrm{i}\omega t} \iint_{C_pC_q} \mathbf{E}_0(p,q,\omega)\mathrm{e}^{\mathrm{i}\tilde{k}(\omega)(px+qy+m\Delta z)} \tilde{k}^2(\omega)\,dp\,dq\right\}, \tag{4.1.48a}$$

$$\mathscr{B}(x,y,z,t)$$

$$= \frac{1}{4\pi^3}\,\mathrm{Re}\left\{\int_{C_+} d\omega \cdot \mathrm{e}^{-\mathrm{i}\omega t} \iint_{C_pC_q} \mathbf{B}_0(p,q,\omega)\mathrm{e}^{\mathrm{i}\tilde{k}(\omega)(px+qy+m\Delta z)} \tilde{k}^2(\omega)\,dp\,dq\right\}, \tag{4.1.48b}$$

where $\Delta z = z - z_0 > 0$ is the normal propagation distance between the observation plane at z and the input plane at $z = z_0$. Here

$$\mathbf{E}_0(p,q,\omega) = \iint\limits_{-\infty}^{\infty} \tilde{\mathbf{E}}_0(x,y,\omega)\mathrm{e}^{-\mathrm{i}\tilde{k}(\omega)(px+qy)}\,dx\,dy\ , \tag{4.1.49a}$$

$$\mathbf{B}_0(p,q,\omega) = \iint\limits_{-\infty}^{\infty} \tilde{\mathbf{B}}_0(x,y,\omega)\mathrm{e}^{-\mathrm{i}\tilde{k}(\omega)(px+qy)}\,dx\,dy\ , \tag{4.1.49b}$$

where

$$\tilde{\mathbf{E}}_0(x,y,\omega) = \int\limits_{-\infty}^{\infty} \mathscr{E}_0(x,y,t)\mathrm{e}^{\mathrm{i}\omega t}\,dt\ , \tag{4.1.50a}$$

$$\tilde{\mathbf{B}}_0(x,y,\omega) = \int\limits_{-\infty}^{\infty} \mathscr{B}_0(x,y,t)\mathrm{e}^{\mathrm{i}\omega t}\,dt\ , \tag{4.1.50b}$$

are the temporal frequency spectra of the initial field vectors on the plane $z = z_0$. Substitution of (4.1.49a) into (4.1.48a) results in

$$\mathscr{E}(x,y,z,t) = \frac{1}{4\pi^3}\mathbb{R}\mathrm{e}\left\{\int\limits_{C_+} d\omega\cdot\mathrm{e}^{-\mathrm{i}\omega t}\iint\limits_{-\infty}^{\infty} dx'\,dy' \cdot\tilde{\mathbf{E}}_0(x',y',\omega)\iint\limits_{C_pC_q} \mathrm{e}^{\mathrm{i}\tilde{k}(\omega)[p(x-x')+q(y-y')+m\Delta z]}\tilde{k}^2(\omega)\,dp\,dq\right\}\ ,$$

while substitution of (4.1.49b) into (4.1.48b) yields

$$\mathscr{B}(x,y,z,t) = \frac{1}{4\pi^3}\mathbb{R}\mathrm{e}\left\{\int\limits_{C_+} d\omega\cdot\mathrm{e}^{-\mathrm{i}\omega t}\iint\limits_{-\infty}^{\infty} dx'\,dy' \cdot\tilde{\mathbf{B}}_0(x',y',\omega)\iint\limits_{C_pC_q} \mathrm{e}^{\mathrm{i}\tilde{k}(\omega)[p(x-x')+q(y-y')+m\Delta z]}\tilde{k}^2(\omega)\,dp\,dq\right\}\ .$$

The *monochromatic spatial impulse response function* for the normal propagation distance $\Delta z = z - z_0$ in the dispersive medium at the angular frequency ω is defined here as

$$h(x-x',y-y';\Delta z,\omega) \equiv \frac{1}{(2\pi)^2}\iint\limits_{C_pC_q} \mathrm{e}^{\mathrm{i}\tilde{k}(\omega)(p(x-x')+q(y-y')+m\Delta z)}\tilde{k}^2(\omega)\,dp\,dq\ . \tag{4.1.51}$$

Notice that this impulse response function is space-invariant (or isopolanatic) and depends upon the angular frequency ω through the complex wave number

$\tilde{k}(\omega)$. With (4.1.19) this expression may be rewritten as

$$h(\xi, \eta; \zeta, \omega) = \frac{1}{(2\pi)^2} \iint_{-\infty}^{\infty} e^{i\gamma(\omega)\zeta} e^{i(k_x \xi + k_y \eta)} \, dk_x \, dk_y$$

$$= \mathscr{F}^{-1}\{e^{i\gamma(\omega)\zeta}\} \, , \tag{4.1.52}$$

where $\gamma(\omega) = [\tilde{k}^2(\omega) - k_x^2 - k_y^2]^{1/2}$. The quantity $e^{i\gamma(\omega)\zeta}$ is seen to be the *spatial transfer function* of the linear dispersive system at the angular frequency ω. With the definition (4.1.51) the preceeding pair of expressions for the propagated field vectors becomes

$$\mathscr{E}(x, y, z, t)$$

$$= \frac{1}{\pi} \mathbb{R}\mathrm{e} \left\{ \int_{C_+} d\omega \cdot e^{-i\omega t} \iint_{-\infty}^{\infty} \tilde{\mathbf{E}}_0(x', y', \omega) h(x - x', y - y'; \Delta z, \omega) \, dx' \, dy' \right\} , \tag{4.1.53a}$$

$$\mathscr{B}(x, y, z, t)$$

$$= \frac{1}{\pi} \mathbb{R}\mathrm{e} \left\{ \int_{C_+} d\omega \cdot e^{-i\omega t} \iint_{-\infty}^{\infty} \tilde{\mathbf{B}}_0(x', y', \omega) h(x - x', y - y'; \Delta z, \omega) \, dx' \, dy' \right\} \tag{4.1.53b}$$

and the spatial part of each propagated field vector is given by the two-dimensional convolution of the spatial part of the corresponding initial field vector on the plane $z = z_0$ with the spatial impulse response function at each value of the angular frequency ω. Notice that the medium dispersion is contained entirely within the spatial impulse response function.

Weyl's integral (3.5.55) expresses a spherical wave in terms of a superposition of plane waves as

$$\frac{e^{i\tilde{k}(\omega)r}}{r} = \frac{i}{2\pi} \iint_{-\infty}^{\infty} \frac{1}{\gamma} e^{i\mathbf{k}^{\pm} \cdot \mathbf{r}} \, dk_x \, dk_y \, , \tag{4.1.54}$$

where $\tilde{\mathbf{k}}^{\pm}(\omega) = \hat{1}_x k_x + \hat{1}_y k_y \pm \hat{1}_z \gamma$, $\mathbf{r} = \hat{1}_x x + \hat{1}_y y + \hat{1}_z z$, and $r = |\mathbf{r}| = (x^2 + y^2 + z^2)^{1/2}$. The positive sign is employed here in the positive half-space $z > 0$, while the negative sign is employed in the negative half-space $z < 0$. From (4.1.51) one has that

$$h(x, y; \Delta z, \omega) = \frac{1}{(2\pi)^2} \iint_{-\infty}^{\infty} e^{i\tilde{\mathbf{k}}^+ \cdot \mathbf{r}^+} \, dk_x \, dk_y = \frac{1}{(2\pi)^2} \iint_{-\infty}^{\infty} e^{i(k_x x + k_y y + \gamma \Delta z)} \, dk_x \, dk_y$$

$$= -\frac{1}{2\pi} \frac{\partial}{\partial z} \left(\frac{i}{2\pi} \iint_{-\infty}^{\infty} \frac{1}{\gamma} e^{i\tilde{\mathbf{k}}^+ \cdot \mathbf{r}^+} \, dk_x \, dk_y \right) , \tag{4.1.55}$$

where $\mathbf{r}^+ = \hat{1}_x x + \hat{1}_y y + \hat{1}_z \Delta z$ with $\Delta z = z - z_0 > 0$. The interchange of the order of the differentiation and integration operations used above is justified by the following argument due to *Lalor* [4.9]. Define the pair of functions

$$f(z) \equiv \iint\limits_{C_p C_q} \frac{1}{m} e^{i\tilde{k}(px+qy+mz)}\, dp\, dq \ ,$$

$$g(z) \equiv i\tilde{k} \iint\limits_{C_p C_q} e^{i\tilde{k}(px+qy+mz)}\, dp\, dq \ ,$$

where it is desired to show that

$$g(z) = \frac{\partial f(z)}{\partial z} \ ,$$

or equivalently, that

$$\int\limits_{z_0}^{z} g(z')\, dz' = f(z) - f(z_0) \ .$$

Consider then the integral

$$\int\limits_{z_0}^{z} g(z')\, dz' = i\tilde{k} \int\limits_{z_0}^{z} dz' \iint\limits_{C_p C_q} e^{i\tilde{k}(px+qy+mz')}\, dp\, dq \ .$$

Since the Lebesgue integrability of the function $\exp[i\tilde{k}(px + qy + mz')]$ is ensured by the existence of the integral $\int_{z_0}^{z} dz' \iint_{C_p C_q} |e^{i\tilde{k}(px+qy+mz')}|\, dp\, dq$ for $z' > 0$, then by Fubini's theorem [4.10], the order of integration in the above integral may be interchanged, so that

$$\int\limits_{z_0}^{z} g(z')\, dz' = \iint\limits_{C_p C_q} \frac{1}{m} [e^{i\tilde{k}(px+qy+mz')}]_{z_0}^{z}\, dp\, dq = f(z) - f(z_0) \ ,$$

as was to be shown.

With the use of Weyl's integral (4.1.54) in (4.1.55) one finally obtains the important result

$$h(x - x', y - y'; \Delta z, \omega) = -\frac{1}{2\pi} \frac{\partial}{\partial z} \left(\frac{e^{i\tilde{k}(\omega)R}}{R} \right) \qquad (4.1.56)$$

where

$$R = [(x - x')^2 + (y - y')^2 + (\Delta z)^2]^{1/2} \ . \qquad (4.1.57)$$

This then identifies the spatial impulse response function defined in (4.1.51) in terms of the normal derivative of the "*free-space*" *Green's function* $G(R) = e^{i\tilde{k}(\omega)R}/R$ that plays a central role in the mathematical embodiment of Huygen's principle, as given by the integral theorem of Helmholtz and Kirchhoff. Substitution of (4.1.56) in the pair of relations given in (4.1.53) then yields

$$\mathscr{E}(x,y,z,t) = -\frac{1}{2\pi^2}\mathrm{Re}\left\{\int_{C_+} d\omega \cdot \mathrm{e}^{-\mathrm{i}\omega t} \iint_{-\infty}^{\infty} \tilde{\mathbf{E}}_0(x',y',\omega)\frac{\partial}{\partial z}\left(\frac{\mathrm{e}^{\mathrm{i}\tilde{k}(\omega)R}}{R}\right) dx'\,dy'\right\}, \tag{4.1.58a}$$

$$\mathscr{B}(x,y,z,t) = -\frac{1}{2\pi^2}\mathrm{Re}\left\{\int_{C_+} d\omega \cdot \mathrm{e}^{-\mathrm{i}\omega t} \iint_{-\infty}^{\infty} \tilde{\mathbf{B}}_0(x',y',\omega)\frac{\partial}{\partial z}\left(\frac{\mathrm{e}^{\mathrm{i}\tilde{k}(\omega)R}}{R}\right) dx'\,dy'\right\}. \tag{4.1.58b}$$

The spatial integrals appearing here are just the *first Rayleigh-Sommerfeld diffraction integrals* [4.11] of classical optics. The solution in terms of the normal derivatives of the initial field vectors, which yields the second Rayleigh-Sommerfeld diffraction integrals, is left as an exercise.

4.2 Polarization Properties of the Freely Propagating Electromagnetic Field

Of considerable interest to the description of the propagation characteristics of a general electromagnetic wave field are the polarization properties of the electric and magnetic field vectors. The used treatment [4.12] of the polarization state of an electromagnetic field is restricted to the special, idealized case of a time-harmonic, plane-wave field. This restriction has, in part, been removed by *Nisbet* and *Wolf* [4.13] for the case of linearly polarized, time-harmonic waves of arbitrary spatial form. A complete extension [4.14] of this treatment to a general, pulsed electromagnetic wave field is then readily accomplished through the use of the angular spectrum of plane waves representation for the freely-propagating field.

From (4.1.15) the angular spectrum of plane waves representation of a freely-propagating electromagnetic field may be written as

$$\mathscr{E}(\mathbf{r},t) = \frac{1}{\pi}\mathrm{Re}\left\{\int_{C_+} \tilde{\mathbf{E}}(\mathbf{r},\omega)\mathrm{e}^{-\mathrm{i}\omega t}\,d\omega\right\}, \tag{4.2.1a}$$

$$\mathscr{B}(\mathbf{r},t) = \frac{1}{\pi}\mathrm{Re}\left\{\int_{C_+} \tilde{\mathbf{B}}(\mathbf{r},\omega)\mathrm{e}^{-\mathrm{i}\omega t}\,d\omega\right\}, \tag{4.2.1b}$$

where the (complex) temporal frequency spectrum of each electromagnetic field vector has the angular spectrum representation

$$\tilde{\mathbf{E}}(\mathbf{r},\omega) = \frac{1}{(2\pi)^2}\iint\limits_{-\infty}^{\infty} \mathbf{E}_0(\mathbf{k}_T,\omega)\mathrm{e}^{\mathrm{i}\tilde{\mathbf{k}}^+\cdot\mathbf{r}^+}\,dk_x\,dk_y$$

$$\equiv \mathbf{p}_e(\mathbf{r},\omega) + \mathrm{i}\mathbf{q}_e(\mathbf{r},\omega)\ , \tag{4.2.2a}$$

$$\tilde{\mathbf{B}}(\mathbf{r},\omega) = \frac{1}{(2\pi)^2}\iint\limits_{-\infty}^{\infty} \mathbf{B}_0(\mathbf{k}_T,\omega)\mathrm{e}^{\mathrm{i}\tilde{\mathbf{k}}^+\cdot\mathbf{r}^+}\,dk_x\,dk_y$$

$$\equiv \mathbf{p}_m(\mathbf{r},\omega) + \mathrm{i}\mathbf{q}_m(\mathbf{r},\omega)\ . \tag{4.2.2b}$$

Here $\mathbf{p}(\mathbf{r},\omega)$ and $\mathbf{q}(\mathbf{r},\omega)$ denote the real and imaginary parts, respectively, of the appropriate complex-valued field vector as indicated by the subscript (e or m). With this substitution the pair of relations given in (4.2.1) becomes

$$\mathscr{E}(\mathbf{r},t) = \frac{1}{\pi}\mathrm{Re}\left\{\int\limits_{C_+} \mathbf{V}_e(\mathbf{r},t;\omega)\,d\omega\right\}\ , \tag{4.2.3a}$$

$$\mathscr{B}(\mathbf{r},t) = \frac{1}{\pi}\mathrm{Re}\left\{\int\limits_{C_+} \mathbf{V}_m(\mathbf{r},t;\omega)\,d\omega\right\}\ , \tag{4.2.3b}$$

where

$$\mathbf{V}_e(\mathbf{r},t;\omega) \equiv [\mathbf{p}_e(\mathbf{r},\omega) + \mathrm{i}q_e(\mathbf{r},\omega)]\mathrm{e}^{-\mathrm{i}\omega t}\ , \tag{3.2.4a}$$

$$\mathbf{V}_m(\mathbf{r},t;\omega) \equiv [\mathbf{p}_m(\mathbf{r},\omega) + \mathrm{i}q_m(\mathbf{r},\omega)]\mathrm{e}^{-\mathrm{i}\omega t}\ , \tag{3.2.4b}$$

are complex vectors that describe the spatial properties of each monochromatic field component appearing in the representation (4.2.3) of the propagated field.

4.2.1 The Polarization Ellipse for the Complex Field Vectors

Following the analysis of *Born* and *Wolf* [4.12], consider the behavior of the monochromatic complex vector field

$$V(\mathbf{r},t) = [\mathbf{p}(\mathbf{r}) + \mathrm{i}\mathbf{q}(\mathbf{r})]\mathrm{e}^{-\mathrm{i}\omega t} \tag{4.2.5}$$

at a specific point $\mathbf{r} = \mathbf{r}_0$ in space. Here $\mathbf{V}(\mathbf{r},t)$ represents either $\mathbf{V}_e(\mathbf{r},t)$ or $\mathbf{V}_m(\mathbf{r},t)$. As time varies the end point of the vector $\mathbf{V}(\mathbf{r}_0,t)$ describes a curve in the plane that is specified by the pair of (real-valued) vectors $\mathbf{p}(\mathbf{r}_0)$ and $\mathbf{q}(\mathbf{r}_0)$. Furthermore, since $\mathbf{V}(\mathbf{r},t)$ is periodic time at any fixedpoint in space, this curve must then be closed. Now let

$$\mathbf{p}(\mathbf{r}_0) + \mathrm{i}\mathbf{q}(\mathbf{r}_0) = (\mathbf{a} + \mathrm{i}\mathbf{b})\mathrm{e}^{i\varphi}\ , \tag{4.2.6}$$

where φ is an arbitrary scalar quantity. The real vectors $\mathbf{a}$ and $\mathbf{b}$ may then be expressed in terms of $\mathbf{p}(\mathbf{r}_0)$, $\mathbf{q}(\mathbf{r}_0)$, and φ as

$$\begin{aligned} \mathbf{a} &= \mathbf{p}(\mathbf{r}_0)\cos\varphi + \mathbf{q}(\mathbf{r}_0)\sin\varphi \ , \\ \mathbf{b} &= -\mathbf{p}(\mathbf{r}_0)\sin\varphi + \mathbf{q}(\mathbf{r}_0)\cos\varphi \ . \end{aligned} \tag{4.2.7}$$

Notice that $\mathbf{a}$, $\mathbf{b}$, and φ are all functions of the position $\mathbf{r}_0$. Consider now choosing the angle φ such that the vectors $\mathbf{a}$ and $\mathbf{b}$ are orthogonal, so that

$$\tan(2\varphi) = \frac{2\mathbf{p}(\mathbf{r}_0)\cdot\mathbf{q}(\mathbf{r}_0)}{\mathbf{p}^2(\mathbf{r}_0) - \mathbf{q}^2(\mathbf{r}_0)} \ . \tag{4.2.8}$$

The parameters that specify the spatial properties of the complex vector field $\mathbf{V}(\mathbf{r}, t)$ at any specific point $\mathbf{r} = \mathbf{r}_0$ may now be considered to be the five independent components of the orthogonal vectors $\mathbf{a}$ and $\mathbf{b}$ and the associated phase factor φ, instead of the six independent components of the vectors $\mathbf{p}$ and $\mathbf{q}$. With (4.2.6) with expression (4.2.5) for the monchromatic complex vector field $\mathbf{V}(\mathbf{r}, t)$ at the point $\mathbf{r} = \mathbf{r}_0$ becomes

$$\begin{aligned} \mathbf{V}(\mathbf{r}_0, t) &= \mathbf{V}^{(r)}(\mathbf{r}_0, t) + \mathrm{i}\mathbf{V}^{(i)}(\mathbf{r}_0, t) \\ &= [\mathbf{a}\cos(\omega t - \varphi) + \mathbf{b}\sin(\omega t - \varphi)] \\ &\quad - \mathrm{i}[\mathbf{a}\sin(\omega t - \varphi) - \mathbf{b}\cos(\omega t - \varphi)] \ . \end{aligned} \tag{4.2.9}$$

If a Cartesian axis system is now defined with origin at the fixed field point $\mathbf{r} = \mathbf{r}_0$ and with the x- and y-coordinate directions chosen along the direction of the vectors $\mathbf{a} = \mathbf{a}(\mathbf{r}_0)$ and $\mathbf{b} = \mathbf{b}(\mathbf{r}_0)$, respectively, then the components of the real part of $\mathbf{V}(\mathbf{r}_0, t)$ with respect to this coordinate system are given by

$$\begin{aligned} V_x^{(r)}(\mathbf{r}_0, t) &= a(\mathbf{r}_0)\cos[\omega t - \varphi(\mathbf{r}_0)] \ , \\ V_y^{(r)}(\mathbf{r}_0, t) &= b(\mathbf{r}_0)\cos[\omega t - \varphi(\mathbf{r}_0)] \ , \\ V_z^{(r)}(\mathbf{r}_0, t) &= 0 \ , \end{aligned} \tag{4.2.10}$$

while the components of the imaginary part of $\mathbf{V}(\mathbf{r}_0, t)$ are

$$\begin{aligned} V_x^{(i)}(\mathbf{r}_0, t) &= a(\mathbf{r}_0)\cos\left[\omega t - \varphi(\mathbf{r}_0) + \frac{\pi}{2}\right] \ , \\ V_y^{(i)}(\mathbf{r}_0, t) &= b(\mathbf{r}_0)\sin\left[\omega t - \varphi(\mathbf{r}_0) + \frac{\pi}{2}\right] \ , \\ V_z^{(i)}(\mathbf{r}_0, t) &= 0 \ , \end{aligned} \tag{4.2.11}$$

where $a(\mathbf{r}_0) = |\mathbf{a}(\mathbf{r}_0)|$ and $b(\mathbf{r}_0) = |\mathbf{b}(\mathbf{r}_0)|$. Both of the above two sets of equations describe an ellipse in time, called the *polarization ellipse*, that is given

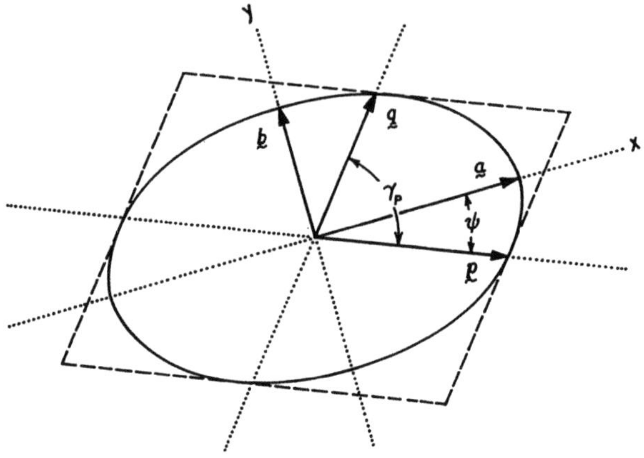

Fig. 4.5. The polarization ellipse.

by

$$\frac{[V_x^{(j)}(\mathbf{r}_0,t)]^2}{a^2}+\frac{[V_y^{(j)}(\mathbf{r}_0,t)]^2}{b^2}=1\ ;\qquad j=r,i\ , \tag{4.2.12}$$

with semi-axes of lengths a and b along the x and y coordinate axes, respectively. Notice that the real and imaginary parts of $\mathbf{V}(\mathbf{r}_0,t)$ are $\pi/2$ out of phase as they trace out the polarization ellipse. The real vectors $\mathbf{p}(\mathbf{r}_0)$ and $\mathbf{q}(\mathbf{r}_0)$ are thus seen to form a pair of conjugate semi-diameters of the polarization ellipse, as illustrated in Fig. 4.5.

The semi-axis lengths a and b of the polarization ellipse described by the complex vector field at the point $\mathbf{r}=\mathbf{r}_0$ are readily obtained from (4.2.7, 8) as

$$a^2=\tfrac{1}{2}\{p^2+q^2+[(p^2-q^2)^2+4(\mathbf{p}\cdot\mathbf{q})^2]^{1/2}\}\ , \tag{4.2.13a}$$

$$b^2=\tfrac{1}{2}\{p^2+q^2-[(p^2-q^2)^2+4(\mathbf{p}\cdot\mathbf{q})^2]^{1/2}\}\ . \tag{4.2.13b}$$

In order to determine the angle ψ between the vectors $\mathbf{a}$ and $\mathbf{p}$ one begins by expressing the equation of the polarization ellipse in parametric form as

$$\left.\begin{aligned}V_x^{(j)}(\mathbf{r}_0,t)&=a\cos\phi_j\\ V_y^{(j)}(\mathbf{r}_0,t)&=b\sin\phi_j\end{aligned}\right\}\ ;\qquad j=r,i\ , \tag{4.2.14}$$

where $\phi_j=\phi_j(\mathbf{r}_0,t)$ denotes the eccentric angle that is depicted in Fig. 4.6 for the case when $a\geqslant b$ (this inequality is assumed to hold throughout the remaining analysis). From the geometry of the figure it is seen that the eccentric angle ϕ_j is related to the polar angle of the point $(V_x^{(j)},V_y^{(j)})$ on the vibration ellipse by

$$\tan\theta=\frac{V_y^{(j)}}{V_x^{(j)}}=\frac{b}{a}\tan\phi_j\ . \tag{4.2.15}$$

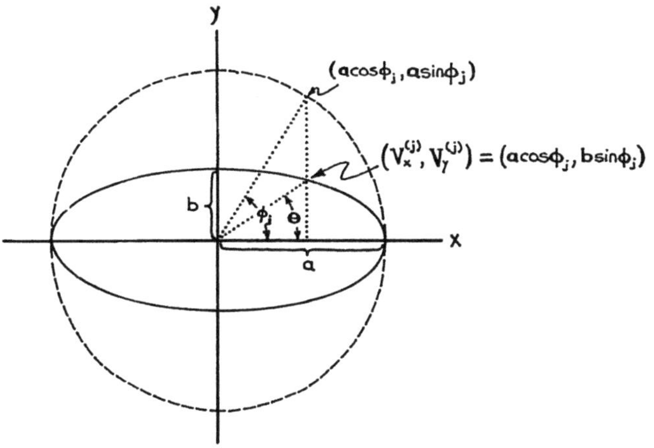

Fig. 4.6. The geometric relation between the eccentric angle ϕ_j and the polar angle θ.

Comparison of (4.2.14) with (4.8.10) shows that for the vibration ellipse of the real part $\mathbf{V}^{(r)}(\mathbf{r}_0, t)$ of the complex field vector

$$\phi_r(\mathbf{r}_0, t) = \omega t - \varphi(\mathbf{r}_0) \ , \tag{4.2.16a}$$

while for the imaginary part

$$\phi_i(\mathbf{r}_0, t) = \omega t - \varphi(\mathbf{r}_0) + \frac{\pi}{2} \ . \tag{4.2.16b}$$

From (4.2.5) it is seen that $\mathbf{V}^{(r)}(\mathbf{r}, t) = \mathbf{p}(\mathbf{r})$ when $t = 0$, while $\mathbf{V}^{(i)}(\mathbf{r}, t) = \mathbf{p}(\mathbf{r})$ when $\omega t = -\pi/2$, so that the eccentric angle of $\mathbf{p}$ is $-\varphi$; from (4.2.15) the angle ψ between the vectors $\mathbf{p}$ and $\mathbf{a}$ is then given by

$$\tan\psi = \frac{b}{a}\tan\varphi \ , \tag{4.2.17}$$

as illustrated in Fig. 4.5 for the polarization ellipse. Finally, if γ_p denotes the angle between the vectors $\mathbf{p}$ and $\mathbf{q}$, and if β is an auxiliary angle defined as

$$\tan\beta \equiv \frac{q}{p} \ , \tag{4.2.18}$$

then (4.2.8) becomes

$$\tan(2\varphi) = \frac{2pq}{p^2 - q^2}\cos\gamma_p = \tan(2\beta)\cos\gamma_p \ . \tag{4.2.19}$$

To summarize these standard results, if the vectors $\mathbf{p} = \mathbf{p}(\mathbf{r}_0)$ and $\mathbf{q} = \mathbf{q}(\mathbf{r}_0)$ are given at the fixed point $\mathbf{r}_0$, where γ_p is the angle between these two vectors and

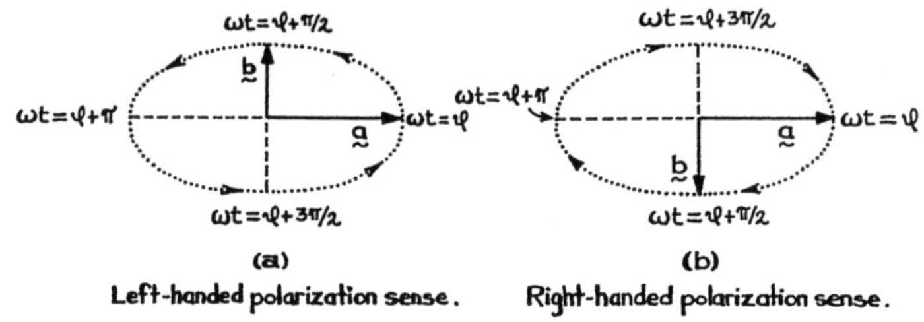

Fig. 4.7. Left- and right-handed polarization ellipses.

β is the auxiliary angle defined in (4.2.18) as the ratio of the magnitudes of these two vectors, then the principal semi-axes of the vibration ellipse at the point $\mathbf{r}_0$ are given by (4.2.13) and the angle ψ which the major axis makes with the vector $\mathbf{p}$ is given by (4.2.17) where the phase factor φ is expressed by (4.2.19).

From (4.2.4–6), each monochromatic field component appearing in the representation (4.2.3) of the propagated field vectors is of the form

$$\mathbf{V}(\mathbf{r}, t; \omega) = [\mathbf{a}(\mathbf{r}, \omega) + \mathrm{i}\mathbf{b}(\mathbf{r}, \omega)]\mathrm{e}^{\mathrm{i}[\varphi(\mathbf{r}, \omega) - \omega t]}$$

which is the complex representation of a time-harmonic wave that is propagating in the direction of $\nabla\varphi$. These are two possible senses in which the polarization ellipse may be described, these corresponding to left-handed and right-handed polarizations. If the sign of the scalar triple product $[\mathbf{a}, \mathbf{b}, \nabla\varphi] = (\mathbf{a} \times \mathbf{b}) \cdot \nabla\varphi$ is positive, then to an observer looking in the direction that is opposite to that in which the electromagnetic wave is propagating, the endpoint of the field vector describes its ellipse in the counterclockwise sense and the polarization is said to be *left-handed*, as illustrated in part (a) of Fig. 4.7. On the other hand, if the sign of the scalar triple product $[\mathbf{a}, \mathbf{b}, \nabla\varphi]$ is negative, then the polarization ellipse is described in the clockwise sense and the polarization is said to be *right-handed*, as illustrated in part (b) of the figure.

Consider now the relation between the initial and propagated polarization properties of the electromagnetic field in the half-space $z \geqslant z_0$. From (4.2.2, 6) the complex representation of the polarization ellipse for the propagated field vectors is seen to be given by

$$[\mathbf{a}_e(\mathbf{r}^+, \omega) + \mathrm{i}\mathbf{b}_e(\mathbf{r}^+, \omega)]\mathrm{e}^{\mathrm{i}\varphi_e(\mathbf{r}^+, \omega)} = \frac{1}{(2\pi)^2} \iint\limits_{-\infty}^{\infty} \mathbf{E}_0(\mathbf{k}_T, \omega)\mathrm{e}^{\mathrm{i}\mathbf{k}^+ \cdot \mathbf{r}^+} \, dk_x \, dk_y \;, \tag{4.2.20a}$$

$$[\mathbf{a}_m(\mathbf{r}^+, \omega) + \mathrm{i}\mathbf{b}_m(\mathbf{r}^+, \omega)]\mathrm{e}^{\mathrm{i}\varphi_m(\mathbf{r}^+, \omega)} = \frac{1}{(2\pi)^2} \iint\limits_{-\infty}^{\infty} \mathbf{B}_0(\mathbf{k}_T, \omega)\mathrm{e}^{\mathrm{i}\mathbf{k}^+ \cdot \mathbf{r}^+} \, dk_x \, dk_y \;, \tag{4.2.20b}$$

where $\mathbf{r}^+ = \hat{1}_x x + \hat{1}_y y + \hat{1}_z(z - z_0)$, $z \geqslant z_0$, is the position vector of the field observation point. Notice that with (4.1.28) these two relations may be expressed solely in terms of either $\mathbf{E}_0(\mathbf{k}_T, \omega)$ or $\mathbf{B}_0(\mathbf{k}_T, \omega)$, if desired. Here

$$\mathbf{E}_0(\mathbf{k}_T, \omega) = \iint_{-\infty}^{\infty} \tilde{\mathbf{E}}_0(\mathbf{r}_T, \omega) e^{-i\mathbf{k}_T \cdot \mathbf{r}_T} dx' \, dy' \,, \tag{4.2.21a}$$

$$\mathbf{B}_0(\mathbf{k}_T, \omega) = \iint_{-\infty}^{\infty} \tilde{\mathbf{B}}_0(\mathbf{r}_T, \omega) e^{-i\mathbf{k}_T \cdot \mathbf{r}_T} dx' \, dy' \,, \tag{4.2.21b}$$

where $\mathbf{k}_T = \hat{1}_x k_x + \hat{1}_y k_y$ and $\mathbf{r}_T = \hat{1}_x x' + \hat{1}_y y'$, with

$$\tilde{\mathbf{E}}_0(\mathbf{r}_T, \omega) = \int_{-\infty}^{\infty} \mathscr{E}_0(\mathbf{r}_T, t) e^{i\omega t} \, dt \,, \tag{4.2.22a}$$

$$\tilde{\mathbf{B}}_0(\mathbf{r}_T, \omega) = \int_{-\infty}^{\infty} \mathscr{B}_0(\mathbf{r}_T, t) e^{i\omega t} \, dt \,. \tag{4.2.22b}$$

As in (4.2.3), the initial field vectors on the $z = z_0$ plane may be written as

$$\mathscr{E}_0(\mathbf{r}_T, t) = \frac{1}{\pi} \mathrm{Re} \left\{ \int_{C_+} \mathbf{V}_e^{(0)}(\mathbf{r}_T, t; \omega) \, d\omega \right\} \,, \tag{4.2.23a}$$

$$\mathscr{B}_0(\mathbf{r}_T, t) = \frac{1}{\pi} \mathrm{Re} \left\{ \int_{C_+} \mathbf{V}_m^{(0)}(\mathbf{r}_T, t; \omega) \, d\omega \right\} \,, \tag{4.2.23b}$$

where

$$\mathbf{V}_e^{(0)}(\mathbf{r}_T, t; \omega) = \tilde{\mathbf{E}}_0(\mathbf{r}_T, \omega) e^{-i\omega t} = [\mathbf{p}_e^{(0)}(\mathbf{r}_T, \omega) + i\mathbf{q}_e^{(0)}(\mathbf{r}_T, \omega)] e^{-i\omega t} \,, \tag{4.2.24a}$$

$$\mathbf{V}_m^{(0)}(\mathbf{r}_T, t; \omega) = \tilde{\mathbf{B}}_0(\mathbf{r}_T, \omega) e^{-i\omega t} = [\mathbf{p}_m^{(0)}(\mathbf{r}_T, \omega) + i\mathbf{q}_m^{(0)}(\mathbf{r}_T, \omega)] e^{-i\omega t} \,, \tag{4.2.24b}$$

are complex vectors that describe the spatial properties of each monochromatic field component that is present in the initial field. With the complex representation (4.2.6) of the polarization ellipse one then has that

$$\begin{aligned} \tilde{\mathbf{E}}_0(\mathbf{r}_T, \omega) &= \mathbf{p}_e^{(0)}(\mathbf{r}_T, \omega) + i\mathbf{q}_e^{(0)}(\mathbf{r}_T, \omega) \\ &= [\mathbf{a}_e^{(0)}(\mathbf{r}_T, \omega) + i\mathbf{b}_e^{(0)}(\mathbf{r}_T, \omega)] e^{i\varphi_e^{(0)}(\mathbf{r}_T, \omega)} \,, \end{aligned} \tag{4.2.25a}$$

$$\begin{aligned} \tilde{\mathbf{B}}_0(\mathbf{r}_T, \omega) &= \mathbf{p}_m^{(0)}(\mathbf{r}_T, \omega) + i\mathbf{q}_m^{(0)}(\mathbf{r}_T, \omega) \\ &= [\mathbf{a}_m^{(0)}(\mathbf{r}_T, \omega) + i\mathbf{b}_m^{(0)}(\mathbf{r}_T, \omega)] e^{i\varphi_m^{(0)}(\mathbf{r}_T, \omega)} \,. \end{aligned} \tag{4.2.25b}$$

Substitution of (4.2.21) with (4.2.25) into (4.2.20) then yields the following general expression that is applicable to either the electric or magnetic polarization ellipse:

$$\begin{aligned} &[\mathbf{a}(\mathbf{r}^+, \omega) + i\mathbf{b}(\mathbf{r}^+, \omega)] e^{i\varphi(\mathbf{r}^+, \omega)} \\ &= \frac{1}{(2\pi)^2} \iint_{-\infty}^{\infty} \left\{ [\mathbf{a}_0(\mathbf{r}_T, \omega) + i\mathbf{b}_0(\mathbf{r}_T, \omega)] e^{i\varphi_0(\mathbf{r}_T, \omega)} \iint_{-\infty}^{\infty} e^{i(\tilde{\mathbf{k}}^+ \cdot \mathbf{r}^+ - \mathbf{k}_T \cdot \mathbf{r}_T)} dk_x \, dk_y \right\} dx' \, dy' \,. \end{aligned} \tag{4.2.26}$$

The integral appearing in the brackets of this expression is just the monochromatic spatial impulse response function, defined in (4.1.51), for the normal propagation distance $\Delta z = z - z_0$ in the dispersive medium at the frequency ω, viz.

$$h(x - x', y - y'; \Delta z, \omega) = \frac{1}{(2\pi)^2} \iint_{-\infty}^{\infty} e^{i(\mathbf{k}^+ \cdot \mathbf{r}^+ - \mathbf{k}_T \cdot \mathbf{r}_T)} dk_x \, dk_y$$

$$= \frac{1}{(2\pi)^2} \iint_{-\infty}^{\infty} e^{i[k_x(x-x') + k_y(y-y') + \gamma \Delta z]} dk_x \, dk_y \, , \qquad (4.2.27)$$

where $\gamma = \gamma(\omega)$ is given by (4.1.10) and should not be confused with the angle defined in (4.2.19) in connection with the polarization ellipse. With this substitution (4.2.26) becomes

$$[\mathbf{a}(\mathbf{r}^+, \omega) + i\mathbf{b}(\mathbf{r}^+, \omega)] e^{i\varphi(\mathbf{r}^+, \omega)}$$
$$= \iint_{-\infty}^{\infty} [\mathbf{a}_0(\mathbf{r}_T, \omega) + i\mathbf{b}_0(\mathbf{r}_T, \omega)] e^{i\varphi_0(\mathbf{r}_T, \omega)} h(x - x', y - y'; \Delta z, \omega) \, dx' \, dy' \, . \qquad (4.2.28)$$

Hence, the polarization properties of each monochromatic component present in the propagated field are given by the convolution of the initial polarization behavior on the plane $z = z_0$ with the spatial impulse response function for the propagation distance $\Delta z = z - z_0$ in the dispersive medium at the frequency ω. As a consequence, if the polarization properties of the initial field vary from point to point over the plane at $z = z_0$, then the polarization properties of the propagated field will also, in general, vary from point to point throughout the half-space $\Delta z > 0$. In a strict sense, the state of polarization refers to the electromagnetic-field vector behavior at a particular point in space and, in general, varies from point to point for each monochromatic component present. Moreover, the frequency dependence allows for the state of polarization to vary in time at any given observation point when the field is pulsed in time.

For a *uniformly polarized field vector* over the plane at $z = z_0$ it is required that

$$[\mathbf{a}_0(\mathbf{r}_T, \omega) + i\mathbf{b}_0(\mathbf{r}_T, \omega)] e^{i\varphi_0(\mathbf{r}_T, \omega)} = W_0(\mathbf{r}_T, \omega)(\hat{\mathbf{a}}_0 + i\hat{\mathbf{b}}_0) e^{i\hat{\varphi}_0} \, , \qquad (4.2.29)$$

where $\hat{\mathbf{a}}_0$ and $\hat{\mathbf{b}}_0$ are both fixed vectors, and $\hat{\varphi}_0$ is a scalar constant; the only spatial variation in the particular field vector at the plane $z = z_0$ appears in the field amplitude. With this substitution (4.2.28) yields

$$[\mathbf{a}(\mathbf{r}^+, \omega) + i\mathbf{b}(\mathbf{r}^+, \omega)] e^{i\varphi(\mathbf{r}^+, \omega)} = W(\mathbf{r}^+, \omega)(\hat{\mathbf{a}}_0 + i\hat{\mathbf{b}}_0) e^{i\hat{\varphi}_0} \, , \qquad (4.2.30)$$

where

$$W(\mathbf{r}^+, \omega) = \iint_{-\infty}^{\infty} W_0(\mathbf{r}_T, \omega) h(x - x', y - y'; \Delta z, \omega) \, dx' \, dy' \, , \qquad (4.2.31)$$

and the polarization state for this field vector remains the same throughout the half-space $z \geqslant z_0$. Notice that the condition specified in (4.2.29) requires that the vectors $\hat{\mathbf{a}}_0$ and $\hat{\mathbf{b}}_0$, and the phase constant $\hat{\varphi}_0$ are all independent of the angular frequency ω. If this is not the case then the polarization state of the field vector considered will in general evolve with time at a given point in space.

Two special cases of considerable interest in regard to the polarization state at a given point are the linearly polarized field and the circularly polarized field. For a *linearly polarized field vector* at the point $\mathbf{r} = \mathbf{r}_0$, the minor axis of the polarization ellipse vanishes so that $b = 0$ and (4.2.13b) then gives

$$p^2 q^2 = (\mathbf{p} \cdot \mathbf{q})^2 \ , \tag{4.2.32}$$

and the angle γ_p between $\mathbf{p}$ and $\mathbf{q}$ is either 0 or π. The complex representation of the field vector at this point is then given by

$$\mathbf{V}(\mathbf{r}_0, t) = \mathbf{a}(\mathbf{r}_0) \mathrm{e}^{\mathrm{i}(\varphi(\mathbf{r}_0) - \omega t)} \ . \tag{4.2.33}$$

For a *circularly polarized field vector* at the point $\mathbf{r} = \mathbf{r}_0$ the vectors $\mathbf{a}$ and $\mathbf{b}$ are indeterminate and consequently the angle φ is also indeterminate. For this to be the case (4.2.8) then requires that

$$\mathbf{p} \cdot \mathbf{q} = p^2 - q^2 = 0 \ , \tag{4.2.34}$$

so that the vectors $\mathbf{p} = \mathbf{p}(\mathbf{r}_0)$ and $\mathbf{q} = \mathbf{q}(\mathbf{r}_0)$ are orthogonal and of equal magnitude. The complex representation of the field vector at this point is then given by

$$\mathbf{V}(\mathbf{r}_0, t) = 2^{1/2} \mathbf{p}(\mathbf{r}_0) \mathrm{e}^{\pm \mathrm{i}\pi/4} \mathrm{e}^{-\mathrm{i}\omega t} \tag{4.2.35}$$

where the sign choice depends upon the polarization sense. From this expression it is seen that any given state of circular polarization may be decomposed into the superposition of two properly phased and orthogonally oriented linearly polarized fields. Furthermore, any given state of elliptic polarization may be decomposed into the superposition of a left-handed and right-handed circularly polarized field of the some angular frequency but with unequal amplitudes.

4.2.2 The Relation Between the Electric and Magnetic Polarization Ellipses

From (4.2.2, 6) the complex representation of the polarization ellipses for the temporal frequency spectra of the electric and magnetic field vectors are given, respectively, by

$$\begin{aligned} \tilde{\mathbf{E}}(\mathbf{r}, \omega) &= \mathbf{p}_e(\mathbf{r}, \omega) + \mathrm{i} q_e(\mathbf{r}, \omega) \\ &= [\mathbf{a}_e(\mathbf{r}, \omega) + \mathrm{i}\mathbf{b}_e(\mathbf{r}, \omega)] \mathrm{e}^{\mathrm{i}\varphi_e(\mathbf{r}, \omega)} = \mathbf{e}(\mathbf{r}, \omega) \mathrm{e}^{\mathrm{i}\varphi_e(\mathbf{r}, \omega)} \ , \end{aligned} \tag{4.2.36a}$$

$$\begin{aligned} \tilde{\mathbf{B}}(\mathbf{r}, \omega) &= \mathbf{p}_m(\mathbf{r}, \omega) + \mathrm{i} q_m(\mathbf{r}, \omega) \\ &= [\mathbf{a}_m(\mathbf{r}, \omega) + \mathrm{i}\mathbf{b}_m(\mathbf{r}, \omega)] \mathrm{e}^{\mathrm{i}\varphi_m(\mathbf{r}, \omega)} = \mathbf{b}(\mathbf{r}, \omega) \mathrm{e}^{\mathrm{i}\varphi_m(\mathbf{r}, \omega)} \ , \end{aligned} \tag{4.2.36b}$$

where the complex vectors defined by

$$\mathbf{e}(\mathbf{r},\omega) \equiv \mathbf{a}_e(\mathbf{r},\omega) + i\mathbf{b}_e(\mathbf{r},\omega) \ , \tag{4.2.37a}$$

$$\mathbf{b}(\mathbf{r},\omega) \equiv \mathbf{a}_m(\mathbf{r},\omega) + i\mathbf{b}_m(\mathbf{r},\omega) \ , \tag{4.2.37b}$$

have been introduced here for notational convenience. Throughout the half-space $z \geqslant z_0$ the temporal frequency spectra of the field vectors satisfy the field equations (2.1.21), viz.

$$\nabla \cdot \tilde{\mathbf{E}}(\mathbf{r},\omega) = 0 \ , \tag{4.2.38a}$$

$$\nabla \times \tilde{\mathbf{E}}(\mathbf{r},\omega) = \left[\!\left[\frac{1}{c}\right]\!\right] i\omega\tilde{\mathbf{B}}(\mathbf{r},\omega) \ , \tag{4.2.38b}$$

$$\nabla \cdot \tilde{\mathbf{B}}(\mathbf{r},\omega) = 0 \ , \tag{4.2.38c}$$

$$\nabla \times \mathbf{B}(\mathbf{r},\omega) = -\left[\!\left[\frac{1}{c}\right]\!\right] i\omega\mu\varepsilon_c(\omega)\tilde{\mathbf{E}}(\mathbf{r},\omega) \ . \tag{4.2.38d}$$

Substitution of the complex representations for the electric and magnetic field vectors given in (4.2.36) into the pair of divergence relations above immediately yields the pair of relations

$$i\mathbf{e}(\mathbf{r},\omega)\cdot\nabla\varphi_e(\mathbf{r},\omega) + \nabla\cdot\mathbf{e}(\mathbf{r},\omega) = 0 \ , \tag{4.2.39a}$$

$$i\mathbf{b}(\mathbf{r},\omega)\cdot\nabla\varphi_m(\mathbf{r},\omega) + \nabla\cdot\mathbf{b}(\mathbf{r},\omega) = 0 \ , \tag{4.2.39b}$$

while substitution into the curl relations gives

$$\begin{aligned}\tilde{\mathbf{B}}(\mathbf{r},\omega) &= \mathbf{b}(\mathbf{r},\omega)e^{i\varphi_m(\mathbf{r},\omega)} \\ &= \frac{[\![c]\!]}{\omega}[\nabla\varphi_e(\mathbf{r},\omega) \times \mathbf{e}(\mathbf{r},\omega) - i\nabla \times \mathbf{e}(\mathbf{r},\omega)]e^{i\varphi_e(\mathbf{r},\omega)} \ ,\end{aligned} \tag{4.2.40a}$$

$$\begin{aligned}\tilde{\mathbf{E}}(\mathbf{r},\omega) &= \mathbf{e}(\mathbf{r},\omega)e^{i\varphi_e(\mathbf{r},\omega)} \\ &= \frac{[\![c]\!]}{\omega\mu\varepsilon_c(\omega)}[-\nabla\varphi_m(\mathbf{r},\omega) \times \mathbf{b}(\mathbf{r},\omega) + i\nabla \times \mathbf{b}(\mathbf{r},\omega)]e^{i\varphi_m(\mathbf{r},\omega)} \ .\end{aligned} \tag{4.2.40b}$$

From this final pair of relations the polarization properties of one field vector can be determined directly from the polarization state of the other field vector at each point of space. For example, if the electric field vector is linearly polarized with $\mathbf{e}(\mathbf{r},\omega) = \mathbf{a}_e(\mathbf{r},\omega)$, where $\mathbf{a}_e$ is a real-valued vector field, then

$$\mathbf{b}(\mathbf{r},\omega)e^{i\varphi_m(\mathbf{r},\omega)} = \frac{[\![c]\!]}{\omega}[\nabla\varphi_e(\mathbf{r},\omega) \times \mathbf{a}_e(\mathbf{r},\omega) - i\nabla \times \mathbf{a}_e(\mathbf{r},\omega)]e^{i\varphi_e(\mathbf{r},\omega)}$$

and the magnetic-field vector will, in general, be elliptically polarized provided that $\mathbf{a}_e$ is spatially dependent; if $\mathbf{a}_e$ is independent of the position vector $\mathbf{r}$ so that the electric field vector is uniformly linearly polarized throughout space, then $\nabla \times \mathbf{a}_e = \mathbf{0}$ and the magnetic field vector is also linearly polarized. As a consequence, the temporal frequency spectra of the electric and magnetic field vectors are not, in general, instantaneously orthogonal to each other; indeed

$$\tilde{\mathbf{E}}(\mathbf{r},\omega)\cdot\tilde{\mathbf{B}}(\mathbf{r},\omega) = -\mathrm{i}\frac{[\![c]\!]}{\omega}\mathbf{e}(\mathbf{r},\omega)\cdot[\nabla \times \mathbf{e}(\mathbf{r},\omega)]\mathrm{e}^{\mathrm{i}2\varphi_e(\mathbf{r},\omega)} , \tag{4.2.41}$$

which does not, in general, vanish. As was previously stated, the state of polarization strictly refers to the electromagnetic field behavior at a particular point in space and, in general, varies from point to point in the field. Even if the electric-field vector is everywhere linearly polarized in the same direction, the magnetic field vector will, in general, be elliptically polarized, as was first shown by *Nisbet* and *Wolf* [4.13]. As a consequence, the temporal frequency spectra of the electric and magnetic field vectors are not, in general, instantaneously orthogonal to each other. However, the long-time average of $\mathbf{E}^{(r)}(\mathbf{r},t;\omega)\cdot\mathbf{B}^{(r)}(\mathbf{r},t;\omega)$, with $\mathbf{E}(\mathbf{r},t;\omega) = \tilde{\mathbf{E}}(\mathbf{r},\omega)\mathrm{e}^{-\mathrm{i}\omega t}$ and $\mathbf{B}(\mathbf{r},t;\omega) = \tilde{\mathbf{B}}(\mathbf{r},\omega)\mathrm{e}^{-\mathrm{i}\omega t}$, does vanish, since

$$\begin{aligned}\langle \mathbf{E}^{(r)}(\mathbf{r},t;\omega)\cdot\mathbf{B}^{(r)}(\mathbf{r},t;\omega)\rangle &= \frac{1}{2}\mathbb{R}\mathrm{e}\{\tilde{\mathbf{E}}(\mathbf{r},\omega)\cdot\tilde{\mathbf{B}}^*(\mathbf{r},\omega)\}\\ &= \frac{[\![c]\!]}{2\omega}\mathbb{R}\mathrm{e}\{\mathbf{e}(\mathbf{r},\omega)\cdot[\nabla\varphi_e(\mathbf{r},\omega)\times\mathbf{e}^*(\mathbf{r},\omega)]\\ &\quad + \mathrm{i}\mathbf{e}(\mathbf{r},\omega)\cdot[\nabla\times\mathbf{e}^*(\mathbf{r},\omega)]\}\\ &= \frac{[\![c]\!]}{2\omega}\mathbb{R}\mathrm{e}\{\mathrm{i}\mathbf{e}(\mathbf{r},\omega)\cdot[\nabla\times\mathbf{e}^*(\mathbf{r},\omega)]\} = 0 .\end{aligned} \tag{4.2.42}$$

The fact that the quantity $\mathbf{e}\cdot(\nabla\times\mathbf{e}^*)$ is real-valued, used in the above proof, directly follows from the vector differential identity $\nabla\cdot(\mathbf{v}\times\mathbf{w}) = \mathbf{w}\cdot(\nabla\times\mathbf{v}) - \mathbf{v}\cdot(\nabla\times\mathbf{w})$. With $\mathbf{v} = \mathbf{e}$ and $\mathbf{w} = \mathbf{e}^*$ as well as the fact that $\mathbf{e}\times\mathbf{e}^* = \mathbf{0}$. this identity yields the result $\mathbf{e}\cdot(\nabla\times\mathbf{e}^*) = \mathbf{e}^*\cdot(\nabla\times\mathbf{e})$, and consequently

$$\begin{aligned}\mathbf{e}\cdot(\nabla\times\mathbf{e}^*) &= \tfrac{1}{2}[\mathbf{e}\cdot(\nabla\times\mathbf{e}^*) + \mathbf{e}^*\cdot(\nabla\times\mathbf{e})]\\ &= \tfrac{1}{2}\{\mathbf{e}\cdot(\nabla\times\mathbf{e}^*) + [\mathbf{e}\cdot(\nabla\times\mathbf{e}^*)]^*\}\\ &= \mathbb{R}\mathrm{e}\{\mathbf{e}\cdot(\nabla\times\mathbf{e}^*)\} .\end{aligned}$$

Thus, *the temporal frequency spectra of the electric and magnetic field vectors are on the average mutually orthogonal.*

$$\{\underset{\sim}{E}(\underset{\sim}{k},\omega)\cdot\underset{\sim}{B}(\underset{\sim}{k},\omega)=0\}$$

$$\Updownarrow\ \mathcal{F}_{\underset{\sim}{k}}$$

$$\left\{\begin{array}{l}\tilde{\underset{\sim}{E}}(\underset{\sim}{r},\omega)\cdot\tilde{\underset{\sim}{B}}(\underset{\sim}{r},\omega)\neq 0\\ \langle\underset{\sim}{E}(\underset{\sim}{r},t;\omega)\cdot\underset{\sim}{B}(\underset{\sim}{r},t;\omega)\rangle=0\end{array}\right\}$$

$$\Updownarrow\ \mathcal{F}_{\omega}$$

$$\left\{\begin{array}{l}\underset{\sim}{\mathcal{E}}(\underset{\sim}{r},t)\cdot\underset{\sim}{\mathcal{B}}(\underset{\sim}{r},t)\neq 0\\ \langle\underset{\sim}{\mathcal{E}}(\underset{\sim}{r},t)\cdot\underset{\sim}{\mathcal{B}}(\underset{\sim}{r},t)\rangle\neq 0\end{array}\right\}$$

Fig. 4.8. Orthogonality properties of the field vectors.

As a summary of these results, indicated in Fig. 4.8, it is first clearly seen from (3.2.16–18) that the spatio-temporal frequency spectra $\mathbf{E}(\mathbf{k},\omega)$, $\mathbf{B}(\mathbf{k},\omega)$ of the electric and magnetic field vectors, respectively, are mutually orthogonal. The temporal frequency spectral wave components $\mathbf{E}(\mathbf{r},t;\omega)=\tilde{\mathbf{E}}(\mathbf{r},\omega)\mathrm{e}^{-\mathrm{i}\omega t}$ and $\mathbf{B}(\mathbf{r},t;\omega)=\tilde{\mathbf{B}}(\mathbf{r},\omega)\mathrm{e}^{-\mathrm{i}\omega t}$ of these field vectors are not, in general, orthogonal, but are mutually orthogonal in the long-time average sense. Finally, from (4.2.1) and the previous results it is seen that the electric $\mathcal{E}(\mathbf{r},t)$ and magnetic $\mathcal{B}(\mathbf{r},t)$ field vectors are not, in general, orthogonal and neither is their long-time average.

4.2.3 The Uniformly Polarized Field

For a *uniformly polarized field in the electric field vector*, (4.2.29 and 36) require that the initial field vectors on the plane $z=z_0$ are given by

$$\tilde{\mathbf{E}}_0(\mathbf{r}_T,\omega)=\tilde{E}_0(\mathbf{r}_T,\omega)(\mathbf{a}_\mathrm{e}+\mathrm{i}\mathbf{b}_\mathrm{e})\mathrm{e}^{\mathrm{i}\varphi_\mathrm{e}}\ , \tag{4.2.43a}$$

$$\tilde{\mathbf{B}}_0(\mathbf{r}_T,\omega)=\tilde{B}_0(\mathbf{r}_T,\omega)[\mathbf{a}_m(\mathbf{r}_T,\omega)+\mathrm{i}\mathbf{b}_m(\mathbf{r}_T,\omega)]\mathrm{e}^{\mathrm{i}\varphi_m(\mathbf{r}_T,\omega)}\ , \tag{4.2.43b}$$

where $\mathbf{a}_\mathrm{e}$ and $\mathbf{b}_\mathrm{e}$ are fixed vectors and where φ_e is a scalar constant. These field vectors are not independent and must be oriented such that (4.2.40) is satisfied. Indeed, (4.2.40a) requires that

$$\tilde{B}_0(\mathbf{r}_T,\omega)[\mathbf{a}_m(\mathbf{r}_T,\omega)+\mathrm{i}\mathbf{b}_m(\mathbf{r}_T,\omega)]\mathrm{e}^{\mathrm{i}\varphi_m(\mathbf{r}_T,\omega)}$$

$$=-\mathrm{i}\frac{[\![c]\!]}{\omega}(\nabla\tilde{E}_0(\mathbf{r}_T,\omega))\times(\mathbf{a}_\mathrm{e}+\mathrm{i}\mathbf{b}_\mathrm{e})\mathrm{e}^{\mathrm{i}\varphi_\mathrm{e}}\ , \tag{4.2.44}$$

from which it is seen that the temporal frequency spectra of the electric and magnetic field vectors are orthogonal. Although not necessary, the complex vectors $(\mathbf{a}_e + i\mathbf{b}_e)$ and $(\mathbf{a}_m + i\mathbf{b}_m)$ may be chosen to be normalized, so that

$$|\mathbf{a} + i\mathbf{b}|^2 = a^2 + b^2 = 1 \ . \tag{4.2.45}$$

The spatio-temporal frequency spectra of the initial field vectors on the plane $z = z_0$ are given by, with (4.2.43–44),

$$\begin{aligned}\mathbf{E}_0(\mathbf{k}_T, \omega) &= (\mathbf{a}_e + i\mathbf{b}_e)e^{i\varphi_e} \iint\limits_{-\infty}^{\infty} \tilde{E}_0(\mathbf{r}_T, \omega)e^{-i\mathbf{k}_T \cdot \mathbf{r}_T} dx\, dy \\ &= (\mathbf{a}_e + i\mathbf{b}_e)e^{i\varphi_e} E_0(\mathbf{k}_T, \omega) \ ,\end{aligned} \tag{4.2.46a}$$

$$\begin{aligned}\mathbf{B}_0(\mathbf{k}_T, \omega) &= i\frac{[\![c]\!]}{\omega}(\mathbf{a}_e + i\mathbf{b}_e)e^{i\varphi_e} \times \iint\limits_{-\infty}^{\infty} [\nabla \tilde{\mathbf{E}}_0(\mathbf{r}_T, \omega)]\, e^{-i\mathbf{k}_T \cdot \mathbf{r}_T} dx\, dy \\ &= \frac{[\![c]\!]}{\omega}\tilde{\mathbf{k}}^+ \times (\mathbf{a}_e + i\mathbf{b}_e)e^{i\varphi_e} E_0(\mathbf{k}_T, \omega) \ ,\end{aligned} \tag{4.2.46b}$$

where

$$E_0(\mathbf{k}_T, \omega) = \iint\limits_{-\infty}^{\infty} \tilde{E}_0(\mathbf{r}_T, \omega)e^{-i\mathbf{k}_T \cdot \mathbf{r}_T} dx\, dy \ . \tag{4.2.47}$$

In addition, the space-time form of the initial field vectors are given by

$$\mathscr{E}_0(\mathbf{r}_T, t) = \frac{1}{\pi}\mathbb{R}\mathrm{e}\left\{(\mathbf{a}_e + i\mathbf{b}_e)e^{i\varphi_e} \int\limits_{C_+} \tilde{E}_0(\mathbf{r}_T, \omega)e^{-i\omega t} d\omega\right\} \ , \tag{4.2.48a}$$

$$\begin{aligned}\mathscr{B}_0(\mathbf{r}_T, t) &= \frac{1}{\pi}\mathbb{R}\mathrm{e}\left\{\int\limits_{C_+} (\mathbf{a}_m(\mathbf{r}_T, \omega) + i\mathbf{b}_m(\mathbf{r}_T, \omega))e^{i\varphi_m(\mathbf{r}_T, \omega)} \tilde{B}_0(\mathbf{r}_T, \omega)e^{-i\omega t} d\omega\right\} \\ &= \frac{[\![c]\!]}{\pi}\mathbb{R}\mathrm{e}\left\{i(\mathbf{a}_e + i\mathbf{b}_e)e^{i\varphi_e} \times \int\limits_{C_+} \frac{1}{\omega}[\nabla \tilde{E}_0(\mathbf{r}_T, \omega)]e^{-i\omega t} d\omega\right\} \ .\end{aligned} \tag{4.2.48b}$$

From (4.1.15) with (4.2.46) the propagated field vectors are found as

$$\mathscr{E}(\mathbf{r}_T, z, t) = \frac{1}{4\pi^3}\mathbb{R}\mathrm{e}\left\{(\mathbf{a}_e + i\mathbf{b}_e)e^{i\varphi_e} \int\limits_{C_+} d\omega \iint\limits_{-\infty}^{\infty} E_0(\mathbf{k}_T, \omega)e^{i(\tilde{\mathbf{k}}^+ \cdot \mathbf{r}^+ - \omega t)} dk_x\, dk_y\right\} \ , \tag{4.2.49a}$$

$$\begin{aligned}\mathscr{B}(\mathbf{r}_T, z, t) = -\frac{[\![c]\!]}{4\pi^3}\mathbb{R}\mathrm{e}\Bigg\{&(\mathbf{a}_e + i\mathbf{b}_e)e^{i\varphi_e} \\ &\times \int\limits_{C_+} \frac{d\omega}{\omega} \iint\limits_{-\infty}^{\infty} \tilde{\mathbf{k}}^+(\omega)E_0(\mathbf{k}_T, \omega)e^{i(\tilde{\mathbf{k}}^+ \cdot \mathbf{r}^+ - \omega t)} dk_x\, dk_y\Bigg\} \ ,\end{aligned} \tag{4.2.49b}$$

Hence, the propagated field vectors are seen to be everywhere orthogonal when the electric field vector is uniformly polarized. However, the magnetic-field vector is not, in general, uniformly polarized because of its directional dependence on the complex wave vector $\tilde{\mathbf{k}}^+(\omega)$ which causes the orientation of the magnetic field polarization ellipse to vary in both space and time. In order that the magnetic-field vector also be uniformly polarized the spectral quantity $\tilde{\mathbf{k}}^+(\omega)E_0(\mathbf{k}_T,\omega)$ must have a fixed direction independent of ω; this in turn implies that the initial field amplitude $E_0(\mathbf{r}_T,\omega)$ must be independent of some coordinate direction.

For a *uniformly polarized field in the magnetic induction field vector* the set of relations given in (4.2.43) are replaced by

$$\tilde{\mathbf{E}}_0(\mathbf{r}_T,\omega) = \tilde{E}_0(\mathbf{r}_T,\omega)[\mathbf{a}_e(\mathbf{r}_T,\omega) + \mathrm{i}\mathbf{b}_e(\mathbf{r}_T,\omega)]\mathrm{e}^{\mathrm{i}\varphi_e(\mathbf{r}_T,\omega)} \,, \tag{4.2.50a}$$

$$\tilde{\mathbf{B}}_0(\mathbf{r}_T,\omega) = \tilde{B}_0(\mathbf{r}_T,\omega)(\mathbf{a}_m + \mathrm{i}\mathbf{b}_m)\mathrm{e}^{\mathrm{i}\varphi_m} \,, \tag{4.2.50b}$$

where $\mathbf{a}_m$ and $\mathbf{b}_m$ are fixed vectors and where φ_m is a scalar constant. The propagated-field vectors in this case are then found to be given by

$$\mathscr{E}(\mathbf{r}_T,z,t) = \frac{[\![c]\!]}{4\pi^3}\mathbb{R}\mathrm{e}(\mathbf{a}_m + \mathrm{i}\mathbf{b}_m)\mathrm{e}^{\mathrm{i}\varphi_m}$$

$$\times \int_{C_+} \frac{d\omega}{\omega\mu\varepsilon_c(\omega)} \iint_{-\infty}^{\infty} \tilde{\mathbf{k}}^+(\omega)B_0(\mathbf{k}_T,\omega)\mathrm{e}^{\mathrm{i}(\tilde{\mathbf{k}}^+\cdot\mathbf{r}^+-\omega t)}\,dk_x\,dk_y \,, \tag{4.2.51a}$$

$$\mathscr{B}(\mathbf{r}_T,z,t) = \frac{1}{4\pi^3}\mathbb{R}\mathrm{e}(\mathbf{a}_m + \mathrm{i}\mathbf{b}_m)\mathrm{e}^{\mathrm{i}\varphi_m}\int_{C_+} d\omega \iint_{-\infty}^{\infty} B_0(\mathbf{k}_T,\omega)\mathrm{e}^{\mathrm{i}(\tilde{\mathbf{k}}^+\cdot\mathbf{r}^+-\omega t)}\,dk_x\,dk_y \,. \tag{4.2.51a}$$

The propagated-field vectors are again seen to be everywhere orthogonal; however, the electric-field vector is not, in general, uniformly polarized.

In order that both field vectors be uniformly polarized, (4.2.49b and 51b) must both be satisfied, so that

$$\mathbb{R}\mathrm{e}\left\{\int_{C_+} d\omega \iint_{-\infty}^{\infty} dk_x\,dk_y\cdot \mathrm{e}^{\mathrm{i}(\tilde{\mathbf{k}}^+\cdot\mathbf{r}^+-\omega t)}\left[B_0(\mathbf{k}_T,\omega)(\mathbf{a}_m + \mathrm{i}\mathbf{b}_m)\mathrm{e}^{\mathrm{i}\varphi_m}\right.\right.$$

$$\left.\left. + \frac{[\![c]\!]}{\omega}E_0(\mathbf{k}_T,\omega)(\mathbf{a}_e + \mathrm{i}\mathbf{b}_e)\mathrm{e}^{\mathrm{i}\varphi_e} \times \tilde{\mathbf{k}}^+(\omega)\right]\right\} = 0 \,,$$

and consequently

$$B_0(\mathbf{k}_T,\omega)(\mathbf{a}_m + \mathrm{i}\mathbf{b}_m)\mathrm{e}^{\mathrm{i}\varphi_m} = -[\![c]\!]\frac{1}{\omega}\tilde{k}(\omega)E_0(\mathbf{k}_T,\omega)(\mathbf{a}_e + \mathrm{i}\mathbf{b}_e) \times \hat{\mathbf{s}}\mathrm{e}^{\mathrm{i}\varphi_e} \,, \tag{4.2.52}$$

where $\hat{s} = \hat{1}_x p + \hat{1}_y q + \hat{1}_z m$ is a real-valued unit vector that is defined by the direction of the complex-valued wave vector as $\tilde{\mathbf{k}}^+(\omega) = \tilde{k}(\omega)\hat{s}$. Notice that (4.2.52) also follows from (4.1.17), viz.

$$\mathbf{B}_0(\mathbf{k}_T, \omega) = [\![c]\!]\frac{1}{\omega}\tilde{\mathbf{k}}^+ \times \mathbf{E}_0(\mathbf{k}_T, \omega)$$

with $\mathbf{E}_0(\mathbf{k}_T, \omega)$ given by (4.2.46a) and $\mathbf{B}_0(\mathbf{k}_T, \omega)$ given by the spatial transform of (4.2.50b). The orientation of the magnetic polarization ellipse is then seen to depend on the orientation of the electric polarization ellipse through the factor $(\mathbf{a}_e + i\mathbf{b}_e) \times \hat{s}$ which depends upon the direction of the complex wave vector $\tilde{\mathbf{k}}^+(\omega) = \tilde{k}(\omega)\hat{s}$. Since it is required that both polarization ellipses are fixed, the direction of the unit vector $\hat{s}$ must then be constrained and this, in turn, constrains the coordinate dependency of the field vectors themselves. For example, if the uniformly polarized electric field is linearly polarized along the x-axis in a nonabsorptive medium so that $\mathbf{a}_e = \hat{1}_x$, $\mathbf{b}_e = \mathbf{0}$, and $\varphi_e = 0$, then

$$\begin{aligned}(\mathbf{a}_m + i\mathbf{b}_m)e^{i\varphi_m} &= [\hat{1}_x p + \hat{1}_y q + \hat{1}_z(1 - p^2 - q^2)^{1/2}] \times \hat{1}_x \\ &= \hat{1}_y(1 - p^2 - q^2)^{1/2} - \hat{1}_z q \ ,\end{aligned}$$

from which it is seen that there are two possibilities in order to maintain the requirement that the magnetic field is uniformly polarized. Either $q = q_0$ is fixed, in which case p may be allowed to vary such that either $p^2 \leqslant 1 - q_0^2$ or $p^2 > 1 - q_0^2$, or else q is allowed to vary in which case p must vary in such a fashion that $m = 0$, i.e. that $p^2 = 1 - q^2$. The situation that results when $p^2 = 1 - q^2$ precludes propagation into the half-space $\Delta z > 0$ (since $m = 0$ for all possible allowed values of p and q) and so is not of interest here. Hence, $q = q_0$ and the field itself must then be independent of the y-coordinate. As another example, let the uniformly polarized electric field be circularly polarized in the xy-plane in a nonabsorptive medium so that $\mathbf{a}_e = \hat{1}_x$, $\mathbf{b}_e = \hat{1}_y$ and $\varphi_e = 0$, in which case

$$\begin{aligned}(\mathbf{a}_m + i\mathbf{b}_m)e^{i\varphi_m} &= [\hat{1}_x p + \hat{1}_y q + \hat{1}_z(1 - p^2 - q^2)^{1/2}] \times (\hat{1}_x + i\hat{1}_y) \\ &= \hat{1}_y(1 - p^2 - q^2)^{1/2} - \hat{1}_z q - i[\hat{1}_x(1 - p^2 - q^2)^{1/2} - \hat{1}_z p]\end{aligned}$$

and p and q must both be fixed (i.e., $p = p_0$, $q = q_0$) in order for the magnetic field polarization ellipse to be uniform. Since $a_m^2 = 1 - p_0^2$ and $b_m^2 = 1 - q_0^2$ the magnetic field is elliptically polarized in general and is circularly polarized if $p_0 = q_0$. In either case the field itself is independent of both the x- and y-coordinates. *In general, the requirement that both field vectors be uniformly polarized demands that the field be independent of some coordinate direction.*

4.3 The Pulsed, Plane-Wave Electromagnetic Field

An important class of fields that arises in many practical situations is that in which either of the field vectors are transverse to some specified propagation direction. These are the transverse electric and transverse magnetic mode fields whose importance arises, for example, in the analysis of reflection and transmission phenomena at a dielectric interface. Common to both of these mode solutions is the plane wave field which also holds a position of fundamental importance in the angular-spectrum representation. Because the analysis of plane-wave pulse propagation through a dispersive medium yields the fundamental dynamics of pulse dispersion that is unencumbered by diffraction effects, this field type is of central importance to the remaining analysis presented in this monograph.

For a transverse electromagnetic mode field with respect to the z-axis it is required that both $\mathcal{E}_z$ and $\mathcal{B}_z$ vanish for all $z \geqslant z_0$. The appropriate solution is then given by either of the equivalent representations given in (4.1.15 or 47) and one obtains a plane wave field that is propagating in the positive z-direction. One may, without any loss of generality, choose the field to be linearly polarized along some convenient direction that is orthogonal to the z-axis. Any other polarization state may then be readily obtained through an appropriate linear superposition of such linearly polarized plane wave fields with suitable orientations of the vibration plane.

Let the electric field vector be linearly polarized along the y-axis so that

$$\mathcal{E}(\mathbf{r},t) = \hat{1}_y \mathcal{E}_y(z,t) \ , \tag{4.3.1a}$$

$$\mathcal{B}(\mathbf{r},t) = \hat{1}_x \mathcal{B}_x(z,t) \ , \tag{4.3.1b}$$

where

$$\mathcal{E}_y(z,t) = \frac{1}{\pi}\mathrm{Re}\left\{\int_{C_+} E_y^{(0)}(\omega) e^{i(\tilde{k}(\omega)\Delta z - \omega t)}\, d\omega\right\} , \tag{4.3.2a}$$

$$\mathcal{B}_x(z,t) = -[\![c]\!]\frac{1}{\pi}\mathrm{Re}\left\{\int_{C_+} \frac{\tilde{\mathbf{k}}(\omega)}{\omega} E_y^{(0)}(\omega) e^{i(\tilde{k}(\omega)\Delta z - \omega t)}\, d\omega\right\} , \tag{4.3.2b}$$

with $\Delta z = z - z_0$ being the propagation distance into the positive half-space $z \geqslant z_0$ from the input plane at $z = z_0$. Let the initial time behavior of the plane wave electric field vector at the plane $z = z_0$ be given by

$$\mathcal{E}_y^{(0)}(t) = E_0 f(t) \tag{4.3.3}$$

with the frequency spectrum

$$E_y^{(0)}(\omega) = E_0 \int_{-\infty}^{\infty} f(t) e^{i\omega t}\, dt = E_0 \tilde{f}(\omega) \ , \tag{4.3.4}$$

where $\tilde{f}(\omega)$ is the Fourier-Laplace transform of $f(t)$. With this substitution the pair of relations appearing in (4.3.2) become

$$\mathscr{E}_y(z,t) = \frac{1}{\pi} E_0 \,\mathrm{Re}\left\{ \int_{ia}^{ia+\infty} \tilde{f}(\omega) e^{i[\tilde{k}(\omega)\Delta z - \omega t]} d\omega \right\} , \tag{4.3.5a}$$

$$\mathscr{B}_x(z,t) = -\frac{[\![c]\!]}{\pi c} E_0 \,\mathrm{Re}\left\{ \int_{ia}^{ia+\infty} n(\omega)\tilde{f}(\omega) e^{i[\tilde{k}(\omega)\Delta z - \omega t]} d\omega \right\} . \tag{4.3.5b}$$

For convenience, this pair of expressions may be rewritten as

$$\mathscr{E}_y(z,t) = \frac{1}{\pi} E_0 \,\mathrm{Re}\left\{ \int_{ia}^{ia+\infty} \tilde{f}(\omega) e^{(\Delta z/c)\phi(\omega,\theta)} d\omega \right\} , \tag{4.3.6a}$$

$$\mathscr{B}_x(z,t) = -\frac{[\![c]\!]}{\pi c} E_0 \,\mathrm{Re}\left\{ \int_{ia}^{ia+\infty} n(\omega)\tilde{f}(\omega) e^{(\Delta z/c)\phi(\omega,\theta)} d\omega \right\} , \tag{4.3.6b}$$

for $\Delta z = z - z_0 \geqslant 0$. The *complex phase function* $\phi(\omega, \theta)$ appearing here is defined as

$$\phi(\omega,\theta) \equiv i\omega[n(\omega) - \theta] , \tag{4.3.7}$$

where $n(\omega) = (c/\omega)\tilde{k}(\omega)$ is the complex index of refraction of the dispersive medium, and where

$$\theta \equiv \frac{ct}{\Delta z} \tag{4.3.8}$$

is a *dimensionless space-time parameter*. Notice that both field vectors given in (4.3.6) may be obtained from the single vector potential

$$\begin{aligned} \mathscr{A}(z,t) &= -\hat{1}_y \mathscr{A}(z,t) \\ &= -\hat{1}_y \frac{[\![c]\!]}{\pi} E_0 \,\mathrm{Re}\left\{ \int_{ia}^{ia+\infty} \frac{1}{\omega} \tilde{f}(\omega) e^{i[\tilde{k}(\omega)\Delta z - \omega t]} d\omega \right\} \end{aligned} \tag{4.3.9}$$

as

$$\mathscr{E}(z,t) = -\left[\!\left[\frac{1}{c}\right]\!\right] \frac{\partial \mathscr{A}(z,t)}{\partial t} , \tag{4.3.10a}$$

$$\mathscr{B}(z,t) = \nabla \times \mathscr{A}(z,t) = \hat{1}_x \frac{\partial \mathscr{A}(z,t)}{\partial z} , \tag{4.3.10b}$$

in agreement with the results of Sect. 3.3.

On the other hand, if the initial time behavior of the plane wave magnetic induction field vector at the plane $z = z_0$ is specified by the dimensionless function $g(t)$ so that

$$\mathscr{B}_x^{(0)}(t) = B_0 g(t) , \tag{4.3.11}$$

then, from (4.1.71),

$$B_0 \tilde{g}(\omega) = [\![c]\!] \frac{\tilde{k}(\omega)}{\omega} E_0 \tilde{f}(\omega) = \frac{[\![c]\!]}{c} n(\omega) E_0 \tilde{f}(\omega) ,$$

and the pair of relations given in (4.3.6) become

$$\mathscr{E}_y(z,t) = \left[\!\left[\frac{1}{c}\right]\!\right] \frac{c}{\pi} B_0 \, \mathrm{Re}\left\{ \int_{ia}^{ia+\infty} \frac{1}{n(\omega)} \tilde{g}(\omega) \mathrm{e}^{(\Delta z/c)\phi(\omega,\theta)} \, d\omega \right\} , \tag{4.3.12a}$$

$$\mathscr{B}_x(z,t) = -\frac{1}{\pi} B_0 \, \mathrm{Re}\left\{ \int_{ia}^{ia+\infty} \tilde{g}(\omega) \mathrm{e}^{(\Delta z/c)\phi(\omega,\theta)} \, d\omega \right\} , \tag{4.3.12b}$$

for the propagated plane-wave field. The corresponding vector potential is then given by $\mathscr{A}(z,t) = -\hat{1}_y \mathscr{A}(z,t)$ with

$$\mathscr{A}(z,t) = \frac{1}{\pi} B_0 \, \mathrm{Re}\left\{ \mathrm{i} \int_{ia}^{ia+\infty} \frac{1}{\tilde{k}(\omega)} \tilde{g}(\omega) \mathrm{e}^{\mathrm{i}(\tilde{k}(\omega)\Delta z - \omega t)} \, d\omega \right\} , \tag{4.3.13}$$

where the field vectors are as in (4.3.10a, b).

Because of the requirement of causality, admissable models for describing the behavior of the complex index of refraction in the complex ω-plane must obey the symmetry relation

$$n(-\omega) = n^*(\omega^*) , \tag{4.3.14}$$

and hence

$$\begin{aligned} \phi(-\omega,\theta) &= -\mathrm{i}\omega[n(-\omega) - \theta] \\ &= \{\mathrm{i}\omega^*[n(\omega^*) - \theta]\}^* = \phi^*(\omega^*,\theta) , \end{aligned} \tag{4.3.15}$$

so that the complex phase function also satisfies the same symmetry relation. Furthermore, for a real-valued initial pulse function $f(t)$ one has that

$$\begin{aligned} \tilde{f}(-\omega) &= \int_{-\infty}^{\infty} f(t) \mathrm{e}^{-\mathrm{i}\omega t} \, dt \\ &= \left[\int_{-\infty}^{\infty} f(t) \mathrm{e}^{\mathrm{i}\omega^* t} \, dt \right]^* = \tilde{f}^*(\omega^*) , \end{aligned} \tag{4.3.16}$$

and since $B_0 g(\omega) = ([\![c]\!]/c)n(\omega)E_0 \tilde{f}(\omega)$ it then follows that (as well as from the fact that $g(t)$ is real-valued)

$$\tilde{g}(-\omega) = \tilde{g}^*(\omega^*) . \tag{4.3.17}$$

Because of these symmetry relations the expression (4.3.6a) may be rewritten as

$$\begin{aligned}\mathscr{E}_y(z,t) &= \frac{1}{2\pi}E_0\left\{\int_{ia}^{ia+\infty} \tilde{f}(\omega)e^{(\Delta z/c)\phi(\omega,\theta)}\,d\omega + \int_{ia}^{ia+\infty} \tilde{f}^*(\omega)e^{(\Delta z/c)\phi^*(\omega,\theta)}\,d\omega^*\right\}\\ &= \frac{1}{2\pi}E_0\left\{\int_{ia}^{ia+\infty} \tilde{f}(\omega)e^{(\Delta z/c)\phi(\omega,\theta)}\,d\omega\right.\\ &\quad\left. + \int_{ia}^{ia+\infty} \tilde{f}(-\omega' + i\omega'')e^{(\Delta z/c)\phi(-\omega'+i\omega'',\theta)}\,d(\omega' - i\omega'')\right\} ,\end{aligned}$$

where $\omega' = \mathrm{Re}\{\omega\}$ and $\omega'' = \mathrm{Im}\{\omega\}$ denote the real and imaginary parts of ω, respectively. Under the transformation $\omega' \to -\omega'$ in the second integral of the above expression there results

$$\begin{aligned}\mathscr{E}_y(z,t) &= \frac{1}{2\pi}E_0\left\{\int_{ia}^{ia+\infty} \tilde{f}(\omega)e^{(\Delta z/c)\phi(\omega,\theta)}\,d\omega + \int_{ia-\infty}^{ia} \tilde{f}(\omega)e^{(\Delta z/c)\phi(\omega,\theta)}\,d\omega\right\}\\ &= \frac{1}{2\pi}E_0 \int_{ia-\infty}^{ia+\infty} \tilde{f}(\omega)e^{(\Delta z/c)\phi(\omega,\theta)}\,d\omega .\end{aligned}$$

A precisely analogous result holds for the magnetic-induction field so that the pair of expressions appearing in (4.3.6) then become

$$\mathscr{E}_y(z,t) = \frac{1}{2\pi}E_0 \int_{ia-\infty}^{ia+\infty} \tilde{f}(\omega)e^{(\Delta z/c)\phi(\omega,\theta)}\,d\omega , \tag{4.3.18a}$$

$$\mathscr{B}_x(z,t) = -\frac{[\![c]\!]}{2\pi c}E_0 \int_{ia-\infty}^{ia+\infty} n(\omega)\tilde{f}(\omega)e^{(\Delta z/c)\phi(\omega,\theta)}\,d\omega . \tag{4.3.18b}$$

On the other hand, if the initial time behavior of the plane-wave magnetic induction field vector is specified at the plane $z = z_0$, as in (4.3.11), then the propagated plane-wave field components are found as, from (4.3.12),

$$\mathscr{E}_y(z,t) = \left[\!\left[\frac{1}{c}\right]\!\right]\frac{c}{2\pi}B_0 \int_{ia-\infty}^{ia+\infty} \frac{1}{n(\omega)}\tilde{g}(\omega)e^{(\Delta z/c)\phi(\omega,\theta)}\,d\omega , \tag{4.3.19a}$$

$$\mathscr{B}_x(z,t) = -\frac{1}{2\pi}B_0 \int_{ia-\infty}^{ia+\infty} \tilde{g}(\omega)e^{(\Delta z/c)\phi(\omega,\theta)}\,d\omega . \tag{4.3.19b}$$

Both field vectors may be obtained through (4.3.10) from the single vector potential $\mathscr{A}(z,t) = -\hat{1}_y\mathscr{A}(z,t)$ with

$$\mathscr{A}(z,t) = \frac{[\![c]\!]}{2\pi}E_0\mathrm{i}\int_{\mathrm{i}a-\infty}^{\mathrm{i}a+\infty}\frac{1}{\omega}\tilde{f}(\omega)\mathrm{e}^{(\Delta z/c)\phi(\omega,\theta)}\,d\omega \tag{4.3.20a}$$

$$= \frac{1}{2\pi}B_0\mathrm{i}\int_{\mathrm{i}a-\infty}^{\mathrm{i}a+\infty}\frac{1}{\tilde{k}(\omega)}\tilde{g}(\omega)\mathrm{e}^{(\Delta z/c)\phi(\omega,\theta)}\,d\omega\ , \tag{4.3.20b}$$

where the first form is appropriate for the representation in (4.3.18), while the second form is appropriate for the representation in (4.3.19).

This final set of expressions for the propagated linearly polarized plane-wave field components in the half-space $z \geqslant z_0$ that is occupied by a homogeneous, isotropic, locally linear, temporally dispersive medium is of the same form as that treated in the classical asymptotic theory of dispersive pulse propagation [4.15–21] provided that either the electric field component alone or the magnetic-field component alone is considered; the classical treatment may also be taken to apply to the single vector potential $\mathscr{A}(z,t) = -\hat{1}_y\mathscr{A}(z,t)$ given in (4.3.20a) with an input pulse spectrum that is given by

$$\tilde{f}_A(\omega) = [\![c]\!]E_0\frac{\mathrm{i}}{\omega}\tilde{f}(\omega)\ .$$

A case of special interest to the analysis of the dynamics of dispersive pulse propagation is that of an input pulse-modulated sinusoidal wave of constant applied *signal frequency* ω_c, given by

$$f(t) = u(t)\sin(\omega_c t + \psi)\ , \tag{4.3.21}$$

where ψ is chosen to be zero for a sine wave while $\psi = \pi/2$ for a cosine wave. Here $u(t)$ is the real-valued initial *envelope function* of the input pulse with temporal spectrum

$$\tilde{u}(\omega) = \int_{-\infty}^{\infty} u(t)\mathrm{e}^{\mathrm{i}\omega t}\,dt\ . \tag{4.3.22}$$

The Fourier spectrum of the input signal given in (4.3.21) is given by

$$\tilde{f}(\omega) = \frac{1}{2\mathrm{i}}[\mathrm{e}^{\mathrm{i}\psi}\tilde{u}(\omega+\omega_c) - \mathrm{e}^{-\mathrm{i}\psi}\tilde{u}(\omega-\omega_c]\ . \tag{4.3.23}$$

For a sine wave $\psi = 0$ and the spectrum is

$$\tilde{f}(\omega) = \frac{1}{2\mathrm{i}}[\tilde{u}(\omega+\omega_c) - \tilde{u}(\omega-\omega_c)]\ ,$$

while for a cosine wave $\psi = \pi/2$ and the spectrum is

$$\tilde{f}(\omega) = \tfrac{1}{2}[\tilde{u}(\omega + \omega_c) + \tilde{u}(\omega - \omega_c)] \ .$$

Substitution of (4.3.23) into (4.3.18) yields the pair of relations

$$\mathscr{E}_y(z,t) = -\frac{1}{4\pi} E_0 \Bigg[\mathrm{i}e^{\mathrm{i}\psi} \int\limits_{ia-\infty}^{ia+\infty} \tilde{u}(\omega + \omega_c) e^{(\Delta z/c)\phi(\omega,\theta)}\, d\omega - \mathrm{i}e^{-\mathrm{i}\psi} \int\limits_{ia-\infty}^{ia+\infty} \tilde{u}(\omega - \omega_c) e^{(\Delta z/c)\phi(\omega,\theta)}\, d\omega \Bigg] , \qquad (4.3.24a)$$

$$\mathscr{B}_x(z,t) = \frac{[\![c]\!]}{4\pi c} E_0 \Bigg[\mathrm{i}e^{\mathrm{i}\psi} \int\limits_{ia-\infty}^{ia+\infty} n(\omega)\tilde{u}(\omega + \omega_c) e^{(\Delta z/c)\phi(\omega,\theta)}\, d\omega - \mathrm{i}e^{-\mathrm{i}\psi} \int\limits_{ia-\infty}^{ia+\infty} n(\omega)\tilde{u}(\omega - \omega_c) e^{(\Delta z/c)\phi(\omega,\theta)}\, d\omega \Bigg] , \qquad (4.3.24b)$$

while substitution into (4.3.6) yields the pair of relations

$$\mathscr{E}_y(z,t) = -\frac{1}{2\pi} E_0\, \mathrm{Re} \Bigg\{ \mathrm{i}e^{\mathrm{i}\psi} \int\limits_{ia}^{ia+\infty} \tilde{u}(\omega + \omega_c) e^{(\Delta z/c)\phi(\omega,\theta)}\, d\omega - \mathrm{i}e^{-\mathrm{i}\psi} \int\limits_{ia}^{ia+\infty} \tilde{u}(\omega - \omega_c) e^{(\Delta z/c)\phi(\omega,\theta)}\, d\omega \Bigg\} , \qquad (4.3.25a)$$

$$\mathscr{B}_x(z,t) = \frac{[\![c]\!]}{2\pi c} E_0\, \mathrm{Re} \Bigg\{ \mathrm{i}e^{\mathrm{i}\psi} \int\limits_{ia}^{ia+\infty} n(\omega)\tilde{u}(\omega + \omega_c) e^{(\Delta z/c)\phi(\omega,\theta)}\, d\omega - \mathrm{i}e^{-\mathrm{i}\psi} \int\limits_{ia}^{ia+\infty} n(\omega)\tilde{u}(\omega - \omega_c) e^{(\Delta z/c)\phi(\omega,\theta)}\, d\omega \Bigg\} . \qquad (4.3.25b)$$

Under the change of variable $\omega \to -\omega$ the first integral appearing in (4.3.25b) may be rewritten as

$$\int\limits_{ia}^{ia+\infty} n(\omega)\tilde{u}(\omega + \omega_c) e^{(\Delta z/c)\phi(\omega,\theta)}\, d\omega = -\int\limits_{-ia}^{-ia-\infty} n^*(\omega^*)\tilde{u}^*(\omega^* - \omega_c) e^{(\Delta z/c)\phi^*(\omega^*,\theta)}\, d\omega \ ,$$

which, under the change of variable $\omega \to \omega^*$, becomes

$$\int\limits_{ia}^{ia+\infty} n(\omega)\tilde{u}(\omega + \omega_c) e^{(\Delta z/c)\phi(\omega,\theta)}\, d\omega = -\int\limits_{ia-\infty}^{ia} n^*(\omega)\tilde{u}^*(\omega - \omega_c) e^{(\Delta z/c)\phi^*(\omega,\theta)}\, d\omega^* \ .$$

Clearly, the same result holds with $n(\omega) = 1$. As a consequence (4.3.25a) may be rewritten as

$$\mathscr{E}_y(z,t) = \frac{E_0}{2\pi}\mathrm{Re}\left\{\int_{ia}^{ia+\infty} ie^{-i\psi}\tilde{u}(\omega - \omega_c)e^{(\Delta z/c)\phi(\omega,\theta)}\,d\omega\right.$$

$$\left. + \int_{ia-\infty}^{ia} [ie^{-i\psi}\tilde{u}(\omega - \omega_c)e^{(\Delta z/c)\phi(\omega,\theta)}\,d\omega]^*\right\}$$

$$= \frac{E_0}{4\pi}\left\{\int_{ia}^{ia+\infty} ie^{-i\psi}\tilde{u}(\omega - \omega_c)e^{(\Delta z/c)\phi(\omega,\theta)}\,d\omega\right.$$

$$+ \int_{ia}^{ia+\infty} [ie^{-i\psi}\tilde{u}(\omega - \omega_c)e^{(\Delta z/c)\phi(\omega,\theta)}\,d\omega]^*$$

$$+ \int_{ia-\infty}^{ia} [ie^{-i\psi}\tilde{u}(\omega - \omega_c)e^{(\Delta z/c)\phi(\omega,\theta)}\,d\omega]^*$$

$$\left. + \int_{ia-\infty}^{ia} ie^{-i\psi}\tilde{u}(\omega - \omega_c)e^{(\Delta z/c)\phi(\omega,\theta)}\,d\omega\right\}$$

$$= \frac{E_0}{4\pi}\left\{\int_{ia-\infty}^{ia+\infty} ie^{-i\psi}\tilde{u}(\omega - \omega_c)e^{(\Delta z/c)\phi(\omega,\theta)}\,d\omega\right.$$

$$\left. + \int_{ia-\infty}^{ia+\infty} [ie^{-i\psi}\tilde{u}(\omega - \omega_c)e^{(\Delta z/c)\phi(\omega,\theta)}\,d\omega]^*\right\}$$

$$= \frac{E_0}{2\pi}\mathrm{Re}\left\{ie^{-i\psi}\int_{ia-\infty}^{ia+\infty} \tilde{u}(\omega - \omega_c)e^{(\Delta z/c)\phi(\omega,\theta)}\,d\omega\right\},$$

and (4.3.25b) may be rewritten in the same manner. The final expressions for the propagated linearly polarized plane-wave field vectors due to the input pulse-modulated sinusoidal wave (4.3.21) are then

$$\mathscr{E}(z,t) = \hat{1}_y\frac{E_0}{2\pi}\mathrm{Re}\left\{ie^{-i\psi}\int_{ia-\infty}^{ia+\infty} \tilde{u}(\omega - \omega_c)e^{(\Delta z/c)\phi(\omega,\theta)}\,d\omega\right\} \quad (4.3.26a)$$

$$\mathscr{B}(z,t) = -\hat{1}_x\frac{[\![c]\!]}{2\pi c}E_0\,\mathrm{Re}\left\{ie^{-i\psi}\int_{ia-\infty}^{ia+\infty} n(\omega)\tilde{u}(\omega - \omega_c)e^{(\Delta z/c)\phi(\omega,\theta)}\,d\omega\right\} \quad (4.3.26b)$$

for $\Delta z \geq 0$. An alternate representation in which the contour of integration is contained in the positive half-space $\omega' \geqslant 0$ is given by, from (4.3.25),

$$\mathscr{E}(z,t) = -\hat{1}_y \frac{E_0}{2\pi} \mathrm{Re}\left\{ \mathrm{i} \int_{ia}^{ia+\infty} [\tilde{u}(\omega + \omega_c)\mathrm{e}^{\mathrm{i}\psi} - \tilde{u}(\omega - \omega_c)\mathrm{e}^{-\mathrm{i}\psi}]\mathrm{e}^{(\Delta z/c)\phi(\omega,\theta)}\, d\omega \right\} \tag{4.3.27a}$$

$$\mathscr{B}(z,t) = \hat{1}_x \frac{[\![c]\!]}{2\pi c} E_0 \mathrm{Re}\left\{ \mathrm{i} \int_{ia}^{ia+\infty} [\tilde{u}(\omega + \omega_c)\mathrm{e}^{\mathrm{i}\psi} - \tilde{u}(\omega - \omega_c)\mathrm{e}^{-\mathrm{i}\psi}] n(\omega)\mathrm{e}^{(\Delta z/c)\phi(\omega,\theta)}\, d\omega \right\} \tag{4.3.27b}$$

for $\Delta z \geqslant 0$. Through a comparison of (4.3.18 and 19) one may readily rewrite the above two representations in terms of the magnetic-induction field vector behavior [with $g(t)$ expressed in the form (4.3.21) for $f(t)$] at the input plane $z = z_0$.

The dynamical evolution of either field vector alone may be analyzed through a study of the scalar plane wave field whose integral representation in the half-space $z \geqslant 0$ (where z_0 is now chosen to be $z_0 = 0$) is given by

$$A(z,t) = \frac{1}{2\pi} \int_C \tilde{f}(\omega)\mathrm{e}^{(z/c)\phi(\omega,\theta)}\, d\omega \tag{4.3.28}$$

where

$$\tilde{f}(\omega) = \int_{-\infty}^{\infty} f(t)\mathrm{e}^{\mathrm{i}\omega t}\, dt \tag{4.3.29}$$

is the temporal Fourier spectrum of the initial pulse $f(t) = A(0,t)$ at the plane $z = 0$. Here $A(z,t)$ represents either the scalar potential or any scalar component of the electric field, magnetic field, Hertz vector, or vector potential field whose spectral amplitude $\tilde{A}(z,\omega)$ satisfies the Helmholtz equation

$$[\nabla^2 + \tilde{k}^2(\omega)]\tilde{A}(z,\omega) = 0 \ . \tag{4.3.30}$$

For an input pulse-modulated sinusoidal wave with signal frequency ω_c and envelope function $u(t)$, as given in (4.3.21), the expression (4.3.28) for the propagated field becomes

$$A(z,t) = \frac{1}{2\pi} \mathrm{Re}\left\{ \mathrm{i}\mathrm{e}^{-i\psi} \int_{ia-\infty}^{ia+\infty} \tilde{u}(\omega - \omega_c)\mathrm{e}^{(z/c)\phi(\omega,\theta)} d\omega \right\} \tag{4.3.31}$$

for $z \geqslant 0$.

Of primary interest in the subsequent analysis of this monograph is the attainment of accurate analytical approximations of the temporal behavior of the scalar plane-wave field $A(z,t)$ due to a particular initial pulse envelope shape $u(t)$ as it propagates through the dispersive medium. To this end, the following specific forms of the initial pulse envelope function $u(t)$ or pulse $f(t)$ are considered.

4.3.1 The Unit Step-Function Modulated Signal

For a unit step-modulated signal, the initial envelope of the pulse is given by the *Heaviside unit step-function*

$$u(t) = \begin{Bmatrix} 0; & \text{for } t < 0 \\ 1; & \text{for } t > 0 \end{Bmatrix} ; \tag{4.3.32}$$

that is, the external current source for the field abruptly begins to radiate harmonically in time at time $t = 0$ and continues indefinitely with a constant amplitude and frequency. The Laplace transform of this particular initial pulse envelope is then given by (for $\mathrm{Im}\{\omega\} > 0$)

$$\tilde{u}(\omega) = \int_0^{\infty} \mathrm{e}^{\mathrm{i}\omega t}\, dt = \frac{\mathrm{i}}{\omega} , \tag{4.3.33}$$

and the integral representation of the propagated disturbance is then

$$A(z,t) = -\frac{1}{2\pi} \mathbb{R}\mathrm{e} \left\{ \int_{ia-\infty}^{ia+\infty} \frac{1}{\omega - \omega_c} \mathrm{e}^{(z/c)\phi(\omega)}\, d\omega \right\} , \tag{4.3.34}$$

for $t > 0$ and is zero for $t < 0$. This field is precisely the same as the signal treated by *Sommerfeld* [4.15] and *Brillouin* [4.16, 17] in their early considerations of this problem in order to give an exact definition of the signal velocity in a dispersive medium. As such, it constitutes the most fundamental canonical problem to be treated in this research.

4.3.2 The Rectangular-Pulse Modulated Signal

For a rectangular pulse-modulated signal of initial time duration T, the initial envelope shape is given by

$$u(t) = \begin{Bmatrix} 0; & \text{for} \quad t < 0 \text{ and } t > T \\ 1; & \text{for} \quad 0 < t < T \end{Bmatrix} ; \tag{4.3.35}$$

that is, the external current source for the field abruptly begins to radiate harmonically in time at time $t = 0$ and continues with a constant amplitude and frequency up to the time T at which it abruptly ceases to radiate. Note that this rectangular envelope can also be represented by the difference between two unit step-functions displaced in time by the amount T. Now, the Laplace transform of this particular initial envelope is

$$\tilde{u}(\omega) = \int_0^T e^{i\omega t}\,dt = \frac{e^{i\omega T} - 1}{i\omega}\,, \tag{4.3.36}$$

and the integral representation of the propagated signal becomes

$$A(z,t) = -\frac{1}{2\pi}\,\mathrm{Re}\left\{\int_{ia-\infty}^{ia+\infty} \frac{1}{\omega - \omega_c} e^{(z/c)\phi(\omega)}\,d\omega - e^{-i\omega_c T}\int_{ia-\infty}^{ia+\infty} \frac{1}{\omega - \omega_c} e^{(z/c)\phi(\omega) + i\omega T}\,d\omega\right\}, \tag{4.3.37}$$

for $t > 0$. By defining a more general phase function, cf. (4.3.7, 8), as

$$\phi_T(\omega) \equiv i\omega[n(\omega) - \theta_T]\,, \tag{4.3.38}$$

$$\theta_T \equiv \frac{c}{z}(t - T)\,, \tag{4.3.39}$$

the field representation (4.3.37) may then be written

$$A(z,t) = -\frac{1}{2\pi}\,\mathrm{Re}\left\{\int_{ia-\infty}^{ia+\infty} \frac{1}{\omega - \omega_c} e^{(z/c)\phi_0(\omega)}\,d\omega - e^{-i\omega_c T}\int_{ia-\infty}^{ia+\infty} \frac{1}{\omega - \omega_c} e^{(z/c)\phi_T(\omega)}\,d\omega\right\}, \tag{4.3.40}$$

for $t > 0$ and is zero for $t < 0$. Note that the first integral in this expression is exactly the same as the integral representation for the unit step-modulated signal (4.3.34), while the second integral in this expression is of the exact same form as the integral appearing in equation (4.3.34), except that the phase function is retarded in time by the initial pulse width T. Consequently, by obtaining the asymptotic behavior for the step-function modulated signal given in (4.3.34), the asymptotic solution for the rectangular pulse-modulated signal (4.3.37) can be readily constructed.

Finally, note that since

$$\lim_{\omega \to \omega_c} \frac{\sin[\frac{1}{2}(\omega - \omega_c)T]}{\omega - \omega_c} = \frac{T}{2}\,,$$

the singularity due to the simple pole at $\omega = \omega_c$ appearing in each of the integrands of (4.3.40) can be removed by simply combining these two integrals as

$$A(z,t) = \frac{1}{\pi}\mathbb{R}\mathrm{e}\left\{\mathrm{i}\mathrm{e}^{-\mathrm{i}\omega_c(T/2)}\int\limits_{ia-\infty}^{ia+\infty}\frac{\sin[\frac{1}{2}(\omega-\omega_c)T]}{\omega-\omega_c}\mathrm{e}^{(z/c)\phi_{T/2}(\omega)}\,d\omega\right\}. \tag{4.3.41}$$

Thus, the spectrum of the initial rectangular pulse shape (4.3.36) is actually analytic for all complex values of ω (i.e., it is entire in the complex ω-plane), as it must be since the initial envelope function $u(t)$ is nonzero only within a finite time domain. As is later seen in Chaps. 8 and 9, the purpose of writing the propagated field $A(z,t)$ in the form (4.3.40) is to facilitate the asymptotic analysis of the signal arrival and the pulse shape in propagating through the dispersive medium.

4.3.3 The Delta-Function Pulse and the Impulse Response of the Model Medium

In order to treat the impulse response of the model medium, one must revert to the more basic integral representation (4.3.28). With the initial wave function

$$f(t) = \delta(t-t_0)\,, \tag{4.3.42}$$

which corresponds to an input delta-function pulse at time $t = t_0 > 0$ whose Laplace transform is

$$\tilde{f}(\omega) = \int\limits_0^\infty \delta(t-t_0)\mathrm{e}^{\mathrm{i}\omega t}\,dt = \mathrm{e}^{\mathrm{i}\omega t_0}\,, \tag{4.3.43}$$

the propagated scalar wave disturbance is then given by the integral representation

$$A(z,t) = \frac{1}{2\pi}\mathbb{R}\mathrm{e}\left\{\int\limits_{ia-\infty}^{ia+\infty}\mathrm{e}^{(z/c)\phi_{t_0}(\omega)}\,d\omega\right\}. \tag{4.3.44}$$

This field is termed the *impulse-response of the model medium.*

4.3.4 The Hyperbolic-Tangent Modulated Signal

In order to study the effects of a finite turn-on time for the unit step-modulated signal, an initial pulse envelope that is nonzero for all finite t is considered. The envelope chosen tends to zero rapidly enough as $t \to -\infty$ so that $A(z,t)$ is still given by (4.3.31) provided that the spectrum $\tilde{u}(\omega)$ is taken to be the Fourier transform of $u(t)$. To that end, consider the following hyperbolic-tangent modu-

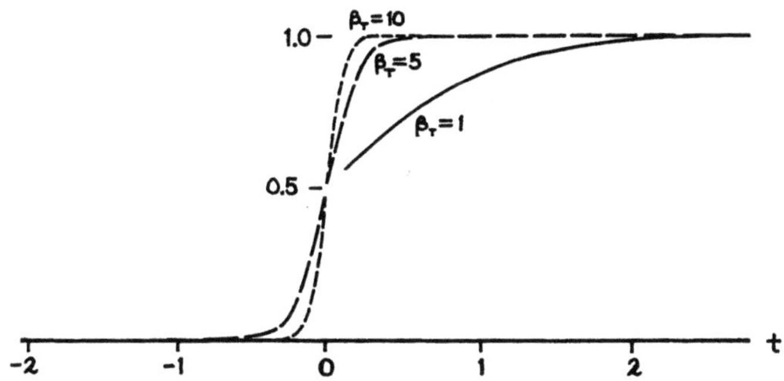

Fig. 4.9. The initial pulse envelope for the hyperbolic-tangent modulated signal. For whatever system of units the time scale t is in, the scale of β_T is given by the inverse so that the product $\beta_T t$ remains dimensionless.

lated signal for the initial pulse envelope

$$u(t) = \frac{1}{2}[1 + \tanh(\beta_T t)] = \frac{1}{1 + e^{-2\beta_T t}} \; ; \qquad \beta_T > 0 \; , \tag{4.3.45}$$

where the parameter β_T, which is indicative of the rapidity of "turn-on" of the signal, is real and positive. Note that in the limit as the time-constant β_T approaches infinity, the initial pulse envelope function (4.3.45) approaches the unit step-modulated envelope (4.3.32). The behavior of this initial envelope as a function of the time t is illustrated in Fig. 4.9 for various values of β_T.

The Fourier transform of this particular initial pulse shape is then given by the integral

$$\begin{aligned}\tilde{u}(\omega) &= \int_{-\infty}^{\infty} \frac{1}{1 + e^{-2\beta_T t}} e^{i\omega t}\, dt \\ &= \frac{1}{\beta_T} \int_0^{\infty} \frac{1}{x(x^2 + 1)} x^{-i(\omega/\beta_T)}\, dx \; ,\end{aligned}$$

where the change of variable $x = e^{-\beta_T t}$ has been performed, and where ω is complex and lies in the upper-half of the complex ω-plane (i.e., $\omega'' = \operatorname{Im}\{\omega\} > 0$) for purposes of convergence.

In order to evaluate the definite integral, consider the following contour integral in the complex z plane

$$\oint \frac{z^{-i\alpha}}{z(z^2 + 1)} dz \; ; \qquad \alpha = \alpha' + i\alpha'' \; , \qquad \alpha'' > 0 \; ,$$

where the integrand has simple poles at $z = 0$, $\pm i$ and a branch point at $z = 0$. Choosing the positive real axis as the branch cut, the contour to be used in

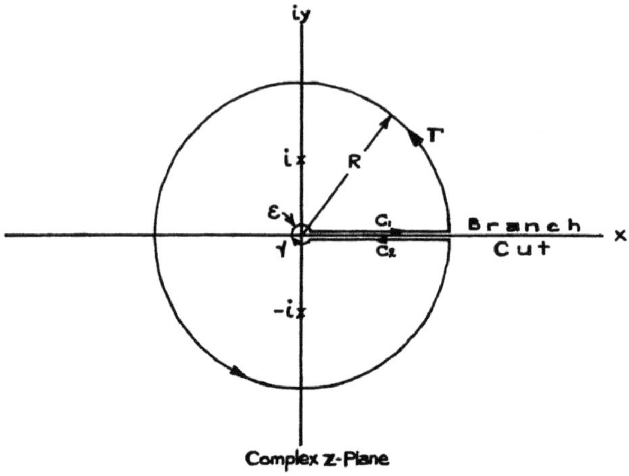

Fig. 4.10. Contour appropriate for the evaluation of the Fourier transform of $u(t) = \frac{1}{2}[1 + \tanh(\beta_T t)]$.

evaluating this integral is illustrated in Fig. 4.10. Along the contour C_1, $z = x$ so that in the limit as R approaches infinity and ε approaches zero

$$\int_{C_1} \frac{z^{-i\alpha}}{z(z^2+1)} dz = \int_0^\infty \frac{x^{-i\alpha}}{x(x^2+1)} dx ,$$

whereas along the contour C_2, $z = xe^{i2\pi}$, so that in the same limits as R approaches infinity and ε approaches zero

$$\int_{C_2} \frac{z^{-i\alpha}}{z(z^2+1)} dz = -e^{2\pi\alpha} \int_0^\infty \frac{x^{-i\alpha}}{x(x^2+1)} dx .$$

Along the outer circular contour $\Gamma, z = Re^{i\theta}$, and the following inequality is obtained

$$\left| \int_\Gamma \frac{z^{-i\alpha}}{z(z^2+1)} dz \right| \leqslant \int_0^{2\pi} \frac{R^{\alpha''}}{R^2+1} e^{\alpha'\theta} \, d\theta ,$$

which goes to zero as R approaches infinity for $\alpha'' < 2$. Furthermore, along the inner circular contour γ, $z = \varepsilon e^{i\theta}$, and the following inequality is obtained

$$\left| \int_\gamma \frac{z^{-i\alpha}}{z(z^2+1)} dz \right| \leqslant \int_0^{2\pi} \frac{\varepsilon^{\alpha''}}{\varepsilon^2+1} e^{\alpha'\theta} \, d\theta ,$$

which goes to zero as ε approaches zero for $\alpha'' > 0$. Finally, by application of the residue theorem, the following result for the entire contour integral is obtained (taking note that only the simple pole singularities at $z = \pm i$ are enclosed by the contour)

$$\oint \frac{z^{-i\alpha}}{z(z^2+1)}dz = 2\pi i\left[\operatorname*{Res}_{z=+i}\left(\frac{z^{-i\alpha}}{z(z+i)(z-i)}\right) + \operatorname*{Res}_{z=-i}\left(\frac{z^{-i\alpha}}{z(z+i)(z-i)}\right)\right]$$

$$= -\pi i(e^{(\pi/2)\alpha} + e^{(3\pi/2)\alpha}) .$$

Combining all of these results together, there results

$$\int_0^\infty \frac{x^{-i\alpha}}{x(x^2+1)}dx = \frac{\pi i}{2}\cdot\frac{1}{\sinh\left(\frac{\pi}{2}\alpha\right)} ; \qquad 0 < \operatorname{Im}\{\alpha\} < 2 ,$$

and hence, the spectrum of the initial pulse shape (4.3.45) is given by

$$\tilde{u}(\omega) = \frac{\pi i}{2\beta_T}\cdot\frac{1}{\sinh\left(\frac{\pi}{2\beta_T}\omega\right)} ; \qquad 0 < \operatorname{Im}\{\omega\} < 2\beta_T . \tag{4.3.46}$$

Since $\sinh(z) = 0$ at $z = \pm n\pi i$ for n an integer, the right-hand side of (4.3.45) possesses simple pole singularities at

$$\omega = \pm 2n\beta_T i ; \qquad n = 0, 1, 2, 3, \ldots , \tag{4.3.47}$$

i.e., there are an infinite number of simple-pole singularities along the imaginary ω''-axis with spacing $2\beta_T$. The inequality $0 < \operatorname{Im}\{\omega\} < 2\beta_T$ requires that the contour of integration in (4.3.31) lies in the upper-half of the complex ω-plane between the real axis and the line parallel to the real axis passing through the first ($n = 1$) simple pole singularity at $\omega = 2\beta_T i$. Note that in the limit as β approaches infinity, the spectrum (4.3.46) approaches the limit

$$\lim_{\beta_T\to\infty} \tilde{u}(\omega) = \frac{i}{\omega} ,$$

which is equal to the spectrum (4.3.33) for the unit step-modulated signal, as was to be expected.

With the spectral function (4.3.46), the propagated scalar wave disturbance is given by the integral representation

$$A(z,t) = -\frac{1}{4\beta}\operatorname{Re}\int_{ia-\infty}^{ia+\infty} \frac{1}{\sinh\left[\frac{\pi}{2\beta_T}(\omega-\omega_c)\right]} e^{(z/c)\phi(\omega)}\,d\omega , \tag{4.3.48}$$

for all time t, with $0 < a < 2\beta_T$. The singularities appearing in the integrand of this expression are simple poles located at

$$\omega = \omega_c \pm 2n\beta_T i ; \qquad n = 0, 1, 2, 3, \ldots , \tag{4.3.49}$$

where ω_c is, by definition, real and positive.

4.4 The Quasimonochromatic Approximation and the Heuristic Theory of Pulse Propagation

Some introductory heuristic results concerning the problem of pulse propagation in a temporally dispersive medium are now presented based upon the quadratic approximation of the dispersion relation [4.22]. It is shown here that such an approximate approach is valid only in the quasimonochromatic case (in regards to the source spectrum). In particular, if the initial pulse rise time is short so that the quasimonochromatic assumption does not apply, the resulting transfer function for the system yields an output before an input is applied, thus violating the principle of causality. However, although the basis of this method depends upon an unfounded approximation of the dispersion relation, and even though this approach does not yield error terms, the results obtained can be useful with regard to the degradation of the slowly-varying envelope shape of the pulse as it propagates through the dispersive medium. In order to obtain the domain of validity for this approximate approach and estimates of the magnitude of its errors, one must necessarily resort to the mathematically rigorous asymptotic approach to dispersive pulse propagation that is developed in this research monograph.

To begin, some introductory concepts arising in dispersive wave propagation are now discussed. Upon taking the angular frequency ω to be real and writing the complex phase function (4.3.7) in the form

$$\frac{z}{c}\phi(\omega) = \mathrm{i}[\tilde{k}(\omega)z - \omega t] \ , \tag{4.4.1}$$

where $\tilde{k}(\omega) = (\omega/c)n(\omega)$, the integral representation of the propagated plane wave field (4.3.31) with $\psi = 0$ may then be written

$$A(z,t) = \frac{1}{2\pi}\mathrm{Re}\left\{\mathrm{i}\int_{-\infty}^{\infty} \tilde{u}(\omega - \omega_c)e^{\mathrm{i}[\tilde{k}(\omega)z-\omega t]}\,d\omega\right\} . \tag{4.4.2}$$

The phase velocity of each monochromatic frequency component comprising the propagated field is then given by the velocity of propagation of the surface of constant phase appearing in the integrand of this expression, viz.

$$v_p = \frac{dz}{dt} = \frac{\omega}{\tilde{k}(\omega)} = \frac{c}{n(\omega)} \ , \tag{4.4.3}$$

so that for a dispersive medium [in which case $n(\omega)$ is replaced by $n_r(\omega) \equiv \mathrm{Re}\{n(\omega)\}$ in the above expression, in addition to $\tilde{k}(\omega)$ being replaced by $\beta(\omega) = \tilde{k}_r(\omega) \equiv \mathrm{Re}\{\tilde{k}(\omega)\}$], the phase velocity is frequency dependent. Consequently, the various different frequency components which comprise the field $A(z,t)$ tend to change phase with respect to one another which leads to a change in shape of the pulse as it propagates through the dispersive medium.

If the initial pulse-envelope shape is slowly varying in time so that the initial spectrum $\tilde{u}(\omega - \omega_c)$ is sharply peaked about the carrier frequency ω_c, then the dominant contribution to the integral representation (4.4.2) will arise from those values of ω which lie about $\omega = \omega_c$. As a consequence, consider expanding the dispersion relation $\tilde{k}(\omega)$ in a Taylor series about the applied signal frequency ω_c as

$$\tilde{k}(\omega) = \tilde{k}(\omega_c) + \left(\frac{\partial \tilde{k}}{\partial \omega}\right)_{\omega_c} (\omega - \omega_c) + \frac{1}{2}\left(\frac{\partial^2 \tilde{k}}{\partial \omega^2}\right)_{\omega_c} (\omega - \omega_c)^2 + \cdots \,. \tag{4.4.4}$$

Indeed, if the spectrum of the initial pulse envelope shape is strongly peaked about the frequency ω_c and is negligible elsewhere, then the major contribution to the integral in (4.4.2) is just due to the immediate vicinity about ω_c, in which case the above series expansion can be terminated after only a few terms. This statement constitutes the essence of the *quasimonochromatic approximation.*

Consider first the case of a *linear dispersion relation* in which the above series expansion is terminated after the second term, so that

$$\tilde{k}(\omega) \approx \tilde{k}(\omega_c) + \tilde{k}'(\omega_c)(\omega - \omega_c) \,. \tag{4.4.5}$$

With this substitution, the integral representation of the propagated field may then be directly evaluated to yield

$$A(z,t) \approx \mathbb{R}\mathrm{e}\{\mathrm{i}u[t - \tilde{k}'(\omega_c)z]\mathrm{e}^{\mathrm{i}[\tilde{k}(\omega_c)z - \omega_c t]}\} \,. \tag{4.4.6}$$

Therefore, although the oscillatory nature of the signal propagates with the phase velocity evaluated at the applied signal frequency ω_c, the pulse-envelope shape itself propagates undistorted in shape with the *group velocity* evaluated at the source frequency ω_c

$$v_g = \frac{1}{\tilde{k}'(\omega_c)} = \left(\frac{1}{\partial \tilde{k}/\partial \omega}\right)_{\omega_c} , \tag{4.4.7}$$

(again, for a dispersive medium $\tilde{k}(\omega)$ is replaced by $\beta(\omega) = \tilde{k}_r(\omega)$ so that v_g is real-valued).

Consider now the more general case (within the quasimonochromatic approximation) of a *quadratic dispersion relation* in which the series expansion (4.4.4) of $\tilde{k}(\omega)$ is terminated after the third term, so that

$$\tilde{k}(\omega) \approx \tilde{k}(\omega_c) + \tilde{k}'(\omega_c)(\omega - \omega_c) + \tfrac{1}{2}\tilde{k}''(\omega_c)(\omega - \omega_c)^2 \,. \tag{4.4.8}$$

With this substitution, the integral representation for the propagated field may then be evaluated to yield

$$A(z,t) \approx \mathbb{R}\mathrm{e}\left\{\frac{1}{[2\pi\tilde{k}''(\omega_c)z]^{1/2}}\mathrm{e}^{\mathrm{i}[\tilde{k}(\omega_c)z - \omega_c t + 3\pi/4]} \cdot \int_{-\infty}^{\infty} u(t')\mathrm{e}^{-\mathrm{i}[\tilde{k}'(\omega_c)z + t' - t]^2/2\tilde{k}''(\omega_c)z}\, dt'\right\} , \tag{4.4.9}$$

where $u(t')$ is the initial pulse envelope shape.

Following the analysis due to *Jones* [4.22], the remaining integral in (4.4.9) is seen to be of the exact same form as the one-dimensional Fresnel-Kirchhoff diffraction integral, which may be written

$$u(x) = \int_{-\infty}^{\infty} s(x')\mathrm{e}^{\mathrm{i}(\pi/R\lambda)(x-x')^2}\,dx' \ .$$

This integral describes the diffracted amplitude in the x-plane due to a transmitting screen in the x'-plane with transmission function $s(x')$, where the separation of these two planes is R and the wavelength of the plane incident light is λ. According to the scalar theory of diffraction, the behavior of $u(x)$ is determined by the scale of variation of the transmission function $s(x')$ and the size of the principal Fresnel zone, which is given by

$$a_0 = \sqrt{\lambda R}$$

If the spatial extent of $s(x')$ is much larger than a_0, then the geometrical optics approximation applies, whereas if the extent of $s(x')$ is much less than a_0, then the Fraunhofer diffraction approximation is appropriate.

On the basis of this close analogy between the pulse propagation problem in the quasimonochromatic approximation and scalar diffraction theory in the Fresnel approximation, *Jones* defined, in analogy to the principal Fresnel zone, the following parameter for the integral appearing in (4.4.9)

$$F \equiv \left[2\pi\left(\frac{\partial^2\tilde{k}}{\partial\omega^2}\right)_{\omega_c} z\right]^{1/2} . \tag{4.4.10}$$

Then, in correspondence to the geometrical optics approximation, if the initial width T of the pulse envelope is much larger than F, the pulse does not broaden significantly. However, for a sufficiently large distance z from the origin of the pulse, the parameter F must always become larger than the initial pulse width T, whereupon the pulse begins to spread in exactly the same manner as light passing through an aperture diverges in the Fraunhofer region. These two limiting cases for the integral expression (4.4.9) are now treated separately.

For $T \gg F$ the scale of variation of $u(t')$ is much larger than F (i.e., over the time interval $\Delta t' = F$, $u(t')$ does not appreciably change), so that over the range of significant values of t' contributing to the integral (4.4.9), namely

$$t - \tilde{k}'(\omega_c)z - F < t' < t - \tilde{k}'(\omega_c)z + F \ , \tag{4.4.11}$$

the pulse envelope shape $u(t')$ does not vary appreciably. The function $u(t')$ may then be regarded as being essentially constant within this region, so that the field (4.4.9) becomes

$$A(z,t) \approx \mathbb{R}\mathrm{e}\{\mathrm{i}u[t - \tilde{k}'(\omega_c)z]\mathrm{e}^{\mathrm{i}[\tilde{k}(\omega_c)z-\omega_c t]}\} \ , \tag{4.4.12}$$

and the pulse envelope propagates undistored in shape with the group velocity, cf. (4.4.6).

When $T \ll F$ the contributions from the quadratic phase terms in the exponential of the integrand in (4.4.9) are negligible due to the fact that the narrowness of the original pulse width effectively confines the integration interval to small values of t'. Consequently, for times t and observation distances z satisfying the inequality

$$|t' - t + \tilde{k}'(\omega_c)z| \ll [2\pi\tilde{k}''(\omega_c)z]^{1/2} , \tag{4.4.13}$$

the approximate integral representation (4.4.9) may be simplified to

$$A(z,t) \approx \mathbb{R}\mathrm{e}\left\{\frac{1}{[2\pi\tilde{k}''(\omega_c)z]^{1/2}} \exp \mathrm{i}\left[\tilde{k}(\omega_c)z - \omega_c t + \frac{3\pi}{4} - \frac{[\tilde{k}'(\omega_c)z - t]^2}{2\tilde{k}''(\omega_c)z}\right]\right.$$
$$\left.\cdot \int_{-\infty}^{\infty} u(t') e^{-\mathrm{i}\{[\tilde{k}'(\omega_c)z - t]/\tilde{k}''(\omega_c)z\}t'} dt'\right\} . \tag{4.4.14}$$

Thus, for a sufficiently large observation distance z from the source and times t such that the inequality (4.4.13) is satisfied, the propagated pulse shape is approximately proportional to the Fourier transform of the original pulse shape $u(t)$.

The quadratic dispersion relation (4.4.8) results in an approximate *effective transfer function* for the propagation problem that is given by

$$\tau(\omega) \approx \exp\{\mathrm{i}[\tilde{k}(\omega_c) + \tilde{k}'(\omega_c)(\omega - \omega_c) + \tfrac{1}{2}\tilde{k}''(\omega_c)(\omega - \omega_c)^2]z\} , \tag{4.4.15}$$

with the output of such a hypothetical system characterized by $\tau(\omega)$ being given by (4.4.9). However, the transfer function is, in actuality, not physically realizable due to its violation of the principle of causality. That is, with the system transfer function according to (4.4.15), an output exists prior to the application of an input, as can easily be seen by considering a delta function pulse input. This quadratic approximation of the dispersion relation assumes that the contribution of the superimposed monochromatic plane waves comprising the pulse is appreciable only for those waves confined to a certain narrow frequency range about the carrier frequency ω_c. Those component waves whose frequencies lie outside this given frequency range are assumed to have negligible spectral amplitudes. As a consequence, these results apply (in the approximate sense) only in the quasimonochromatic case. The difficulty then is that if the initial rise time of the pulse is short, this narrow-band frequency assumption is no longer valid and the transfer function for the system under this quasimonochromatic approximation yields an output before an input is applied, in violation of the principle of causality. It is therefore expected that the result (4.4.9) will, to some extent, approximately represent the propagated field $A(z,t)$ for input pulse envelope shapes which do not turn on too abruptly, such as, for example, the slowly-varying gaussian and hyperbolic tangent waveforms, which are present for all times. Furthermore, for pulse shapes which turn on at time $t = 0$ and off at time $t = T$, the representation (4.4.9) should again hold for times $t \gg z/c$ to some unspecified degree of approximation. However, the accuracy of these approximations cannot be determined within the context of this approximate analysis, and their validity must be verified by the use of mathematically rigorous asymptotic expansions of the exact integral representation of the propagated field.

As a final note, this heuristic approach to the problem of pulse propagation in dispersive media has been extended by a number of researchers [4.23–32] so as to include higher-order contributions in the expansion (4.4.4) of the medium dispersion relation. However, such an approach still relies upon the quasimonochromatic assumption and will continue to yield erroneous results for input pulse envelope shapes which turn on or off too abruptly. What is lacking in these approaches is a consideration of the total medium response to the input field, not just the medium response about the applied carrier frequency of the signal. That is, it is important from a physical standpoint to obtain solutions of the integral representation of the propagated field which accurately depict the behavior of the exact solution for all times t. In order to accomplish this, the dispersive properties of the medium must be accurately taken into account over the entire frequency spectrum. As is shown in this monograph, such considerations are taken into full account by the uniform asymptotic approach to this problem.

5 Advanced Saddle-Point Methods for the Asymptotic Evaluation of Single Contour Integrals

The integral representation developed in the previous chapters is an exact, formal solution to the problem of electromagnetic pulse propagation in homogeneous, isotropic, temporally dispersive media filling all space. However, an exact analytic evaluation of the resulting integral is not possible for realistic initial pulse shapes. Consequently, an approximate evaluation of the integral representation for a given initial pulse envelope is necessary in order to determine the behavior of the phenomena of primary interest, e.g., the properties of the precursor fields, the arrival of the main signal, and the evolution of the pulse. There are two possible approaches to accomplish this approximate evaluation. The first is a direct numerical evaluation of the integral representation; such a numerical integration is difficult to accurately do, however, and can be done for only one initial pulse envelope and carrier frequency and one set of medium parameters at a time. Since the dependance of the propagation characteristics on these parameters is complicated, such an approach would require many calculations in order to obtain a general knowledge of dispersive pulse-propagation phenomena. The alternate approach is an asymptotic analysis of the integral representation for any given initial pulse envelope. This approach is also difficult to accomplish, but it yields analytic approximations for the propagated field that display clearly all of the basic features of the propagation phenomena as a function of the medium parameters and the applied signal frequency. Such an asymptotic approach yields a very accurate approximation to the actual field behavior in the regions of primary interest, and the analysis gives insight into the resultant dynamic behavior not given by the numerical approach. Consequently, the asymptotic analysis approach to the problem of dispersive pulse propagation is taken in this treatise.

The generic form of the basic integral representation which remains to be evaluated by asymptotic techniques is, c.f. (4.3.31),

$$I(z) = \int_{ia-\infty}^{ia+\infty} q(\omega) e^{zp(\omega)} d\omega \ .$$

There exists two standard approaches for obtaining the asymptotic expansion of such integrals for large real values of the parameter z. The simplest is the method of stationary phase. However, that approach is applicable to only those integrals in which the argument of the exponential term appearing in the integrand is purely imaginary. Since dispersive media must necessarily be absorptive due to the Plemelj (dispersion) relations [5.1], the method of stationary phase is not applicable to this problem. The other approach is the method of steepest descent, which relies upon a factor that decays exponentially with the parameter z in the integrand. Consequently, that approach is applicable to the analysis of the asymptotic behavior of the integrals to be treated in this research. However, simple application of the method of steepest descent leads to discontinuous results in the asymptotic approximation of the integral if the order of the relevant saddle point changes abruptly at some critical time or if another contribution to the asymptotic behavior of the integral becomes dominant over the saddle-point contribution. Such seemingly discontinuous changes can be accounted for in a continuous fashion with modern *uniform asymptotic approaches* based upon extensions of the method of steepest descent. In this research, analytic approximations of the field behavior are obtained based upon this modern theory of asymptotic analysis.

In this chapter, a summary of the basic results of the modern theory of asymptotic analysis required for the research reported in the remainder of the book is presented. For purposes of conciseness, the results are stated in theorem form. The theorems are then followed by a discussion and analysis that permit their application to the asymptotic analysis of the integrals of interest. The methods presented are sufficient to obtain the complete uniform asymptotic behavior of a variety of optical pulses in Lorentz media. Throughout the chapter, a knowledge of the basic theory of asymptotic expansions is assumed. The required background material may be found in the second chapter of [5.2].

The starting point of the theory to be reviewed has its origin in the results developed by *Olver* [5.3]. *Olver*'s method, to be discussed in Sect. 5.1, is an alternative to the method of steepest descents, which is desirable in the present research since it is less stringent in its requirements on the deformation of the contour of integration through the saddle point. *Olver*'s method is used in this work to obtain the basic asymptotic nature of the propagated field in regions away from certain critical transition points at which the approach breaks down. The same approach is used to obtain the transition in the asymptotic behavior of the integral when there are two relevant isolated saddle points through which the contour of integration may be deformed and the relative importance of these two saddle points changes with time. That analysis, to be presented in Sect. 5.2.1, is necessary to obtain the transition from the first to the second precursor field.

The uniform asymptotic expansion due to two nearby first-order saddle points [5.2, 4–5] is next reviewed in Sect. 5.2.2. When two saddle points approach one another and coalesce, a single saddle point of higher order results; such is the case during the evolution of the second precursor field. The change in form of the asymptotic expansion (obtained by *Olver*'s method) due to the abrupt change in order of the relevant saddle point (or points) makes it necessary to have an asymptotic representation that is valid uniformly in a neighborhood of the exceptional value of the parameter describing the order of the saddle point. That analysis is necessary to obtain a complete, continuous evolution of the second precursor field.

Next, the uniform asymptotic expansion valid when the saddle point is near a simple pole singularity of the integrand [5.4, 6–7] is considered in Sect. 5.3. In this case, the uniform asymptotic expansion provides a continuous transition as the deformed contour of integration crosses the simple pole singularity and a residue is contributed to (or subtracted from) the integral. Such a uniform asymptotic analysis is necessary to describe the continuous transition from the precursor field to the main signal.

Finally, the uniform asymptotic expansion valid when the saddle point is at infinity [5.8] is considered in Sect. 5.4. Such a uniform expansion is necessary to describe the arrival and initial evolution of the first precursor field. As the saddle point moves into the finite complex plane, this uniform asymptotic expansion reduces to the expansion obtained by Olver's method.

5.1 The Saddle-Point Method due to Olver

5.1.1 Peak Value of the Integrand at the Endpoint of Integration

Consider the Contour Integral

$$I(z) = \int_P q(\omega) e^{zp(\omega)} d\omega \tag{5.1.1}$$

along a given contour P from ω_1 to ω_2 in the complex ω-plane. Both $p(\omega)$ and $q(\omega)$ are holomorphic (regular analytic) functions of the complex variable ω in a domain D containing the contour P. What is desired here is to obtain the asymptotic expansion of $I(z)$ for large absolute values of the real or complex parameter z that is uniform with respect to the phase of z. The case to be considered at present is the one in which the exponential factor $[zp(\omega)]$ attains its maximum real value along the contour P at the starting point ω_1. The angle of slope of the contour P at ω_1 is given by

$$\bar{\alpha} = \lim_{\substack{\omega \to \omega_1 \\ P}} \arg(\omega - \omega_1) \ , \tag{5.1.2}$$

where, as is indicated, this limit is taken along the contour of integration.

The assumptions made by *Olver* [5.3] in his analysis are as follows:

(a) The functions $p(\omega)$ and $q(\omega)$ are both independent of the parameter z, and are both single-valued and holomorphic in an open domain D in the complex ω-plane.

(b) The contour of integration P is independent of the parameter z, $|\omega_1|$ is finite, $|\omega_2|$ may be either finite or infinite, and the entire path P from ω_1 to ω_2 lies in the domain of holomorphicity D with the possible exception of the endpoints which may be boundary points of D.

(c) In the neighborhood of the endpoint ω_1, the functions $p(\omega)$ and $q(\omega)$ can be expanded in convergent series of the form

$$p(\omega) = p(\omega_1) + \sum_{s=0}^{\infty} p_s(\omega - \omega_1)^{s+\mu} \ , \tag{5.1.3}$$

$$q(\omega) = \sum_{s=0}^{\infty} q_s(\omega - \omega_1)^{s+\lambda-1} \ , \tag{5.1.4}$$

where $p_0 \neq 0$, $\mathrm{Re}\{\lambda\} > 0$, and μ is real and positive. When μ and λ are not integers (and this can occur only when ω_1 is a boundary point of D), the branches of $(\omega - \omega_1)^\mu$ and $(\omega - \omega_1)^\lambda$ are determined by the limiting forms

$$(\omega - \omega_1)^\mu = |\omega - \omega_1|^\mu e^{i\mu\bar{\alpha}} \ , \tag{5.1.5}$$

$$(\omega - \omega_1)^\lambda = |\omega - \omega_1|^\lambda e^{i\lambda\bar{\alpha}} \ , \tag{5.1.6}$$

as ω approaches ω_1 along P, and by continuity elsewhere on the contour P.

(d) The parameter z ranges along a ray or over a subsector $\Theta_1 \leqslant \Theta \leqslant \Theta_2$ in the complex z-plane and $|z| \geqslant Z$, where $\Theta \equiv \arg(z)$, $\Theta_2 - \Theta_1 < \pi$, and $Z > 0$. Furthermore, it is assumed that the integral $I(z)$ converges at its endpoint ω_2 both absolutely and uniformly with respect to z.

(e) Finally, it is assumed that the quantity $\mathrm{Re}\{e^{i\Theta}[p(\omega_1) - p(\omega)]\}$ is positive for all values of ω along the contour P, except at the lower endpoint ω_1 of P, and that this quantity is bounded away from zero uniformly with respect to Θ as ω approaches ω_2 along P. This implies that the real part of the exponential factor $[zp(\omega)]$ appearing in the integrand of (5.1.1) attains its maximum value along the contour P at the starting point ω_1.

Note that neither $\bar{\alpha}$ nor Θ need be confined to their principal range $(-\pi, \pi]$, provided that consistency is maintained throughout their usage. Finally, the following convention is introduced: the value of the angle $\bar{\alpha}_0 \equiv \arg(-p_0)$ is not necessarily its principal value, but rather is chosen so as to satisfy the inequality

$$|\bar{\alpha}_0 + \Theta + \mu\bar{\alpha}| \leqslant \frac{\pi}{2} \ , \tag{5.1.7}$$

and furthermore, this "branch" of $\arg(-p_0)$ is to be used in constructing all of the fractional powers of $(-p_0)$ which may occur. Since Θ is restricted to lie within an interval less than π, the value of $\bar{\alpha}_0$ which satisfies the inequality given in (5.1.7) is independent of $\Theta = \arg(z)$.

With these assumptions and conventions, the asymptotic behavior of the contour integral (5.1.1) is given by the following theorem due to *Olver* [5.3].

Theorem 5.1: *Subject to the conditions* (a) *through* (e),

$$\int_P q(\omega) e^{zp(\omega)}\, d\omega \sim e^{zp(\omega_1)} \sum_{s=0}^{\infty} \Gamma\left(\frac{s+\lambda}{\mu}\right) \frac{a_s}{z^{(s+\lambda)/\mu}} , \tag{5.1.8}$$

as $|z| \to \infty$ *uniformly with respect to* $\Theta = \arg(z)$ *for* $\Theta_1 \leqslant \Theta \leqslant \Theta_2$. *Here* $\Gamma(\zeta)$ *denotes the gamma function. The branch of* $z^{(s+\lambda)/\mu}$ *to be employed in* (5.1.8) *has phase* $(s+\lambda)\Theta/\mu$, *and the first three coefficients* a_s *are given by*

$$a_0 = \frac{q_0}{\mu(-p_0)^{\lambda/\mu}} , \tag{5.1.9a}$$

$$a_1 = \left(\frac{q_1}{\mu} - \frac{(\lambda+1)p_1 q_0}{\mu^2 p_0}\right) \frac{1}{(-p_0)^{(\lambda+1)/\mu}} , \tag{4.1.9b}$$

$$a_2 = \left\{\frac{q_2}{\mu} - \frac{(\lambda+2)p_1 q_1}{\mu^2 p_0} + [(\lambda+\mu+2)p_1^2 - 2\mu p_0 p_2] \frac{(\lambda+2)q_0}{2\mu^3 p_0^2}\right\} \frac{1}{(-p_0)^{(\lambda+2)/\mu}} . \tag{5.1.9c}$$

The infinite series appearing in the expression (5.1.8) need not be convergent; it is termed an asymptotic expansion. By definition [5.2], (5.1.8) states that for an arbitrary positive integer N,

$$\int_P q(\omega) e^{zp(\omega)}\, d\omega = e^{zp(\omega_1)} \left[\sum_{s=0}^{N-1} \Gamma\left(\frac{s+\lambda}{\mu}\right) \frac{a_s}{z^{(s+\lambda)/\mu}} + R_N\right] , \tag{5.1.10}$$

where

$$R_N = \mathcal{O}(z^{-(N+\lambda)/\mu}) , \tag{5.1.11}$$

as $|z| \to \infty$. The statement that (5.1.8) is satisfied uniformly with respect to Θ for $\Theta_1 \leqslant \Theta \leqslant \Theta_2$ means that (5.1.11) for the remainder R_N is satisfied uniformly with respect to Θ for $\Theta_1 \leqslant \Theta \leqslant \Theta_2$. The finite sum in (5.1.10) is termed an asymptotic approximation of the integral, and the first term of the series is called the dominant term of the asymptotic expansion.

If the complex phase function $p(\omega)$ and the contour of integration P appearing (5.1.8) are continuous functions of a parameter θ that varies continuously over a given domain $\mathcal{R}$ (i.e., $p(\omega) = p(\omega, \theta)$ and $P = P(\theta)$), then the asymptotic behavior of the integral can change discontinuously as θ varies over $\mathcal{R}$ even when the conditions of Theorem 5.1 are satisfied for all values of θ in $\mathcal{R}$. This can happen, for example, when either of the parameters λ or μ changes discontinuously as θ changes, see (5.1.3) and (5.1.4). If for all $\theta \in \mathcal{R}$, however, conditions (a) through (e) are satisfied with both λ and μ independent of θ, and the path P

moves in the complex ω-plane in a continuous fashion as θ varies continuously, then the asymptotic expansion given in (5.1.8) is uniform with respect to the parameter θ for all $\theta \in \mathscr{R}$ and the asymptotic behavior of the integral varies continuously with θ.

5.1.2 Peak Value of the Integrand at an Interior Point of the Path of Integration

Consider the contour integral (5.1.1) with the lower limit ω_1 replaced by ω_0. Both $p(\omega)$ and $q(\omega)$ are regular analytic functions in an open domain D containing the contour $(\omega_0, \omega_2)_P$, and $\Theta \equiv \arg(z)$ is either fixed or ranges over an interval $[\Theta_1, \Theta_2]$ such that $\Theta_2 - \Theta_1 < \pi$. However, suppose now that the maximum value of the expression $\mathrm{Re}\{e^{i\Theta}p(\omega)\}$ occurs at some point ω_1 interior to the path P, so that $\omega_1 \in (\omega_0, \omega_2)_P$ and is independent of Θ. Accordingly, the contour may be subdivided at ω_1 so that

$$I(z) = \int_{P^+} q(\omega)e^{zp(\omega)}\,d\omega - \int_{P^-} q(\omega)e^{zp(\omega)}\,d\omega \ , \tag{5.1.12}$$

where P^+ is that portion of the original contour P extending from ω_1 to ω_2 and P^- is that portion of P extending from ω_1 to ω_0. Note that P^+ is traversed in the same sense as P, whereas P^- is traversed in the opposite direction of P. The results of Theorem 5.1 apply to each of the above two contour integrals for large $|z|$ subject to the conditions (a) through (e). Since $p(\omega)$ and $q(\omega)$ are now both analytic about the point ω_1 contained within the domain D, the parameter μ appearing in (5.1.3) is an integer and the expansion (5.1.3) is a Taylor series of $p(\omega)$ about the point ω_1 with coefficients p_s given by $p_s = p^{(s+\mu)}(\omega_1)/[(s+\mu)!]$, where $p^{(n)}(\omega_1)$ denotes the n'th derivative of $p(\omega)$ with respect to ω evaluated at ω_1.

Consider first the case in which $p^{(1)}(\omega_1)$ does not vanish, so that $\mu = 1$. The condition (e) that the quantity $\mathrm{Re}\{e^{i\Theta}p(\omega)\}$ be a maximum at $\omega = \omega_1$ then gives (upon taking the derivative of $p(\omega)$ along the contour P and setting the result to zero at $\omega = \omega_1$)

$$\cos(\bar{\alpha}_0 + \Theta + \bar{\alpha}) = 0 \ ; \qquad p^{(1)}(\omega_1) \neq 0 \ , \tag{5.1.13}$$

where $\bar{\alpha}_0 = \arg(-p_0) = \arg(-p^{(1)}(\omega_1))$ and $\bar{\alpha}$ is the angle of slope of the contour P at ω_1 along which the derivative is evaluated. Since the two possible values of $\bar{\alpha}$ differ by π, the quantity $(\bar{\alpha}_0 + \Theta + \bar{\alpha})$ is $\frac{1}{2}\pi$ for one integral and $-\frac{1}{2}\pi$ for the other. As a result, the value of $\bar{\alpha}_0$ and hence, the coefficients a_s, $s = 0, 1, 2, 3, \ldots$, are exactly the same for both of the contour integrals taken over P^+ and P^-. Consequently, the asymptotic expansions of the two contour integrals appearing in (5.1.12) are the same, and all that remains upon substitution of (5.1.10) into (5.1.12) is an error term $R'_N = \mathcal{O}\{z^{-(N+\lambda)}e^{zp(\omega_1)}\}$, N being an arbitrary positive integer. Hence, the method does not give an asymptotic approximation to the integral in this case.

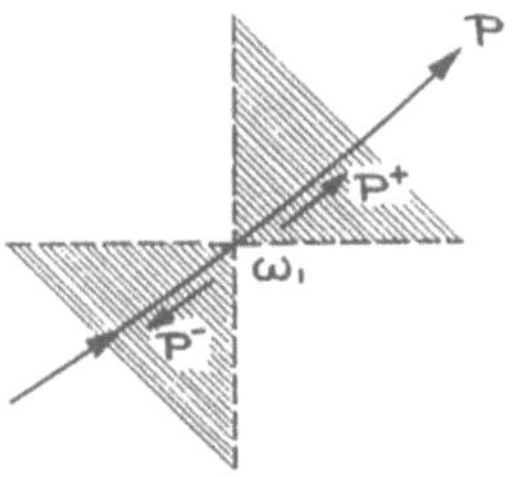

Fig. 5.1. An interior first-order saddle point.

Alternatively, if ω_1 is a saddle point so that $p^{(1)}(\omega_1)$ vanishes identically, then by condition (c) and the expansion (5.1.3) with the Taylor series coefficients p_s, it is seen that the parameter μ is an integer such that $\mu \geqslant 2$ (note that the quantity $(\mu - 1)$ specifies the order of the saddle point). Thus, $\mu\bar{\alpha}$ differs by $\mu\pi$ for the two integrals appearing in (5.1.12), causing the values of $\bar{\alpha}_0$ which satisfy the inequality (5.1.7) to differ by either $\mu\pi$ or $(\mu - 1)\pi$, according as to whether μ is even and odd, respectively. Consequently, different branches are used for the quantity $(-p_0)^{1/\mu}$ in constructing the coefficients a_s, and the asymptotic expansions of these two integrals no longer cancel upon substitution into (5.1.12). In this case, an asymptotic expansion of the integral $I(z)$ is obtained with the application of *Olver*'s method.

As an example, consider the case in which the contour of integration P passes through a single isolated saddle point of the first order, as illustrated in Fig. 5.1. In this diagram, the hatched area indicates the region wherein the quantity $\mathbb{R}\mathrm{e}\{e^{i\theta}[p(\omega_1) - p(\omega)]\}$ is positive, i.e., the region of the complex ω-plane within which the contour of integration must lie in order that condition (e) is satisfied. It is further assumed that conditions (a) through (e) are satisfied. According to (5.1.12), the integral $I(z)$ taken over the contour P may be expressed as the difference between the two integrals $I^+(z)$ and $I^-(z)$ taken over the contours P^+ and P^-, respectively, both of which start at opposite sides of the saddle point ω_1 and progress away from it, P^+ being taken in the same sense as the original contour P and P^- being taken in the opposite sense. Since $\mu = 2$, $\bar{\alpha}_0^+$ and $\bar{\alpha}_0^-$ differ by 2π and the coefficients a_s for the asymptotic expansions of these two integrals are related by

$$a_s^- = (-1)^{s+1} a_s^+ , \qquad s = 0, 1, 2, 3, \dots .$$

Consequently, the even-order coefficients add whereas the odd-order coefficients cancel with each other in the asymptotic expansion of $I(z)$. Therefore, according to Theorem 5.1, the first two terms in the asymptotic expansion of the integral (5.1.12) are given by

$$\begin{aligned} I(z) &= I^+(z) - I^-(z) \\ &= 2e^{zp(\omega_1)}\left[\Gamma\left(\frac{\lambda}{2}\right)\frac{a_0^+}{z^{\lambda/2}} + \Gamma\left(1 + \frac{\lambda}{2}\right)\frac{a_2^+}{z^{1+\lambda/2}} + \mathcal{O}(z^{-(2+\lambda/2)})\right], \end{aligned} \tag{5.1.14}$$

where the coefficients a_0^+ and a_2^+ are as given in (5.1.9). With λ equal to unity, this result reduces directly to the well-known result obtained by the method of steepest descent.

5.1.3 The Application of Olver's Method

The integrals to be evaluated asymptotically in this research, as given in (4.4.28, 31), have the same form as the integral appearing in (5.1.1), but the path of integration does not necessarily pass through a saddle point of the integrand. As a result, *Olver*'s method cannot be applied directly to obtain the asymptotic expansion of the desired integral. The first step in the asymptotic analysis is to apply Cauchy's residue theorem to change the path of integration so that *Olver*'s method can be applied to the resulting deformed contour integral.

When the contour of integration P passes through a saddle point ω_1 in such a way that the integral $I(z)$ can be written as in (5.1.12) with each of the component integrals over P^+ and P^- satisfying conditions (a) through (e) of *Olver*'s theorem, then P is called in this research an *Olver-type path* with respect to the saddle point ω_1. Furthermore, if Cauchy's residue theorem can be applied to express the integral in (5.1.1) taken over the contour P' as the sum of the same integral over a contour P plus the contributions of any pole singularities of the function $q(\omega)$, then the contour P' is said to be deformable to the contour P, and vice-versa. In particular, the contour P' is not deformable to P if the difference between the two contour integrals includes contributions due to any integral along arcs at $|\omega| = \infty$ or along branch cuts of $p(\omega)$ or $q(\omega)$.

In order to apply *Olver*'s method to obtain an asymptotic expansion of a given integral $I(z)$ taken over a contour P', the first step is to try to find an Olver-type path P with respect to a saddle point ω_1 of the integrand to which the contour P' may be deformed. Even when such a path exists, the task of finding one can be formidable when the function $p(\omega)$ appearing in the exponential of the integrand is complicated. Nevertheless, that task is usually much simpler than the one of determining a path of steepest descent through the saddle point to which P' may be deformed, as is required to apply the method of steepest descent. The essential feature of an Olver-type path is that the real part of $[zp(\omega)]$ is larger at the saddle point than at any other point along the contour P. This condition is much more general than the constraint on the path of steepest descent through the saddle point. There is a finite domain in the complex ω-plane that has the property that any path in this domain that passes through the saddle point ω_1 is an Olver-type path with respect to ω_1. Since the path of steepest descent is one of these paths, the method of steepest descent is seen to be a special case of *Olver*'s method. The only significance of the path of steepest descent is that it permits the determination of the smallest upper bound on the estimate of the magnitude of the remainder term that results when the asymptotic series is terminated after a finite number of terms [5.3].

Additional complications can arise when the complex phase function $p(\omega)$ is also a function $p(\omega, \theta)$ of a parameter θ that varies over some domain of interest $\mathscr{R}$, as is the case for the integrals appearing in (4.3.28, 31). Suppose that $p(\omega, \theta)$ is a continuous function of θ and that for each value of θ, there is an Olver-type path $P(\theta)$, to which the original contour of integration P' may be deformed, that moves in a continuous fashion in the complex ω-plane as θ varies continuously over $\mathscr{R}$. Even with such simplications, the asymptotic approximation of the integral $I(z)$ obtained by using *Olver*'s method for each value of $\theta \in \mathscr{R}$ can change discontinuously as θ varies. These discontinuities can arise from a variety of complications. In particular, *Brillouin*'s [5.9, 5.10] asymptotic approximations of signal propagation in a Lorentz medium, based upon the method of steepest descent, exhibit three discontinuities with θ that arise from three different sources. The same discontinuities result when *Olver*'s method is applied in place of the method of steepest descent, but with the former method, the sources of these discontinuities are made more transparent.

The discontinuous nature of *Brillouin*'s results are an artifact of the asymptotic analysis known as "Stokes' phenomena" [5.11]. For fixed but large values of z, the integrals being analyzed are actually continuous functions of the parameter θ. In order to obtain asymptotic approximations of the integrals that provide their true functional behavior as θ varies over the region of interest, it is necessary to apply uniform asymptotic techniques. Three different techniques are needed in order to deal with the three different sources of discontinuous behavior mentioned. A review of the required techniques is presented in the remaining sections of this chapter.

5.2 The Uniform Asymptotic Expansion for Two First-Order Saddle Points

5.2.1 The Uniform Asymptotic Expansion for Two Isolated First-Order Saddle Points

Consider a contour integral of the form given in (5.1.1) taken over a path of integration P' which extends from $|\omega| = \infty$ through the finite complex ω-plane and back to $|\omega| = \infty$ without forming a closed contour. Let the complex phase function $p(\omega, \theta)$ be a continuous function of a real parameter θ that varies over a domain $\mathscr{R}$. Furthermore, let $\omega_1(\theta)$ and $\omega_2(\theta)$ denote two isolated[1] first order saddle points of $p(\omega, \theta)$ such that $\mathrm{Re}\{p(\omega_1, \theta)\} > \mathrm{Re}\{p(\omega_2, \theta)\}$ for all θ in the range $\theta < \theta_s$, $\mathrm{Re}\{p(\omega_1, \theta)\} < \mathrm{Re}\{p(\omega_2, \theta))$ for all θ in the range $\theta > \theta_s$, and

[1] The term "isolated", as used here, means that the distance between the two points is bounded away from zero for all $\theta \in \mathscr{R}$.

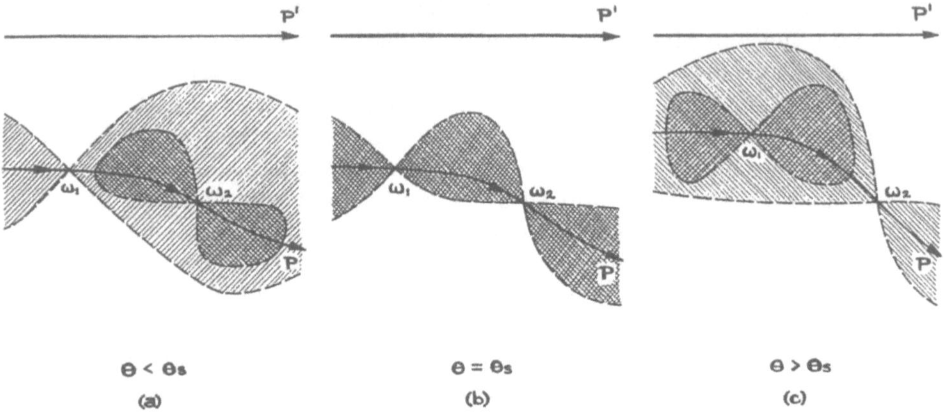

Fig. 5.2. Interaction of two isolated first-order saddle points. The hatched area indicates the region of the complex ω-plane wherein the inequality $\mathbb{R}\{p(\omega,\theta)\} < \mathbb{R}\mathrm{e}\{p(\omega_1,\theta)\}$ is satisfied, and the hatched area indicates the region of the complex ω-plane wherein the inequality $\mathbb{R}\mathrm{e}\{p(\omega,\theta)\} < \mathbb{R}\mathrm{e}\{p(\omega_2,\theta)\}$ is satisfied.

$\mathbb{R}\mathrm{e}\{p(\omega_1, \theta_s)\} = \mathbb{R}\mathrm{e}\{p(\omega_2, \theta_s\}$ for $\theta_s \in \mathscr{R}$, as illustrated in Fig. 5.2. In this case, ω_1 is called the dominant saddle point for $\theta < \theta_s$, and ω_2 is said to be dominant for $\theta > \theta_s$.

Let the original contour of integration P' be deformable to a path $P(\theta)$ that, for all θ in $\mathscr{R}$, passes through both of the saddle points $\omega_1(\theta)$ and $\omega_2(\theta)$ and has the following properties. For all $\theta \in \mathscr{R}$, the contour $P(\theta)$ moves continuously in the complex ω-plane as θ varies over $\mathscr{R}$ continuously. Moreover, $P(\theta)$ can be divided into two parts $P_1(\theta)$ and $P_2(\theta)$ such that $P(\theta) = P_1(\theta) + P_2(\theta)$, where (for $i = 1, 2$) $P_i(\theta)$ passes through the saddle point $\omega_i(\theta)$ and is an Olver-type path with respect to $\omega_i(\theta)$. The integral $I(z, \theta)$ over the contour P can then be expressed as

$$I(z,\theta) = I_1(z,\theta) + I_2(z,\theta) \ , \tag{5.2.1}$$

where

$$I_i(z,\theta) = \int_{P_i} q(\omega) e^{zp(\omega,\theta)} \, d\omega \ , \qquad i = 1, 2 \ . \tag{5.2.2}$$

It follows from the constraints imposed on the contours $P_1(\theta)$ and $P_2(\theta)$ that $P(\theta)$ is an Olver-type path for the integral $I(z, \theta)$ with respect to ω_1 when $\theta < \theta_s$ and with respect to ω_2 when $\theta > \theta_s$. Hence, according to Theorem 5.1 and the results of Sect. 5.1.2, the asymptotic expansion of $I(z, \theta)$ as $|z| \to \infty$ is given by

$$I(z,\theta) \sim 2e^{zp(\omega_i,\theta)} \sum_{s=0}^{\infty} \Gamma(s + \lambda/2) \frac{a_{2s}^{(i)}}{z^{s+\lambda/2}} \ , \tag{5.2.3}$$

where $\omega_i = \omega_1$ for $\theta < \theta_s$ and $\omega_i = \omega_2$ for $\theta > \theta_s$, and the coefficients $a_{2s}^{(i)}$ are calculated with respect to the saddle point ω_i according (5.1.9, 3, and 4). The discontinuous nature at $\theta = \theta_s$ of this asymptotic approximation of $I(z, \theta)$ as a function of θ for fixed z is obvious. Furthermore, at $\theta = \theta_s$ *Olver*'s method cannot be applied to obtain an asymptotic expansion of the integral $I(z, \theta_s)$ since condition (e) of Theorem 5.1 is not satisfied.

The discontinuity can be avoided and an expansion at $\theta = \theta_s$ can be obtained, however, by applying *Olver*'s method to $I_i(z, \theta)$ for $i = 1, 2$ instead of just applying it to $I(z, \theta)$. The asymptotic expansion of $I_i(z, \theta)$ as $|z| \to \infty$ is given uniformly with respect to θ by the right-hand side of (5.2.3) for all $\theta \in \mathscr{R}$. Application of (5.2.1) consequently yields

$$I(z,\theta) = 2e^{zp(\omega_1,\theta)}\left[\sum_{s=0}^{N-1} \Gamma\left(s+\frac{\lambda}{2}\right)\frac{a_{2s}^{(1)}}{z^{s+\lambda/2}} + \mathcal{O}(z^{-(N+\lambda/2)})\right]$$

$$+ 2e^{zp(\omega_2,\theta)}\left[\sum_{s=0}^{M-1} \Gamma\left(s+\frac{\lambda}{2}\right)\frac{a_{2s}^{(2)}}{z^{s+\lambda/2}} + \mathcal{O}(z^{-(M+\lambda/2)})\right], \tag{5.2.4}$$

as $|z| \to \infty$ uniformly with respect to θ for all $\theta \in \mathscr{R}$, where N and M are arbitrary positive integers.

For sufficiently large values of $|z|$ and for fixed $\theta \neq \theta_s$, the second term in (5.2.4) is negligible in comparison to the first term when $\theta < \theta_s$, and the first term is negligible in comparison to the second term when $\theta > \theta_s$. As a result, (5.2.4) is equivalent to (5.2.3) under these conditions. As $|\theta - \theta_s|$ tends to zero, however, $|z|$ must be increased without bound in order for (5.2.3) to give a good asymptotic approximation of $I(z, \theta)$ with a finite fixed number of terms N. Equation (5.2.4) does not have this difficulty. For sufficiently large fixed $|z|$, (5.2.4) can be used with fixed values of M and N to obtain an asymptotic approximation of $I(z, \theta)$ that is uniformly valid for all $\theta \in \mathscr{R}$. The result is a continuous function of θ.

The difficulty with the nonuniform expansion (5.2.3) arises from the error term associated with that asymptotic expansion truncated after N terms, c.f. (5.1.10). This error term satisfies the inequality

$$|R_N| \leqslant A|z|^{-(N+\lambda/2)} + A'e^{\mathbb{R}\mathrm{e}\{z[p(\omega_j,\theta)-p(\omega_i,\theta)]\}}, \tag{5.2.5}$$

where ω_i is the dominant saddle point and ω_j is the other saddle point. Since ω_i gives the dominant contribution, the second term in equation (5.2.5) is negligible in comparison to the first for large enough $|z|$, and hence $|R_N| = \mathcal{O}(z^{-(N+\lambda/2)})$. However, the closer $\mathbb{R}\mathrm{e}\{p(\omega_j, \theta)\}$ is to $\mathbb{R}\mathrm{e}\{p(\omega_i, \theta)\}$ (i.e., the closer θ is to θ_s), the larger $|z|$ must be made in order for the second term to be negligible in comparison to the first term. The error terms in the uniform expansion (5.2.4) do not have this additional second term because the contour P_i for the integral I_i does not pass through the saddle point ω_j.

5.2.2 The Uniform Asymptotic Expansion for Two Neighboring First-Order Saddle Points

Consider again a contour integral of the form given in (5.1.1) take over a path of integration P' which extends from $|\omega| = \infty$ through the finite complex ω-plane and back to $|\omega| = \infty$ without forming a closed contour. Let the complex phase function $p(\omega, \theta)$ be a continuous function of a real parameter θ that varies over a domain $\mathscr{R}$. Furthermore, let $\omega_1(\theta)$ and $\omega_2(\theta)$ denote two first-order saddle points of $p(\omega, \theta)$ which coalesce into a single saddle point ω_s of second order when $\theta = \theta_s \in \mathscr{R}$. Let the original contour of integration P' be deformable to a path $P(\theta)$ that, for all θ in $\mathscr{R}$, passes either through one of the saddle points or through both of them. When the deformed contour $P(\theta)$ passes through one of the saddle points, it is an Olver-type path with respect to that saddle point. When it passes through both saddle points, it may be expressed as $P(\theta) = P_1(\theta) + P_2(\theta)$, where $P_i(\theta)$ is an Olver-type path with respect to the saddle point $\omega_i(\theta)$, for $i = 1, 2$. Furthermore, for all $\theta \in \mathscr{R}$, the contour $P(\theta)$ moves continuously in the complex ω-plane as θ varies continuously over $\mathscr{R}$.

As an example, the situation that is encountered in this research is illustrated in Fig. 5.3. In this case, for $\theta < \theta_s$, the path P' is not deformable to an Olver-type path with respect to the saddle point ω_2, and the deformed contour passes through the saddle point $\omega_1(\theta)$ only. At $\theta = \theta_s$ the deformed contour $P(\theta)$ passes through the second-order saddle point $\omega_1(\theta_s) = \omega_2(\theta_s) = \omega_s$, and for $\theta > \theta_s$, $P(\theta)$ passes through both of the saddle points $\omega_1(\theta)$ and $\omega_2(\theta)$.

Although *Olver*'s method can be applied to obtain an asymptotic approximation of the integral valid for large $|z|$ for each value of θ in $\mathscr{R}$, the result is a

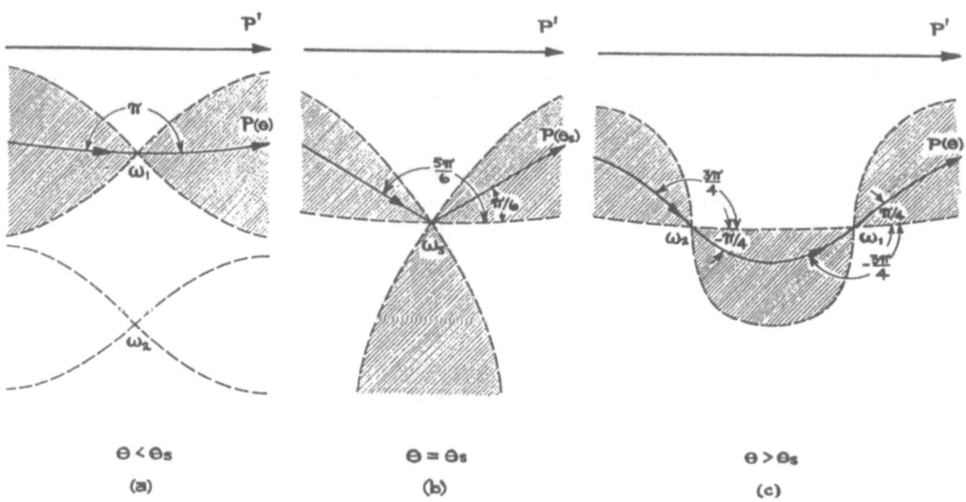

Fig. 5.3. Interaction of two neighboring first-order saddle points. The hatched area indicates the region of the complex ω-plane wherein $\mathrm{Re}\{p(\omega, \theta)\}$ is smaller than that at the relevant saddle point or points.

discontinuous function of θ at $\theta = \theta_s$. The source of the discontinuity is the fact that the parameter μ (which describes the order of the saddle points) is equal to 2 for $\theta \neq \theta_s$ and is equal to 3 at $\theta = \theta_s$. The resulting asymptotic approximation obtained for $\theta \neq \theta_s$ remains useful as θ approaches θ_s only if $|z|$ increases without bound as $|\theta - \theta_s|$ becomes arbitrarily small.

In order to obtain an asymptotic approximation of the integral for large fixed $|z|$ that is a continuous function of θ for all θ in $\mathscr{R}$, it is necessary to apply uniform asymptotic techniques. The required result is stated most simply if z is taken to be real and positive and if $P(\theta)$ [or $P_i(\theta)$ for $i = 1, 2$ when $P(\theta) = P_1(\theta) + P_2(\theta)$] is taken to be the path of steepest descent with respect to the relevant saddle point. Extensions of the result to include other paths $P(\theta)$ and complex z are discussed later. For simplicity, only the dominant term in the asymptotic expansion is considered. The result can be stated as follows [5.4, 5].

Theorem 5.2: *In the integrand of the contour integral*

$$I(z,\theta) = \int\limits_{P(\theta)} q(\omega) e^{zp(\omega,\theta)}\, d\omega \ ,$$

let the functions $p(\omega,\theta)$ and $q(\omega)$ both be holomorphic in a domain containing the two first-order saddle points $\omega_1(\theta)$ and $\omega_2(\theta)$ of $p(\omega,\theta)$, which vary in position in the complex ω-plane as the parameter θ varies over a domain $\mathscr{R}$ in such a way that as θ approaches some critical value $\theta_s \in \mathscr{R}$, the two saddle points coalesce into a single saddle point of second order [i.e., $p^{(1)}(\omega_1) = p^{(1)}(\omega_2) = 0$ and $p^{(2)}(\omega_1) \neq 0$, $p^{(2)}(\omega_2) \neq 0$ for $\theta \neq \theta_s$, while at $\theta = \theta_s$, $\omega_1 = \omega_2 = \omega_s$, $p^{(1)}(\omega_s) = p^{(2)}(\omega_s) = 0$ and $p^{(3)}(\omega_s) \neq 0$]. The path $P(\theta)$ [or $P_i(\theta)$ for $i = 1, 2$ when $P(\theta) = P_1(\theta) + P_2(\theta)$] is the path of steepest descent with respect to the relevant saddle point or points, and the parameter z is real and positive. Both $p(\omega,\theta)$ and $P(\theta)$ are continuous functions of θ for all $\theta \in \mathscr{R}$. Then the asymptotic behavior of the integral $I(z,\theta)$ is given uniformly for all $\theta \in \mathscr{R}$ as $z \to \infty$ by

$$I(z,\theta) = e^{\alpha_0(\theta)z}\left(\frac{2\pi i}{z^{1/3}}\mathscr{C}(\alpha_1(\theta)z^{2/3})\left\{\frac{1}{2}[q(\omega_1)h_1(\theta) + q(\omega_2)h_2(\theta)] + \mathscr{O}\left(\frac{1}{z}\right)\right\}\right.$$
$$\left. + \frac{2\pi i}{z^{2/3}}\mathscr{C}^{(1)}(\alpha_1(\theta)z^{2/3})\left\{\frac{1}{2\alpha_1^{1/2}(\theta)}[q(\omega_1)h_1(\theta) - q(\omega_2)h_2(\theta)] + \mathscr{O}\left(\frac{1}{z}\right)\right\}\right) , \tag{5.2.6}$$

where the function $\mathscr{C}(\zeta)$ is given by

$$\mathscr{C}(\zeta) = \frac{1}{2\pi i}\int\limits_L e^{\zeta v - (1/3)v^3}\, dv \ . \tag{5.2.7}$$

The path of integration L appearing in (5.2.7) is the path mapped into the complex v-plane as ω traverses the contour $P(\theta)$ by the root of the equation

$$\tfrac{1}{3}v^3 - \alpha_1(\theta)v - \alpha_0(\theta) + p(\omega,\theta) = 0 \tag{5.2.8}$$

that satisfies

$$\left.\frac{dv}{d\omega}\right|_{\omega=\omega_s} = \frac{1}{h_s(\theta_s)} \tag{5.2.9}$$

when $\theta = \theta_s$. *The coefficients appearing in* (5.2.6, 8, *and* 9) *are given by*

$$\alpha_0(\theta) = \frac{1}{2}[p(\omega_1, \theta) + p(\omega_2, \theta)] \ , \tag{5.2.10}$$

$$\alpha_1^{1/2}(\theta) = \left\{\frac{3}{4}[p(\omega_1, \theta) - p(\omega_2, \theta)]\right\}^{1/3} \ , \tag{5.2.11}$$

$$h_1(\theta) = \left(-\frac{2\alpha_1^{1/2}(\theta)}{p^{(2)}(\omega_1, \theta)}\right)^{1/2} \ , \qquad h_2(\theta) = \left(\frac{2\alpha_1^{1/2}(\theta)}{p^{(2)}(\omega_2, \theta)}\right)^{1/2} \ , \tag{5.2.12}$$

for $\theta \neq \theta_s$. *At the critical value* θ_s *of* θ *when the two saddle points coalesce, the coefficients in the expansion* (4.2.6) *take on the limiting values*

$$\lim_{\theta \to \theta_s} h_i(\theta) = \left(-\frac{2}{p^{(3)}(\omega_s)}\right)^{1/3} \equiv h_s(\theta_s) \ ; \qquad i = 1, 2 \ , \tag{5.2.13}$$

$$\lim_{\theta \to \theta_s} \frac{1}{2}[q(\omega_1)h_1(\theta) + q(\omega_2)h_2(\theta)] = q(\omega_s)h_s(\theta_s) \ , \tag{5.2.14}$$

$$\lim_{\theta \to \theta_s} \frac{1}{2\alpha_1^{1/2}(\theta)}[q(\omega_1)h_1(\theta) - q(\omega_2)h_2(\theta)] = h_s^2(\theta_s)q^{(1)}(\omega_s) \ , \tag{5.2.15}$$

and $\alpha_0(\theta_s)$ *is given by* (5.2.10). *The cube root in* (5.2.13) *is made single-valued by the requirement that*

$$\arg[h_s(\theta_s)] = \bar{\alpha}_s \ , \tag{5.2.16}$$

where $\bar{\alpha}_s$ *is the angle of slope of the path* $P(\theta_s)$ *as it leaves the second-order saddle point at* $\omega = \omega_s$ *for* $\theta = \theta_s$. *The cube root in* (5.2.11) *is made single-valued by the requirement that*

$$\lim_{\theta \to \theta_s} \left(\frac{\alpha_1^{1/2}(\theta)}{\omega_1(\theta) - \omega_2(\theta)}\right) = \frac{1}{2h_s(\theta_s)} \ . \tag{5.2.17}$$

The square roots in (5.2.12) *are made single-valued by the requirement that* $\lim_{\theta \to \theta_s} h_i(\theta) = h_s(\theta_s)$ *for* $i = 1, 2$, *with the argument of* $h_s(\theta_s)$ *specified by* (5.2.16).

The requirement on the argument of $h_s(\theta_s)$, as given in (5.2.16), is the same as is obtained when the more general condition used in *Olver*'s method, c.f. (5.1.7), is applied to the special case of z real and positive and $P(\theta)$ equal to the path of steepest descent for $\theta = \theta_s$ (at which $\mu = 3$). Theorem 5.2 can be extended to complex z and arbitrary Olver-type paths by applying (5.1.7) instead of (5.2.16). The theorem is sufficient as stated, however, for the problems treated in this research.

Consider now the argument of the coefficient $\alpha_1^{1/2}(\theta)$. If the saddle point ω_2 encircles ω_1 once as θ varies over $\mathscr{R}$, the argument of $\alpha_1^{3/2}(\theta)$ varies over a range of 6π and the argument of $\alpha_1^{1/2}(\theta)$ varies over a range of 2π. Hence, the cube root in (5.2.11) is not confined to a single branch of the cube root of $\alpha_1^{3/2}(\theta)$ as would be obtained by using a branch cut to restrict the argument of $\alpha_1^{3/2}(\theta)$ to a range of less tha 2π. To determine the argument of $\alpha_1^{1/2}(\theta)$ implied by (5.2.17), it is useful to apply the following geometrical concepts. Let $\bar{\alpha}_{12}$ be the angle of slope of the vector from ω_2 to ω_1 in the complex ω-plane. Then, according to (5.2.16, 17),

$$\lim_{\theta\to\theta_s} \arg[\alpha_1^{1/2}(\theta)] = \bar{\alpha}_{12} - \bar{\alpha}_s + 2\pi n \ , \tag{5.2.18}$$

where n is an arbitrary integer. Hence, as ω_1 approaches ω_2 along a straight line, the argument of $\alpha_1^{1/2}(\theta)$ approaches $2\pi n$ plus the angle that that line makes with the vector tangent to the path of steepest descent as it leaves the saddle point ω_s when $\theta = \theta_s$. If desired, the integer n can be chosen so that the argument of $\alpha_1^{1/2}(\theta)$ lies within the principal range $(-\pi, \pi)$ for all $\theta \in \mathscr{R}$. As is shown later, however, it is most convenient to set $n = 0$. In the example illustrated in Fig. 5.3, the argument of $\alpha_1^{1/2}(\theta)$ is approximately $\pi/3$ for small positive $\theta_s - \theta$ and is approximately $-\pi/6$ for small positive $\theta - \theta_s$.

Consider now the argument of the coefficient $h_i(\theta)$, $i = 1, 2$. In order to determine the argument of $h_i(\theta)$ as specified by the theorem, consider the argument of $h_i^2(\theta)$. It can be shown that [5.4]

$$\lim_{\theta\to\theta_s} \frac{-p^{(2)}(\omega_1)}{\omega_1 - \omega_2} = \lim_{\theta\to\theta_s} \frac{p^{(2)}(\omega_2)}{\omega_1 - \omega_2} = -\frac{1}{2} p^{(3)}(\omega_s) \ . \tag{5.2.19}$$

Hence, it follows from (5.2.13, 16, and 19) that

$$\lim_{\theta\to\theta_s} \arg\left(-\frac{\omega_1 - \omega_2}{p^{(2)}(\omega_1)}\right) = \lim_{\theta\to\theta_s} \arg\left(\frac{\omega_1 - \omega_2}{p^{(2)}(\omega_2)}\right) = 3\bar{\alpha}_s \ . \tag{5.2.20}$$

Consequently, application of (5.2.12, 18, and 20) yields

$$\lim_{\theta\to\theta_s} \arg[h_i^2(\theta)] = 2\bar{\alpha}_s + 2\pi n \ ; \qquad i = 1, 2 \ , \tag{5.2.21}$$

where n is the arbitrary integer appearing in (5.2.18). It is most convenient to take $n = 0$ since in that case, it follows from the requirement that $\lim_{\theta\to\theta_s} h_i(\theta) = h_s(\theta_s)$, $i = 1, 2$, and from (5.2.16, 21) that $\arg[h_i(\theta)]$ is given by

$$\arg[h_i(\theta)] = \tfrac{1}{2}\arg[h_i^2(\theta)] \ ; \qquad i = 1, 2 \ . \tag{5.2.22}$$

Consideration is given finally to the determination of the contour L into which the path $P(\theta)$ is mapped under the cubic transformation (5.2.8). Equation (5.2.8) maps the path $P(\theta)$ into three contours in the complex v-plane. It follows from (5.2.19, 16) that the path of integration L appearing in (5.2.7) is the one contour of the three that satisfies

$$\left.\frac{dv}{d\omega}\right|_{\omega=\omega_s} = -\bar{\alpha}_s \tag{5.2.23}$$

when $\theta = \theta_s$.

Since the integrand in (5.2.7) is an entire function of complex v, the only features of the path L that effect the value of $\mathscr{C}(\zeta)$ are the endpoints of the path. For the integral treated in this research, the endpoints are at infinity since $|p(\omega, \theta)| \to \infty$ as $|\omega| \to \infty$ in each direction along $P(\theta)$. In order for the integral $I(z, \theta)$ to be convergent, $p(\omega, \theta)$ must tend to infinity in the left half of the complex plane as $|\omega| \to \infty$ in each direction along $P(\theta)$. Consequently, it follows from (5.2.8) that the arguments of the endpoints of L can lie only in the following regions:

$$\text{Region 1:} \quad -\frac{\pi}{6} < \arg(v) < \frac{\pi}{6}\,, \tag{5.2.24a}$$

$$\text{Region 2:} \quad \frac{\pi}{2} < \arg(v) < \frac{5\pi}{6}\,, \tag{5.2.24b}$$

$$\text{Region 3:} \quad -\frac{5\pi}{6} < \arg(v) < -\frac{\pi}{2}\,, \tag{5.2.24c}$$

as illustrated in Fig. 5.4 below.

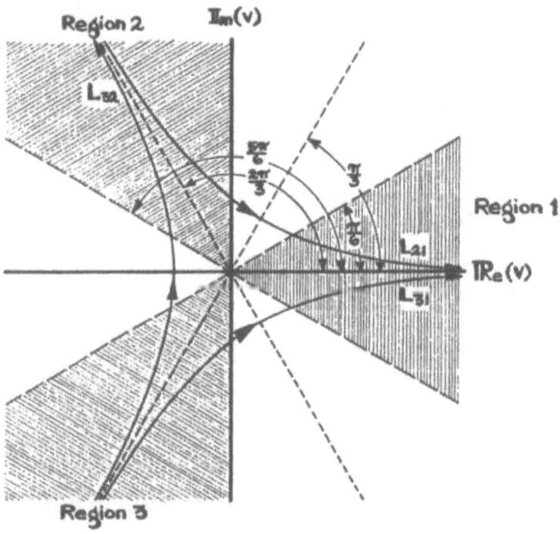

Fig. 5.4. Possible contours of integration L_{ij} in the complex v-plane into which the path $P(\theta)$ may be mapped.

Any contour that originates in Region i and terminates in Region j is labeled L_{ij}. Three such contours are illustrated in Fig. 5.4. If L is taken to be an L_{ij} contour, the value of $\mathscr{C}(\zeta)$ is determined completely by specifying i and j independent of any other details of the path L_{ij}. In particular, it can easily be shown that [5.2–4]

$$\mathscr{C}(\zeta) = \mathrm{Ai}(\zeta) \; ; \qquad \text{for } i = 3, j = 2 \; , \tag{5.2.25a}$$

$$\mathscr{C}(\zeta) = \mathrm{e}^{-\mathrm{i}(2\pi/3)} A_i(\zeta \mathrm{e}^{-\mathrm{i}(2\pi/3)}) \; ; \qquad \text{for } i = 2, j = 1 \; , \tag{5.2.25b}$$

$$\mathscr{C}(\zeta) = \mathrm{e}^{-\mathrm{i}(\pi/3)} A_i(\zeta \mathrm{e}^{\mathrm{i}(2\pi/3)}) \; ; \qquad \text{for } i = 3, j = 1 \; , \tag{5.2.25c}$$

$$\mathscr{C}(\zeta) = 0 \; ; \qquad \text{for } i = j \; , \tag{5.2.25d}$$

where $\mathrm{Ai}(\zeta)$ is the *Airy function*.

If the transformation given in (5.2.8) [combined with the requirement given in (5.2.9)] maps the path $P(\theta)$ into an L_{ij} path for one value of $\theta \in \mathscr{R}$, then that transformation maps $P(\theta)$ into an L_{ij} path with fixed values of i and j for all θ in $\mathscr{R}$. Hence, in order to determine the particular values of i and j for all $\theta \in \mathscr{R}$, it suffices to determine the arguments of the endpoints of L for one value of $\theta \in \mathscr{R}$. For that purpose, it is most convenient to choose $\theta = \theta_s$ since $P(\theta_s)$ is mapped into two straight lines, one passing from infinity in Region i to the origin and the other passing from the origin to infinity in Region j. Hence, the values of i and j can be determined most simply by examining the slopes of L at the origin for the special case $\theta = \theta_s$, where, according to equation (5.2.23)

$$\arg(\Delta v) = -\bar{\alpha}_s + \arg(\Delta\omega) \; , \tag{5.2.26}$$

at $v = 0$ for $\theta = \theta_s$. Here $\Delta\omega$ denotes the change in ω along the contour $P(\theta_s)$ taken in the direction leading away from ω_s (i.e., $\Delta\omega = \omega - \omega_s$ with ω on $P(\theta_s)$) and Δv denotes the corresponding change in v along the path L taken in the direction leading away from $v = 0$. This relationship is also approximately valid for values of $\theta \neq \theta_s$ such that $|\theta - \theta_s|$ is small.

To illustrate the determination of the path L, consider the example depicted in Fig. 5.3. According to Fig. 5.3b, the angle of slope of the path $P(\theta_s)$ as it leaves the second order saddle point at $\omega = \omega_s$ for $\theta = \theta_s$ is given by $\bar{\alpha}_s = \pi/6$. If $\Delta\omega$ in (5.2.26) is taken to lie along the portion of $P(\theta_s)$ that approaches ω_s from the left in Fig. 5.3b, then, according to the figure, $\arg(\Delta\omega) = 5\pi/6$. Hence, (5.2.26) states that the argument of Δv that lies along the corresponding portion of the transformed contour is $\arg(\Delta v) = 2\pi/3$, showing that the transformed contour originates in Region 2. If $\Delta\omega$ is taken to lie along the portion of $P(\theta_s)$ that departs from ω_s towards the right in Fig. 5.3b, then, according to the figure, $\arg(\Delta\omega) = \pi/6$. Hence, (5.2.26) states that the argument of Δv that lies along the corresponding portion of the transformed contour is $\arg(\Delta v) = 0$, showing that the transformed contour terminates in Region 1. Consequently, the transformation (5.2.8) combined with (5.2.23) transforms the contour $P(\theta)$ into an L_{21} path so that the function $\mathscr{C}(\zeta)$ occurring in (5.2.6) is given by (5.2.25b).

As an aside, it is of peripheral interest to consider the behavior of the contour $P(\theta)$ illustrated in Fig. 5.3 under the cubic transformation (5.2.8). This extra material (which is not necessary to apply the theorem) is presented only to help clarify the physical picture of what is going on. Upon application of the cubic transformation (5.2.8), the saddle points $\omega_1(\theta)$ and $\omega_2(\theta)$ map into the two points $v_{1,2} = \pm\alpha_1^{1/2}(\theta)$ for all $\theta \in \mathscr{R}$. For values of θ such that $(\theta_s - \theta)$ is small and positive, $\arg(\Delta\omega) = \pi$ along the contour $P(\theta)$ approaching ω_1 in Fig. 5.3a so that [since (5.2.26) is approximately valid for $\theta < \theta_s$] $\arg(\Delta v) \cong 5\pi/6$ along the transformed contour approaching the saddle point $v_1 = \alpha_1^{1/2}(\theta)z^{1/3}$ of $\mathscr{C}(\alpha_1(\theta)z^{2/3})$, and $\arg(\Delta\omega) = 0$ along the contour leaving ω_1 so that $\arg(\Delta v) \cong -\pi/6$ along the transformed contour leaving $v_1 = \alpha_1^{1/2}(\theta)z^{1/3}$, as illustrated in Fig. 5.5a. For $\theta = \theta_s$, $\arg(\Delta\omega) = 5\pi/6$ along the contour $P(\theta_s)$ approaching ω_s so that $\arg(\Delta v) = 2\pi/3$ along the transformed contour approaching the saddle point $v = 0$, and $\arg(\Delta\omega) = \pi/6$ along the contour $P(\theta_s)$ leaving ω_s so that $\arg(\Delta v) = 0$ along the transformed contour leaving $v = 0$, as illustrated in Fig. 5.5b. For values of θ such that $(\theta - \theta_s)$ is small and positive, $\arg(\Delta\omega) = 3\pi/4$ along the contour $P(\theta)$ approaching the saddle point ω_2 so that $\arg(\Delta v) \cong 7\pi/12$ along the transformed contour approaching the saddle point $v_2 = -\alpha_1^{1/2}(\theta)z^{1/3}$, $\arg(\Delta\omega) = -\pi/4$ along the contour leaving ω_2 so that $\arg(\Delta v) \cong -5\pi/12$ along the transformed contour leaving $v = -\alpha_1^{1/2}(\theta)z^{1/3}$, $\arg(\Delta\omega) = -3\pi/4$ along the contour approaching the saddle point ω_1 so that $\arg(\Delta v) \cong -11\pi/12$ along the transformed contour approaching the saddle point $v_1 = \alpha_1^{1/2}(\theta)z^{1/3}$, and $\arg(\Delta\omega) = \pi/4$ along the contour leaving ω_1 so that $\arg(\Delta v) \cong \pi/12$ along the transformed contour leaving the saddle point $v_1 = \alpha_1^{1/2}(\theta)z^{1/3}$, as illustrated in Fig. 5.5c. As a consequence of this analysis, it is seen that the approaching and departing slopes of the contour $P_v(\theta)$ at the saddle point or points are discontinuous at $\theta = \theta_s$. This discontinuous behavior, however, is just an anomally of the coalescence of the two saddle points for $\theta = \theta_s$. As θ approaches closer to θ_s, the path of steepest descent $P_v(\theta)$ exterior to a small neighborhood of the saddle point it passes through deviates more sharply from the tangent line to the appropriate slope at the relevant saddle point, such that at $\theta = \theta_s$, the path $P_v(\theta)$ has evolved in a continuous fashion into $P_v(\theta_s)$, as depicted in Fig. 5.5. Consequently, the transformed contour of integration $P_v(\theta)$ and its preimage $P(\theta)$ evolve in a continuous fashion with θ for all $\theta \in \mathscr{R}$, as is required in Theorem 5.2.

For sufficiently large but fixed finite values of z, (5.2.6) provides an asymptotic approximation for the integral $I(z, \theta)$ that is a continuous function of θ for all $\theta \in \mathscr{R}$, a feature that does not result when *Olver*'s method is applied directly. In return for this continuous behavior, however, (5.2.6) is more complicated than the results of *Olver*'s method due to the presence of the Airy function and its first derivative. To obtain the simpler results that follow from *Olver*'s method, (5.2.6) is now reduced for the situation depicted in Fig. 5.3.

Case I. $\theta < \theta_s$ and $|\alpha_1(\theta)|z^{2/3} \gg 1$

For values of $\theta < \theta_s$, the angle of slope of the vector from ω_2 to ω_1 is given by (Fig. 5.3a) $\bar{\alpha}_{12} = \pi/2$. Then, according to (5.2.18) with $n = 0$ and $\bar{\alpha}_s = \pi/6$, $\arg(\alpha_1^{1/2}(\theta)) \cong \pi/3$ for $\theta \simeq \theta_s$, and, according to (5.2.13, 16), $\arg(h_i(\theta)) \cong \pi/6$ for

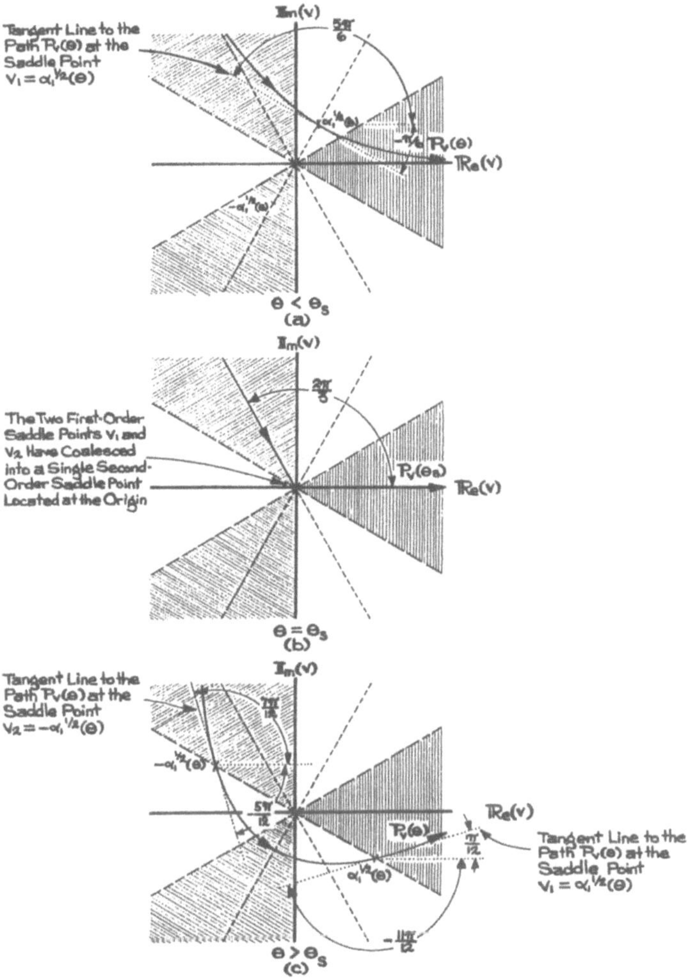

Fig. 5.5. The continuous evolution with θ of the transformed contour of integration which is the image of the contour of integration $P(\theta)$ illustrated in Fig. 5.3.

$\theta \cong \theta_s$ with $i = 1, 2$. Although these results for $\arg[\alpha_1^{1/2}(\theta)]$ and $\arg[h_i(\theta)]$ have been obtained only in the limit as θ approaches θ_s, they serve to fix the proper phase of these two functions for all $\theta < \theta_s$ such that $\theta \in \mathscr{R}$. For the situation encountered in this research, these limiting values are valid for all $\theta < \theta_s$. In this case, the argument of the Airy function (5.2.25b) appearing in the uniform expansion (5.2.6) is then given by

$$\alpha_1(\theta) z^{2/3} e^{-i(2\pi/3)} = |\alpha_1(\theta)| z^{2/3} \ ,$$

which is positive and real for z positive. For $|\alpha_1(\theta)| z^{2/3} \gg 1$, the large argument asymptotic expansion of the Airy function and its first derivative with a positive real argument may be employed in the expansion (5.2.6), where [5.3]

$$\mathrm{A_i}(|\alpha_1(\theta)|z^{2/3}) = \frac{1}{2\sqrt{\pi}|\alpha_1(\theta)|^{1/4}z^{1/6}} \mathrm{e}^{-(2/3)|\alpha_1(\theta)|^{3/2}z}\left[1 + \mathcal{O}\left(\frac{1}{|\alpha_1(\theta)|^{3/2}z}\right)\right],$$

$$\mathrm{A_i^{(1)}}(|\alpha_1(\theta)|z^{2/3}) = -\frac{1}{2\sqrt{\pi}}|\alpha_1(\theta)|^{1/4}z^{1/6}\mathrm{e}^{-(2/3)|\alpha_1(\theta)|^{3/2}z}\left[1 + \mathcal{O}\left(\frac{1}{|\alpha_1(\theta)|^{3/2}z}\right)\right],$$

as $|\alpha_1(\theta)|z^{2/3} \to \infty$. Upon substitution of these results into (5.2.6), there results

$$I(z,\theta) = \left(-\frac{2\pi}{zp^{(2)}(\omega_1,\theta)}\right)^{1/2} q(\omega_1)\mathrm{e}^{zp(\omega_1)} + \mathrm{e}^{zp(\omega_1)}\mathcal{O}[(\alpha_1(\theta)z)^{-3/2}], \quad (5.2.27)$$

as $|\alpha_1(\theta)|z^{2/3} \to \infty$. This expression (5.2.27) is the same result as is obtained when *Olver*'s method is applied to the case of a single isolated first order saddle point at ω_1, except that the dependence of the remainder term on the separation between the two saddle points is displayed explicitly through the factor $\alpha_1(\theta)$ appearing in (5.2.27).

Case II. $\theta = \theta_s$

When $\theta = \theta_s$, at which $\omega_1(\theta_s) = \omega_2(\theta_s) = \omega_s$ and $\alpha_1(\theta_s)$ identically vanishes, the limiting expressions (5.2.13) through (5.2.15) must be employed in the expansion (5.2.6). In particular, the Airy function and its first derivative at this critical value are given by

$$\mathrm{A_i}(0) = \frac{1}{2\pi}3^{-1/6}\Gamma\left(\frac{1}{3}\right), \qquad \mathrm{A_i^{(1)}}(0) = -\frac{1}{2\pi}3^{1/6}\Gamma\left(\frac{2}{3}\right).$$

Substitution of these results into (5.2.6) yields

$$I(z,\theta_s) = \mathrm{e}^{zp(\omega_s,\theta_s)}\left[\frac{\Gamma(\frac{1}{3})}{3^{1/6}}\left(-\frac{2}{zp^{(3)}(\omega_s,\theta_s)}\right)^{1/3} \mathrm{e}^{-i(\pi/6)}q(\omega_s) + \mathcal{O}(z^{-4/3})\right], \quad (5.2.28)$$

as $z \to \infty$. This expression is the same result as that obtained when *Olver*'s method is applied to the case of a single isolated second-order saddle point at ω_s.

Case III. $\theta > \theta_s$ and $|\alpha_1(\theta)|z^{2/3} \gg 1$

For values of $\theta > \theta_s$, the angle of slope of the vector from ω_2 to ω_1 is given by (Fig. 5.3c) $\bar{\alpha}_{12} = 0$. Then, according to (5.2.18) with $n = 0$ and $\bar{\alpha}_s = \pi/6$, $\arg[\alpha_1^{1/2}(\theta)] \cong -\pi/6$ for $\theta \cong \theta_s$, and, according to (5.2.13, 16), $\arg[h_i(\theta)] \cong \pi/6$ for $\theta \cong \theta_s$ with $i = 1, 2$. Although these results for $\arg[\alpha_1^{1/2}(\theta)]$ and $\arg[h_i(\theta)]$ have been obtained only in the limit as θ approaches θ_s from above, they serve to fix the proper phase of these two functions for all $\theta > \theta_s$ such that $\theta \in \mathcal{R}$. For the situation encountered in this research, these limiting values are valid for all $\theta > \theta_s$. In this case, the argument of the Airy function (5.2.25b) appearing in the uniform expansion (5.2.6) is then given by

$$\alpha_1(\theta)z^{2/3}e^{-i(2\pi/3)} = -|\alpha_1(\theta)|z^{2/3} ,$$

which is real and negative for z positive. For $|\alpha_1(\theta)|z^{2/3} \gg 1$, the large argument asymptotic expansion of the Airy function and its first derivative with a negative real argument may be employed in the expansion (5.2.6), where [5.3]

$$\mathrm{A_i}(-|\alpha_1(\theta)|z^{2/3}) = \frac{\sin\left[\frac{2}{3}|\alpha_1(\theta)|^{3/2}z + \frac{\pi}{4}\right]}{\sqrt{\pi}|\alpha_1(\theta)|^{1/4}z^{1/6}}\left[1 + \mathcal{O}\left(\frac{1}{|\alpha_1(\theta)|^{3/2}z}\right)\right] ,$$

$$\mathrm{A_i^{(1)}}(-|\alpha_1(\theta)|z^{2/3}) = -\frac{1}{\sqrt{\pi}}|\alpha_1(\theta)|^{1/4}z^{1/6}\cos\left[\frac{2}{3}|\alpha_1(\theta)|^{3/2}z + \frac{\pi}{4}\right] \times\left[1 + \mathcal{O}\left(\frac{1}{|\alpha_1(\theta)|^{3/2}z}\right)\right] ,$$

as $|\alpha_1(\theta)|z^{2/3} \to \infty$. Substitution of these results into (5.2.6) yields

$$I(z,\theta) = \mathrm{e}^{zp(\omega_1,\theta)}\left[q(\omega_1)\left(-\frac{2\pi}{zp^{(2)}(\omega_1,\theta)}\right)^{1/2} + \mathcal{O}(\alpha_1^{-3/2}(\theta)z^{-3/2})\right\} + \mathrm{e}^{zp(\omega_2,\theta)}\left[q(\omega_2)\left(-\frac{2\pi}{zp^{(2)}(\omega_2,\theta)}\right)^{1/2} + \mathcal{O}(\alpha_1^{-3/2}(\theta)z^{-3/2})\right] , \quad (5.2.29)$$

as $|\alpha_1(\theta)|z^{2/3} \to \infty$. This expression is the same result as that obtained when *Olver*'s method is applied to the case of two isolated first-order saddle points at ω_1 and ω_2 that are of equal importance, except that the dependence of the remainder term on the separation between the two saddle points is displayed explicitly through the factor $\alpha_1(\theta)$ appearing in (5.2.29).

5.3 The Uniform Asymptotic Expansion for a First-Order Saddle Point and a Simple-Pole Singularity of the Integrand

Consider the contour integral

$$I(z,\theta) = \int_{P'} q(\omega)\mathrm{e}^{zp(\omega,\theta)}\,d\omega , \quad (5.3.1)$$

where the path of integration P' extends from $|\omega| = \infty$ through the finite complex ω-plane and back to $|\omega| = \infty$ without forming a closed contour. Let the complex phase function $p(\omega,\theta)$ be a continuous function of a real parameter θ

that varies over a domain $\mathscr{R}$. The complex amplitude function $q(\omega)$ and the contour P' are independent of θ. Let $P(\theta)$ be an Olver-type path with respect to a saddle point $\omega_{sp}(\theta)$ of $p(\omega, \theta)$ and let the original contour P' be deformable to the path $P(\theta)$ which moves continuously in the complex ω-plane as θ varies continuously over $\mathscr{R}$. If no pole singularities of the function $q(\omega)$ are crossed when P' is deformed to $P(\theta)$, then the integral $I(z, \theta)$ is equal to the integral $I_{sp}(z, \theta)$ defined by

$$I_{sp}(z,\theta) = \int_{P(\theta)} q(\omega) e^{zp(\omega,\theta)} d\omega \ . \tag{5.3.2}$$

Moreover, if any pole singularities of $q(\omega)$ are crossed when P' is deformed to $P(\theta)$, then the value of the original integral $I(z,\theta)$ can be obtained from the value of $I_{sp}(z,\theta)$ by adding the appropriate residue contributions of the pole singularities according to Cauchy's residue theorem [5.2].

In particular, suppose that for all values of $\theta \in \mathscr{R}$ less than a critical value θ_s, the path P' is deformable to $P(\theta)$ without crossing any poles of $q(\omega)$, and that for $\theta > \theta_s$, a single, simple pole singularity of $q(\omega)$ located at $\omega = \omega_c$ is crossed when P' is deformed to $P(\theta)$ such that the pole is encircle in the clockwise sense. The situation is depicted in Fig. 5.6. In this diagram, the shaded area indicates the region of the complex ω-plane wherein the inequality $\mathbb{R}\mathrm{e}\{p(\omega,\theta)\} < \mathbb{R}\mathrm{e}\{p(\omega_{sp}(\theta),\theta)\}$ is satisfied, i.e., the region within which the contour $P(\theta)$ must lie in order to be an Olver-type path with respect to the saddle point $\omega_{sp}(\theta)$. Then, the integral $I(z,\theta)$ is given in terms of $I_{sp}(z,\theta)$ by

$$I(z,\theta) = I_{sp}(z,\theta) \ ; \qquad \text{for } \theta < \theta_s \ , \tag{5.3.3a}$$

$$I(z,\theta) = I_{sp}(z,\theta) - \pi i \gamma e^{zp(\omega_c,\theta_s)} \ ; \qquad \text{for } \theta = \theta_s \ , \tag{5.3.3b}$$

$$I(z,\theta) = I_{sp}(z,\theta) - 2\pi i \gamma e^{zp(\omega_c,\theta)} \ ; \qquad \text{for } \theta > \theta_s \ , \tag{5.3.3c}$$

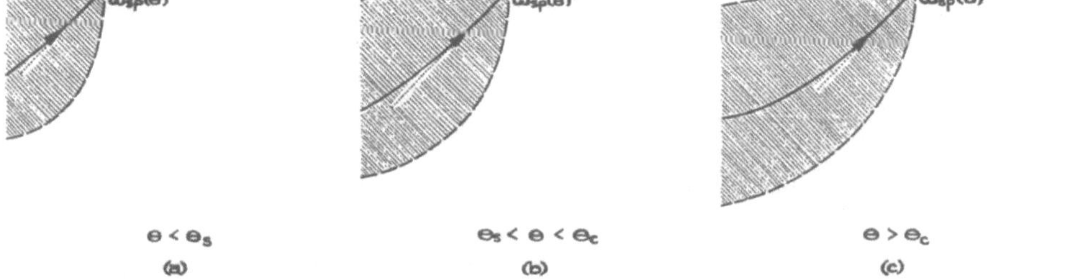

Fig. 5.6. Interaction of a first-order saddle point with a simple pole at ω_c. The dotted line indicates the tangent line to the path of steepest descent at the saddle point located at $\omega = \omega_{sp}(\theta)$

where

$$\gamma = \lim_{\omega \to \omega_c} [(\omega - \omega_c)q(\omega)] \tag{5.3.4}$$

is the residue of the simple pole singularity at $\omega = \omega_c$.

If *Olver*'s method is applied to obtain an asymptotic approximation of the "saddle-point integral" $I_{sp}(z,\theta)$, the result is a continuous function of θ for fixed finite values of z of sufficiently large magnitude. In particular, this result is continuous at $\theta = \theta_s$. However, the resulting approximation for the integral $I(z,\theta)$, obtained from (5.3.3), is a discontinuous function of θ at $\theta = \theta_s$. The discontinuity is of no consequence for fixed values of $|z|$ larger than some positive constant Z, however, because the saddle-point integral $I_{sp}(z,\theta)$ varies exponentially as $e^{zp(\omega_{sp},\theta)}$ which dominates the exponential behavior of the last term of (5.3.3b and c) so long as ω_c lies within the shaded region in Fig. 5.6. Since ω_c lies on the Olver-type path $P(\theta)$ at $\theta = \theta_s$, the singularity must lie within the shaded region at $\theta = \theta_s$ when the discontinuity occurs. The minimum value of Z depends upon the particular Olver-type path chosen. The closer $P(\theta)$ is to the path of steepest descent through the saddle point ω_{sp}, the larger is the difference $|p(\omega_{sp}(\theta_s),\theta_s) - p(\omega_c,\theta_s)|$, and hence, the smaller is the minimum value of $|z|$ for which the discontinuity can be ignored.

When the pole singularity lies outside of the shaded region, as depicted in Fig. 5.6c, the last term of (5.3.3c) is exponentially dominant over the first term. Let $\mathbb{R}\mathrm{e}\{p(\omega_c,\theta)\} > \mathbb{R}\mathrm{e}\{p(\omega_{sp}(\theta),\theta)\}$ for $\theta > \theta_c > \theta_s$ and $\mathbb{R}\mathrm{e}\{p(\omega_c,\theta)\} < \mathbb{R}\mathrm{e}\{p(\omega_{sp}(\theta),\theta)\}$ for $\theta_s < \theta < \theta_c$. Then the asymptotic behavior of $I(z,\theta)$ is the same as that of $I_{sp}(z,\theta)$ for $\theta < \theta_c$ and it is given by the last term of (5.3.3c) when $\theta > \theta_c$ (Fig. 5.6c). Hence, the asymptotic behavior of $I(z,\theta)$ changes abruptly as θ crosses from $\theta < \theta_c$ to $\theta > \theta_c$. For fixed finite values of $|z| > Z$, however, the asymptotic approximation of $I(z,\theta)$ obtained by substituting an asymptotic approximation of $I_{sp}(z,\theta)$ in (5.3.3c) is a continuous function of θ for all $\theta > \theta_s$ as long as the last term in (5.3.3c) is retained for all $\theta > \theta_s$.

An additional complication arises when the saddle point $\omega_{sp}(\theta)$ approaches close to the pole singularity ω_c at $\theta = \theta_s$. As the quantity $|\omega_{sp}(\theta_s) - \omega_c|$ becomes small, so does $|p(\omega_{sp}(\theta_s),\theta_s) - p(\omega_c,\theta_s)|$ and hence, Z becomes large. For $|z| < Z$, the discontinuity displayed in (5.3.3) can be significant. To avoid this discontinuous behavior when the saddle point is near the pole, it is necessary to apply a technique known as subtraction of the pole. This method yields an asymptotic expansion of the integral $I_{sp}(z,\theta)$ as $|z| \to \infty$ that is uniform for all $\theta \in \mathscr{R}$. The resulting asymptotic approximation of the integral $I(z,\theta)$ is a continuous function of $\theta \in \mathscr{R}$ for fixed finite z.

The asymptotic behavior of the saddle point integral $I_{sp}(z,\theta)$ can be stated most simply if z is taken to be real and positive and if $P(\theta)$ is taken to be a path of steepest descent defined by the equation $\mathbb{I}\mathrm{m}\{p(\omega,\theta)\} = \mathbb{I}\mathrm{m}\{p(\omega_{sp}(\theta),\theta)\}$. The result for more general Olver-type paths is deferred until later. For simplicity, only the dominant term in the asymptotic expansion is considered. The result can then be stated as follows [5.4, 6, 7].

Theorem 5.3: *In the integral $I_{sp}(z,\theta)$ given in (5.3.2), the contour of integration $P(\theta)$ is a path of steepest descent through a first-order saddle point $\omega_{sp}(\theta)$ isolated from any other saddle points of $p(\omega,\theta)$ and z is real and positive. All of the conditions required in order for $P(\theta)$ to be an Olver-type path with respect to $\omega_{sp}(\theta)$ are satisfied for all $\theta \in \mathscr{R}$ except that the function $q(\omega)$ exhibits a single first order pole singularity at $\omega = \omega_c$ in the domain D wherein $p(\omega,\theta)$ is analytic. The complex phase function $p(\omega,\theta)$ is a continuous function of θ for all $\theta \in \mathscr{R}$, whereas $q(\omega)$ and z are independent of θ. Then, the saddle point integral $I_{sp}(z,\theta)$ satisfies*

$$I_{sp}(z,\theta) = q(\omega_{sp})\left(-\frac{2\pi}{zp^{(2)}(\omega_{sp},\theta)}\right)^{1/2} e^{zp(\omega_{sp},\theta)} + \gamma\left[\pm i\pi\,\mathrm{erfc}(\mp i\Delta(\theta)\sqrt{z})e^{zp(\omega_c,\theta)} + \frac{1}{\Delta(\theta)}\sqrt{\frac{\pi}{z}}e^{zp(\omega_{sp},\theta)}\right] + R_1 e^{zp(\omega_{sp},\theta)} \; ; \qquad \mathrm{Im}\{\Delta(\theta)\} \gtrless 0 \; , \tag{5.3.5a}$$

$$I_{sp}(z,\theta) = q(\omega_{sp})\left(-\frac{2\pi}{zp^{(2)}(\omega_{sp},\theta)}\right)^{1/2} e^{zp(\omega_{sp},\theta)} + \gamma\left[i\pi\,\mathrm{erfc}(-i\Delta(\theta)\sqrt{z})e^{zp(\omega_c,\theta)} + \frac{1}{\Delta(\theta)}\sqrt{\frac{\pi}{z}}e^{zp(\omega_{sp},\theta)}\right] - i\pi\gamma e^{zp(\omega_c,\theta)} + R_1 e^{zp(\omega_{sp},\theta)} \; ; \quad \mathrm{Im}\{\Delta(\theta)\} = 0 \; , \quad \Delta(\theta) \neq 0 \; , \tag{5.3.5b}$$

$$I_{sp}(z,\theta) = \left(-\frac{2\pi}{zp^{(2)}(\omega_{sp},\theta)}\right)^{1/2} e^{zp(\omega_{sp},\theta)}\left\{\left[q(\omega_{sp}) - \frac{\gamma}{\omega_{sp}-\omega_c}\right] - \gamma\frac{p^{(3)}(\omega_{sp},\theta)}{6p^{(2)}(\omega_{sp},\theta)}\right\} + R_1 e^{zp(\omega_{sp},\theta)} \; ; \qquad \Delta(\theta) = 0 \; , \tag{5.3.5c}$$

where

$$R_1 = \mathcal{O}(z^{-3/2}) \; , \qquad \text{as} \quad z \to \infty \tag{5.3.6}$$

uniformly with respect to θ for all $\theta \in \mathscr{R}$, and where γ is as defined in (5.3.4), and

$$\Delta(\theta) \equiv [p(\omega_{sp}(\theta),\theta) - p(\omega_c,\theta)]^{1/2} \; . \tag{5.3.7}$$

The argument of the quantity $[-zp^{(2)}(\omega_{sp}(\theta),\theta)]^{1/2}$ is defined to be equal to $\arg(d\omega)_{\omega_{sp}}$, *where $d\omega$ is an element along the path of steepest descent through the saddle point $\omega_{sp}(\theta)$, and the argument of $\Delta(\theta)$ is defined so that*

$$\lim_{\omega_c \to \omega_{sp}(\theta)} [\Delta(\theta)] = [\omega_c - \omega_{sp}(\theta)]\left(-\frac{p^{(2)}(\omega_{sp}(\theta),\theta)}{2}\right)^{1/2} \; . \tag{5.3.8}$$

Finally, the function $\mathrm{erfc}(\mathscr{S})$ *is the complementary error function, defined by*

$$\mathrm{erfc}(\zeta) \equiv \frac{2}{\sqrt{\pi}} \int_{\zeta}^{\infty} e^{-\xi^2} d\xi \ . \tag{5.3.9}$$

The asymptotic behavior of the saddle-point integral $I_{sp}(z, \theta)$ is given by (5.3.5a) with the upper signs when the contour $P(\theta)$ lies on one side of the pole and with the lower signs when $P(\theta)$ lies on the other side of the pole with respect to the original path P'. When the pole lies on the contour $P(\theta)$, $I_{sp}(z, \theta)$ satisfies (5.3.5b), and when the saddle point coalesces with the pole, $I_{sp}(z, \theta)$ satisfies (5.3.5c). Since (5.3.6) for the error term is satisfied uniformly with respect to θ for all $\theta \in \mathscr{R}$, the apparent discontinuities in the asymptotic behavior of $I_{sp}(z, \theta)$ exhibited by (5.3.5) are real. In particular, when the path $P(\theta)$ passes from one side of the pole to the other, the discontinuous jump in $I_{sp}(z, \theta)$ due to the change in sign of $\mathrm{Im}\{\Delta(\theta)\}$ in (5.3.5a) is $2\pi i \gamma e^{zp(\omega_c, \theta)}$. This discontinuity in $I_{sp}(z, \theta)$ exactly cancels the discontinuity in $I(z, \theta)$ introduced by the contribution of the simple-pole singularity when Cauchy's residue theorem is applied to deform the original contour P' to the path of steepest descent $P(\theta)$, as seen in the set of relations given in (5.3.3). As a result, the asymptotic behavior of $I(z, \theta)$ is a continuous function of θ for all $\theta \in \mathscr{R}$ for fixed finite values of z.

If $P(\theta)$ is an Olver-type path other than the path of steepest descent, then Theorem 5.3 remains valid if $P(\theta)$ is deformable to the path of steepest descent without crossing the pole singularity. If the pole is crossed when $P(\theta)$ is deformed to the path of steepest descent, then the set of relations given in (5.3.5) are changed by the addition or subtraction [5.12] of a term $2\pi i \gamma e^{zp(\omega_c, \theta)}$. Since the change in the expression for $I_{sp}(z, \theta)$ is equal but with opposite sign to the change introduced between $I(z, \theta)$ and $I_{sp}(z, \theta)$ when Cauchy's residue theorem is applied to change the contour of integration from the path of steepest descent to the new Olver-type path $P(\theta)$, the resulting asymptotic expression for $I(z, \theta)$ remains unchanged. Hence, the asymptotic approximation obtained for $I(z, \theta)$ is independent of the particular Olver-type path chosen. Nevertheless, in order to apply Theorem 5.3 to obtain a uniform asymptotic approximation of $I(z, \theta)$, it is still necessary to determine the path of steepest descent relative to the position of the pole in order to determine whether or not the contribution of the pole should be added to the right-hand side of (5.3.5)

If the pole encircles the saddle point once as θ varies over $\mathscr{R}$, the argument of $\Delta^2(\theta)$ varies over a range of 4π and the argument of $\Delta(\theta)$ varies over a range of 2π. Hence, $\Delta(\theta)$ is not confined to a single branch of the square root of $\Delta^2(\theta)$ as would be obtained by using a branch cut to restrict the argument of $\Delta^2(\theta)$ to a range of less than 2π. To determine the argument of $\Delta(\theta)$ that is implied by (5.3.8), it is useful to apply the following geometrical concepts. Let $\bar{\alpha}_c$ be the angle of slope of the vector from $\omega_{sp}(\theta)$ to ω_c in the complex ω-plane. Then, according to (5.3.8),

$$\lim_{\omega_c \to \omega_{sp}(\theta)} \arg[\Delta(\theta)] = \bar{\alpha}_c + \arg\{[-p^{(2)}(\omega_{sp}(\theta), \theta)]^{1/2}\} + 2\pi n \ , \tag{5.3.10}$$

where the limit is taken along the straight line with slope $\bar{\alpha}_c$ and where n is an arbitrary integer. According to the definition given in the theorem,

$$\arg\{[-p^{(2)}(\omega_{sp}(\theta),\theta)]^{1/2}\} = -\bar{\alpha}_{SD} \ , \tag{5.3.11}$$

where $\bar{\alpha}_{SD}$ is the angle of slope of a vector tangent to the path of steepest descent at the saddle point, as defined in (5.1.2) with P the path of steepest descent and $\omega_1 = \omega_{sp}(\theta)$. Substitution of (5.3.11) into (5.3.10) yields

$$\lim_{\omega_c \to \omega_{sp}(\theta)} \arg[\Delta(\theta)] = \bar{\alpha}_c - \bar{\alpha}_{SD} + 2\pi n \ . \tag{5.3.12}$$

Hence, as the pole approaches the saddle point along a straight line, the argument of $\Delta(\theta)$ approaches $2\pi n$ plus the angle that that line makes with the vector tangent to the path of steepest descent at the saddle point $\omega_{sp}(\theta)$. The integer n can be chosen so that the argument of $\Delta(\theta)$ for all $\theta \in \mathscr{R}$ lies within the principal range $(-\pi, \pi]$. In the example illustrated in Fig. 5.6, the limit of $\arg[\Delta(\theta)]$ is a small negative angle in part (a) and a small positive angle in parts (b) and (c).

In order to understand the uniform asymptotic nature of $I_{sp}(z,\theta)$ as given in (5.3.5), it is necessary to understand the properties of the complementary error function. Since $e^{-\xi^2}$ is entire, both the *error function* $\operatorname{erf}(\zeta)$ and the *complementary error function* $\operatorname{erfc}(\zeta) = 1 - \operatorname{erf}(\zeta)$ are entire functions of the complex argument ζ as seen in (5.3.9). According to (5.3.9), for ζ along the real axis, $\operatorname{erfc}(\zeta)$ approaches 2 as ζ goes to negative infinity, is equal to unity at the origin, and approaches 0 as ζ goes to positive infinity. Along the imaginary axis where $\zeta = \rho e^{\pm i(\pi/2)}$, with ρ positive or zero, $\operatorname{erfc}(\zeta)$ is given by

$$\operatorname{erfc}(\rho e^{\pm i(\pi/2)}) = 1 \mp i \frac{2}{\sqrt{\pi}} \int_0^{\rho} e^{t^2}\, dt \ , \tag{5.3.13}$$

which is related to *Dawson's integral* $F(\rho) = e^{-\rho^2} \int_0^{\rho} e^{t^2}\, dt$. Consequently, $\operatorname{erfc}(\rho e^{\pm i(\pi/2)})$ approaches $1 \mp i\infty$ as ρ approaches infinity for positive ρ. Finally, along the diagonal axes where $\zeta = \rho e^{\pm i(\pi/4)}$, with ρ real, $\operatorname{erfc}(\zeta)$ is given by

$$\operatorname{erfc}(\rho e^{\pm i(\pi/4)}) = 1 - \sqrt{2} e^{\pm i(\pi/4)} \left[\mathscr{C}\left(\sqrt{\frac{2}{\pi}}\rho\right) \mp i \mathscr{S}\left(\sqrt{\frac{2}{\pi}}\rho\right)\right] , \tag{5.3.14}$$

where $\mathscr{C}(z)$ and $\mathscr{S}(z)$ are the *cosine* and *sine Fresnel integrals*, respectively, defined by

$$\mathscr{C}(z) = \int_0^z \cos(\tfrac{1}{2}\pi t^2)\, dt \ , \tag{5.3.15a}$$

$$\mathscr{S}(z) = \int_0^z \sin(\tfrac{1}{2}\pi t^2)\, dt \ . \tag{5.3.15b}$$

Since $\mathscr{C}(\infty) = \mathscr{S}(\infty) = \frac{1}{2}$ and $\mathscr{C}(-\infty) = \mathscr{S}(-\infty) = -\frac{1}{2}$, it is seen that $\operatorname{erfc}(\rho e^{\pm i(\pi/4)})$ approaches 0 for positive real values of ρ and 2 for negative real values of ρ as ρ approaches infinity.

The asymptotic behavior of the complementary error function is given by [Ref. 5.2, pp. 82–87]

$$\operatorname{erfc}(\zeta) = \frac{1}{\sqrt{\pi}} e^{-\zeta^2} \sum_{n=0}^{\infty} \frac{(-1)^n \Gamma(n+\frac{1}{2})}{\Gamma(\frac{1}{2})\zeta^{2n+1}} \tag{5.3.16}$$

as $|\zeta| \to \infty$ uniformly in $|\arg(\zeta)| < \pi/2$. By use of the identity

$$\operatorname{erfc}(\zeta) + \operatorname{erfc}(-\zeta) = 2 \tag{5.3.17}$$

the asymptotic behavior of $\operatorname{erfc}(\zeta)$ in the half-plane $\mathrm{Re}\{\zeta\} < 0$ may be obtained directly from (5.3.16). As pointed out by *Copson* [5.2], the region of validity of (5.3.16) can be extended to include purely imaginary values of ζ so that $\arg(\zeta) = \pm\pi/2$. It is more convenient, however, to apply

$$\operatorname{erfc}(\zeta) = 1 + \frac{1}{\sqrt{\pi}} e^{-\zeta^2} \sum_{n=0}^{\infty} \frac{(-1)^n \Gamma(n+\frac{1}{2})}{\Gamma(\frac{1}{2})\zeta^{2n+1}} \tag{5.3.18}$$

as $|\zeta| \to \infty$ with $\arg(\zeta) = \pm\pi/2$. Equations (5.3.16 and 18) are equivalent for $\arg(\zeta) = \pm\pi/2$ because the first term on the right-hand side of (5.3.18) is asymptotically negligible in comparison to the second term as $|\zeta| \to \infty$. Both equations give the correct asymptotic behavior of $\operatorname{erfc}(\zeta)$ for $\arg(\zeta) = \pm\pi/2$. Equation (5.3.18) has the advantage, however, that it gives the correct asymptotic behavior of the real and imaginary parts of $\operatorname{erfc}(\zeta)$ separately as can be seen by comparison with (5.3.13) and noting that the second term on the right-hand side of (5.3.18) is purely imaginary. Finally, from (5.3.16–18) it is seen that the asymptotic behavior of $\operatorname{erfc}(\zeta)$ is exponentially attenuated for $-\pi/4 < \arg(\zeta) < \pi/4$ and $3\pi/4 < \arg(\zeta) < 5\pi/4$, whereas it is exponentially amplified for $\pi/4 < \arg(\zeta) < 3\pi/4$ and $-3\pi/4 < \arg(\zeta) < -\pi/4$.

When the pole is far enough away from the saddle point that the quantity $|\Delta(\theta)|\sqrt{z}$ is large, then the dominant term in the corresponding asymptotic expansion of the complementary error function with large argument can be substituted into (5.3.5). If the integer n in (5.3.12) is chosen so that $\arg[\Delta(\theta)]$ lies within the principal range $(-\pi, \pi]$, and if the pole does not lie on the path of steepest descent, then the argument $(\mp i\Delta(\theta)\sqrt{z})$ of the complementary error function appearing in (5.3.5a) has phase that satisfies

$$|\arg[\mp i\Delta(\theta)\sqrt{z}]| < \frac{\pi}{2}\,; \qquad \mathrm{Im}\{\Delta(\theta)\} \gtrless 0\,. \tag{5.3.19}$$

Hence, (5.3.16) is applicable in (5.3.5a). According to (5.3.16), the dominant term in the expansion of the complementary error function appearing in (5.3.5a) is given by

$$\operatorname{erfc}[\mp \mathrm{i}\Delta(\theta)\sqrt{z}] = \frac{1}{\mp \mathrm{i}\sqrt{\pi z}\Delta(\theta)} \mathrm{e}^{z[p(\omega_{sp},\theta)-p(\omega_c,\theta)]} + \mathcal{O}\{[\Delta(\theta)\sqrt{z}]^{-3}\}\ , \tag{5.3.20}$$

as $|\Delta(\theta)|\sqrt{z} \to \infty$ uniformly with respect to $\arg[\Delta(\theta)]$ for $-\pi/2 < \arg[\mp \mathrm{i}\Delta(\theta)\sqrt{z}] < \pi/2$. Substitution of (5.3.20) into (5.3.5a) yields

$$I_{sp}(z,\theta) = q(\omega_{sp})\left(-\frac{2\pi}{zp^{(2)}(\omega_{sp},\theta)}\right)^{1/2} \mathrm{e}^{zp(\omega_{sp},\theta)} + \mathcal{O}\{[\Delta(\theta)\sqrt{z}]^{-3}\}\ , \tag{5.3.21}$$

as $|\Delta(\theta)|\sqrt{z} \to \infty$ uniformly with respect to $\arg[\Delta(\theta)]$ for $0 < \arg[\Delta(\theta)] < \pi$. This equation (5.3.21) is the same result as is obtained by direct application of *Olver*'s method except that the dependence of the remainder term on the separation between the pole and the saddle point is displayed in (5.3.21) through the function $\Delta(\theta)$.

If the pole lies on the path of steepest descent ($\theta = \theta_s$) but remains far from the saddle point, then the argument $-\mathrm{i}\Delta(\theta_s)\sqrt{z}$ of the complementary error function appearing in (5.3.5b) has phase given by

$$\arg[-\mathrm{i}\Delta(\theta_s)\sqrt{z}] = -\frac{\pi}{2}\ ; \qquad \operatorname{Im}\{\Delta(\theta_s)\} = 0\ , \qquad \Delta(\theta_s) \neq 0\ . \tag{5.3.22}$$

Hence, (5.3.18) is applicable to (5.3.5b). According to (5.3.18), the complementary error function appearing in (5.3.5b) satisfies

$$\operatorname{erfc}[-\mathrm{i}\Delta(\theta_s)\sqrt{z}] = 1 + \frac{1}{-\mathrm{i}\sqrt{\pi z}\Delta(\theta_s)} \mathrm{e}^{z[p(\omega_{sp},\theta_s)-p(\omega_c,\theta_s)]} + \mathcal{O}\{[\Delta(\theta_s)\sqrt{z}]^{-3}\}\ , \tag{5.3.23}$$

as $|\Delta(\theta_s)|\sqrt{z} \to \infty$ with $\arg[\Delta(\theta_s)\sqrt{z}] = 0$. Substitution of (5.3.23) into (5.3.5b) yields

$$I_{sp}(z,\theta_s) = q(\omega_{sp})\left(-\frac{2\pi}{zp^{(2)}(\omega_{sp},\theta_s)}\right)^{1/2} \mathrm{e}^{zp(\omega_{sp},\theta_s)} + \mathcal{O}\{[\Delta(\theta_s)\sqrt{z}]^{-3}\}\ , \tag{5.3.24}$$

as $|\Delta(\theta_s)|\sqrt{z} \to \infty$ with $\arg[\Delta(\theta_s)] = 0$. This equation (5.3.24) is the same result as that obtained by direct application of *Olver*'s method except that the dependence of the remainder term on the separation between the pole and the saddle point is displayed in (5.3.24) through the factor $\Delta(\theta_s)$.

To illustrate the application of Theorem 5.3, the example illustrated in Fig. 5.6 is now continued. Substitution of (5.3.5) into (5.3.3) yields

$$I(z,\theta) = q(\omega_{sp})\left(-\frac{2\pi}{zp^{(2)}(\omega_{sp},\theta)}\right)^{1/2} e^{zp(\omega_{sp},\theta)}$$
$$+ \gamma\left\{i\pi\,\mathrm{erfc}[-i\Delta(\theta)\sqrt{z}]e^{zp(\omega_c,\theta)} + \frac{1}{\Delta(\theta)}\sqrt{\frac{\pi}{z}}e^{zp(\omega_{sp},\theta)}\right\}$$
$$+ R_1 e^{zp(\omega_{sp},\theta)}\ ; \qquad \theta < \theta_s\ , \tag{5.3.25a}$$

$$I(z,\theta_s) = q(\omega_{sp})\left(-\frac{2\pi}{zp^{(2)}(\omega_{sp},\theta_s)}\right)^{1/2} e^{zp(\omega_{sp},\theta_s)}$$
$$+ \gamma\left\{i\pi\,\mathrm{erfc}[-i\Delta(\theta_s)\sqrt{z}]e^{zp(\omega_c,\theta_s)} + \frac{1}{\Delta(\theta_s)}\sqrt{\frac{\pi}{z}}e^{zp(\omega_{sp},\theta_s)}\right\}$$
$$- 2i\pi\gamma e^{zp(\omega_c,\theta_s)} + R_1 e^{zp(\omega_{sp},\theta_s)}\ ; \qquad \theta = \theta_s\ , \tag{5.3.25b}$$

$$I(z,\theta) = q(\omega_{sp})\left(-\frac{2\pi}{zp^{(2)}(\omega_{sp},\theta)}\right)^{1/2} e^{zp(\omega_{sp},\theta)}$$
$$+ \gamma\left\{i\pi\,\mathrm{erfc}[i\Delta(\theta)\sqrt{z}]e^{zp(\omega_c,\theta)} + \frac{1}{\Delta(\theta)}\sqrt{\frac{\pi}{z}}e^{zp(\omega_{sp},\theta)}\right\}$$
$$- 2i\pi\gamma e^{zp(\omega_c,\theta)} + R_1 e^{zp(\omega_{sp},\theta)}\ ; \qquad \theta > \theta_s\ , \tag{5.3.25c}$$

where $R_1 = \mathcal{O}\{[\Delta(\theta)\sqrt{z}]^{-3}\}$ as $|\Delta(\theta)|\sqrt{z} \to \infty$ uniformly with respect to θ for all $\theta \in \mathcal{R}$. Now, at $\theta = \theta_s$, $\arg[-i\Delta(\theta_s)\sqrt{z}] = -\pi/2$ for z real and positive. Consequently, the complementary error function appearing in (5.3.25b) may be replaced by (5.3.13), so that

$$I(z,\theta_s) = q(\omega_{sp})\left(-\frac{2\pi}{zp^{(2)}(\omega_{sp},\theta_s)}\right)^{1/2} e^{zp(\omega_{sp},\theta_s)}$$
$$+ \gamma\left\{-2\sqrt{\pi}F[\Delta(\theta_s)\sqrt{z}]e^{zp(\omega_{sp},\theta_s)} + \frac{1}{\Delta(\theta_s)}\sqrt{\frac{\pi}{z}}e^{zp(\omega_{sp},\theta_s)}\right\}$$
$$- i\pi\gamma e^{zp(\omega_c,\theta_s)} + R_1 e^{zp(\omega_{sp},\theta_s)}\ ; \qquad \theta = \theta_s\ , \tag{5.3.26}$$

where $F(\zeta)$ is Dawson's integral. Finally, at $\theta = \theta_c$, $\arg[i\Delta(\theta_c)\sqrt{z}] = \pi/4$ for z real and positive. Consequently, in this case the complementary error function appearing in (5.3.25c) may be replaced by (5.3.14), so that

$$I(z,\theta_c) = q(\omega_{sp})\left(-\frac{2\pi}{zp^{(2)}(\omega_{sp},\theta_c)}\right)^{1/2} e^{zp(\omega_{sp},\theta_c)}$$
$$+ \gamma\left\{i\sqrt{2}\pi e^{i(\pi/4)}\left[\mathscr{C}\left(\sqrt{\frac{2z}{\pi}}|\Delta(\theta_c)|\right) - \mathscr{S}\left(\sqrt{\frac{2z}{\pi}}|\Delta(\theta_c)|\right)\right]e^{zp(\omega_c,\theta_c)}\right.$$
$$\left. + \frac{1}{\Delta(\theta_c)}\sqrt{\frac{\pi}{z}}e^{zp(\omega_{sp},\theta_c)}\right\}$$
$$- 3i\pi\gamma e^{zp(\omega_c,\theta_c)} + R_1 e^{zp(\omega_{sp},\theta_c)}\ ; \qquad \theta = \theta_c\ , \tag{5.3.27}$$

where $\mathscr{C}(\zeta)$ and $\mathscr{S}(\zeta)$ are the cosine and sine Fresnel integrals, respectively. With these two results, the behavior of the integral $I(z,\theta)$ (and, in particular, the pole contribution) at the two critical values $\theta = \theta_s$ and $\theta = \theta_c$ is given as a function of the variable z in terms of well known real-valued functions.

When the pole is far enough away from the saddle point that (5.3.21, 24) can be applied in (5.3.3), then (5.3.25) becomes

$$I(z,\theta) = q(\omega_{sp})\left(-\frac{2\pi}{zp^{(2)}(\omega_{sp},\theta)}\right)^{1/2} e^{zp(\omega_{sp},\theta)} + R_1 e^{zp(\omega_{sp},\theta)} \ ; \qquad \theta < \theta_s \ , \tag{5.3.28a}$$

$$I(z,\theta_s) = q(\omega_{sp})\left(-\frac{2\pi}{zp^{(2)}(\omega_{sp},\theta_s)}\right)^{1/2} e^{zp(\omega_{sp},\theta_s)} - i\pi\gamma e^{zp(\omega_c,\theta_s)} + R_1 e^{zp(\omega_{sp},\theta_s)} \ ; \qquad \theta = \theta_s \ , \tag{5.3.28b}$$

$$I(z,\theta) = q(\omega_{sp})\left(-\frac{2\pi}{zp^{(2)}(\omega_{sp},\theta)}\right)^{1/2} e^{zp(\omega_{sp},\theta)} - 2i\pi\gamma e^{zp(\omega_c,\theta)} + R_1 e^{zp(\omega_{sp},\theta)} \ ; \qquad \theta > \theta_s \ , \tag{5.3.28c}$$

where $R_1 = \mathcal{O}\{[\Delta(\theta)\sqrt{z}]^{-3}\}$ as $|\Delta(\theta)|\sqrt{z} \to \infty$ uniformly with respect to θ for all $\theta \in \mathscr{R}$. These results are the same as obtained when *Olver*'s method is applied directly, except that the dependence of the remainder term on the separation between the pole and the saddle point is displayed explicitly through the factor $\Delta(\theta)$ in (5.3.28). They yield a continuous asymptotic approximation of $I(z,\theta)$ for fixed large z as θ varies over $\mathscr{R}$ as long as the quantity $z|p(\omega_{sp},\theta_s) - p(\omega_c,\theta_s)|$ is large enough so that the discontinuity at $\theta = \theta_s$ is negligible. When $z|p(\omega_{sp},\theta_s) - p(\omega_c,\theta_s)|$ is small enough that the discontinuity is significant, then (5.3.25) must be employed.

Finally, consider the case in which there is a simple pole singularity of the function $q(\omega)$ at $\omega = \omega_c$ and there are two relevant saddle points of the complex phase function $p(\omega,\theta)$ which remain isolated from each other over the entire range $\mathscr{R}$ of values of θ. As in Sect. 5.2.1, the Olver-type path $P(\theta)$ can then be divided into two parts $P_1(\theta)$ and $P_2(\theta)$ such that $P(\theta) = P_1(\theta) + P_2(\theta)$, where (for $i = 1, 2$) $P_i(\theta)$ passes through the saddle point $\omega_i(\theta)$ and is an Olver-type path with respect to the saddle point ω_i. The integral $I_{sp}(z,\theta)$ over the contour $P(\theta)$ can then be expressed as the sum of the the integrals $I_{sp_1}(z,\theta)$ and $I_{sp_2}(z,\theta)$ over the paths $P_1(\theta)$ and $P_2(\theta)$, respectively, as in (5.2.1). The results of Theorem 5.3 are then applicable to whichever one of the contour integrals taken over the path $P_i(\theta)$ crosses the simple pole singularity at $\omega = \omega_c$ when $\theta = \theta_s$. For example, let $P_1(\theta)$ cross the simple pole singularity. Then the uniform asymptotic behavior of $I_{sp_1}(z,\theta)$ is given by Theorem 5.3, whereas $I_{sp_2}(z,\theta)$ is given by direct application of *Olver*'s method. The value of θ_s is then determined by the value of θ at which the path of steepest descent through the saddle point $\omega_1(\theta)$ crosses the pole singularity. However, the value of θ_c is given by the value of θ at which $\mathrm{Re}\{p(\omega_c,\theta)\}$ is greater than both $\mathrm{Re}\{p(\omega_1,\theta)\}$ and $\mathrm{Re}\{p(\omega_2,\theta)\}$. Finally, if the two saddle points remain well removed from the pole singularity for all $\theta \in \mathscr{R}$, the absolute value of the parameter $\Delta(\theta)$ remains large for all $\theta \in \mathscr{R}$, so that the complementary error function appearing in the asymptotic expansion may be replaced by its individual asymptotic expansion, as in (5.3.28). In this case, the results are the same as are obtained by direct application of *Olver*'s method.

5.4 The Uniform Asymptotic Expansion for Two First-Order Saddle Points at Infinity

This section presents an asymptotic expression for certain integrals of the form given in (5.1.1) that is valid uniformly as the relevant saddle points tend towards infinity in the complex ω-plane. Let the complex phase function $p(\omega,\theta)$ appearing in (5.1.1) have two first-order saddle points $\omega_\pm(\theta)$ with equal imaginary parts and with real parts that approach $\pm\infty$, respectively, as θ approaches unity from above. Let the contour of integration P' be deformable to a continuous path $P(\theta) = P^+(\theta) + P^-(\theta)$ where $P^\pm(\theta)$ is a path of steepest descent through $\omega_\pm(\theta)$ with one endpoint satisfying $\mathrm{Re}\{\omega\} = \pm\infty$. Furthermore, let the function $\psi(\omega,\theta)$ defined by

$$\psi(\omega,\theta) \equiv -ip(\omega,\theta) \tag{5.4.1}$$

have a Laurent series expansion of the form

$$\psi(\omega,\theta) = \omega(1-\theta) + \sum_{n=0}^{\infty} a_n(\theta)\omega^{-n} \tag{5.4.2}$$

for all ω such that $|\omega| \geqslant R_1$ and for all θ in the closed interval $[1,\theta']$, where R_1 is a finite positive constant and $\theta' > 1$ is a positive constant. All other saddle points of $p(\omega,\theta)$ (if any) are finite in number and are confined to some bounded region in the complex ω-plane such that $|\omega| \leqslant R_2 < R_1$ for all $\theta \in [1,\theta']$. Moreover, the amplitude function $q(\omega)$ may be written in the form

$$q(\omega) = \omega^{-(1+\nu)}\tilde{q}(\omega)\ , \tag{5.4.3}$$

for large $|\omega|$ with real $\nu > 0$, where the function $\tilde{q}(\omega)$ has a Laurent series expansion which is convergent for $|\omega| \geqslant R_1$ and is such that $\lim_{|\omega|\to\infty}[\tilde{q}(\omega)] \neq 0$. When $\nu < 0$, the uniform asymptotic approximation presented in Theorem 5.4 is still applicable for all values of $\theta \in [1,\theta']$ provided that its limiting value as θ approaches unity is finite [5.8].

Since the two saddle points $\omega_\pm(\theta)$ are located at infinity for $\theta = 1$, condition (b) of Theorem 5.1 is not satisfied and *Olver*'s method becomes inapplicable in this limit. An asymptotic approximation of the integral (5.1.1) is desired, therefore, which is uniform in the parameter θ as θ approaches the critical value of unity. For simplicity, only the dominant term in the asymptotic expansion is considered. This uniform asymptotic approximation is given by the following theorem due to *Handelsman* and *Bleistein* [5.8].

Theorem 5.4: *In the integrand of the contour integral*

$$I(z,\theta) = \int_{P(\theta)} q(\omega) e^{zp(\omega,\theta)} d\omega$$

with real z, let the function $p(\omega,\theta)$ possess two first-order saddle points $\omega_\pm(\theta)$ with equal imaginary parts and whose real parts approach $\pm\infty$, respectively, as the parameter θ approaches unity. Let the contour $P(\theta) = P^+(\theta) + P^-(\theta)$ be a continuous function of θ, where $P^\pm(\theta)$ is a path of steepest descent through $\omega_\pm(\theta)$ with one endpoint satisfying $\mathrm{Re}\{\omega\} = \pm\infty$, and let the function $p(\omega,\theta)$ satisfy (5.4.1). *Then, subject to the conditions stated in connection with* (5.4.2, 3), *the integral $I(z,\theta)$ satisfies*

$$\begin{aligned} I(z,\theta) = &-2\pi i e^{-iz\beta(\theta)}[2\alpha(\theta)e^{-i(\pi/2)}]^\nu \\ &\cdot[\gamma_0 J_\nu(\alpha(\theta)z) + 2\alpha(\theta)e^{-i(\pi/2)}\gamma_1 J_{\nu+1}(\alpha(\theta)z)] \\ &+ R(z,\theta)\ , \end{aligned} \tag{5.4.4}$$

where (with K a positive real constant independent of θ and z)

$$|R(z,\theta)| \leqslant K\frac{2^{\nu+1}|\alpha(\theta)|^{\nu+1}}{z}[|J_{\nu+1}(\alpha(\theta)z)| + |J_{\nu+2}(\alpha(\theta)z)|] \tag{5.4.5}$$

for $z \geqslant Z > 0$ and $\theta \in [1,\theta']$. This error term is small for large z independent of $\alpha(\theta)$. The coefficients appearing in (5.4.4, 5) *are given by*

$$\alpha(\theta) = -\frac{\psi(\omega_+,\theta) - \psi(\omega_-,\theta)}{2}\ , \tag{5.4.6}$$

$$\beta(\theta) = -\frac{\psi(\omega_+,\theta) + \psi(\omega_-,\theta)}{2}\ , \tag{5.4.7}$$

$$\begin{aligned} \gamma_0(\theta) = \frac{1}{2}\Bigg[&q(\omega_+)\left(\frac{1}{2\alpha(\theta)}\right)^{1+\nu}\left(-\frac{4\alpha^3(\theta)}{\psi^{(2)}(\omega_+,\theta)}\right)^{1/2} \\ &+ q(\omega_-)\left(-\frac{1}{2\alpha(\theta)}\right)^{1+\nu}\left(\frac{4\alpha^3(\theta)}{\psi^{(2)}(\omega_-,\theta)}\right)^{1/2}\Bigg]\ , \end{aligned} \tag{5.4.8}$$

$$\begin{aligned} \gamma_1(\theta) = \frac{1}{4\alpha(\theta)}\Bigg\{ &q(\omega_+)\left(\frac{1}{2\alpha(\theta)}\right)^{1+\nu}\left(-\frac{4\alpha^3(\theta)}{\psi^{(2)}(\omega_+,\theta)}\right)^{1/2} \\ &- q(\omega_-)\left(-\frac{1}{2\alpha(\theta)}\right)^{1+\nu}\left(\frac{4\alpha^3(\theta)}{\psi^{(2)}(\omega_-,\theta)}\right)^{1/2}\Bigg\}\ . \end{aligned} \tag{5.4.9}$$

The argument of the quantity $[\mp\psi^{(2)}(\omega_\pm,\theta)]$ *is chosen so as to satisfy the inequality* (5.1.7) *with* $\Theta = \arg(\mathrm{i}z) = \pi/2$ *for z real and positive and* $\bar{\alpha}^+$ *being the angle of slope of the contour* $P^+(\theta)$ *leading away from the saddle point* ω_+ *in the limit as* $\theta \to 1$. *For values of* θ *close to unity, these coefficients reduce to*

$$\alpha(\theta) = -2\sqrt{-a_1(\theta)(\theta-1)} + o(\sqrt{\theta-1}) \ , \tag{5.4.10}$$

$$\beta(\theta) = a_0(\theta) + o(\sqrt{\theta-1}) \ , \tag{5.4.11}$$

and the coefficients $\gamma_0(\theta)$ *and* $\gamma_1(\theta)$ *are both* $\mathcal{O}(1)$ *for* $\theta \in [1,\theta']$. *The proper branch of* $\alpha^{1/2}(\theta)$ *is chosen so as to satisfy the relation*

$$\left(\mp\frac{4\alpha^3(\theta)}{\psi^{(2)}(\omega_\pm,\theta)}\right)^{1/2} = \sqrt{2}a_1(\theta)[1+\mathcal{O}(\sqrt{\theta-1})] \tag{5.4.12}$$

as $\theta \to 1$ *for all* $\theta \in [1,\theta']$.

In order to illustrate the application of Theorem 5.4 to obtain the uniform asymptotic expansion of the integral (5.1.1), consider the case in which $v = 0$ and the complex-valued phase function $p(\omega,\theta) \equiv X(\omega,\theta) + \mathrm{i}Y(\omega,\theta)$ satisfies $p(-\omega,\theta) = p^*(\omega^*,\theta)$, so that with $\omega = \omega' + \mathrm{i}\omega''$

$$X(\omega'+\mathrm{i}\omega'',\theta) = X(-\omega'+\mathrm{i}\omega'',\theta) \ , \tag{5.4.13a}$$

$$Y(\omega'+\mathrm{i}\omega'',\theta) = -Y(-\omega'+\mathrm{i}\omega'',\theta) \ . \tag{5.4.13b}$$

The saddle points of $p(\omega,\theta)$ are then symmetrically located in the complex ω-plane about the imaginary axis. In particular, let $p(\omega,\theta)$ posses two first-order saddle points $\omega_\pm(\theta)$ which tend to $\pm\infty + \mathrm{i}\omega''$ as θ approaches unity from above. The phase function $\psi(\omega,\theta)$, defined by (5.4.1), is then given by

$$\psi(\omega,\theta) = Y(\omega,\theta) - \mathrm{i}X(\omega,\theta) \ , \tag{5.4.14}$$

and the coefficients appearing in the expansion (5.4.4) are

$$\alpha(\theta) = -Y(\omega_+,\theta) \ , \tag{5.4.15}$$

$$\beta(\theta) = \mathrm{i}X(\omega_+,\theta) \ , \tag{5.4.16}$$

$$\gamma_0(\theta) = \frac{1}{2}Y^{1/2}(\omega_+,\theta)\left[-q(\omega_+)\left(\frac{1}{\mathrm{i}p^{(2)}(\omega_+,\theta)}\right)^{1/2} + q(\omega_-)\left(-\frac{1}{\mathrm{i}p^{(2)}(\omega_-,\theta)}\right)^{1/2}\right] , \tag{5.4.17}$$

$$\gamma_1(\theta) = \frac{1}{4Y^{1/2}(\omega_+,\theta)}\left[-q(\omega_+)\left(\frac{1}{\mathrm{i}p^{(2)}(\omega_+,\theta)}\right)^{1/2} - q(\omega_-)\left(-\frac{1}{\mathrm{i}p^{(2)}(\omega_-,\theta)}\right)^{1/2}\right] . \tag{5.4.18}$$

Upon substitution of these expressions into the uniform asymptotic approximation (5.4.4) there results

$$
\begin{aligned}
I(z,\theta) = -2\pi \mathrm{i} e^{zX(\omega_+,\theta)} \Bigg\{ & \frac{1}{2} Y^{1/2}(\omega_+,\theta) \left[-q(\omega_+) \left(\frac{1}{\mathrm{i}p^{(2)}(\omega_+,\theta)} \right)^{1/2} \right. \\
& \left. + q(\omega_-) \left(-\frac{1}{\mathrm{i}p^{(2)}(\omega_-,\theta)} \right)^{1/2} \right] \mathrm{J}_0[-Y(\omega_+,\theta)z] \\
& + \frac{1}{2} e^{-\mathrm{i}(\pi/2)} Y^{1/2}(\omega_-,\theta) \left[-q(\omega_+) \left(\frac{1}{\mathrm{i}p^{(2)}(\omega_+,\theta)} \right)^{1/2} \right. \\
& \left. - q(\omega_-,\theta) \left(-\frac{1}{\mathrm{i}p^{(2)}(\omega_-,\theta)} \right)^{1/2} \right] \mathrm{J}_1[-Y(\omega_+,\theta)z] \Bigg\} \\
+ R_1(z,\theta) \ , \qquad & \qquad (5.4.19)
\end{aligned}
$$

which is uniformly valid in the parameter θ as θ approaches the critical value of unity. At $\theta = 1$, $Y(\omega_+, 1) = 0$, see (5.4.15 and 10), and this result identically vanishes.

For large values of the argument $|Y(\omega_+,\theta)|z$ (i.e., for values of θ bounded away from unity), the Bessel functions appearing in (5.4.19) may be replaced by their asymptotic approximations given by

$$
\mathrm{J}_\nu(\zeta) = \sqrt{\frac{2}{\pi\zeta}} \cos(\zeta - \tfrac{1}{2}\nu\pi - \tfrac{1}{4}\pi) + \mathcal{O}(\zeta^{-1}) e^{|\mathrm{Im}(\zeta)|}
$$

for $|\arg(\zeta)| < \pi$. In this limit, with $|\arg(-Y(\omega_+,\theta)z)| < \pi$, the uniform expansion given in (5.4.19) becomes

$$
\begin{aligned}
I(z,\theta) = \sqrt{\frac{2\pi}{z}} e^{zX(\omega_+,\theta)} \Bigg\{ & \left[q(\omega_+) \left(-\frac{1}{p^{(2)}(\omega_+,\theta)} \right)^{1/2} e^{-\mathrm{i}(\pi/4)} \right. \\
& \left. + q(\omega_-) \left(-\frac{1}{p^{(2)}(\omega_-,\theta)} \right)^{1/2} e^{\mathrm{i}(\pi/4)} \right] \cos\left[-Y(\omega_+,\theta)z - \frac{\pi}{4} \right] \\
& - \mathrm{i} \left[q(\omega_+) \left(-\frac{1}{p^{(2)}(\omega_+,\theta)} \right)^{1/2} e^{-\mathrm{i}(\pi/4)} \right. \\
& \left. - q(\omega_-) \left(-\frac{1}{p^{(2)}(\omega_-,\theta)} \right)^{1/2} e^{\mathrm{i}(\pi/4)} \right] \sin\left[-Y(\omega_+,\theta)z - \frac{\pi}{4} \right] \Bigg\} \\
& + \mathcal{O}[(Y(\omega_+,\theta)z)^{-1}] \\
= q(\omega_+) & \left(-\frac{2\pi}{zp^{(2)}(\omega_+,\theta)} \right)^{1/2} e^{zp(\omega_+,\theta)} \\
& + q(\omega_-) \left(-\frac{2\pi}{zp^{(2)}(\omega_-,\theta)} \right)^{1/2} e^{zp(\omega_-,\theta)} + \mathcal{O}[(Y(\omega_+,\theta)z)^{-1}] \qquad (5.4.20)
\end{aligned}
$$

as $|Y(\omega_+,\theta)z| \to \infty$. This is the same result as that obtained by application of *Olver*'s method except that the dependence of the remainder term on the location of ω_+ is displayed in (5.4.20) through the factor $Y(\omega_+,\theta)$. Since it follows from (5.4.10, 15) that $Y(\omega_+,\theta) \to 0$ as $\theta \to 1$, the estimate of the remainder term in (5.4.20) is not useful when $\mathbb{R}\mathrm{e}\{\omega_+\} \to \infty$ for fixed z.

6 Analysis of the Phase Function and Its Saddle Points

In preparation for the asymptotic analysis of the integral representation given in either equation (4.3.31) or (4.3.28) for the propagated field in a temporally dispersive medium, it is necessary to first determine the topography of the real part $X(\omega, \theta)$ of the complex phase function $\phi(\omega, \theta)$, defined in (4.37), in the complex ω-plane. In particular, the location of the saddle points of $\phi(\omega, \theta)$, the value of $\phi(\omega, \theta)$ at these points, and the regions of the complex ω-plane wherein $X(\omega, \theta)$ is less than the value of $X(\omega, \theta)$ at the dominant saddle point for a given value of θ are all required. This chapter is devoted to that purpose. The analysis presented here is involved because the behavior of the complex-valued phase function is complicated. Specifically, for a Lorentz medium with a single-resonance frequency, $\phi(\omega, \theta)$ possesses four saddle points and four branch points, and its topography evolves with the space-time parameter $\theta = ct/z$ in a complicated manner.

By studying the structure of $\phi(\omega, \theta)$ in special regions of the complex ω-plane where its behavior is relatively simple (e.g., along the real and imaginary axes, in the vicinities of the branch points and the origin, and for large values of $|\omega|$), *Brillouin* [6.1] was able to determine a rough picture of the topography of $X(\omega)$. He could not provide that topography with much detail, however, because it was necessary to resort to numerical methods without the aid of modern electronic computers. This chapter begins with the presentation of accurate, detailed, computer-generated contour plots of the topography of $X(\omega, \theta)$ for several different values of the parameter θ for the same medium as was investigated by *Brillouin*. These results show that the most important features are essentially the same as was originally presented by *Brillouin*. These computer-generated contour plots are used to determine acceptable deformed contours of integration that can be used in applying both Olver's saddle point method and the uniform asymptotic techniques to obtain the required asymptotic approximations. A discussion of how the evolution of the topography of $X(\omega, \theta)$ with θ affects this asymptotic analysis and a comparison with *Brillouin's* approach is also given.

The next step in this analysis is to obtain analytic approximations for the locations of the relevant saddle points and the behavior of the complex phase at the saddle points. The approximations obtained by *Brillouin* are crude and break down for much of the time domain of interest. Most of this chapter is devoted to developing higher-order approximations of this important behavior. These higher-order approximations are then used to determine analytically which relevant saddle point is the dominant one at any given time.

During the course of the analysis presented in this chapter, approximations are occasionally made without complete justifications. The justifications are provided in the final section which presents a numerical evaluation of the saddle-point locations and the behavior of the complex phase function $\phi(\omega, \theta)$ at these locations based on both the exact expressions and on the approximate expressions. These numerical results show that the new, higher-order approximations give excellent results, (much improved over *Brillouin's* results), over the entire time range of interest, resulting in much more accurate results concerning the dynamic evolution of the precursor fields and the time of arrival of the main signal.

6.1 The Behavior of the Phase in the Complex ω-Plane

The behavior of the complex phase function to be considered here is given by (4.3.7), e.g.

$$\phi(\omega, \theta) = i\omega[n(\omega) - \theta] \ , \tag{6.1.1}$$

where $\theta = ct/z$. This behavior is dictated by the analytic form of the complex index of refraction $n(\omega)$, which is taken in this research to be given by the Lorentz model for a homogeneous, isotropic, locally linear, temporally dispersive medium characterized by a single resonance frequency ω_0, viz. (Sect. 2.4)

$$n(\omega) = \left(1 - \frac{b^2}{\omega^2 - \omega_0^2 + 2\delta i\omega}\right)^{1/2} \ . \tag{6.1.2}$$

Here, $b^2 = ([\![4\pi]\!]/\varepsilon_0)(Ne^2/m)$ is the square of the plasma frequency of the medium, N is the number density of electrons of charge e and mass m bound with the resonant frequency ω_0, and δ is the associated phenomenological damping constant of this optical electron (Sect. 2.4). The values of these parameters as chosen by *Brillouin* in his analysis for a medium possessing a single ultraviolet resonance frequency are

$$\omega_0 = 4.0 \times 10^{16}/\mathrm{s} \ ,$$

$$b^2 = 20.0 \times 10^{32}/\mathrm{s}^2 \ ,$$

$$\delta = 0.28 \times 10^{16}/\mathrm{s} \ .$$

As will be seen, this choice of the medium parameters corresponds to an extremely absorptive medium, but it is used in many of the numerical examples presented in this research to facilitate a comparison with *Brillouin's* results. A more realistic set of parameters describing a less absorptive medium which is also used in some of the examples presented in this research is

$$\omega_0 = 2.0 \times 10^{16}/\mathrm{s} \ ,$$

$$b^2 = 0.4 \times 10^{32}/\mathrm{s}^2 \ ,$$

$$\delta = 0.00017 \times 10^{16}/\mathrm{s} \ .$$

That both of these sets of medium parameters possess resonance frequencies in the near ultraviolet region of the spectrum is just a matter of choice and is not indicative of any limitations of the analysis presented here. The analysis and asymptotic expansions presented in this research are equally applicable to media which possess resonance frequencies in the visible and infrared regions of the electromagnetic spectrum.

It is important in the present analysis to accurately describe the behavior of the complex phase function $\phi(\omega, \theta)$ in the complex ω-plane. This analysis is simplified by the following result; since, by (4.3.15)

$$\phi(-\omega, \theta) = \phi^*(\omega^*, \theta) \ ,$$

it then follows directly that the real part $X(\omega, \theta)$ of the complex phase function $\phi(\omega, \theta)$ satisfies

$$X(-\omega' + i\omega'', \theta) = X(\omega' + i\omega'', \theta) \ , \tag{6.1.3}$$

whereas the imaginary part $Y(\omega, \theta)$ of $\phi(\omega, \theta)$ satisfies

$$Y(-\omega' + i\omega'', \theta) = -Y(\omega' + i\omega'', \theta) \ , \tag{6.1.4}$$

where $\omega' \equiv \mathrm{Re}\{\omega\}$ and $\omega'' \equiv \mathrm{Im}\{\omega\}$ are the real and imaginary parts of the complex variable ω, respectively. Consequently, $X(\omega, \theta) \equiv \mathrm{Re}\{\phi(\omega, \theta)\}$ is symmetric about the imaginary axis whereas $Y(\omega, \theta) \equiv \mathrm{Im}\{\phi(\omega, \theta)\}$ is antisymmetric about the imaginary axis. Hence, one need only examine the right-half of the complex ω-plane in regards to the behavior of the real and imaginary parts of the complex phase function $\phi(\omega, \theta)$.

The branch points for $n(\omega)$, and hence, for $\phi(\omega, \theta)$, can be directly determined by rewriting the complex index of refraction as

$$n(\omega) = \left(\frac{\omega^2 - \omega_1^2 + 2\delta i\omega}{\omega^2 - \omega_0^2 + 2\delta i\omega}\right)^{1/2} = \left(\frac{(\omega - \omega'_+)(\omega - \omega'_-)}{(\omega - \omega_+)(\omega - \omega_-)}\right)^{1/2} \ , \tag{6.1.5}$$

where

$$\omega_1^2 \equiv \omega_0^2 + b^2 \ . \tag{6.1.6}$$

The branch point locations are then given by

$$\omega'_\pm = \pm\sqrt{\omega_1^2 - \delta^2} - \delta i \ , \tag{6.1.7a}$$

$$\omega_\pm = \pm\sqrt{\omega_0^2 - \delta^2} - \delta i \ , \tag{6.1.7b}$$

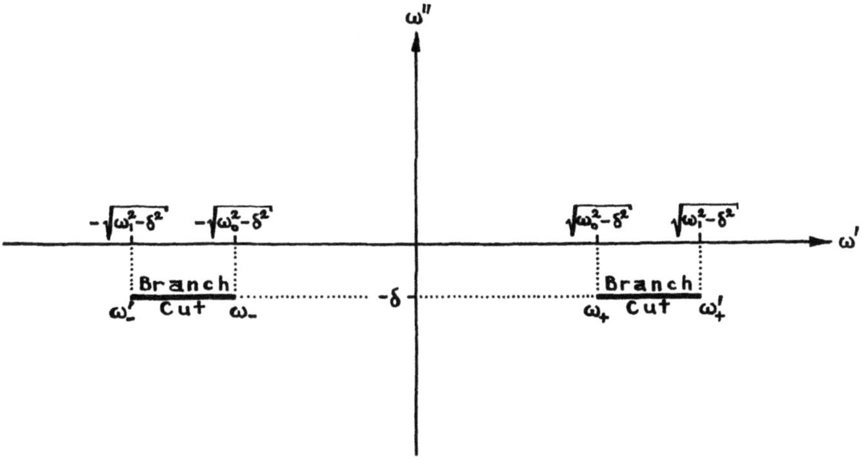

Fig. 6.1. Branch points and branch cuts for the single resonance Lorentz model dielectric.

and lie along the line $\omega'' = -\delta$ symmetrically located about the imaginary ω''-axis (it is assumed here that $\omega_0 > \delta$). The branch cuts chosen here are the line segments $\omega'_-\omega_-$ and $\omega_+\omega'_+$, as shown in Fig. 6.1. The complex index of refraction $n(\omega)$ and the complex phase function $\phi(\omega, \theta)$ are then analytic in the complex ω-plane, except, of course, at the branch points $\omega'_\pm$ and $\omega_\pm$.

6.1.1 Brillouin's Analysis

Some general aspects of the behavior of $X(\omega, \theta)$ in specific regions of the complex ω-plane, as discussed originally by *Brillouin* [6.1, 2], are reviewed in this section for background material. New results with much more detail are presented in the next subsection. According to the results of Sect. 2.3, the complex index of refraction may be written as

$$n(\omega) = n_r(\omega) + in_i(\omega) ,$$

where $n_r(\omega)$ is the real index of refraction of the medium and $n_i(\omega)$ is related to the coefficient of absorption of the medium; both $n_r(\omega)$ and $n_i(\omega)$ are real-valued functions of ω. The phase function $\phi(\omega, \theta)$ may then be written as

$$\begin{aligned}\phi(\omega, \theta) &= \mathrm{i}(\omega' + \mathrm{i}\omega'')[n_r(\omega) - \theta + \mathrm{i}n_i(\omega)] \\ &= -\{\omega''[n_r(\omega) - \theta] + \omega' n_i(\omega)\} \\ &\quad + \mathrm{i}\{\omega'[n_r(\omega) - \theta] - \omega'' n_i(\omega)\} ,\end{aligned} \tag{6.1.8}$$

so that

$$X(\omega, \theta) \equiv \mathrm{Re}\{\phi(\omega, \theta)\} = -\{\omega''[n_r(\omega) - \theta] + \omega' n_i(\omega)\} , \tag{6.1.9}$$

and

$$Y(\omega,\theta) \equiv \mathrm{Im}\{\phi(\omega,\theta)\} = \omega'[n_r(\omega) - \theta] - \omega'' n_i(\omega) \ . \tag{6.1.10}$$

The real and imaginary parts of the complex index of refraction must now be considered. From (6.1.2), it follows that

$$n^2(\omega) = \frac{\omega^2 - \omega_1^2 + 2\delta i\omega}{\omega^2 - \omega_0^2 + 2\delta i\omega}$$
$$= \frac{[(\omega'^2-\omega''^2-\omega_0^2-2\delta\omega'')^2 - b^2(\omega'^2-\omega''^2-\omega_0^2-2\delta\omega'') + 4\omega'^2(\omega''+\delta)^2 + 2i\omega'(\omega''+\delta)b^2]}{(\omega'^2-\omega''^2-\omega_0^2-2\delta\omega'')^2 + 4\omega'^2(\omega''+\delta)^2} \ . \tag{6.1.11}$$

The magnitude and phase of this complex-valued expression are given by

$$|n^2(\omega)|^2 = 1 + \frac{b^4 - 2b^2(\omega'^2 - \omega''^2 - \omega_0^2 - 2\delta\omega'')}{(\omega'^2 - \omega''^2 - \omega_0^2 - 2\delta\omega'')^2 + 4\omega'^2(\omega''+\delta)^2} \ , \tag{6.1.12}$$

$$\zeta(\omega) \equiv \arg[n^2(\omega)]$$
$$= \tan^{-1}\left(\frac{2\omega'(\omega''+\delta)b^2}{(\omega'^2 - \omega''^2 - \omega_0^2 - 2\delta\omega'')^2 - b^2(\omega'^2 - \omega''^2 - \omega_0^2 - 2\delta\omega'') + 4\omega'^2(\omega''+\delta)^2}\right) \ , \tag{6.1.13}$$

where the principal branch is to be chosen in this last expression, viz. $-\pi \leqslant \zeta < \pi$. As a consequence, the complex index of refraction may be written in the phasor form

$$n(\omega) = \left(1 + \frac{b^4 - 2b^2(\omega'^2 - \omega''^2 - \omega_0^2 - 2\delta\omega'')}{(\omega'^2 - \omega''^2 - \omega_0^2 - 2\delta\omega'')^2 + 4\omega'^2(\omega''+\delta)^2}\right)^{1/4} e^{i(\zeta/2)} \ . \tag{6.1.14}$$

With these results, the analytic behavior of $X(\omega, \theta)$ may now be examined analytically in specific regions of the complex ω-plane and numerically throughout the entire complex plane.

Along the real axis, $\omega'' = 0$ and the real and imaginary parts of the complex index of refraction are then given by the two expressions

$$n_r(\omega') = \left(1 + \frac{b^4 + 2b^2(\omega_0^2 - \omega'^2)}{(\omega'^2 - \omega_0^2)^2 + 4\delta^2\omega'^2}\right)^{1/4} \cos\left[\frac{1}{2}\zeta(\omega')\right] \ , \tag{6.1.15}$$

and

$$n_i(\omega') = \left(1 + \frac{b^4 + 2b^2(\omega_0^2 - \omega'^2)}{(\omega'^2 - \omega_0^2)^2 + 4\delta^2\omega'^2}\right)^{1/4} \sin\left[\frac{1}{2}\zeta(\omega')\right] \ , \tag{6.1.16}$$

respectively. The spectral regions wherein the real index of refraction $n_r(\omega')$ increases with ω' (i.e., are of positive slope) are termed *normally dispersive*, while the region wherein $n_r(\omega')$ decreases with increasing ω' (i.e., are of negative slope) is said to exhibit *anomolous dispersion.* The real index of refraction $n_r(\omega')$ varies rapidly within the region of anomolous dispersion, and this region essentially coincides with the region of strong absorption of the medium. According to (6.1.9), the behavior of $X(\omega')$ along the real ω'-axis is given by

$$X(\omega') = -\omega' n_i(\omega') \,. \tag{6.1.17}$$

Since $n_i(\omega') \geqslant 0$ for $\omega' \geqslant 0$ and $n_i(\omega') \leqslant 0$ for $\omega' \leqslant 0$, it then follows that $X(\omega') \leqslant 0$ for all values of ω'. Note that $X(\omega')$ is independent of the parameter θ, and hence is independent of both z and t. The absolute value of $X(\omega')$ is large in the region of anomolous dispersion about the characteristic frequency ω_0 of the medium, is vanishingly small for very small ($|\omega'| \ll \omega_0$) or very large ($|\omega'| \gg \omega_0$) absolute frequencies $|\omega'|$, and vanishes identically at the origin and at infinity.

Consider now determining the approximate value of $\omega' = \omega_{\min}$ along the real axis at which $X(\omega')$ attains its minimum value. As is shown in Chap. 9, the signal velocity attains a minimum value for the applied signal frequency $\omega_c = \omega_{\min}$. Since $X(\omega')$ is an even function along the real axis, it is necessary to consider only the behavior exhibited along the positive real axis. For δ small in comparison to the other medium parameters it is seen that the minimum in $X(\omega')$ is attained very near to the resonance frequency ω_0 of the medium. Consequently, for ω' close to ω_0 the approximation

$$\zeta(\omega') \cong \tan^{-1}\left(\frac{b^2}{2\delta\omega'}\right) = \frac{\pi}{2} - \frac{2\delta\omega'}{b^2} + \frac{8\delta\omega'^3}{3b^6} - \cdots \tag{6.1.18}$$

may be employed, from which it is seen that $\zeta(\omega')$ varies comparatively slowly in the region about $\omega' = \omega_0$. The approximate behavior of $X(\omega')$ in the vicinity of ω_0 may then be written

$$X(\omega') \cong -\omega'\left(1 + \frac{b^4 - 4b^2\omega_0(\omega' - \omega_0)}{4\omega_0^2(\omega' - \omega_0)^2 + 4\delta^2\omega_0^2}\right)^{1/4} \sin\left[\frac{1}{2}\zeta(\omega_0)\right] . \tag{6.1.19}$$

Differentiating this approximate expression with respect to ω' and setting the result equal to zero, there results

$$\begin{aligned}
&16[\omega_0^2(\omega' - \omega_0)^2 + \delta^2\omega_0^2]^2 \\
&+ 4b^2[b^2 - 4\omega_0(\omega' - \omega_0)][\omega_0^2(\omega' - \omega_0)^2 + \delta^2\omega_0^2] \\
&+ b^2\omega'[4\omega_0^3(\omega' - \omega_0)^2 - 2b^2\omega_0^2(\omega' - \omega_0) - 4\delta^2\omega_0^3] \\
&= 0
\end{aligned}$$

at $\omega' = \omega_{\min}$. In order to solve this quartic equation in ω' for the root $\omega_{\min}$, let $\omega' = \omega_{\min} = \omega_0 + \varepsilon$, where ε is assumed small. Then retaining only terms of order ε and lower, the following approximate solution is obtained

$$\omega_{\min} \cong \omega_0 \left[1 + \frac{2\delta^2}{\omega_0^2} \left(1 - \frac{\omega_0^2}{b^2} \right) \right] . \tag{6.1.20}$$

In order to obtain a more accurate approximation of $\omega_{\min}$ than that given above, the dependency of the phase angle ζ on ω', c.f. (6.1.18), must be taken into account. However, for the present research, (6.1.20) is of sufficient accuracy.

Consider next the behavior of $X(\omega, \theta)$ in the limit as ω approaches infinity in an arbitrary direction. According to (6.1.12, 13), at $|\omega| = \infty$, $n(\omega)$ is real and equal to unity, viz

$$\lim_{|\omega| \to \infty} [n(\omega)] = 1 . \tag{6.1.21}$$

Therefore, by equation (6.1.9) there results

$$\lim_{|\omega| \to \infty} [X(\omega, \theta)] = \omega''(\theta - 1) , \tag{6.1.22}$$

and the following behavior for $X(\omega, \theta)$ at $|\omega| = \infty$ is obtained: For $\theta < 1$, $X(\omega, \theta)$ is equal to $-\infty$ in the upper-half of the complex ω-plane, zero at the real ω'-axis, and $+\infty$ in the lower-half plane; at $\theta = 1$, $X(\omega, 1)$ is equal to zero everywhere at $|\omega| = \infty$; and for $\theta > 1$, $X(\omega, \theta)$ is equal to $+\infty$ in the upper-half plane, zero at the real ω'-axis, and $-\infty$ in the lower half of the complex ω-plane.

Consider next the behavior of $X(\omega, \theta)$ along the straight line $\omega'' = -\delta$, along which lie the branch points $\omega'_{\pm}$ [where $n(\omega)$ vanishes] and $\omega_{\pm}$ [where $n(\omega)$ becomes infinite], and also the behavior of $X(\omega, \theta)$ in the neighborhood of the branch cuts $\omega'_{-}\omega_{-}$ and $\omega_{+}\omega'_{+}$ [the coordinates of these points are given in (6.1.7a, b)]. Along the straight line $\omega = \omega' - i\delta$, the complex index of refraction is given by

$$n(\omega' - i\delta) = \left(1 + \frac{b^2}{\omega_0^2 - \delta^2 - \omega'^2} \right)^{1/2} . \tag{6.1.23}$$

Consequently, in the right half of complex ω-plane (and, symmetrically, in the left half of the ω-plane), $n(\omega' - i\delta)$ is real when $\omega' \geqslant \sqrt{\omega_1^2 - \delta^2}$ and when $0 \leqslant \omega' \leqslant \sqrt{\omega_0^2 - \delta^2}$. Thus, $n(\omega' - i\delta)$ is real and positive except for along the two branch cuts $\omega'_{-}\omega_{-}$ and $\omega_{+}\omega'_{+}$ wherein it is purely imaginary. Now, by (6.1.9), on the line $\omega'' = -\delta$, $X(\omega, \theta)$ is given by

$$X(\omega' - i\delta, \theta) = \delta[n_r(\omega') - \theta] - \omega' n_i(\omega') \tag{6.1.24}$$

so that, by (6.1.23) and the above results, the following behavior is obtained:

$$X(\omega' - \mathrm{i}\delta, \theta) = \delta\left[\left(1 + \frac{b^2}{\omega_0^2 - \delta^2 - \omega'^2}\right)^{1/2} - \theta\right] ;$$

when either $\omega' < \mathrm{Re}\{\omega'_-\}$, or
$\mathrm{Re}\{\omega_-\} < \omega' < \mathrm{Re}\{\omega_+\}$, or $\mathrm{Re}\{\omega'_+\} < \omega'$, (6.1.25a)

$$X(\omega' - \mathrm{i}\delta, \theta) = -\delta\theta - \omega'\left(\frac{b^2}{\omega'^2 + \delta^2 - \omega_0^2} - 1\right)^{1/2} ;$$

when either $\mathrm{Re}\{\omega'_-\} \leqslant \omega' \leqslant \mathrm{Re}\{\omega_-\}$, or
$\mathrm{Re}\{\omega_+\} \leqslant \omega' \leqslant \mathrm{Re}\{\omega'_+\}$. (6.1.25b)

The behavior of $n(\omega)$ about the two branch points ω'_+ and ω_+ is now considered. In the region about ω'_+, ω may be written as

$$\omega = \sqrt{\omega_1^2 - \delta^2} - \mathrm{i}\delta + r\mathrm{e}^{\mathrm{i}\alpha} ,$$

where r and α are the polar coordinates about the point $\omega'_+ = \sqrt{\omega_1^2 - \delta^2} - \mathrm{i}\delta$. The square of the complex index of refraction (6.1.1) can then be expressed as

$$n^2(\omega) = \frac{(\omega - \omega'_+)(\omega - \omega'_-)}{(\omega - \omega_+)(\omega - \omega_-)}$$

$$= \frac{[\omega - (\sqrt{\omega_1^2 - \delta^2} - \delta\mathrm{i})][\omega - (-\sqrt{\omega_1^2 - \delta^2} - \delta\mathrm{i})]}{[\omega - (\sqrt{\omega_0^2 - \delta^2} - \delta\mathrm{i})][\omega - (-\sqrt{\omega_0^2 - \delta^2} - \delta\mathrm{i})]} ,$$

so that

$$n^2(r, \alpha)$$

$$= \frac{r\mathrm{e}^{\mathrm{i}\alpha}(2\sqrt{\omega_1^2 - \delta^2} + r\mathrm{e}^{\mathrm{i}\alpha})}{(\sqrt{\omega_1^2 - \delta^2} - \sqrt{\omega_0^2 - \delta^2} + r\mathrm{e}^{\mathrm{i}\alpha})(\sqrt{\omega_1^2 - \delta^2} + \sqrt{\omega_0^2 - \delta^2} + r\mathrm{e}^{\mathrm{i}\alpha})}$$

$$\cong \frac{2\sqrt{\omega_1^2 - \delta^2}}{b^2} r\mathrm{e}^{\mathrm{i}\alpha} ,$$

for r small. Consequently, in this same limit of small r, the complex index of refraction in the region of the complex ω-plane about the branch point ω'_+ is given by

$$n(r, \alpha) \cong \frac{[2\sqrt{\omega_1^2 - \delta^2} r]^{1/2}}{b} \mathrm{e}^{\mathrm{i}(\alpha/2)} . \tag{6.1.26}$$

Similarly, in the region about ω_+, ω may be written as

$$\omega = \sqrt{\omega_0^2 - \delta^2} - \mathrm{i}\delta + R\mathrm{e}^{\mathrm{i}\xi}$$

where R and ξ are the polar coordinates about the branch point $\omega_+ = \sqrt{\omega_0^2-\delta^2} - i\delta$. The square of the complex index of refraction can then be expressed as follows:

$$n^2(R,\xi) = \frac{(\sqrt{\omega_0^2-\delta^2} - \sqrt{\omega_1^2-\delta^2} + \mathrm{R}e^{i\xi})(\sqrt{\omega_0^2-\delta^2} + \sqrt{\omega_1^2-\delta^2} + \mathrm{R}e^{i\xi})}{\mathrm{R}e^{i\xi}(2\sqrt{\omega_0^2-\delta^2} + \mathrm{R}e^{i\xi})}$$

$$\cong \frac{-b^2}{2\sqrt{\omega_0^2-\delta^2}R} e^{-i\xi} = \frac{b^2}{2\sqrt{\omega_0^2-\delta^2}R} e^{i(\pi-\xi)} ,$$

for R small. Consequently, in this same limit of small R, the complex index of refraction in the region of the complex ω-plane about the branch point ω_+ is given by

$$n(R,\xi) \cong \frac{b}{(2\sqrt{\omega_0^2-\delta^2}R)^{1/2}} e^{i((\pi-\xi)/2)} . \tag{6.1.27}$$

The behavior of $n(\omega)$ at the branch points ω'_+ and ω_+, as described by (6.1.26 and 27), respectively, is illustrated in Fig. 6.2 below. Similar results hold for the behavior of the complex index of refraction about the branch points ω'_- and ω_- in the left half of the complex ω-plane.

From these results the behavior of $X(\omega,\theta)$ about the branch points ω'_+ and ω_+ is easily determined by the use of (6.1.24). Thus, in the immediate region of the complex ω-plane about the branch point ω'_+, there results

$$X(r,\alpha,\theta) \cong \delta\left(\frac{(2\sqrt{\omega_1^2-\delta^2}r)^{1/2}}{b}\cos\left(\frac{\alpha}{2}\right) - \theta\right) - \sqrt{\omega_1^2-\delta^2}\frac{(2\sqrt{\omega_1^2-\delta^2}r)^{1/2}}{b}\sin\left(\frac{\alpha}{2}\right) , \tag{6.1.28}$$

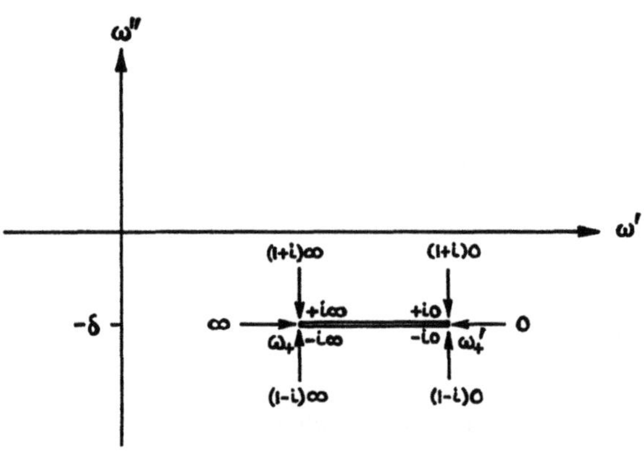

Fig. 6.2. The behavior of $n(\omega)$ in the immediate regions about the branch points ω_+ and ω'_+.

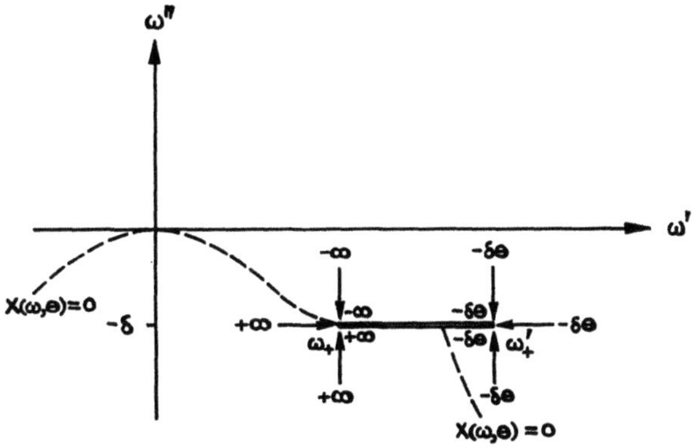

Fig. 6.3. Behavior of $X(\omega, \theta)$ in the immediate regions about the branch points ω_+ and ω'_+. The dashed line indicates the contour $X(\omega, \theta) = 0$ for $\theta > 0$.

whereas in the immediate region of the complex ω-plane about the branch point ω_+, there results

$$X(R, \xi, \theta) \cong \delta\left(\frac{b}{(2\sqrt{\omega_0^2 - \delta^2}R)^{1/2}} \cos\left(\frac{\pi - \xi}{2}\right) - \theta\right)$$

$$- \sqrt{\omega_0^2 - \delta^2} \frac{b}{(2\sqrt{\omega_0^2 - \delta^2}R)^{1/2}} \sin\left(\frac{\pi - \xi}{2}\right). \tag{6.1.29}$$

Hence, $X(\omega, \theta)$ is negative on both sides of the branch cut near ω'_+ for $\theta > 0$, is zero at $\theta = 0$, and is positive for $\theta < 0$. Near ω_+, however, $X(\omega, \theta)$ is negative on the upper side of the branch cut and positive on the lower side for all θ. From the behavior of $X(\omega, \theta)$ in the region of the complex ω-plane about ω_+, it is seen that the contour $X(\omega, \theta) = 0$ must pass through the branch point ω_+ and continue on from the lower side of the branch cut between ω_+ and ω'_+ for $\theta > 0$, as illustrated in Fig. 6.3. For $\theta < 0$, the contour $X(\omega, \theta) = 0$ continues on from the upper side of the branch cut between ω_+ and ω'_+, and for $\theta = 0$, the contour continues on from the branch point ω'_+.

6.1.2 Numerical Results

These simple results concerning the behavior of $X(\omega, \theta) = \mathbb{R}\text{e}\{\phi(\omega, \theta)\}$ in specific regions of the complex ω-plane are supplemented in this section with computer-generated contour plots of $X(\omega, \theta)$ in the right half of the complex ω-plane. These plots are needed to give a more complete picture of the topography of $X(\omega, \theta)$ in order to determine the number of saddle points of $\phi(\omega, \theta)$, their approximate locations, the approximate locations of the Olver-type paths which pass through these saddle points, and the way they move in the complex ω-plane with changing θ. Utilizing *Brillouin*'s choice of the medium parameters

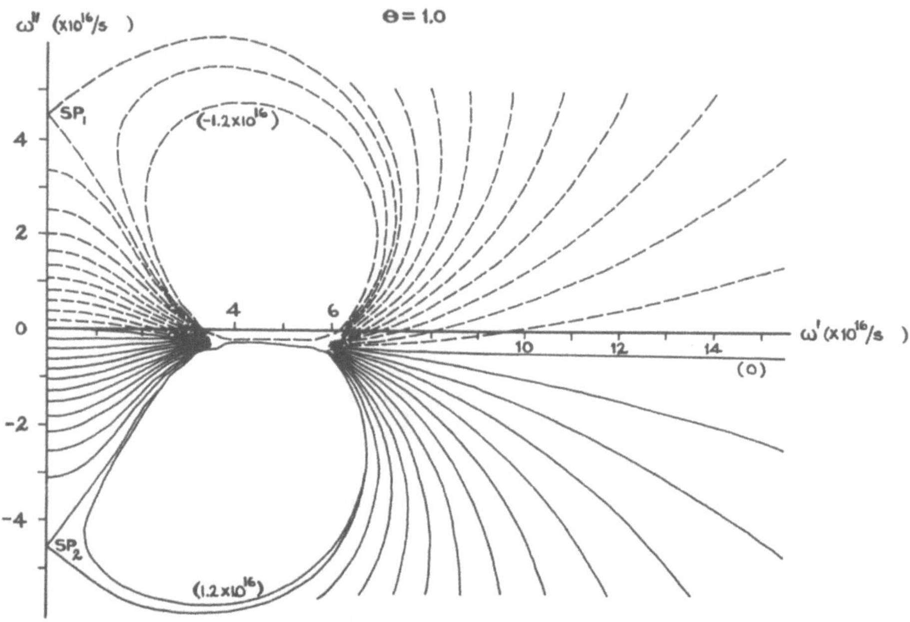

Fig. 6.4. Contours of the real phase behavior $X(\omega, \theta)$ in the right half of the complex ω-plane for $\theta = 1$. Note that $X(\omega, 1) < 0$ in the upper half plane, and the dominant distant saddle points are located in the lower half plane at $\omega_{sp_D^\pm} = \pm\infty - 2\delta i$ at which $X(\omega_{sp_D^\pm}, 1) = 0$.

(viz. $\omega_0 = 4.0 \times 10^{16}$/s, $b^2 = 20.0 \times 10^{32}/\text{s}^2$, and $\delta = 0.28 \times 10^{16}$/s), specific contours of the real phase function $X(\omega, \theta)$ in the complex ω-plane [including the contours which pass through the saddle points of $\phi(\omega, \theta)$] have been plotted by computer techniques and are reproduced in Figs. 6.4 through 6.9 for various values of the parameter θ increasing away from unity. Those contours of $X(\omega, \theta)$ which are greater than or equal to the value at the dominant saddle point are indicated by solid lines, whereas those which are less than the value at the dominant saddle point are indicated by short dashed lines. The values of the maximum and minimum contours of $X(\omega, \theta)$ illustrated in each figure are indicated by the numbers in parenthesis adjacent to the appropriate contour, as is the contour $X(\omega, \theta) = 0$ in each figure.

As was predicted by *Brillouin* [6.1, 2], for all values of θ except one, there are four first-order saddle points of the complex phase function $\phi(\omega, \theta)$ symmetrically located with respect to the imaginary axis. Two of the saddle points lie in the region near the origin, whereas the other two lie in the region removed from the origin. As can be seen from these contour plots, the two saddle points near the origin lie along the imaginary axis for $\theta = 1$ and approach one another along the imaginary axis as θ increases, coalescing into a single second-order saddle point when $\theta = \theta_1$, where the value of θ_1 for *Brillouin's* choice of the medium parameters is just slightly larger than 1.501 (Fig. 6.7). For larger values of θ, these saddle points separate symmetrically with respect to the imaginary

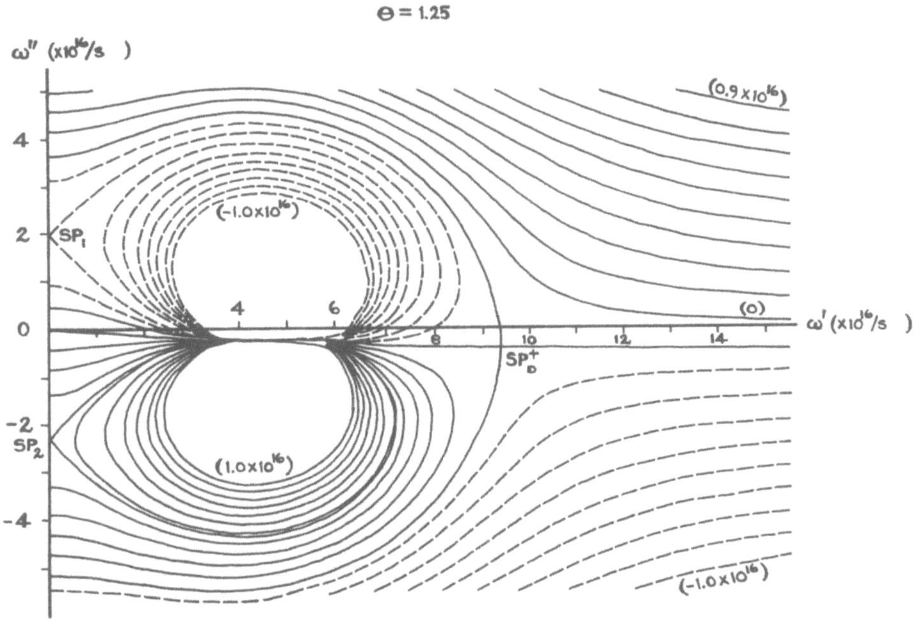

Fig. 6.5. Contours of the real phase behavior $X(\omega, \theta)$ in the right half of the complex ω-plane for $\theta = 1.25$. Note that the distant saddle point has moved in from infinity and is still dominant over the upper near saddle point SP_1.

axis and approach the branch points ω_+ and ω_-, respectively, as θ approaches infinity. On the other hand, the two distant saddle points are located in the lower-half of the complex ω-plane and are at $\pm\infty - 2\delta i$ for $\theta = 1$. As θ increases, these saddle points move in from infinity and approach the branch points ω'_+ and ω'_-, respectively, as θ approaches infinity. Initially, the distant saddle points have less exponential decay associated with them than does the upper near saddle point labeled SP_1 in Fig. 6.4 and 6.5. Since the original path of integration in the integral of interest in this research is not deformable into an Olver-type path through the lower near saddle point SP_2, that saddle point is irrelevant for the present analysis for all θ below and bounded away from θ_1. For $\theta = \theta_{SB} \cong 1.33425$, illustrated in Fig. 6.6, the upper near saddle point SP_1 has precisely the same exponential decay associated with it as do the two distant saddle points. Consequently, for $\theta = \theta_{SB}$, those three saddle points are of equal importance in the asymptotic analysis of the propagated field. The remaining figures show that for values of θ within the range $1 \leqslant \theta < \theta_{SB}$, the two distant saddle points are dominant over the upper near saddle point SP_1, for values of θ within the range $\theta_{SB} < \theta \leqslant \theta_1$, the upper near saddle point SP_1 is dominant over the distant saddle points, and for all later values of the parameter θ, the two near saddle points are dominant over the two distant saddle points.

The primary purpose of the remainder of this chapter is to obtain approximate analytic expressions which accurately describe the locations of the four

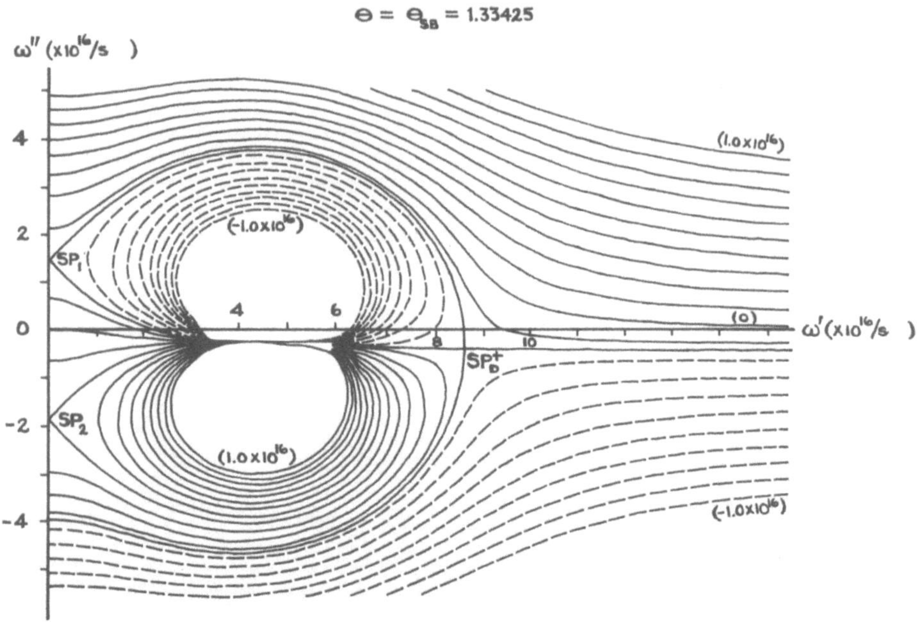

Fig. 6.6. Contours of the real phase behavior $X(\omega, \theta)$ in the right half of the complex ω-plane for $\theta = \theta_{SB} = 1.33425$, at which the distant saddle points and upper near saddle point SP_1 are all of equal importance.

saddle points of $\phi(\omega, \theta)$ and the phase behavior at those locations. In addition, analytic expressions are obtained for the values of the parameter θ at which the upper near and distant saddle points are of equal importance and at which the two near saddle points coalesce into a single higher-order saddle point.

6.2 The Location of the Saddle Points and the Approximation of the Phase

In order to obtain the asymptotic expansion of the propagated field $A(z, t)$ for large values of the observation distance z, the saddle points of the complex phase function $\phi(\omega, \theta)$ must be located in the complex ω-plane and the behavior of $\phi(\omega, \theta)$ at these points determined. The condition that $\phi(\omega, \theta)$ be stationary at a saddle point is

$$\phi'(\omega, \theta) = \mathrm{i}[n(\omega) - \theta] + \mathrm{i}\omega n'(\omega) \equiv 0 \ ,$$

where the prime denotes differentiation with respect to ω, so that

$$n(\omega) + \omega n'(\omega) - \theta = 0 \ . \tag{6.2.1}$$

The roots of this equation then give the desired saddle-point locations.

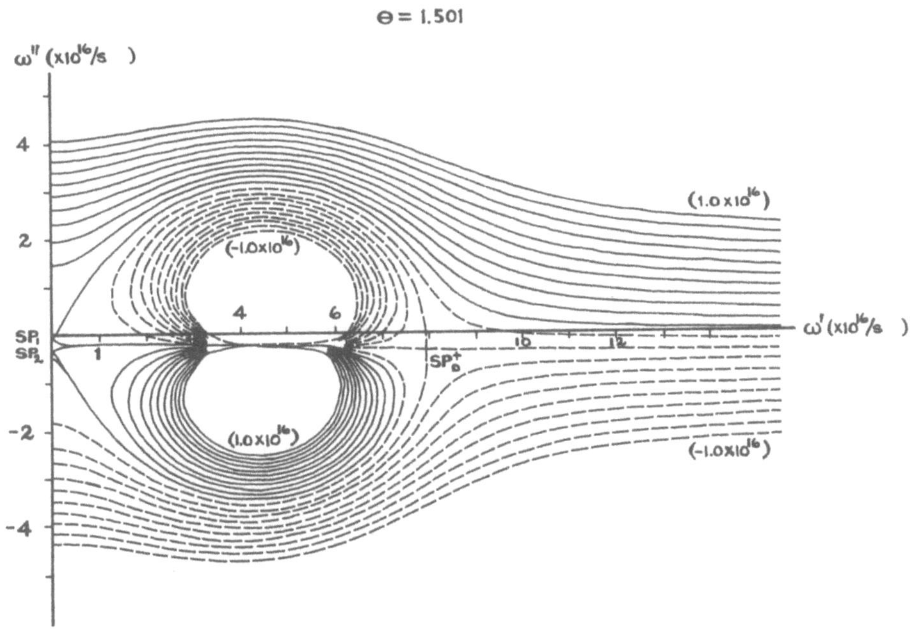

Fig. 6.7. Contours of the real phase behavior $X(\omega,\theta)$ in the right half of the complex ω-plane for $\theta = 1.501$ just prior to the coalescence of the two near first order saddle points SP_1 and SP_2 into a single second order saddle point. Note that the upper near saddle point SP_1 is dominant over the distant saddle points.

An exact analytic expression for the location of the saddle points of $\phi(\omega,\theta)$ is considered first. The complex index of refraction is given by (6.5.9), viz.

$$n(\omega) = \left(1 - \frac{b^2}{\omega^2 - \omega_0^2 + 2\delta i\omega}\right)^{1/2} , \tag{6.2.2}$$

so that the first derivative of $n(\omega)$ is

$$n'(\omega) = \frac{1}{2}\left(1 - \frac{b^2}{\omega^2 - \omega_0^2 + 2\delta i\omega}\right)^{-1/2} \cdot \frac{b^2(2\omega + 2\delta i)}{(\omega^2 - \omega_0^2 + 2\delta i\omega)^2} .$$

Hence, from (6.2.1), the saddle point locations are given exactly by

$$\begin{aligned}&\left(1 - \frac{b^2}{\omega^2 - \omega_0^2 + 2\delta i\omega}\right)^{1/2} \\ &+ \frac{b^2\omega(\omega + \delta i)}{(\omega^2 - \omega_0^2 + 2\delta i\omega)^2\left(1 - \dfrac{b^2}{\omega^2 - \omega_0^2 + 2\delta i\omega}\right)^{1/2}} - \theta \\ &= 0 .\end{aligned} \tag{6.2.3}$$

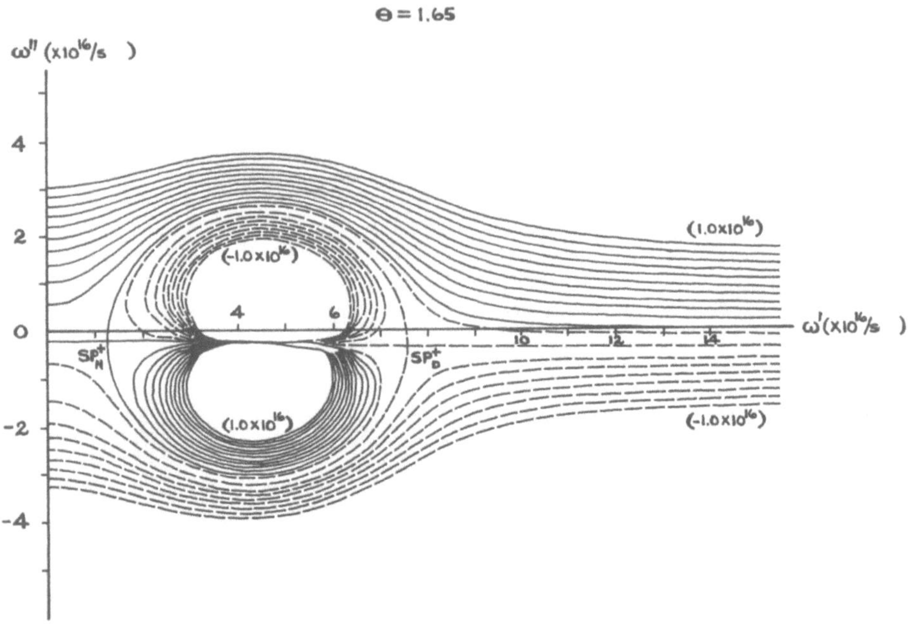

Fig. 6.8. Contours of the real phase behavior $X(\omega, \theta)$ in the right half of the complex ω-plane for $\theta = 1.65 > \theta_1$, at which the dominant near saddle points have moved off of the imaginary axis and into the lower half of the complex ω-plane.

In order to eliminate the square roots appearing in this expression, it may be rewritten as

$$\frac{\omega^2 - \omega_1^2 + 2\delta i\omega}{\omega^2 - \omega_0^2 + 2\delta i\omega} + \frac{b^2\omega(\omega + \delta i)}{(\omega^2 - \omega_0^2 + 2\delta i\omega)^2} = \theta\left(\frac{\omega^2 - \omega_1^2 + 2\delta i\omega}{\omega^2 - \omega_0^2 + 2\delta i\omega}\right)^{1/2} ,$$

where $\omega_1^2 = \omega_0^2 + b^2$. Squaring both sides of this equation yields

$$\left(\omega^2 - \omega_1^2 + 2\delta i\omega + \frac{b^2\omega(\omega + \delta i)}{\omega^2 - \omega_0^2 + 2\delta i\omega}\right)^2$$
$$= \theta^2(\omega^2 - \omega_1^2 + 2\delta i\omega)(\omega^2 - \omega_0^2 + 2\delta i\omega) \ . \tag{6.2.4}$$

Since both of the expressions given in (6.2.3, 4) are complicated functions of the complex variable ω, it is difficult (if not impossible) to determine exact general analytic expressions for the locations of the saddle points as a function of θ. However, from the computer plots of the topography of $X(\omega, \theta)$ in the complex ω-plane, given in Sect. 6.1.2, it is found that there are in general two saddle points in the region near the origin and two saddle points in the region well removed from the origin. These two regions may then be considered separately in order to determine the approximate locations of the saddle points. The resulting approximate solutions of (6.2.3 or 4) can then be utilized as initial values

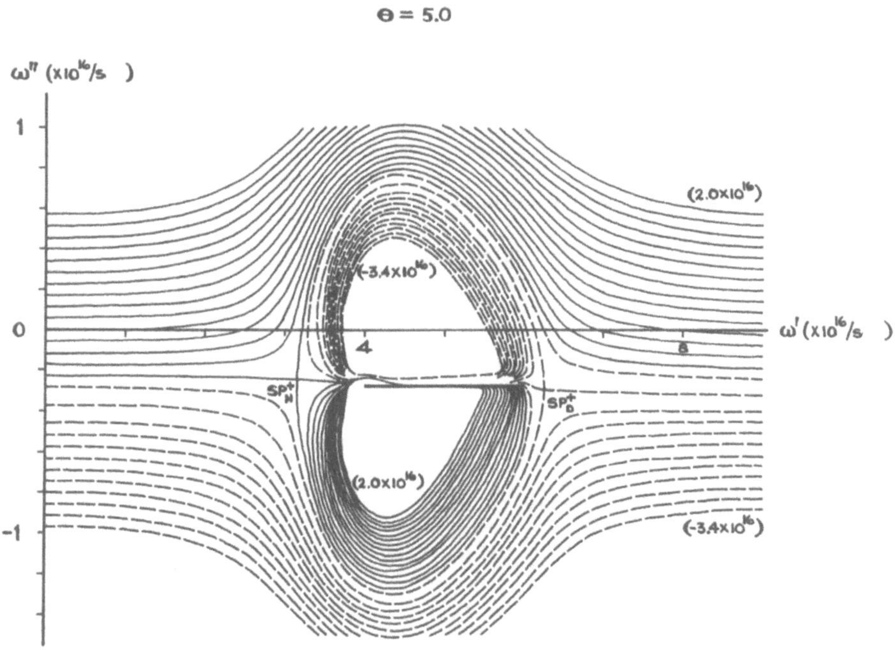

Fig. 6.9. Contours of the real phase behavior $X(\omega, \theta)$ in the right half of the complex ω-plane for $\theta = 5.0$. Note the approach of the dominant near saddle point towards the branch point $\omega_+ = (3.99 - 0.28i) \times 10^{16}$/s and the approach of the distant saddle point towards the branch point $\omega' = (5.99 - 0.28i) \times 10^{16}$/s.

in a numerical solution of the saddle-point equaton in order to test their accuracy and also to numerically determine more accurately the roots of the equation.

Beforc proceeding with this approximate analysis, note that two exact roots of the saddle-point equation (6.2.3) are readily obtained in the limit as θ approaches infinity. In that limiting case, either $(\omega^2 - \omega_0^2 + 2\delta i\omega) = 0$, yielding the roots

$$\omega = \pm\sqrt{\omega_0^2 - \delta^2} - \delta i \ ,$$

or $n(\omega) = 0$, yielding the roots

$$\omega = \pm\sqrt{\omega_1^2 - \delta^2} - \delta i \ .$$

Thus, in the limit as θ approaches infinity, the saddle points move onto the branch points ω_+ and ω'_+ in the right-half plane and ω_- and ω'_- in the left-half plane.

Moreover, an exact polynomial equation describing the location of the saddle points can be obtained. To that end, (6.2.3) is rewritten to eliminate the square roots as follows:

$$\theta\left(\frac{\omega^2-\omega_1^2+2\delta i\omega}{\omega^2-\omega_0^2+2\delta i\omega}\right)^{1/2}(\omega^2-\omega_0^2+2\delta i\omega)^2$$
$$=\omega^4+4\delta i\omega^3-2(\omega_0^2+2\delta^2)\omega^2-\delta i(4\omega_0^2+3b^2)\omega+\omega_1^2\omega_0^2\ .$$

Squaring both sides of this equation yields

$$\theta^2(\omega^2-\omega_1^2+2\delta i\omega)(\omega^2-\omega_0^2+2\delta i\omega)^3$$
$$=[\omega^4+4\delta i\omega^3-2(\omega_0^2+2\delta^2)\omega^2-\delta i(4\omega_0^2+3b^2)\omega+\omega_1^2\omega_0^2]^2\ ,$$

and after a bit of algebraic manipulation, there results finally the following exact polynomial equation for the saddle point locations:

$$\begin{aligned}&(\theta^2-1)\omega^8+8\delta i(\theta^2-1)\omega^7-[4(\omega_0^2+6\delta^2)(\theta^2-1)+b^2\theta^2]\omega^6\\&-2\delta i(12\omega_0^2+3b^2+16\delta^2)(\theta^2-1)\omega^5\\&+[(6\omega_0^4+48\delta^2\omega_0^2+2b^2\omega_0^2+12b^2\delta^2+16\delta^4)(\theta^2-1)\\&+b^2(\omega_0^2\theta^2-12\delta^2)]\omega^4\\&+4\delta i[(6\omega_0^4+4b^2\omega_0^2+8\delta^2\omega_0^2+4\delta^2b^2)(\theta^2-1)\\&+(\omega_0^2+2\delta^2)(2\omega_0^2\theta^2-b^2)]\omega^3\\&-[\omega_0^2(4\omega_0^4+3b^2\omega_0^2+24\delta^2\omega_0^2+12\delta^2b^2)(\theta^2-1)\\&-b^2(\omega_0^4+20\delta^2\omega_0^2+9\delta^2b^2)]\omega^2\\&-2\delta i\omega_0^2(4\omega_0^2+3b^2)[\omega_0^2(\theta^2-1)-b^2]\omega\\&+\omega_0^4(\omega_0^2+b^2)[\omega_0^2(\theta^2-1)-b^2]\\&=0\ .\end{aligned}\tag{6.2.5}$$

Since (6.2.5) is extremely formidable and difficult to approximate, the approximate solution of the saddle-point equation as given in (6.2.3 or 4) is now undertaken for the two separate classes of saddle points, i.e., the saddle points in the region removed from the origin and the saddle points in the region near the origin.

6.2.1 The Region Removed from the Origin

(a) *The First Approximation*

In order to permit comparison of the new results obtained below (called herein the second approximation) with the results previously available, this subsection reviews the analysis given by *Brillouin* [6.1, 2] to obtain an approximate expression (called herein the first approximation) for the location of the distant saddle points. For large values of $|\omega|$ the parameter ω_0^2 can be neglected in comparison to ω^2 in the expression for the complex refractive index. To this first approximation then, (6.2.2) becomes

$$n^2(\omega) \cong 1 - \frac{b^2}{\omega(\omega + 2\delta i)} , \tag{6.2.6}$$

for $|\omega|^2 \gg \omega_0^2$. Since the second term in this expression is small in comparison to unity for $|\omega|^2 \gg b^2$, the complex index of refraction is then given by

$$n(\omega) \cong 1 - \frac{b^2}{2\omega(\omega + 2\delta i)} . \tag{6.2.7}$$

Differentiation of this expression with respect to ω yields

$$n'(\omega) \cong \frac{b^2(\omega + \delta i)}{(\omega^2 + 2\delta i\omega)^2} . \tag{6.2.8}$$

Therefore, according to (6.2.1), the saddle point locations in the region removed from the origin are given, to a first approximation, by the expression

$$1 - \frac{b^2}{2\omega(\omega + 2\delta i)} + \frac{b^2(\omega + \delta i)}{\omega(\omega + 2\delta i)^2} - \theta = 0 ,$$

whose solutions are

$$\omega_{sp_D^\pm}(\theta) \cong \pm\frac{b}{\sqrt{2(\theta - 1)}} - 2\delta i . \tag{6.2.9}$$

This result was first obtained by *Brillouin* [6.1, 2] and is herein called the *first approximation to the location of the saddle points in the region removed from the origin.* Thus, there are two saddle points symmetrically located about the imaginary axis, lying on the straight line $\omega = -2\delta i$. At $\theta = 1$ these two saddle points are at infinity on the line $\omega = -2\delta i$, and as θ increases from unity they move in symmetrically towards the imaginary ω''-axis (i.e., $\mathrm{Re}\{\omega_{sp_D^\pm}\}$ approaches 0 in the limit as θ tends to ∞). However, when θ gets large enough so that $|\omega_{sp_D^\pm}|^2$ is no longer large in comparison to both ω_0^2 and b^2 and the second term in the expression (6.2.6) for $n^2(\omega)$ is no longer small in comparison to unity (i.e., when the saddle points come close to the branch cuts), then the above approximation loses its validity and, as is shown in the second approximation, the saddle points $\omega_{sp_D^\pm}(\theta)$ leave the line $\omega = -2\delta i$ and approach the branch points ω'_+ and ω'_-, respectively, in the limit as θ approaches infinity.

(b) *The Second Approximation*

In order to obtain a more accurate description of the locations of the distant saddle points, particularly for large values of θ, the exact saddle point equation (6.2.4) may be rewritten as

$$\frac{1}{\theta^2(\omega^2-\omega_0^2+2\delta i\omega)}\left(\omega^2-\omega_1^2+2\delta i\omega+\frac{b^2\omega(\omega+\delta i)}{\omega^2-\omega_0^2+2\delta i\omega}\right)^2$$
$$=\omega^2-\omega_1^2+2\delta i\omega\ . \tag{6.2.10}$$

In this particular form, it is seen that in the limit as θ approaches infinity, the right-hand side of the equation must approach zero, and hence

$$\lim_{\theta\to\infty}[\omega_{sp_D^\pm}(\theta)]=\pm\sqrt{\omega_1^2-\delta^2}-\delta i\ .$$

With this limit in mind, the rational function appearing in the squared term of (6.2.10) may be approximated by

$$\frac{b^2\omega(\omega+\delta i)}{\omega^2-\omega_0^2+2\delta i\omega}\cong b^2\frac{\omega+\delta i}{\omega+2\delta i}=b^2\frac{1+\delta i/\omega}{1+2\delta i/\omega}$$
$$\cong b^2\left[1-\frac{\delta i}{\omega}+\mathcal{O}\left(\frac{1}{\omega^2}\right)\right], \tag{6.2.11}$$

for $|\omega|$ large in comparison to δ.

As a first approximation (for comparison purposes to the previous first approximation), the very approximate result that the above expression (6.2.11) is equal to b^2 is taken. Under this approximation, the saddle point equation (6.2.10) becomes

$$\frac{1}{\theta^2(\omega^2-\omega_0^2+2\delta i\omega)}(\omega^2-\omega_0^2+2\delta i\omega)^2=\omega^2-\omega_1^2+2\delta i\omega\ ,$$

the solution of which is

$$\omega_{sp_D^\pm}(\theta)\cong\pm\sqrt{\omega_0^2-\delta^2+\frac{b^2\theta^2}{\theta^2-1}}-\delta i\ . \tag{6.2.12}$$

This result is to be compared with *Brillouin's* result given in (6.2.9). Thus, although the saddle points in this first approximation (6.2.12) now lie along the line $\omega=-\delta i$ symmetrically located about the imaginary axis and are at infinity at $\theta=1$, in the limit as θ approaches infinity the two saddle-points approach the branch points $\omega'_\pm$, in agreement with the exact result.

Consider now the second approximation to the location of the distant saddle points, where (6.2.11) is approximated as

$$\frac{b^2\omega(\omega+\delta i)}{\omega^2-\omega_0^2+2\delta i\omega}\cong b^2-\frac{b^2\delta i}{\omega}\ .$$

In this case the saddle-point equation (6.2.10) becomes

$$(\omega^2 - \omega_0^2 + 2\delta i\omega)^2 - 2i\frac{\delta b^2}{\omega}(\omega^2 - \omega_0^2 + 2\delta i\omega) - \frac{\delta^2 b^4}{\omega^2}$$
$$= \theta^2(\omega^2 - \omega_0^2 + 2\delta i\omega)(\omega^2 - \omega_1^2 + 2\delta i\omega) \ .$$

The term $\delta^2 b^4/\omega^2$ may be neglected in comparison to the other two terms on the left-hand side of this equaton, resulting in the equation

$$\omega^3 + 2\delta i\omega^2 - \left(\omega_0^2 + \frac{b^2\theta^2}{\theta^2 - 1}\right)\omega + \frac{2i\delta b^2}{\theta^2 - 1} = 0 \ . \tag{6.2.13}$$

In order to obtain analytically the roots of this cubic equation, the procedure outlined in [6.3], is employed. Accordingly, if one defines the coefficients

$$a_1 = 2\delta i \ ,$$
$$b_1 = -\left(\omega_0^2 + \frac{b^2\theta^2}{\theta^2 - 1}\right) ,$$
$$c_1 = \frac{2i\delta b^2}{\theta^2 - 1} ,$$

and performs the change of variable

$$\omega = \xi - \frac{a_1}{3} = \xi - \frac{2}{3}\delta i \ , \tag{6.2.14}$$

the cubic equation (6.2.13) is then reduced to the form

$$\xi^3 + a_2\xi + b_2 = 0 \ , \tag{6.2.15}$$

where

$$a_2 = b_1 - \frac{1}{3}a_1^2 = -\omega_0^2 - \frac{b^2\theta^2}{\theta^2 - 1} + \frac{4}{3}\delta^2 \ ,$$
$$b_2 = \frac{2}{27}a_1^3 - \frac{1}{3}a_1 b_1 + c_1 = \frac{2}{3}\delta i\left(\omega_0^2 - \frac{8}{9}\delta^2 + \frac{b^2(\theta^2 + 3)}{\theta^2 - 1}\right) .$$

For the solution of this reduced cubic equation (6.2.15), let

$$A = \left(-\frac{b_2}{2} + \sqrt{\frac{b_2^2}{4} + \frac{a_2^3}{27}}\right)^{1/3} ,$$
$$B = \left(-\frac{b_2}{2} - \sqrt{\frac{b_2^2}{4} + \frac{a_2^3}{27}}\right)^{1/3} ,$$

where

$$\sqrt{\frac{b_2^2}{4}+\frac{a_2^3}{27}}=\frac{\mathrm{i}}{3\sqrt{3}}\left\{\omega_0^4(\omega_0^2-\delta^2)\right.$$

$$+\frac{b^2}{\theta^2-1}[(3\omega_0^2-2\delta^2)\omega_0^2\theta^2+2\delta^2(9\omega_0^2-8\delta^2)]$$

$$+\frac{b^4}{(\theta^2-1)^2}[(3\omega_0^2-\delta^2)\theta^4+9\delta^2(2\theta^2+3)]$$

$$\left.+\frac{b^6\theta^6}{(\theta^2-1)^3}\right\}^{1/2} .$$

The algebraic expression under the square-root operation in this expression is positive for $\theta \geqslant 1$, which is the only case that need be considered here. Finally, if one defines the real parameters

$$\beta_1 \equiv \frac{1}{3}\delta\left(\omega_0^2-\frac{8}{9}\delta^2+\frac{b^2(\theta^2+3)}{\theta^2-1}\right) , \tag{6.2.16a}$$

$$\beta_2 \equiv \frac{1}{3\sqrt{3}}\left\{\omega_0^4(\omega_0^2-\delta^2)\right.$$

$$+\frac{b^2}{\theta^2-1}[(3\omega_0^2-2\delta^2)\omega_0^2\theta^2+2\delta^2(9\omega_0^2-8\delta^2)]$$

$$+\frac{b^4}{(\theta^2-1)^2}[(3\omega_0^2-\delta^2)\theta^4+9\delta^2(2\theta^2+3)]$$

$$\left.+\frac{b^6\theta^6}{(\theta^2-1)^3}\right\}^{1/2} , \tag{6.2.16b}$$

there results

$$A=(-\mathrm{i}\beta_1+\mathrm{i}\beta_2)^{1/3}=(-\mathrm{i})^{1/3}(\beta_1-\beta_2)^{1/3}=\mathrm{i}(\beta_1-\beta_2)^{1/3} ,$$

$$B=(-\mathrm{i}\beta_1-\mathrm{i}\beta_2)^{1/3}=\mathrm{i}(\beta_1+\beta_2)^{1/3} ,$$

where the principal branch $0 \leqslant \arg(x) < 2\pi$ has been chosen for these cube root expressions. The three solutions of the reduced cubic equation (6.2.15) are then given by [6.3]

$$\xi=A+B ; \qquad -\frac{A+B}{2}+\frac{A-B}{2}\sqrt{3}\mathrm{i} ; \qquad -\frac{A+B}{2}-\frac{A-B}{2}\sqrt{3}\mathrm{i} .$$

Since there are only two saddle points which are symmetrically located with respect to the imaginary axis, it is easily seen that the last two solutions are of the proper form. Hence, the two sought-after solutions are given by

$$\begin{aligned}\xi^{\pm} &= -\frac{A+B}{2} \pm \frac{A-B}{2}\sqrt{3}\mathrm{i}\\ &= \pm\frac{\sqrt{3}}{2}[(\beta_1+\beta_2)^{1/3}-(\beta_1-\beta_2)^{1/3}]\\ &\quad -\frac{\mathrm{i}}{2}[(\beta_1-\beta_2)^{1/3}+(\beta_1+\beta_2)^{1/3}] \ .\end{aligned} \tag{6.2.17}$$

Therefore, by (6.2.14), the locations of the two distant saddle points are approximately given by

$$\begin{aligned}\omega_{sp_D^{\pm}}(\theta) &\cong \pm\frac{\sqrt{3}}{2}[(\beta_1+\beta_2)^{1/3}-(\beta_1-\beta_2)^{1/3}]\\ &\quad -\mathrm{i}\left\{\frac{2}{3}\delta+\frac{1}{2}[(\beta_1+\beta_2)^{1/3}+(\beta_1-\beta_2)^{1/3}]\right\} \ .\end{aligned} \tag{6.2.18}$$

This complicated expression is rather formidable to work with, however, and a more simplified version which is of greater accuracy than either of the first order approximations (6.2.9 or 12) is sought. Since for all values of $\theta \geqslant 1$ it is seen that $\beta_2 \gg \beta_1$, the following approximations can then be made

$$(\beta_1+\beta_2)^{1/3} = \beta_2^{1/3}\left(1+\frac{\beta_1}{\beta_2}\right)^{1/3} \cong \beta_2^{1/3}+\frac{\beta_1}{3\beta_2^{2/3}} \ , \tag{6.2.19a}$$

$$(\beta_1-\beta_2)^{1/3} = \beta_2^{1/3}\left(\frac{\beta_1}{\beta_2}-1\right)^{1/3} \cong -\beta_2^{1/3}+\frac{\beta_1}{3\beta_2^{2/3}} \ . \tag{6.2.19b}$$

Under this additional approximation, (6.2.18) becomes

$$\omega_{sp_D^{\pm}}(\theta) \cong \pm\sqrt{3}\beta_2^{1/3} - \mathrm{i}\left[\frac{2}{3}\delta+\frac{\beta_1}{3\beta_2^{2/3}}\right] \ . \tag{6.2.20}$$

Finally, the quantity $\beta_2^{1/3}$ may be approximated as follows

$$\begin{aligned}\sqrt{3}\beta_2^{1/3} = &\left\{\frac{b^6\theta^6}{(\theta^2-1)^3}+\frac{b^4}{(\theta^2-1)^2}[(3\omega_0^2-\delta^2)\theta^4+9\delta^2(2\theta^2+3)]\right.\\ &+\frac{b^2}{\theta^2-1}[(3\omega_0^2-2\delta^2)\omega_0^2\theta^2+2\delta^2(9\omega_0^2-8\delta^2)]\\ &\left.+\,\omega_0^4(\omega_0^2-\delta^2)\right\}^{1/6}\end{aligned}$$

$$\cong \left[\frac{b^2\theta^2}{\theta^2 - 1} + \omega_0^2 - \frac{\delta^2}{3} + \left(\omega_0^2 - \frac{2}{3}\delta^2\right)\frac{\omega_0^2}{b^2} + \frac{\omega_0^4}{3b^4}(\omega_0^2 - \delta^2) \right]^{1/2}$$

$$\cong \sqrt{\omega_0^2 - \delta^2 + \frac{b^2\theta^2}{\theta^2 - 1}} ,$$

to a fair degree of approximation in order to obtain agreement with the real part of the first approximation (6.2.12). Consequently, with this approximate result one obtains the expression

$$\frac{\beta_1}{3\beta_2^{2/3}} \cong \frac{1}{3}\delta \left\{ 1 + \frac{\frac{1}{9}\delta^2 + \frac{3b^2}{\theta^2 - 1}}{\omega_0^2 - \delta^2 + \frac{b^2\theta^2}{\theta^2 - 1}} \right\} .$$

Therefore, with these approximate results, the *second approximation to the locations of the distant saddle points* is given by

$$\omega_{sp_D^\pm}(\theta) \cong \pm\xi(\theta) - \delta i[1 + \eta(\theta)] , \tag{6.2.21}$$

where, for convenience, the following two functions have been defined here

$$\xi(\theta) \equiv \sqrt{\omega_0^2 - \delta^2 + \frac{b^2\theta^2}{\theta^2 - 1}} , \tag{6.2.22}$$

$$\eta(\theta) \equiv \frac{\frac{1}{27}\delta^2 + \frac{b^2}{\theta^2 - 1}}{\xi^2(\theta)} . \tag{6.2.23}$$

It is easily seen that this second approximation (6.2.21) properly describes the behavior of the distant saddle points. For values of θ close to unity,

$$\xi(\theta) \cong \frac{b}{\sqrt{2(\theta - 1)}}$$

and

$$\eta(\theta) \cong 1 ,$$

so that in the limit as θ approaches unity from above

$$\lim_{\theta \to 1} [\omega_{sp_D^\pm}(\theta)] = \lim_{\theta \to 1} \left[\pm\frac{b}{\sqrt{2(\theta - 1)}} - 2\delta i \right] = \pm\infty - 2\delta i ,$$

in agreement with the first approximation (6.2.9). Alternatively, for very large values of θ,

$$\xi(\theta) \cong \sqrt{\omega_1^2 - \delta^2}$$

and

$$\eta(\theta) \cong \frac{\delta^2}{27(\omega_1^2 - \delta^2)} \cong 0 \ ,$$

so that in the limit as θ approaches infinity,

$$\lim_{\theta \to \infty} [\omega_{sp_D^\pm}(\theta)] \cong \pm\sqrt{\omega_1^2 - \delta^2} - \delta \mathrm{i} = \omega'_\pm \ ,$$

in agreement with the first approximation (6.2.12). Notice that this result is not exactly valid since $\eta(\theta)$ does not identically vanish as θ approaches infinity. However, this small discrepency is of little importance and is shown to be of minor (if not null) consequence to the remaining analysis either concerned with or dependent upon the locations of the distant saddle points.

An analytic approximation of the phase behavior $\phi(\omega, \theta)$ valid in that region of the complex ω-plane which is traversed by the distant saddle points is now derived. For the present analysis, the complex index of refraction as given by the approximate expression (6.2.7) is sufficiently accurate, so that the phase behavior in the region of the complex ω-plane removed from the origin can be approximated by

$$\phi(\omega, \theta) \cong \mathrm{i}\omega(1 - \theta) - \mathrm{i}\frac{b^2}{2(\omega + 2\delta \mathrm{i})} \ , \tag{6.2.24}$$

for both $|\omega| \gg \omega_0$ and $|\omega| \gg b$. In order to obtain the behavior of $X(\omega, \theta) = \mathbb{R}\mathrm{e}\{\phi(\omega, \theta)\}$ in the vicinity of the distant saddle point in the right-half of the complex ω-plane, let

$$\begin{aligned} \omega &= \omega_{sp_D^+}(\theta) + r\mathrm{e}^{\mathrm{i}\varphi} \\ &= \xi(\theta) - \delta \mathrm{i}[1 + \eta(\theta)] + r\mathrm{e}^{\mathrm{i}\varphi} \ . \end{aligned}$$

With this substitution, $\phi(\omega, \theta)$ as given by (6.2.24) becomes

$$\begin{aligned} &\phi(\theta, r, \varphi) \\ &\quad \cong (1 - \theta)\{\mathrm{i}\xi(\theta) + \delta[1 + \eta(\theta)] + \mathrm{i}r\mathrm{e}^{\mathrm{i}\varphi}\} - \mathrm{i}\frac{b^2}{2} \\ &\quad \cdot \frac{\xi(\theta) - \delta \mathrm{i}[1 - \eta(\theta)] + r\mathrm{e}^{-\mathrm{i}\varphi}}{\xi^2(\theta) + \delta^2[1 - \eta(\theta)]^2 + 2\xi(\theta) r \cos\varphi + 2\delta[1 - \eta(\theta)] r \sin\varphi + r^2} \ . \end{aligned} \tag{6.2.25}$$

The real part of this equation is

$$X(\theta, r, \varphi)$$

$$\cong (1-\theta)\{\delta[1+\eta(\theta)] - r\sin\varphi\} - \frac{b^2}{2}$$

$$\cdot \frac{\delta[1-\eta(\theta)] + r\sin\varphi}{\xi^2(\theta) + \delta^2[1-\eta(\theta)]^2 + 2\xi(\theta)r\cos\varphi + 2\delta[1-\eta(\theta)]r\sin\varphi + r^2}, \tag{6.2.26}$$

from which it is seen that $X(\theta, r, \varphi)$ attains its maximum variation with r for $\varphi = \pi/4$, $3\pi/4$, $5\pi/4$, and $7\pi/4$. Consequently, in the right-half plane the lines of steepest descent through the distant saddle point are at the angles $\varphi = 3\pi/4$ and $7\pi/4$, while the lines of steepest ascent are at $\varphi = \pi/4$ and $5\pi/4$. Due to the symmetric behavior of $X(\omega, \theta)$ about the imaginary axis, the reverse holds true for the distant saddle point in the left half of the complex ω-plane. These two distant saddle points are illustrated in Fig. 6.10, where the arrows indicate the direction of ascent along the lines of steepest descent and ascent through the saddle points which are at angles of 45° to the coordinate axes, and the hatched area indicates the region of the complex ω-plane wherein the inequality $X(\omega) < X(\omega_{sp_D^\pm}, \theta)$ is satisfied and in which the path of descent from the saddle point lies. These general features are in complete agreement with the numerical results presented in Figs. 6.4 through 6.9.

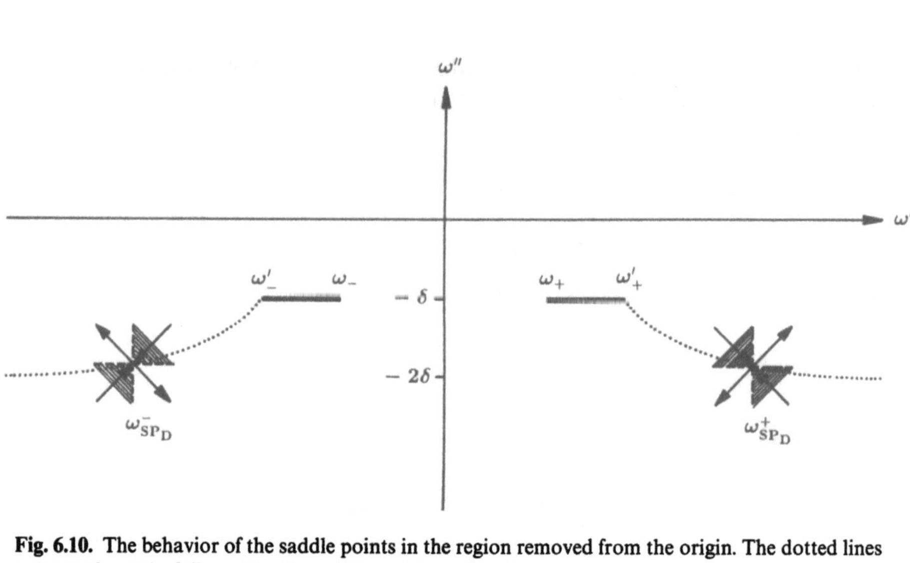

Fig. 6.10. The behavior of the saddle points in the region removed from the origin. The dotted lines indicate the paths followed by the saddle points as θ varies.

6.2.2 The Region Near the Origin

(a) *The First Approximation*

As in Sect. 6.2.1a, this subsection reviews the analysis given by *Brillouin* [6.1, 2] to obtain the first approximation for the location of the saddle points in order to permit comparison with the new, more accurate second approximation obtain in Sect. 6.2.2b. In order to locate the saddle points in the region of the complex ω-plane about the origin, the complex index of refraction $n(\omega)$ is expanded in an ascending power series in ω. To accomplish this, the complex index of refraction (6.2.2) is written in the form

$$\begin{aligned} n(\omega) &= \left(\frac{\omega^2 - \omega_1^2 + 2\delta i\omega}{\omega^2 - \omega_0^2 + 2\delta i\omega}\right)^{1/2} \\ &= \left(\frac{\omega_1^2 - \varepsilon}{\omega_0^2 - \varepsilon}\right)^{1/2} , \end{aligned} \tag{6.2.27}$$

where $\omega_1^2 = \omega_0^2 + b^2$ and

$$\varepsilon = \omega^2 + 2\delta i\omega \ . \tag{6.2.28}$$

Since $|\omega|$ is small in the region about the origin, $|\varepsilon|$ is small in comparison with both ω_0^2 and ω_1^2, and the following expansion is justified

$$\begin{aligned} \frac{\omega_1^2 - \varepsilon}{\omega_0^2 - \varepsilon} &= \frac{\omega_1^2}{\omega_0^2}\left(\frac{1 - \varepsilon/\omega_1^2}{1 - \varepsilon/\omega_0^2}\right) \\ &\cong \frac{\omega_1^2}{\omega_0^2}\left(1 + \varepsilon\frac{b^2}{\omega_1^2\omega_0^2} + \varepsilon^2\frac{b^2}{\omega_1^2\omega_0^4}\right) . \end{aligned}$$

With this result employed in the expression for the complex index of refraction, the square root may then be expanded in a Taylor series, resulting in

$$n(\omega) \cong \frac{\omega_1}{\omega_0}\left(1 + \frac{\varepsilon}{2}\frac{b^2}{\omega_1^2\omega_0^2} + \frac{\varepsilon^2}{8}\frac{b^2(4\omega_1^2 - b^2)}{\omega_1^4\omega_0^4}\right) .$$

Substitution of the expression for ε, given by (6.2.28), results in

$$n(\omega) \cong \frac{\omega_1}{\omega_0} + \frac{b^2}{2\omega_1\omega_0^3}\omega(\omega + 2\delta i) - \frac{\delta^2 b^2(4\omega_1^2 - b^2)}{2\omega_1^3\omega_0^5}\omega^2 , \tag{6.2.29}$$

where this final expression for the complex index of refraction is correct to $\mathcal{O}(\omega^2)$. Differentiation of this result with respect to ω yields

$$n'(\omega) \cong \frac{b^2}{\omega_1\omega_0^3}(\omega + \delta i) - \frac{\delta^2 b^2(4\omega_1^2 - b^2)}{\omega_1^3\omega_0^5}\omega .$$

Therefore, within the above approximations, the locations of the near saddle points are given by the roots of the equation, c.f. (6.2.1),

$$\omega^2 + \frac{4}{3}\mathrm{i}\frac{\delta}{\alpha}\omega + \frac{2}{3}\frac{\omega_0^3\omega_1}{\alpha b^2}(\theta_0 - \theta) = 0 \ , \tag{6.2.30}$$

where the following parameters have been introduced

$$\theta_0 \equiv \frac{\omega_1}{\omega_0} = \sqrt{1 + \frac{b^2}{\omega_0^2}} = n(0) \ , \tag{6.2.31}$$

$$\alpha \equiv 1 - \frac{\delta^2(4\omega_1^2 - b^2)}{\omega_0^2\omega_1^2} \ , \tag{6.2.32}$$

which depend only upon the medium properties. The roots of (6.2.30), where terms of order ω^3 have been neglected, then give the first approximate locations of the near saddle points. The general roots of this equaton are

$$\omega_{sp_N^\pm} \cong \pm\frac{1}{3}\left(6\frac{\theta_0\omega_0^4}{\alpha b^2}(\theta - \theta_0) - 4\frac{\delta^2}{\alpha^2}\right)^{1/2} - \frac{2}{3}\mathrm{i}\frac{\delta}{\alpha} \ . \tag{6.2.33}$$

This result is called the *first approximation to the location of the saddle points in the region near the origin. Brillouin*'s result is obtained with α, defined in (6.2.32), approximated by unity.

In order to analyze the behavior of the near saddle points as specified by this first approximation (6.2.33), it is necessary to determine the algebraic sign of the argument of the square root. Consider then determining the value of $\theta \equiv \theta_1$ at which this argument vanishes, i.e.

$$6\frac{\theta_0\omega_0^4}{\alpha b^2}(\theta_1 - \theta_0) - 4\frac{\delta^2}{\alpha^2} = 0 \ .$$

Hence, θ_1 is given by

$$\theta_1 = \theta_0 + \frac{2\delta^2 b^2}{3\alpha\theta_0\omega_0^4} \ . \tag{6.2.34}$$

According to the first approximation (6.2.33), the two near saddle points lie along the imaginary axis, symmetrically located about the point $\omega = -(2/3)\mathrm{i}(\delta/\alpha)$ for $1 \leqslant \theta < \theta_1$; for $\theta = \theta_1$ the two saddle points coalesce into a single saddle point of higher order at $\omega = -(2/3)\mathrm{i}(\delta/\alpha)$; for $\theta > \theta_1$ the two saddle points move off the imaginary axis and into the complex ω-plane along the line $\omega = -(2/3)\mathrm{i}(\delta/\alpha)$, symmetrically located about the imaginary axis, and as θ approaches infinity the two saddle points approach $\pm\infty$, respectively, along this line. However, when $\theta > \theta_1$ gets large enough so that $|\varepsilon| = |\omega^2 + 2\delta\mathrm{i}\omega|$ is no longer small in comparison to ω_0^2, then the above approximate locations of the near saddle points is no longer valid. As is shown below in the second approximation, the near saddle points leave the line $\omega = -(2/3)\mathrm{i}(\delta/\alpha)$ for $\theta > \theta_1$ and approach the branch points ω_+ and ω_- in the limit as θ approaches infinity.

(b) *The Second Approximation*

In order to obtain a more accurate description of the locations of the near saddle points, particularly for values of θ greater than θ_1, the exact saddle-point equation (6.2.4) is again employed. Equation (6.2.4) may be rewritten in the form

$$\theta^2(\omega^2 - \omega_0^2 + 2\delta\mathrm{i}\omega) = \omega^2 - \omega_1^2 + 2\delta\mathrm{i}\omega + 2b^2 \frac{\omega(\omega + \delta\mathrm{i})}{\omega^2 - \omega_0^2 + 2\delta\mathrm{i}\omega} + b^4 \frac{\omega^2(\omega + \delta\mathrm{i})^2}{(\omega^2 - \omega_1^2 + 2\delta\mathrm{i}\omega)(\omega^2 - \omega_0^2 + 2\delta\mathrm{i}\omega)^2}, \tag{5.2.35}$$

where, up to this point, this equation is exact. Now, for $|\omega|$ small in comparison to ω_0, the following two expansions are useful:

$$\frac{\omega(\omega + \delta\mathrm{i})}{\omega^2 - \omega_0^2 + 2\delta\mathrm{i}\omega} = -\frac{1}{\omega_0^2} \cdot \frac{\omega^2 + \delta\mathrm{i}\omega}{1 - \dfrac{2\delta\mathrm{i}}{\omega_0^2}\omega - \dfrac{\omega^2}{\omega_0^2}}$$

$$\cong -\frac{1}{\omega_0^2}\left[\delta\mathrm{i}\omega + \left(1 - \frac{2\delta^2}{\omega_0^2}\right)\omega^2 + \frac{\delta\mathrm{i}}{\omega_0^2}\left(3 - 4\frac{\delta^2}{\omega_0^2}\right)\omega^3\right],$$

and

$$\frac{\omega^2(\omega + \delta\mathrm{i})^2}{(\omega^2 - \omega_1^2 + 2\delta\mathrm{i}\omega)(\omega^2 - \omega_0^2 + 2\delta\mathrm{i}\omega)^2} \cong \frac{\omega^2(\omega^2 + 2\delta\mathrm{i}\omega - \delta^2)}{-\omega_1^2\omega_0^4 + 2\delta\mathrm{i}\omega_0^2(2\omega_1^2 + \omega_0^2)\omega}$$

$$\cong -\frac{1}{\omega_1^2\omega_0^4}\omega^2\left[-\delta^2 + 2\delta\mathrm{i}\omega - \frac{2\delta^3\mathrm{i}}{\omega_1^2\omega_0^2}(2\omega_1^2 + \omega_0^2)\omega\right].$$

With these approximate expressions, the saddle-point equation (6.2.35) for the near saddle points becomes

$$\theta^2(\omega^2 - \omega_0^2 + 2\delta\mathrm{i}\omega) \cong -\frac{2\delta\mathrm{i}b^2}{\omega_0^4}\left[3 + \frac{b^2}{\omega_1^2} - \frac{4\delta^2}{\omega_0^2} - \frac{\delta^2 b^2}{\omega_1^4\omega_0^2}(2\omega_1^2 + \omega_0^2)\right]\omega^3$$

$$+ \left[1 - \frac{b^2}{\omega_0^2}\left(2 - \frac{4\delta^2}{\omega_0^2} - \frac{\delta^2 b^2}{\omega_1^2\omega_0^2}\right)\right]\omega^2$$

$$+ 2\delta\mathrm{i}\left(1 - \frac{b^2}{\omega_0^2}\right)\omega - \omega_0^2 - b^2,$$

from which it is easily seen that the coefficient of the cubic term is very small in comparison to the other terms. Hence, this term may be neglected to obtain the equation

$$\left[1-\theta^2-\frac{b^2}{\omega_0^2}\left(2-\frac{4\delta^2}{\omega_0^2}-\frac{\delta^2 b^2}{\omega_1^2\omega_0^2}\right)\right]\omega^2+2\delta\mathrm{i}\left(1-\theta^2-\frac{b^2}{\omega_0^2}\right)\omega$$
$$-(1-\theta^2)\omega_0^2-b^2$$
$$=0 .$$

The coefficients appearing in this equation can be rewritten as follows:

$$\omega_0^2(1-\theta^2)-b^2=\omega_0^2(\theta_0^2-\theta^2) ,$$

$$1-\theta^2-\frac{b^2}{\omega_0^2}=\theta_0^2-\theta^2-\frac{2b^2}{\omega_0^2} ,$$

$$1-\theta^2-\frac{b^2}{\omega_0^2}\left(2-\frac{4\delta^2}{\omega_0^2}-\frac{\delta^2 b^2}{\omega_1^2\omega_0^2}\right)=\theta_0^2-\theta^2-\frac{3b^2}{\omega_0^2}\alpha ,$$

where the parameter α has been redefined here as, c.f. (6.2.32),

$$\alpha\equiv 1-\frac{\delta^2}{3\omega_0^2\omega_1^2}(4\omega_1^2+b^2) . \tag{6.2.36}$$

With these substitutions, the second approximate saddle-point equation becomes

$$\omega^2+2\delta\mathrm{i}\frac{\theta^2-\theta_0^2+\dfrac{2b^2}{\omega_0^2}}{\theta^2-\theta_0^2+\dfrac{3b^2}{\omega_0^2}\alpha}\omega-\frac{\omega_0^2(\theta^2-\theta_0^2)}{\theta^2-\theta_0^2+\dfrac{3b^2}{\omega_0^2}\alpha}=0 . \tag{6.2.37}$$

Before studying the exact solution to this equation, consider first its approximate behavior for small values of θ. For $\theta\cong\theta_0$, the coefficients in this equation reduce to

$$\left.\frac{\theta^2-\theta_0^2+\dfrac{2b^2}{\omega_0^2}}{\theta^2-\theta_0^2+\dfrac{3b^2}{\omega_0^2}\alpha}\right|_{\theta\cong\theta_0}=\frac{2}{3\alpha} ,$$

$$\left.\frac{\omega_0^2(\theta^2-\theta_0^2)}{\theta^2-\theta_0^2+\dfrac{3b^2}{\omega_0^2}\alpha}\right|_{\theta\cong\theta_0}=\frac{2}{3}\frac{\omega_0^3\omega_1}{b^2\alpha}(\theta-\theta_0) ,$$

so that the saddle-point equation (6.2.37) becomes

$$\omega^2 + \frac{4}{3}\frac{\delta i}{\alpha}\omega - \frac{2}{3}\frac{\omega_0^3\omega_1}{b^2\alpha}(\theta - \theta_0) = 0 \ ,$$

for $\theta \cong \theta_0$. This is the exact same equation as was obtained in the first approximation, c.f. (6.2.30), if one identifies the value of the parameter α as given by (6.2.32) as being an approximation of the more accurate value given by (6.2.36). As an immediate consequence of this result, it is seen that the first approximate locations of the near saddle points are valid only for $\theta \cong \theta_0$, as was pointed out earlier.

Consider now the second approximate locations of the near saddle points which are given by using the solution of (6.2.37), namely

$$\omega_{sp_N^\pm}(\theta) \cong \pm\psi(\theta) - \tfrac{2}{3}i\delta\zeta(\theta) \ , \tag{6.2.38}$$

where, for convenience, the following two functions have been defined:

$$\zeta(\theta) \equiv \frac{3}{2}\cdot\frac{\theta^2 - \theta_0^2 + \dfrac{2b^2}{\omega_0^2}}{\theta^2 - \theta_0^2 + \dfrac{3b^2}{\omega_0^2}\alpha} \ , \tag{6.2.39}$$

$$\psi(\theta) \equiv \left[\frac{\omega_0^2(\theta^2 - \theta_0^2)}{\theta^2 - \theta_0^2 + \dfrac{3b^2}{\omega_0^2}\alpha} - \delta^2\left(\frac{\theta^2 - \theta_0^2 + \dfrac{2b^2}{\omega_0^2}}{\theta^2 - \theta_0^2 + \dfrac{3b^2}{\omega_0^2}\alpha}\right)^2\right]^{1/2} . \tag{6.2.40}$$

Here, the positive square root is to be taken and α is expressed by (6.2.36). The expression for $\omega_{sp_N^\pm}(\theta)$ given in (6.2.38) is called *the second approximation to the location of the saddle points in the region near the origin.*

For values of θ approximately equal to θ_0, the functions (6.2.39 and 40) simplify to

$$\zeta(\theta)|_{\theta\cong\theta_0} \cong \frac{1}{\alpha} \ ,$$

$$\psi(\theta)|_{\theta\cong\theta_0} \cong \left[\frac{2\theta_0\omega_0^4}{3b^2\alpha}(\theta - \theta_0) - \frac{4\delta^2}{9\alpha^2}\right]^{1/2} \ ,$$

and (6.2.38) reduces to the first approximate solution (6.2.33). Furthermore, in the limit as θ approaches infinity, the second approximate solution yields

$$\lim_{\theta\to\infty}[\omega_{sp_N^\pm}(\theta)] = \pm\sqrt{\omega_0^2 - \delta^2} - \delta i = \omega_\pm \ , \tag{6.2.41}$$

so that the two near saddle-points approach the branch points $\omega_\pm$, as expected.

In order to analyze the behavior of the near saddle points as specified by the second approximation (6.2.38), it is again necessary to first determine the algebraic sign of the argument of the square root in this expression. This amounts to determining a more accurate value of $\theta \equiv \theta_1$ at which that argument vanishes, i.e.

$$\frac{\omega_0^2(\theta_1^2 - \theta_0^2)}{\theta_1^2 - \theta_0^2 + \dfrac{3b^2}{\omega_0^2}\alpha} - \delta^2 \left[\frac{\theta_1^2 - \theta_0^2 + \dfrac{2b^2}{\omega_0^2}}{\theta_1^2 - \theta_0^2 + \dfrac{3b^2}{\omega_0^2}\alpha}\right]^2 = 0 ,$$

which simplifies to

$$(\omega_0^2 - \delta^2)(\theta_1^2 - \theta_0^2)^2 + b^2\left(3\alpha - \frac{4\delta^2}{\omega_0^2}\right)(\theta_1^2 - \theta_0^2) - \frac{4\delta^2 b^4}{\omega_0^4} = 0 .$$

Since θ_1 is greater than θ_0 for nonzero values of δ, c.f. (6.2.34), the corresponding solution of this binomial equation is then

$$\theta_1^2 = \theta_0^2 + \frac{b^2}{2}\frac{3\alpha - \dfrac{4\delta^2}{\omega_0^2}}{\omega_0^2 - \delta^2}\left(\sqrt{1 + 16\frac{\delta^2(\omega_0^2 - \delta^2)}{(3\alpha\omega_0^2 - 4\delta^2)^2}} - 1\right) ,$$

so that

$$\theta_1 = \left[\theta_0^2 + \frac{b^2}{2}\frac{3\alpha - \dfrac{4\delta^2}{\omega_0^2}}{\omega_0^2 - \delta^2}\left(\sqrt{1 + 16\frac{\delta^2(\omega_0^2 - \delta^2)}{(3\alpha\omega_0^2 - 4\delta^2)^2}} - 1\right)\right]^{1/2} , \tag{6.2.42}$$

where the positive values of both square roots in this expression are to be taken.

In order to compare the more accurate expression of the critical parameter θ_1 given in (6.2.42) with the first approximate expression (6.2.34), the above expression for θ_1^2 may be expanded as follows:

$$\theta_1^2 - \theta_0^2 \cong \frac{b^2}{2}\frac{3\alpha - \dfrac{4\delta^2}{\omega_0^2}}{\omega_0^2 - \delta}\left(1 + 8\frac{\delta^2(\omega_0^2 - \delta^2)}{(3\alpha\omega_0^2 - 4\delta^2)^2} - 1\right)$$

$$= \frac{4\delta^2 b^2}{\omega_0^2(3\alpha\omega_0^2 - 4\delta^2)} , \tag{6.2.43a}$$

and hence

$$\theta_1 \cong \left(\theta_0^2 + \frac{4\delta^2 b^2}{\omega_0^2(3\alpha\omega_0^2 - 4\delta^2)}\right)^{1/2} \cong \theta_0 + \frac{2\delta^2 b^2}{\theta_0\omega_0^2(3\alpha\omega_0^2 - 4\delta^2)} , \tag{6.2.43b}$$

which differs only slightly from (6.2.34). The approximate expression (6.2.43a) for θ_1^2 is of much simpler form than that obtained from (6.2.42) and greatly simplifies subsequent calculations concerning the behavior of the saddle points at that critical value.

An analytic approximation of the phase behavior $\phi(\omega, \theta)$ valid in that region of the complex ω-plane which is traversed by the near saddle points is now considered. For the present analysis, the complex index of refraction as given by the approximate expression (6.2.29) is sufficiently accurate. By (6.2.32), this expression may be rewritten as

$$n(\omega) \cong \theta_0 + \frac{b^2}{2\theta_0 \omega_0^4} \omega(\alpha\omega + 2\delta \mathrm{i}) \ , \tag{6.2.44}$$

so that, to a good approximation, $\phi(\omega, \theta)$ becomes

$$\phi(\omega, \theta) \cong \mathrm{i}\omega(\theta_0 - \theta) + \frac{b^2}{2\theta_0 \omega_0^4} \omega^2(\mathrm{i}\alpha\omega - 2\delta) \ . \tag{6.2.45}$$

(c) *Behavior of the Second Approximation*

The behavior of the near saddle points in the complex ω-plane as given by (6.2.38) is now considered for the three separate cases $1 \leqslant \theta < \theta_1$, $\theta = \theta_1$, and $\theta > \theta_1$.

For $1 \leqslant \theta < \theta_1$, the near saddle point locations are determined by

$$\omega_{sp_N^\pm} \cong \mathrm{i}[\pm|\psi(\theta)| - \tfrac{2}{3}\delta\zeta(\theta)] \ , \tag{6.2.46}$$

where the quantity inside the brackets is real. Hence, the near saddle points are located on the imaginary axis and are symmetrically situated about the point $-\frac{2}{3}\delta\zeta(\theta)\mathrm{i}$, where $\zeta(\theta)$ varies slowly for $1 \leqslant \theta < \theta_1$, as is seen from (6.2.39).

In order to obtain the behavior of $X(\omega, \theta)$ (and hence, the paths of steepest ascent and descent) in the vicinity of either of the near saddle points for $1 \leqslant \theta < \theta_1$, ω is expressed in polar coordinates r, φ about the particular saddle point as

$$\begin{aligned} \omega &= \omega_{sp_N^\pm} + r e^{\mathrm{i}\varphi} \\ &= \mathrm{i}\omega'' + r e^{\mathrm{i}\varphi} \ , \end{aligned}$$

where

$$\omega'' = \pm|\psi(\theta)| - \tfrac{2}{3}\delta\zeta(\theta) \ .$$

With this result and (6.2.45), the approximate phase behavior in the vicinity of the near saddle points becomes

$$\begin{aligned}\phi(\theta,r,\varphi) \cong (-\omega'' + \mathrm{i}r\mathrm{e}^{\mathrm{i}\varphi})(\theta_0 - \theta) + \frac{b^2}{2\theta_0\omega_0^4}[\alpha\omega''^3 + 2\delta\omega''^2 \\ - \mathrm{i}(3\alpha\omega''^2 + 4\delta\omega'')r\mathrm{e}^{\mathrm{i}\varphi} - (3\alpha\omega'' + 2\delta)r^2\mathrm{e}^{\mathrm{i}2\varphi} \\ + \mathrm{i}\alpha r^3\mathrm{e}^{\mathrm{i}3\varphi}] \ . \end{aligned} \tag{6.2.47}$$

Hence, taking the real part of this equation yields

$$\begin{aligned}X(\theta,r,\varphi) \cong (\omega'' + r\sin\varphi)(\theta - \theta_0) + \frac{b^2}{2\theta_0\omega_0^4}[\alpha\omega''^3 + 2\delta\omega''^2 \\ + (3\alpha\omega''^2 + 4\delta\omega'')r\sin\varphi - (3\alpha\omega'' + 2\delta)r^2\cos 2\varphi \\ - \alpha r^3\sin 3\varphi] \ , \end{aligned} \tag{6.2.48}$$

from which it is seen that $X(\theta,r,\varphi)$ attains its maximum variation for $\varphi = 0, \pi/2$, π, and $3\pi/2$. That is, the lines of steepest descent and ascent are parallel to the coordinate axes, as illustrated in Fig. 6.11, where the arrows indicate ascent along these lines, and where the hatched areas indicate the regions of the complex ω-plane wherein $X(\omega)$ is below that value at the saddle points. The saddle point labeled SP_1 is located at $\omega_{sp_N^+}$ and the saddle point labeled SP_2 is located at $\omega_{sp_N^-}$. A further consideration of (6.2.48) for values of θ within the interval $1 \leqslant \theta < \theta_1$ and the associated values of ω'' shows that for the upper near saddle point SP_1, the paths of steepest descent are at $\varphi = 0$ and π, whereas for the

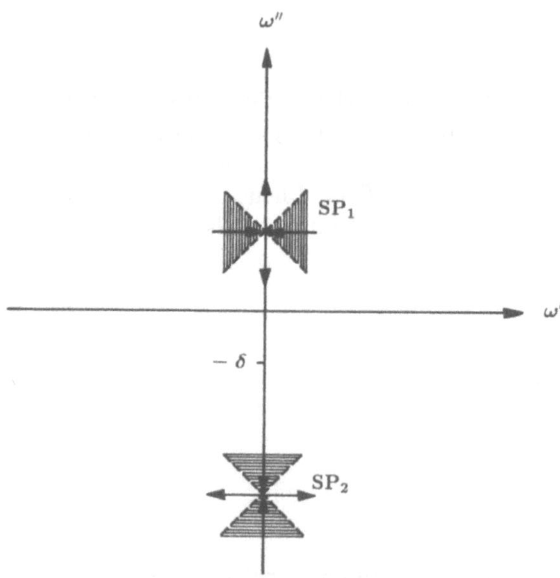

Fig. 6.11. Behavior of the saddle points in the region near the origin for $1 \leqslant \theta < \theta_1$.

lower near saddle point SP_2, the paths of steepest descent are at $\varphi = \pi/2$ and $3\pi/2$, as is indicated in Fig. 6.11.

For the special value $\theta = \theta_0$ (where $1 < \theta < \theta_0$), (6.2.38) yields the roots

$$\omega_{sp_N^+}(\theta_0) = 0 \ , \tag{6.2.49a}$$

$$\omega_{sp_N^-}(\theta_0) \cong -\frac{4}{3}\frac{\delta}{\alpha}\mathrm{i} \ , \tag{6.2.49b}$$

where it is to be noted that the root $\omega_{sp_N^+}(\theta_0)$ is exact. At the upper near saddle point SP_1, not only does $\phi'(\omega_{sp_N^+}, \theta_0)$ vanish, but also $\phi(\omega_{sp_N^+}, \theta_0)$ vanishes by consideration of (6.2.47) for $\theta = \theta_0$ and $\omega'' = 0$ [this vanishing of $\phi(\omega_{sp_N^+}, \theta_0)$ is also exact].

At the critical value $\theta = \theta_1$, (6.2.38) yields only one saddle point which is located at

$$\begin{aligned} \omega_{sp_N}(\theta_1) &\cong -\frac{2}{3}\mathrm{i}\delta\zeta(\theta_1) \\ &= -\delta\mathrm{i}\frac{\theta_1^2 - \theta_0^2 + \dfrac{2b^2}{\omega_0^2}}{\theta_1^2 - \theta_0^2 + \dfrac{3b^2}{\omega_0^2}\alpha} \qquad (6.2.50a) \\ &\cong -\frac{2}{3}\frac{\delta}{\alpha}\mathrm{i} \ , \qquad (6.2.50b) \end{aligned}$$

where the final expression (6.2.50b) is obtained upon use of the approximate expression (6.2.43a) for $\theta_1^2 - \theta_0^2$ and the fact that $\delta \ll \omega_0$. At this critical value of θ, not only does $\phi'(\omega_{sp_N}, \theta_1)$ vanish, but also

$$\phi''(\omega_{sp_N}, \theta_1) \cong \frac{b^2}{\theta_0\omega_0^4}[3\mathrm{i}\alpha\omega_{sp_N}(\theta_1) - 2\delta] = 0 \ . \tag{6.2.51}$$

Hence, at $\theta = \theta_1$ the two saddle points along the imaginary axis coalesce into a single second-order saddle point at which both $\phi'(\omega_{sp_N}, \theta_1)$ and $\phi''(\omega_{sp_N}, \theta_1)$ vanish. By (6.2.48) with $\omega'' = -(2/3)(\delta/\alpha)$, the behavior of $X(\omega, \theta_1)$ in the vicinity about this saddle point is given by

$$X(\theta_1, r, \varphi) \cong -\frac{b^2}{\theta_0\omega_0^4}\left(\frac{4\delta^3}{27\alpha^2} + \frac{1}{2}\alpha r^3 \sin 3\varphi\right) \ . \tag{6.2.52}$$

Hence, $X(\theta_1, r, \varphi)$ attains its maximum variation for $\varphi = \pi/6$, $\pi/2$, $5\pi/6$, $7\pi/6$, $3\pi/2$, and $11\pi/6$; the lines of steepest descent from the saddle point are at $\varphi = \pi/6$, $5\pi/6$, and $3\pi/2$, while the lines of steepest ascent are at $\varphi = \pi/2$, $7\pi/6$, and $11\pi/6$. Figure 6.12 illustrates this special saddle point of higher order [again, the

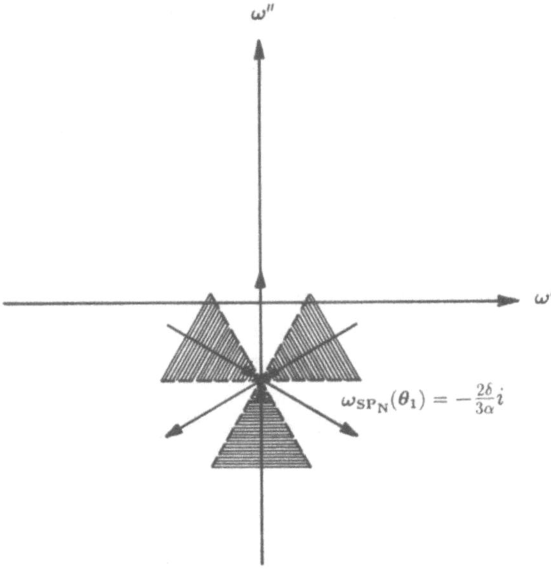

Fig. 6.12. The two saddle points in the region near the origin have coalesced into a single special saddle point of second order for $\theta = \theta_1$.

arrows indicate ascent along the lines of steepest descent and ascent through the saddle point, and the hatched areas indicate the regions of the complex ω-plane wherein $X(\omega)$ is less than that at the saddle point].

Consider finally the near saddle point behavior for $\theta > \theta_1$. In this case, by (6.2.38), the near saddle point locations are given approximately by

$$\omega_{sp_N^\pm}(\theta) \cong \pm\psi(\theta) - \tfrac{2}{3}i\delta\zeta(\theta) \ , \qquad (6.2.53)$$

where both $\psi(\theta)$ and $\zeta(\theta)$ are real. Thus, for $\theta > \theta_1$ the two near saddle points move off of the imaginary axis into the lower-half of the complex ω-plane, and are symmetrically located about the imaginary axis. In the limit as θ approaches infinity, these two saddle points approach the branch points ω_+ and ω_-, respectively.

In order to obtain the behavior of $X(\omega, \theta)$ in the vicinity of the near saddle points, ω is again expressed in polar coordinates r, φ about the particular saddle point. Since $X(\omega, \theta)$ is symmetric about the imaginary axis, only the saddle point in the right-half plane need be considered. Hence, let

$$\begin{aligned}\omega &= \omega_{sp_N^+} + re^{i\varphi} \\ &= \psi(\theta) - \tfrac{2}{3}i\delta\zeta(\theta) + re^{i\varphi} \ .\end{aligned}$$

With this expression substituted into (6.2.45), the approximate phase behavior in the vicinity of the near saddle point $\omega_{sp_N^+}(\theta)$ for $\theta > \theta_1$ becomes

$$\phi(\theta, r, \varphi) \cong \left[\frac{2}{3}\delta\zeta(\theta) + \mathrm{i}\psi(\theta) + \mathrm{i}r\mathrm{e}^{\mathrm{i}\varphi}\right](\theta_0 - \theta)$$

$$+ \frac{b^2}{2\theta_0\omega_0^4}\left[\psi(\theta) - \frac{2}{3}\mathrm{i}\delta\zeta(\theta) + r\mathrm{e}^{\mathrm{i}\varphi}\right]^2$$

$$\cdot \left[\frac{2}{3}\alpha\delta\zeta(\theta) - 2\delta + \mathrm{i}\alpha\psi(\theta) + \mathrm{i}\alpha r\mathrm{e}^{\mathrm{i}\varphi}\right]$$

$$= \left[\frac{2}{3}\delta\zeta(\theta) + \mathrm{i}\psi(\theta) + \mathrm{i}r\mathrm{e}^{\mathrm{i}\varphi}\right](\theta_0 - \theta)$$

$$+ \frac{b^2}{2\theta_0\omega_0^4}\left[\frac{8}{9}\delta^3\zeta^2(\theta)\left[1 - \frac{1}{3}\alpha\zeta(\theta)\right]\right.$$

$$+ 2\delta\psi^2(\theta)[\alpha\zeta(\theta) - 1]$$

$$+ \mathrm{i}\left\{\frac{4}{3}\delta^2\zeta(\theta)\psi(\theta)[2 - \alpha\zeta(\theta)] + \alpha\psi^3(\theta)\right\}$$

$$+ \left(4\delta\psi(\theta)[\alpha\zeta(\theta) - 1]\right.$$

$$\left. + \mathrm{i}\left\{3\alpha\psi^2(\theta) + \frac{4}{3}\delta^2\zeta(\theta)[2 - \alpha\zeta(\theta)]\right\}\right) r\mathrm{e}^{\mathrm{i}\varphi}$$

$$\left. + \{2\delta[\alpha\zeta(\theta) - 1] + 3\mathrm{i}\alpha\psi(\theta)\}r^2\mathrm{e}^{\mathrm{i}2\varphi} + \mathrm{i}\alpha r^3\mathrm{e}^{\mathrm{i}3\varphi}\right]. \tag{6.2.54}$$

Taking the real part of this equation yields

$$X(\theta, r, \varphi) \cong \left[r\sin\varphi - \frac{2}{3}\delta\zeta(\theta)\right](\theta - \theta_0)$$

$$+ \frac{b^2}{2\theta_0\omega_0^4}\left(\frac{8}{9}\delta^3\zeta^2(\theta)\left[1 - \frac{1}{3}\alpha\zeta(\theta)\right]\right.$$

$$+ 2\delta\psi^2(\theta)[\alpha\zeta(\theta) - 1] + 4\delta\psi(\theta)[\alpha\zeta(\theta) - 1]r\cos\varphi$$

$$- \left\{3\alpha\psi^2(\theta) + \frac{4}{3}\delta^2\zeta(\theta)[2 - \alpha\zeta(\theta)]\right\}r\sin\varphi$$

$$+ 2\delta[\alpha\zeta(\theta) - 1]r^2\cos 2\varphi - 3\alpha\psi(\theta)r^2\sin 2\varphi$$

$$\left. - \alpha r^3\sin 3\varphi\right), \tag{6.2.55}$$

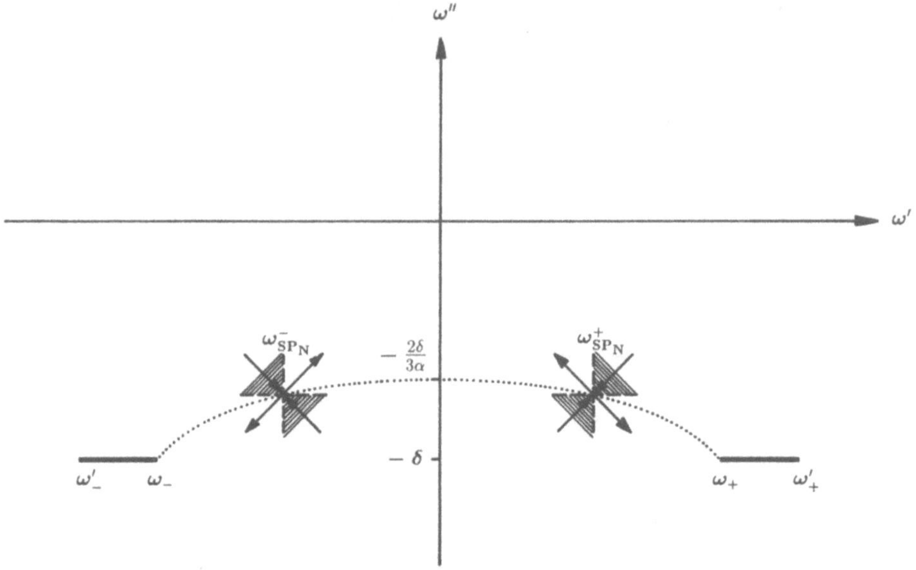

Fig. 6.13. Behavior of the saddle points in the region near the origin for $\theta > \theta_1$. The dotted lines indicate the paths followed by the saddle points as θ varies.

from which it is seen that $X(\theta, r, \varphi)$ attains its maximum variation for $\varphi = \pi/4$, $3\pi/4$, $5\pi/4$, and $7\pi/4$. In the right-half plane, the lines of steepest descent through the near saddle point are at $\varphi = \pi/4$ and $5\pi/4$, while the lines of steepest ascent are at $\varphi = 3\pi/4$ and $7\pi/4$. In the left-half plane the lines of steepest ascent and descent are reversed due to the symmetricity of $X(\omega)$ about the imaginary axis. These two near saddle points are illustrated in Fig. 6.13, where the arrows indicate ascent along the lines of steepest descent and ascent through the saddle points, and where the hatched areas indicate the regions of the complex ω-plane wherein $X(\omega)$ is less than that at the saddle points.

Thus, in summary, the two saddle points in the region near the origin, which are at first on the imaginary axis symmetrically located about the point $\omega'' = -\frac{2}{3}\delta\zeta(\theta)$, move closer together with increasing θ; at $\theta = \theta_0$ the phase function $\phi(\omega, \theta_0)$ vanishes identically at the upper near saddle point $\omega_{sp^+_N}(\theta_0) = 0$; at $\theta = \theta_1$ these two saddle points coalesce into a single special saddle point of second order at $\omega_{sp_N} \cong -(2\delta/3\alpha)i$, at which both $\phi'(\omega_{sp_N}, \theta_1)$ and $\phi''(\omega_{sp_N}, \theta_1)$ vanish; and for $\theta > \theta_1$ they move symmetrically off of the imaginary axis, approaching the branch points $\omega_\pm$ in the limit as θ approaches infinity.

6.3 Analytic Determination of the Dominant Saddle Point

In order to obtain the asymptotic approximation of the integral representation of the propagated field $A(z, t)$, it is necessary to determine the saddle point (called the dominant saddle point) that gives the least exponential decay in the integrand. That task is accomplished in this section with the aid of the expressions obtained in Sect. 6.2 for the complex phase behavior at the saddle points. Except for the lower near saddle point SP_2, the results of Sect. 6.2 show that the values of $X(\omega, \theta)$ at the remaining saddle points are always negative or zero for $\theta \geqslant 1$. The values of $X(\omega, \theta)$ at SP_2 are positive at first, and then become negative as θ approaches θ_1 from below. However, the original contour of integration is not deformable into an Olver-type path through SP_2 for $1 \leqslant \theta < \theta_1$, since that would require deforming the contour across the branch cuts $\omega_+\omega'_+$ and $\omega_-\omega'_-$. Consequently, the lower near saddle point plays no role in the asymptotic analysis and may be ignored for $1 \leqslant \theta < \theta_1$ in the following analysis. Since the asymptotic expansion of the field $A(z, t)$ is to be carried out for large positive distances z, the dominant saddle point is that one for which $X(\omega_{sp}, \theta) \equiv \mathrm{Re}\{\phi(\omega_{sp}, \theta)\}$ is least negative or zero. Since $X(\omega, \theta)$ is symmetric about the imaginary axis, only the behavior in the right-half of the complex ω-plane need be considered.

Consider first the saddle point in the region removed from the origin, which, by (6.2.21), is located at

$$\omega_{sp_D^+} \cong \xi(\theta) - \delta\mathrm{i}[1 + \eta(\theta)] \ , \tag{6.3.1}$$

where the functions $\xi(\theta)$ and $\eta(\theta)$ are as defined in (6.2.22 and 23). According to (6.2.26), the real part of the phase behavior at this saddle point is given to a good approximation by

$$X(\omega_{sp_D}, \theta) \cong \delta[1 + \eta(\theta)](1 - \theta) - \frac{b^2}{2} \frac{\delta[1 - \eta(\theta)]}{\xi^2(\theta) + \delta^2[1 - \eta(\theta)]^2} \ , \tag{6.3.2}$$

where this result is valid for both of the distant saddle points. From this equation and the expressions for $\xi(\theta)$ and $\eta(\theta)$ it is easily seen that at $\theta = 1$, $X(\omega_{sp_D}, 1) = 0$. Furthermore, this vanishing of $X(\omega_{sp_D}, \theta)$ at $\theta = 1$ is satisfied exactly, c.f. (6.1.22). Finally, as θ increases away from unity, $X(\omega_{sp_D}, \theta)$ is seen to grow negative monotonically.

Consider now the saddle points in the region about the origin, which, by (6.2.38), are approximately located at

$$\omega_{sp_N^\pm} \cong \mathrm{i}[\pm|\psi(\theta)| - \tfrac{2}{3}\delta\zeta(\theta)] \equiv \mathrm{i}\omega'' \ ; \qquad 1 \leqslant \theta \leqslant \theta_1 \ , \tag{6.3.3}$$

$$\omega_{sp_N^+} \cong \psi(\theta) - \tfrac{2}{3}\mathrm{i}\delta\zeta(\theta) \ ; \qquad \theta \geqslant \theta_1 \ , \tag{6.3.4}$$

where the functions $\zeta(\theta)$ and $\psi(\theta)$ are as defined in (6.2.39 and 40). According to (6.2.48–55), the real part of the phase behavior at these saddle points is given to a good approximation by the pair of expressions

$$X(\omega_{sp_N^\pm}, \theta) \cong \omega''(\theta - \theta_0) + \frac{b^2}{2\theta_0 \omega_0^4} \omega''^2(\alpha\omega'' + 2\delta) \; ; \qquad 1 \leqslant \theta \leqslant \theta_1 \; , \tag{6.3.5}$$

$$X(\omega_{sp_N}, \theta) \cong -\frac{2}{3}\delta\zeta(\theta)(\theta - \theta_0) + \frac{b^2}{2\theta_0 \omega_0^4}\left\{\frac{8}{9}\delta^3\zeta^2(\theta)\left[1 - \frac{1}{3}\alpha\zeta(\theta)\right] + 2\delta\psi^2(\theta)[\alpha\zeta(\theta) - 1]\right\} \; ; \qquad \theta \geqslant \theta_1 \; , \tag{6.3.6}$$

where (6.3.6) is valid for both of the near saddle points for $\theta \geqslant \theta_1$. From these equations and the expressions for $\zeta(\theta)$ and $\psi(\theta)$, it is seen that at the upper near saddle point SP_1 [$\omega'' = |\psi(\theta)| - \frac{2}{3}\delta\zeta(\theta)$ in (6.3.5)], $X(\omega_{sp_N^+}, \theta)$ is at first negative for $1 \leqslant \theta < \theta_0$, vanishes identically at $\theta = \theta_0$, and grows negative monotonically for $\theta_0 < \theta \leqslant \theta_1$. Finally, for $\theta \geqslant \theta_1$, during which the two near saddle points have first coalesced and have then moved off into the complex ω-plane symmetrically located about the imaginary axis, it is seen from (6.3.6) that $X(\omega_{sp_N}, \theta)$ continues to decrease monotonically with increasing θ.

Attention is now turned to determining the domains of θ over which either the distant or near saddle points are dominant. According to the above discussion, the following progression of events is obtained, as illustrated in Fig. 6.14. At the distant saddle points, $X(\omega_{sp_D}, \theta)$ vanishes identically at $\theta = 1$, and decreases monotonically as θ increases away from unity. At the upper near saddle point SP_1, $X(\omega_{sp_N^+}, \theta)$ is at first negative for $1 \leqslant \theta < \theta_0$, vanishes identically at $\theta = \theta_0$, and decreases monotonically as θ increases away from θ_0, but, as is found in Sect. 6.4, is less negative than $X(\omega_{sp_D}, \theta)$ for all $\theta \geqslant \theta_0$. Consequently,

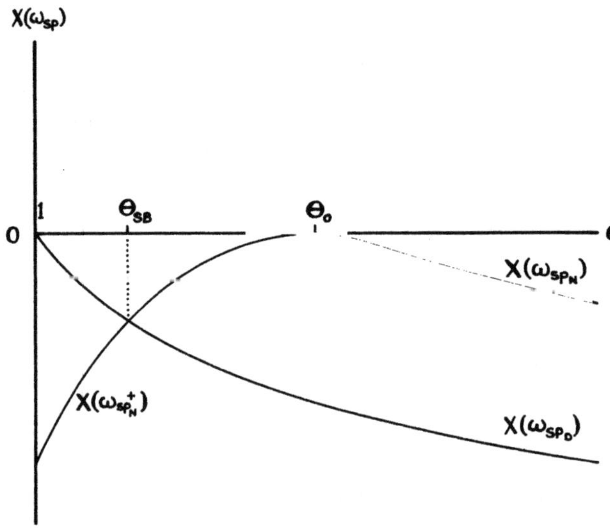

Fig. 6.14. Behavior of $X(\omega, \theta)$ at the relevant saddle points.

at first, the distant saddle points are dominant, but for some value of θ between unity and θ_0, which shall be called θ_{SB}, the near saddle point becomes dominant, and remains dominant for all later values of θ. It is therefore known that

$$1 < \theta_{SB} < \theta_0 \ .$$

All of these results are verified numerically in Sect. 6.4.

All that remains now to be determined is an analytic expression for the critical value θ_{SB} at which the upper near saddle point becomes dominant over the initially dominant distant saddle points. As is shown in Chap. 9, the quantity θ_{SB} is the value of θ at which the propagated field evolves from the first precursor over to the second precursor. The value of θ_{SB} is obtained by equating the real phase behavior $X(\omega_{sp_D}, \theta)$ at the distant saddle points with the real phase behavior $X(\omega_{sp_N^+}, \theta)$ at the upper near saddle point SP_1, i.e.

$$X(\omega_{sp_D}, \theta_{SB}) \equiv X(\omega_{sp_N^+}, \theta_{SB}) \ . \tag{6.3.7}$$

The resulting algebraic equation is then solved for θ_{SB}. If the second approximate expressions for $X(\omega_{sp_D}, \theta)$ and $X(\omega_{sp_N^+}, \theta)$ given by (6.3.2 and 5), respectively, are employed in (6.3.7) at $\theta = \theta_{SB}$, the resulting algebraic expression for θ_{SB} is intractable to solve. In order to obtain a manageable algebraic expression for θ_{SB}, the first approximate expressions for $X(\omega_{sp_D}, \theta)$ and $X(\omega_{sp_N^+}, \theta)$ are employed in (6.3.7). The numerical results of Sect. 6.4 show this to be a good approximation. (Figure 6.18 shows that the first approximate expressions yield a value of θ_{SB} which is 2.2% lower than the actual value, while the second approximate expressions yield a value of θ_{SB} which is 1% lower than the actual value). If a value of θ_{SB} is needed which is more accurate than that obtained when the first approximate expressions of $X(\omega_{sp_D}, \theta)$ and $X(\omega_{sp_N^+}, \theta)$ are used in (6.3.7), then a numerical solution of this equation using the second approximations of $X(\omega_{sp_D}, \theta)$ and $X(\omega_{sp_N^+}, \theta)$ could be employed with the first approximate solution for θ_{SB} used as an initial value.

From (6.3.2), the real phase behavior at the distant saddle points is given to a good approximation by

$$X(\omega_{sp_D}, \theta) \cong -2\delta(\theta - 1) \ ,$$

for $1 \leqslant \theta < \theta_0$. At the upper near saddle point SP_1, ω'' is given to a good approximation by

$$\omega'' \cong -\frac{2}{3}\frac{\delta}{\alpha} + \frac{1}{3}\sqrt{4\frac{\delta^2}{\alpha^2} + 6\frac{\theta_0\omega_0^4}{\alpha b^2}(\theta_0 - \theta)} \ ,$$

for $1 \leqslant \theta < \theta_1$, so that, from (6.3.5), the real part of the phase behavior at SP_1 is given by

$$X(\omega_{sp_N^+},\theta) \cong \omega''\left[\theta - \theta_0 + \frac{b^2}{2\theta_0\omega_0^4}(\alpha\omega''^2 + 2\delta\omega'')\right]$$

$$= -\frac{2}{27}\left\{\left[2\frac{\delta^2b^2}{\alpha\theta_0\omega_0^4} + 3(\theta_0 - \theta)\right]\sqrt{4\frac{\delta^2}{\alpha^2} + 6\frac{\theta_0\omega_0^4}{\alpha b^2}(\theta_0 - \theta)}\right.$$

$$\left. - 9\frac{\delta}{\alpha}(\theta_0 - \theta) - 4\frac{\delta^3b^2}{\alpha^2\theta_0\omega_0^4}\right\} .$$

These two expressions are then equated at $\theta = \theta_{SB}$, so that the following approximate equation for θ_{SB} is obtained:

$$\left[2\frac{\delta^2b^2}{\alpha\theta_0\omega_0^4} + 3(\theta_0 - \theta_{SB})\right]\sqrt{4\frac{\delta^2}{\alpha^2} + 6\frac{\theta_0\omega_0^4}{\alpha b^2}(\theta_0 - \theta_{SB})}$$

$$\cong -9\delta\left(3 - \frac{1}{\alpha}\right)(\theta_0 - \theta_{SB}) + 27\delta(\theta_0 - 1) + 4\frac{\delta^3b^2}{\alpha^2\theta_0\omega_0^4} . \tag{6.3.8}$$

This equation can be simplified considerably by noting that $\alpha \cong 1$, and that the last term on the right is negligible in comparison to the preceeding term. Under these approximations, the above equation becomes

$$\left[2\frac{\delta^2b^2}{\theta_0\omega_0^4} + 3(\theta_0 - \theta_{SB})\right]\sqrt{4\delta^2 + 6\frac{\theta_0\omega_0^4}{b^2}(\theta_0 - \theta_{SB})}$$

$$\cong -18\delta(\theta_0 - \theta_{SB}) + 27\delta(\theta_0 - 1) . \tag{6.3.9}$$

Upon squaring both sides of this equation and neglecting terms in δ^4 and higher, the following approximate cubic equation in $(\theta_0 - \theta_{SB})$ is obtained:

$$(\theta_0 - \theta_{SB})^3 - 4\frac{\delta^2b^2}{\theta_0\omega_0^4}(\theta_0 - \theta_{SB})^2 + 18\frac{\delta^2b^2(\theta_0 - 1)}{\theta_0\omega_0^4}(\theta_0 - \theta_{SB})$$

$$-\frac{27\delta^2b^2(\theta_0 - 1)^2}{2\theta_0\omega_0^4}$$

$$\cong 0 . \tag{6.3.10}$$

In order to solve this cubic equation by the standard technique, the following change of variable is performed

$$\theta_{SB} = \theta_0 - \frac{4\delta^2b^2}{3\theta_0\omega_0^4} - \xi , \tag{6.3.11}$$

where ξ is the real solution of the reduced cubic equation

$$\xi^3 + a\xi + c = 0 , \tag{6.3.12}$$

whose coefficients are given by

$$a = 18\frac{\delta^2 b^2(\theta_0 - 1)}{\theta_0\omega_0^4} - \frac{16\delta^4 b^4}{3\theta_0^2\omega_0^8}$$

$$\cong 18\frac{\delta^2 b^2(\theta_0 - 1)}{\theta_0\omega_0^4},$$

$$c = -\frac{128\delta^6 b^6}{27\theta_0^3\omega_0^{12}} + 24\frac{\delta^4 b^4(\theta_0 - 1)}{\theta_0^2\omega_0^8} - \frac{27\delta^2 b^2(\theta_0 - 1)^2}{2\theta_0\omega_0^4}$$

$$\cong -\frac{27\delta^2 b^2(\theta_0 - 1)^2}{2\theta_0\omega_0^4},$$

where terms in δ^4 and higher have been neglected. In order to obtain the solution of the reduced cubic equation (6.3.12) whose coefficients are given by the above approximate expressions, let

$$A = \left(-\frac{c}{2} + \sqrt{\frac{c^2}{4} + \frac{a^3}{27}}\right)^{1/3}$$

$$\cong \left[\frac{27\delta^2 b^2(\theta_0 - 1)^2}{4\theta_0\omega_0^4}\left(\sqrt{1 + \frac{64\delta^2 b^2}{27\theta_0(\theta_0 - 1)\omega_0^4}} + 1\right)\right]^{1/3},$$

$$B = \left(-\frac{c}{2} - \sqrt{\frac{c^2}{4} + \frac{a^3}{27}}\right)^{1/3}$$

$$\cong -\left[\frac{27\delta^2 b^2(\theta_0 - 1)^2}{4\theta_0\omega_0^4}\left(\sqrt{1 + \frac{64\delta^2 b^2}{27\theta_0(\theta_0 - 1)\omega_0^4}} - 1\right)\right]^{1/3},$$

where that branch of the cube root is chosen so that both A and B are real. The real-valued solution of (6.3.12) is then given by $\xi = A + B$, and hence, by (6.3.11), the critical value θ_{SB} is given approximately by the complicated expression

$$\theta_{SB} \cong \theta_0 - \frac{4\delta^2 b^2}{3\theta_0\omega_0^4} - \left(\frac{27\delta^2 b^2(\theta_0 - 1)^2}{4\theta_0\omega_0^4}\right)^{1/3} \cdot\left[\left(\sqrt{1 + \frac{\delta^2 b^2}{27\theta_0(\theta_0 - 1)\omega_0^4}} + 1\right)^{1/3} - \left(\sqrt{1 + \frac{\delta^2 b^2}{27\theta_0(\theta_0 - 1)\omega_0^4}} - 1\right)^{1/3}\right]. \tag{6.3.13}$$

A useful, simple estimate of the value of θ_{SB} is provided by the first two terms on the right-hand side of (6.3.13).

A related quantity of interest is the value of the real frequency $\omega_{SB} \geqslant \sqrt{\omega_1^2 - \delta^2}$ defined by the equation

$$X(\omega_{SB}) = X(\omega_{sp_D^+}, \theta_{SB}) \ . \tag{6.3.14}$$

The importance of this value arises in Chap. 9, wherein it is shown that for input signals with applied signal frequencies ω_c in the spectral range $0 \leqslant \omega_c < \omega_{SB}$, the two precursor fields evolve prior to the main signal arrival, whereas for applied signal frequencies ω_c in the range $\omega_c \geqslant \omega_{SB}$, only the first precursor field evolves prior to the signal arrival, while the second precursor field evolution occurs during the signal evolution. According to (6.3.14), the frequency ω_{SB} is the real coordinate value at which the contour $X(\omega_{sp_D^+}, \theta)$ at the distant saddle point in the right half of the complex ω-plane crosses the positive real axis when $\theta = \theta_{SB}$. The solution of (6.3.14) for ω_{SB} is an extremely formidable, if not impossible task. However, since the contour of $X(\omega_{sp_D^+}, \theta)$ through the distant saddle point at the angle $\pi/2$ to the real axis remains at essentially this angle when it intersects the real axis for not too large values of θ, as is evident in Figs. 6.5 through 6.9, an approximate expression for ω_{SB} is given by the real coordinate value of the distant saddle point in the right half of the complex ω-plane at $\theta = \theta_{SB}$. Consequently, according to (6.2.22) and the first two terms on the right-hand side of (6.3.13), this approximate expression for ω_{SB} is given by

$$\begin{aligned} \omega_{SB} &\cong \xi(\theta_{SB}) = \left(\omega_0^2 - \delta^2 + b^2 \frac{\theta_{SB}^2}{\theta_{SB}^2 - 1}\right)^{1/2} \\ &\cong \omega_0 \left(2 + \frac{b^2}{\omega_0^2} + \frac{5\delta^2}{3\omega_0^2}\right)^{1/2} , \end{aligned} \tag{6.3.15}$$

where terms in δ^4 and higher have been neglected.

6.4 Numerical Determination of the Saddle-Point Locations and the Associated Phase Behavior at the Saddle Points

This section presents a numerical evaluation of the exact saddle point locations and the exact behavior of the complex phase behavior $\phi(\omega, \theta)$ at these locations. These exact results are compared to both the first and second approximations of the saddle-point locations and the approximate behavior of the complex phase function $\phi(\omega, \theta)$ at these locations, as described in Sects. 6.2.1, 2. This comparison of the approximate and exact solutions is done over the entire range of values of θ of importance so that the range of values of θ over which a given approximation closely describes the exact behavior may be ascertained. *Brillouin*'s choice of the Lorentz medium parameters (viz., $\omega_0 = 4.0 \times 10^{16}$/s, $b^2 = 20.0 \times 10^{32}/\text{s}^2$, and $\delta = 0.28 \times 10^{16}$/s) are used in all of the numerical evaluations presented in this section.

As was discussed at the beginning of Sect. 6.2, with the complex index of refraction given by (6.2.2), the exact locations of the saddle points of $\phi(\omega, \theta)$ are given by the roots of (6.2.3), which may be written more simply (and more suitably for the purposes of numerical analysis) as

$$F(\omega, \theta) = n(\omega) + \frac{b^2 \omega(\omega + \delta i)}{(\omega^2 - \omega_0^2 + 2\delta i\omega)^2 n(\omega)} - \theta \equiv 0 \ . \tag{6.4.1}$$

What is necessary first then is a numerical solution of the exact saddle-point equation (6.4.1) as a function of the dimensionless parameter θ as it increases away from unity. This can be accomplished most easily by employing Newton's method at each value of θ with the second approximate solution at that value of θ as a starting point. Suppose then that ω is an approximate solution of equation (6.4.1) at some fixed value of θ, and let $\Delta\omega$ be a small correction that is to be determined such that $F(\omega + \Delta\omega) = 0$. If this equaton is expanded in a Taylor series about the approximate solution ω that is truncated after the first-order term, there results

$$\Delta\omega = -\frac{F(\omega)}{dF(\omega)/d\omega} , \tag{6.4.2}$$

from which the next approximation is obtained and the process repeated, if necessary. That is, let ω_1 denote the second approximate solution of the desired saddle point location at a given fixed value of θ, and let $\Delta\omega_1$, as determined from (6.4.2), denote the associated correction factor. The second trial solution of (6.4.1) is then $\omega_2 = \omega_1 + \Delta\omega_1$, and after j such iterations, the $(j + 1)$th solution is given by

$$\omega_{j+1} = \omega_j - \frac{F(\omega_j)}{dF(\omega)/d\omega|_{\omega=\omega_j}} . \tag{6.4.3}$$

This iteration procedure is terminated at a value ω_K such that $|F(\omega_K)|$ is less than some small positive number ε. The closer ε is to zero, the more accurate is the resulting numerically determined saddle point location. For all of the calculations presented here, $\varepsilon = 1.0 \times 10^{-11}$. Furthermore, the exact expression for $dF(\omega)/d\omega$ appearing in (6.4.3), given by

$$\frac{dF(\omega)}{d\omega} = \frac{b^2}{(\omega^2 - \omega_0^2 + 2\delta i\omega)^2 n(\omega)} \cdot \left[3\omega + 2\delta i - \omega(\omega + \delta i)^2 \cdot \frac{4n(\omega) + \dfrac{b^2}{(\omega^2 - \omega_0^2 + 2\delta i\omega)n(\omega)}}{(\omega^2 - \omega_0^2 + 2\delta i\omega)n(\omega)} \right] \tag{6.4.4}$$

was employed in the calculations of the saddle-point locations. Once the numerically determined exact saddle-point location at a given value of θ is obtained, the values of $X(\omega, \theta)$ and $Y(\omega, \theta)$ at this point are calculated according to the exact expressions (6.1.9 and 10), respectively.

Employing *Brillouin's* choice of the medium parameters, the following critical values of θ, as given by (6.2.64, 31, and 42) are obtained

$$\theta_{SB} \cong 1.295 \ ,$$

$$\theta_0 = 1.500 \ ,$$

$$\theta_1 \cong 1.503 \ .$$

The numerical results of the saddle point locations as a function of θ are illustrated graphically in Figs. 6.15 through 6.17; the first approximate locations (originally due to *Brillouin* [6.1, 2]) are indicated by the short-dashed lines, the second approximate locations (due to *Oughstun* and *Sherman* [6.4]) are indicated by the alternating long an short dashed lines, and the exact locations (obtained numerically using *Newton's* method) are indicated by the solid lines.

Consider first the locations of the distant saddle points in the right-half of the complex ω-plane as a function of the continuous parameter θ, as is illustrated in Fig. 6.15. In this diagram, the values of θ corresponding to the saddle-point locations marked by the crosses along the respective path are indicated

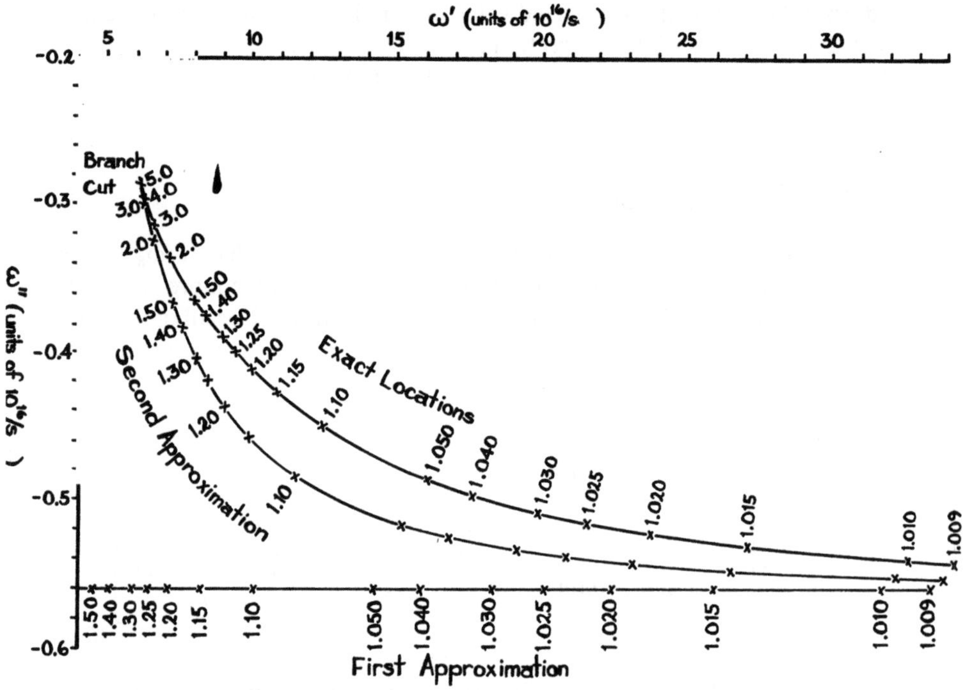

Fig. 6.15. Distant saddle-point locations in the right half of the complex ω-plane as a function of θ.

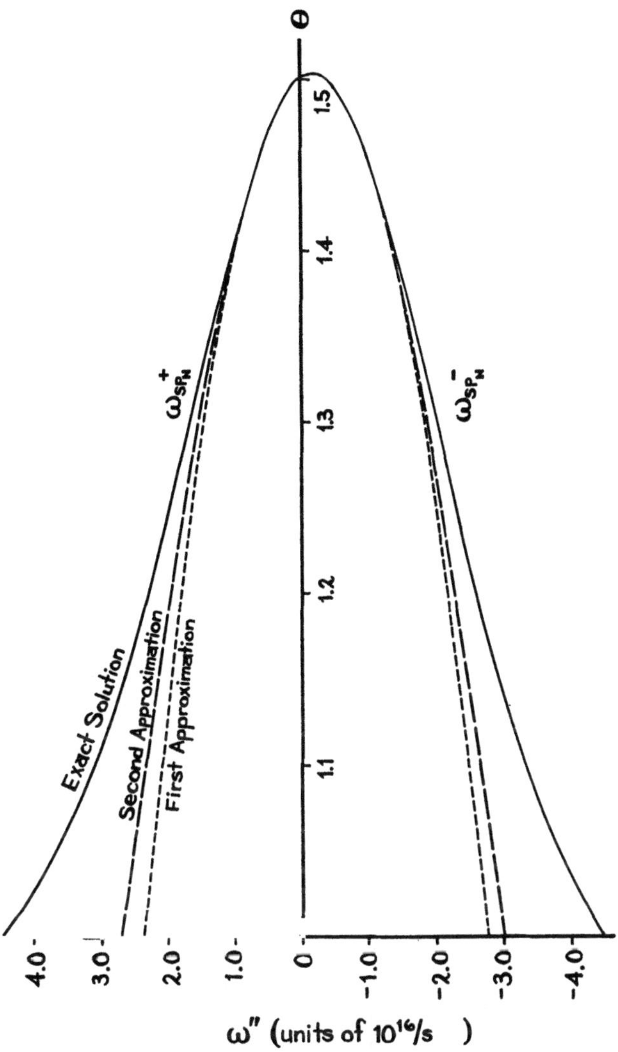

Fig. 6.16. Near saddle-point locations along the imaginary axis as a function of θ for $1 \leqslant \theta \leqslant \theta_1$, where $\theta_1 \cong 1.503$.

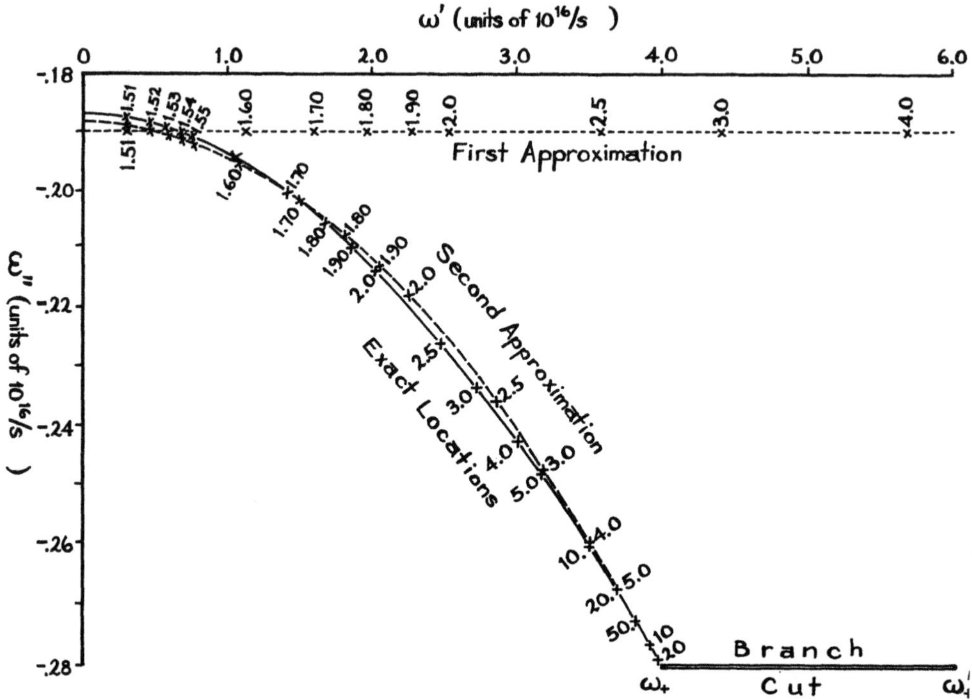

Fig. 6.17. Near saddle-point locations in the right half of the complex ω-plane as a function of θ for $\theta \geqslant \theta_1$.

along the various curves. As can easily be seen, the second approximate locations of this distant saddle point are in close agreement with the exact locations over the entire range of values of θ. Furthermore, it is seen that these second approximate locations move at roughly the same rate as do the exact locations. As regards this rate of motion of the saddle points with θ, it is observed for both the exact and second approximations that for θ very close to unity, a small increase in the value of θ produces a large change in the saddle-point location, and as θ increases away from unity, the change in location of the saddle point with θ steadily diminishes, so that as the saddle points $\omega_{sp_D^\pm}$ approach ever closer to the branch points $\omega'_\pm$, respectively, a large increase in the value of θ produces a small change in the locations of the two distant saddle points. The second approximate saddle point locations are further seen to be a marked improvement over the first approximate locations, which are seen to fail rapidly as $\omega' = \mathbb{R}\mathrm{e}\{\omega\}$ approaches near to the real part of the branch point ω'_+, corresponding roughly to values of θ greater than 1.05. Consequently, the second approximation (6.2.21) for the location of the saddle points in the region removed from the origin accurately describes the exact behavior over the entire range of values of θ, whereas the first approximation (6.2.9) holds only for values of θ very close to unity.

Consider next the locations of the near saddle points in the complex ω-plane as a function of θ. According to the results of Sect. 6.2.2, for $1 \leqslant \theta \leqslant \theta_1$ the approximate locations of the near saddle points are along the imaginary axis, approximately symmetric about the value $\omega'' = -(2/3)\delta/\alpha$, as illustrated in Fig. 6.16. The actual locations of the two near saddle points SP_1 and SP_2 in this case are also along the imaginary axis, and they are approximately symmetric about the value $\omega'' = -(2/3)\delta/\alpha$. They are further out from this value, however, than are the first and second approximations, both of which become less accurate as θ approaches unity. Note also that the second approximation yields only slightly more accurate locations in this case than does the first approximation. The exact and approximate locations are in excellent agreement, however, for values of θ in the range $1.3 \leqslant \theta \leqslant \theta_1$. The critical value of $\theta = \theta_1$ is defined as that value of θ when the two near first order saddle points SP_1 and SP_2 coalesce into a single saddle point of second order. According to the first approximation, see (6.2.33, 34), this coalescence occurs at $\omega'' = -(2/3)\delta/\alpha$ when $\theta_1 \cong 1.50277$, whereas by the second approximation, see (6.2.38, 42), this coalescence occurs very near to this same point, see (6.2.50a, b) when $\theta_1 \cong 1.50275$. From Fig. 6.17, which illustrates the behavior of the exact and approximate locations of the near saddle points in the right half of the complex ω-plane, it is seen that the actual value of θ_1 lies somewhere between the values 1.502752 and 1.503. At $\theta =$ 1.502752 the actual locations of the two near saddle points are still separated by a small amount along the imaginary axis. Consequently, for values of θ within the range $1.3 \leqslant \theta \leqslant \theta_1$, both the first and second approximate locations accurately depict the exact behavior of the near saddle points along the imaginary axis. For $\theta > \theta_1$, however, during which the near saddle points move off symmetrically about the imaginary axis into the complex ω-plane, the first approximation is seen in Fig. 6.17 to fail rapidly since $|\omega_{sp_N^\pm}(\theta)|$ is no longer small in comparison to both ω_0 and ω_1. The second approximate locations do, however, follow closely along with the exact locations as they approach the branch points ω_+ in the right-half plane and ω_- in the left-half plane. Note that the path traced out by the second approximate locations lies closely adjacent to the path traced out by the exact locations, but that the positions predicted by the second approximation lie slightly ahead of the actual positions. Furthermore, the second approximate locations are seen to move at roughly the same rate as do the exact locations over the entire range of θ. Again, as in the case of the distant saddle points, this rate of motion is rapid at first, but as the saddle points $\omega_{sp_N^\pm}(\theta)$ approach closer to the branch points $\omega_\pm$, respectively, their rate of motion with θ rapidly decreases. Consequently, the second approximation (6.2.38) for the locations of the saddle points in the region about the origin accurately describes the exact behavior over the entire range of values $\theta \geqslant 1.3$, whereas the first approximation (6.2.33) is accurate only for values of θ within the very limited range $1.3 \leqslant \theta \leqslant \theta_1$.

Finally, consider the behavior of the complex phase function $\phi(\omega, \theta)$ at these saddle points as a function of θ, as illustrated in Figs. 6.18 through 6.20, the first two of which illustrate the real phase behavior $X(\omega, \theta)$ at the saddle points, and

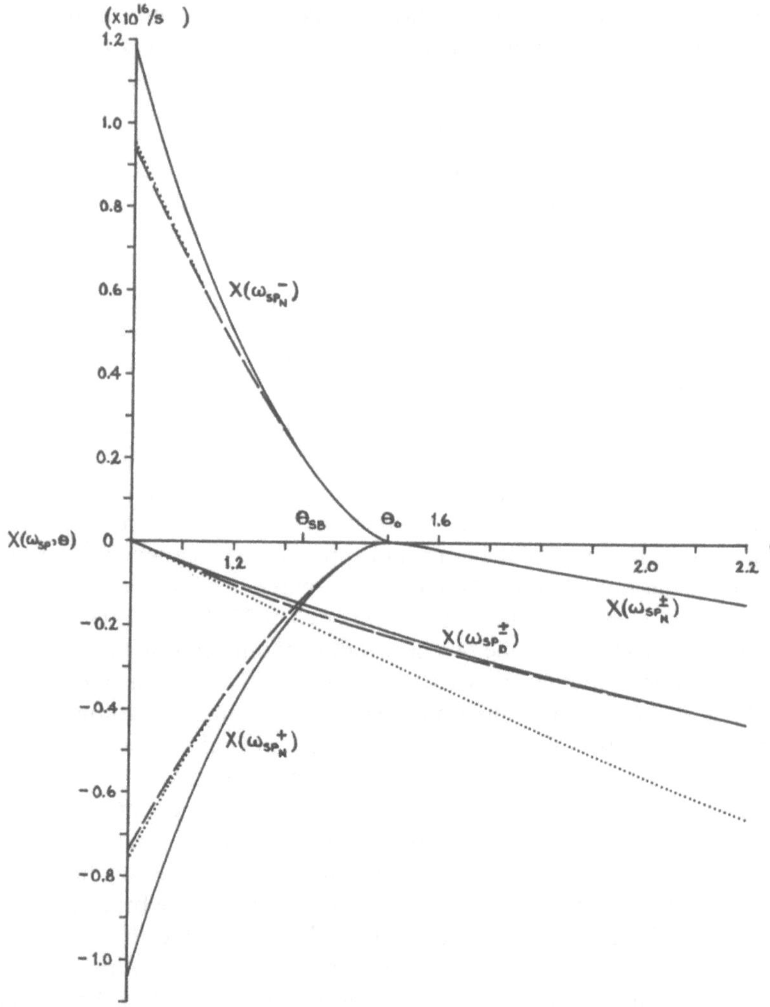

Fig. 6.18. Behavior of the real part of the phase, $X(\omega, \theta) = \mathbb{R}\mathrm{e}\{\phi(\omega, \theta)\}$, at the saddle points as a function of θ for $1 \leqslant \theta \leqslant 1.9$.

the third of which illustrates the imaginary phase behavior $Y(\omega, \theta)$ at the saddle points. The behavior predicted by the first approximation is indicated by the dotted lines, that due to the second approximation by the long dashed lines, and the exact behavior, as determined numerically, is indicated by the solid lines.

Consider first the real phase behavior at the saddle points as a function of θ. At the distant saddle points, $X(\omega_{sp_D^\pm}, \theta)$ is seen to vanish identically at θ equals unity, and then to monotonically decrease as θ increases away from unity. The first approximation to $X(\omega_{sp_D^\pm}, \theta)$ is seen to diverge rapidly away from the exact

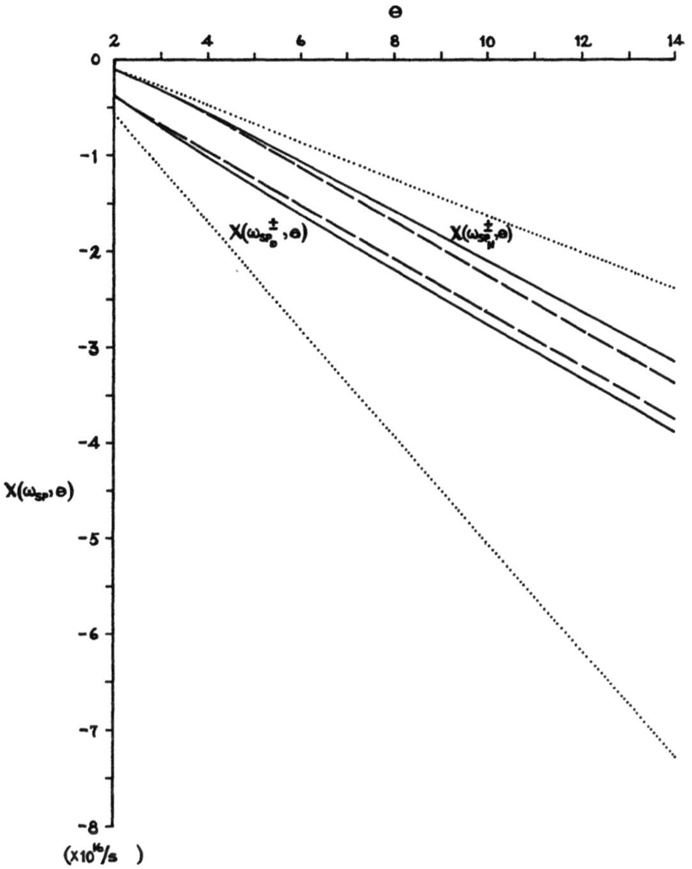

Fig. 6.19. Behavior of the real part of the phase, $X(\omega, \theta) = \mathbb{R}\mathrm{e}\{\phi(\omega_{sp}, \theta)\}$, at the saddle points as a function of θ for $2 \leqslant \theta \leqslant 14$.

behavior for values of θ above approximately $\theta_0 = 1.5$, whereas the second approximation closely follows the actual behavior over the entire range of values of θ. However, since the distant saddle points are dominant over the upper near saddle point SP_1 only for values of θ within the range $1 \leqslant \theta \leqslant \theta_{SB} < \theta_0$, either of the first or second approximate expressions for $X(\omega_{sp_D^\pm}, \theta)$ may be employed in the asymptotic description of the first precursor field. Indeed, it is for this reason that the first approximate expression for $X(\omega_{sp_D^\pm}, \theta)$ is employed in Sect. 6.3 for the analytic determination of θ_{SB}.

At the upper near saddle point SP_1, for $1 \leqslant \theta \leqslant \theta_1$, both the first and second approximations to $X(\omega_{sp_N^+}, \theta)$ lie very close together and are in fair agreement with the exact behavior, the agreement becoming excellent for values of θ roughly above a value of 1.3. However, it is seen that for values of θ up to $\theta_{SB} > 1.3$, the two distant saddle points are dominant over the upper near saddle

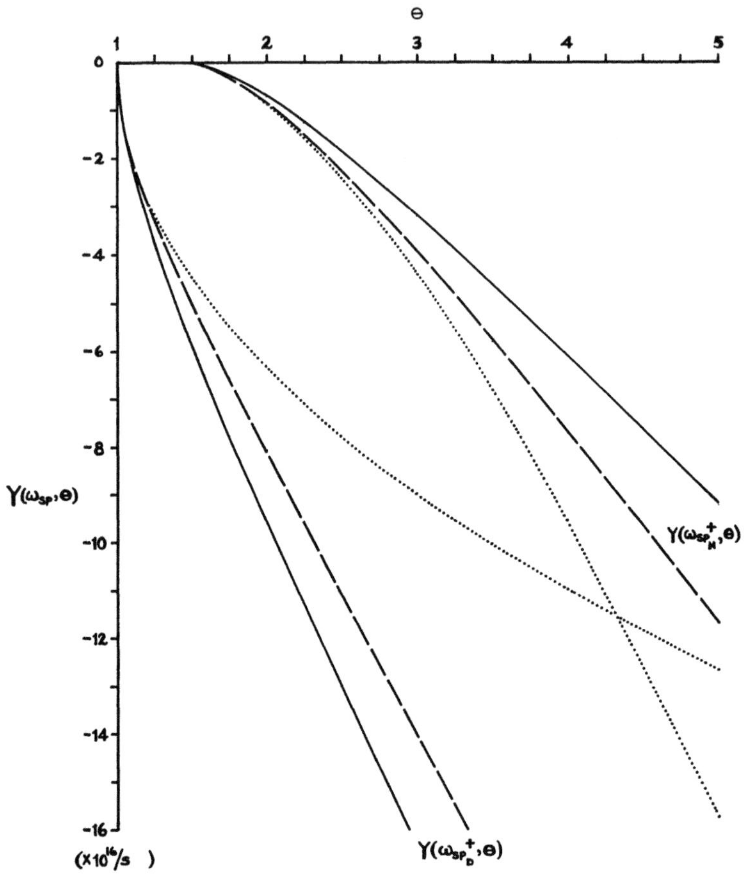

Fig. 6.20. Behavior of the imaginary part of the phase, $Y(\omega, \theta) = \mathrm{Im}\{\phi(\omega, \theta)\}$, at the saddle points in the right half of the complex ω-plane as a function of θ.

point SP_1. From Fig. 6.18, the exact value of θ_{SB} at which the upper near saddle point becomes dominant over the distant saddle points is $\theta_{SB} = 1.334$, at which $X(\omega_{sp_D^\pm}, \theta_{SB}) = X(\omega_{sp_N^+}, \theta_{SB}) = -0.1451 \times 10^{16}/\mathrm{s}$. The first approximate curves yield a value of $\theta_{SB} \simeq 1.305$, which is 2.2% off from the actual value, and the second approximate curves yield a value of 1.320, which is 1% off from the actual value. The approximate expression (6.3.12) yields a value of $\theta_{SB} \cong 1.295$, which is sufficiently accurate for the purposes of this research. Consequently, over the range of values of θ given by $1 \leqslant \theta \leqslant \theta_{SB}$ wherein the second approximation does not sufficiently describe the exact behavior of $X(\omega_{sp_D^+}, \theta)$ at the upper near saddle point, the far saddle points are dominant, and for these two distant saddle points, both the first and second approximations of $X(\omega_{sp_D^\pm}, \theta)$ accurately describe the actual behavior for $1 \leqslant \theta \leqslant \theta_{SB}$. For $\theta \geqslant \theta_{SB}$, during which the near saddle point (or points for $\theta > \theta_1$) remains dominant over the distant saddle points, the second approximation of $X(\omega_{sp_N^+}, \theta)$ for $\theta_{SB} \leqslant \theta \leqslant \theta_1$ and $X(\omega_{sp_N^\pm}, \theta)$ for $\theta \geqslant \theta_1$ accurately describes the exact behavior. For values of θ in the region about $\theta_0 = 1.5$, the first approximation is also seen to accurately describe the actual behavior at the near saddle points, as was to be expected, but as θ increases further and further away from θ_0, the accuracy of the first approximation is seen to steadily diminish. Finally, note that for values of θ sufficiently larger than θ_0, $X(\omega_{sp}, \theta)$ at both the near and distant saddle points decreases steadily in a nearly linear relationship to θ, with $X(\omega_{sp_N^\pm}, \theta) > X(\omega_{sp_D^\pm}, \theta)$.

Consider finally the imaginary part of the complex phase behavior at the saddle points in the right-half of the complex ω-plane as a function of the continuous parameter θ, as illustrated in Fig. 6.20. Since $Y(-\omega' + i\omega'', \theta) = -Y(\omega' + i\omega'', \theta)$, just the negative of this behavior is obtained in the left-half of the complex ω-plane. As is evident from this figure, the first approximate behavior at the distant saddle points diverges rapidly away from the actual behavior as θ increases away from unity, and the first approximate behavior at the near saddle points diverges rapidly away from the actual behavior as θ increases away from θ_1 [for $1 \leqslant \theta \leqslant \theta_1$, $Y(\omega_{sp_N^\pm}, \theta) = 0$]. The second approximations in both of these cases, however, are seen to be a significant improvement over their respective first approximations in depicting the exact behavior.

In conclusion, it has been shown here that, as regards the location and behavior of the saddle points, the first approximation fails to accurately describe the actual behavior over the range of values of θ necessary to describe the asymptotic behavior of optical pulse propagation in temporally dispersive Lorentz media. The second approximation, however, accurately describes the actual behavior over the entire range of values of θ needed. The second approximation to the phase behavior at the near saddle points accurately depicts the actual behavior for all $\theta \geqslant \theta_{SB}$ during which the near saddle points are dominant, whereas the first approximation is accurate only within the restricted range $\theta_{SB} \leqslant \theta \leqslant \theta_1$, which is insufficient for the purposes of this research. However, both the first and second approximations to the phase behavior at the distant saddle points accurately depict the actual behavior for all θ within the range $1 \leqslant \theta \leqslant \theta_{SB}$, during which these two saddle points are dominant. However, since the first approximate depiction of the distant saddle point locations is accurate only for values of θ close to unity, and is widely divergent from the actual behavior for larger values of θ, these first approximate expressions are inadequate for the purposes of the research, since they will be needed in evaluating $q(\omega_{sp_D^\pm})$ in the asymptotic analysis of the following two chapters. Furthermore, the second approximate expressions for both the saddle point locations and the complex phase behavior at them are continuous functions of θ for all $\theta \geqslant 1$. Thus, the second approximation for both the near and distant saddle points is employed in the asymptotic analysis in the following chapters. As a consequence of this, much more accurate expressions describing the dynamic behavior of the two precursor fields and the main signal than could be obtained by use of the first approximation are obtained.

6.5 Procedure for the Asymptotic Analysis of the Field $A(z,t)$

Equipped with the knowledge of the topography of the phase function $\phi(\omega,\theta)$ and the dynamics of its saddle points, it is now possible to formulate a procedure to perform the asymptotic analysis of the field $A(z,t)$ as given by either (4.3.28 or 31). This final section describes such a procedure based on the modern asymptotic methods described in Chap. 5.

The first step in the asymptotic analysis of the field $A(z,t)$ is to express the integral representation of $A(z,t)$ in terms of an integral $I(z,\theta)$ with the same integrand but with a new contour of integration $P(\theta)$ to which the original contour of integration may be deformed. In the present application, it is found that any poles of the spectral function $\tilde{u}(\omega-\omega_c)$ that are crossed when the original contour is deformed to $P(\theta)$ are encircled in the process in the clockwise sense[1]. Hence, according to *Cauchy's* residue theorem [6.5], the integral representation of $A(z,t)$ and the integral $I(z,\theta)$ are related by

$$A(z,t) = I(z,\theta) - \mathrm{Re}\{2\pi \mathrm{i} \Lambda(\theta)\} \ , \tag{6.5.1}$$

where $\Lambda(\theta) = \sum_p \mathrm{Res}_{\omega=\omega_p}\{(1/2\pi)\mathrm{i}\tilde{u}(\omega-\omega_c)\mathrm{e}^{(z/c)\phi(\omega,\theta)}\}$ is the sum of the residues of the poles that were crossed[2] and $I(z,\theta)$ is defined by

$$I(z,\theta) = \frac{1}{2\pi}\mathrm{Re}\left\{\mathrm{i}\int_{P(\theta)} \tilde{u}(\omega-\omega_c)\mathrm{e}^{(z/c)\phi(\omega,\theta)}\,d\omega\right\} . \tag{6.5.2}$$

Because the distant saddle points are dominant for some values of θ and the near saddle points are dominant for other values of θ, there is no path $P(\theta)$ that is an Olver-type path with respect to a single saddle point and that evolves continuously with θ for all $\theta \geqslant 1$. Consequently, the method of Sect. 5.2.1 is required in order to obtain an asymptotic representation that remains uniformly valid for all $\theta \geqslant 1$. To apply that method, the contour $P(\theta)$ must evolve continuously for all $\theta \geqslant 1$ and, in the vicinity of $\theta = \theta_{SB}$ (when the saddle point dominance changes), the path must pass through both the dominant and nondominant saddle points. Moreover, the path must be divisible into the sum of subpaths, each of which is an Olver-type path with respect to one of the saddle points.

For values of θ in the range $1 \leqslant \theta < \theta_1$ during which the two near saddle points are on the imaginary axis, the lower near saddle point SP_2 is dominant over all of the other saddle points. That saddle point is not useful, however, because the Olver-type paths with respect to it are not deformable to the origi-

---

[1] Recall that it is assumed throughout this research that the spectral function $\tilde{u}(\omega-\omega_c)$ is an analytic function of the complex variable ω, regular in the entire complex ω-plane except at a countable number of isolated points where $\tilde{u}(\omega-\omega_c)$ may exhibit poles.

[2] Note that $\Lambda(\theta)$ changes discontinuously with the parameter θ as the path $P(\theta)$ crosses over the poles. These discontinuities are cancelled by the corresponding discontinuous changes in $I(z,\theta)$.

nal contour (and vice versa) due to the presence of the branch points. For this reason, the saddle point SP_2 is not included in the subsequent discussion about which saddle point is dominant for $1 \leqslant \theta < \theta_1$.

There are, however, many paths having the required properties that pass through both the upper near saddle point SP_1 and the distant saddle points $SP_D^{\pm}$ for values of θ within the range $1 \leqslant \theta < \theta_1$. There also are many paths having the required properties that pass through both of the near saddle points $SP_N^{\pm}$ and the distant saddle points $SP_D^{\pm}$ for $\theta > \theta_1$. Finally, there are many paths having the desired properties that pass through the single second-order near saddle point $SP_1 = SP_2$ and both the distant saddle points $SP_D^{\pm}$ for $\theta = \theta_1$. As a result, the contour $P(\theta)$ can be chosen so that it passes through the upper near saddle point SP_1 and the distant saddle points $SP_D^{\pm}$ for $1 \leqslant \theta \leqslant \theta_1$ (remembering that $SP_1 = SP_2$ at $\theta = \theta_1$) and through all four saddle points for $\theta > \theta_1$ so that it evolves in a continuous fashion as θ varies over $\theta \geqslant 1$ and can be divided into the desired subpaths.

An example of such a path $P(\theta)$ and its component subpaths is illustrated in Fig. 6.21. For values of θ in the range $1 \leqslant \theta \leqslant \theta_1$, the component subpaths are $P_D^-(\theta)$, $P_1(\theta)$, and $P_D^+(\theta)$, and for $\theta > \theta_1$, the component subpaths are $P_D^-(\theta)$, $P_N^-(\theta)$, $P_N^+(\theta)$, and $P_D^+(\theta)$. The subpaths $P_D^{\pm}(\theta)$ and $P_N^{\pm}(\theta)$ are Olver-type paths with respect to the saddle points $SP_D^{\pm}$ and $SP_N^{\pm}$, respectively, and the subpath $P_1(\theta)$ is an Olver-type path with respect to the saddle point SP_1.

Provided that the path $P(\theta)$ and its component subpaths satisfy the above constraints, it is unimportant which particular paths are used; the asymptotic results are independent of the choice. Some particular choices may be more convenient, however, in that they reduce the computation required. In the analysis to follow, the paths are sometimes taken to follow the path of steepest descent in the vicinity of the saddle points in order to simplify the determination of the appropriate values of the multivalued functions appearing in the asymptotic approximations.

The deformed contour employed by *Brillouin* [6.1, 2] followed along the entire paths of steepest descent through the distant saddle points and the entire steepest-descent paths through SP_1 for $1 \leqslant \theta \leqslant \theta_1$ and through $SP_N^{\pm}$ for $\theta > \theta_1$ with the various paths connected along the branch cuts. Although it is perfectly valid, that path is unnecessarily complicated. It is avoided in this analysis by applying *Olver*'s method instead of the method of steepest descent.

In accordance with the method described in Sect. 5.2.1, the integral $I(z,\theta)$ is expressed as the sum of integrals with the same integrand over the various subpaths, viz.

$$I(z,\theta) = I_D^-(z,\theta) + I_1(z,\theta) + I_D^+(z,\theta) \; ; \qquad \text{for } 1 \leqslant \theta \leqslant \theta_1 \; , \tag{6.5.3a}$$

$$I(z,\theta) = I_D^-(z,\theta) + I_N^-(z,\theta) + I_N^+(z,\theta) + I_D^+(z,\theta) \; ; \qquad \text{for } \theta_1 < \theta \; , \tag{6.5.3b}$$

where $I_D^{\pm}(\theta)$ and $I_N^{\pm}(\theta)$ denote the contour integrals taken over the paths $P_D^{\pm}(\theta)$ and $P_N^{\pm}(\theta)$, respectively, and where $I_1(z,\theta)$ denotes the contour integral taken over the path $P_1(\theta)$. In order to obtain an asymptotic approximation of the

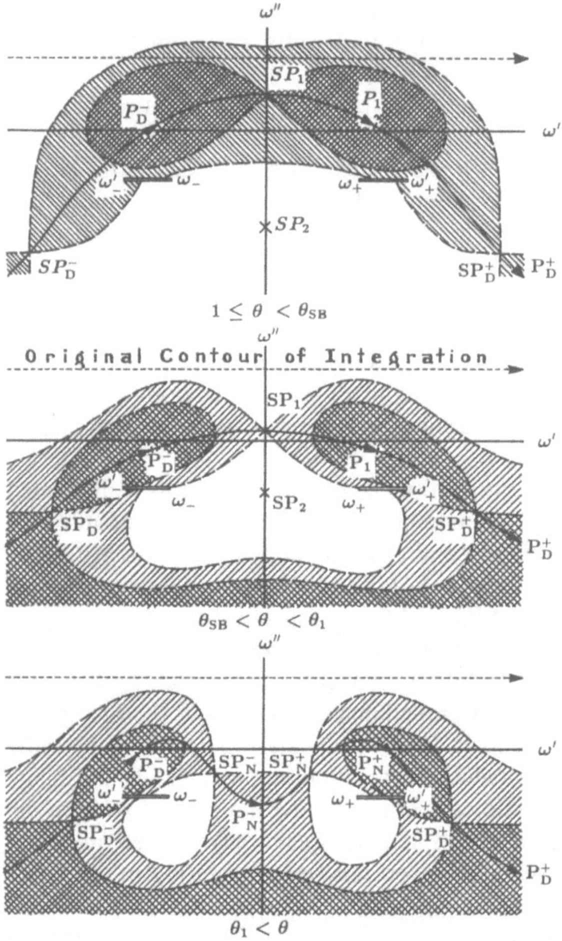

Fig. 6.21. Olver-type paths through the saddle points of a Lorentz dielectric. The hatched area //// indicates the region of the complex ω-plane wherein the inequality $X(\omega, \theta) < X(\omega_{sp_N}, \theta)$ is satisfied, and the hatched area \\\\ indicates the region wherein the inequality $X(\omega, \theta) < X(\omega_{sp_D}, \theta)$.

integral representation of the field $A(z, t)$, it then remains to obtain asymptotic approximations of the integrals appearing on the right-hand side of (6.5.3). A description of this procedure is given in the following paragraphs.

If the distant saddle points $\mathrm{SP}_D^{\pm}$ do not pass too near to any poles of the spectral function $\tilde{u}(\omega - \omega_c)$, then the results of Sect. 5.4 can be applied to obtain an asymptotic approximation of the quantity $(I_D^-(z, \theta) + I_D^+(z, \theta))$ in the form

$$I_D^-(z, \theta) + I_D^+(z, \theta) = A_S(z, t) + R(z, \theta) , \qquad (6.5.4)$$

where $A_S(z, t)$ is obtained from (5.4.19) and an estimate of the remainder as $z \to \infty$ is given by (5.4.5). Equation (6.5.4) is uniformly vaid for all $\theta \geqslant 1$ so long as both of the distant saddle points remain isolated from any poles of $\tilde{u}(\omega - \omega_c)$. For values of θ bounded away from unity from above, (6.5.4) reduces to the result obtained by application of *Olver*'s method directly to $I_D^-(z, \theta)$ and $I_D^+(z, \theta)$ and adding the results.

If the near saddle points SP_1 and $SP_N^\pm$ do not pass too close to any poles of the spectral function $\tilde{u}(\omega - \omega_c)$, then the results of Sect. 5.2.2 can be applied to obtain asymptotic approximations of $I_1(z, \theta)$ and $(I_N^-(z, \theta) + I_N^+(z, \theta))$ in the form

$$I_1(z, \theta) = A_B(z, t) + R'(z, \theta) \; ; \qquad \text{for } 1 \leqslant \theta \leqslant \theta_1 \; , \tag{6.5.5a}$$

$$I_N^-(z, \theta) + I_N^+(z, \theta) = A_B(z, t) + R'(z, \theta) \; ; \qquad \text{for } \theta > \theta_1 \; , \tag{6.5.5b}$$

where the expression for $A_B(z, t)$ and an estimate of the remainder $R'(z, \theta)$ as $z \to \infty$ are obtained from (5.2.6). Taken together, (6.5.5a, b) yields an asymptotic approximation of $I(z, \theta) - I_D^-(z, \theta) - I_D^+(z, \theta)$ that is valid uniformly for all $\theta \geqslant 1$ so long as the near saddle points remain isolated from the poles of $\tilde{u}(\omega - \omega_c)$. For values of θ bounded away from θ_1 from below, (6.5.5a) reduces to the result that would be obtained by applying *Olver*'s method directly to obtain the asymptotic approximation of $I_1(z, \theta)$. Similarly, for values of θ bounded away from θ_1 from above, (6.5.5b) reduces to the result that would be obtained by applying *Olver*'s method directly to obtain the asymptotic approximations of $I_N^-(z, \theta)$ and $I_N^+(z, \theta)$ and summing the results.

Consider now the case when either one of the distant saddle points $SP_D^\pm$ approaches (as θ varies) a pole of $\tilde{u}(\omega - \omega_c)$ located in a region of the complex ω-plane bounded away from the limiting values $\pm\infty - 2\delta i$ approached by $SP_D^\pm$ as $\theta \to 1^+$. Then the methods of Sects. 5.3, 4 can be applied to obtain an asymptotic approximation of $(I_D^-(z, \theta) + I_D^+(z, \theta))$ that is of the form

$$I_D^-(z, \theta) + I_D^+(z, \theta) = A_S(z, t) + C_D^\pm(z, t) + R(z, \theta) \tag{6.5.6}$$

with $A_S(z, t)$ given by the same expression as before. Since the pole is bounded away from the limiting values of $\omega_{sp_D^\pm}(\theta)$ as $\theta \to 1^+$, the saddle point and pole do not interact for values of θ near unity. Hence, the asymptotic approximation of $(I_D^-(z, \theta) + I_D^+(z, \theta))$ is determined by applying the methods of Sect. 5.4, $C_D^\pm(z, t)$ is asymptotically negligible, and (6.5.6) reduces to (6.5.4) for values of θ near 1. For values of θ bounded away from unity from above, the results of Sect. 5.3 are applicable, and (6.5.6) is obtained from (5.3.5a) with $A_S(z, t)$ being given by the first term and $C_D^\pm(z, t)$ being given by the second term. The resulting expression for $A_S(z, t)$ is the same as before. Hence, $A_S(z, t)$ in equation (6.5.6) is given by the same expression as is $A_S(z, t)$ in (6.5.4) for all $\theta \geqslant 1$. The quantity $C_D^\pm(z, t)$ is asymptotically negligible if either of the distant saddle points $SP_D^\pm$ does not approach a pole of $\tilde{u}(\omega - \omega_c)$ so that in that case (6.5.6) reduces (6.5.4).

Similarly, let either one of the near saddle points SP_1 or $SP_N^\pm$ approach (as θ varies) a pole of the spectral function $\tilde{u}(\omega - \omega_c)$ with location bounded away from the point where the two near saddle points coalesce for $\theta = \theta_1$ to form a single second-order saddle point. Then, the methods of Sects. 5.2.2 and 5.3

can be applied to obtain an asymptotic approximation of $I_1(z,\theta)$ or $(I_N^-(z,\theta) + I_N^+(z,\theta))$ of the form

$$I_1(z,\theta) = A_B(z,t) + C_1(z,t) + R'(z,\theta) \tag{6.5.7a}$$

for $1 \leqslant \theta \leqslant \theta_1$, or

$$I_N^-(z,\theta) + I_N^+(z,\theta) = A_B(z,t) + C_N^{\pm}(z,t) + R'(z,\theta) \tag{6.5.7b}$$

for $\theta > \theta_1$. In both cases, the expression for $A_B(z,t)$ is the same as before. The quantity $C_1(z,t)$ or $C_N^{\pm}(z,t)$ is asymptotically negligible if the corresponding saddle point does not approach a pole so that in that case, the corresponding equation (6.5.7) reduces to the appropriate equation (6.5.5).

Combination of (6.5.1–7) leads to the following final general expression for the asymptotic approximation of the integral representation of the field $A(z,t)$:

$$A(z,t) = A_S(z,t) + A_B(z,t) + A_C(z,t) + R''(z,\theta) \ , \tag{6.5.8}$$

which is uniformly valid for all $\theta \geqslant 1$ provided that all the poles of the spectral function $\tilde{u}(\omega - \omega_c)$ are bounded away from the limiting locations taken by $\mathrm{SP}_D^{\pm}$ as $\theta \to 1^+$ and by SP_1 as $\theta \to \theta_1$ from below. The contribution $A_C(z,t)$ is obtained by adding all of the terms that involve the poles, viz.

$$A_C(z,t) = -\mathbb{R}\mathrm{e}\{2\pi \mathrm{i} \Lambda(\theta)\} + C_D^-(z,t) + C_D^+(z,t) + C_1(z,t) \ ;$$
$$\text{for } 1 \leqslant \theta \leqslant \theta_1 \ , \tag{6.5.9a}$$

$$A_C(z,t) = -\mathbb{R}\mathrm{e}\{2\pi \mathrm{i} \Lambda(\theta)\} + C_D^-(z,t) + C_D^+(z,t) + C_N^-(z,t) + C_N^+(z,t) \ ;$$
$$\text{for } \theta > \theta_1 \ . \tag{6.5.9b}$$

An estimate of the remainder term $R''(z,\theta)$ as $z \to \infty$ is obtained by taking the largest estimate of the remainder terms appearing in (6.5.4–7).

An important feature of (6.5.8) is that the asymptotic behavior of the propagated field $A(z,t)$ is expressed as the sum of three terms which are essentially uncoupled so that they can be treated independently of one another. The dynamic behavior of $A_S(z,t)$ is determined by the dynamics of the distant saddle points and the value of the integrand at those saddle points. Since the distant saddle points are dominant first, $A_S(z,t)$, describes the dynamic behavior of the *first (Sommerfeld's) precursor field.* It is negligible during most of the evolution of the second precursor and the main signal.

The dynamic behavior of $A_B(z,t)$ is determined by the dynamics of the near saddle points and the value of the integrand at those saddle points. Hence, $A_B(z,t)$ describes the dynamic behavior of the *second (Brillouin's) precursor field.* It is negligible during most of the evolution of the first precursor and the main signal.

The dynamic behavior of $A_C(z,t)$ is determined by the poles of the spectral function $\tilde{u}(\omega - \omega_c)$ and the dynamics of the saddle points that pass near these poles. The contribution $A_C(z,t)$ is nonzero only if $\tilde{u}(\omega - \omega_c)$ [or $\tilde{f}(\omega)$, see (4.3.28)] has poles. If the envelope function $u(t)$, defined in (4.3.21), of the field $A(0,t)$ on the plane $z = 0$ is bounded for all time t, then $\tilde{u}(\omega - \omega_c)$ can have poles only if $u(t)$ does not tend to zero too fast as $t \to \infty$.[3] Hence, the implication of nonzero $A_C(z,t)$ is that the field $A(z,t)$ oscillates with angular frequency ω_c for positive time t on the plane $z = 0$ and will tend to do the same at larger values of z for large enough t. As a result, the contribution $A_C(z,t)$ describes the dynamic behavior of the *main signal oscillating with angular frequency* ω_c. This contribution is negligible during most of the evolution of the precursor fields.

For most values of θ, only one of the terms $A_S(z,t)$, $A_B(z,t)$, and $A_C(z,t)$ appearing in (6.5.8) is important at a time. There are short intevals of θ, however, during which two or more of these terms are significant for fixed values of z. These intervals mark the transition periods when the field is changing its character from one form to another and the presence of both terms in the expression leads to a continuous transition in the behavior of the field. As a result, (6.5.8) displays the entire evolution of the field through its various forms in a continuous manner.

The chapters that follow are devoted to evaluating the three terms $A_S(z,t)$, $A_B(z,t)$, and $A_C(z,t)$ and to examining their contribution to the total propagated field $A(z,t)$. The *Sommerfeld* and *Brillouin* precursor fields are evaluated in Chap. 7 and the contribution of the poles $A_C(z,t)$ is evaluated in Chap. 8. These three contributions are combined in Chap. 9 to determine the total evolution of the propagated field $A(z,t)$ and the signal velocity of the pulse.

---

[3] If $u(t)$ is bounded and tends to zero rapidly enough so that the Fourier integral of $u(t)$ converges uniformly for all real ω, then $\tilde{u}(\omega - \omega_c)$ is an entire function of complex ω.

7 Evolution of the Precursor Fields

Based upon the foundational analyses just completed, the asymptotic behavior of pulse propagation in a Lorentz medium may now be developed. The analysis presented in this chapter begins with an examination of the exact field behavior for times t such that $\theta = ct/z < 1$, for a fixed observation distance z. By applying the method *Sommerfeld* [7.1] used to treat the step-function modulated signal, it is shown here [7.2] that for fields with the initial envelopes $u(t)$ that are zero for times $t < 0$, the propagated field is identically zero for all values of $\theta < 1$, in agreement with the relativistic principle of causality [7.3]. The remainder of the chapter is devoted to determining the evolutionary properties of the precursor fields associated with the propagating pulse. The analysis follows the approach used by *Brillouin* [7.4, 5] in his treatment of the step-function modulated signal except that the new approximations developed in Chap. 6 for the saddle-point locations and the behavior of the complex phase function $\phi(\omega, \theta)$ at the saddle points are used and the advanced asymptotic techniques reviewed in Chap. 5 are applied. Application of the new approximations from Chap. 6 yields approximate expressions describing the dynamic behavior of the precursor fields which are much more accurate than any results available previously. Application of the advanced saddle-point methods of Chap. 5 yields approximations which have functional forms very different from previously available results and which display, for the first time, the continuous evolution with time of the total precursor field. In addition to the step-function modulated signal analyzed by *Brillouin*, the entire precursor fields are obtained also for the rectangular-pulse modulated signal, the delta-function pulse, and the hyperbolic-tangent modulated pulse defined in Sect. 4.3. In this chapter, the precursors are studied without reference to the main signal. The arrival and evolution of the main signal is studied in Chap. 8 and its interaction with the precursor fields is discussed in Chap. 9.

In the asymptotic analysis that follows in both this chapter and Chap. 8, the expression $A \sim B$ is used to mean that B is an approximation of the dominant term in the asymptotic expansion of A valid as $z \to \infty$ with fixed θ. The reason that B is not equal to the dominant term exactly is that approximations for the saddle point locations and the phase function have been used in determining B. When the exact saddle point locations and phase behavior are employed so that B is equal to the dominant term exactly, then the asymptotic relation is written $A = B + \mathcal{O}(z^{-3/2})$ as $z \to \infty$.

7.1 The Field Behavior for $\theta < 1$

If the initial envelope $u(t)$ of the field on the plane $z = 0$ (as defined in Sect. 4.3) is zero for times $t < 0$, then the field $A(z,t)$ can be zero for values of $\theta \equiv ct/z < 1$ only if the field propagates with a velocity greater than the speed c of light in vacuum, in violation of the special theory of relativity [or the principle of relativistic causality [7.3]]. *Sommerfeld* [7.1] proved that for the step-function modulated signal, the propagated field is identically zero for all $\theta < 1$. For completeness, his argument is extended in this section to show that the same result holds for all fields with initial envelopes $u(t)$ on the plane $z = 0$ that are zero for all times $t < 0$. In addition, a series representation of $A(z,t)$, valid for all $\theta < 1$, is obtained for the hyperbolic-tangent modulated signal, which is the one particular field discussed in this research that has an initial envelope $u(t)$ that is not zero for $t < 0$.

The integral representation of the field $A(z,t)$ to be analyzed in the remaining chapters is given by (4.3.31), viz.

$$A(z,t) = \frac{1}{2\pi}\mathrm{Re}\left\{ \mathrm{i} \int_{ia-\infty}^{ia+\infty} \tilde{u}(\omega - \omega_c) e^{(z/c)\phi(\omega,\theta)}\, d\omega \right\} . \tag{7.1.1}$$

For fields with an envelope function $u(t)$ that is zero for times $t < 0$, the spectral amplitude function appearing in the integrand of (7.1.1) is given by

$$\tilde{u}(\omega) = \int_0^\infty u(t) e^{\mathrm{i}\omega t}\, dt . \tag{7.1.2}$$

If $u(t)$ is bounded for all t, then it is easy to show by direct differentiation of (7.1.2) that $\tilde{u}(\omega)$ is an analytic function of complex ω for $\omega'' > 0$ (where, as before, ω' and ω'' are, respectively, the real and imaginary parts of ω). In addition, if $du(t)/dt$ is bounded for all t, then it is easy to show by integration of the integral in (7.1.2) by parts that $\tilde{u}(\omega)$ tends to zero uniformly with respect to $\psi \equiv \arg(\Omega)$ for $0 \leqslant \psi \leqslant \pi$ as $|\Omega| \to \infty$, where $\Omega = \omega - ia$ with $a > 0$. Since ω_c is real, the same results apply to $\tilde{u}(\omega - \omega_c)$.

With the knowledge that the spectral amplitude function $\tilde{u}(\omega - \omega_c)$ satisfies the above conditions, it is now possible to express $A(z,t)$ by (7.1.1) with the change that the integration is taken over the closed contour that encircles the region $\omega'' > a > 0$ of the complex ω-plane. All that is required is to show that $I(z,\theta,|\Omega|) \to 0$ uniformly with respect to z and θ for $z \geqslant Z$ and $\theta \leqslant 1 - \varepsilon$ as $|\Omega| \to \infty$ for arbitrary $Z > 0$ and ε such that $0 < \varepsilon < 1$, where $I(z,\theta,|\Omega|)$ is defined by

$$I(z,\theta,|\Omega|) = \int_C \tilde{u}(\omega - \omega_c) e^{(z/c)\phi(\omega,\theta)}\, d\omega , \tag{7.1.3}$$

with C being the contour $0 \leqslant \psi \leqslant \pi$ for fixed $|\Omega|$. The proof makes use of the proof of Jordan's lemma. It follows from (7.1.3) that the following inequality is satisfied

$$|I(z,\theta,|\Omega|)| \leqslant \int_C |\tilde{u}(\omega-\omega_c)|\,\mathrm{e}^{(z/c)X(\omega,\theta)}|d\omega| \;, \tag{7.1.4}$$

where $X(\omega,\theta)$ is the real part of $\phi(\omega,\theta)$ as given by (6.1.9), viz.

$$X(\omega,\theta) = -\omega''[n_r(\omega)-\theta] - \omega' n_i(\omega) \;, \tag{7.1.5}$$

and where $n_r(\omega)$ and $n_i(\omega)$ are, respectively, the real and imaginary parts of the complex index of refraction of the medium. It follows from (6.1.13, 14) that there exists a positive constant Ω_0 such that for $|\Omega| > \Omega_0$, $\omega' n_i(\omega) > 0$. Hence, for $|\Omega| > \Omega_0$ and $0 \leqslant \psi \leqslant \pi$,

$$X(\omega,\theta) < -\omega''[n_r(\omega)-\theta] \;. \tag{7.1.6}$$

Furthermore, it follows from (6.1.13, 14) that there is a positive constant such that for $|\Omega|$ larger than that constant, $n_r(\omega) \geqslant 1$. Henceforth, Ω_0 is taken to be larger than that constant. It then follows that

$$X(\omega,\theta) \leqslant -\omega''(1-\theta) \tag{7.1.7}$$

for $|\Omega| > \Omega_0$ with $0 \leqslant \psi \leqslant \pi$, [the inequality appearing in (7.1.7) also follows directly from (6.2.24)]. Consequently, for $\theta \leqslant 1-\varepsilon$ with $0 < \varepsilon < 1$, the combination of the inequalities appearing in (7.1.4 and 7) yields

$$|I(z,\theta,|\Omega|)| \leqslant \int_C |\tilde{u}(\omega-\omega_c)|\,\mathrm{e}^{-(z/c)\varepsilon\omega''}|d\omega| \;, \tag{7.1.8}$$

for $|\Omega| > \Omega_0$. From here on, the proof follows exactly the proof of Jordan's lemma as given in [Ref. 7.6, Sect. 6.222]. Hence, $I(z,\theta,|\Omega|) \to 0$ uniformly with respect to both z and θ for $z \geqslant Z$ and $\theta \leqslant 1-\varepsilon$ as $|\Omega| \to \infty$ for arbitrary $Z > 0$ and ε such that $0 < \varepsilon < 1$. Consequently, that integral can be added to the integral in (7.1.1) to express $A(z,t)$ as an integral over a closed contour that encircles the region $\omega'' > a$ of the complex ω-plane.

Since the integrand of the closed contour integral is a regular analytic function of complex ω for $\omega'' > a$, with a positive, it then follows from Cauchy's residue theorem [7.6, 7] that the integral is identically zero for $z \geqslant Z$ and $\theta \leqslant 1-\varepsilon$ for arbitrary $Z > 0$ and arbitrarily small $\varepsilon > 0$. Hence, $A(z,t) = 0$ for $z > 0$ and $\theta < 1$ when the initial envelope function $u(t)$ is zero for $t < 0$.

Consider now the special case of the hyperbolic-tangent modulated signal, which is determined by (7.1.1) with $\tilde{u}(\omega)$ given by (4.3.46), viz.

$$\tilde{u}(\omega) = \frac{\pi\mathrm{i}}{2\beta}\cdot\frac{1}{\sinh\left(\dfrac{\pi}{2\beta}\omega\right)} \;, \tag{7.1.9}$$

with a satisfying the inequality $0 < a < 2\beta$. Since the envelope function $u(t)$ does not vanish for times $t < 0$ for this field, $\tilde{u}(\omega)$ need not satisfy the limiting properties and the analyticity properties previously described for the spectra of the other signals that have been presented. In fact, the spectrum $\tilde{u}(\omega)$, as given in (7.1.9), has an infinite number of simple pole singularities located at $\omega = 2n\beta\mathrm{i}$ with $n = 0, \pm 1, \pm 2, \ldots$. Moreover, the spectrum (7.1.9) does not tend to zero as $|\Omega| \to \infty$ for all $0 \leqslant \psi \leqslant \pi$, where $\psi \equiv \arg(\Omega)$ and $\Omega = \omega - \mathrm{i}a$.

Nevertheless, in the limit as $|\Omega| \to \infty$, the spectrum $\tilde{u}(\omega)$ does tend to zero uniformly with respect to $\psi = \arg(\Omega)$ for $0 \leqslant \psi \leqslant \Psi$ and for $\pi - \Psi \leqslant \psi \leqslant \pi$ for small enough $\Psi > 0$. Hence, the proof given previously to show that $I(z, \theta, |\Omega|) \to 0$ as $|\Omega| \to 0$, can be applied in this case to show that the same result holds when $\tilde{u}(\omega)$ is given by (7.1.9) if the contour C in (7.13) is taken to either be the arc $0 \leqslant \psi \leqslant \Psi$ or the arc $\pi - \Psi \leqslant \psi \leqslant \pi$. Furthermore, the same integral over the arc $\Psi \leqslant \psi \leqslant \pi - \Psi$ also vanishes as $|\Omega| \to \infty$ due to the exponential decay of the factor $e^{(z/c)\phi(\omega, \theta)}$ on that arc for $\theta < 1$. Hence, the integral over the entire arc $0 \leqslant \psi \leqslant \pi$ tends to zero uniformly with respect to both z and θ for $z \geqslant Z$ and $\theta \leqslant 1 - \varepsilon$ as $|\Omega| \to \infty$ for arbitrary $Z > 0$ and ε such that $0 < \varepsilon < 1$. Consequently, that integral can be added to the integral in (7.1.1) with $\tilde{u}(\omega)$ given by (7.1.9) to express $A(z, t)$ as an integral over a closed contour that encircles the region $\omega'' > a$ exactly as before.

In the present case, however, the closed contour integral expressing $A(z, t)$ is not zero as it was before because the integrand now has an infinite number of simple-pole singularities located within the closed contour. Application of *Cauchy*'s residue theorem to evaluate the integral yields

$$A(z,t) = -\frac{1}{4\beta}\mathrm{Re}\left\{2\pi\mathrm{i}\sum_{n=1}^{\infty}\operatorname*{Res}_{\omega=\omega_c+2n\beta\mathrm{i}}\left[\frac{e^{(z/c)\phi(\omega,\theta)}}{\sinh\left[\frac{\pi}{2\beta}(\omega-\omega_c)\right]}\right]\right\}$$

$$= -\mathrm{Re}\left\{\mathrm{i}\sum_{n=1}^{\infty}(-1)^n e^{(z/c)\phi(\omega_c+2n\beta\mathrm{i},\theta)}\right\}, \tag{7.1.10}$$

for $\theta < 1$, where this expression is exact. For sufficiently large values of $|\omega_c + 2\beta\mathrm{i}| = +\sqrt{\omega_c^2 + 4\beta^2}$, the complex phase function $\phi(\omega, \theta)$ is, to a good approximation, given by (6.2.24), so that

$$\phi(\omega_c + 2n\beta\mathrm{i}, \theta) \cong \mathrm{i}\omega_c(1-\theta) - 2n\beta(1-\theta) - \frac{b^2}{2}\cdot\frac{2(\delta + n\beta) + \mathrm{i}\omega_c}{\omega_c^2 + 4(\delta + n\beta)^2}.$$

With this approximate expression substituted into (7.1.10) there then results

$$A(z,t) \cong \sum_{n=1}^{\infty}(-1)^n \exp\left\{-\frac{z}{c}\left[2n\beta(1-\theta) + b^2\frac{\delta + n\beta}{\omega_c^2 + 4(\delta + n\beta)^2}\right]\right\}$$

$$\cdot\sin\left[\frac{z}{c}\omega_c\left(1 - \theta - \frac{b^2}{2}\cdot\frac{1}{\omega_c^2 + 4(\delta + n\beta)^2}\right)\right], \tag{7.1.11}$$

for $\theta < 1$. This result approaches zero exponentially as θ approaches negative infinity. Note also that this result approaches zero for all values of $\theta < 1$ in the limit as β approaches infinity, which limit corresponds to an instantaneous turn-on time (i.e., the limiting case is a step-function modulated signal).

Consequently, the exact behavior of the propagated field $A(z,t)$ for values of $\theta < 1$ is readily obtained by the method of completing the contour of integration at infinity in the upper half of the complex ω-plane. This approach works by virtue of the fact that for $\theta < 1$, the real phase function $X(\omega,\theta) \equiv \mathbb{R}\mathrm{e}\{\phi(\omega,\theta)\}$ tends to negative infinity in the limit as $|\omega|$ approaches infinity in the upper half of the complex ω-plane. For values of $\theta \geqslant 1$, however, the real phase function $X(\omega,\theta)$ does not tend to negative infinity in the limit as $|\omega| \to \infty$ in the upper half of the complex ω-plane. According to (6.1.22), for $\theta = 1$, $X(\omega, 1)$ is equal to zero everywhere at $|\omega| = \infty$, and for $\theta > 1$, $X(\omega,\theta)$ is equal to $+\infty$ in the upper half of the complex ω-plane, zero at the real ω'-axis, and $-\infty$ in the lower half of the complex ω-plane. Consequently, the method used for $\theta < 1$ can not be applied to evaluate the integral for $A(z,t)$ for $\theta \geqslant 1$. Moreover, attempts to find other methods for exactly evaluating the integral for $\theta \geqslant 1$ have not proved fruitful. Since all of the features of pulse propagation of interest occur in the regime $\theta \geqslant 1$, analytic approximations for $A(z,t)$ are required. The remainder of this research is devoted to this approximate evaluation of the integral representation of $A(z,t)$. That analysis is begun in this chapter by evaluating the first and second precursor fields.

7.2 The First Precursor Field (Sommerfeld's Precursor)

The contributions of the two distant saddle points to the asymptotic behavior of the field $A(z,t)$ for sufficiently large values of the observation distance z yield the dynamic evolution of the first precursor field. This contribution to the total asymptotic behavior of the field $A(z,t)$ is denoted by $A_S(z,t)$, c.f. (6.5.4, 8), and is dominant over the second precursor field for all values of θ in the range $1 \leqslant \theta < \theta_{SB}$. Since the two distant saddle points $SP_D^{\pm}$ remain isolated from one another and do not change their order, a straightforward asymptotic analysis based upon *Olver*'s method (Sect. 5.1) is applied in Sect. 7.2.1 in order to obtain the first precursor evolution for $\theta > 1$. However, as the parameter θ approaches unity from above, the two distant saddle points approach infinity and *Olver*'s method fails. In order to describe the initial behavior of the first precursor field for values of θ in a neighborhood of unity, the uniform asymptotic expansion presented in Sect. 5.4 is employed in Sect. 7.2.2. This uniform expansion is valid for all values of $\theta \geqslant 1$ and reduces to *Olver*'s result for all θ bounded away from unity.

7.2.1 The Nonuniform Approximation

The expression for the first precursor field that results from a direct application of *Olver*'s method is derived in this subsection. As is pointed out in Chap. 5,

Olver's method is applicable only when the parameter θ is bounded away from unity from above. If the first approximation for the saddle point locations were employed in *Olver*'s method, the result for the step-modulated signal would be the same as that obtained by *Brillouin* [7.4, 5]. As is pointed out in Chap. 6, however, the first approximation is accurate only for values of θ equal to or very close to unity; although the first approximation gives a good approximation of the complex phase $\phi(\omega, \theta)$ at the distant saddle points for $1 \leqslant \theta \leqslant \theta_{SB}$, it does not give a good approximation of the coefficients $a_0(\omega_{sp_D^\pm})$ that appear in the asymptotic expansion of $A_S(z, t)$ except near $\theta = 1$. Hence, in this subsection, the second approximation of the distant saddle point locations is used in *Olver*'s method to obtain an asymptotic approximation of the first precursor field $A_S(z, t)$. The result is a good approximation of $A_S(z, t)$ for positive values of $(\theta - 1)$ bounded away from zero.

The asymptotic behavior of the first precursor field $A_S(z, t)$ for a particular initial pulse envelope is obtained from the asymptotic expansion of the integral representation of the propagated field (4.3.31), viz.

$$A(z,t) = \frac{1}{2\pi}\,\mathrm{Re}\left\{\mathrm{i}\int_{ia-\infty}^{ia+\infty} \tilde{u}(\omega - \omega_c)e^{(z/c)\phi(\omega,\theta)}\,d\omega\right\} \tag{7.2.1}$$

about the two distant saddle points $SP_D^\pm$. The distant saddle point locations are given by

$$\omega_{sp_D^\pm} \cong \pm\xi(\theta) - \delta\mathrm{i}[1 + \eta(\theta)] \ , \tag{7.2.2a}$$

where, for convenience, the following functions have been defined, c.f. (6.2.22, 3):

$$\xi(\theta) \equiv \sqrt{\omega_0^2 - \delta^2 + \frac{b^2\theta^2}{\theta^2 - 1}} \ , \tag{7.2.2b}$$

$$\eta(\theta) \equiv \frac{\dfrac{1}{27}\delta^2 + \dfrac{b^2\theta^2}{\theta^2 - 1}}{\xi^2(\theta)} \ . \tag{7.2.2c}$$

In the limit as θ approaches unity these functions attain the limiting forms used by *Brillouin* [7.4, 5]

$$\lim_{\theta \to 1^+} \xi(\theta) = \frac{b}{\sqrt{2(\theta - 1)}} \ ,$$

$$\lim_{\theta \to 1^+} \eta(\theta) = 1 \ ,$$

which are valid only for values of θ in the immediate vicinity of unity. The second approximate expressions (7.2.2b and c) developed in Chap. 6, which accurately depict the distant saddle point behavior for all values of $\theta \geqslant 1$, are used here.

For the examples considered in this research, both the spectral amplitude function $\tilde{u}(\omega - \omega_c)$ and the phase function $\phi(\omega, \theta)$ appearing in the integrand of $A(z, t)$ are analytic about these two distant saddle points. For values of θ within the range $1 < \theta \leqslant \theta_1$ the conditions of *Olver*'s Theorem 5.1 are satisfied at the two distant saddle points $\mathrm{SP}_D^{\pm}$ when the original contour of integration is deformed to the path $P(\theta) = P_D^-(\theta) + P_1(\theta) + P_D^+(\theta)$, and for values of θ within the range $\theta > \theta_1$ the conditions of *Olver*'s theorem are satisfied at the two distant saddle points when the original contour of integration is deformed to the path $P(\theta) = P_D^-(\theta) + P_N^-(\theta) + P_N^+(\theta) + P_D^+(\theta)$, where $P_D^-(\theta)$ is an Olver-type path with respect to the distant saddle point SP_D^- and $P_D^+(\theta)$ is an Olver-type path with respect to the distant saddle point SP_D^+ (Fig. 6.21). Consequently, (5.1.14) applies for each of the two distant saddle points with (5.1.3, 4) taken to be Taylor series expansions about the saddle points. Since $\mathrm{SP}_D^{\pm}$ are first order saddle points, the parameter μ is 2, and since $\tilde{u}(\omega - \omega_c)$ is regular at the saddle points, the parameter λ is 1. Hence, according to Theorem 5.1 and the results of Sect. 6.5, the integral (7.2.1) taken over the two contours $P_D^{\pm}(\theta)$ yields the first precursor $A_S(z, t)$, which satisfies

$$A_S(z,t) = \frac{1}{2\pi}\,\mathrm{Re}\,\mathrm{i}\left\{2\mathrm{e}^{(z/c)\phi(\omega_{sp_D^+},\theta)}\sqrt{\pi}\,\frac{a_0(\omega_{sp_D^+})}{(z/c)^{1/2}}[1 + \mathcal{O}(z^{-1})]\right.$$
$$\left. + 2\mathrm{e}^{(z/c)\phi(\omega_{sp_D^-},\theta)}\sqrt{\pi}\,\frac{a_0(\omega_{sp_D^-})}{(z/c)^{1/2}}[1 + \mathcal{O}(z^{-1})]\right\}, \tag{7.2.3}$$

as $z \to \infty$ uniformly for $\theta \geqslant 1 + \varepsilon$, where $\varepsilon > 0$.

In order to evaluate the coefficients $a_0(\omega_{sp_D^{\pm}})$ appearing in (7.2.3), the first two coefficients in the Taylor series expansion (5.1.3) of $\phi(\omega, \theta)$ about the distant saddle points are now obtained. According to (6.2.24), the approximate phase behavior in the region of the complex ω-plane far removed from the origin is

$$\phi(\omega, \theta) \cong \mathrm{i}\omega(1 - \theta) - \mathrm{i}\frac{b^2}{2}\cdot\frac{1}{\omega + 2\delta\mathrm{i}},$$

so that differentiation of this expression twice with respect to ω yields

$$\phi^{(1)}(\omega, \theta) \cong \mathrm{i}(1 - \theta) + \mathrm{i}\frac{b^2}{2}\cdot\frac{1}{(\omega + 2\delta\mathrm{i})^2},$$

$$\phi^{(2)}(\omega, \theta) \cong -\mathrm{i}\frac{b^2}{(\omega + 2\delta\mathrm{i})^3}.$$

Thus, at $\omega = \omega_{sp_D^{\pm}}(\theta)$, there results

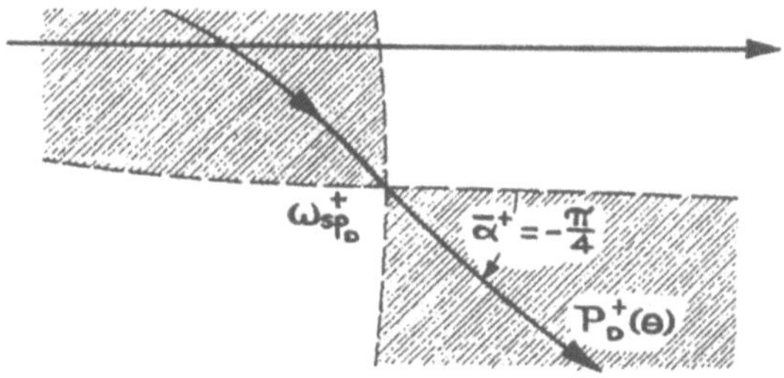

Fig. 7.1. Olver-type path through the distant saddle point SP_D^+. The hatched area indicates the region of the complex ω-plane wherein the inequality $X(\omega) < X(\omega_{sp_D^+}, \theta)$ is satisfied.

$$\phi(\omega_{sp_D^\pm}, \theta) \cong -\delta\left\{[1 + \eta(\theta)](\theta - 1) + \frac{b^2}{2}\cdot\frac{1 - \eta(\theta)}{\xi^2(\theta) + \delta^2[1 - \eta(\theta)]^2}\right\}$$

$$\mp \mathrm{i}\xi(\theta)\left(\theta - 1 + \frac{b^2}{2}\cdot\frac{1}{\xi^2(\theta) + \delta^2[1 - \eta(\theta)]^2}\right), \qquad (7.2.4)$$

and

$$\phi^{(2)}(\omega_{sp_D^\pm}, \theta) \cong -\mathrm{i}\frac{b^2}{\{\pm\xi(\theta) + \delta\mathrm{i}[1 - \eta(\theta)]\}^3}, \qquad (7.2.5)$$

so that the second coefficient in the Taylor series expansion (5.1.3) of $\phi(\omega, \theta)$ is

$$p_0(\omega_{sp_D^\pm}, \theta) = \frac{\phi^{(2)}(\omega_{sp_D^\pm}, \theta)}{2!} \cong -\mathrm{i}\frac{b^2}{2}\cdot\frac{1}{\{\pm\xi(\theta) + \delta\mathrm{i}[1 - \eta(\theta)]\}^3}. \qquad (7.2.6)$$

Furthermore, the first coefficient in the Taylor series expansion (5.1.4) of the spectrum is given by

$$q_0(\omega_{sp_D^\pm}) = \tilde{u}(\omega_{sp_D^\pm} - \omega_c), \qquad (7.2.7)$$

the specific form of which depends upon the particular initial pulse envelope function $u(t)$.

The proper value of $\arg[-p_0(\omega_{sp_D^\pm}, \theta)]$ must now be ascertained according to the convention set forth in *Olver*'s analysis, c.f. (4.1.7). For simplicity, the path $P_D^+(\theta)$ through the distant saddle point SP_D^+ in the right-half of the complex ω-plane is taken along the path of steepest descent in the vicinity of $\omega_{sp_D^+}(\theta)$, as illustrated in Fig. 7.1. The angle of slope of the contour at the saddle point, c.f. (6.2.26), is then $\bar{\alpha}^+ = -\pi/4$. Therefore, since $\Theta \equiv \arg(z) = 0$, the proper value of $\bar{\alpha}_0^+ \equiv \arg[-p_0(\omega_{sp_D^+}, \theta)]$ as determined by the inequality (5.1.7) is $\bar{\alpha}_0^+ \cong \pi/2$. By a similar argument, the proper value of $\bar{\alpha}_0^- \equiv \arg[-p_0(\omega_{sp_D^-}, \theta)]$ is $\bar{\alpha}_0^- \cong -\pi/2$.

According to (5.1.9a), the coefficients $a_0(\omega_{sp_D^\pm})$ appearing in (7.2.3) are given by

$$\begin{aligned}
a_0(\omega_{sp_D^\pm}) &= \frac{q_0(\omega_{sp_D^\pm})}{\mu[-p_0(\omega_{sp_D^\pm}, \theta)]^{\lambda/\mu}} \\
&\cong \tilde{u}(\omega_{sp_D^\pm} - \omega_c)\cdot\frac{1}{2}\left(\frac{2}{ib^2}\{\pm\xi(\theta) + \delta i[1-\eta(\theta)]\}^3\right)^{1/2} \\
&\cong \tilde{u}(\omega_{sp_D^\pm} - \omega_c)\frac{1}{\sqrt{2}b}e^{\mp i(\pi/4)}\left\{\xi^{3/2}(\theta) \pm \frac{3}{2}\delta i[1-\eta(\theta)]\xi^{1/2}(\theta)\right\},
\end{aligned} \tag{7.2.8}$$

where the facts that $0 < 1 - \eta(\theta) < 1$ and $\xi(\theta) \gg \delta$ for $1 \leqslant \theta \leqslant \infty$ have been employed in expanding the square root expression appearing in this equation. Substitution of (7.2.4 and 8) into the expansion (7.2.3) for the first precursor field then yields

$$\begin{aligned}
A_S(z,t) \sim{}& \frac{1}{b}\sqrt{\frac{c\xi(\theta)}{2\pi z}}\exp\left(-\delta\frac{z}{c}\left\{[1+\eta(\theta)](\theta-1)\right.\right. \\
&\left.\left.+\frac{b^2}{2}\cdot\frac{1-\eta(\theta)}{\xi^2(\theta)+\delta^2[1-\eta(\theta)]^2}\right\}\right) \\
&\cdot \mathbb{R}\mathrm{e}\, i\left\{\tilde{u}(\omega_{sp_D^+} - \omega_c)\left\{\xi(\theta) + \frac{3}{2}\delta i[1-\eta(\theta)]\right\}\right. \\
&\cdot\exp\left\{-i\left[\frac{z}{c}\xi(\theta)\left(\theta - 1 + \frac{b^2}{2}\cdot\frac{1}{\xi^2(\theta)+\delta^2[1-\eta(\theta)]^2}\right)+\frac{\pi}{4}\right]\right. \\
&+\tilde{u}(\omega_{sp_D^-} - \omega_c)\left\{\xi(\theta) - \frac{3}{2}\delta i[1-\eta(\theta)]\right\} \\
&\left.\cdot\exp\left\{i\left[\frac{z}{c}\xi(\theta)\left(\theta - 1 + \frac{b^2}{2}\cdot\frac{1}{\xi^2(\theta)+\delta^2[1-\eta(\theta)]^2}\right)+\frac{\pi}{4}\right]\right\}\right\},
\end{aligned} \tag{7.2.9}$$

as $z \to \infty$ uniformly for $\theta \geqslant 1 + \varepsilon$, where $\varepsilon > 0$. This expression constitutes the second approximate expression of the nonuniform asymptotic behavior of the *Sommerfeld* precursor field.

7.2.2 The Uniform Approximation

As the parameter θ is allowed to approach unity from above, the asymptotic representation (7.2.9) of the first precursor field $A_S(z,t)$ is no longer valid and must be replaced by the uniform asymptotic representation presented in Sect. 5.4. *Handelsman* and *Bleistein* [7.8] have performed the required analysis for the step-function modulated signal using the first approximation for the distant saddle-point locations. They obtained a result derived earlier by *Sommerfeld* [7.1]. As *Handelsman* and *Bleistein* pointed out, however, the result is not truely uniform because the first approximation is useful only for very small positive values of $\theta - 1$ and hence the asymptotic approximation of $A_S(z,t)$ is valid only for values of θ in the vicinity of $\theta = 1$. In this subsection, the second approximation of the distant saddle point locations is employed to obtain a uniform asymptotic approximation of the first precursor $A_S(z,t)$ that is valid uniformly for all $\theta \geqslant 1$.

According to Theorem 5.4 and the results of Sect. 6.5, the integral (7.2.1) taken over the two contours $P_D^+(\theta)$ and $P_D^-(\theta)$ yields the first or *Sommerfeld* precursor field $A_S(z,t)$, which satisfies

$$A_S(z,t) = \mathrm{Re}\left\{ e^{-i(z/c)\beta(\theta)}[2\alpha(\theta)e^{-i(\pi/2)}]^{\nu}\left[\gamma_0 J_\nu\left(\frac{z}{c}\alpha(\theta)\right) + 2\alpha(\theta)e^{-i(\pi/2)}\gamma_1 J_{\nu+1}\left(\frac{z}{c}\alpha(\theta)\right)\right]\right\} + R_1(z,\theta) \ , \tag{7.2.10}$$

for all $\theta \geqslant 1$. The remainder term $R_1(z,\theta)$ is bounded by the inequality (5.4.5) for $\theta \geqslant 1$ with the constant K independent of θ. Consequently, (7.2.10) provides the dominant term in the asymptotic expansion of the first precursor field that is valid uniformly for all $\theta \geqslant 1$. The real parameter ν is defined by (5.4.3). Furthermore, according to the defining relations (5.4.6) through (5.4.9) and (7.2.4, 5), the coefficients appearing in this uniform expansion are given by

$$\alpha(\theta) = i\frac{\phi(\omega_{sp_D^+},\theta) - \phi(\omega_{sp_D^-},\theta)}{2} \cong \xi(\theta)\left(\theta - 1 + \frac{b^2}{2}\cdot\frac{1}{\xi^2(\theta) + \delta^2[1-\eta(\theta)]^2}\right) , \tag{7.2.11}$$

$$\beta(\theta) = i\frac{\phi(\omega_{sp_D^+},\theta) + \phi(\omega_{sp_D^-},\theta)}{2} \cong -i\delta\left\{[1+\eta(\theta)](\theta-1) + \frac{b^2}{2}\cdot\frac{1-\eta(\theta)}{\xi^2(\theta) + \delta^2[1-\eta(\theta)]^2}\right\} , \tag{7.2.12}$$

$$\gamma_0(\theta) = \frac{1}{2}\left[\tilde{u}(\omega_{sp_D^+} - \omega_c)\left(\frac{1}{2\alpha(\theta)}\right)^{(1+\nu)}\left(\frac{4\alpha^3(\theta)}{i\phi^{(2)}(\omega_{sp_D^+},\theta)}\right)^{1/2} + \tilde{u}(\omega_{sp_D^-} - \omega_c)\left(-\frac{1}{2\alpha(\theta)}\right)^{(1+\nu)}\left(-\frac{4\alpha^3(\theta)}{i\phi^{(2)}(\omega_{sp_D^-},\theta)}\right)^{1/2}\right]$$

$$\cong \left(\frac{1}{2}\right)^{(1+\nu)} \left[\xi(\theta) \left(\theta - 1 + \frac{b^2}{2} \cdot \frac{1}{\xi^2(\theta) + \delta^2[1-\eta(\theta)]^2} \right) \right]^{((1/2)-\nu)}$$
$$\cdot \left[\tilde{u}(\omega_{sp_D^+} - \omega_c) \left(\frac{\{\xi(\theta) + \delta \mathrm{i}[1-\eta(\theta)]\}^3}{b^2} \right)^{1/2} \right.$$
$$\left. + (-1)^{(1+\nu)} \tilde{u}(\omega_{sp_D^-} - \omega_c) \left(-\frac{\{-\xi(\theta) + \delta \mathrm{i}[1-\eta(\theta)]\}^3}{b^2} \right)^{1/2} \right]$$
$$\cong \left(\frac{1}{2}\right)^{(1+\nu)} \frac{1}{b} \xi^{1/2}(\theta) \left[\xi(\theta) \left(\theta - 1 \right.\right.$$
$$\left.\left. + \frac{b^2}{2} \cdot \frac{1}{\xi^2(\theta) + \delta^2[1-\eta(\theta)]^2} \right) \right]^{((1/2)-\nu)}$$
$$\cdot \left(\tilde{u}(\omega_{sp_D^+} - \omega_c) \left\{ \xi(\theta) + \frac{3}{2} \delta \mathrm{i}[1-\eta(\theta)] \right\} \right.$$
$$\left. + (-1)^{(1+\nu)} \tilde{u}(\omega_{sp_D^-} - \omega_c) \left\{ \xi(\theta) - \frac{3}{2} \delta \mathrm{i}[1-\eta(\theta)] \right\} \right), \tag{7.2.13}$$

$$\gamma_1(\theta) = \frac{1}{4\alpha(\theta)} \left[\tilde{u}(\omega_{sp_D^+} - \omega_c) \left(\frac{1}{2\alpha(\theta)} \right)^{(1+\nu)} \left(\frac{4\alpha^3(\theta)}{\mathrm{i}\phi^{(2)}(\omega_{sp_D^+}, \theta)} \right)^{1/2} \right.$$
$$\left. - \tilde{u}(\omega_{sp_D^-} - \omega_c) \left(-\frac{1}{2\alpha(\theta)} \right)^{(1+\nu)} \left(-\frac{4\alpha^3(\theta)}{\mathrm{i}\phi^{(2)}(\omega_{sp_D^-}, \theta)} \right)^{1/2} \right]$$
$$\cong \left(\frac{1}{2}\right)^{(2+\nu)} \left[\xi(\theta) \left(\theta - 1 + \frac{b^2}{2} \cdot \frac{1}{\xi^2(\theta) + \delta^2[1-\eta(\theta)]^2} \right) \right]^{-((1/2)+\nu)}$$
$$\cdot \left[\tilde{u}(\omega_{sp_D^+} - \omega_c) \left(\frac{\{\xi(\theta) + \delta \mathrm{i}[1-\eta(\theta)]\}^3}{b^2} \right)^{1/2} \right.$$
$$\left. - (-1)^{(1+\nu)} \tilde{u}(\omega_{sp_D^-} - \omega_c) \left(-\frac{\{-\xi(\theta) + \delta \mathrm{i}[1-\eta(\theta)]\}^3}{b^2} \right)^{1/2} \right]$$
$$\cong \left(\frac{1}{2}\right)^{(2+\nu)} \frac{1}{b} \xi^{1/2}(\theta) \left[\xi(\theta) \left(\theta - 1 \right.\right.$$
$$\left.\left. + \frac{b^2}{2} \cdot \frac{1}{\xi^2(\theta) + \delta^2[1-\eta(\theta)]^2} \right) \right]^{-((1/2)+\nu)}$$
$$\cdot \left(\tilde{u}(\omega_{sp_D^+} - \omega_c) \left\{ \xi(\theta) + \frac{3}{2} \delta \mathrm{i}[1-\eta(\theta)] \right\} \right.$$
$$\left. - (-1)^{(1+\nu)} \tilde{u}(\omega_{sp_D^-} - \omega_c) \left\{ \xi(\theta) - \frac{3}{2} \delta \mathrm{i}[1-\eta(\theta)] \right\} \right). \tag{7.2.14}$$

The branch of the square root expression $(\pm 4\alpha^3(\theta)/i\phi^{(2)}(\omega_{sp_D^\pm},\theta))^{1/2}$ appearing in the defining equations for the coefficients $\gamma_0(\theta)$ and $\gamma_1(\theta)$ has been determined by the limiting relation (5.4.12) as follows. In accordance with Theorem 5.4, the argument of the quantity $[\pm i\phi^{(2)}(\omega_{sp_D^\pm},\theta)] \cong b^2/\{\xi(\theta) \pm \delta i[1-\eta(\theta)]\}^3$ as determined by the inequality (5.1.7) with $\Theta \equiv \arg(iz) = \pi/2$ and $\bar{\alpha}^+ = -\pi/4$ is approximately zero [since $\xi(\theta) \gg \delta(1-\eta(\theta))$ for all $\theta \geqslant 1$]. Then, according to (5.4.12), $\arg[\alpha(\theta)] = \frac{2}{3}\arg(a_1)$ in the limit as θ approaches unity from above, where, cf. (5.4.2), $a_1 = b^2/2$, which is real and positive. Hence, $\arg(a_1) = 0$, so that $\arg(\alpha(\theta)) = 0$ in the limit as $\theta \to 1^+$. By continuity, this result is, to a good approximation, valid for all $\theta \geqslant 1$. This branch requirement, in addition to the fact that $\xi(\theta) \gg \delta[1-\eta(\theta)]$ for all $\theta \geqslant 1$, has been used in obtaining the final approximate expressions for the coefficients $\gamma_0(\theta)$ and $\gamma_1(\theta)$ in (7.2.13 and 14), respectively. All of the square root expressions appearing in (7.2.13 and 14) are real and positive.

Substitution of these expressions into (7.2.10) then yields

$$
\begin{aligned}
A_S(z,t) \sim \frac{1}{2b}\xi(\theta)\left(\theta - 1 + \frac{b^2}{2}\cdot\frac{1}{\xi^2(\theta)+\delta^2[1-\eta(\theta)]^2}\right)^{1/2} \\
\cdot\exp\left(-\delta\frac{z}{c}\left\{[1+\eta(\theta)](\theta-1) + \frac{b^2}{2}\cdot\frac{1-\eta(\theta)}{\xi^2(\theta)+\delta^2[1-\eta(\theta)]^2}\right\}\right) \\
\cdot\mathrm{Re}\left\{e^{-i(\pi/2)\nu}\left[\left(\tilde{u}(\omega_{sp_D^+}-\omega_c)\left\{\xi(\theta)+\frac{3}{2}\delta i[1-\eta(\theta)]\right\}\right.\right.\right. \\
\left. + (-1)^{(1+\nu)}\tilde{u}(\omega_{sp_D^-}-\omega_c)\left\{\xi(\theta)-\frac{3}{2}\delta i[1-\eta(\theta)]\right\}\right) \\
\cdot J_\nu\left[\frac{z}{c}\xi(\theta)\left(\theta-1+\frac{b^2}{2}\cdot\frac{1}{\xi^2(\theta)+\delta^2[1-\eta(\theta)]^2}\right)\right] \\
+ e^{-i(\pi/2)}\left(\tilde{u}(\omega_{sp_D^+}-\omega_c)\left\{\xi(\theta)+\frac{3}{2}\delta i[1-\eta(\theta)]\right\}\right. \\
\left. - (-1)^{(1+\nu)}\tilde{u}(\omega_{sp_D^-}-\omega_c)\left\{\xi(\theta)-\frac{3}{2}\delta i[1-\eta(\theta)]\right\}\right) \\
\left.\left.\cdot J_{\nu+1}\left[\frac{z}{c}\xi(\theta)\left(\theta-1+\frac{b^2}{2}\cdot\frac{1}{\xi^2(\theta)+\delta^2[1-\eta(\theta)]^2}\right)\right]\right]\right\},
\end{aligned}
\tag{7.2.15}
$$

as $z \to \infty$ uniformly for all $\theta \geqslant 1$. This expression constitutes the second approximate expression of the uniform asymptotic behavior of the *Sommerfeld* precursor field. As shown in Sect. 5.4, this result reduces to the nonuniform expansion given in (7.2.9) for values of $\theta > 1$ bounded away from unity.

Consider finally the limiting behavior of the uniform expansion (7.2.15) as θ goes to unity from above. In this limit, the functions $\xi(\theta)$ and $\eta(\theta)$ attain the limiting forms

$$\lim_{\theta\to 1^+} \xi(\theta) = \frac{b}{\sqrt{2(\theta-1)}} ,$$

$$\lim_{\theta\to 1^+} \eta(\theta) = 1 ,$$

so that the argument of the Bessel functions appearing in (7.2.15) becomes

$$\lim_{\theta\to 1^+} \left[\frac{z}{c}\xi(\theta)\left(\theta - 1 + \frac{b^2}{2}\cdot\frac{1}{\xi^2(\theta) + \delta^2[1-\eta(\theta)]^2}\right)\right] = \frac{z}{c} b\sqrt{2(\theta-1)} .$$

Consequently, for values of θ very close to unity the argument of the Bessel functions is very small so that the small argument limiting form of the Bessel functions appearing in (7.2.15) may be employed, where (for integer values of the order ν)

$$\mathrm{J}_\nu(\zeta) \sim \frac{(\frac{1}{2}\zeta)^\nu}{\Gamma(\nu+1)} ,$$

as $\zeta \to 0$ with ν fixed and nonnegative. For negative values of the order ν, the relation $\mathrm{J}_{-\nu}(\zeta) = (-1)^\nu \mathrm{J}_\nu(\zeta)$ may be employed in conjunction with the above limiting expression to obtain

$$\mathrm{J}_{-\nu}(\zeta) \sim (-1)^\nu \cdot \frac{(\frac{1}{2}\zeta)^\nu}{\Gamma(\nu+1)} ,$$

as $\zeta \to 0$ with ν fixed and nonnegative.

For integer $\nu \geqslant 0$, substitution of the above results into equation (7.2.15) yields the following limiting behavior of the first precursor field for values of θ very close to 1:

$$\begin{aligned}\lim_{\theta\to 1^+} A_S(z,t) \sim \frac{1}{2\sqrt{2}} e^{-2(\delta z/c)(\theta-1)} \cdot \mathbb{R}\mathrm{e}\, e^{-i(\pi/2)\nu} \Bigg\{ & [\tilde{u}(\omega_{sp_D^+} - \omega_c) \\ & + (-1)^{(1+\nu)}\tilde{u}(\omega_{sp_D^-} - \omega_c)] \frac{1}{\Gamma(\nu+1)} \\ & \cdot \frac{b}{\sqrt{2(\theta-1)}} \left(\frac{bz}{2c}\sqrt{2(\theta-1)}\right)^\nu + e^{-i(\pi/2)}[\tilde{u}(\omega_{sp_D^+} - \omega_c) \\ & - (-1)^{(1+\nu)}\tilde{u}(\omega_{sp_D^-} - \omega_c)] \frac{1}{\Gamma(\nu+2)} \\ & \cdot \frac{b}{\sqrt{2(\theta-1)}} \left(\frac{bz}{2c}\sqrt{2(\theta-1)}\right)^{\nu+1} \Bigg\} . \end{aligned} \tag{7.2.16}$$

Since the initial envelope function $u(t)$ is real so that its spectrum satisfies $\tilde{u}(-\omega) = \tilde{u}^*(\omega^*)$, and since $\mathrm{Re}\{\omega_{sp_D^+}\} = -\mathrm{Re}\{\omega_{sp_D^-}\}$ and $\mathrm{Im}\{\omega_{sp_D^+}\} = \mathrm{Im}\{\omega_{sp_D^-}\}$, this expression (7.2.16) identically vanishes at $\theta = 1$ for finite values of ω_c. However, for $\theta = 1 + \varepsilon$ with $\varepsilon > 0$ arbitrarily small, this expression is, in general, nonzero. Consequently, *for integer* $v \geqslant 0$ *the front of the first precursor field travels with the velocity of light c in vacuum.*

For $v = -1$ the limiting behavior of the first precursor field for values of θ very close to 1 is given by

$$\begin{aligned}\lim_{\theta\to 1^+} A_S(z,t) \sim \frac{1}{2\sqrt{2}} \mathrm{e}^{-2(\delta z/c)(\theta-1)} \cdot \mathrm{Re}\Bigg\{ &-\mathrm{e}^{\mathrm{i}(\pi/2)}[\tilde{u}(\omega_{sp_D^+} - \omega_c) \\ &+ \tilde{u}(\omega_{sp_D^-} - \omega_c)]\frac{b^2 z}{2c} + [\tilde{u}(\omega_{sp_D^+} - \omega_c) \\ &- \tilde{u}(\omega_{sp_D^-} - \omega_c)]\frac{b}{\sqrt{2(\theta-1)}}\Bigg\} ,\end{aligned} \tag{7.2.17}$$

which diverges as $(\theta - 1)^{-1/2}$ as $\theta \to 1^+$, while it is finite for any $\theta = 1 + \varepsilon$ with $\varepsilon > 0$ arbitrarily small. Consequently, *for* $v = -1$ *the front of the first precursor field is singular and this point travels with the velocity of light c in vacuum.* The uniform expansion is then valid only for $\theta \geqslant 1 + \varepsilon$, $\varepsilon > 0$. For smaller values of v (viz., $v = -2, -3, -4, \ldots$), the spectrum $\tilde{u}(\omega)$ does not remain finite in the limit as $|\omega| \to \infty$. Such initial pulse envelope functions $u(t)$ are excluded from this analysis.

7.2.3 The Instantaneous Angular Frequency

It is convenient to define the instantaneous angular frequency of oscillation of the precursor field as the time derivative of the oscillatory phase [7.4, 5]. Note that the oscillatory phase terms appearing in (7.2.9, 15) are identical. Since $d\theta/dt = c/z$ and

$$\frac{d\xi(\theta)}{dt} = -\frac{cb^2}{z\xi(\theta)} \cdot \frac{\theta}{(\theta^2-1)^2} ,$$

$$\frac{d\eta(\theta)}{dt} = \frac{2b^2c}{z} \cdot \frac{\theta}{(\theta^2-1)^2} \cdot \frac{\eta(\theta)-1}{\xi^2(\theta)} ,$$

then the instantaneous angular frequency of oscillation of the first precursor field is given by

$$\begin{aligned}\omega_s &\equiv \frac{d}{dt}\left[\frac{z}{c}\xi(\theta)\left(\theta - 1 + \frac{b^2}{2} \cdot \frac{1}{\xi^2(\theta) + \delta^2[1-\eta(\theta)]^2}\right)\right] \\ &= \xi(\theta) + \frac{b^2}{2\xi(\theta)} \cdot \frac{\theta}{(\theta^2-1)^2}\left[b^2 \frac{\xi^2(\theta) - 5\delta^2[1-\eta(\theta)]^2}{\{\xi^2(\theta) + \delta^2[1-\eta(\theta)]^2\}^2} - 2(\theta - 1)\right]\end{aligned} \tag{7.2.18a}$$

$$\cong \xi(\theta) . \tag{7.2.18b}$$

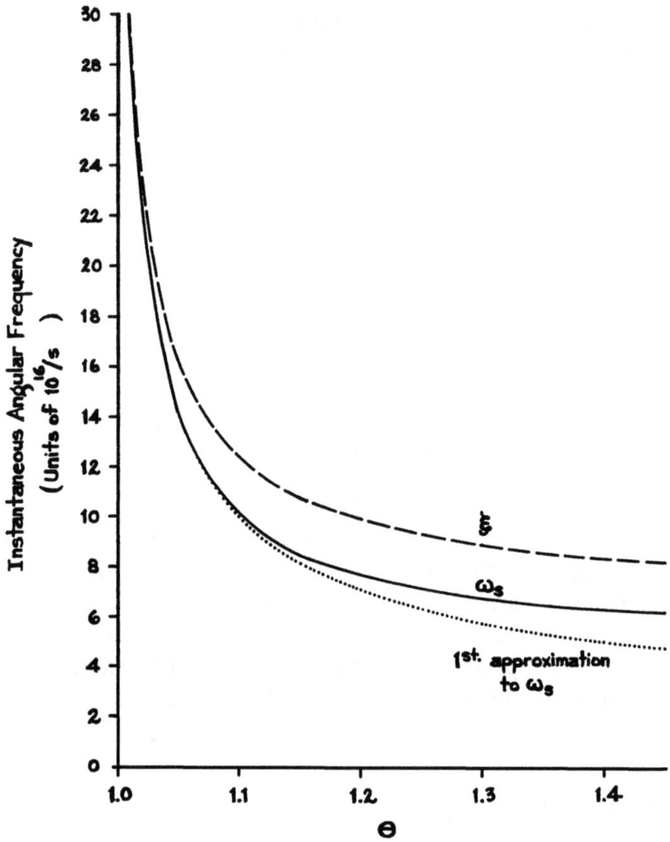

Fig. 7.2. The instantaneous angular frequency of oscillation ω_s of the first precursor field as a function of θ for $\omega_0 = 4.0 \times 10^{16}$/s, $\delta = 0.28 \times 10^{16}$/s, and $b^2 = 20.0 \times 10^{32}/\text{s}^2$.

A graph of the instantaneous angular frequency of oscillation of the first precursor field as a function of θ according to (7.2.18a) is illustrated in Fig. 7.2 along with the corresponding behavior of $\xi(\theta)$ for *Brillouin*'s choice of the medium parameters. It is easily seen that, for values of θ very close to unity, ω_s is given by $\xi(\theta)$ to a high degree of accuracy [at $\theta = 1$, the second term on the right in (7.2.18a) identically vanishes and $\omega_s = \xi(1) = \infty$/s]. However, the approximate expression (7.2.18b) becomes less accurate for increasing values of θ. Finally, for large values of θ, the second term on the right in (7.2.18a) again approaches zero and the instantaneous angular frequency ω_s approaches the value $\xi(\infty) = \sqrt{\omega_1^2 - \delta^2}$ asymptotically from above. As a final point, notice that the instantaneous angular frequency is, strictly speaking, only a heuristic mathematical concept which, at times, may yield completely erroneous results [7.9]. For the present situation, however, the instantaneous frequency monotonically decreases with increasing θ; thus it does yield proper results (i.e., results that describe correctly the dynamical evolution of the field under consideration).

7.2.4 The Unit Step-Function Modulated Signal

For the unit step-function modulated signal, the spectrum of the initial field envelope is given by (4.3.33), so that

$$\tilde{u}(\omega_{sp_D^\pm} - \omega_c) = \frac{i}{\omega_{sp_D^\pm} - \omega_c} \cong \frac{i}{\pm\xi(\theta) - \omega_c - \delta i[1 + \eta(\theta)]} \,. \tag{7.2.19}$$

With $\tilde{u}(\omega_{sp_D^\pm} - \omega_c)$ given by (7.2.19), v is equal to zero and the uniform asymptotic expansion (7.2.15) of the first precursor field in this case becomes

$$\begin{aligned} A_S(z,t) \sim{}& \frac{1}{2b}\xi(\theta)\left(\theta - 1 + \frac{b^2}{2}\frac{1}{\xi^2(\theta) + \delta^2[1 - \eta(\theta)]^2}\right)^{1/2} \\ &\cdot \exp\left(-\delta\frac{z}{c}\left\{[1 + \eta(\theta)](\theta - 1) + \frac{b^2}{2}\cdot\frac{1 - \eta(\theta)}{\xi^2(\theta) + \delta^2[1 - \eta(\theta)]^2}\right\}\right) \\ &\cdot\left\{\frac{\delta}{2}\left(\frac{\xi(\theta)[5 - \eta(\theta)] + 3\omega_c[1 - \eta(\theta)]}{[\xi(\theta) + \omega_c]^2 + \delta^2[1 + \eta(\theta)]^2}\right.\right. \\ &\left. - \frac{\xi(\theta)[5 - \eta(\theta)] - 3\omega_c[1 - \eta(\theta)]}{[\xi(\theta) - \omega_c]^2 + \delta^2[1 + \eta(\theta)]^2}\right) \\ &\cdot J_0\left[\frac{z}{c}\xi(\theta)\left(\theta - 1 + \frac{b^2}{2}\cdot\frac{1}{\xi^2(\theta) + \delta^2[1 - \eta(\theta)]^2}\right)\right] \\ &+ \left(\frac{\xi(\theta)[\xi(\theta) - \omega_c] - \frac{3}{2}\delta^2[1 - \eta^2(\theta)]}{[\xi(0) - \omega_c]^2 + \delta^2[1 + \eta(\theta)]^2}\right. \\ &\left. - \frac{\xi(\theta)[\xi(\theta) + \omega_c] - \frac{3}{2}\delta^2[1 - \eta^2(\theta)]}{[\xi(\theta) + \omega_c]^2 + \delta^2[1 + \eta(\theta)]^2}\right) \\ &\left.\cdot J_1\left[\frac{z}{c}\xi(\theta)\left(\theta - 1 + \frac{b^2}{2}\cdot\frac{1}{\xi^2(\theta) + \delta^2[1 - \eta(\theta)]^2}\right)\right]\right\}, \end{aligned} \tag{7.2.20}$$

as $z \to \infty$ uniformly for all $\theta \geqslant 1$. It is evident from this result for the unit step-function modulated signal that $A_S(z,t)$, and consequently $A(z,t)$, "asymptotically" vanish at $\theta = 1$, but are nonzero for $\theta = 1 + \varepsilon$ where $\varepsilon > 0$ can be arbitrarily small. Consequently, the front of the first precursor field in this case travels with the velocity of light c in vacuum.

For values of $\theta > 1$ bounded away from unity, the two Bessel functions appearing in (7.2.20) may be replaced by their asymptotic approximations given by

$$J_\nu(\zeta) = \sqrt{\frac{2}{\pi\zeta}}\cos\left(\zeta - \frac{1}{2}\nu\pi - \frac{\pi}{4}\right) + \mathcal{O}(\zeta^{-1})e^{|\mathrm{Im}\{\zeta\}|}$$

as $\zeta \to \infty$ for $|\arg(\zeta)| < \pi$. With the substitution of this expression into (7.2.20), the nonuniform asymptotic expansion of the first precursor field is obtained, viz.

$$\begin{aligned} A_S(z,t) \sim -\sqrt{\frac{c\xi(\theta)}{2\pi z}}\frac{1}{b}\exp\Bigg(-\delta\frac{z}{c}\Bigg\{&[1+\eta(\theta)](\theta-1) \\ &+\frac{b^2}{2}\cdot\frac{1-\eta(\theta)}{\xi^2(\theta)+\delta^2[1-\eta(\theta)]^2}\Bigg\}\Bigg) \\ \cdot\Bigg\{\Bigg(&\frac{\xi(\theta)[\xi(\theta)-\omega_c]-\frac{3}{2}\delta^2[1-\eta^2(\theta)]}{[\xi(\theta)-\omega_c]^2+\delta^2[1+\eta(\theta)]^2} \\ &-\frac{\xi(\theta)[\xi(\theta)+\omega_c]-\frac{3}{2}\delta^2[1-\eta^2(\theta)]}{[\xi(\theta)+\omega_c]^2+\delta^2[1+\eta(\theta)]^2}\Bigg) \\ &\cdot\cos\left[\frac{z}{c}\xi(\theta)\left(\theta-1+\frac{b^2}{2}\cdot\frac{1}{\xi^2(\theta)+\delta^2[1-\eta(\theta)]^2}\right)+\frac{\pi}{4}\right] \\ &+\frac{\delta}{2}\left[\frac{\xi(\theta)[5-\eta(\theta)]-3\omega_c[1-\eta(\theta)]}{[\xi(\theta)-\omega_c]^2+\delta^2[1+\eta(\theta)]^2}\right. \\ &\left.-\frac{\xi(\theta)[5-\eta(\theta)]+3\omega_c[1-\eta(\theta)]}{[\xi(\theta)+\omega_c]^2+\delta^2[1+\eta(\theta)]^2}\right] \\ &\cdot\sin\left[\frac{z}{c}\xi(\theta)\left(\theta-1+\frac{b^2}{2}\cdot\frac{1}{\xi^2(\theta)+\delta^2[1-\eta(\theta)]^2}\right)+\frac{\pi}{4}\right]\Bigg\}, \end{aligned} \tag{7.2.21}$$

as $z \to \infty$ for $\theta \geqslant 1 + \varepsilon$ with $\varepsilon > 0$. The same result is obtained with the substitution of (7.2.19) into (7.2.9).

As pointed out earlier, the expression in (7.2.21) is not a valid approximation of the first precursor field in the limit as $\theta \to 1^+$ for fixed values of the observation distance z. In order to establish contact with *Brillouin*'s result, however, the value of this expression in that limit is now examined. For values of θ approximately equal to but greater than unity, (7.2.21) simplifies to

$$A_S(z,t) \sim -\sqrt{\frac{bc}{2\pi z}}\cdot\frac{1}{[2(\theta-1)]^{3/4}}\,e^{-2\delta(z/c)(\theta-1)}$$

$$\cdot\left\{\left[\frac{\dfrac{b}{\sqrt{2(\theta-1)}}-\omega_c}{\left(\dfrac{b}{\sqrt{2(\theta-1)}}-\omega_c\right)^2+4\delta^2}-\frac{\dfrac{b}{\sqrt{2(\theta-1)}}+\omega_c}{\left(\dfrac{b}{\sqrt{2(\theta-1)}}+\omega_c\right)^2+4\delta^2}\right]\cos\left[\frac{bz}{c}\sqrt{2(\theta-1)}+\frac{\pi}{4}\right]\right.$$

$$\left.+2\delta\left[\frac{1}{\left(\dfrac{b}{\sqrt{2(\theta-1)}}-\omega_c\right)^2+4\delta^2}-\frac{1}{\left(\dfrac{b}{\sqrt{2(\theta-1)}}+\omega_c\right)^2+4\delta^2}\right]\sin\left[\frac{bz}{c}\sqrt{2(\theta-1)}+\frac{\pi}{4}\right]\right\},\tag{7.2.22}$$

as $z\to\infty$ with $\theta>1$ and $\theta\cong 1$. This equation is the result which would be obtained by using the first approximate expressions for the locations of the distant saddle points, c.f. (6.2.9), in the nonuniform expansion (7.2.3). The expression can be further simplified by noting that for values of θ very close to unity, any finite applied signal frequency ω_c will become negligible in comparison to the quantity $b[2(\theta-1)]^{-1/2}$. Hence, in the limit as θ approaches unity from above, one may ignore ω_c in comparison to $b[2(\theta-1)]^{-1/2}$ in the denominators of (7.2.22), and this equation reduces to

$$A_S(z,t)\sim\sqrt{\frac{2bc}{\pi z}}\cdot\frac{\omega_c[2(\theta-1)]^{1/4}}{b^2+8\delta^2(\theta-1)}e^{-2\delta(z/c)(\theta-1)}\cos\left[\frac{bz}{c}\sqrt{2(\theta-1)}+\frac{\pi}{4}\right],\tag{7.2.23}$$

as $z\to\infty$ and $\theta\to 1^+$. Equation (7.2.23) is *Brillouin*'s result [Ref. 7.5, p. 73]. Hence, *Brillouin*'s expression for the first precursor field is an approximation, valid for θ near 1, of an expression, (7.2.21), that is not valid for θ near 1. Consequently, *Brillouin*'s expression for the asymptotic behavior of the first precursor field for the unit step-function modulated signal is not applicable.

The large argument asymptotic approximations of the Bessel functions appearing in the uniform expansion (7.2.20) can be employed in numerical calcula-

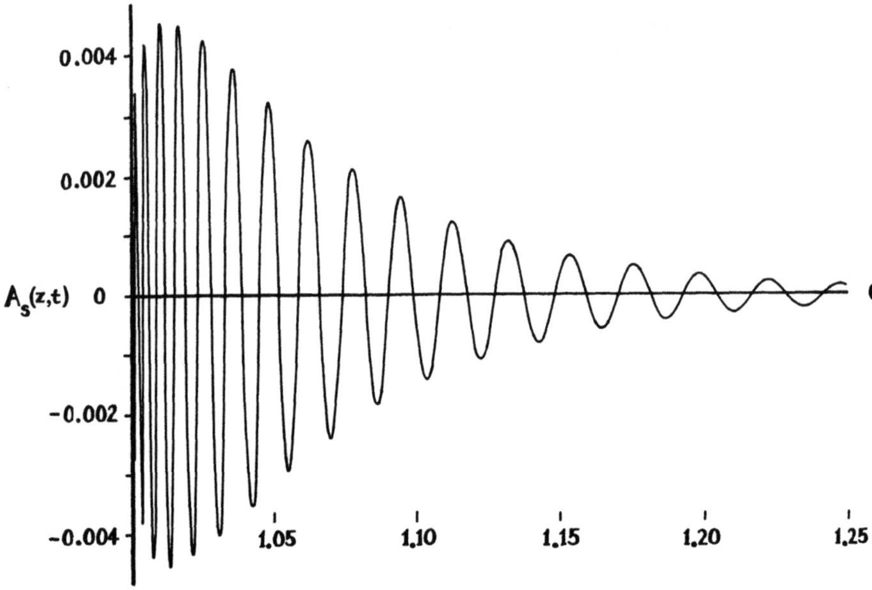

Fig. 7.3. Dynamic behavior of the Sommerfeld precursor field $A_S(z,t)$ for the unit step-function modulated signal with applied signal frequency $\omega_c = 1 \times 10^{16}$/s at an observation distance of $z = 1 \times 10^{-4}$ cm, where *Brillouin*'s choice of the medium parameters are employed.

tions of the field behavior for values of θ which satisfy the inequality $(z/c)|\alpha(\theta)| = (z/c)|Y(\omega_{sp_D^\pm}, \theta)| > 1$. That is, for values of θ such that $|Y(\omega_{sp_D^\pm}, \theta)| > c/z$, the nonuniform asymptotic expansion (7.2.21) may be employed in describing the behavior of the first precursor field, whereas for values of θ such that $|Y(\omega_{sp_D^\pm}, \theta)| \leqslant c/z$, the uniform asymptotic expansion (7.2.20) must be employed. For example, for $z = 1 \times 10^{-4}$ cm, this inequality requires that $|Y(\omega_{sp_D^\pm}, \theta)| > 3 \times 10^{14}$/s, and this inequality is satisfied for all $\theta > 1.00005$ for *Brillouin*'s choice of the medium parameters (Fig. 6.20). For larger observation distances z, the difference $\theta - 1$ becomes smaller.

The dynamic evolution with θ of the first precursor field $A_S(z,t)$ for the unit step-function modulated signal, as given by (7.2.20), is illustrated in Fig. 7.3 at an observation distance of $z = 1 \times 10^{-4}$ cm and in Fig. 7.4 at an observation distance of $z = 1 \times 10^{-3}$ cm. In both cases, *Brillouin*'s choice for the medium parameters are employed [viz., $\omega_0 = 4 \times 10^{16}$/s, $b^2 = 20 \times 10^{32}/\text{s}^2$, and $\delta = 0.28 \times 10^{16}$/s] and the applied signal frequency is given by $\omega_c = 1 \times 10^{16}$/s, which is in the near-ultraviolet region of the electromagnetic spectrum. The rapid build-up of the amplitude of the first precursor field from its initially zero value at $\theta = 1$ to a maximum value is evident from the figures. For larger values of θ, the amplitude damps out exponentially with increasing θ. The instantaneous angular frequency of oscillation of the first precursor field is also seen to rapidly decrease as θ increases away from unity, this decrease becoming less and less rapid as θ continues to increase. As the distance of observation z is increased

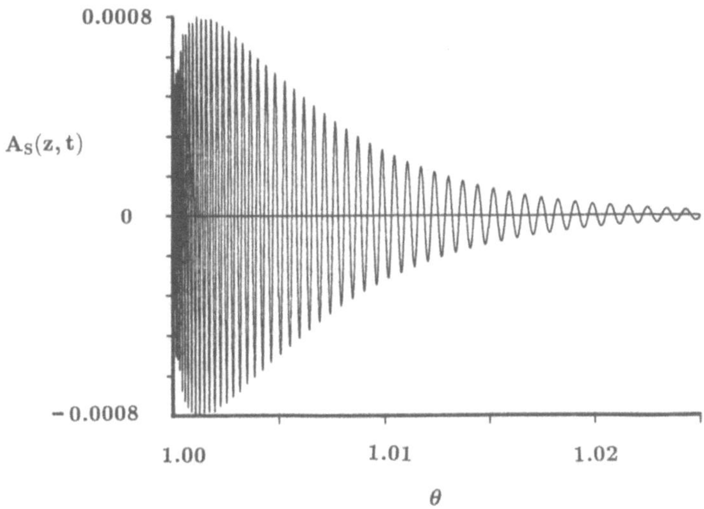

Fig. 7.4. Dynamic behavior of the Sommerfeld precursor field $A_S(z, t)$ for the unit step-function modulated signal with applied signal frequency $\omega_c = 1 \times 10^{16}$/s at an observation distance of $z = 1 \times 10^{-3}$ cm, where *Brillouin*'s choice of the medium parameters are employed.

away from the plane $z = 0$, the amplitude of the first precursor will increasingly diminish, and its maximum amplitude will shift to smaller values of θ [for the example illustrated here for $z = 1 \times 10^{-4}$ cm this maximum occurs at $\theta \cong 1.0134$, while for $z = 1 \times 10^{-3}$ cm it occurs at $\theta \cong 1.0013$].

7.2.5 The Rectangular-Pulse Modulated Signal

According to the results of Sect. 4.3.2 for a unit rectangular pulse-modulated signal of initial time duration T and frequency ω_c, the integral representation for the propagated disturbance may be represented by the difference between two unit step-function modulated signals displaced in time by the amount T, viz.

$$A(z,t) = -\frac{1}{2\pi}\mathrm{Re}\left\{\int_{ia-\infty}^{ia+\infty} \frac{1}{\omega - \omega_c} e^{(z/c)\phi_0(\omega,\theta)}\, d\omega \right.$$

$$\left. - e^{-i\omega_c T}\int_{ia-\infty}^{ia+\infty} \frac{1}{\omega - \omega_c} e^{(z/c)\phi_T(\omega,\theta_T)}\, d\omega \right\} ,$$

where the initial field begins to radiate time-harmonically at time $t = 0$, and where

$$\phi_T(\omega, \theta_T) = i\omega[n(\omega) - \theta_T] \; ,$$

$$\theta_T = \frac{c}{z}(t - T) \; .$$

Each of these integrals has associated with it a first and second precursor field, and the total precursor evolution is given by the difference of these two sets of fields, which difference is strongly dependent upon the initial pulse duration T. For example, for a sufficiently long initial temporal width T, the first and second precursor fields associated with the leading edge $A_I(z,t)$ of the pulse will completely evolve prior to the arrival of the precursor fields associated with the trailing edge $A_{II}(z,t)$ of the pulse. For shorter initial pulse widths T such that $(c/z)T$ is still somewhat greater than θ_{SB}, the first precursor field associated with the leading edge of the pulse will evolve completely undisturbed. During the second precursor evolution, however, the arrival and evolution of the precursor fields associated with the trailing edge of the pulse will occur. Finally, for initial pulse widths T such that $(c/z)T$ lies within the range $1 < (c/z)T \leqslant \theta_{SB}$, there will be a complete overlap of these two sets of precursor fields.

The Sommerfeld precursor field associated with the first integral $A_I(z,t)$ in the above expression for $A(z,t)$ is the same as $A_S(z,t)$ given in (7.2.20, 21). Since the second integral $A_{II}(z,t)$ appearing in the expression for the propagated field is of exactly the same form as the first, except that it is retarded in θ by the amount $-(c/z)T$ and is multiplied by the additional phase factor $\mathrm{e}^{-\mathrm{i}\omega_c T}$ (which depends only upon the duration and applied frequency of the input pulse), the first precursor field associated with this integral may then be easily constructed from the analysis of Sect. 7.2.4 to obtain

$$
\begin{aligned}
&A_{IIS}(z,t)\\
&\sim \frac{1}{2b}\xi(\theta_T)\left(\theta_T - 1 + \frac{b^2}{2}\cdot\frac{1}{\xi^2(\theta_T) + \delta^2[1-\eta(\theta_T)]^2}\right)^{1/2}\\
&\quad\cdot\exp\left(-\delta\frac{z}{c}\left\{[1+\eta(\theta_T)](\theta_T - 1) + \frac{b^2}{2}\cdot\frac{1-\eta(\theta_T)}{\xi^2(\theta_T) + \delta^2[1-\eta(\theta_T)]^2}\right\}\right)\\
&\quad\cdot\left[\left(\frac{-\frac{1}{2}\delta\{\xi(\theta_T)[5-\eta(\theta_T)] - 3\omega_c[1-\eta(\theta_T)]\}\cos(\omega_c T) + \{\xi(\theta_T)[\xi(\theta_T)+\omega_c] - \frac{3}{2}\delta^2[1-\eta^2(\theta_T)]\}\sin(\omega_c T)}{(\xi(\theta_T)-\omega_c)^2 + \delta^2(1+\eta(\theta_T))^2}\right.\right.\\
&\quad\left.+\frac{\frac{1}{2}\delta\{\xi(\theta_T)[5-\eta(\theta_T)] + 3\omega_c[1-\eta(\theta_T)]\}\cos(\omega_c T) + \{\xi(\theta_T)[\xi(\theta_T)+\omega_c] - \frac{3}{2}\delta^2[1-\eta^2(\theta_T)]\}\sin(\omega_c T)}{[\xi(\theta_T)+\omega_c]^2 + \delta^2[1+\eta(\theta_T)]^2}\right]\\
&\quad\cdot J_0\left[\frac{z}{c}\xi(\theta_T)\left(\theta_T - 1 + \frac{b^2}{2}\cdot\frac{1}{\xi^2(\theta_T) + \delta^2[1-\eta(\theta_T)]^2}\right)\right]\\
&\quad+\left[\frac{\{\xi(\theta_T)[\xi(\theta_T)-\omega_c] - \frac{3}{2}\delta^2[1-\eta^2(\theta_T)]\}\cos(\omega_c T) + \frac{1}{2}\delta\{\xi(\theta_T)[5-\eta(\theta_T)] - 3\omega_c[1-\eta(\theta_T)]\}\sin(\omega_c T)}{[\xi(\theta_T)-\omega_c]^2 + \delta^2[1+\eta(\theta_T)]^2}\right.\\
&\quad\left.-\frac{\{\xi(\theta_T)[\xi(\theta_T)+\omega_c] - \frac{3}{2}\delta^2[1-\eta^2(\theta_T)]\}\cos(\omega_c T) - \frac{1}{2}\delta\{\xi(\theta_T)[5-\eta(\theta_T)] + 3\omega_c[1-\eta(\theta_T)]\}\sin(\omega_c T)}{[\xi(\theta_T)+\omega_c]^2 + \delta^2[1+\eta(\theta_T)]^2}\right)\\
&\quad\left.\cdot J_1\left[\frac{z}{c}\xi(\theta_T)\left(\theta_T - 1 + \frac{b^2}{2}\cdot\frac{1}{\xi^2(\theta_T) + \delta^2[1-\eta(\theta_T)]^2}\right)\right]\right],
\end{aligned}
\tag{7.2.24}
$$

as $z \to \infty$ uniformly for all $\theta_T \geqslant 1$. This expression vanishes at $\theta_T = 1$, and reduces to (7.2.20) for $\omega_c T = 2n\pi$, $n = 0, 1, 2, 3, \ldots$. For values of $\theta_T > 1$ bounded away from unity, the two Bessel functions appearing in (7.2.24) may be replaced by their large argument asymptotic approximations. With this substitution, equation (7.2.24) becomes

$$
\begin{aligned}
A_{IIS}(z,t) \sim & -\sqrt{\frac{c\xi(\theta_T)}{2\pi z}}\cdot\frac{1}{b}\exp\left\{-\delta\frac{z}{c}\left[[1+\eta(\theta_T)](\theta_T-1)\right.\right.\\
&\left.\left.+\frac{b^2}{2}\cdot\frac{1-\eta(\theta_T)}{\xi^2(\theta_T)+\delta^2[1-\eta(\theta_T)]^2}\right]\right\}\\
&\cdot\left\{\frac{\xi(\theta_T)[\xi(\theta_T)-\omega_c]-\frac{3}{2}\delta^2[1-\eta^2(\theta_T)]}{[\xi(\theta_T)-\omega_c]^2+\delta^2[1+\eta(\theta_T)]^2}\cos\left[\frac{z}{c}\xi(\theta_T)\right.\right.\\
&\left.\cdot\left(\theta_T-1+\frac{b^2}{2}\cdot\frac{1}{\xi^2(\theta_T)+\delta^2[1-\eta(\theta_T)]^2}\right)+\frac{\pi}{4}+\omega_c T\right]\\
&-\frac{\xi(\theta_T)[\xi(\theta_T)+\omega_c]-\frac{3}{2}\delta^2[1-\eta^2(\theta_T)]}{[\xi(\theta_T)+\omega_c]^2+\delta^2[1+\eta(\theta_T)]^2}\cos\left[\frac{z}{c}\xi(\theta_T)\right.\\
&\left.\cdot\left(\theta_T-1+\frac{b^2}{2}\cdot\frac{1}{\xi^2(\theta_T)+\delta^2[1-\eta(\theta_T)]^2}\right)+\frac{\pi}{4}-\omega_c T\right]\\
&+\frac{\delta}{2}\cdot\frac{\xi(\theta_T)[5-\eta(\theta_T)]-3\omega_c[1-\eta(\theta_T)]}{[\xi(\theta_T)-\omega_c]^2+\delta^2[1+\eta(\theta_T)]^2}\sin\left[\frac{z}{c}\xi(\theta_T)\right.\\
&\left.\cdot\left(\theta_T-1+\frac{b^2}{2}\cdot\frac{1}{\xi^2(\theta_T)+\delta^2[1-\eta(\theta_T)]^2}\right)+\frac{\pi}{4}+\omega_c T\right]\\
&-\frac{\delta}{2}\cdot\frac{\xi(\theta_T)[5-\eta(\theta_T)]+3\omega_c[1-\eta(\theta_T)]}{[\xi(\theta_T)+\omega_c]^2+\delta^2[1+\eta(\theta_T)]^2}\sin\left[\frac{z}{c}\xi(\theta_T)\right.\\
&\left.\left.\cdot\left(\theta_T-1+\frac{b^2}{2}\cdot\frac{1}{\xi^2(\theta_T)+\delta^2[1-\eta(\theta_T)]^2}\right)+\frac{\pi}{4}-\omega_c T\right]\right\},
\end{aligned}
\tag{7.2.25}
$$

as $z \to \infty$ for $\theta_T \geqslant 1 + \varepsilon$ with $\varepsilon > 0$. This equation reduces to (7.2.21) for $\omega_c T = 2n\pi$, $n = 0, 1, 2, 3, \ldots$.

7.2.6 The Delta-Function Pulse

According to (4.3.44), the integral representation for the propagated disturbance due to an input delta-function pulse at time $t = t_0$ is given by

$$A(z,t) = \frac{1}{2\pi}\mathrm{Re}\left\{\int_{ia-\infty}^{ia+\infty} e^{(z/c)\phi_{t_0}(\omega,\theta_{t_0})}\,d\omega\right\} ,$$

where

$$\phi_{t_0}(\omega,\theta_{t_0}) = \mathrm{i}\omega[n(\omega) - \theta_{t_0}] ,$$

$$\theta_{t_0} = \frac{c}{z}(t - t_0) .$$

It is necessary only to consider an input delta-function pulse at the time $t_0 = 0$, since other such pulses at arbitrary times t_0 can easily be constructed by a linear shift in the time coordinate. Upon comparison of (4.3.28) with (6.3.31), it is seen that

$$\tilde{u}(\omega - \omega_c) = -\mathrm{i} . \tag{7.2.26}$$

With the spectrum given by (7.2.26), it is seen that $\nu = -1$ in this case, c.f. (5.4.3). As was pointed out in Sect. 5.4, the uniform asymptotic expansion stated in Theorem 4.4 is still applicable for all values of $\theta \geqslant 1$ when $\nu < 0$ provided that its limiting behavior as θ approaches unity is finite [7.8]. Upon substitution of (7.2.26) into (7.2.15) with $\nu = -1$, there results

$$\begin{aligned} A_S(z,t) \sim \frac{1}{2b}\xi(\theta)\left(\theta - 1 + \frac{b^2}{2}\cdot\frac{1}{\xi^2(\theta) + \delta^2[1-\eta(\theta)]^2}\right)^{1/2} \\ \cdot\exp\left(-\delta\frac{z}{c}\left\{[1+\eta(\theta)](\theta-1) + \frac{b^2}{2}\cdot\frac{1-\eta(\theta)}{\xi^2(\theta)+\delta^2[1-\eta(\theta)]^2}\right\}\right) \\ \cdot\left\{2\xi(\theta)\mathrm{J}_{-1}\left[\frac{z}{c}\xi(\theta)\left(\theta - 1 + \frac{b^2}{2}\cdot\frac{1}{\xi^2(\theta)+\delta^2[1-\eta(\theta)]^2}\right)\right]\right. \\ \left. + 3\delta(1-\eta(\theta))\mathrm{J}_0\left[\frac{z}{c}\xi(\theta)\left(\theta - 1 + \frac{b^2}{2}\cdot\frac{1}{\xi^2(\theta)+\delta^2[1-\eta(\theta)]^2}\right)\right]\right\} . \end{aligned} \tag{7.2.27}$$

Since this expression diverges in the limit as $\theta \to 1^+$, (7.2.27) is a valid approximation of the first precursor field as $z \to \infty$ for all $\theta > 1$. Consequently, the front of the first precursor field for a delta-function input propagates at the velocity of light in vacuum.

With the substitution of (7.2.26) into the nonuniform expansion (7.2.9) [or, equivalently, with the substitution of the large argument asymptotic expansions of the Bessel functions $\mathrm{J}_0(\zeta)$ and $\mathrm{J}_{-1}(\zeta)$ into (7.2.27)], there results

$$A_S(z,t) \sim \sqrt{\frac{c\xi(\theta)}{2\pi z}} \cdot \frac{1}{b} \exp\left(-\delta\frac{z}{c}\left\{[1+\eta(\theta)](\theta-1)\right.\right.$$
$$\left.\left.+\frac{b^2}{2}\cdot\frac{1-\eta(\theta)}{\xi^2(\theta)+\delta^2[1-\eta(\theta)]^2}\right\}\right)\cdot\left\{2\xi(\theta)\cos\left[\frac{z}{c}\xi(\theta)\left(\theta-1\right.\right.\right.$$
$$\left.\left.+\frac{b}{2}\cdot\frac{1-\eta(\theta)}{\xi^2(\theta)+\delta^2[1-\eta(\theta)]^2}\right)+\frac{\pi}{4}\right]+3\delta[1-\eta(\theta)]\sin\left[\frac{z}{c}\xi(\theta)\right.$$
$$\left.\left.\cdot\left(\theta-1+\frac{b^2}{2}\cdot\frac{1-\eta(\theta)}{\xi^2(\theta)+\delta^2[1-\eta(\theta)]^2}\right)+\frac{\pi}{4}\right]\right\}, \tag{7.2.28}$$

as $z \to \infty$ for $\theta \geqslant 1 + \varepsilon$ with $\varepsilon > 0$. In the limit as θ approaches unity from above, the nonuniform asymptotic approximation (7.2.28) must be replaced by the uniform expansion (7.2.27). Indeed, $A_S(z,t)$ as given in (7.2.28) tends to ∞ as $\theta \to 1^+$.

7.2.7 The Hyperbolic-Tangent Modulated Signal

For the hyperbolic-tangent modulated signal, the spectrum of the initial field envelope is given by (4.3.46), so that

$$\tilde{u}(\omega_{sp_D^\pm} - \omega_c)$$
$$= \frac{\dfrac{\pi i}{2\beta_T}}{\sinh\left[\dfrac{\pi}{2\beta_T}(\omega_{sp_D^\pm} - \omega_c)\right]}$$
$$\cong \frac{\pi i}{2\beta_T}$$
$$\cdot\frac{\sinh\left\{\frac{\pi}{2\beta_T}[\pm\xi(\theta)-\omega_c]\right\}\cos\left\{\frac{\pi}{2\beta_T}\delta[1+\eta(\theta)]\right\}+i\cosh\left\{\frac{\pi}{2\beta_T}[\pm\xi(\theta)-\omega_c]\right\}\sin\left\{\frac{\pi}{2\beta_T}\delta[1+\eta(\theta)]\right\}}{\sinh^2\left\{\frac{\pi}{2\beta_T}[\pm\xi(\theta)-\omega_c]\right\}+\sin^2\left\{\frac{\pi}{2\beta_T}\delta[1+\eta(\theta)]\right\}}. \tag{7.2.29}$$

It is of interest in the present analysis to consider only large values of the parameter β_T, which correspond to rapid turn-on times for the initial signal[1].

---

[1] In the limit as β approaches zero, the parametric family of functions $(1/4\beta)\{\sinh[(\pi/2\beta)(\omega - \omega_c)]\}^{-1}$ approaches a delta-function at $\omega = \omega_c$. Consequently, for small values of the parameter β approaching zero, a monochromatic field is approached, which is not of much interest here.

Since

$$\frac{1}{\sinh(z)} = \frac{1}{z}\left(1 + 2z^2 \sum_{n=1}^{\infty} \frac{(-1)^n}{(n\pi)^2 + z^2}\right)$$

and since the quantity in brackets is convergent for large enough $|z|$ and does not vanish in the limit as $|z| \to \infty$, then, according to (5.4.3), $\nu = 0$. With this identification, substitution of (7.2.29) into the uniform expansion (7.2.15) for the first precursor field yields

$$\begin{aligned}
A_S(z,t) &\sim \frac{\pi}{4\beta_T b}\xi(\theta)\left(\theta - 1 + \frac{b^2}{2}\cdot\frac{1}{\xi^2(\theta)+\delta^2[1-\eta(\theta)]^2}\right)^{1/2} \exp\left(-\delta\frac{z}{c}\left\{[1+\eta(\theta)](\theta-1) + \frac{b^2}{2}\cdot\frac{1-\eta(\theta)}{\xi^2(\theta)+\delta^2[1-\eta(\theta)]^2}\right\}\right)\\
&\cdot\Bigg\{\Bigg[\frac{\xi(\theta)\cosh\left\{\frac{\pi}{2\beta_T}[\xi(\theta)-\omega_c]\right\}\sin\left\{\frac{\pi}{2\beta_T}\delta[1+\eta(\theta)]\right\} + \frac{3}{2}\delta[1-\eta(\theta)]\sinh\left\{\frac{\pi}{2\beta_T}[\xi(\theta)-\omega_c]\right\}\cos\left\{\frac{\pi}{2\beta_T}\delta[1+\eta(\theta)]\right\}}{\sinh^2\left(\frac{\pi}{2\beta_T}(\xi(\theta)-\omega_c)\right) + \sin^2\left(\frac{\pi}{2\beta_T}\delta(1+\eta(\theta))\right)}\\
&+ \frac{\xi(\theta)\cosh\left\{\frac{\pi}{2\beta_T}[\xi(\theta)+\omega_c]\right\}\sin\left\{\frac{\pi}{2\beta_T}\delta[1+\eta(\theta)]\right\} + \frac{3}{2}\delta[1-\eta(\theta)]\sinh\left\{\frac{\pi}{2\beta_T}[\xi(\theta)+\omega_c]\right\}\cos\left\{\frac{\pi}{2\beta_T}\delta[1+\eta(\theta)]\right\}}{\sinh^2\left(\frac{\pi}{2\beta_T}(\xi(\theta)+\omega_c)\right) + \sin^2\left(\frac{\pi}{2\beta_T}\delta(1+\eta(\theta))\right)}\Bigg]\\
&\cdot J_0\left[\frac{z}{c}\xi(\theta)\left(\theta - 1 + \frac{b^2}{2}\cdot\frac{1}{\xi^2(\theta)+\delta^2[1-\eta(\theta)]^2}\right)\right]\\
&+\Bigg[\frac{\xi(\theta)\sinh\left\{\frac{\pi}{2\beta_T}[\xi(\theta)-\omega_c]\right\}\cos\left\{\frac{\pi}{2\beta_T}\delta[1+\eta(\theta)]\right\} - \frac{3}{2}\delta[1-\eta(\theta)]\cosh\left\{\frac{\pi}{2\beta_T}[\xi(\theta)-\omega_c]\right\}\sin\left\{\frac{\pi}{2\beta_T}\delta[1+\eta(\theta)]\right\}}{\sinh^2\left(\frac{\pi}{2\beta_T}(\xi(\theta)-\omega_c)\right) + \sin^2\left(\frac{\pi}{2\beta_T}\delta(1+\eta(\theta))\right)}\\
&+ \frac{-\xi(\theta)\sinh\left\{\frac{\pi}{2\beta_T}[\xi(\theta)+\omega_c]\right\}\cos\left\{\frac{\pi}{2\beta_T}\delta[1+\eta(\theta)]\right\} + \frac{3}{2}\delta[1-\eta(\theta)]\cosh\left\{\frac{\pi}{2\beta_T}[\xi(\theta)+\omega_c]\right\}\sin\left\{\frac{\pi}{2\beta_T}\delta[1+\eta(\theta)]\right\}}{\sinh^2\left(\frac{\pi}{2\beta_T}(\xi(\theta)+\omega_c)\right) + \sin^2\left(\frac{\pi}{2\beta_T}\delta(1+\eta(\theta))\right)}\Bigg]\\
&\cdot J_1\left[\frac{z}{c}\xi(\theta)\left(\theta - 1 + \frac{b^2}{2}\cdot\frac{1}{\xi^2(\theta)+\delta^2[1-\eta(\theta)]^2}\right)\right]\Bigg\}, \qquad (7.2.30)
\end{aligned}$$

as $z \to \infty$ for all $\theta \geqslant 1$. In the limit as $\beta_T \to \infty$ (i.e. in the limit of an instantaneous turn-on time for the initial field envelope), equation (7.2.30) reduces to the uniform expansion (7.2.20) of the first precursor field for the unit step-function modulated signal.

With the substitution of (7.2.29) into the nonuniform expansion (7.2.9) [or, equivalently, with the substitution of the large argument asymptotic expansions of the Bessel functions $J_0(\zeta)$ and $J_1(\zeta)$ into (7.2.30)], there results

$$
\begin{aligned}
A_S(z,t) & \\
\sim -\frac{1}{2\beta_T b}\sqrt{\frac{\pi c\xi(\theta)}{2z}}\exp\Bigg(-\delta\frac{z}{c}\bigg\{[1+\eta(\theta)](\theta-1)+\frac{b^2}{2}\cdot\frac{1}{\xi^2(\theta)+\delta^2[1-\eta(\theta)]^2}\bigg\}\Bigg) & \\
\cdot\Bigg\{\Bigg[\frac{\xi(\theta)\sinh\left\{\frac{\pi}{2\beta_T}[\xi(\theta)-\omega_c]\right\}\cos\left\{\frac{\pi}{2\beta_T}\delta[1+\eta(\theta)]\right\}-\frac{3}{2}\delta[1-\eta(\theta)]\cosh\left\{\frac{\pi}{2\beta_T}[\xi(\theta)-\omega_c]\right\}\sin\left\{\frac{\pi}{2\beta_T}\delta[1+\eta(\theta)\right\}}{\sinh^2\left\{\frac{\pi}{2\beta_T}[\xi(\theta)-\omega_c]\right\}+\sin^2\left\{\frac{\pi}{2\beta_T}\delta[1+\eta(\theta)]\right\}} & \\
-\frac{\xi(\theta)\sinh\left\{\frac{\pi}{2\beta_T}[\xi(\theta)+\omega_c]\right\}\cos\left\{\frac{\pi}{2\beta_T}\delta[1+\eta(\theta)]\right\}-\frac{3}{2}\delta[1-\eta(\theta)]\cosh\left\{\frac{\pi}{2\beta_T}[\xi(\theta)+\omega_c]\right\}\sin\left\{\frac{\pi}{2\beta_T}\delta[1+\eta(\theta)]\right\}}{\sinh^2\left(\frac{\pi}{2\beta_T}(\xi(\theta)+\omega_c)\right)+\sin^2\left(\frac{\pi}{2\beta_T}\delta(1+\eta(\theta))\right)}\Bigg] & \\
\cdot\cos\left[\frac{z}{c}\xi(\theta)\left(\theta-1+\frac{b^2}{2}\cdot\frac{1}{\xi^2(\theta)+\delta^2[1+\eta(\theta)]^2}\right)+\frac{\pi}{4}\right] & \\
+\Bigg[\frac{\xi(\theta)\cosh\left\{\frac{\pi}{2\beta_T}[\xi(\theta)-\omega_c]\right\}\sin\left\{\frac{\pi}{2\beta_T}\delta[1+\eta(\theta)]\right\}+\frac{3}{2}\delta[1-\eta(\theta)]\sinh\left\{\frac{\pi}{2\beta_T}[\xi(\theta)-\omega_c]\right\}\cos\left\{\frac{\pi}{2\beta_T}\delta[1+\eta(\theta)]\right\}}{\sinh^2\left\{\frac{\pi}{2\beta_T}[\xi(\theta)-\omega_c]\right\}+\sin^2\left\{\frac{\pi}{2\beta_T}\delta[1+\eta(\theta)]\right\}} & \\
-\frac{\xi(\theta)\cosh\left\{\frac{\pi}{2\beta_T}[\xi(\theta)+\omega_c]\right\}\sin\left\{\frac{\pi}{2\beta_T}\delta[1+\eta(\theta)]\right\}+\frac{3}{2}\delta[1-\eta(\theta)]\sinh\left\{\frac{\pi}{2\beta_T}[\xi(\theta)+\omega_c]\right\}\cos\left\{\frac{\pi}{2\beta_T}\delta[1+\eta(\theta)]\right\}}{\sinh^2\left(\frac{\pi}{2\beta_T}(\xi(\theta)+\omega_c)\right)+\sin^2\left(\frac{\pi}{2\beta_T}\delta(1+\eta(\theta))\right)}\Bigg] & \\
\cdot\sin\left(\frac{z}{c}\xi(\theta)\left(\theta-1+\frac{b^2}{2}\cdot\frac{1}{\xi^2(\theta)+\delta^2[1+\eta(\theta)]^2}\right)+\frac{\pi}{4}\right]\Bigg\}, &
\end{aligned}
\tag{7.2.31}
$$

as $z \to \infty$ for $\theta \geqslant 1 + \varepsilon$ with $\varepsilon > 0$. In the limit as $\beta_T \to \infty$, (7.2.31) reduces to the nonuniform expansion (7.2.21) of the first precursor field for the unit step-function modulated signal.

7.3 The Second Precursor Field (Brillouin's Precursor)

The contributions of the near saddle points to the asymptotic behavior of the field $A(z,t)$ for sufficiently large values of the observation distance z yield the dynamic evolution of the second precursor field. This contribution to the total asymptotic behavior of the field $A(z,t)$ is denoted by $A_B(z,t)$, c.f. (6.5.5, 8), and is dominant over the first precursor field $A_S(z,t)$ for all $\theta > \theta_{SB}$. According to the results of Sect. 6.2.2, the two near first-order saddle points, which are initially isolated from each other at $\theta = 1$, approach one another along the imaginary axis as θ approaches θ_1 and coalesce into a single second-order saddle point at $\theta = \theta_1$, after which they again separate into two first-order saddle points and move away from each other in the complex ω-plane. A straightforward asymptotic analysis based upon Olver's method is employed in Sect. 6.3.1 for values of θ in the three ranges $1 \leqslant \theta < \theta_1$, $\theta = \theta_1$, and $\theta > \theta_1$. However, these results are nonuniform in a neighborhood of the critical value $\theta = \theta_1$, and one must necessarily turn to the uniform asymptotic analysis described in Sect. 5.2 to obtain the complete continuous evolution of the second precursor field for all $\theta \geqslant 1$. This is done in Sect. 7.3.2.

7.3.1 The Nonuniform Approximation

The expression for the second precursor field that results from direct application of Olver's method is derived in this subsection. If the first approximate expressions for the near saddle point locations were applied in Olver's method, the result (for the step-function modulated signal) would be the same as that obtained by *Brillouin* [7.4, 5]. As is pointed out in Chap. 6, however, the first approximation is accurate only for values of θ within the restricted range $\theta_{SB} \leqslant \theta \leqslant \theta_1$. Hence, in this subsection, the second approximation of the near saddle point locations is used. The results are good asymptotic approximations that are valid for $\theta \geqslant 1$ except in the transition region in the vicinity of $\theta = \theta_1$. The uniform asymptotic approximation valid for all $\theta \geqslant 1$ is obtained in the following subsection.

The asymptotic behavior of the second precursor field $A_B(z,t)$ for a particular initial pulse envelope is obtained from the asymptotic expansion of the integral representation of the propagated field (4.3.31), viz.

$$A(z,t) = \frac{1}{2\pi}\mathrm{Re}\left\{\mathrm{i}\int_{ia-\infty}^{ia+\infty} \tilde{u}(\omega-\omega_c)\mathrm{e}^{(z/c)\phi(\omega,\theta)}\,d\omega\right\} \tag{7.3.1}$$

about the near saddle points. The near saddle point locations are given by

$$\omega_{sp_1} \cong \mathrm{i}[|\psi(\theta)| - \tfrac{2}{3}\delta\zeta(\theta)] \ , \tag{7.3.2a}$$

$$\omega_{sp_2} \cong -\mathrm{i}[|\psi(\theta)| + \tfrac{2}{3}\delta\zeta(\theta)] \ , \tag{7.3.2b}$$

for $1 \leqslant \theta \leqslant \theta_1$,

$$\omega_{sp_1} = \omega_{sp_2} \cong -\mathrm{i}\tfrac{2}{3}\delta\zeta(\theta_1) \ , \tag{7.3.2c}$$

for $\theta = \theta_1$, and

$$\omega_{sp_N^\pm} \cong \pm\psi(\theta) - \mathrm{i}\tfrac{2}{3}\delta\zeta(\theta) \ , \tag{7.3.2d}$$

for $\theta \geqslant \theta_1$, where, for convenience, the following functions have been defined, c.f.

(6.2.39, 40):

$$\zeta(\theta) \equiv \frac{3}{2} \cdot \frac{\theta^2 - \theta_0^2 + \dfrac{2b^2}{\omega_0^2}}{\theta^2 - \theta_0^2 + \dfrac{3b^2}{\omega_0^2}\alpha}, \tag{7.3.3a}$$

$$\psi(\theta) \equiv \left[\frac{\omega_0^2(\theta^2 - \theta_0^2)}{\theta^2 - \theta_0^2 + \dfrac{3b^2}{\omega_0^2}\alpha} - \frac{4}{9}\delta^2 \zeta^2(\theta) \right]^{1/2} . \tag{7.3.3b}$$

The square root in (7.3.3b) is taken to be positive, and α is given by (6.2.36). For values of θ approximately equal to θ_0, these functions simplify to the approximate forms used by *Brillouin* [7.4, 5] (with $\alpha = 1$)

$$\zeta(\theta) \cong \frac{1}{\alpha},$$

$$\psi(\theta) \cong \left[\frac{2\theta_0 \omega_0^4}{3\alpha b^2}(\theta - \theta_0) - \frac{4\delta^2}{9\alpha^2} \right]^{1/2},$$

which are valid only for values of θ in the vicinity of θ_0. The relations (7.3.2 and 3) represent the second approximate locations of the near saddle points which accurately depict the near saddle point behavior for all values of $\theta \geqslant 1$.

For the examples considered in this research, both the spectral amplitude function $\tilde{u}(\omega - \omega_c)$ and the phase function $\phi(\omega, \theta)$ appearing in the integrand of $A(z, t)$ are analytic about the two near saddle points. For values of θ within the range $1 \leqslant \theta < \theta_1$ the conditions of Olver's Theorem 5.1 are satisfied at the upper near saddle point SP_1 when the original contour of integration is deformed to the path $P(\theta) = P_D^-(\theta) + P_1(\theta) + P_D^+(\theta)$, where $P_1(\theta)$ is an Olver-type path with respect to the upper near first order saddle point SP_1 (Fig. 6.21a and b). Consequently, (5.1.14) applies with $\mu = 2$ and $\lambda = 1$ and with (5.1.3, 4) taken to be Taylor series expansions about the near saddle point SP_1. Hence, according to Theorem 5.1 and the results of Section 6.5, the integral (7.3.1) taken over the contour $P_1(\theta)$ yields the second precursor field $A_B(z, t)$, which satisfies

$$A_B(z, t) = \frac{1}{2\pi} \mathbb{R}\mathrm{e}\, \mathrm{i} \left\{ 2 e^{(z/c)\phi(\omega_{sp_1}, \theta)} \sqrt{\pi} \frac{a_0(\omega_{sp_1})}{(z/c)^{1/2}} [1 + \mathcal{O}(z^{-1})] \right\}, \tag{7.3.4}$$

as $z \to \infty$ uniformly for $1 \leqslant \theta \leqslant \theta_1 - \varepsilon$ with $0 < \varepsilon < \theta_1 - 1$.

In order to evaluate the coefficient $a_0(\omega_{sp_1})$ appearing in (7.3.4), the first two coefficients in the Taylor series expansion (5.1.3) of $\phi(\omega, \theta)$ about the upper near saddle point SP_1 are now obtained. According to (6.2.45), the approximate phase behavior in the region of the complex ω-plane about the origin is given by

$$\phi(\omega,\theta) \cong \mathrm{i}\omega(\theta_0 - \theta) + \frac{b^2}{2\theta_0\omega_0^4}\omega^2(\mathrm{i}\alpha\omega - 2\delta) \; . \tag{7.3.5a}$$

Upon differentiation of this expression twice with respect to ω, there results

$$\phi^{(1)}(\omega,\theta) \cong \mathrm{i}(\theta_0 - \theta) + \frac{b^2}{2\theta_0\omega_0^4}(3\mathrm{i}\alpha\omega^2 - 4\delta\omega) \; , \tag{7.3.5b}$$

$$\phi^{(2)}(\omega,\theta) \cong \frac{b^2}{\theta_0\omega_0^4}(3\mathrm{i}\alpha\omega - 2\delta) \; , \tag{7.3.5c}$$

so that at $\omega = \omega_{sp_1}$, there results

$$\begin{aligned}\phi(\omega_{sp_1},\theta) &\cong \frac{1}{3}[2\delta\zeta(\theta) - 3|\psi(\theta)|](\theta_0 - \theta) \\ &\quad + \frac{b^2}{54\theta_0\omega_0^4}[2\delta\zeta(\theta) - 3|\psi(\theta)|]^2\{2\delta[3 - \alpha\zeta(\theta)] \\ &\quad + 3\alpha|\psi(\theta)|\} \; ,\end{aligned} \tag{7.3.6}$$

and

$$\phi^{(2)}(\omega_{sp_1},\theta) \cong -\frac{b^2}{\theta_0\omega_0^4}\{2\delta[1 - \alpha\zeta(\theta)] + 3\alpha|\psi(\theta)|\} \; . \tag{7.3.7}$$

The second coefficient in the Taylor series expansion (5.1.3) of $\phi(\omega,\theta)$ about ω_{sp_1} is then given by

$$p_0(\omega_{sp_1},\theta) = \frac{\phi^{(2)}(\omega_{sp_1},\theta)}{2!} \cong -\frac{b^2}{2\theta_0\omega_0^4}\{2\delta[1 - \alpha\zeta(\theta)] + 3\alpha|\psi(\theta)|\} \; . \tag{7.3.8}$$

Furthermore, the first coefficient in the Taylor series expansion (5.1.4) of the spectrum is

$$q_0(\omega_{sp_1}) = \tilde{u}(\omega_{sp_1} - \omega_c) \; , \tag{7.3.9}$$

the specific form of which depends upon the particular initial pulse envelope function $u(t)$.

The proper value of $\bar{\alpha}_0 \equiv \arg[-p_0(\omega_{sp_1},\theta)]$ must now be ascertained according to the convention set forth in Olver's analysis, c.f. (5.1.7). For simplicity, the deformed contour of integration is taken along the path of steepest descent through the upper near saddle point SP_1, as shown in Fig. 7.5. The angle of slope of the contour $P_1(\theta)$ at the saddle point is then $\bar{\alpha} = 0$, c.f. (6.2.48). Therefore, since $\Theta \equiv \arg(z) = 0$, the proper value of $\bar{\alpha}_0 \equiv \arg[-p_0(\omega_{sp_1},\theta)]$ as determined by the inequality (5.1.7) is $\bar{\alpha}_0 = 0$.

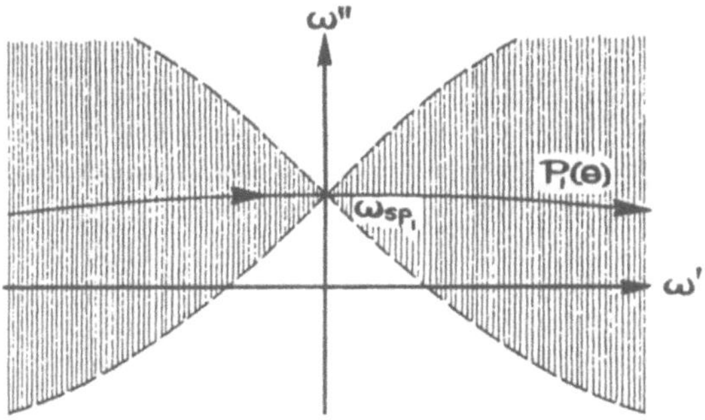

Fig. 7.5. Olver-type path through the upper near saddle point SP_1. The hatched area indicates the region of the complex ω-plane wherein the inequality $X(\omega) < X(\omega_{sp_1}, \theta)$ is satisfied.

According to (5.1.9a), the coefficient $a_0(\omega_{sp_1})$ appearing in (6.3.4) is given by

$$a_0(\omega_{sp_1}) = \frac{q_0(\omega_{sp_1})}{\mu[-p_0(\omega_{sp_1}, \theta)]^{\lambda/\mu}}$$

$$\cong \tilde{u}(\omega_{sp_1} - \omega_c) \cdot \frac{\omega_0^2}{b} \left(\frac{\theta_0}{4\delta[1 - \alpha\zeta(\theta)] + 6\alpha|\psi(\theta)|} \right)^{1/2}, \tag{7.3.10}$$

where the positive value of the square root in this expression is to be taken. Upon substitution of (7.3.6 and 10) into (7.3.4) for the second precursor field, there results

$$A_B(z,t) \sim \frac{\omega_0^2}{b} \left(\frac{\theta_0 c}{\pi z \{4\delta[1 - \alpha\zeta(\theta)] + 6\alpha|\psi(\theta)|\}} \right)^{1/2} \cdot \mathbb{R}\mathrm{e}\{i\tilde{u}(\omega_{sp_1} - \omega_c)\}$$
$$\cdot \exp\left[\frac{z}{3c} [2\delta\zeta(\theta) - 3|\psi(\theta)|] \left(\theta_0 - \theta + \frac{b^2}{18\theta_0\omega_0^4} [2\delta\zeta(\theta) - 3|\psi(\theta)|] \{2\delta[3 - \alpha\zeta(\theta)] + 3\alpha|\psi(\theta)|\} \right) \right] \tag{7.3.11}$$

as $z \to \infty$ uniformly for $1 \leqslant \theta \leqslant \theta_1 - \varepsilon$ with $0 < \varepsilon < \theta_1 - 1$.

Consider now the algebraic sign of the exponential argument appearing in (7.3.11). Since the quantity $[2\delta\zeta(\theta) - 3|\psi(\theta)|]$ is negative for $1 \leqslant \theta < \theta_0$, vanishes at $\theta = \theta_0$, and is positive for $\theta_0 < \theta < \theta_1$, and since the quantity $\{2\delta[3 - \alpha\zeta(\theta)] + 3\alpha|\psi(\theta)|\}$ is positive for $1 \leqslant \theta < \theta_1$, and since the inequality

$$|\theta_0 - \theta| \geqslant \frac{b^2}{18\theta_0\omega_0^4} |2\delta\zeta(\theta) - 3|\psi(\theta)|| \{2\delta[3 - \alpha\zeta(\theta)] + 3\alpha|\psi(\theta)|\}$$

is satisfied for all θ in the range $1 \leqslant \theta < \theta_1$, with the equality holding only at $\theta = \theta_0$ when both sides of this equation vanish, then the exponential argument is negative for $1 \leqslant \theta < \theta_0$, vanishes identically at $\theta = \theta_0$, and is again negative for $\theta_0 < \theta < \theta_1$. Consequently, the second precursor field $A_B(z, t)$ first grows with θ as the exponential argument decreases for increasing values of θ in the range $1 \leqslant \theta < \theta_0$ and becomes exponentially dominant over the first precursor field for $\theta > \theta_{SB}$ (where $1 < \theta_{SB} < \theta_0$). At $\theta = \theta_0$ the exponential argument vanishes and the field (7.3.11) varies only as $z^{-1/2}$, making this particular case strikingly different from the field observed at any other value of θ. Finally, for increasing values of θ in the range $\theta_0 < \theta < \theta_1$ the exponential argument decreases and the field is again exponentially attenuated. Note that in this range of values of θ the second precursor field is nonoscillatory.

At $\theta = \theta_1 = ct_1/z$, the conditions of *Olver*'s Theorem 5.1 are satisfied at the near saddle point SP_N when the original contour of integration is deformed to the path $P(\theta) = P_D^-(\theta) + P_1(\theta) + P_D^+(\theta)$, where $P_1(\theta)$ is an Olver-type path with resect to the second order near saddle point SP_N, located at

$$\omega_{sp_N}(\theta_1) \cong -\frac{2\delta}{3\alpha}\mathrm{i} \ . \tag{7.3.12}$$

Since both the spectral amplidude and phase functions appearing in the integrand of (7.3.1) are analytic about this saddle point, (5.1.3, 4) are Taylor series about that point with $\mu = 3$ and $\lambda = 1$. Since $\mu = 3$, the argument leading up to (5.1.14), which is valid only for $\mu = 2$, must be changed.

Substitution of (7.3.12) into (7.3.5a) yields

$$\phi(\omega_{sp_N}, \theta_1) \cong \frac{2\delta}{3\alpha}\left(\theta_0 - \theta_1 + \frac{8\delta^2 b^2}{27\alpha\theta_0\omega_0^4}\right) , \tag{7.3.13a}$$

and differentiation of (7.3.5c) with respect to ω yields

$$\phi^{(3)}(\omega, \theta) \cong 3\mathrm{i}\frac{\alpha b^2}{\theta_0\omega_0^4} , \tag{7.3.13b}$$

which is independent of both θ and the saddle point location. The second coefficient in the Taylor series expansion (5.1.3) of $\phi(\omega, \theta_1)$ is then given by

$$p_0(\omega_{sp_N}, \theta_1) = \frac{\phi^{(3)}(\omega_{sp_N}, \theta_1)}{3!} \cong \mathrm{i}\frac{\alpha b^2}{2\theta_0\omega_0^4} , \tag{7.3.14}$$

and the first coefficient in the Taylor series expansion (5.1.4) of the spectrum is given by

$$q_0(\omega_{sp_N}) = \tilde{u}(\omega_{sp_N} - \omega_c) , \tag{7.3.15}$$

the specific form of which depends upon the particular initial pulse envelope function $u(t)$.

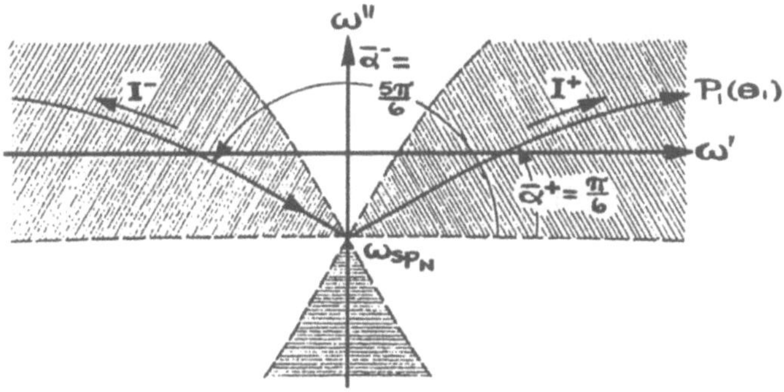

Fig. 7.6. Olver-type path through the second order near saddle point at $\theta = \theta_1$. The hatched area indicates the region of the complex ω-plane wherein the inequality $X(\omega) < X(\omega_{sp_N}, \theta_1)$ is satisfied.

The proper value of $\bar{\alpha}_0 \equiv \arg[-p_0(\omega_{sp_N}, \theta_1)] \cong \arg(-\mathrm{i})$ must now be determined according to the convention prescribed in *Olver*'s analysis. For simplicitly, the deformed contour of integration is taken along the path of steepest descent through the second order saddle point SP_N, as shown in Fig. 7.6 above. Along this deformed contour of integration, the asymptotic behavior of the integral $A(z, t)$ may then be written

$$A(z, t_1) = A_B(z, t_1) = I^+ - I^- \ ,$$

where I^+ and I^- denote the contour integrals both of whose directions of integration lead away from the saddle point, as indicated in Fig. 7.6. The angle of slope of the contours at the saddle point are $\bar{\alpha}^+ = \pi/6$ and $\bar{\alpha}^- = 5\pi/6$, c.f. (6.2.52), as indicated in the diagram. Consequently, since $\Theta \equiv \arg(z) = 0$, the proper values of $\bar{\alpha}_0$ for these contours is determined by the inequality (5.1.7) to be $\bar{\alpha}_0^+ = -\pi/2$ for I^+ and $\bar{\alpha}_0^- = -5\pi/2$ for I^-. The phase difference between the coefficients $a_0^+(\omega_{sp_N})$ and $a_0^-(\omega_{sp_N})$ then yields the factor

$$\mathrm{e}^{-\mathrm{i}\bar{\alpha}_0^+/3} - \mathrm{e}^{-\mathrm{i}\bar{\alpha}_0^-/3} = \mathrm{e}^{\mathrm{i}(\pi/6)} - \mathrm{e}^{\mathrm{i}(5\pi/6)} = 2\cos\left(\frac{\pi}{6}\right) = \sqrt{3} \ .$$

According to (5.1.9a) and the above specified phase angles, the coefficient $a_0^+(\omega_{sp_N})$ is given by

$$a_0^+(\omega_{sp_N}) = \frac{q_0(\omega_{sp_N})}{\mu[-p_0(\omega_{sp_N}, \theta_1)]^{\lambda/\mu}}$$

$$\cong \frac{1}{3}\tilde{u}(\omega_{sp_N} - \omega_c)\left(\frac{2\theta_0\omega_0^4}{\alpha b^2}\right)^{1/3} \mathrm{e}^{\mathrm{i}(\pi/6)} \ . \tag{7.3.16}$$

Thus, asymptotically as $z \to \infty$ with fixed $\theta = \theta_1 = ct_1/z$, the following asymptotic behavior for the second precursor field is obtained:

$$
\begin{aligned}
A_B(z, t_1) &\\
&= \frac{1}{2\pi} \mathrm{Re}\, \mathrm{i} \left\{ e^{(z/c)\phi(\omega_{sp_N}, \theta_1)} \Gamma\left(\frac{1}{3}\right) [a_0^+(\omega_{sp_N}) - a_0^-(\omega_{sp_N})] \left(\frac{c}{z}\right)^{1/3} \right. \\
&\quad \left. \cdot [1 + \mathcal{O}(z^{-1/3})] \right\} \\
&\sim \frac{\omega_0}{2\pi\sqrt{3}} \Gamma\left(\frac{1}{3}\right) \left(\frac{2\theta_0 \omega_0 c}{\alpha b^2 z}\right)^{1/3} \exp\left[\frac{2\delta z}{3\alpha c}\left(\theta_0 - \theta_1 + \frac{4\delta^2 b^2}{9\alpha\theta_0\omega_0^4}\right)\right] \\
&\quad \cdot \mathrm{Re}\{\mathrm{i}\tilde{u}(\omega_{sp_N} - \omega_c)\} \; . \qquad (7.3.17)
\end{aligned}
$$

This result represents the asymptotic behavior of the second precursor field $A_B(z, t_1)$ at the critical value $\theta = \theta_1$ when the two near first order saddle points have coalesced into a single second order saddle point. According to (6.2.43b), the exponential argument appearing in (7.3.17) satisfies the inequality

$$
\theta_0 - \theta_1 + \frac{8\delta^2 b^2}{27\alpha\theta_0\omega_0^4} < -\frac{20\delta^2 b^2}{81\alpha\theta_0\omega_0^4} \; ,
$$

and hence, as was to be expected, the second precursor field is exponentially attenuated at $\theta = \theta_1$.

For values of θ within the range $\theta > \theta_1$, the conditions of *Olver*'s Theorem 5.1 are satisfied at the near saddle points $\mathrm{SP}_N^\pm$ when the original contour of integration is deformed to the path $P(\theta) = P_D^-(\theta) + P_N^-(\theta) + P_N^+(\theta) + P_D^+(\theta)$, where $P_N^-(\theta)$ is an Olver-type path with respect to the near saddle point SP_N^- and $P_N^+(\theta)$ is an Olver-type path with respect to the near saddle point SP_N^+ (Fig. 6.21c). Consequently, (5.1.14) applies for each of the two near saddle points with (5.1.3 and 4) taken to be Taylor series expansions about the two saddle points. Since $SP_N^\pm$ are first order saddle points, the parameter μ is 2, and since $\tilde{u}(\omega - \omega_c)$ is regular at the saddle points, the parameter λ is 1. Hence, according to Theorem 5.1 and the results of Sect. 6.5, the integral (7.2.1) taken over the two contours $P_N^\pm(\theta)$ yields the second precursor field $A_B(z, t)$, which satisfies

$$
\begin{aligned}
A_B(z, t) = \frac{1}{2\pi} \mathrm{Re}\, \mathrm{i} &\left\{ 2e^{(z/c)\phi(\omega_{sp_N^+}, \theta)} \sqrt{\pi} \frac{a_0(\omega_{sp_N^+})}{(z/c)^{1/2}} [1 + \mathcal{O}(z^{-1})] \right. \\
&\left. + 2e^{(z/c)\phi(\omega_{sp_N^-}, \theta)} \sqrt{\pi} \frac{a_0(\omega_{sp_N^-})}{(z/c)^{1/2}} [1 + \mathcal{O}(z^{-1})] \right\} , \qquad (7.3.18)
\end{aligned}
$$

as $z \to \infty$ uniformly for $\theta \geqslant \theta_1 + \varepsilon$ for positive ε arbitrarily small.

According to (7.3.5a, c), the phase function and its second derivative at the two near saddle point locations are given by

$$\phi(\omega_{sp_N^\pm},\theta) \cong -\delta\left(\frac{2}{3}\zeta(\theta)(\theta-\theta_0) + \frac{b^2}{\theta_0\omega_0^4}\left\{[1-\alpha\zeta(\theta)]\psi^2(\theta)\right.\right.$$
$$\left.\left. + \frac{4}{9}\delta^2\zeta^2(\theta)\left[\frac{1}{3}\alpha\zeta(\theta)-1\right]\right\}\right) \pm i\psi(\theta)\left(\theta_0-\theta\right.$$
$$\left. + \frac{b^2}{2\theta_0\omega_0^4}\left\{\frac{4}{3}\delta^2\zeta(\theta)[2-\alpha\zeta(\theta)] + \alpha\psi^2(\theta)\right\}\right) , \tag{7.3.19a}$$

$$\phi^{(2)}(\omega_{sp_N^\pm},\theta) \cong \frac{b^2}{\theta_0\omega_0^4}\{2\delta[\alpha\zeta(\theta)-1] \pm 3i\alpha\psi(\theta)\} \tag{7.3.19b}$$

The second coefficient in the Taylor series expansion (5.1.3) of $\phi(\omega,\theta)$ is therefore given by

$$p_0(\omega_{sp_N^\pm},\theta) = \frac{\phi^{(2)}(\omega_{sp_N^\pm},\theta)}{2!} \cong \frac{b^2}{2\theta_0\omega_0^4}\{2\delta[\alpha\zeta(\theta)-1] \pm 3i\alpha\psi(\theta)\} , \tag{7.3.20}$$

and the first coefficient in the Taylor series expansion (5.1.4) of the spectrum is given by

$$q_0(\omega_{sp_N^\pm}) = \tilde{u}(\omega_{sp_N^\pm} - \omega_c) , \tag{7.3.21}$$

the specific form of which depends upon the particular initial pulse envelope function $u(t)$.

The proper value of $\bar{\alpha}_0^\pm \equiv \arg[-p_0(\omega_{sp_N^\pm},\theta)]$ must now be determined according to the convention set forth in Olver's analysis, c.f. (5.1.7). For simplicity, the deformed contour of integration through the near saddle point SP_N^+ in the right half of the complex ω-plane is taken along the path of steepest descent in the vicinity of $\omega_{sp_N^+}$, as shown in Fig. 7.7. The angle of slope of the contour $P_N^+(\theta)$ at the saddle point SP_N^+, c.f. (6.2.55), is $\bar{\alpha}^+ = \pi/4$, as indicated in the diagram.

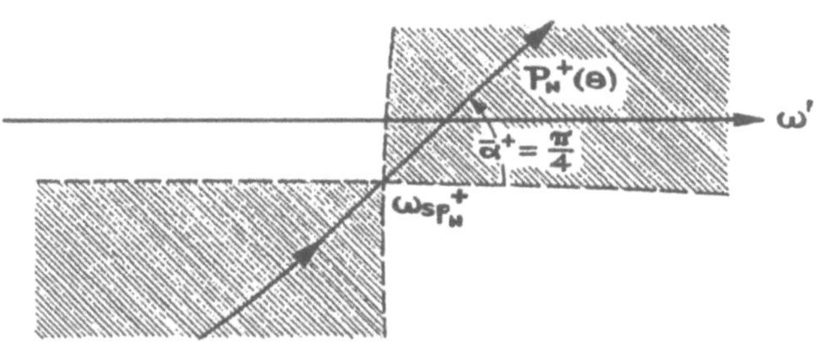

Fig. 7.7. Olver-type path through the near saddle point SP_N^+. The hatched area indicates the region of the complex ω-plane wherein the inequality $X(\omega) < X(\omega_{sp_N^+},\theta)$ is satisfied.

Therefore, since $\Theta \equiv \arg(z) = 0$, the proper value of $\bar{\alpha}_0^+ \equiv \arg[-p_0(\omega_{sp_N^+}, \theta)]$, as determined by the inequality (5.1.7), is $\bar{\alpha}_0^+ \cong -\pi/4$. By a similar argument, the proper value of $\bar{\alpha}_0^- \equiv \arg[-p_0(\omega_{sp_N^-}, \theta)]$ is $\bar{\alpha}_0^- \cong +\pi/4$.

According to (5.19a), the coefficients $a_0(\omega_{sp_N^\pm})$ appearing in (6.3.18) are given by

$$a_0(\omega_{sp_N^\pm}) = \frac{q_0(\omega_{sp_N^\pm})}{\mu[-p_0(\omega_{sp_N^\pm}, \theta)]^{\lambda/\mu}}$$

$$\cong \frac{1}{2}\tilde{u}(\omega_{sp_N^\pm} - \omega_c)\frac{\omega_0^2}{b}\left(\frac{-2\theta_0}{2\delta[\alpha\zeta(\theta) - 1] \pm 3i\alpha\psi(\theta)}\right)^{1/2} .$$

For values of θ slightly larger than θ_1, the term $2\delta[\alpha\zeta(\theta) - 1]$ becomes negligible in comparison to $3\alpha\psi(\theta)$, and remains so for all larger values of θ. Hence, for values of θ sufficiently larger than θ_1, this term may be neglected, in which case the above expression may be simplified somewhat to

$$a_0(\omega_{sp_N^\pm}) \cong \tilde{u}(\omega_{sp_N^\pm} - \omega_c)\frac{\omega_0^2}{2b}\left(\frac{2\theta_0}{3\alpha\psi(\theta)}\right)^{1/2} e^{\pm i(\pi/4)} , \tag{7.3.22}$$

provided that θ is not too close to θ_1 [when θ approaches θ_1 from above, this analysis must be replaced by the uniform expansion presented in Sect. 5.2.2, so that the approximation (7.3.22) is valid for the nonuniform asymptotic analysis presented here]. Substitution of (7.3.19a and 22) into (7.3.18) for the second precursor field yields the result

$$\begin{aligned}
A_B(z,t) \sim \frac{\omega_0^2}{2b}\left(\frac{2\theta_0 c}{3\pi\alpha\psi(\theta)z}\right)^{1/2} \exp\Bigg[-\delta\frac{z}{c}\Bigg(\frac{2}{3}\zeta(\theta)(\theta - \theta_0)\\
+ \frac{b^2}{\theta_0\omega_0^4}\left\{[1 - \alpha\zeta(\theta)]\psi^2(\theta) + \frac{4}{9}\delta^2\zeta^2(\theta)\left[\frac{1}{3}\alpha\zeta(\theta) - 1\right]\right\}\Bigg)\Bigg]\\
\cdot \mathrm{Re}\, i\Bigg\{\tilde{u}(\omega_{sp_N^+} - \omega_c)\exp\Bigg[i\psi(\theta)\frac{z}{c}\Bigg(\theta_0 - \theta\\
+ \frac{b^2}{2\theta_0\omega_0^4}\left\{\frac{4}{3}\delta^2\zeta(\theta)[2 - \alpha\zeta(\theta)] + \alpha\psi^2(\theta)\right\}\Bigg) + i\frac{\pi}{4}\Bigg]\\
+ \tilde{u}(\omega_{sp_N^-} - \omega_c)\exp\Bigg[-i\psi(\theta)\frac{z}{c}\Bigg(\theta_0 - \theta\\
+ \frac{b^2}{2\theta_0\omega_0^4}\left\{\frac{4}{3}\delta^2\zeta(\theta)[2 - \alpha\zeta(\theta)] + \alpha\psi^2(\theta)\right\}\Bigg) - i\frac{\pi}{4}\Bigg]\Bigg\} ,
\end{aligned} \tag{7.3.23}$$

as $z \to \infty$ for $\theta > \theta_1$. As can easily be seen, the second precursor field for $\theta > \theta_1$ is oscillatory and exponentially attenuated with the propagation distance z. The attenuation coefficient increases monotonically with increasing $\theta > \theta_1$ so that the second precursor field is increasingly attenuated as θ increases with fixed z.

7.3.2 The Uniform Approximation

Equations (7.3.11, 17, and 23) represent the nonuniform asymptotic approximation of the second precursor field for values of θ within the ranges $1 \leqslant \theta < \theta_1$, $\theta = \theta_1$, and $\theta > \theta_1$, respectively. The results are discontinuous at the critical transition point $\theta = \theta_1$ at which the two near first-order saddle points have coalesced into a single second-order saddle point. In order to obtain a continuous transition in the behavior of the second precursor field as θ is allowed to vary across the point $\theta = \theta_1$, the uniform asymptotic approximation presented in Sect. 5.2.2 is employed in this subsection.

The required uniform asymptotic approximation is obtained by direct application of Theorem 5.2. Since the behavior of the saddle points and the path of integration in the present case is the same as in the example treated in Sect. 5.2.2, if follows from the discussion of that example that the path of integration L appearing in (5.2.7) is an L_{21} contour (Fig. 7.4) so that the function $\mathscr{C}(\zeta)$ occuring in (4.2.6) is given by (4.2.25b). Although Theorem 4.2 is directly applicable to the present problem for all $\theta \geqslant 1$, it is necessary to treat the two cases $1 \leqslant \theta \leqslant \theta_1$ and $\theta \geqslant \theta_1$ separately because the approximate expressions for the locations of the near saddle points differ in the two cases. Nonetheless, the results for the two cases combined are continuous at $\theta = \theta_1$ and constitute an asymptotic approximation of the second precursor field $A_B(z, t)$ that is uniformly valid for all $\theta \geqslant 1$.

Consider first the uniform asymptotic behavior of the second precursor field for values of θ within the range $1 \leqslant \theta \leqslant \theta_1$. In this case, the two near saddle point locations are given by (6.3.2a, b), so that, according to (6.2.45), the approximate phase behavior at those locations is given by

$$\phi(\omega_{sp_1}, \theta) \cong \left[\frac{2}{3}\delta\zeta(\theta) - |\psi(\theta)|\right](\theta_0 - \theta) - \frac{b^2}{2\theta_0\omega_0^4}\left[-\frac{2}{3}\delta\zeta(\theta) + |\psi(\theta)|\right]^2\left[\frac{2}{3}\alpha\delta\zeta(\theta) - \alpha|\psi(\theta)| - 2\delta\right], \tag{7.3.24a}$$

$$\phi(\omega_{sp_2}, \theta) \cong \left[\frac{2}{3}\delta\zeta(\theta) + |\psi(\theta)|\right](\theta_0 - \theta) - \frac{b^2}{2\theta_0\omega_0^4}\left[\frac{2}{3}\delta\zeta(\theta) + |\psi(\theta)|\right]^2\left[\frac{2}{3}\alpha\delta\zeta(\theta) - \alpha|\psi(\theta)| - 2\delta\right]. \tag{7.3.24b}$$

Application of Theorem 5.2 therefore yields the asymptotic expansion of the second precursor field $A_B(z,t)$ given by

$$A_B(z,t) = -\mathbb{R}\mathrm{e}\, e^{(z/c)\alpha_0(\theta)}\left\{\left(\frac{c}{z}\right)^{1/3} e^{-i(2\pi/3)} A_i\left[\alpha_1(\theta)e^{-i(2\pi/3)}\left(\frac{z}{c}\right)^{2/3}\right]\right.$$
$$\cdot\left\{\frac{1}{2}[\tilde{u}(\omega_{sp_1}-\omega_c)h_1(\theta)+\tilde{u}(\omega_{sp_2}-\omega_c)h_2(\theta)]+\mathcal{O}\left(\frac{1}{z}\right)\right\}$$
$$+\left(\frac{c}{z}\right)^{2/3} e^{-i(4\pi/3)} A_i^{(1)}\left[\alpha_1(\theta)e^{-i(2\pi/3)}\left(\frac{z}{c}\right)^{2/3}\right]$$
$$\left.\cdot\left\{\frac{1}{2\alpha_1^{1/2}(\theta)}[\tilde{u}(\omega_{sp_1}-\omega_c)h_1(\theta)-\tilde{u}(\omega_{sp_2}-\omega_c)h_2(\theta)]+\mathcal{O}\left(\frac{1}{z}\right)\right\}\right\}, \tag{7.3.25}$$

as $z\to\infty$ uniformly for $1\leqslant\theta\leqslant\theta_1$. According to (5.2.10–12) and (7.3.24), the coefficients appearing in this expression are given by

$$\alpha_0(\theta) = \frac{1}{2}[\phi(\omega_{sp_1},\theta)+\phi(\omega_{sp_2},\theta)]$$
$$\cong -\frac{2}{3}\delta\zeta(\theta)(\theta-\theta_0) - \frac{\delta b^2}{\theta_0\omega_0^4}\left\{|\psi(\theta)|^2[\alpha\zeta(\theta)-1]\right.$$
$$\left.+\frac{4}{9}\delta^2\zeta^2(\theta)\left[\frac{1}{3}\alpha\zeta(\theta)-1\right]\right\}, \tag{7.3.26}$$

$$\alpha_1^{1/2}(\theta) = \left\{\frac{3}{4}[\phi(\omega_{sp_1},\theta)-\phi(\omega_{sp_2},\theta)]\right\}^{1/3}$$
$$\cong\left(|\psi(\theta)|\left\{\frac{3}{2}(\theta-\theta_0)+\frac{b^2}{\theta_0\omega_0^4}\left[\frac{3}{4}\alpha|\psi(\theta)|^2+\alpha\delta^2\zeta^2(\theta)\right.\right.\right.$$
$$\left.\left.\left.-2\delta^2\zeta(\theta)\right]\right\}\right)^{1/3}, \tag{7.3.27}$$

and, upon use of (7.3.5c),

$$h_{1,2}(\theta) = \left(\mp\frac{2\alpha_1^{1/2}(\theta)}{\phi^{(2)}(\omega_{sp_{1,2}},\theta)}\right)^{1/2}$$
$$\cong\left(\left(\frac{2\theta_0\omega_0^4}{b^2\{3\alpha|\psi(\theta)|\pm 2\delta[1-\alpha\zeta(\theta)]\}}\right)^3\cdot|\psi(\theta)|\left\{\frac{3}{2}(\theta-\theta_0)\right.\right.$$
$$\left.\left.+\frac{b^2}{\theta_0\omega_0^4}\left[\frac{3}{4}\alpha|\psi(\theta)|^2+\alpha\delta^2\zeta^2(\theta)-2\delta^2\zeta(\theta)\right]\right\}\right)^{1/6}, \tag{7.3.28}$$

for all values of θ in the range $1 \leqslant \theta \leqslant \theta_1$. In (7.3.28), the upper sign corresponds to $h_1(\theta)$ and the lower sign corresponds to $h_2(\theta)$. Finally, according to (5.2.13 and 7.3.13b), this expression reduces in the limit as θ approaches θ_1 from below to

$$\lim_{\theta \to \theta_1^-} h_{1,2}(\theta) = \left(-\frac{2}{\phi^{(3)}(\omega_{sp_N}, \theta_1)}\right)^{1/3}$$

$$\cong \left(-\frac{2\theta_0\omega_0^4}{3i\alpha b^2}\right)^{1/3} \equiv h(\theta_1) \ . \tag{7.3.29}$$

The proper values of the multivalued functions appearing in (7.3.27–29) are determined by the conditions presented in Sect. 5.2.2. In particular, the phase of $h_{1,2}(\theta)$ is specified by (5.2.16), i.e., in the notation of the present chapter,

$$\lim_{\theta \to \theta_1^-} \arg[h_{1,2}(\theta)] = \bar{\alpha}^+ \ , \tag{7.3.30}$$

where $\bar{\alpha}^+$ is the angle of slope of the path of steepest descent as it leaves the second-order saddle point at $\omega_{sp_N} = \omega_{sp_1}(\theta_1) = \omega_{sp_2}(\theta_1)$ for $\theta = \theta_1$. From Fig. 7.6, it is seen that $\bar{\alpha}^+ = \pi/6$. Hence, (7.3.30) shows that the argument of $h(\theta_1)$ is $\pi/6$. Moreover, since the quantity within the brackets ( ) in equation (7.3.28) is real and negative for all θ in the range $1 \leqslant \theta \leqslant \theta_1$, the argument of $h_{1,2}(\theta)$ is independent of θ over that range. Hence, $\arg[h_{1,2}(\theta)] = \pi/6$, and (7.3.28) can be written

$$h_{1,2}(\theta) \cong \left| \left(\left(\frac{2\theta_0\omega_0^4}{b^2\{3\alpha|\psi(\theta)| \pm 2\delta[1-\alpha\zeta(\theta)]\}} \right)^3 \cdot |\psi(\theta)| \left\{ \frac{3}{2}(\theta-\theta_0) + \frac{b^2}{\theta_0\omega_0^4}\left[\frac{3}{4}\alpha|\psi(\theta)|^2 + \alpha\delta^2\zeta^2(\theta) - 2\delta^2\zeta(\theta)\right]\right\}\right)^{1/6} \right| e^{i(\pi/6)} \ , \tag{7.3.31}$$

for $1 \leqslant \theta \leqslant \theta_1$.

The proper value of the phase of $\alpha_1^{1/2}(\theta)$ is determined by (5.2.18) with $n = 0$, i.e., in the notation of the present chapter,

$$\lim_{\theta \to \theta_1^-} \arg[\alpha_1^{1/2}(\theta)] = \bar{\alpha}_{12} - \bar{\alpha}^+ \ , \tag{7.3.32}$$

where $\bar{\alpha}_{12}$ is the angle of slope of the vector from the saddle point SP_2 to the saddle point SP_1 in the complex ω-plane and $\bar{\alpha}^+ = \pi/6$. Since $\bar{\alpha}_{12} = \pi/2$ for $1 \leqslant \theta < \theta_1$, (7.3.32) yields

$$\lim_{\theta \to \theta_1^-} \arg[\alpha_1^{1/2}(\theta)] = \frac{\pi}{3} \ . \tag{7.3.33}$$

Moreover, since the quantity in the brackets { } in equation (7.3.27) is real and negative for all θ in the range $1 \leqslant \theta \leqslant \theta_1$, the argument of $\alpha_1^{1/2}(\theta)$ is independent of θ in that range. Hence, $\arg[\alpha_1^{1/2}(\theta)] = \pi/3$, and (7.3.27) can be written

$$\alpha_1^{1/2}(\theta) \cong \left| \left(|\psi(\theta)| \left\{ \frac{3}{2}(\theta - \theta_0) + \frac{b^2}{\theta_0 \omega_0^4} \left[\frac{3}{4}\alpha |\psi(\theta)|^2 + \alpha\delta^2\zeta^2(\theta) - 2\delta^2\zeta(\theta) \right] \right\} \right)^{1/3} \right| e^{i(\pi/3)}, \tag{7.3.34}$$

for $1 \leqslant \theta \leqslant \theta_1$.

Since $\arg[\alpha_1(\theta)] = 2\pi/3$, the argument of the Airy function and its first derivative appearing in (7.3.25) is real and nonnegative for $1 \leqslant \theta \leqslant \theta_1$. Application of the values of the arguments of $h_{1,2}(\theta)$ and $\alpha_1(\theta)$ just found simplify equation (7.3.25) to read

$$\begin{aligned} A_B(z,t) \sim e^{(z/c)\alpha_0(\theta)} \Bigg(\frac{1}{2}\left(\frac{c}{z}\right)^{1/3} & \mathbb{R}\mathrm{e}\{ i[\tilde{u}(\omega_{sp_1} - \omega_c)|h_1(\theta)| \\ & + \tilde{u}(\omega_{sp_2} - \omega_c)|h_2(\theta)|]\} A_i\left[|\alpha_1(\theta)| \left(\frac{z}{c}\right)^{2/3} \right] \\ & - \frac{1}{2|\alpha_1(\theta)|^{1/2}} \left(\frac{c}{z}\right)^{2/3} \mathbb{R}\mathrm{e}\{ i[\tilde{u}(\omega_{sp_1} - \omega_c)|h(\theta_1)| \\ & - \tilde{u}(\omega_{sp_2} - \omega_c)|h_2(\theta)|]\} A_i^{(1)}\left[|\alpha_1(\theta)| \left(\frac{z}{c}\right)^{2/3} \right] \Bigg) \end{aligned} \tag{7.3.35}$$

as $z \to \infty$ uniformly for all θ in the range $1 \leqslant \theta \leqslant \theta_1$.

As θ approaches the critical value θ_1 from below, the argument of the Airy function and its first derivative in (7.3.35) tends to zero and the amplitude coefficients in that equation tend to indeterminate forms. The determinate form of (7.3.35) in this limit is found from (5.2.13–15) and (7.3.29) to be

$$\begin{aligned} A_B(z,t_1) \sim \frac{\omega_0}{2\pi\sqrt{3}} \Gamma\left(\frac{1}{3}\right) \left(\frac{2\theta_0\omega_0 c}{\alpha b^2 z} \right)^{1/3} & \mathbb{R}\mathrm{e}\{ i\tilde{u}(\omega_{sp_N} - \omega_c) \\ & \cdot \exp\left[\frac{2\delta z}{3\alpha c}\left(\theta_0 - \theta_1 + \frac{8\delta^2 b^2}{27\alpha\theta_0\omega_0^4} \right) \right] \}, \end{aligned} \tag{7.3.36}$$

as $z \to \infty$ for $\theta = \theta_1 = ct_1/z$. Note that this result is identical with that given in (7.3.17) by Olver's method.

For a numerical evaluation of the second precursor field $A_B(z,t)$ for values of θ in the range $1 \leqslant \theta \leqslant \theta_1$, (7.3.35) is useless for $\theta_1 - \theta > 0$ too small because of the numerical instabilities of that equation that result from the indeterminate form the equation takes as θ approaches θ_1 from below. For values of θ such that (7.3.35) is not useful because $\theta_1 - \theta > 0$ is too small and (7.3.36) is not accurate enough because $\theta_1 - \theta$ is too large, an alternative expression is required for a numerical evaluation of $A_B(z,t)$. As θ approaches the critical value θ_1 from below, the limiting forms (5.2.13–15) may be employed in the uniform expansion (7.3.25). Moreover, for values of θ close to θ_1, the second term in (7.3.25) is negligible in comparison to the first, and hence, may be neglected. With these substitutions, the following asymptotic behavior for the second precursor field is obtained

$$A_B(z,t) \sim e^{(z/c)\alpha_0(\theta)} \frac{1}{2}\left(\frac{c}{z}\right)^{1/3} |h(\theta_1)| \,\mathrm{Re}\{i[\tilde{u}(\omega_{sp_1} - \omega_c) + \tilde{u}(\omega_{sp_2} - \omega_c)]\} A_i\left[|\alpha_1(\theta)|\left(\frac{z}{c}\right)^{2/3}\right], \tag{7.3.37}$$

as $z \to \infty$ and $\theta \to \theta_1^-$. At $\theta = \theta_1$, this expression reduces to that in (7.3.36).

It now remains to determine the range of values of θ prior to θ_1 over which the asymptotic expression (7.3.37) may be employed in place of (7.3.35) for a numerical calculation of the field behavior. This is accomplished by noting that the limiting form (7.3.37) of the asymptotic approximation (7.3.35) is due primarily to the change in form of the function $h_{1,2}(\theta)$ from that given in (7.3.31) to that in (7.3.29) as θ approaches the critical value θ_1 from below. In order to obtain a continuous transformation in the variable θ as $h_{1,2}(\theta)$ changes its form, the value of θ at which the expression (7.3.31) reduces to that in (7.3.29) is thus desired. Now, at $\theta = \theta_0$ one has that $|\psi(\theta_0)| = 2\delta/3\alpha$ and $\zeta(\theta_0) = 1/\alpha$, so that with these substitutions Eq. (7.3.31) becomes

$$h_{1,2}(\theta_0) \cong \left|\left[\frac{\theta_0\omega_0^4}{\delta b^2}\left(-\frac{4\delta^3 b^2}{9\alpha^2\theta_0\omega_0^4}\right)^{1/3}\right]^{1/2}\right| e^{i(\pi/6)} = \left|\left(\frac{2\theta_0\omega_0^4}{3\alpha b^2}\right)^{1/3}\right| e^{i(\pi/6)} ,$$

which is precisely the expression in (7.3.29). Consequently, for numerical calculations, (7.3.35 and 37) will yield a uniform, continuous evolution of the second precursor field for all values of θ in the range $1 \leqslant \theta \leqslant \theta_1$ when (7.3.35) is restricted to the range $1 \leqslant \theta \leqslant \theta_0$ and (7.3.37) is restricted to the range $\theta_0 \leqslant \theta \leqslant \theta_1$.

Equations (7.3.35 and 36) express the asymptotic behavior of the second precursor field valid uniformly for all θ in the range $1 \leqslant \theta \leqslant \theta_1$. Since the argument of the Airy function and its first derivative is real and positive in this range of values of θ, the second precursor field is nonoscillatory. As shown in Sect. 5.2.2, Eq. (7.3.35) reduces to the nonuniform result obtained in Sect. 7.3.1 for $1 \leqslant \theta \leqslant \theta_1$ when θ is sufficiently bounded away from θ_1 for fixed z so that the large argument asymptotic behavior of the Airy function and its first derivative can be applied.

Consider now the uniform asymptotic behavior of the second precursor field for values of θ within the range $\theta \geqslant \theta_1$. In this case, the two near saddle point locations are given by (7.3.2d) and the approximate phase behavior at these points is given by (7.3.19). Application of Theorem 5.2 therefore yields the asymptotic expansion of the second precursor field $A_B(z,t)$

$$A_B(z,t) = -\mathbb{R}\mathrm{e}\, e^{(z/c)\alpha_0(\theta)}\left\{\left(\frac{c}{z}\right)^{1/3} e^{-i(2\pi/3)} A_i\left(\alpha_1(\theta)e^{-i(2\pi/3)}\left(\frac{z}{c}\right)^{2/3}\right)\right.$$
$$\cdot\left\{\frac{1}{2}(\tilde{u}(\omega_{sp_N^+} - \omega_c)h^+(\theta) + \tilde{u}(\omega_{sp_N^-} - \omega_c)h^-(\theta)] + \mathcal{O}\left(\frac{1}{z}\right)\right\}$$
$$+ \left(\frac{c}{z}\right)^{2/3} e^{-i(4\pi/3)} A_i^{(1)}\left(\alpha_1(\theta)e^{-i(2\pi/3)}\left(\frac{z}{c}\right)^{2/3}\right)$$
$$\left.\cdot\left\{\frac{1}{2\alpha_1^{1/2}(\theta)}[\tilde{u}(\omega_{sp_N^+} - \omega_c)h^+(\theta) - \tilde{u}(\omega_{sp_N^-} - \omega_c)h^-(\theta)] + \mathcal{O}\left(\frac{1}{z}\right)\right\}\right\}, \tag{7.3.38}$$

as $z \to \infty$ uniformly for $\theta \geqslant \theta_1$. According to (5.2.10–12) and (7.3.19), the coefficients appearing in this expression are

$$\alpha_0(\theta) = \frac{1}{2}[\phi(\omega_{sp_N^+},\theta) + \phi(\omega_{sp_N^-},\theta)]$$
$$\cong -\delta\left(\frac{2}{3}\zeta(\theta)(\theta-\theta_0) + \frac{b^2}{\theta_0\omega_0^4}\left\{[1-\alpha\zeta(\theta)]\psi^2(\theta)\right.\right.$$
$$\left.\left. + \frac{4}{9}\delta^2\zeta^2(\theta)\left[\frac{1}{3}\alpha\zeta(\theta) - 1\right]\right\}\right), \tag{7.3.39}$$

$$\alpha_1^{1/2}(\theta) = \left\{\frac{3}{4}[\phi(\omega_{sp_N^+},\theta) - \phi(\omega_{sp_N^-},\theta)]\right\}^{1/3}$$
$$\cong \left[-\frac{3}{2}i\psi(\theta)\left(\theta - \theta_0 - \frac{b^2}{2\theta_0\omega_0^4}\left\{\frac{4}{3}\delta^2\zeta(\theta)[2-\alpha\zeta(\theta)]\right.\right.\right.$$
$$\left.\left.\left. + \alpha\psi^2(\theta)\right\}\right)\right]^{1/3}, \tag{7.3.40}$$

and

$$h^{\pm}(\theta) = \left(\mp\frac{2\alpha_1^{1/2}(\theta)}{\phi^{(2)}(\omega_{sp_N^{\pm}},\theta)}\right)^{1/2}$$
$$\cong \left[\left(i\frac{2\theta_0\omega_0^4}{b^2\{3\alpha\psi(\theta) \mp 2\delta i[\alpha\zeta(\theta)-1]\}}\right)^3\left(-\frac{3}{2}i\psi(\theta)\right)\right.$$
$$\left.\cdot\left(\theta - \theta_0 - \frac{b^2}{2\theta_0\omega_0^4}\left\{\frac{4}{3}\delta^2\zeta(\theta)[2-\alpha\zeta(\theta)] + \alpha\psi^2(\theta)\right\}\right)\right]^{1/6}$$

$$\cong \left[\left(\mathrm{i}\frac{2\theta_0\omega_0^4}{3\alpha b^2\psi(\theta)}\right)^3\left[-\frac{3}{2}\mathrm{i}\psi(\theta)\right]\left(\theta-\theta_0\right.\right.$$
$$\left.\left.-\frac{b^2}{2\theta_0\omega_0^4}\left\{\frac{4}{3}\delta^2\zeta(\theta)[2-\alpha\zeta(\theta)]+\alpha\psi^2(\theta)\right\}\right)\right]^{1/6}, \tag{7.3.41}$$

for all θ in the range $\theta \geqslant \theta_1$. The approximation made in obtainining the final form of (7.3.41) [viz., $\{3\alpha\psi(\theta) \mp 2\delta\mathrm{i}[\alpha\zeta(\theta)-1]\} \cong 3\alpha\psi(\theta)$] is valid for all $\theta \geqslant \theta_1$. In the limit as θ approaches θ_1 from above, this expression (7.3.41) is replaced by its limiting form given in (5.2.13), so that

$$\lim_{\theta\to\theta_1^+} h^{\pm}(\theta) = \left(-\frac{2}{\phi^{(3)}(\omega_{sp_N},\theta_1)}\right)^{1/3}$$
$$\cong \left(-\frac{2\theta_0\omega_0^4}{3\mathrm{i}\alpha b^2}\right)^{1/3} \equiv h(\theta_1)\ . \tag{7.3.42}$$

The proper values of the multivalued functions appearing in (7.3.39–42) are determined by the conditions presented in Sect. 5.2.2. In particular, the phase of $h^{\pm}(\theta)$ is specified by (5.2.16), i.e., in the notation of the present chapter

$$\lim_{\theta\to\theta_1^+} \arg[h^{\pm}(\theta)] = \bar{\alpha}^+ \tag{7.3.43}$$

where $\bar{\alpha}^+$ is the angle of slope of the path of steepest descent as it leaves the second order saddle point at $\omega_{sp_N} = \omega_{sp_N^\pm}(\theta_1)$ for $\theta = \theta_1$. From Fig. 7.6, it is seen that $\bar{\alpha}^+ = \pi/6$. Hence, (7.3.43) shows that the argument of $h(\theta_1)$ is $\pi/6$. Moreover, since the quantity within the outer brackets in (7.3.41) is real and negative for all θ in the range $\theta \geqslant \theta_1$, the argument of $h^{\pm}(\theta)$ is independent of θ over that range. Hence, $\arg[h^{\pm}(\theta)] = \pi/6$ and (7.3.41) can be written

$$h^{\pm}(\theta) \cong \left|\left[\left(\frac{2\theta_0\omega_0^4}{3\alpha b^2\psi(\theta)}\right)^3 \frac{3}{2}\psi(\theta)\left(\theta-\theta_0\right.\right.\right.$$
$$\left.\left.\left.-\frac{b^2}{2\theta_0\omega_0^4}\left\{\frac{4}{9}\delta^2\zeta(\theta)[2-\alpha\zeta(\theta)]+\alpha\psi^2(\theta)\right\}\right)\right]^{1/6}\right| \mathrm{e}^{\mathrm{i}(\pi/6)}\ , \tag{7.3.44}$$

for $\theta \geqslant \theta_1$.

The proper value of the phase of $\alpha_1^{1/2}(\theta)$ is determined by (5.2.18) with $n = 0$, i.e., in the notation of the present chapter

$$\lim_{\theta\to\theta_1^+} \arg[\alpha_1^{1/2}(\theta)] = \bar{\alpha}_{12} - \bar{\alpha}^+\ , \tag{7.3.45}$$

where $\bar{\alpha}_{12}$ is the angle of slope of the vector from the saddle point SP_N^- to the saddle point SP_N^+ in the complex ω-plane and $\bar{\alpha}^+ = \pi/6$. Since $\bar{\alpha}_{12} = 0$ for $\theta \geqslant \theta_1$,

(7.3.45) yields

$$\lim_{\theta \to \theta_1^+} \arg[\alpha_1^{1/2}(\theta)] = -\frac{\pi}{6} . \tag{7.3.46}$$

Moreover, since the quantity in the brackets ( ) in equation (7.3.40) is negative imaginary for all θ in the range $\theta \geqslant \theta_1$, the argument of $\alpha_1^{1/2}(\theta)$ is independent of θ in that range. Hence $\arg[\alpha_1^{1/2}(\theta)] = -\pi/6$ and (7.3.40) can be written

$$\alpha_1^{1/2}(\theta) \cong \left| \left[\frac{3}{2}\psi(\theta)\left(\theta - \theta_0 - \frac{b^2}{2\theta_0\omega_0^4}\left\{\frac{4}{3}\delta^2\zeta(\theta)[2 - \alpha\zeta(\theta)] + \alpha\psi^2(\theta)\right\}\right)\right]^{1/3} \right| e^{-i(\pi/6)} , \tag{7.3.47}$$

for $\theta \geqslant \theta_1$.

Since $\arg[\alpha_1(\theta)] = -\pi/3$, the argument of the Airy function and its first derivative appearing in (7.3.38) is real and nonpositive for $\theta \geqslant \theta_1$. Application of the values of the arguments of $h^{\pm}(\theta)$ and $\alpha_1(\theta)$ just found simplify equation (7.3.38) to read

$$\begin{aligned} A_B(z,t) \sim e^{(z/c)\alpha_0(\theta)} \left(\frac{1}{2}\left(\frac{c}{z}\right)\right)^{1/3} & \mathrm{Re}\{i[\tilde{u}(\omega_{sp_N^+} - \omega_c)|h^+(\theta)| \\ & + \tilde{u}(\omega_{sp_N^-} - \omega_c)|h^-(\theta)|]\} A_i\left(-|\alpha_1(\theta)|\left(\frac{z}{c}\right)^{2/3}\right) \\ & + \frac{1}{2|\alpha_1(\theta)|^{1/2}}\left(\frac{c}{z}\right)^{2/3} \mathrm{Re}\{\tilde{u}(\omega_{sp_N^+} - \omega_c)|h^+(\theta)| \\ & - \tilde{u}(\omega_{sp_N^-} - \omega_c)|h^-(\theta)|\} A_i^{(1)}\left(-|\alpha_1(\theta)|\left(\frac{z}{c}\right)^{2/3}\right)\bigg) , \end{aligned} \tag{7.3.48}$$

as $z \to \infty$ uniformly for all θ in the range $\theta \geqslant \theta_1$.

As θ approaches the critical value θ_1 from above, the argument of the Airy function and its first derivative in (7.3.48) tends to zero and the amplitude coefficients in that equation tend to indeterminate forms. The determinate form of (7.3.48) in this limit is found from (5.2.13–15) and (7.3.42) to be precisely that given in (7.3.17, 36) as $z \to \infty$ for $\theta = \theta_1 = ct_1/z$.

For a numerical evaluation of the second precursor field $A_B(z,t)$ for values of θ in the range $\theta \geqslant \theta_1$, (7.3.48) is useless for $\theta - \theta_1 > 0$ too small because of the numerical instabilities of that equation that result from the indeterminate form the equation takes as θ approaches θ_1 from above. For values of θ such that (7.3.48) is not useful because $\theta - \theta_1 > 0$ is too small and (7.3.36) is not accurate enough because $\theta - \theta_1$ is too large, an alternative expression is required for a numerical evaluation of $A_B(z,t)$. As θ approaches the critical value θ_1 from above, the limiting forms (5.2.13 to 15) may be employed in the uniform expansion (7.3.38). Moreover, for values of θ close to θ_1, the second term in (7.3.38) is negligible in comparison to the first, and hence, may be neglected. With these substitutions, the following asymptotic behavior for the second precursor field is obtained

$$A_B(z,t) \sim e^{(z/c)\alpha_0(\theta)} \frac{1}{2}\left(\frac{c}{z}\right)^{1/3} |h(\theta_1)| \,\mathrm{Re}\{i(\tilde{u}(\omega_{sp_N^+} - \omega_c) + \tilde{u}(\omega_{sp_N^-} - \omega_c)\} A_i\left(-|\alpha_1(\theta)|\left(\frac{z}{c}\right)^{2/3}\right), \tag{7.3.49}$$

as $z \to \infty$ and $\theta \to \theta_1^+$. At $\theta = \theta_1$, this expression reduces to that in (7.3.36).

It now remains to determine the range of values of θ after θ_1 over which the asymptotic expression (7.3.49) may be employed in place of (7.3.48) for a numerical calculation of the field behavior. As in the previous case, this is accomplished by noting that the limiting form (7.3.49) of the asymptotic approximation (7.3.48) is due primarily to the change in form of the function $h^\pm(\theta)$ from that given in (7.3.44) to that in (7.3.42) as θ approaches the critical value θ_1 from above. In order to obtain a continuous transformation in the variable θ as $h^\pm(\theta)$ changes its form, the value of $\theta \equiv \theta_2$ at which the expression (7.3.44) reduces to that given in (7.3.42) is thus desired. In the previous case for $\theta \leqslant \theta_1$, the corresponding value of θ was shown to be equal to θ_0. Since $|\psi(\theta)|$ is symmetric in θ about the critical value θ_1, it is then expected that $\theta_2 = \theta_1 + (\theta_1 - \theta_0) = 2\theta_1 - \theta_0$, at which value $\psi(\theta_2) \cong 2\delta/3\alpha$ and $\zeta(\theta_2) \cong 1/\alpha$. With these substitutions, (7.3.44) becomes

$$h^\pm(\theta_2) \cong \left|\left[\frac{\theta_0\omega_0^4}{\delta b^2}\left(\frac{4\delta^3 b^2}{9\alpha^2\theta_0\omega_0^4}\right)^{1/3}\right]^{1/2}\right| e^{i(\pi/6)} = \left|\left(\frac{2\theta_0\omega_0^4}{3\alpha b^2}\right)^{1/3}\right| e^{i(\pi/6)},$$

which is precisely the expression in (7.3.42). Consequently, for numerical calculations, (7.3.48 and 49) will yield a uniform, continuous evolution of the second precursor field for all values of θ in the range $\theta \geqslant \theta_1$ when (7.3.49) is restricted to the range $\theta_1 \leqslant \theta \leqslant \theta_2$ and (7.3.48) is restricted to the range $\theta \geqslant \theta_2$. Upon the use of (6.2.43b), the value of θ_2 is given by

$$\theta_2 = 2\theta_1 - \theta_0 \cong \theta_0 + \frac{4\delta^2 b^2}{\theta_0\omega_0^2(3\alpha\omega_0^2 - 4\delta^2)} \tag{7.3.50}$$

to a good approximation.

Equations (7.3.36 and 48) express the asymptotic behavior of the second precursor field valid uniformly for all θ in the range $\theta \geqslant \theta_1$. Since the argument of the Airy function and its first derivative is real and negative for all θ in the range $\theta > \theta_1$, the second precursor field is oscillatory in this range of values of θ. As shown in Sect. 5.2.2, (7.3.48) reduces to the nonuniform result obtained in Sect. 7.3.1 for $\theta > \theta_1$ when θ is sufficiently bounded away from θ_1 for fixed z so that the large argument asymptotic behavior of the Airy function and its first derivative can be applied.

The asymptotic approximation of $A_B(z,t)$ uniformly valid for all $\theta \geqslant 1$ is given by (7.3.26, 29, 31, 34–36, 39, 42, 44, 47, and 48). The results are collected here for convenient reference. As $z \to \infty$, the asymptotic behavior of the second precursor field is given by

$$A_B(z,t) \sim e^{(z/c)\alpha_0(\theta)} \left\{ \frac{1}{2}\left(\frac{c}{z}\right)^{1/3} \mathbb{R}\mathrm{e}\{i[\tilde{u}(\omega_{sp_1} - \omega_c)|h_1(\theta)| + \tilde{u}(\omega_{sp_2} - \omega_c)|h_2(\theta)|]\} A_i\left(|\alpha_1(\theta)|\left(\frac{z}{c}\right)^{2/3}\right) - \frac{1}{2|\alpha_1(\theta)|^{1/2}}\left(\frac{c}{z}\right)^{2/3} \mathbb{R}\mathrm{e}\{i[\tilde{u}(\omega_{sp_1} - \omega_c)|h_1(\theta)| - \tilde{u}(\omega_{sp_2} - \omega_c)|h_2(\theta)|]\} A_i^{(1)}\left(|\alpha_1(\theta)|\left(\frac{z}{c}\right)^{2/3}\right) \right\}, \tag{7.3.51}$$

uniformly for $1 \leqslant \theta \leqslant \theta_1$, and by

$$A_B(z,t) \sim e^{(z/c)\alpha_0(\theta)} \left(\frac{1}{2}\left(\frac{c}{z}\right)^{1/3} \mathbb{R}\mathrm{e}\{i[\tilde{u}(\omega_{sp_N^+} - \omega_c)|h^+(\theta)| + \tilde{u}(\omega_{sp_N^-} - \omega_c)|h^-(\theta)|]\} A_i\left(-|\alpha_1(\theta)|\left(\frac{z}{c}\right)^{2/3}\right) + \frac{1}{2|\alpha_1(\theta)|^{1/2}}\left(\frac{c}{z}\right)^{2/3} \mathbb{R}\mathrm{e}\{\tilde{u}(\omega_{sp_N^+} - \omega_c)|h^+(\theta)| - \tilde{u}(\omega_{sp_N^-} - \omega_c)|h^-(\theta)|\} A_i^{(1)}\left(-|\alpha_1(\theta)|\left(\frac{z}{c}\right)^{2/3}\right) \right), \tag{7.3.52}$$

uniformly for $\theta \geqslant \theta_1$. The functional parameters appearing in (7.3.51–52) are given by

$$\alpha_0(\theta) \cong -\frac{2}{3}\delta\zeta(\theta)(\theta - \theta_0) - \frac{\delta b^2}{\theta_0\omega_0^4}\left\{|\psi(\theta)|^2[\alpha\zeta(\theta) - 1] + \frac{4}{9}\delta^2\zeta^2(\theta)\left[\frac{1}{3}\alpha\zeta(\theta) - 1\right]\right\}, \tag{7.3.53a}$$

$$|\alpha_1(\theta)| \cong \left| |\psi(\theta)| \left\{ \frac{3}{2}(\theta - \theta_0) + \frac{b^2}{\theta_0\omega_0^4}\left[\frac{3}{4}\alpha|\psi(\theta)|^2 + \alpha\delta^2\zeta^2(\theta) - 2\delta^2\zeta(\theta)\right]\right\}\right|^{2/3}, \tag{7.3.53b}$$

$$|h_{1,2}(\theta)| \cong \left| \frac{2\theta_0\omega_0^4}{b^2\{3\alpha|\psi(\theta)| \pm 2\delta[1-\alpha\zeta(\theta)]\}} \cdot \left(|\psi(\theta)| \left\{ \frac{3}{2}(\theta-\theta_0) + \frac{b^2}{\theta_0\omega_0^4}\left[\frac{3}{4}\alpha|\psi(\theta)|^2 + \alpha\delta^2\zeta^2(\theta) - 2\delta^2\zeta(\theta)\right]\right\}\right)^{1/3} \right|^{1/2}, \quad (7.3.53c)$$

$$|h(\theta_1)| \cong \left| \frac{2\theta_0\omega_0^4}{3\alpha b^2} \right|^{1/3}, \quad (7.3.53d)$$

for $\theta \leqslant \theta_1$, and

$$\alpha_0(\theta) \cong -\delta\left(\frac{2}{3}\zeta(\theta)(\theta-\theta_0) + \frac{b^2}{\theta_0\omega_0^4}\left\{[1-\alpha\zeta(\theta)]\psi^2(\theta) + \frac{4}{9}\delta^2\zeta^2(\theta)\left[\frac{1}{3}\alpha\zeta(\theta) - 1\right]\right\}\right), \quad (7.3.54a)$$

$$|\alpha_1(\theta)| \cong \left| \frac{3}{2}\psi(\theta)\left(\theta - \theta_0 - \frac{b^2}{2\theta_0\omega_0^4}\left\{\frac{4}{3}\delta^2\zeta(\theta)[2-\alpha\zeta(\theta)] + \alpha\psi^2(\theta)\right\}\right)\right|^{2/3}, \quad (7.3.54b)$$

$$|h^{\pm}(\theta)| \cong \left| \frac{2\theta_0\omega_0^4}{3\alpha b^2\psi(\theta)}\left[\frac{3}{2}\psi(\theta)\left(\theta-\theta_0 - \frac{b^2}{2\theta_0\omega_0^4}\left\{\frac{4}{3}\delta^2\zeta(\theta)[2-\alpha\zeta(\theta)] + \alpha\psi^2(\theta)\right\}\right)\right]^{1/3}\right|^{1/2}, \quad (7.3.54c)$$

$$|h(\theta_1)| \cong \left| \frac{2\theta_0\omega_0^4}{3\alpha b^2} \right|^{1/3}, \quad (7.3.54d)$$

for $\theta \geqslant \theta_1$. Although the asymptotic behavior of $A_B(z,t)$ is given by different expressions in the different regions $1 \leqslant \theta \leqslant \theta_1$ and $\theta \geqslant \theta_1$, the different expressions approach the same expression in the limit as $\theta \to \theta_1$, i.e.

$$A_B(z,t_1) \sim \frac{\omega_0}{2\pi\sqrt{3}}\Gamma\left(\frac{1}{3}\right)\left(\frac{2\theta_0\omega_0 c}{\alpha b^2 z}\right)^{1/3} \mathbb{R}\mathrm{e}\{i\tilde{u}(\omega_{sp_N} - \omega_c)\} \cdot \exp\left[\frac{2\delta z}{3\alpha c}\left(\theta_0 - \theta_1 + \frac{8\delta^2 b^2}{27\alpha\theta_0\omega_0^4}\right)\right]. \quad (7.3.55)$$

Consequently, (7.3.51 and 52) taken together give an asymptotic approximation of the second precursor field $A_B(z,t)$ that is valid uniformly for all $\theta \geqslant 1$.

For a numerical evaluation of $A_B(z,t)$ (7.3.51 and 52) are unstable and are not useful for small $|\theta - \theta_1|$. For that purpose, they should be replaced by

$$A_B(z,t) \sim e^{(z/c)\alpha_0(\theta)} \frac{1}{2}\left(\frac{c}{z}\right)^{1/3} |h(\theta_1)| \operatorname{Re}\{i[\tilde{u}(\omega_{sp_1} - \omega_c) + \tilde{u}(\omega_{sp_2} - \omega_c)]\} A_i\left(|\alpha_1(\theta)|\left(\frac{z}{c}\right)^{2/3}\right), \quad (7.3.56)$$

for $\theta_0 \leqslant \theta \leqslant \theta_1$, and by

$$A_B(z,t) \sim e^{(z/c)\alpha_0(\theta)} \frac{1}{2}\left(\frac{c}{z}\right)^{1/3} |h(\theta_1)| \operatorname{Re}\{i[\tilde{u}(\omega_{sp_N^+} - \omega_c) + \tilde{u}(\omega_{sp_N^-} - \omega_c)]\} A_i\left(-|\alpha_1(\theta)|\left(\frac{z}{c}\right)^{2/3}\right), \quad (7.3.57)$$

for $\theta_1 \leqslant \theta \leqslant \theta_2$, where

$$\theta_0 = \sqrt{1 + \frac{b^2}{\omega_0^2}}, \quad (7.3.58a)$$

$$\theta_1 \cong \theta_0 + \frac{2\delta^2 b^2}{\theta_0 \omega_0^2(3\alpha\omega_0^2 - 4b^2)}, \quad (7.3.58b)$$

$$\theta_2 \cong \theta_0 + \frac{4\delta^2 b^2}{\theta_0 \omega_0^2(3\alpha\omega_0^2 - 4\delta^2)}. \quad (7.3.58c)$$

7.3.3 The Instantaneous Angular Frequency

The instantaneous angular frequency of oscillation of the second precursor field $A_B(z,t)$ is defined as minus the time derivative of the oscillatory phase [7.4, 5], where the minus is included so as to obtain a positive-valued angular frequency. Since $A_B(z,t)$ is not oscillatory for values of θ prior and equal to θ_1, the instantaneous angular frequency then identically vanishes for values of θ within this range:

$$\omega_B = 0, \quad (7.3.59)$$

for $1 \leqslant \theta \leqslant \theta_1$. For values of θ greater than θ_1, $A_B(z,t)$ is oscillatory [note that the oscillatory phase terms appearing in both the nonuniform and uniform expressions (7.3.23 and 52) are identical], and ω_B increases away from zero. Consequently, since $d\theta/dt = c/z$, application of

$$\frac{d\zeta(\theta)}{dt} = \frac{d}{dt}\left[\frac{3}{2} \cdot \frac{\theta^2 - \theta_0^2 + \dfrac{2b^2}{\omega_0^2}}{\theta^2 - \theta_0^2 + \dfrac{3b^2}{\omega_0^2}\alpha}\right] \cong \frac{3b^2\theta c}{\omega_0^2 z} \cdot \frac{1}{\left(\theta^2 - \theta_0^2 + \dfrac{3b^2}{\omega_0^2}\right)^2},$$

and

$$\frac{d\psi(\theta)}{dt} = \frac{d}{dt}\left[\frac{\omega_0^2(\theta^2-\theta_0^2)}{\theta^2-\theta_0^2+\frac{3b^2}{\omega_0^2}\alpha} - \delta^2\left\{\frac{\theta^2-\theta_0^2+\frac{2b^2}{\omega_0^2}}{\theta^2-\theta_0^2+\frac{3b^2}{\omega_0^2}\alpha}\right\}^2\right]^{1/2}$$

$$\cong \frac{b^2\theta c}{z\psi(\theta)}\cdot\frac{3\left(\theta^2-\theta_0^2+\frac{3b^2}{\omega_0^2}\right)-2\frac{\delta^2}{\omega_0^2}\left(\theta^2-\theta_0^2+\frac{2b^2}{\omega_0^2}\right)}{\left(\theta^2-\theta_0^2+\frac{3b^2}{\omega_0^2}\right)^3},$$

(where the approximation that α is essentially equal to unity has been employed), yields the instantaneous angular frequency of the second precursor field for values of θ greater than θ_1, given by

$$\omega_B \equiv \frac{d}{dt}\left[\frac{z}{c}\psi(\theta)\left(\theta-\theta_0-\frac{b^2}{2\theta_0\omega_0^4}\left\{\frac{4}{3}\delta^2\zeta(\theta)[2-\zeta(\theta)]+\psi^2(\theta)\right\}\right)+\frac{\pi}{4}\right]$$

$$\cong \psi(\theta)\left[1-\frac{3b^4\theta}{2\theta_0\omega_0^4}\cdot\frac{3\left(\theta^2-\theta_0^2+\frac{3b^2}{\omega_0^2}\right)-2\frac{\delta^2}{\omega_0^2}\left(\theta^2-\theta_0^2+\frac{2b^2}{\omega_0^2}\right)}{\left(\theta^2-\theta_0^2+\frac{3b^2}{\omega_0^2}\right)^3}\right.$$

$$\left.-\frac{4\delta^2b^4\theta}{\theta_0\omega_0^6}\cdot\frac{1-\zeta(\theta)}{\left(\theta^2-\theta_0^2+\frac{3b^2}{\omega_0^2}\right)^2}\right]$$

$$+\frac{b^2\theta}{\psi(\theta)}\cdot\frac{3\left(\theta^2-\theta_0^2+\frac{3b^2}{\omega_0^2}\right)-2\frac{\delta^2}{\omega_0^2}\left(\theta^2-\theta_0^2+\frac{2b^2}{\omega_0^2}\right)}{\left(\theta^2-\theta_0^2+\frac{3b^2}{\omega_0^2}\right)^3}$$

$$\cdot\left\{\theta-\theta_0-\frac{2\delta^2b^2}{3\theta_0\omega_0^4}\zeta(\theta)[2-\zeta(\theta)]\right\}. \tag{7.3.60}$$

Now, according to (6.2.43b), one has that

$$\theta-\theta_0-\frac{2\delta^2b^2}{3\theta_0\omega_0^4}\zeta(\theta)[2-\zeta(\theta)] \cong \theta-\theta_1,$$

for all values of θ within the range $\theta > \theta_1$. Consequently, since

$$\frac{3b^4\theta}{2\theta_0\omega_0^4}\cdot\frac{3\left(\theta^2-\theta_0^2+\dfrac{3b^2}{\omega_0^2}\right)-2\dfrac{\delta^2}{\omega_0^2}\left(\theta^2-\theta_0^2+\dfrac{2b^2}{\omega_0^2}\right)}{\left(\theta^2-\theta_0^2+\dfrac{3b^2}{\omega_0^2}\right)^3}$$

$$+\frac{4\delta^2 b^4\theta}{\theta_0\omega_0^6}\cdot\frac{1-\zeta(\theta)}{\left(\theta^2-\theta_0^2+\dfrac{3b^2}{\omega_0^2}\right)^2}$$

$$=\frac{b^4\theta}{\omega_0^4\theta_0}\cdot\frac{\left(\dfrac{9}{2}+8\dfrac{\delta^2}{\omega_0^2}\right)\left(\theta^2-\theta_0^2+\dfrac{3b^2}{\omega_0^2}\right)-15\dfrac{\delta^2}{\omega_0^2}\left(\theta^2-\theta_0^2+\dfrac{2b^2}{\omega_0^2}\right)}{\left(\theta^2-\theta_0^2+\dfrac{3b^2}{\omega_0^2}\right)^3},$$

it is seen that (7.3.60) reduces to

$$\omega_B \cong \psi(\theta)\left[1-\frac{b^4\theta}{\omega_0^4\theta_0}\cdot\frac{\left(\dfrac{9}{2}+8\dfrac{\delta^2}{\omega_0^2}\right)\left(\theta^2-\theta_0^2+\dfrac{3b^2}{\omega_0^2}\right)-15\dfrac{\delta^2}{\omega_0^2}\left(\theta^2-\theta_0^2+\dfrac{2b^2}{\omega_0^2}\right)}{\left(\theta^2-\theta_0^2+\dfrac{3b^2}{\omega_0^2}\right)^3}\right]$$

$$+\frac{b^2\theta}{\psi(\theta)}(\theta-\theta_1)\cdot\frac{3\left(\theta^2-\theta_0^2+\dfrac{3b^2}{\omega_0^2}\right)-2\dfrac{\delta^2}{\omega_0^2}\left(\theta^2-\theta_0^2+\dfrac{2b^2}{\omega_0^2}\right)}{\left(\theta^2-\theta_0^2+\dfrac{3b^2}{\omega_0^2}\right)^3} \qquad (7.3.61a)$$

$$\cong \psi(\theta) \qquad (7.3.61b)$$

A graph of the instantaneous angular frequency of the second precursor field as a function of θ according to (7.3.61a) is illustrated in Fig. 7.8 along with the behavior of $\psi(\theta)$ for *Brillouin's* choice of the medium parameters. It is easily seen that for values of θ very close to θ_1, ω_B is given by $\psi(\theta)$ to a high degree of accuracy, as it is for very large values of θ. However, the approximation stated in (7.3.61b) becomes less accurate for intermediate values of θ. Finally, as θ increases, the instantaneous angular frequency ω_B approaches the value $\psi(\infty) = \sqrt{\omega_0^2-\delta^2}$ asymptotically from below.

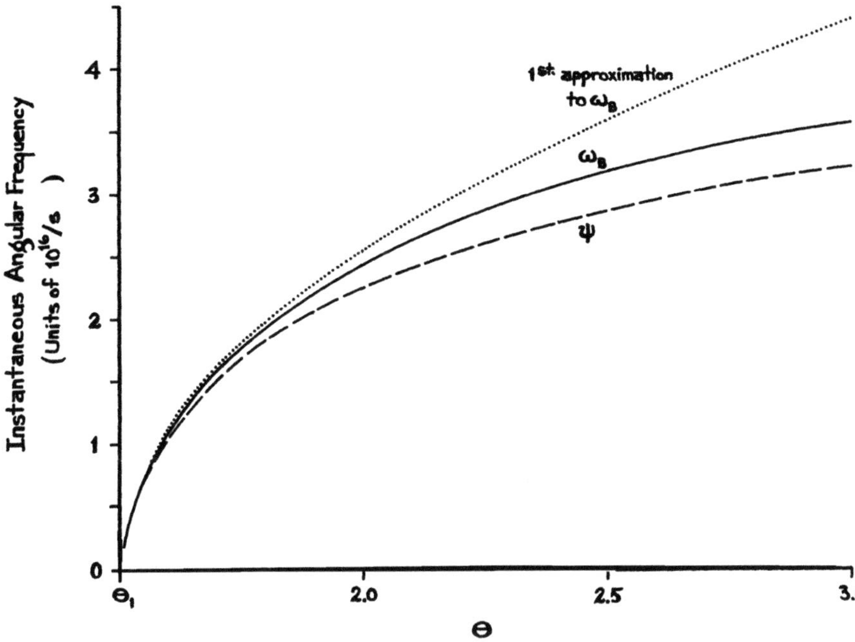

Fig. 7.8. The instantaneous angular frequency of oscillation ω_B of the second precursor field as a function of θ for $\omega_0 = 4.0 \times 10^{16}$/s, $\delta = 0.28 \times 10^{16}$/s, and $b^2 = 20.0 \times 10^{32}/\text{s}^2$.

7.3.4 The Unit Step-Function Modulated Signal

For the unit step-function modulated signal, the spectrum of the initial field envelope is given by (4.3.33), so that

$$\tilde{u}(\omega_{sp_1} - \omega_c) = \frac{i}{\omega_{sp_1} - \omega_c} \cong \frac{\frac{1}{3}[3|\psi(\theta)| - 2\delta\zeta(\theta)] - i\omega_c}{\omega_c^2 + \frac{1}{9}[3|\psi(\theta)| - 2\delta\zeta(\theta)]^2}, \tag{7.3.62a}$$

$$\tilde{u}(\omega_{sp_2} - \omega_c) = \frac{i}{\omega_{sp_2} - \omega_c} \cong \frac{-\frac{1}{3}[3|\psi(\theta)| + 2\delta\zeta(\theta)] - i\omega_c}{\omega_c^2 + \frac{1}{9}[3|\psi(\theta)| + 2\delta\zeta(\theta)]^2}, \tag{7.3.62b}$$

for $1 \leqslant \theta \leqslant \theta_1$, and

$$\tilde{u}(\omega_{sp_N^\pm} - \omega_c) = \frac{i}{\omega_{sp_N^\pm} - \omega_c} \cong \frac{-\frac{2}{3}\delta\zeta(\theta) + i[\pm\psi(\theta) - \omega_c]}{[\pm\psi(\theta) - \omega_c]^2 + \frac{4}{9}\delta^2\zeta^2(\theta)}, \tag{7.3.63}$$

for $\theta \geqslant \theta_1$. At $\theta = \theta_1$, these equations reduce to

$$\tilde{u}(\omega_{sp_N} - \omega_c) \cong \frac{-\frac{2}{3}\frac{\delta}{\alpha} - i\omega_c}{\omega_c^2 + \frac{4}{9}\frac{\delta^2}{\alpha^2}}. \tag{7.3.64}$$

For values of θ in the range $1 \leqslant \theta \leqslant \theta_1$, the uniform asymptotic behavior of the second precursor field is given by (7.3.51). Substitution of (7.3.62) into that equation yields

$$
\begin{aligned}
A_B(z,t) \sim \frac{\omega_0^2\omega_c}{2b}\left(\frac{c}{z}\right)^{1/3} e^{(z/c)\alpha_0(\theta)}
&\cdot\Bigg\{\Bigg[\frac{1}{\omega_c^2 + \frac{1}{9}[3|\psi(\theta)| - 2\delta\zeta(\theta)]^2}\left(\frac{2\theta_0}{3\alpha|\psi(\theta)| + 2\delta[1-\alpha\zeta(\theta)]}\right)^{1/2} \\
&+ \frac{1}{\omega_c^2 + \frac{1}{9}[3|\psi(\theta)| + 2\delta\zeta(\theta)]^2}\left(\frac{2\theta_0}{3\alpha|\psi(\theta)| - 2\delta[1-\alpha\zeta(\theta)]}\right)^{1/2}\Bigg] \\
&\cdot|\alpha_1(\theta)|^{1/4}\mathrm{A_i}\left(|\alpha_1(\theta)|\left(\frac{z}{c}\right)^{2/3}\right) \\
&-\Bigg[\frac{1}{\omega_c^2 + \frac{1}{9}[3|\psi(\theta)| - 2\delta\zeta(\theta)]^2}\left(\frac{2\theta_0}{3\alpha|\psi(\theta)| + 2\delta[1-\alpha\zeta(\theta)]}\right)^{1/2} \\
&- \frac{1}{\omega_c^2 + \frac{1}{9}[3|\psi(\theta)| + 2\delta\zeta(\theta)]^2}\left(\frac{2\theta_0}{3\alpha|\psi(\theta)| - 2\delta[1-\alpha\zeta(\theta)]}\right)^{1/2}\Bigg] \\
&\cdot\frac{1}{|\alpha_1(\theta)|^{1/4}}\left(\frac{c}{z}\right)^{1/3}\mathrm{A_i^{(1)}}\left(|\alpha_1(\theta)|\left(\frac{z}{c}\right)^{2/3}\right)\Bigg\},
\end{aligned}
\tag{7.3.65}
$$

as $z \to \infty$ uniformly for all θ in the range $1 \leqslant \theta \leqslant \theta_1$. The functions $\alpha_0(\theta)$ and $|\alpha_1(\theta)|$ appearing in (7.3.65) are given by (7.3.53a and b), respectively. For values of θ in the range $\theta \geqslant \theta_1$, the uniform asymptotic behavior of the second precursor field is given by (7.3.52). Substitution of (7.3.63) into that equation yields

$$
\begin{aligned}
A_B(z,t) \sim -\frac{\omega_0^2}{b}\left(\frac{\theta_0}{6\alpha\psi(\theta)}\right)^{1/2}\left(\frac{c}{z}\right)^{1/3} e^{(z/c)\alpha_0(\theta)}
&\cdot\Bigg[\frac{2}{3}\delta\zeta(\theta)|\alpha_1(\theta)|^{1/4}\Bigg(\frac{1}{[\psi(\theta)+\omega_c]^2 + \frac{4}{9}\delta^2\zeta^2(\theta)} \\
&- \frac{1}{[\psi(\theta)-\omega_c]^2 + \frac{4}{9}\delta^2\zeta^2(\theta)}\Bigg)\mathrm{A_i}\left(-|\alpha_1(\theta)|\left(\frac{z}{c}\right)^{2/3}\right) \\
&- \frac{1}{|\alpha_1(\theta)|^{1/4}}\Bigg(\frac{\psi(\theta)-\omega_c}{[\psi(\theta)-\omega_c]^2 + \frac{4}{9}\delta^2\zeta^2(\theta)} \\
&- \frac{\psi(\theta)+\omega_c}{[\psi(\theta)+\omega_c]^2 + \frac{4}{9}\delta^2\zeta^2(\theta)}\Bigg)\left(\frac{c}{z}\right)^{1/3}\mathrm{A_i^{(1)}}\left(-|\alpha_1(\theta)|\left(\frac{z}{c}\right)^{2/3}\right)\Bigg],
\end{aligned}
\tag{7.3.66}
$$

as $z \to \infty$ uniformly for all θ in the range $\theta \geqslant \theta_1$. The functions $\alpha_0(\theta)$ and $|\alpha_1(\theta)|$ appearing in (7.3.66) are given by (7.3.54a and b), respectively. Equations (7.3.65 and 66) taken together give an asymptotic approximation of the second precursor field for the unit step-function modulated signal that is valid uniformly for all $\theta \geqslant 1$. At $\theta = \theta_1$, both of these expressions reduce to

$$A_B(z, t_1) \sim \frac{1}{2\pi\sqrt{3}} \Gamma\left(\frac{1}{3}\right) \frac{\omega_0 \omega_c}{\omega_c^2 + \dfrac{4\delta^2}{9\alpha^2}} \left(\frac{2\theta_0 \omega_0 c}{\alpha b^2 z}\right)^{1/3} \cdot \exp\left[\frac{2\delta z}{3\alpha c}\left(\theta - \theta_1 + \frac{8\delta^2 b^2}{27\alpha\theta_0\omega_0^4}\right)\right], \tag{7.3.67}$$

as $z \to \infty$ for $\theta = \theta_1 = \dfrac{ct_1}{z}$.

For a numerical evaluation of $A_B(z,t)$, (7.3.65 and 66) are unstable and are not useful for small values of $|\theta - \theta_1|$. For that purpose, (7.3.56 and 57) are employed. Substitution of (7.3.62) into (7.3.56) yields

$$A_B(z, t) \sim \frac{\omega_0 \omega_c}{\omega_c^2 + \dfrac{4\delta^2}{9\alpha^2}} \left(\frac{2\theta_0 \omega_0 c}{3\alpha b^2 z}\right)^{1/3} \mathrm{A_i}\left(|\alpha_1(\theta)| \left(\frac{z}{c}\right)^{2/3}\right) \cdot \exp\left[\frac{2\delta z}{3\alpha c}\left(\theta_0 + \frac{8\delta^2 b^2}{27\alpha\theta_0\omega_0^4} - \theta\right)\right], \tag{7.3.68}$$

as $z \to \infty$ for $\theta_0 \leqslant \theta \leqslant \theta_1$. Here, $|\alpha_1(\theta)|$ is given by (7.3.53b), and the approximate behavior of $\alpha_0(\theta)$, obtained from (7.3.53a) for values of θ close to θ_1, is employed in this approximate expression. Substitution of (7.3.63) into (7.3.57) yields

$$A_B(z, t) \sim \frac{\omega_0 \omega_c}{\omega_c^2 + \dfrac{4\delta^2}{9\alpha^2}} \left(\frac{2\theta_0 \omega_0 c}{3\alpha b^2 z}\right)^{1/3} \mathrm{A_i}\left(-|\alpha_1(\theta)| \left(\frac{z}{c}\right)^{2/3}\right) \cdot \exp\left[-\frac{2\delta z}{3\alpha c}\left(\theta - \theta_0 - \frac{8\delta^2 b^2}{27\alpha\theta_0\omega_0^4}\right)\right] \tag{7.3.69}$$

as $z \to \infty$ for $\theta_1 \leqslant \theta \leqslant \theta_2$. Here, $|\alpha_1(\theta)|$ is given by (7.3.54b), and the approximate behavior of $\alpha_0(\theta)$, obtained from (7.3.54a) for values of θ close to θ_1, is employed in this approximate expression.

For values of θ in the range $1 \leqslant \theta < \theta_1$ with θ bounded away from θ_1, substitution of (7.3.62a) into the nonuniform approximation (7.3.11) [or, equivalently, with the substitution of the large argument asymptotic behavior of the

Airy function and its first derivative in (7.3.65)] yields

$$A_B(z,t) \sim \frac{\omega_0^2\omega_c}{b\{\omega_c^2 + \frac{1}{9}[3|\psi(\theta)| - 2\delta\zeta(\theta)]^2\}} \cdot \left(\frac{\theta_0 c}{\pi z\{4\delta[1-\alpha\zeta(\theta)] + 6\alpha|\psi(\theta)|\}}\right)^{1/2} \cdot \exp\left[\frac{z}{3c}[2\delta\zeta(\theta) - 3|\psi(\theta)|]\left(\theta_0 - \theta + \frac{b^2}{18\theta_0\omega_0^4}[2\delta\zeta(\theta) - 3|\psi(\theta)|]\{2\delta[3-\alpha\zeta(\theta)] - 3\alpha|\psi(\theta)|\}\right)\right], \tag{7.3.70}$$

as $z \to \infty$. For values of θ approximately equal to θ_0, this nonuniform asymptotic approximation of the second precursor field simplifies somewhat to

$$A_B(z,t) \sim \frac{\omega_0^2\omega_c}{b\left\{\omega_c^2 + \left[\chi(\theta) - \frac{2\delta}{3\alpha}\right]^2\right\}}\left(\frac{\theta_0 c}{6\pi\alpha z\chi(\theta)}\right)^{1/2} \cdot \exp\left(\frac{z}{c}\left(\frac{2\delta}{3\alpha} - \chi(\theta)\right)\left\{\theta_0 - \theta + \frac{b^2}{2\theta_0\omega_0^4}\left[\frac{2\delta}{3\alpha} - \chi(\theta)\right] \cdot \left[\alpha\chi(\theta) + \frac{4}{3}\delta\right]\right\}\right), \tag{7.3.71}$$

as $z \to \infty$, where

$$\chi(\theta) = \left[\frac{4\delta^2}{9\alpha^2} + \frac{2\theta_0\omega_0^4}{3\alpha b^2}(\theta_0 - \theta)\right]^{1/2}.$$

The nonuniform asymptotic approximation (7.3.71) of the second precursor field for the unit step-function modulated signal is the result obtained by *Brillouin* for $1 \leqslant \theta < \theta_1$ [Ref. 7.5, p. 66]. It is seen to be accurate only for values of θ close to θ_0, becoming invalid as θ approaches θ_1. In the special case when $\theta = \theta_0$ [at which $\psi(\theta_0) = \chi(\theta_0) = 2\delta/3\alpha$], both of the above expressions (7.3.70, 71) reduce to

$$A_B(z,t_0) \sim \frac{\omega_0^2}{b\omega_c}\left(\frac{\theta_0 c}{4\pi\delta z}\right)^{1/2}, \tag{7.3.72}$$

and the second precursor field varies only as $z^{-1/2}$. The same result is obtained from the uniform asymptotic approximation (7.3.65) for sufficiently large values of z.

For values of θ in the range $\theta > \theta_1$ with θ bounded away from θ_1, substitution of (7.3.63) into the nonuniform approximation (7.3.23) [or, equivalently, with the substitution of the large argument asymptotic behavior of the Airy function and its first derivative in (7.3.66)] yields

$$A_B(z,t) \sim \frac{\omega_0^2 \omega_c}{b\{[\psi(\theta) - \omega_c]^2 + \frac{4}{9}\delta^2\zeta^2(\theta)\}\{[\psi(\theta) + \omega_c]^2 + \frac{4}{9}\delta^2\zeta^2(\theta)\}}$$
$$\cdot \left(\frac{2\theta_0 c}{3\pi\alpha\psi(\theta)z}\right)^{1/2} \exp\left[-\frac{\delta z}{c}\left(\frac{2}{3}(\theta - \theta_0)\zeta(\theta)\right.\right.$$
$$\left.\left. + \frac{b^2}{\theta_0\omega_0^4}\left\{[1 - \alpha\zeta(\theta)]\psi^2(\theta) + \frac{4}{9}\delta^2\zeta^2(\theta)\left[\frac{1}{3}\alpha\zeta(\theta) - 1\right]\right\}\right)\right]$$
$$\cdot \left\{\left[\omega_c^2 + \frac{4}{9}\delta^2\zeta^2(\theta) - \psi^2(\theta)\right]\cos\left[\frac{z}{c}\psi(\theta)\left(\theta_0 - \theta\right.\right.\right.$$
$$\left.\left. + \frac{b^2}{2\theta_0\omega_0^4}\left\{\frac{4}{3}\delta^2\zeta(\theta)[2 - \alpha\zeta(\theta)] + \alpha\psi^2(\theta)\right\}\right) + \frac{\pi}{4}\right]$$
$$+ \frac{4}{3}\delta\zeta(\theta)\psi(\theta)\sin\left[\frac{z}{c}\psi(\theta)\left(\theta_0 - \theta + \frac{b^2}{2\theta_0\omega_0^4}\left\{\frac{4}{3}\delta^2\zeta(\theta)[2 - \alpha\zeta(\theta)]\right.\right.\right.$$
$$\left.\left.\left. + \alpha\psi^2(\theta)\right\}\right) + \frac{\pi}{4}\right]\right\}, \tag{7.3.73}$$

as $z \to \infty$. This nonuniform expression reduces to *Brillouin*'s result [Ref. 7.5, p. 71] if $\psi(\theta)$ is replaced by $|\chi(\theta)|$ and $\zeta(\theta)$ is replaced by $1/\alpha \cong 1$. However, $\psi(\theta)$ can be replaced by $|\chi(\theta)|$ and $\zeta(\theta)$ can be replaced by $1/\alpha$ only for values of θ not too distant from θ_0. Hence, *Brillouin*'s expression for the second precursor field for $\theta > \theta_1$ is an approximation, valid for θ near θ_1, of an cxpression, (7.3.73), that is not valid for θ near θ_1. Consequently, Brillouin's expression for the asymptotic behavior of the second precursor field for the unit step-function modulated signal for $\theta > \theta_1$ is not applicable.

The dynamic evolution of the second precursor field $A_B(z,t)$ for the unit step-function modulated signal, as described by the uniform asymptotic representation given in (7.3.65, 69), is illustrated in Fig. 7.9 as a function of θ. This field was calculated at an observation distance of $z = 1.0 \times 10^{-3}$ cm for an applied signal frequency $\omega_c = 1.0 \times 10^{16}$/s, where Brillouin's choice of the medium parameters have again been employed. Consequently, the critical values of θ in this case are given by

$$\theta_0 = 1.50 \ ,$$

$$\theta_1 \cong 1.50275 \ ,$$

$$\theta_2 \cong 1.50549 \ .$$

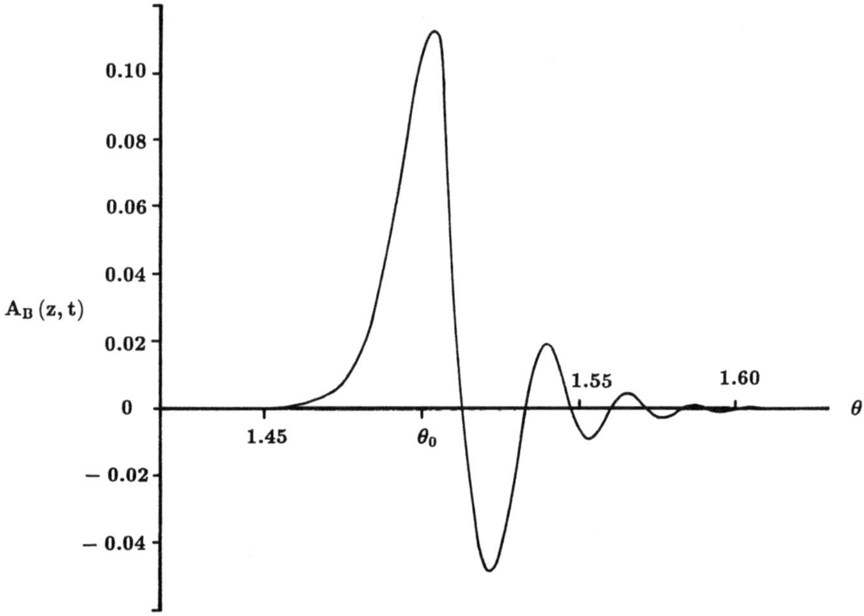

Fig. 7.9. Dynamic behavior of the *Brillouin* precursor field $A_B(z,t)$ for the unit step-function modulated signal with applied signal frequency $\omega_c = 1.0 \times 10^{16}$/s at an observation distance of $z = 1 \times 10^{-3}$ cm, where *Brillouin*'s choice of the medium parameters are employed.

As can be seen, the second precursor field builds up rapidly as θ approaches θ_0 from below and then decays with increasing $\theta > \theta_0$. The instantaneous angular frequency of oscillation of the second precursor field is also seen to increase monotonically with θ. The first precursor field behavor for the same case that is illustrated here is depicted in Fig. 7.4. Comparison of Figs. 7.4 and 7.9 shows that, for this case, the maximum amplitude obtained by the second precursor field is two orders of magnitude larger than that of the first precursor field.

7.3.5 The Rectangular-Pulse Modulated Signal

According to the results of Section 4.3.2, the propagated disturbance due to an initial unit rectangular pulse modulated signal of time duration T and frequency ω_c may be represented by the difference between two unit step-function modulated signals displaced in time by the amount T, viz.

$$A(z,t) = -\frac{1}{2\pi}\mathrm{Re}\left\{\int_{ia-\infty}^{ia+\infty} \frac{1}{\omega-\omega_c} e^{(z/c)\phi_0(\omega,\theta)}\,d\omega - e^{-i\omega_c T}\int_{ia-\infty}^{ia+\infty} \frac{1}{\omega-\omega_c} e^{(z/c)\phi_T(\omega,\theta_T)}\,d\omega\right\}$$

$$\equiv A_I(z,t) - A_{II}(z,t) \ ,$$

where the initial field begins to radiate time-harmonically at time $t = 0$, and where

$$\phi_T(\omega, \theta_T) = \mathrm{i}\omega[n(\omega) - \theta_T] \ ,$$

$$\theta_T = \frac{c}{z}(t - T) \ .$$

Since the first term $A_I(z,t)$ in the above expression for the propagated field is the same as the field treated in Sect. 7.3.4, the uniform asymptotic behavior of the second precursor field associated with that term is given by (7.3.65 and 66). Since the second term $A_{II}(z,t)$ in the above expression for the propagated field is of exactly the same form as the first, except that it is retarded in θ by the amount $-(c/z)T$ and is multiplied by the additional phase factor $\mathrm{e}^{-\mathrm{i}\omega_c T}$, the second precursor field associated with that term may then be easily constructed from the same equations. Hence, the uniform asymptotic approximation of the second precursor field of $A_{II}(z,t)$ for $1 \leqslant \theta_T \leqslant \theta_1$ is given by

$$\begin{aligned}
A_{IIB}(z,t) \sim \frac{\omega_0^2}{2b}\left(\frac{c}{z}\right)^{1/3} \mathrm{e}^{(z/c)\alpha_0(\theta_T)} \Bigg\{ \omega_c \cos(\omega_c T) \Bigg[\Bigg(& \frac{1}{\omega_c^2 + (|\psi(\theta_T)| - \frac{2}{3}\delta\zeta(\theta_T))^2} \\
\cdot \left(\frac{2\theta_0}{3\alpha|\psi(\theta_T)| + 2\delta(1 - \alpha\zeta(\theta_T))}\right)^{1/2} & + \frac{1}{\omega_c^2 + (|\psi(\theta_T)| + \frac{2}{3}\delta\zeta(\theta_T))^2} \\
\cdot \left(\frac{2\theta_0}{3\alpha|\psi(\theta_T)| - 2\delta(1 - \alpha\zeta(\theta_T))}\right)^{1/2} \Bigg) & |\alpha_1(\theta_T)|^{1/4} \\
\cdot \mathrm{A_i}\left(|\alpha_1(\theta_T)|\left(\frac{z}{c}\right)^{2/3}\right) - \Bigg(& \frac{1}{\omega_c^2 + (|\psi(\theta_T)| - \frac{2}{3}\delta\zeta(\theta_T))^2} \\
\cdot \left(\frac{2\theta_0}{3\alpha|\psi(\theta_T)| + 2\delta(1 - \alpha\zeta(\theta_T))}\right)^{1/2} & - \frac{1}{\omega_c^2 + (|\psi(\theta_T)| + \frac{2}{3}\delta\zeta(\theta_T))^2} \\
\cdot \left(\frac{2\theta_0}{3\alpha|\psi(\theta_T)| - 2\delta(1 - \alpha\zeta(\theta_T))}\right)^{1/2} \Bigg) & \frac{1}{|\alpha_1(\theta_T)|^{1/4}}\left(\frac{c}{z}\right)^{1/3} \\
\cdot \mathrm{A_i^{(1)}}\left(|\alpha_1(\theta_T)|\left(\frac{z}{c}\right)^{2/3}\right)\Bigg] + \sin(\omega_c T)\Bigg[\Bigg(& \frac{|\psi(\theta_T)| - \frac{2}{3}\delta\zeta(\theta_T)}{\omega_c^2 + (|\psi(\theta_T)| - \frac{2}{3}\delta\zeta(\theta_T)^2} \\
\cdot \left(\frac{2\theta_0}{3\alpha|\psi(\theta_T)| + 2\delta(1 - \alpha\zeta(\theta_T))}\right)^{1/2} & - \frac{|\psi(\theta_T)| + \frac{2}{3}\delta\zeta(\theta_T)}{\omega_c^2 + (|\psi(\theta_T)| + \frac{2}{3}\delta\zeta(\theta_T))^2} \\
\cdot \left(\frac{2\theta_0}{3\alpha|\psi(\theta_T)| - 2\delta(1 - \alpha\zeta(\theta_T))}\right)^{1/2} \Bigg) & |\alpha_1(\theta_T)|^{1/4} \\
\cdot \mathrm{A_i}\left(|\alpha_1(\theta_T)|\left(\frac{z}{c}\right)^{2/3}\right) - \Bigg(& \frac{|\psi(\theta_T)| - \frac{2}{3}\delta\zeta(\theta_T)}{\omega_c^2 + (|\psi(\theta_T)| - \frac{2}{3}\delta\zeta(\theta_T)^2}
\end{aligned}$$

$$\cdot\left(\frac{2\theta_0}{3\alpha|\psi(\theta_T)| + 2\delta(1-\alpha\zeta(\theta_T))}\right)^{1/2} + \frac{|\psi(\theta_T)| + \frac{2}{3}\delta\zeta(\theta_T)}{\omega_c^2 + (|\psi(\theta_T)| + \frac{2}{3}\delta\zeta(\theta_T))^2}$$

$$\cdot\left(\frac{2\theta_0}{3\alpha|\psi(\theta_T)| - 2\delta(1-\alpha\zeta(\theta_T))}\right)^{1/2}\Bigg)\frac{1}{|\alpha_1(\theta_T)|^{1/4}}\left(\frac{c}{z}\right)^{1/3}$$

$$\cdot \mathrm{A}_\mathrm{i}^{(1)}\left(|\alpha_1(\theta_T)|\left(\frac{z}{c}\right)^{2/3}\right)\Bigg]\Bigg\}, \tag{7.3.74}$$

as $z \to \infty$, where $\alpha_0(\theta_T)$ and $|\alpha_1(\theta_T)|$ are as given in (7.3.53a and 53b), respectively. Furthermore, the uniform asymptotic approximation of the second precursor field of $A_{II}(z,t)$ for $\theta_T \geqslant \theta_1$ is given by

$$A_{IIB}(z,t) \sim -\frac{\omega_0^2}{b}\left[\frac{\theta_0}{6\alpha\psi(\theta_T)}\right]^{1/2} e^{(z/c)\alpha_0(\theta_T)} \cdot \Bigg\{\cos(\omega_c T)\Bigg[\frac{2}{3}\delta\zeta(\theta_T)|\alpha_1(\theta_T)|^{1/4}$$

$$\cdot\left(\frac{1}{(\psi(\theta_T)+\omega_c)^2 + \frac{4}{9}\delta^2\zeta^2(\theta_T)} - \frac{1}{(\psi(\theta_T)-\omega_c)^2 + \frac{4}{9}\delta^2\zeta^2(\theta_T)}\right)$$

$$\cdot \mathrm{A}_\mathrm{i}\left(-|\alpha_1(\theta_T)|\left(\frac{z}{c}\right)^{2/3}\right) - \frac{1}{|\alpha_1(\theta_T)|^{1/4}}\left(\frac{\psi(\theta_T)-\omega_c}{(\psi(\theta_T)-\omega_c)^2 + \frac{4}{9}\delta^2\zeta^2(\theta_T)}\right.$$

$$\left. - \frac{\psi(\theta_T)+\omega_c}{(\psi(\theta_T)+\omega_c)^2 + \frac{4}{9}\delta^2\zeta^2(\theta_T)}\right)\left(\frac{c}{z}\right)^{1/3} \mathrm{A}_\mathrm{i}^{(1)}\left(-|\alpha_1(\theta_T)|\left(\frac{z}{c}\right)^{2/3}\right)\Bigg]$$

$$+ \sin(\omega_c T)\Bigg[|\alpha_1(\theta_T)|^{1/4}\left(\frac{\psi(\theta_T)-\omega_c}{(\psi(\theta_T)-\omega_c)^2 + \frac{4}{9}\delta^2\zeta^2(\theta_T)}\right.$$

$$\left. + \frac{\psi(\theta_T)+\omega_c}{(\psi(\theta_T)+\omega_c)^2 + \frac{4}{9}\delta^2\zeta^2(\theta_T)}\right)\mathrm{A}_\mathrm{i}\left(-|\alpha_1(\theta_T)|\left(\frac{z}{c}\right)^{2/3}\right)$$

$$- \frac{2}{3}\delta\zeta(\theta_T)\frac{1}{|\alpha_1(\theta_T)|^{1/4}}\left(\frac{1}{(\psi(\theta_T)-\omega_c)^2 + \frac{4}{9}\delta^2\zeta^2(\theta_T)}\right.$$

$$\left. + \frac{1}{(\psi(\theta_T)+\omega_c)^2 + \frac{4}{9}\delta^2\zeta^2(\theta_T)}\right)\left(\frac{c}{z}\right)^{1/3}$$

$$\cdot \mathrm{A}_\mathrm{i}^{(1)}\left(-|\alpha_1(\theta_T)|\left(\frac{z}{c}\right)^{2/3}\right)\Bigg]\Bigg\}, \tag{7.3.75}$$

as $z \to \infty$, where $\alpha_0(\theta_T)$ and $|\alpha_1(\theta_T)|$ are as given in (7.3.54a and 54b), respectively. Equations (7.3.74 and 75) taken together give an asymptotic approximation of the second precursor field of $A_{II}(z,t)$ that is valid uniformly for all $\theta_T \geqslant 1$. Finally, note that (7.3.74) reduces (7.3.65) and (7.3.75) reduces to (7.3.66) for $\omega_c T = 2n\pi$, $n = 0, 1, 2, 3, \ldots$.

For a numerical evaluation of $A_{IIB}(z,t)$, (7.3.74 and 75) are unstable and are not useful for small $|\theta - \theta_1|$. For that purpose they should be replaced by

$$A_{IIB}(z,t) \sim \frac{\omega_0}{\omega_c^2 + \dfrac{4\delta^2}{9\alpha^2}}\left(\omega_0\cos(\omega_c T) - \frac{2\delta}{3\alpha}\sin(\omega_c T)\right)\left[\frac{2\theta_0\omega_0 c}{3\alpha b^2 z}\right]^{1/3}$$
$$\cdot \mathrm{A_i}\left(|\alpha_1(\theta_T)|\left(\frac{z}{c}\right)^{2/3}\right)\exp\left\{\frac{2\delta z}{3\alpha c}\left[\theta_0 + \frac{8\delta^2 b^2}{27\alpha\theta_0\omega_0^4} - \theta_T\right]\right\}, \quad (7.3.76)$$

for $\theta_0 \leqslant \theta_T \leqslant \theta_1$, and by

$$A_{IIB}(z,t) \sim \frac{\omega_0}{\omega_c^2 + \dfrac{4\delta^2}{9\alpha^2}}\left(\omega_c\cos(\omega_c T) - \frac{2\delta}{3\alpha}\sin(\omega_c T)\right)\left[\frac{2\theta_0\omega_0 c}{3\alpha b^2 z}\right]^{1/3}$$
$$\cdot \mathrm{A_i}\left(-|\alpha_1(\theta_T)|\left(\frac{z}{c}\right)^{2/3}\right)\exp\left\{-\frac{2\delta z}{3\alpha c}\left[\theta_T - \theta_0 - \frac{8\delta^2 b^2}{27\alpha\theta_0\omega_0^4}\right]\right\}, \quad (7.3.77)$$

for $\theta_1 \leqslant \theta_T \leqslant \theta_T$. In (7.3.76), the quantity $|\alpha_1(\theta_T)|$ is given by (7.3.53b), whereas in (7.3.77), $|\alpha_1(\theta_T)|$ is given by (7.3.54b). Note that (7.3.76) reduces to (7.3.68) and (7.3.77) reduces to (7.3.69) for $\omega_c T = 2n\pi$, $n = 0, 1, 2, 3, \ldots$.

7.3.6 The Delta-Function Pulse

For an input delta-function pulse at time $t = 0$, the initial field envelope spectrum is given by

$$\tilde{u}(\omega - \omega_c) = -\mathrm{i}\ . \quad (7.3.78)$$

For values of θ in the range $1 \leqslant \theta \leqslant \theta_1$, the uniform asymptotic behavior of the second precursor field is given by (7.3.51). Substitution of (7.3.78) into that equation yields

$$A_B(z,t) \sim \frac{\omega_0^2}{2b}\left(\frac{c}{z}\right)^{1/3} e^{(z/c)\alpha_0(\theta)}\left\{\left[\left(\frac{2\theta_0}{3\alpha|\psi(\theta)| + 2\delta(1 - \alpha\zeta(\theta))}\right)^{1/2}\right.\right.$$
$$\left. + \left(\frac{2\theta_0}{3\alpha|\psi(\theta)| - 2\delta(1 - \alpha\zeta(\theta))}\right)^{1/2}\right]|\alpha_1(\theta)|^{1/4}\mathrm{A_i}\left(|\alpha_1(\theta)|\left(\frac{z}{c}\right)^{2/3}\right)$$
$$- \left[\left(\frac{2\theta_0}{3\alpha|\psi(\theta)| + 2\delta(1 - \alpha\zeta(\theta))}\right)^{1/2}\right.$$
$$\left. - \left(\frac{2\theta_0}{3\alpha|\psi(\theta)| - 2\delta(1 - \alpha\zeta(\theta))}\right)^{1/2}\right]\frac{1}{|\alpha_1(\theta)|^{1/4}}\left(\frac{c}{z}\right)^{1/3}$$
$$\left. \cdot \mathrm{A_i^{(1)}}\left(|\alpha_1(\theta)|\left(\frac{z}{c}\right)^{2/3}\right)\right\}, \quad (7.3.79)$$

as $z \to \infty$ uniformly for all θ in the range $1 \leqslant \theta \leqslant \theta_1$. The functions $\alpha_0(\theta)$ and $|\alpha_1(\theta)|$ appearing in this expression are given by equations (7.3.53a) and (7.3.53b), respectively. For values of θ in the range $\theta \geqslant \theta_1$, the uniform asymptotic behavior of the second precursor field is given by equation (7.3.52). Substitution of equation (7.3.78) into that equation yields

$$A_B(z,t) \sim \frac{\omega_0^2}{b}\left[\frac{2\theta_0}{3\alpha\psi(\theta)}\right]^{1/2}\left(\frac{c}{z}\right)^{1/3}|\alpha_1(\theta)|^{1/4}e^{(z/c)\alpha_0(\theta)}\mathrm{A_i}\left(-|\alpha_1(\theta)|\left(\frac{z}{c}\right)^{2/3}\right), \tag{7.3.80}$$

as $z \to \infty$ uniformly for all θ in the range $\theta \geqslant \theta_1$. The functions $\alpha_0(\theta)$ and $|\alpha_1(\theta)|$ appearing in this expression are given by equations (7.3.54a) and (7.3.54b), respectively. Equations (7.3.79 and 80) taken together give an asymptotic approximation of the second precursor field for an input delta-function pulse that is valid uniformly for all $\theta \geqslant 1$. At $\theta = \theta_1$, at which $\alpha_1(\theta_1)$ identically vanishes, both of these expressions reduce to

$$A_B(z,t_1) \sim \frac{1}{2\pi\sqrt{3}}\left[\frac{2\theta_0\omega_0^4 c}{\alpha b^2 z}\right]^{1/3}\Gamma\left(\frac{1}{3}\right)\exp\left\{\frac{2\delta z}{3\alpha c}\left[\theta_0 + \frac{8\delta^2 b^2}{27\alpha\theta_0\omega_0^4} - \theta_1\right]\right\}, \tag{7.3.81}$$

as $z \to \infty$ for $\theta = \theta_1 = ct_1/z$.

For a numerical evaluation of $A_B(z,t)$, (7.3.79 and 80) are unstable and are not useful for small values of $|\theta - \theta_1|$. For that purpose, (7.3.56 and 57) are employed. Substitution of (7.3.78) into (7.3.56) yields

$$A_B(z,t) \sim \left[\frac{2\theta_0\omega_0^4 c}{3\alpha b^2 z}\right]^{1/3}\mathrm{A_i}\left(|\alpha_1(\theta)|\left(\frac{z}{c}\right)^{2/3}\right)\exp\left\{\frac{2\delta z}{3\alpha c}\left[\theta_0 + \frac{8\delta^2 b^2}{27\alpha\theta_0\omega_0^4} - \theta\right]\right\}, \tag{7.3.82}$$

as $z \to \infty$ for $\theta_0 \leqslant \theta \leqslant \theta_1$, where $|\alpha_1(\theta)|$ is given by (7.3.53b). Substitution of (7.3.78) into (7.3.57) yields

$$A_B(z,t) \sim \left[\frac{2\theta_0\omega_0^4 c}{3\alpha b^2 z}\right]^{1/3}\mathrm{A_i}\left(-|\alpha_1(\theta)|\left(\frac{z}{c}\right)^{2/3}\right)\exp\left\{-\frac{2\delta z}{3\alpha c}\left[\theta - \theta_0 - \frac{8\delta^2 b^2}{27\alpha\theta_0\omega_0^4}\right]\right\}, \tag{7.3.83}$$

as $z \to \infty$ for $\theta_1 \leqslant \theta \leqslant \theta_2$, where $|\alpha_1(\theta)|$ is given by (7.3.54b).

For values of θ in the range $1 \leqslant \theta < \theta_1$ with θ bounded away from θ_1, substitution of (7.3.78) into the nonuniform approximation (7.3.11) (or, equivalently, substitution of the large argument asymptotic behavior of the Airy function and its first derivative in (7.3.79)) yields

$$A_B(z,t) \sim \frac{\omega_0^2}{b}\left[\frac{\theta_0 c}{2\pi z[2\delta(1-\delta\zeta(\theta)) + 3\alpha|\psi(\theta)|]}\right]^{1/2} \cdot \exp\left\{\frac{z}{3c}(2\delta\zeta(\theta) - 3|\psi(\theta)|)\right)\left[\theta_0 - \theta + \frac{b^2}{18\theta_0\omega_0^4}(2\delta\zeta(\theta) - 3|\psi(\theta)|)(2\delta(3-\alpha\zeta(\theta)) + 3\alpha|\psi(\theta)|)\right]\right\}, \tag{7.3.84}$$

as $z \to \infty$. For values of θ in the range $\theta > \theta_1$ with θ bounded away from θ_1, substitution of (7.3.78) into the nonuniform approximation (7.3.23) (or, equivalently, substitution of the large argument asymptotic behavior of the Airy function and its first derivative in (7.3.80)) yields

$$A_B(z,t) \sim \frac{\omega_0^2}{b}\left[\frac{2\theta_0 c}{3\pi\alpha\psi(\theta)z}\right]^{1/2} \cdot \exp\left\{-\delta\frac{z}{c}\left[\frac{2}{3}\zeta(\theta)(\theta-\theta_0) + \frac{b^2}{\theta_0\omega_0^4}\left((1-\alpha\zeta(\theta))\psi^2(\theta) + \frac{4}{9}\delta^2\zeta^2(\theta)\left(\frac{1}{3}\alpha\zeta(\theta) - 1\right)\right)\right]\right\} \cdot \cos\left\{\frac{z}{c}\psi(\theta)\left[\theta_0 - \theta + \frac{b^2}{2\theta_0\omega_0^4}\left(\frac{4}{3}\delta^2\zeta(\theta)(2-\alpha\zeta(\theta)) + \alpha\psi^2(\theta)\right)\right] + \frac{\pi}{4}\right\}, \tag{7.3.85}$$

as $z \to \infty$.

7.3.7 The Hyperbolic-Tangent Modulated Signal

For the hyperbolic-tangent modulated signal, the spectrum of the initial field envelope is given by (4.3.46), so that

$$\tilde{u}(\omega_{sp_1} - \omega_c) = \frac{\dfrac{\pi i}{2\beta_T}}{\sinh\left(\dfrac{\pi}{2\beta_T}(\omega_{sp_1} - \omega_c)\right)} \cong -\frac{i\pi}{2\beta_T} \cdot \frac{\sinh\left(\frac{\pi}{2\beta_T}\omega_c\right)\cos\left(\frac{\pi}{2\beta_T}\left(|\psi(\theta)| - \frac{2}{3}\delta\zeta(\theta)\right)\right) + i\cosh\left(\frac{\pi}{2\beta_T}\omega_c\right)\sin\left(\frac{\pi}{2\beta_T}\left(|\psi(\theta)| - \frac{2}{3}\delta\zeta(\theta)\right)\right)}{\sinh^2\left(\frac{\pi}{2\beta_T}\omega_c\right) + \sin^2\left(\frac{\pi}{2\beta_T}\left(|\psi(\theta)| - \frac{2}{3}\delta\zeta(\theta)\right)\right)}, \tag{7.3.86a}$$

$$\tilde{u}(\omega_{sp_2} - \omega_c) = \frac{\dfrac{\pi i}{2\beta_T}}{\sinh\left(\dfrac{\pi}{2\beta_T}(\omega_{sp_2} - \omega_c)\right)}$$

$$\cong \frac{i\pi}{2\beta_T} \cdot \frac{-\sinh\left(\dfrac{\pi}{2\beta_T}\omega_c\right)\cos\left(\dfrac{\pi}{2\beta_T}\left(|\psi(\theta)| + \dfrac{2}{3}\delta\zeta(\theta)\right)\right) + i\cosh\left(\dfrac{\pi}{2\beta_T}\omega_c\right)\sin\left(\dfrac{\pi}{2\beta_T}\left(|\psi(\theta)| + \dfrac{2}{3}\delta\zeta(\theta)\right)\right)}{\sinh^2\left(\dfrac{\pi}{2\beta_T}\omega_c\right) + \sin^2\left(\dfrac{\pi}{2\beta_T}\left(|\psi(\theta)| + \dfrac{2}{3}\delta\zeta(\theta)\right)\right)}, \tag{7.3.86b}$$

for $1 \leqslant \theta \leqslant \theta_1$, and

$$\tilde{u}(\omega_{sp_N^\pm} - \omega_c) = \frac{\dfrac{\pi i}{2\beta_T}}{\sinh\left(\dfrac{\pi}{2\beta_T}(\omega_{sp_N^\pm} - \omega_c)\right)}$$

$$\cong \frac{i\pi}{2\beta_T} \cdot \frac{\sinh\left(\dfrac{\pi}{2\beta_T}(\pm\psi(\theta) - \omega_c)\right)\cos\left(\dfrac{\pi}{3\beta_T}\delta\zeta(\theta)\right) + i\cosh\left(\dfrac{\pi}{2\beta_T}(\pm\psi(\theta) - \omega_c)\right)\sin\left(\dfrac{\pi}{3\beta_T}\delta\zeta(\theta)\right)}{\sinh^2\left(\dfrac{\pi}{2\beta_T}(\pm\psi(\theta) - \omega_c)\right) + \sin^2\left(\dfrac{\pi}{2\beta_T}\delta\zeta(\theta)\right)}, \tag{7.3.87}$$

for $\theta \geqslant \theta_1$. At $\theta = \theta_1$, these equations reduce to the result

$$\tilde{u}(\omega_{sp_N} - \omega_c) \cong -\frac{i\pi}{2\beta_T} \cdot \frac{\sinh\left(\dfrac{\pi}{2\beta_T}\omega_c\right)\cos\left(\dfrac{\pi}{3\beta_T}\delta\zeta(\theta_1)\right) - i\cosh\left(\dfrac{\pi}{2\beta_T}\omega_c\right)\sin\left(\dfrac{\pi}{3\beta_T}\delta\zeta(\theta_1)\right)}{\sinh^2\left(\dfrac{\pi}{2\beta_T}\omega_c\right) + \sin^2\left(\dfrac{\pi}{3\beta_T}\delta\zeta(\theta_1)\right)}. \tag{7.3.88}$$

For values of θ in the range $1 \leqslant \theta \leqslant \theta_1$, the uniform asymptotic behavior of the second precursor field is given by (7.3.51). Substitution of (7.3.86) into that equation yields

$$
\begin{aligned}
A_B(z,t) \sim{}& \frac{\pi\omega_0^2}{4\beta_T b}\left(\frac{c}{z}\right)^{1/3} e^{(z/c)\alpha_0(\theta)} \sinh\left(\frac{\pi}{2\beta_T}\omega_c\right) \\
&\cdot\Bigg\{\Bigg[\frac{\cos\left(\frac{\pi}{2\beta_T}\left(|\psi(\theta)| - \frac{2}{3}\delta\zeta(\theta)\right)\right)}{\sinh^2\left(\frac{\pi}{2\beta_T}\omega_c\right) + \sin^2\left(\frac{\pi}{2\beta_T}\left(|\psi(\theta)| - \frac{2}{3}\delta\zeta(\theta)\right)\right)} \\
&\cdot\left(\frac{2\theta_0}{3\alpha|\psi(\theta)| + 2\delta(1-\alpha\zeta(\theta))}\right)^{1/2} \\
&+ \frac{\cos\left(\frac{\pi}{2\beta_T}\left(|\psi(\theta)| + \frac{2}{3}\delta\zeta(\theta)\right)\right)}{\sinh^2\left(\frac{\pi}{2\beta_T}\omega_c\right) + \sin^2\left(\frac{\pi}{2\beta_T}\left(|\psi(\theta)| + \frac{2}{3}\delta\zeta(\theta)\right)\right)} \\
&\cdot\left(\frac{2\theta_0}{3\alpha|\psi(\theta)| - 2\delta(1-\alpha\zeta(\theta))}\right)^{1/2}\Bigg] |\alpha_1(\theta)|^{1/4} \mathrm{A_i}\left(|\alpha_1(\theta)|\left(\frac{z}{c}\right)^{2/3}\right) \\
&- \Bigg[\frac{\cos\left(\frac{\pi}{2\beta_T}\left(|\psi(\theta)| - \frac{2}{3}\delta\zeta(\theta)\right)\right)}{\sinh^2\left(\frac{\pi}{2\beta_T}\omega_c\right) + \sin^2\left(\frac{\pi}{2\beta_T}\left(|\psi(\theta)| - \frac{2}{3}\delta\zeta(\theta)\right)\right)} \\
&\cdot\left(\frac{2\theta_0}{3\alpha|\psi(\theta)| + 2\delta(1-\alpha\zeta(\theta))}\right)^{1/2} \\
&- \frac{\cos\left(\frac{\pi}{2\beta_T}\left(|\psi(\theta)| + \frac{2}{3}\delta\zeta(\theta)\right)\right)}{\sinh^2\left(\frac{\pi}{2\beta_T}\omega_c\right) + \sin^2\left(\frac{\pi}{2\beta_T}\left(|\psi(\theta)| + \frac{2}{3}\delta\zeta(\theta)\right)\right)} \\
&\cdot\left(\frac{2\theta_0}{3\alpha|\psi(\theta)| - 2\delta(1-\alpha\zeta(\theta))}\right)^{1/2}\Bigg] \frac{1}{|\alpha_1(\theta)|^{1/4}}\left(\frac{c}{z}\right)^{1/3} \\
&\cdot \mathrm{A_i^{(1)}}\left(|\alpha_1(\theta)|\left(\frac{z}{c}\right)^{2/3}\right)\Bigg\}, \qquad (7.3.89)
\end{aligned}
$$

as $z \to \infty$ uniformly for all θ in the range $1 \leqslant \theta \leqslant \theta_1$. The functions $\alpha_0(\theta)$ and $|\alpha_1(\theta)|$ appearing in this expression are given by (7.3.53a and 53b), respectively. For values of θ in the range $\theta \geqslant \theta_1$, the uniform asymptotic behavior of the second precursor field is given by (7.3.52). Substitution of (7.3.87) into that equation yields

$$A_B(z,t) \sim \frac{\pi}{4\beta_T}\left[\frac{2\theta_0\omega_0^4}{3\alpha b^2\psi(\theta)}\right]^{1/2}\left(\frac{c}{z}\right)^{1/3} e^{(z/c)\alpha_0(\theta)} \cdot \left\{ |\alpha_1(\theta)|^{1/4}\cos\left(\frac{\pi}{3\beta_T}\delta\zeta(\theta)\right) \right.$$
$$\cdot\left[\frac{\sinh\left(\frac{\pi}{2\beta_T}(\psi(\theta)+\omega_c)\right)}{\sinh^2\left(\frac{\pi}{2\beta_T}(\psi(\theta)+\omega_c)\right)+\sin^2\left(\frac{\pi}{3\beta_T}\delta\zeta(\theta)\right)} - \frac{\sinh\left(\frac{\pi}{2\beta_T}(\psi(\theta)-\omega_c)\right)}{\sinh^2\left(\frac{\pi}{2\beta_T}(\psi(\theta)-\omega_c)\right)+\sin^2\left(\frac{\pi}{3\beta_T}\delta\zeta(\theta)\right)}\right]$$
$$\cdot \mathrm{A_i}\left(-|\alpha_1(\theta)|\left(\frac{z}{c}\right)^{2/3}\right) - \frac{1}{|\alpha_1(\theta)|^{1/4}}\sin\left(\frac{\pi}{3\beta_T}\delta\zeta(\theta)\right)$$
$$\cdot\left[\frac{\cosh\left(\frac{\pi}{2\beta_T}(\psi(\theta)-\omega_c)\right)}{\sinh^2\left(\frac{\pi}{2\beta_T}(\psi(\theta)-\omega_c)\right)+\sin^2\left(\frac{\pi}{3\beta_T}\delta\zeta(\theta)\right)} - \frac{\cosh\left(\frac{\pi}{2\beta_T}(\psi(\theta)+\omega_c)\right)}{\sinh^2\left(\frac{\pi}{2\beta_T}(\psi(\theta)+\omega_c)\right)+\sin^2\left(\frac{\pi}{3\beta_T}\delta\zeta(\theta)\right)}\right]\left(\frac{c}{z}\right)^{1/3}$$
$$\left.\cdot \mathrm{A_i^{(1)}}\left(-|\alpha_1(\theta)|\left(\frac{z}{c}\right)^{2/3}\right)\right\}, \tag{7.3.90}$$

as $z \to \infty$ uniformly for all θ in the range $\theta \geqslant \theta_1$. The functions $\alpha_0(\theta)$ and $|\alpha_1(\theta)|$ appearing in this expression are given by (7.3.54a and 54b), respectively. Equations (7.3.89 and 90) taken together give an asymptotic approximation of the second precursor field for the hyperbolic-tangent modulated signal that is valid uniformly for all $\theta \geqslant 1$. In the limit as $\beta_T \to \infty$ (i.e. in the limit of an instantaneous turn-on time for the initial field envelope), (7.3.89 and 90) reduce to the uniform expansion of the second precursor field for the unit step-function modulated signal given by (7.3.65 and 66), respectively. At $\theta = \theta_1$, both (7.3.89 and 90) reduce to

$$A_B(z,t_1) \sim \frac{\Gamma\left(\frac{1}{3}\right)}{4\sqrt{3}\beta_T}\left[\frac{2\theta_0\omega_0^4 c}{\alpha b^2 z}\right]^{1/3} \cdot \frac{\sinh\left(\frac{\pi}{2\beta_T}\omega_c\right)\cos\left(\frac{\pi\delta}{3\beta_T\alpha}\right)}{\sinh^2\left(\frac{\pi}{2\beta_T}\omega_c\right)+\sin^2\left(\frac{\pi\delta}{3\beta_T\alpha}\right)} \cdot \exp\left\{\frac{2\delta z}{3\alpha c}\left[\theta_0 + \frac{8\delta^2 b^2}{27\alpha\theta_0\omega_0^4} - \theta_1\right]\right\}, \tag{7.3.91}$$

as $z \to \infty$ for $\theta = \theta_1 = ct_1/z$.

For a numerical evaluation of $A_B(z,t)$, (7.3.89 and 90) are unstable and are not useful for small values of $|\theta - \theta_1|$. For that purpose, (7.3.56 and 57) are employed. Substitution of (7.3.86) into (7.3.56) yields

$$A_B(z,t) \sim \frac{\pi}{2\beta_T}\left[\frac{2\theta_0\omega_0^4 c}{3\alpha b^2 z}\right]^{1/3} \cdot \frac{\sinh\left(\frac{\pi}{2\beta_T}\omega_c\right)\cos\left(\frac{\pi\delta}{3\beta_T\alpha}\right)}{\sinh^2\left(\frac{\pi}{2\beta_T}\omega_c\right)+\sin^2\left(\frac{\pi\delta}{3\beta_T\alpha}\right)} \cdot \mathrm{A_i}\left(|\alpha_1(\theta)|\left(\frac{z}{c}\right)^{2/3}\right)\exp\left\{\frac{2\delta z}{3\alpha c}\left[\theta_0 + \frac{8\delta^2 b^2}{27\alpha\theta_0\omega_0^4} - \theta\right]\right\}, \tag{7.3.92}$$

as $z \to \infty$ for $\theta_0 \leqslant \theta \leqslant \theta_1$, where $|\alpha_1(\theta)|$ is given by (7.3.53b). Substitution of (7.3.87) into (7.3.57) yields

$$A_B(z,t) \sim \frac{\pi}{2\beta_T}\left[\frac{2\theta_0\omega_0^4 c}{3\alpha b^2 z}\right]^{1/3} \cdot \frac{\sinh\left(\frac{\pi}{2\beta_T}\omega_c\right)\cos\left(\frac{\pi\delta}{3\beta_T\alpha}\right)}{\sinh^2\left(\frac{\pi}{2\beta_T}\omega_c\right)+\sin^2\left(\frac{\pi\delta}{3\beta_T\alpha}\right)} \cdot \mathrm{A_i}\left(-|\alpha_1(\theta)|\left(\frac{z}{c}\right)^{2/3}\right)\exp\left\{-\frac{2\delta z}{3\alpha c}\left[\theta - \theta_0 - \frac{8\delta^2 b^2}{27\alpha\theta_0\omega_0^4}\right]\right\}, \tag{7.3.93}$$

as $z \to \infty$ for $\theta_1 \leqslant \theta \leqslant \theta_2$, where $|\alpha_1(\theta)|$ is given by (7.3.54b).

For values of θ in the range $1 \leqslant \theta < \theta_1$ with θ bounded away from θ_1, substitution of (7.3.86a) into the nonuniform approximation (7.3.11) (or, equivalently, substitution of the large argument asymptotic behavior of the Airy function and its first derivative in (7.3.89)) yields

$$A_B(z,t) \sim \frac{\omega_0^2}{4\beta_T b}\left[\frac{2\pi\theta_0 c}{z[2\delta(1-\alpha\zeta(\theta)) + 3\alpha|\psi(\theta)|]}\right]^{1/2}$$

$$\cdot \frac{\sinh\left(\frac{\pi}{2\beta_T}\omega_c\right)\cos\left(\frac{\pi}{2\beta_T}\left(|\psi(\theta)| - \frac{2}{3}\delta\zeta(\theta)\right)\right)}{\sinh^2\left(\frac{\pi}{2\beta_T}\omega_c\right) + \sin^2\left(\frac{\pi}{2\beta_T}\left(|\psi(\theta)| - \frac{2}{3}\delta\zeta(\theta)\right)\right)}$$

$$\cdot \exp\left\{\frac{z}{3c}(2\delta\zeta(\theta) - 3|\psi(\theta)|)\left[\theta_0 - \theta + \frac{b^2}{18\theta_0\omega_0^4}\right.\right.$$

$$\left.\left.\cdot(2\delta\zeta(\theta) - 3|\psi(\theta)|)(2\delta(3 - \alpha\zeta(\theta)) + 3\alpha|\psi(\theta)|)\right]\right\} \qquad (7.3.94)$$

as $z \to \infty$. For values of θ in the range $\theta > \theta_1$ with θ bounded away from θ_1, substitution of (7.3.87) into the nonuniform approximation (7.3.23) (or, equivalently, substitution of the large argument asymptotic behavior of the Airy function and its first derivative in equation (7.3.90)) yields

$$A_B(z,t) \sim -\frac{1}{4\beta_T}\left[\frac{2\pi\theta_0\omega_0^4 c}{3\alpha b^2\psi(\theta)z}\right]^{1/2}\exp\left\{-\delta\frac{z}{c}\left[\frac{2}{3}\zeta(\theta)(\theta - \theta_0)\right.\right.$$

$$\left.\left.+\frac{b^2}{\theta_0\omega_0^4}\left((1 - \alpha\zeta(\theta))\psi^2(\theta) + \frac{4}{9}\delta^2\zeta^2(\theta)\left(\frac{1}{3}\alpha\zeta(\theta) - 1\right)\right)\right]\right\}$$

$$\cdot\left\{\cos\left(\frac{\pi}{3\beta_T}\delta\zeta(\theta)\right)\left[\frac{\sinh\left(\frac{\pi}{2\beta_T}(\psi(\theta) + \omega_c)\right)}{\sinh^2\left(\frac{\pi}{2\beta_T}(\psi(\theta) + \omega_c)\right) + \sin^2\left(\frac{\pi}{3\beta_T}\delta\zeta(\theta)\right)}\right.\right.$$

$$\left. - \frac{\sinh\left(\frac{\pi}{2\beta_T}(\psi(\theta) - \omega_c)\right)}{\sinh^2\left(\frac{\pi}{2\beta_T}(\psi(\theta) - \omega_c)\right) + \sin^2\left(\frac{\pi}{3\beta_T}\delta\zeta(\theta)\right)}\right]$$

$$\cdot\cos\left[\frac{z}{c}\psi(\theta)\left(\theta_0 - \theta + \frac{b^2}{2\theta_0\omega_0^4}\left(\frac{4}{3}\delta^2\zeta(\theta)(2 - \alpha\zeta(\theta))\right.\right.\right.$$

$$\left.\left.\left.+\alpha\psi^2(\theta)\right)\right) + \frac{\pi}{4}\right]$$

$$+\sin\left(\frac{\pi}{3\beta_T}\delta\zeta(\theta)\right)\left[\frac{\cosh\left(\frac{\pi}{2\beta_T}(\psi(\theta) - \omega_c)\right)}{\sinh^2\left(\frac{\pi}{2\beta_T}(\psi(\theta) - \omega_c)\right) + \sin^2\left(\frac{\pi}{3\beta_T}\delta\zeta(\theta)\right)}\right.$$

$$- \frac{\cosh\left(\frac{\pi}{2\beta_T}(\psi(\theta) + \omega_c)\right)}{\sinh^2\left(\frac{\pi}{2\beta_T}(\psi(\theta) + \omega_c)\right) + \sin^2\left(\frac{\pi}{3\beta_T}\delta\zeta(\theta)\right)}\Bigg]$$

$$\cdot \sin\left[\frac{z}{c}\psi(\theta)\left(\theta_0 - \theta + \frac{b^2}{2\theta_0\omega_0^4}\left(\frac{4}{3}\delta^2\zeta(\theta)(2 - \alpha\zeta(\theta))\right.\right.\right.$$

$$\left.\left.\left. + \alpha\psi^2(\theta)\right)\right) + \frac{\pi}{4}\right]\Bigg\},$$

as $z \to \infty$.

8 Evolution of the Main Signal

This chapter obtains the contribution $A_c(z,t)$ to the asymptotic behavior of the propagated field $A(z,t)$ that is due to the presence of any simple pole singularities of the spectral function $\tilde{u}(\omega - \omega_c)$. The field $A_c(z,t)$ is associated with any long term signal being propagated. The velocity of propagation of the signal and the transition from the precursor field to the signal field are determined by the relative importance of the component fields $A_s(z,t)$, $A_B(z,t)$, and $A_c(z,t)$. Hence, discussion of those topics is deferred to Chap. 9 which deals with the evolution of the total field $A(z,t)$ by combining the results of this chapter with those of Chap. 7.

The first section of this chapter presents the nonuniform asymptotic analysis based on the direct application of Olver's method and the Cauchy Residue Theorem. Even though the nonuniform result exhibits a discontinuous change in behavior as the parameter $\theta = ct/z$ varies, it is a useful expression for $A_c(z,t)$ in the final expression for the total field $A(z,t)$ for all θ provided that the dominant saddle point remains isolated from the pole, since in that case, $A_c(z,t)$ is asymptotically negligible in comparison to the precursor field at the time when the discontinuity occurs. When applied to the unit step-function modulated signal, the nonuniform expression for $A_c(z,t)$ is the same as the one obtained by *Brillouin* [8.1,2] except for the value of the time at which the discontinuous jump occurs. For those cases in which the nonuniform result is useful, the difference in values of the time at which the discontinuity occurs is of no consequence in the final expression for the total field $A(z,t)$ because $A_c(z,t)$ is negligible during a time period that includes both values. Although Brillouin attaches physical significance to the value of $\theta = \theta_s$ at which the discontinuous jump occurs, calling $t_s = (z/c)\theta_s$ the time of arrival of the signal, it is shown in Chap. 9 that the time of the jump has no physical significance and that the signal arrives at a later time.

The nonuniform expression for $A_c(z,t)$ is not useful in cases in which the dominant saddle point passes too near the pole singularity because $A_c(z,t)$ is not then negligible at the time of the discontinuous jump. For such cases, it is necessary to apply the technique presented in Sect. 5.3 to obtain a uniform asymptotic expression for $A_c(z,t)$. That analysis is given in Sect. 8.2. During times when the saddle point is near the pole, the results are very different from those of *Brillouin*, but they reduce to *Brillouin*'s results in the case of the unit step-function modulated signal for all other times.

The explicit forms of the pole contributions for the special pulses considered in this thesis are treated in the final section of the chapter. In particular, for the hyperbolic-tangent modulated signal, the number of poles which must be treated increases as the parameter $\beta_T > 0$ decreases[1]. The results are obtained for all finite applied signal frequencies ω_c.

8.1 The Nonuniform Asymptotic Approximation

This section obtains the nonuniform asymptotic approximation of the field $A_c(z,t)$ due to the contribution of any pole singularities appearing in the integrand of the propagated scalar wave field (4.3.31). For that purpose, let ω_p denote a simple pole singularity of the spectral amplitude function $\tilde{u}(\omega - \omega_c)$ with residue γ_p given by

$$\gamma_p = \lim_{\omega \to \omega_p} [(\omega - \omega_p)\tilde{u}(\omega - \omega_c)] \ . \tag{8.1.1}$$

In accordance with the results of Sect. 6.5, the pole contribution is given by (6.5.9). Furthermore, according to the results of Sect. 5.3, all of the functions $C_{\bar{D}}^{\pm}(z,t)$, $C_1(z,t)$, and $C_{\bar{N}}^{\pm}(z,t)$ appearing in (6.5.9) are zero in the nonuniform asymptotic approximation, so that

$$A_c(z,t) = -\mathbb{R}\mathrm{e}[2\pi i \Lambda(\theta)] \ , \tag{8.1.2}$$

where

$$\begin{aligned} \Lambda(\theta) &= \sum_p \operatorname*{Res}_{\omega=\omega_p} \left\{ \frac{1}{2\pi} \mathrm{i}\tilde{u}(\omega - \omega_c) \mathrm{e}^{(z/c)\phi(\omega,\theta)} \right\} \\ &= \sum_p \frac{1}{2\pi} \mathrm{i}\gamma_p \mathrm{e}^{(z/c)\phi(\omega_p,\theta)} \end{aligned} \tag{8.1.3}$$

is the sum of the residues of the poles that are crossed when the original contour of integration is deformed to the path $P(\theta)$ that is specified in Sect. 6.5 (see Fig. 6.21). For reasons of simplicity, it is assumed here that the deformed contour of integration $P(\theta)$ is near only one pole at a time, and attention is restricted to obtaining the nonuniform asymptotic contribution due to that single pole alone. The results obtained are easily generalized to account for several individual pole contributions.

The contribution of the simple pole singularity at $\omega = \omega_p$ occurs when the original contour of integration, extending along the straight line from $ia - \infty$ to $ia + \infty$ in the upper-half of the complex ω-plane, is deformed across the pole to

[1] Recall that the parameter β_T is indicative of the turn-on time of the initial signal envelope and that the larger the value of β_T, the more rapid is that initial turn-on time.

$P(\theta)$. For concreteness, assume that there is only one pole and let $P(\theta)$ and the original contour of integration lie on the same side of the pole for $\theta < \theta_s$ and on opposite sides for $\theta > \theta_s$. Then, according to the definition of $\Lambda(\theta)$,

$$\Lambda(\theta) = 0 \ ; \qquad \text{for } \theta < \theta_s \ , \tag{8.1.4a}$$

$$\Lambda(\theta_s) = \frac{1}{2}\cdot\frac{1}{2\pi}\mathrm{i}\gamma_p \mathrm{e}^{(z/c)\phi(\omega_p,\theta_s)} \ ; \qquad \text{for } \theta = \theta_s \ , \tag{8.1.4b}$$

$$\Lambda(\theta) = \frac{1}{2\pi}\mathrm{i}\gamma_p \mathrm{e}^{(z/c)\phi(\omega_p,\theta)} \ ; \qquad \text{for } \theta > \theta_s \ . \tag{8.1.4c}$$

Upon substitution of (8.1.4) into (8.1.2), the following nonuniform expression for the pole contribution at $\omega = \omega_p$ is obtained:

$$A_c(z,t) = 0 \ ; \qquad \theta < \theta_s \ , \tag{8.1.5a}$$

$$A_c(z,t_s) = \frac{1}{2}\mathrm{e}^{(z/c)X(\omega_p,\theta_s)}\left[\gamma_p' \cos\left(\frac{z}{c}Y(\omega_p,\theta_s)\right) - \gamma_p'' \sin\left(\frac{z}{c}Y(\omega_p,\theta_s)\right)\right] ; \qquad \theta = \theta_s \ , \tag{8.1.5b}$$

$$A_c(z,t) = \mathrm{e}^{(z/c)X(\omega_p,\theta)}\left[\gamma_p' \cos\left(\frac{z}{c}Y(\omega_p,\theta)\right) - \gamma_p'' \sin\left(\frac{z}{c}Y(\omega_p,\theta)\right)\right] ; \qquad \theta > \theta_s \ , \tag{8.1.5c}$$

where $\gamma_p' \equiv \mathrm{Re}(\gamma_p)$ and $\gamma_p'' \equiv \mathrm{Im}(\gamma_p)$.

A special case of interest is that for which the simple pole singularity of the spectral amplitude function $\tilde{u}(\omega - \omega_c)$ is real and positive. In that case, according to (6.1.9 and 10), there results

$$X(\omega_p) = -\omega_p n_i(\omega_p) \ , \tag{8.1.6a}$$

$$Y(\omega_p,\theta) = \omega_p[n_r(\omega_p) - \theta] \ . \tag{8.1.6b}$$

Consequently, the *amplitude attenuation coefficient* at the real frequency ω_p appearing in (8.1.5) is given by

$$\alpha(\omega_p) \equiv -\frac{1}{c}X(\omega_p) = \frac{\omega_p}{c}n_i(\omega_p) \ . \tag{8.1.7}$$

Furthermore, the oscillatory phase term appearing in the nonuniform expansion (8.1.5) may, in this case, be written

$$\begin{aligned}\frac{z}{c}Y(\omega_p,\theta) &= \frac{z}{c}\omega_p\left[n_r(\omega_p) - \frac{ct}{z}\right] \\ &= \beta(\omega_p)z - \omega_p t \ ,\end{aligned} \tag{8.1.8}$$

where the *propagation factor* $\beta(\omega_p)$ at the real frequency ω_p is given by

$$\beta(\omega_p) = \frac{\omega_p}{c} n_r(\omega_p) \ . \tag{8.1.9}$$

With these results, the pole contribution (8.1.5) may be written

$$A_c(z,t) = 0 \ ; \qquad \theta < \theta_s \ , \tag{8.1.10a}$$

$$A_c(z,t_s) = \tfrac{1}{2} e^{-z\alpha(\omega_p)}[\gamma' \cos(\beta(\omega_p)z - \omega_p t) - \gamma'' \sin(\beta(\omega_p)z - \omega_p t)] \ ; \qquad \theta = \theta_s \ , \tag{8.1.10b}$$

$$A_c(z,t) = e^{-z\alpha(\omega_p)}[\gamma' \cos(\beta(\omega_p)z - \omega_p t) - \gamma'' \sin(\beta(\omega_p)z - \omega_p t)] \ ; \qquad \theta > \theta_s \ , \tag{8.1.10c}$$

for ω_p real and positive. As can be seen, in this special case the pole contribution has an angular frequency of ω_p and is attenuated with propagation distance z with a constant attenuation coefficient $\alpha(\omega_c)$ that is independent of the time t (compare with the general expression in (8.1.5)).

The integral representation (3.3.31) of the propagated field $A(z,t)$ is a continuous function of θ, and, in particular, is continuous at $\theta = \theta_s$. However, the resulting asymptotic approximation of $A(z,t)$ is a discontinuous function of θ at $\theta = \theta_s$[2] when the pole contribution is given by the nonuniform expression (8.1.5). The discontinuity is of no consequence for fixed values of z larger than some positive constant Z, however, because the contribution to the field from the dominant saddle point at ω_{sp} varies exponentially as $e^{(z/c)X(\omega_{sp},\theta)}$ which dominates the exponential behavior of the pole contribution given by (8.1.5b) at $\theta = \theta_s$[3]. Hence, at $\theta = \theta_s$ when the discontinuity in the asymptotic behavior of the propagated field occurs, the pole contribution is negligible in comparison to the saddle point contribution, and, as a consequence, the discontinuous behavior itself is negligible. For that reason, the particular value of θ_s at which the pole crossing occurs is of little or no importance to the asymptotic behavior of the propagated field.

8.2 The Uniform Asymptotic Approximation

If the saddle point at $\omega_{sp}(\theta)$ approaches close to the pole singularity ω_p at $\theta = \theta_s$ so that the quantity $|\omega_{sp}(\theta) - \omega_p|$ becomes small, then the quantity

[2] The value of θ_s depends upon which Olver-type path is chosen for $P(\theta)$. If that path is taken to lie along the path of steepest descent through the saddle point nearest the pole, then the value of θ_s is given by the expression $Y(\omega_{sp}, \theta_s) = Y(\omega_p, \theta_s)$.

[3] The dominance of the saddle point contribution over the pole contribution for $\theta = \theta_s$ is guaranteed by the fact that $P(\theta)$ is an Olver-type path.

$|\phi(\omega_{sp}(\theta_s), \theta_s) - \phi(\omega_p, \theta_s)|$ also becomes small. In that case, the positive constant Z mentioned at the end of Sect. 8.1 becomes increasingly large. As a result, it becomes impractical to take $z > Z$ in order to make $A_c(z, t)$ asymptotically negligible at $\theta = \theta_s$. In order to avoid the discontinuous behavior of $A_c(z, t)$ at $\theta = \theta_s$ when the saddle point is near the pole, the uniform asymptotic approximation presented in Sect. 5.3 must be applied. That technique is employed in this Section to obtain the uniform asymptotic behavior of the pole contribution.

In accordance with the results of Sect. 6.5, the pole contribution is given by (cf. (6.5.9))

$$A_c(z, t) = -\mathrm{Re}[2\pi \mathrm{i} \Lambda(\theta)] + C_D^-(z, t) + C_D^+(z, t) + C_1(z, t) \; ; \qquad 1 \leqslant \theta \leqslant \theta_1 \; , \tag{8.2.1a}$$

$$A_c(z, t) = -\mathrm{Re}[2\pi \mathrm{i} \Lambda(\theta)] + C_D^-(z, t) + C_D^+(z, t) + C_N^-(z, t) + C_N^+(z, t) \; ; \qquad \theta > \theta_1 \; , \tag{8.2.1b}$$

where $\Lambda(\theta)$, which is given by (8.1.3), is the sum of the residues of the poles that are crossed when the original contour of integration is deformed to the path $P(\theta)$. For reasons of simplicity, it is again assumed that the deformed contour of integration $P(\theta)$ is near only one pole at a time, and attention is restricted to obtaining the uniform asymptotic contribution due to that single pole alone. The results obtained are easily generalized to account for several individual pole contributions.

By Theorem 5.3, the C-functions appearing in (8.2.1) are given by

$$C(z, t) = \frac{1}{2\pi} \mathrm{Re}\, \mathrm{i}\gamma \left\{ \pm \mathrm{i}\pi \,\mathrm{erfc}\left(\mp \mathrm{i}\Delta(\theta) \sqrt{\frac{z}{c}} \right) \mathrm{e}^{(z/c)\phi(\omega_p, \theta)} + \frac{1}{\Delta(\theta)} \sqrt{\frac{c\pi}{z}} \mathrm{e}^{(z/c)\phi(\omega_{sp}, \theta)} \right\} \; ; \qquad \mathrm{Im}[\Delta(\theta)] \gtrless 0 \; , \tag{8.2.2a}$$

$$C(z, t) = \frac{1}{2\pi} \mathrm{Re}\, \mathrm{i}\gamma \left\{ \mathrm{i}\pi \,\mathrm{erfc}\left(-\mathrm{i}\Delta(\theta) \sqrt{\frac{z}{c}} \right) \mathrm{e}^{(z/c)\phi(\omega_p, \theta)} - \mathrm{i}\pi \mathrm{e}^{(z/c)\phi(\omega_p, \theta)} + \frac{1}{\Delta(\theta)} \sqrt{\frac{c\pi}{z}} \mathrm{e}^{(z/c)\phi(\omega_{sp}, \theta)} \right\} \; ; \qquad \begin{array}{l} \mathrm{Im}[\Delta(\theta)] = 0 \\ \Delta(\theta) \neq 0 \end{array} \; , \tag{8.2.2b}$$

$$C(z, t) = -\frac{1}{2\pi} \mathrm{Re}\, \mathrm{i}\gamma \left\{ \left[-\frac{2\pi c}{z\phi^{(2)}(\omega_{sp}, \theta)} \right]^{1/2} \left[\frac{1}{\omega_{sp} - \omega_c} + \frac{\phi^{(3)}(\omega_{sp}, \theta)}{6\phi^{(2)}(\omega_{sp}, \theta)} \right] \right\} \mathrm{e}^{(z/c)\phi(\omega_{sp}, \theta)} \; ; \qquad \Delta(\theta) = 0 \tag{8.2.2c}$$

The particular form of the C-function to be employed depends upon the sign of the imaginary part of the quantity $\Delta(\theta)$, which is given by (cf. (5.3.7))

$$\Delta(\theta) \equiv [\phi(\omega_{sp}, \theta) - \phi(\omega_p, \theta)]^{1/2} \; . \tag{8.2.3}$$

Here ω_{sp} denotes the location of the interacting saddle point. The proper argument of this expression is determined by (5.3.8 and 12), so that

$$\lim_{\omega_p \to \omega_{sp}(\theta)} \arg[\Delta(\theta)] = \bar{\alpha}_c - \bar{\alpha}_{SD} + 2n\pi \ , \tag{8.2.4}$$

where $\bar{\alpha}_c$ is the angle of slope of the vector from ω_{sp} to ω_p in the complex ω-plane, $\bar{\alpha}_{SD}$ is the angle of slope of the tangent vector to the path of steepest descent at the interacting saddle point, and n is an integer which can be chosen so that the argument of $\Delta(\theta)$ lies within the principal range $(-\pi, \pi]$ for all $\theta \geqslant 1$.

Consider the case in which either one of the distant saddle points $SP_D^{\pm}$ approaches the pole singularity at $\omega = \omega_p$ located in a region of the complex ω-plane bounded away from the limiting values $\pm\infty - 2\delta i$ approach by $SP_D^{\pm}$ as $\theta \to 1^+$, respectively. To be specific, let SP_D^+ approach $\omega = \omega_p$ which is assumed here to be the only pole singularity of the spectral amplitude function $\tilde{u}(\omega - \omega_c)$. Then $C_D^+(z,t)$ is given by (8.2.2) with $\omega_{sp_D^+}$ substituted for ω_{sp} throughout, and the remaining C-functions appearing in (8.2.1) are negligible in comparison to $C_D^+(z,t)$ and the residue contribution. Similarly, if the distant saddle point SP_D^- approaches the simple pole singularity at $\omega = \omega_p$, then $C_D^-(z,t)$ is given by (8.2.2) with $\omega_{sp_D^-}$ substituted for ω_{sp} throughout, and the remaining C-functions appearing in (8.2.1) are negligible in comparison to $C_D^-(z,t)$ and the residue contribution.

Consider next the case in which one of the near saddle point SP_1 or $SP_N^{\pm}$ approaches the pole singularity at $\omega = \omega_p$. To be specific, let SP_1 approach ω_p, which is again assumed to be the only pole singularity of the spectral amplitude function $\tilde{u}(\omega - \omega_c)$. Then $C_1(z,t)$ is given by (8.2.2) with ω_{sp_1} substituted for ω_{sp} throughout, and the remaining C-functions appearing in (8.2.1a) are asymptotically negligible in comparison to $C_1(z,t)$ and the residue contribution. If the near saddle point SP_N^+ approaches the simple pole singularity at $\omega = \omega_p$, then $C_N^+(z,t)$ is given by (8.2.2) with $\omega_{sp_N^+}$ substituted for ω_{sp} throughout, and the remaining C-functions appearing in (8.2.1b) are negligible in comparison to $C_N^+(z,t)$ and the residue contribution. Similarly, if the near saddle point SP_N^- approaches the simple pole singularity at $\omega = \omega_p$, then $C_N^-(z,t)$ is given by (8.2.2) with $\omega_{sp_N^-}$ substituted for ω_{sp} throughout, and the remaining C-functions appearing in (8.2.1b) are asymptotically negligible in comparison to $C_N^-(z,t)$ and the residue contribution.

Consider finally the case in which none of the saddle points approaches the pole singularity at $\omega = \omega_p$. In that case, all of the C-functions appearing in (8.2.1) are asymptotically negligible in comparison to the residue contribution for all θ, and the uniform asymptotic approximation for the pole contribution reduces to the nonuniform approximation given in Sect. 8.1.

A special case of interest is that for which the pole singularity ω_p of the spectral amplitude function $\tilde{u}(\omega - \omega_c)$ is real and positive, and hence, lies along the positive real axis of the complex ω-plane. The complex phase behavior at the simple pole singularity is then given by (6.1.8), viz

$$\phi(\omega_p, \theta) = -\omega_p n_i(\omega_p) + i\omega_p[n_r(\omega_p) - \theta] \ , \tag{8.2.5}$$

so that

$$X(\omega_p) = -\omega_p n_i(\omega_p) \ , \tag{8.2.6a}$$

$$Y(\omega_p, \theta) = \omega_p(n_r(\omega_p) - \theta] \ , \tag{8.2.6b}$$

where $n_r(\omega_p)$ and $n_i(\omega_p)$ are the real and imaginary parts of the complex index of refraction and are given respectively by (6.1.15 and 16). As can easily be seen, the saddle points SP_D^- and SP_N^- located in the left-half of the complex ω-plane do not interact with the pole at $\omega = \omega_p$, and hence, their contributions C_D^- and C_N^- are negligible. Furthermore, since the real coordinate location of the near saddle point SP_N^+ in the right-half plane lies within the range from 0 to $+\sqrt{\omega_0^2 - \delta^2}$ while the real coordinate location of the distant saddle point SP_D^+ in the right-half plane is greater than or equal to $+\sqrt{\omega_1^2 - \delta^2}$ for all $\theta \geqslant 1$, this problem separates naturally into the following three cases. For values of ω_p in the range $0 \leqslant \omega_p \leqslant \sqrt{\omega_0^2 - \delta^2}$, either of the near saddle points SP_1 or SP_N^+ will interact with the simple pole singularity at $\omega = \omega_p$. In that case, either (or both) of the functions $C_1(z, t)$ or $C_N^+(z, t)$ appearing (8.2.1) may be significant, while the remaining functions $C_D^\pm(z, t)$ and $C_N^-(z, t)$ are negligible. For values of ω_p in the range $\omega_p \geqslant \sqrt{\omega_1^2 - \delta^2}$, the distant saddle point SP_D^+ interacts with the simple pole singularity at $\omega = \omega_p$. In that case, the function $C_D^+(z, t)$ appearing in (8.2.1) may be significant, while the remaining functions $C_1(z, t)$, $C_N^\pm(z, t)$, and $C_D^-(z, t)$ are all negligible. Finally, for values of ω_p in the range $\sqrt{\omega_0^2 - \delta^2} < \omega_p < \sqrt{\omega_1^2 - \delta^2}$, neither the near nor distant saddle points approaches the simple pole singularity located at $\omega = \omega_p$. In that case, all of the C-functions appearing in (8.2.1) are asymptotically negligible in comparison to the residue contribution, and the nonuniform expression results. These three special cases are now treated in detail.

8.2.1 Frequencies ω_p in the Range $0 \leqslant \omega_p \leqslant \sqrt{\omega_0^2 - \delta^2}$

For real frequencies ω_p in the range $0 \leqslant \omega_p \leqslant \sqrt{\omega_0^2 - \delta^2}$, which is below the absorption band of the medium, it is the near saddle point in the right-half of the complex ω-plane which interacts with the simple pole singularity at $\omega = \omega_p$. The results of (8.2.1 and 2) then apply for either $C_1(z, t)$ or $C_N^+(z, t)$ with ω_{sp} denoting the location of the upper near saddle point SP_1 for $1 < \theta < \theta_1$, the second-order near saddle point SP_N at $\theta = \theta_1$, and the near saddle point SP_N^+ for $\theta > \theta_1$. The quantity $\Delta(\theta)$ is then given by

$$\Delta(\theta) = [\phi(\omega_{sp_1}, \theta) - \phi(\omega_c, \theta)]^{1/2} \ ; \qquad \text{for } 1 < \theta < \theta_1 \ , \tag{8.2.7a}$$

$$\Delta(\theta_1) = [\phi(\omega_{sp_N}, \theta_1) - \phi(\omega_c, \theta_1)]^{1/2} \ ; \qquad \text{for } \theta = \theta_1 \ , \tag{8.2.7b}$$

$$\Delta(\theta) = [\phi(\omega_{sp_N^+}, \theta) - \phi(\omega_c, \theta)]^{1/2} \ ; \qquad \text{for } \theta > \theta_1 \ , \tag{8.2.7c}$$

where (from (7.4.6, 12a, and 18a), respectively)

$$\phi(\omega_{sp_1},\theta) \cong \frac{1}{3}(2\delta\zeta(\theta) - 3|\psi(\theta)|)(\theta_0 - \theta) + \frac{b^2}{54\theta_0\omega_0^4}(2\delta\zeta(\theta) - 3|\psi(\theta)|)^2$$

$$\cdot(2\delta(3 - \alpha\zeta(\theta)) + 3\alpha|\psi(\theta)|) \; ; \qquad 1 < \theta < \theta_1 \; , \qquad (8.2.8a)$$

$$\phi(\omega_{sp_N},\theta_1) \cong \frac{2\delta}{3\alpha}\left(\theta_0 - \theta_1 + \frac{8\delta^2 b^2}{27\alpha\theta_0\omega_0^4}\right) \; ; \qquad \theta = \theta_1 \; , \qquad (8.2.8b)$$

$$\phi(\omega_{sp_N^+},\theta) \cong -\delta\left\{\frac{2}{3}\zeta(\theta)(\theta - \theta_0) + \frac{b^2}{\theta_0\omega_0^4}\left[(1 - \alpha\zeta(\theta)\psi^2(\theta)\right.\right.$$

$$\left.\left. + \frac{4}{9}\delta^2\zeta^2(\theta)\left(\frac{1}{3}\alpha\zeta(\theta) - 1\right)\right]\right\}$$

$$+ i\psi(\theta)\left\{\theta_0 - \theta + \frac{b^2}{2\theta_0\omega_0^4}\left[\frac{4}{3}\delta^2\zeta(\theta)(2 - \alpha\zeta(\theta)) + \alpha\psi^2(\theta)\right]\right\} \; ;$$

$$\theta > \theta_1 \; . \qquad (8.2.8c)$$

The argument of $\Delta(\theta)$, as given by (8.2.7), must now be determined according to (8.2.4) in the limit as ω_c approaches the saddle point location, with the integer n chosen so that this argument lies within the principal range $(-\pi, \pi]$. A depiction of the interaction of the near saddle point with a simple pole singularity at $\omega = \omega_p$ bounded away from the origin along the positive real axis is illustrated in Fig. 8.1. As can be seen from this diagram, $-\pi/2 \leqslant \bar{\alpha}_c < \pi$ for all values of ω_p for all $\theta > 1$, where $\bar{\alpha}_c$ increases monotonically with increasing θ. Furthermore, according to the uniform asymptotic approximation of the second precursor field $A_B(z,t)$ presented in Sect. 7.3, the angle of slope $\bar{\alpha}_{SD}$ of the path of steepest descent through the upper near saddle point SP_1 is equal to 0 as θ increases from unity to θ_1, at $\theta = \theta_1$, $\bar{\alpha}_{SD} = \pi/6$, and the angle of slope of the path of steepest descent through the near saddle point SP_N^+ is equal to $\pi/4$ as θ increases without bound from θ_1. Note that although the value of $\bar{\alpha}_{SD}$ changes abruptly at $\theta = \theta_1$, the path of steepest descent varies in a continuous fashion with θ for all θ in the range $\theta > 1$ (see the example treated in Sect. 5.2.2). Substitution of these results into (8.2.4) yields $\arg(\Delta(\theta)) = -\pi/2$ for $1 < \theta < \theta_s$, $\arg(\Delta(\theta_s)) = 0$ at $\theta = \theta_s$, and $\arg(\Delta(\theta)) = 3\pi/4$ for $\theta > \theta_s$. Note that for all values of ω_p in the range $0 < \omega_p \leqslant \sqrt{\omega_0^2 - \delta^2}$, $\theta_s > \theta_0$, while for $\omega_p = 0$, $\theta_s = \theta_0$.

Upon application of (8.2.1 and 2), the uniform asymptotic contribution of the simple pole singularity at $\omega = \omega_p$, where $0 < \omega_p \leqslant \sqrt{\omega_0^2 - \delta^2}$, is given as follows. For $\theta < \theta_s$, $\mathrm{Im}(\Delta(\theta)) < 0$ and (8.2.1 and 2a) yield

$$A_c(z,t) \sim \frac{1}{2\pi}\mathrm{Re}\, i\gamma\left\{-i\pi\,\mathrm{erfc}\left(i\Delta(\theta)\sqrt{\frac{z}{c}}\right)e^{(z/c)\phi(\omega_p,\theta)}\right.$$

$$\left. + \frac{1}{\Delta(\theta)}\sqrt{\frac{\pi c}{z}}e^{(z/c)\phi(\omega_{sp},\theta)}\right\} \; ; \qquad \theta < \theta_s \; , \qquad (8.2.9)$$

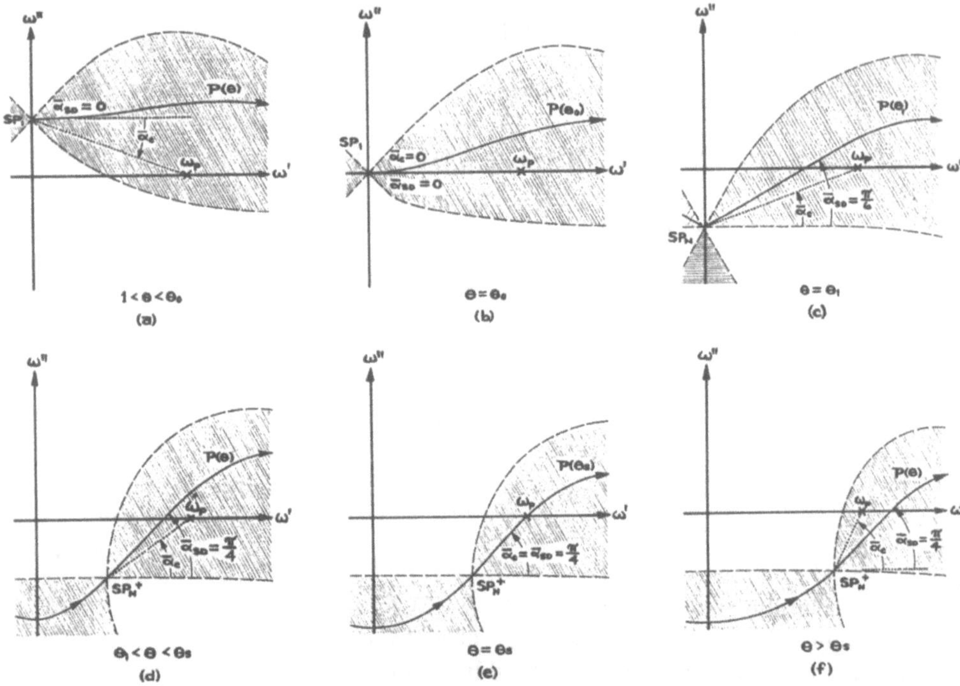

Fig. 8.1. The interaction of the near saddle point with the simple pole singularity located at $\omega = \omega_p$ within the upper end of the range $0 < \omega_p < \sqrt{\omega_0^2 - \delta^2}$. The hatched area indicates the region of the complex ω-plane wherein the inequality $X(\omega_{sp}, \theta) > X(\omega, \theta)$ is satisfied, where ω_{sp} denotes ω_{sp_1} for $1 < \theta < \theta_1$, ω_{sp_N} at $\theta = \theta_1$, and $\omega_{sp_D^+}$ for $\theta > \theta_1$.

as $z \to \infty$, where ω_{sp} denotes ω_{sp_1} and $\Delta(\theta)$ is given by (8.2.7a) for $1 \leqslant \theta < \theta_1$, ω_{sp} denotes ω_{sp_N} and $\Delta(\theta)$ is given by (8.2.7b) at $\theta = \theta_1$, and ω_{sp} denotes $\omega_{sp_N^+}$ and $\Delta(\theta)$ is given by (8.2.7c) for $\theta > \theta_1$. At $\theta = \theta_s$, $\mathrm{Im}(\Delta(\theta)) = 0$[4] and (8.2.1 and 2b) yield for $\omega_p \neq 0$

$$
\begin{aligned}
A_c(z, t_s) &\sim \frac{1}{2\pi} \mathrm{Re}\, \mathrm{i}\gamma \left\{ \mathrm{i}\pi \,\mathrm{erfc}\left(-\mathrm{i}\Delta(\theta_s) \sqrt{\frac{z}{c}} \right) \mathrm{e}^{(z/c)\phi(\omega_p, \theta_s)} + \frac{1}{\Delta(\theta_s)} \sqrt{\frac{\pi c}{z}} \mathrm{e}^{(z/c)\phi(\omega_{sp}, \theta_s)} \right\} \\
&\quad + \mathrm{Re}\left[\gamma \mathrm{e}^{(z/c)\phi(\omega_p, \theta_s)}\right] \\
&= \frac{1}{2\pi} \mathrm{Re}\, \mathrm{i}\gamma \left\{ -2\sqrt{\pi} \mathrm{e}^{(z/c)\phi(\omega_{sp}, \theta_s)} F\left(|\Delta(\theta_s)| \sqrt{\frac{z}{c}} \right) + \frac{1}{\Delta(\theta_s)} \sqrt{\frac{\pi c}{z}} \mathrm{e}^{(z/c)\phi(\omega_{sp}, \theta_s)} \right\} \\
&\quad + \frac{1}{2} \mathrm{Re}\left[\gamma \mathrm{e}^{(z/c)\phi(\omega_p, \theta_s)}\right] ; \qquad \begin{matrix} \theta = \theta_s \\ \omega_c \neq 0 \end{matrix} ,
\end{aligned}
\tag{8.2.10}
$$

[4] For all finite nonzero values of ω_p, $\Delta(\theta_s) \neq 0$ for dispersive and absorptive media. If $\Delta(\theta_s) = 0$ for finite $\omega_p \neq 0$, it then follows that $\delta = 0$, which in turn implies that the medium is nonabsorbing if either $0 \leqslant \omega_p \leqslant \omega_0$ or $\omega_1 \leqslant \omega_p$. However, if $\omega_0 < \omega_p < \omega_1$ then the field component of that frequency is cut off.

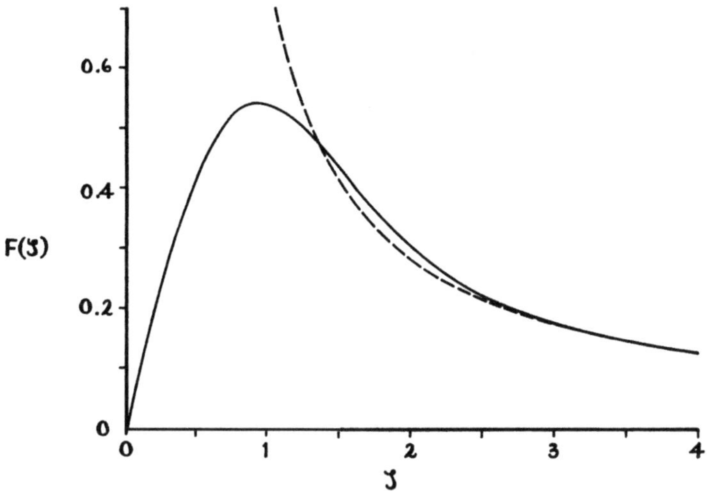

Fig. 8.2. The functional dependence of *Dawson*'s integral $F(\zeta) = e^{-\zeta^2}\int_0^\zeta e^{t^2}\,dt$ for real ζ. The exact behavior is illustrated by the solid line and the behavior of the first two terms in the asymptotic expansion of $F(\zeta)$ as $\zeta \to \infty$ is illustrated by the dashed line.

as $z \to \infty$, where ω_{sp} denotes ω_{sp_1} and $\Delta(\theta_s)$ is given by (8.2.7a) if $\theta_0 < \theta_s < \theta_1$, ω_{sp} denotes ω_{sp_N} and $\Delta(\theta_s)$ is given by (8.2.7b) if $\theta_s = \theta_1$, and ω_{sp} denotes $\omega_{sp_N^+}$ and $\Delta(\theta_s)$ is given by (8.2.7c) if $\theta_s > \theta_1$. Here $F(\zeta)$ denotes *Dawson's integral*, given by

$$F(\zeta) \equiv e^{-\zeta^2}\int_0^\zeta e^{t^2}\,dt \ .$$

The exact behavior (as determined numerically) of Dawson's integral as a function of real ζ is illustrated by the solid line in Fig. 8.2. The dashed curve in this figure illustrates the behavior of the first two terms in the asymptotic expansion of $F(\zeta)$ as $\zeta \to \infty$, where

$$F(\zeta) = \frac{1}{2}\sum_{n=0}^{N-1}\frac{\Gamma(n+\frac{1}{2})}{\Gamma(\frac{1}{2})\zeta^{2n+1}} + \mathcal{O}\left(\frac{1}{\zeta^{2N+1}}\right) ; \qquad \text{as } \zeta \to \infty \ .$$

For $\theta > \theta_s$, $\mathrm{Im}(\Delta(\theta)) > 0$ and (8.2.1 and 2a) yield

$$A_c(z,t) \sim \frac{1}{2\pi}\mathrm{Re}\,\mathrm{i}\gamma\left\{\mathrm{i}\pi\,\mathrm{erfc}\left(-\mathrm{i}\Delta(\theta)\sqrt{\frac{z}{c}}\right)e^{(z/c)\phi(\omega_p,\theta)} + \frac{1}{\Delta(\theta)}\sqrt{\frac{\pi c}{z}}e^{(z/c)\phi(\omega_{sp},\theta)}\right\}$$
$$+ \mathrm{Re}[\gamma e^{(z/c)\phi(\omega_p,\theta)}] \ ; \qquad \theta > \theta_s \ , \tag{8.2.11}$$

as $z \to \infty$, where ω_{sp} denotes ω_{sp_1} and $\Delta(\theta)$ is given by (8.2.7a) for $\theta_0 < \theta < \theta_1$, ω_{sp} denotes ω_{sp_N} and $\Delta(\theta)$ is given by (8.2.7b) at $\theta = \theta_1$, and ω_{sp} denotes $\omega_{sp_N^+}$ and $\Delta(\theta)$ is given by (8.2.7c) for $\theta > \theta_1$.

For the special case when $\omega_p = 0$, the upper near saddle point SP_1 and the simple pole singularity coalesce when $\theta = \theta_0$, so that $\theta_s = \theta_0$ and $\Delta(\theta_0) = 0$. For $\theta < \theta_s$, $\mathrm{Im}(\Delta(\theta)) < 0$ and the uniform asymptotic approximation of the pole contribution is given by (8.2.9), where ω_{sp} denotes ω_{sp_1} and $\Delta(\theta)$ is given by (8.2.7a). At $\theta = \theta_s = \theta_0$, $\Delta(\theta_0) = 0$ and (8.2.2c) applys for $C_1(z,t)$. Since the path $P(\theta_0)$ crosses over the pole at $\omega_p = 0$, then $\Lambda(\theta_0) = 1/2 \cdot (1/2\pi)\mathrm{i}\gamma e^{(z/c)\phi(0,\theta_0)}$ (cf. (8.1.4b)). Hence, in this special case one obtains

$$A_c(z,t_0) \sim -\frac{1}{2\pi}\mathrm{Re}\,\mathrm{i}\gamma\left\{\left[-\frac{2\pi c}{z\phi^{(2)}(\omega_{sp_1},\theta_0)}\right]^{1/2}\left[\frac{1}{\omega_{sp_1}(\theta_0)} + \frac{\phi^{(3)}(\omega_{sp_1},\theta_0)}{6\phi^{(2)}(\omega_{sp_1},\theta_0)}\right]\right\}$$

$$\cdot\, e^{(z/c)\phi(\omega_{sp_1},\theta_0)} + \frac{1}{2}\mathrm{Re}[\gamma e^{(z/c)\phi(0,\theta_0)}] \; ; \qquad \begin{matrix} \theta = \theta_s = \theta_0 \\ \omega_p = 0 \end{matrix} \; ,$$

as $z \to \infty$. In accordance with (7.4.2 and 5), one has that

$$\phi(\omega_{sp_1},\theta_0) = \phi(0,\theta_0) = 0 \; ,$$

$$\phi^{(2)}(\omega_{sp_1},\theta_0) \cong -\frac{2\delta b^2}{\theta_0\omega_0^4} \; ,$$

$$\phi^{(3)}(\omega_{sp_1},\theta_0) \cong \frac{3\mathrm{i}\alpha b^2}{\theta_0\omega_0^4} \; .$$

Substitution of these results into the above equation then gives the result[5]

$$A_c(z,t_0) \sim \frac{\omega_0^2}{2b}\left[\frac{\theta_0 c}{\pi\delta z}\right]^{1/2} \mathrm{Re}\,\mathrm{i}\gamma\left\{-\frac{1}{\omega_{sp_1}(\theta_0)} + \frac{\mathrm{i}\alpha}{4\delta}\right\} + \frac{1}{2}\mathrm{Re}(\gamma) \; ;$$

$$\begin{matrix} \theta = \theta_s = \theta_0 \\ \omega_p = 0 \end{matrix} \; , \qquad (8.2.12)$$

as $z \to \infty$, where $t_0 = (z/c)\theta_0$. For $\theta > \theta_s$, $\mathrm{Im}(\Delta(\theta)) > 0$ and the uniform asymptotic approximation of the pole contribution is given by (8.2.11) where ω_{sp} denotes ω_{sp_1} and $\Delta(\theta)$ is given by (8.2.7a) for $\theta_0 < \theta < \theta_1$, ω_{sp} denotes ω_{sp_N} and $\Delta(\theta)$ is given by (8.2.7b) at $\theta = \theta_1$, and ω_{sp} denotes $\omega_{sp_N^+}$ and $\Delta(\theta)$ is given by (8.2.7c) for $\theta > \theta_1$.

Equations (8.2.9, 10, and 11) taken together constitute the uniform asymptotic approximation of the pole contribution at $\omega = \omega_p$ with $0 < \omega_p \leqslant \sqrt{\omega_0^2 - \delta^2}$, whereas (8.2.9, 12, and 11) taken together constitute the uniform asymptotic approximation of the pole contribution at $\omega = \omega_p = 0$. For values

[5] It is to be noted that even though the term $1/\omega_{sp_1}(\theta_0)$ is singular since $\omega_{sp_1}(\theta_0) = 0$, this result for $A_c(z,t_0)$ when combined with the asymptotic behavior of the second precursor field yields a uniform asymptotic approximation of the total field $A(z,t)$ that is well-behaved at $\theta = \theta_0$. This point is examined in detail in Sect. 9.2.

of $\theta < \theta_s$ and sufficiently large observation distances z such that quantity $|\Delta(\theta)|\sqrt{z/c}$ is large, the dominant term in the asymptotic expansion of $\mathrm{erfc}(\mathrm{i}\Delta(\theta)\sqrt{z/c})$ may be substituted into (8.2.9) with the result that the first and second terms in that equation identically cancel. Hence, for values of θ sufficiently less than θ_s, there is no contribution to the asymptotic behavior of the total field from the simple pole singularity. For values of $\theta > \theta_s$ and sufficiently large observation distances z such that the quantity $|\Delta(\theta)|\sqrt{z/c}$ is large, the dominant term in the asymptotic expansion of $\mathrm{erfc}(-\mathrm{i}\Delta(\theta)\sqrt{z/c})$ may be substituted into (8.2.11) with the result

$$\begin{aligned} A_c(z,t) &\sim \mathbb{R}\mathrm{e}[\gamma e^{(z/c)\phi(\omega_p,\theta)}] \\ &= e^{-z\alpha(\omega_p)}[\gamma' \cos(\beta(\omega_p)z - \omega_p t) - \gamma'' \sin(\beta(\omega_p)z - \omega_p t)] \; ; \\ &\qquad \theta \gg \theta_s \; , \end{aligned} \tag{8.2.13}$$

as $z \to \infty$ with θ bounded away from θ_s. This result is in agreement with the nonuniform approximation given in (8.1.10c), where the amplitude attenuation coefficient $\alpha(\omega_p)$ is given by (8.1.7) and the propagation constant $\beta(\omega_p)$ is given by (8.1.9).

8.2.2 Frequencies ω_p in the Range $\omega_p \geqslant \sqrt{\omega_1^2 - \delta^2}$

For real frequencies ω_p in the range $\omega_p \geqslant \sqrt{\omega_1^2 - \delta^2}$, which is above the absorption band of the medium, it is the distant saddle point SP_D^+ in the right-half of the complex ω-plane which interacts with the sample pole singularity at $\omega = \omega_p$. The results of (8.2.1 and 2) then apply for $C_D^+(z,t)$ with $\omega_{sp} = \omega_{sp_D^+}$ for all $\theta \geqslant 1$. The quantity $\Delta(\theta)$ is then given by

$$\Delta(\theta) = [\phi(\omega_{sp_D^+},\theta) - \phi(\omega_p,\theta)]^{1/2} \; , \tag{8.2.14}$$

where (from (7.3.4))

$$\begin{aligned} \phi(\omega_{sp_D^+},\theta) \cong &-\delta\left\{[1+\eta(\theta)](\theta-1) + \frac{b^2}{2}\cdot\frac{1-\eta(\theta)}{\xi^2(\theta)+\delta^2(1-\eta(\theta))^2}\right\} \\ &- \mathrm{i}\xi(\theta)\left\{\theta - 1 + \frac{b^2}{2}\cdot\frac{1}{\xi^2(\theta)+\delta^2(1-\eta(\theta))^2}\right\} \; ; \qquad \theta \geqslant 1 \; , \end{aligned} \tag{8.2.15}$$

and where $\phi(\omega_p,\theta)$ is given by (8.2.5).

The argument of $\Delta(\theta)$, as given in (8.2.14), must now be determined according to (8.2.4) in the limit as ω_p approaches the saddle point $\omega_{sp_D^+}$ with the integer n chosen so that this argument lies within the principal range $(-\pi,\pi]$. A depiction of the interaction of the distant saddle point SP_D^+ with a simple pole singularity located at $\omega = \omega_p \geqslant \sqrt{\omega_1^2 - \delta^2}$ is illustrated in Fig. 8.3. In accor-

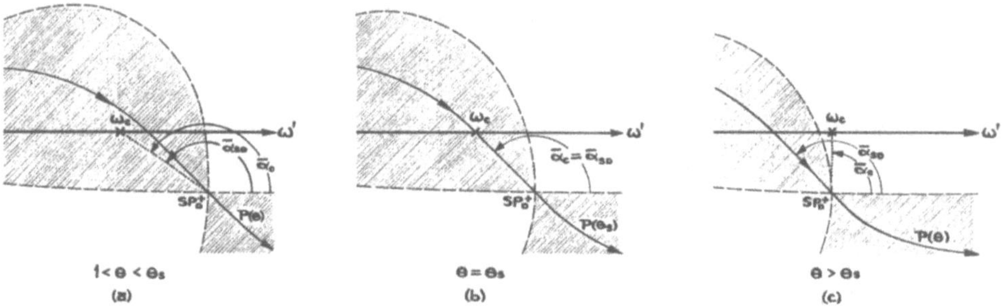

Fig. 8.3. The interaction of the distant saddle point with the simple pole singularity located at $\omega = \omega_p$ within the range $\omega_p \geqslant \sqrt{\omega_1^2 - \delta^2}$ with finite ω_p. The hatched area indicates the region of the complex ω-plane wherein the inequality $X(\omega_{sp_D^+}, \theta) > X(\omega, \theta)$ is satisfied.

dance with (6.2.26), the angle of slope of the tangent vector to the path of steepest descent at the saddle point is $\bar{\alpha}_{SD} = 3\pi/4$ for all $\theta \geqslant 1$. Furthermore, as can be seen from Fig. 8.3, the angle of slope $\bar{\alpha}_c$ of the vector from $\omega_{sp_D^+}$ to ω_p in the complex ω-plane decreases monotonically with increasing $\theta \geqslant 1$ and lies within the range $\pi \geqslant \bar{\alpha}_c \geqslant 0$. Upon substitution of these results into (8.2.4) with $n = 0$, it then follows that $\pi/4 \geqslant \arg(\Delta(\theta)) > 0$ for $1 \leqslant \theta < \theta_s$, $\arg(\Delta(\theta_s)) = 0$ at $\theta = \theta_s$, and $0 > \arg(\Delta(\theta)) \geqslant -3\pi/4$ for $\theta > \theta_s$, where $\arg(\Delta(\theta))$ decreases monotonically with increasing θ.

Upon application of (8.2.1 and 2), the uniform asymptotic approximation of the contribution due to the simple pole singularity at $\omega = \omega_p$, where $\omega_p \geqslant \sqrt{\omega_1^2 - \delta^2}$, is given as follows[6]. For $1 \leqslant \theta < \theta_s$, $\mathrm{Im}(\Delta(\theta)) > 0$ and (8.2.2a and 1) yield

$$A_c(z,t) \sim \frac{1}{2\pi}\,\mathrm{Re}\,\mathrm{i}\gamma\left\{\mathrm{i}\pi\,\mathrm{erfc}\left(-\mathrm{i}\Delta(\theta)\sqrt{\frac{z}{c}}\right)\mathrm{e}^{(z/c)\phi(\omega_p,\theta)} + \frac{1}{\Delta(\theta)}\sqrt{\frac{\pi c}{z}}\,\mathrm{e}^{(z/c)\phi(\omega_{sp_D^+},\theta)}\right\}\,; \qquad 1 \leqslant \theta < \theta_s\,, \tag{8.2.16}$$

as $z \to \infty$. At $\theta = \theta_s$, $\mathrm{Im}(\Delta(\theta_s)) = 0$ with $\Delta(\theta_s) \neq 0$ and (8.2.2b) yields

$$A_c(z,t_s) \sim \frac{1}{2\pi}\,\mathrm{Re}\,\mathrm{i}\gamma\left\{\mathrm{i}\pi\,\mathrm{erfc}\left(-\mathrm{i}\Delta(\theta_s)\sqrt{\frac{z}{c}}\right)\mathrm{e}^{(z/c)\phi(\omega_p,\theta_s)} + \frac{1}{\Delta(\theta_s)}\sqrt{\frac{\pi c}{z}}\,\mathrm{e}^{(z/c)\phi(\omega_{sp_D^+},\theta_s)}\right\} + \mathrm{Re}\left[\gamma\mathrm{e}^{(z/c)\phi(\omega_p,\theta_s)}\right]$$

[6] The nonphysical case of $\omega_p = \infty$ is not considered here.

$$= \frac{1}{2\pi} \mathrm{Re}\, \mathrm{i}\gamma \left\{ -2\sqrt{\pi} e^{(z/c)\phi(\omega_{sp_D^+}, \theta_s)} F\left(|\Delta(\theta_s)| \sqrt{\frac{z}{c}} \right) \right.$$

$$\left. + \frac{1}{\Delta(\theta_s)} \sqrt{\frac{\pi c}{z}} e^{(z/c)\phi(\omega_{sp_D^+}, \theta_s)} \right\}$$

$$+ \frac{1}{2} \mathrm{Re}[\gamma e^{(z/c)\phi(\omega_c, \theta_s)}] \; ; \qquad \theta = \theta_s = \frac{ct_s}{z} \, , \tag{8.2.17}$$

as $z \to \infty$, where $F(\zeta)$ denotes *Dawson*'s integral (depicted in Fig. 8.2). For $\theta > \theta_s$, $\mathrm{Im}(\Delta(\theta)) < 0$ and (8.2.1 and 2a) yield

$$A_c(z,t) \sim \frac{1}{2\pi} \mathrm{Re}\, \mathrm{i}\gamma \left\{ -\mathrm{i}\pi \,\mathrm{erfc}\left(\mathrm{i}\Delta(\theta) \sqrt{\frac{z}{c}} \right) e^{(z/c)\phi(\omega_p, \theta)} \right.$$

$$\left. + \frac{1}{\Delta(\theta)} \sqrt{\frac{\pi c}{z}} e^{(z/c)\phi(\omega_{sp_D^+}, \theta)} \right\} + \mathrm{Re}[\gamma e^{(z/c)\phi(\omega_p, \theta)}] \; ; \qquad \theta > \theta_s \, , \tag{8.2.18}$$

as $z \to \infty$.

Taken together, (8.2.16) through (8.2.18) constitute the uniform asymptotic approximation of the pole contribution at $\omega = \omega_p$ with finite $\omega_p \geqslant \sqrt{\omega_1^2 - \delta^2}$. For fixed values of $\theta < \theta_s$ and sufficiently large observation distances z such that the quantity $|\Delta(\theta)|\sqrt{z/c}$ is large, the dominant term in the asymptotic expansion of $\mathrm{erfc}(-\mathrm{i}\Delta(\theta)\sqrt{z/c})$ may be substituted into (8.2.16) with the result that the first and second terms in that equation identically cancel. Hence, for values of θ sufficiently less than θ_s, there is no contribution to the asymptotic behavior of the total field from the simple pole singularity. For values of $\theta > \theta_s$ and sufficiently large observation distances z such that the quantity $|\Delta(\theta)|\sqrt{z/c}$ is large, the dominant term in the asymptotic expansion of $\mathrm{erfc}(\mathrm{i}\Delta(\theta)\sqrt{z/c})$ may be substituted into (8.2.19) with the result

$$A_c(z,t) \sim \mathrm{Re}[\gamma e^{(z/c)\phi(\omega_p, \theta)}]$$

$$= e^{-z\alpha(\omega_p)}[\gamma' \cos(\beta(\omega_p)z - \omega_p t) - \gamma'' \sin(\beta(\omega_p)z - \omega_p t)] \; ;$$

$$\theta \gg \theta_p \, , \tag{8.2.19}$$

as $z \to \infty$ with θ bounded away from θ_s. This result is in agreement with the nonuniform approximation given in (8.1.10c).

8.2.3 Frequencies ω_p in the Range $\sqrt{\omega_0^2 - \delta^2} < \omega_p < \sqrt{\omega_1^2 - \delta^2}$

For values of ω_p in the range $\sqrt{\omega_0^2 - \delta^2} < \omega_p < \sqrt{\omega_1^2 - \delta^2}$, which is within the absorption band of the medium, neither the near nor distant saddle points comes within close proximity of the simple pole singularity located at $\omega = \omega_p$.

The results of (5.3.28) then apply in this case with the following results. For $\theta < \theta_s$ there is no pole contribution (cf. (5.3.28a)) to the asymptotic behavior of the total field from the simple pole singularity. At $\theta = \theta_s$ (5.3.28b) yields

$$A_c(z, t_s) \sim \frac{1}{2}\mathrm{Re}[\gamma e^{(z/c)\phi(\omega_p, \theta_s)}]$$

$$= \frac{1}{2}e^{-z\alpha(\omega_p)}[\gamma' \cos(\beta(\omega_p)z - \omega_p t_s) - \gamma'' \sin(\beta(\omega_p)z - \omega_p t_s)] ;$$

$$\theta = \theta_s = \frac{ct_s}{z} , \quad (8.2.20)$$

as $z \to \infty$, and for $\theta > \theta_s$ (5.3.28c) yields

$$A_c(z, t) \sim \mathrm{Re}[\gamma e^{(z/c)\phi(\omega_p, \theta)}]$$

$$= e^{-z\alpha(\omega_p)}[\gamma' \cos(\beta(\omega_p)z - \omega_p t) - \gamma'' \sin(\beta(\omega_p)z - \omega_p t)] ;$$

$$\theta > \theta_s , \quad (8.2.21)$$

as $z \to \infty$. These results are the same as the nonuniform approximation given in Sect. 8.1 (cf. (8.1.10)).

8.3 Special Pulses

The uniform asymptotic approximation of the pole contribution (or contributions if there are several pole singularities) $A_c(z, t)$ for the special pulses considered in this research are obtained in this section. The results are based upon the analysis given in Sect. 8.2.

8.3.1 The Unit Step-Function Modulated Signal

For the unit step-function modulated signal, whose initial spectrum $\tilde{u}(\omega - \omega_c)$ is given by (4.3.33) and possesses a single simple pole singularity at the applied signal frequency $\omega_p = \omega_c$, the value of the residue at that point is given by (cf. (8.1.1))

$$\gamma = \lim_{\omega \to \omega_c} \left[(\omega - \omega_c)\frac{i}{\omega - \omega_c}\right] = i . \quad (8.3.1)$$

Substitution of this result into (8.2.9) through (8.2.12) (with $\omega_p = \omega_c$) yields the following set of equations describing the uniform asymptotic behavior of the pole contribution for applied signal frequencies in the range $0 \leqslant \omega_c \leqslant$

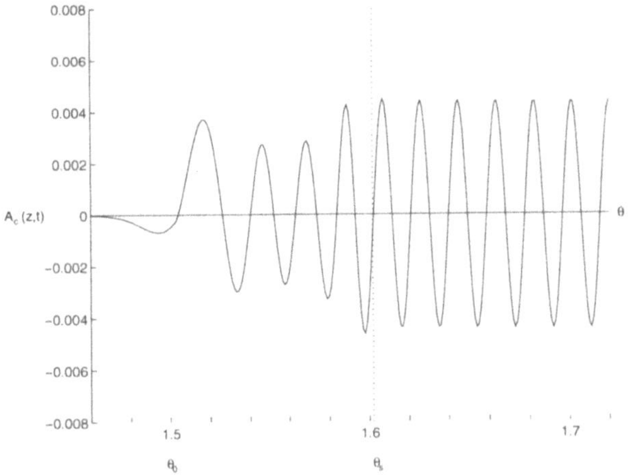

Fig. 8.4. Dynamic behavior of the main signal $A_c(z,t)$ for the unit step-function modulated signal with applied signal frequency $\omega_c = 1.0 \times 10^{16}$/s at an observation distance of $z = 1 \times 10^{-3}$ cm, where *Brillouin*'s choice of the medium parameters are employed.

$\sqrt{\omega_0^2 - \delta^2}$:

$$A_c(z,t) \sim \frac{1}{2\pi}\mathrm{Re}\left\{\mathrm{i}\pi\,\mathrm{erfc}\left(\mathrm{i}\Delta(\theta)\sqrt{\frac{z}{c}}\right)\mathrm{e}^{(z/c)\phi(\omega_c,\theta)} - \frac{1}{\Delta(\theta)}\sqrt{\frac{\pi c}{z}}\,e^{(z/c)\phi(\omega_{sp},\theta)}\right\}\ ; \quad \theta < \theta_s\ , \tag{8.3.2a}$$

$$A_c(z,t_s) \sim \frac{1}{2\pi}\mathrm{Re}\left\{2\sqrt{\pi}\mathrm{e}^{(z/c)\phi(\omega_{sp},\theta_s)}F\left(|\Delta(\theta_s)|\sqrt{\frac{z}{c}}\right) - \frac{1}{\Delta(\theta_s)}\sqrt{\frac{\pi c}{z}}\mathrm{e}^{(z/c)\phi(\omega_{sp},\theta_s)}\right\} - \frac{1}{2}\mathrm{e}^{-z\alpha(\omega_c)}\sin(\beta(\omega_c)z - \omega_c t_s)\ ; \quad \begin{matrix}\theta = \theta_s = \dfrac{ct_s}{z}\\ \omega_c \neq 0\end{matrix}\ ; \tag{8.3.2b}$$

$$A_c(z,t_s) \sim \frac{\omega_0^2}{2b}\left[\frac{\theta_0 c}{\pi\delta z}\right]^{1/2}\frac{1}{\omega_{sp_1}(\theta_0)}\ ; \quad \begin{matrix}\theta = \theta_s = \theta_0\\ \omega_c = 0\end{matrix}\ , \tag{8.3.2c}$$

$$A_c(z,t) \sim \frac{1}{2\pi}\mathrm{Re}\left\{-\mathrm{i}\pi\,\mathrm{erfc}\left(-\mathrm{i}\Delta(\theta)\sqrt{\frac{z}{c}}\right)\mathrm{e}^{(z/c)\phi(\omega_c,\theta)} - \frac{1}{\Delta(\theta)}\sqrt{\frac{\pi c}{z}}\mathrm{e}^{(z/c)\phi(\omega_{sp},\theta)}\right\} - \mathrm{e}^{-z\alpha(\omega_c)}\sin(\beta(\omega_c)z - \omega_c t)\ ; \quad \theta > \theta_s\ , \tag{8.3.2d}$$

as $z \to \infty$. Here ω_{sp} denotes ω_{sp_1} and $\Delta(\theta)$ is given by (8.2.7a) for $1 \leqslant \theta \leqslant \theta_1$, ω_{sp} denotes ω_{sp_N} and $\Delta(\theta)$ is given by (8.2.7b) for $\theta = \theta_1$, and ω_{sp} denotes $\omega_{sp_N^+}$ and $\Delta(\theta)$ is given by (8.2.7c) for $\theta > \theta_1$.

The dynamic behavior of the pole contribution as described by (8.3.2a, 2b, and 2d) for the unit step-function modulated signal is illustrated in Fig. 8.4

as a function of θ. This field was calculated at an observation distance of $z = 1 \times 10^{-3}$ cm. for an applied signal frequency $\omega_c = 1.0 \times 10^{16}$/sec., where Brillouin's choice of the medium parameters have again been employed. The first precursor field and second precursor field behavior for the same case as illustrated here is given in Fig. 7.4 and 7.9, respectively. Comparison of Fig. 7.4, 7.9, and 8.4 shows that, for this particular case, the maximum amplitude attained by the pole contribution is an order of magnitude larger than that of the first precursor field and an order of magnitude less than that of the second precursor field.

Substitution of (8.3.1) into (8.2.16) through (8.2.18) (with $\omega_p = \omega_c$) yields the following set of equations describing the uniform asymptotic behavior of the pole contribution for finite applied signal frequencies ω_c in the range $\omega_c \geqslant \sqrt{\omega_1^2 - \delta^2}$:

$$A_c(z,t) \sim \frac{1}{2\pi}\,\mathrm{Re}\left\{-\mathrm{i}\pi\,\mathrm{erfc}\left(-\mathrm{i}\Delta(\theta)\sqrt{\frac{z}{c}}\right)\mathrm{e}^{(z/c)\phi(\omega_c,\theta)} - \frac{1}{\Delta(\theta)}\sqrt{\frac{\pi c}{z}}\,\mathrm{e}^{(z/c)\phi(\omega_{sp_D^+},\theta)}\right\}; \qquad 1 \leqslant \theta < \theta_s \tag{8.3.3a}$$

$$\begin{aligned} A_c(z,t_s) \sim \frac{1}{2\pi}\,\mathrm{Re}&\left\{2\sqrt{\pi}\,\mathrm{e}^{(z/c)\phi(\omega_{sp_D^+},\theta_s)}F\left(|\Delta(\theta_s)|\sqrt{\frac{z}{c}}\right) - \frac{1}{\Delta(\theta_s)}\sqrt{\frac{\pi c}{z}}\,\mathrm{e}^{(z/c)\phi(\omega_{sp_D^+},\theta_s)}\right\} \\ &-\frac{1}{2}\mathrm{e}^{-z\alpha(\omega_c)}\sin(\beta(\omega_c)z - \omega_c t); \qquad \theta = \theta_s = \frac{ct_s}{z}, \end{aligned} \tag{8.3.3b}$$

$$\begin{aligned} A_c(z,t) \sim \frac{1}{2\pi}\,\mathrm{Re}&\left\{\mathrm{i}\pi\,\mathrm{erfc}\left(\mathrm{i}\Delta(\theta)\sqrt{\frac{z}{c}}\right)\mathrm{e}^{(z/c)\phi(\omega_c,\theta)} - \frac{1}{\Delta(\theta)}\sqrt{\frac{\pi c}{z}}\,\mathrm{e}^{(z/c)\phi(\omega_{sp_D^+},\theta)}\right\} \\ &- \mathrm{e}^{-z\alpha(\omega_c)}\sin(\beta(\omega_c)z - \omega_c t); \qquad \theta > \theta_s, \end{aligned} \tag{8.3.3c}$$

as $z \to \infty$. Here $\Delta(\theta)$ is given by (8.2.14).

Finally, for applied signal frequencies ω_c in the range $\sqrt{\omega_0^2 - \delta^2} < \omega_c < \sqrt{\omega_1^2 - \delta^2}$, which is within the absorption band of the medium, substitution of (8.3.1) into (8.2.20) and (8.2.21) (with $\omega_p = \omega_c$) yields the following set of equations describing the uniform asymptotic behavior of the pole contribution at $\omega_p = \omega_c$:

$$A_c(z,t) = 0; \qquad \theta < \theta_s, \tag{8.3.4a}$$

$$A_c(z,t_s) \sim -\frac{1}{2}\mathrm{e}^{-z\alpha(\omega_c)}\sin(\beta(\omega_c)z - \omega_c t); \qquad \theta = \theta_s = \frac{ct_s}{z}, \tag{8.3.4b}$$

$$A_c(z,t) \sim -\mathrm{e}^{-z\alpha(\omega_c)}\sin(\beta(\omega_c)z - \omega_c t); \qquad \theta > \theta_s, \tag{8.3.4c}$$

as $z \to \infty$.

Equations (8.3.4) are the same as the expressions that *Brillouin* [8.1, 2] obtained for the pole contribution for all positive values of ω_c. In the limit of large values of the quantity $|\Delta(\theta)|\sqrt{z/c}$, (8.3.2 and 3) also reduce to the same expressions. Even the value of the time at which the discontinuous jump in the behavior of $A_c(z,t)$ occurs is the same as that obtained by Brillouin because θ_s is taken in the uniform asymptotic analysis of this section to be the value of θ at which the path of steepest descent crosses the simple pole singularity located at $\omega_p = \omega_c$. The nonuniform analysis presented in Sect. 8.1 yields the same expressions for $A_c(z,t)$, but the value of θ_s can be different since θ_s is then the value of θ at which an arbitrary Olver-type path crosses the pole. The difference in values obtained for θ_s is of no consequence, however, since $A_c(z,t)$ is asymptotically negligible in the final expression for the total field $A(z,t)$ for values of θ in a range that includes all possible values of θ_s. Although *Brillouin* associates θ_s with the time of arrival of the main signal, that interpretation is shown to be incorrect in Chap. 9.

8.3.2 The Rectangular-Pulse Modulated Signal

In accordance with the results of Sect. 4.3.2, the propagated disturbance due to an initial unit rectangular-modulated field of time duration T and angular frequency ω_c may be represented by the difference between two unit step-function modulated signals with angular-frequency ω_c and displaced in time by the amount T, viz

$$A(z,t) = -\frac{1}{2\pi}\mathrm{Re}\left\{\int_{ia-\infty}^{ia+\infty}\frac{1}{\omega-\omega_c}e^{(z/c)\phi_0(\omega,\theta)}\,d\omega\right.$$

$$\left.-\,e^{-\omega_c T}\int_{ia-\infty}^{ia+\infty}\frac{1}{\omega-\omega_c}e^{(z/c)\phi_T(\omega,\theta)}\,d\omega\right\}$$

$$\equiv A_I(z,t) - A_{II}(z,t)\ , \tag{8.3.5a}$$

where the initial field begins to radiate time-harmonically at time $t = 0$, and where

$$\phi_T(\omega,\theta_T) = i\omega[n(\omega) - \theta_T]\ , \tag{8.3.5b}$$

$$\theta_T = \frac{c}{z}(t - T)\ . \tag{8.3.5c}$$

In order to avoid confusion for $T = 0$, the subscript is dropped from the parameter θ_T in that case, where $\phi_0(\omega,\theta) = \phi(\omega,\theta)$.

Since the first term $A_I(z,t)$ in the above expression (8.3.5a) for the propagated field is the same as the field treated in Sect. 8.3.1, the uniform asymptotic approximation of the pole contribution associated with that term is given by the set of equations (8.3.2) for values of ω_c in the range $0 \leqslant \omega_c \leqslant \sqrt{\omega_0^2 - \delta^2}$, the set of equations (8.3.3) for $\omega_c \geqslant \sqrt{\omega_1^2 - \delta^2}$, and the set of equations (8.3.4) for $\sqrt{\omega_0^2 - \delta^2} < \omega_c < \sqrt{\omega_1^2 - \delta^2}$. Furthermore, since the second term $A_{II}(z, t)$ in (8.3.5a) is of exactly the same form as the first, except that it is retarded in θ by the amount $-(c/z)T$ and is multiplied by the additional phase factor $e^{-i\omega_c T}$, the uniform asymptotic approximation of the pole contribution associated with that term may then be easily constructed from the same equations. Accordingly, the uniform asymptotic approximation of the pole contribution to $A_{II}(z, t)$ for applied signal frequencies ω_c in the range $0 \leqslant \omega_c \leqslant \sqrt{\omega_0^2 - \delta^2}$ is given by (cf. (8.3.2))

$$A_{IIc}(z,t) \sim \frac{1}{2\pi}\mathbb{R}\mathrm{e}\, \mathrm{e}^{i\omega_c T}\left\{ i\pi\, \mathrm{erfc}\left(i\Delta(\theta_T)\sqrt{\frac{z}{c}}\right) \mathrm{e}^{(z/c)\phi_T(\omega_c,\theta_T)} - \frac{1}{\Delta(\theta_T)}\sqrt{\frac{\pi c}{z}}\mathrm{e}^{(z/c)\phi_T(\omega_{sp},\theta_T)}\right\}; \qquad \theta_T < \theta_s\,, \tag{8.3.6a}$$

$$A_{IIc}(z,t_{Ts}) \sim \frac{1}{2\pi}\mathbb{R}\mathrm{e}\, \mathrm{e}^{i\omega_c T}\left\{ 2\sqrt{\pi}\mathrm{e}^{(z/c)\phi_T(\omega_{sp},\theta_s)} F\left(|\Delta(\theta_s)|\sqrt{\frac{z}{c}}\right) - \frac{1}{\Delta(\theta_s)}\sqrt{\frac{\pi c}{z}}\mathrm{e}^{(z/c)\phi_T(\omega_{sp},\theta_s)}\right\}$$
$$- \frac{1}{2}\mathrm{e}^{-z\alpha(\omega_c)}\sin(\beta(\omega_c)z - \omega_c t_{Ts})\,; \qquad \begin{matrix}\theta_T = \theta_s = \frac{c}{z}t_s \\ \omega_c \neq 0\end{matrix}\,, \tag{8.3.6b}$$

$$A_{IIc}(z,t_{Ts}) \sim \frac{\omega_0^2}{2b}\left[\frac{\theta_0 c}{\pi\delta z}\right]^{1/2}\left[\frac{\cos(\omega_c T)}{\omega_{sp_1}(\theta_s)} + \frac{\alpha}{4\delta}\sin(\omega_c T)\right] - \frac{1}{2}\sin(\omega_c T)\,; \qquad \begin{matrix}\theta_T = \theta_s = \theta_0 \\ \omega_c = 0\end{matrix}\,, \tag{8.3.3c}$$

$$A_{IIc}(z,t) \sim \frac{1}{2\pi}\mathbb{R}\mathrm{e}\, \mathrm{e}^{i\omega_c T}\left\{ -i\pi\, \mathrm{erfc}\left(-i\Delta(\theta_T)\sqrt{\frac{z}{c}}\right) \mathrm{e}^{(z/c)\phi_T(\omega_c,\theta_T)} - \frac{1}{\Delta(\theta_T)}\sqrt{\frac{\pi c}{z}}\mathrm{e}^{(z/c)\phi_T(\omega_{sp},\theta_T)}\right\}$$
$$- \mathrm{e}^{-z\alpha(\omega_c)}\sin(\beta(\omega_c)z - \omega_c t)\,; \qquad \theta_T > \theta_s\,, \tag{8.3.6d}$$

as $z \to \infty$, where $t_{Ts} = t_s + T$ and $t_s = (z/c)\theta_s$. Here ω_{sp} denotes ω_{sp_1} and $\Delta(\theta_T)$ is given by (8.2.7a) with $\phi(\omega, \theta)$ replaced by $\phi_T(\omega, \theta_T)$ for $1 \leqslant \theta_T < \theta_1$, ω_{sp} denotes ω_{sp_N} and $\Delta(\theta_T)$ is given by (8.2.7b) with $\phi(\omega, \theta)$ replaced by $\phi_T(\omega, \theta_T)$ for $\theta_T = \theta_1$, and ω_{sp} denotes $\omega_{sp_N^+}$ and $\Delta(\theta_T)$ is given by (8.2.7c) with $\phi(\omega, \theta)$ replaced by $\phi_T(\omega, \theta_T)$ for $\theta_T > \theta_1$. For applied signal frequencies in the range $\omega_c \geqslant$

$\sqrt{\omega_1^2 - \delta^2}$, the uniform asymptotic approximation of the pole contribution to $A_{IIc}(z,t)$ is given by (cf. (8.3.3))

$$A_{IIc}(z,t) \sim \frac{1}{2\pi} \mathbb{R}\mathrm{e}\, \mathrm{e}^{\mathrm{i}\omega_c T} \left\{ -\mathrm{i}\pi \operatorname{erfc}\left(-\mathrm{i}\Delta(\theta_T) \sqrt{\frac{z}{c}} \right) \mathrm{e}^{(z/c)\phi(\omega_c,\theta_T)} \right.$$
$$\left. - \frac{1}{\Delta(\theta_T)} \sqrt{\frac{\pi c}{z}} \mathrm{e}^{(z/c)\phi(\omega_{sp_D^+},\theta_T)} \right\} ; \qquad 1 \leqslant \theta_T < \theta_s , \quad (8.3.7a)$$

$$A_{IIc}(z,t_{Ts}) \sim \frac{1}{2\pi} \mathbb{R}\mathrm{e}\, \mathrm{e}^{\mathrm{i}\omega_c T} \left\{ 2\sqrt{\pi} \mathrm{e}^{(z/c)\phi(\omega_{sp_D^+},\theta_s)} F\left(|\Delta(\theta_s)| \sqrt{\frac{z}{c}} \right) \right.$$
$$\left. - \frac{1}{\Delta(\theta_s)} \sqrt{\frac{\pi c}{z}} \mathrm{e}^{(z/c)\phi(\omega_{sp_D^+},\theta_s)} \right\}$$
$$- \frac{1}{2} \mathrm{e}^{-z\alpha(\omega_c)} \sin(\beta(\omega_c)z - \omega_c t_{Ts}) ; \qquad \theta_T = \theta_s = \frac{c}{z} t_s , \quad (8.3.7b)$$

$$A_{IIc}(z,t) \sim \frac{1}{2\pi} \mathbb{R}\mathrm{e}\, \mathrm{e}^{\mathrm{i}\omega_c T} \left\{ \mathrm{i}\pi \operatorname{erfc}\left(\mathrm{i}\Delta(\theta_T) \sqrt{\frac{z}{c}} \right) \mathrm{e}^{(z/c)\phi(\omega_c,\theta_T)} \right.$$
$$\left. - \frac{1}{\Delta(\theta_T)} \sqrt{\frac{\pi c}{z}} \mathrm{e}^{(z/c)\phi(\omega_{sp_D^+},\theta_T)} \right\}$$
$$- \mathrm{e}^{-z\alpha(\omega_c)} \sin(\beta(\omega_c)z - \omega_c t) ; \qquad \theta_T > \theta_s , \quad (8.3.7c)$$

as $z \to \infty$, where $t_{Ts} = t_s + T$. Here $\Delta(\theta_T)$ is given by (8.2.14) with $\phi(\omega, \theta)$ replaced by $\phi_T(\omega, \theta_T)$. Finally, for applied signal frequencies ω_c in the range $\sqrt{\omega_0^2 - \delta^2} < \omega_c < \sqrt{\omega_1^2 - \delta^2}$, which is within the absorption band of the medium, the uniform asymptotic approximation of the pole contribution to $A_{IIc}(z,t)$ is given by (cf. (8.3.4))

$$A_{IIc}(z,t) = 0 ; \qquad \theta_T < \theta_s , \quad (8.3.8a)$$

$$A_{IIc}(z,t_{Ts}) \sim -\frac{1}{2} \mathrm{e}^{-z\alpha(\omega_c)} \sin(\beta(\omega_c)z - \omega_c t_{Ts}) ; \qquad \theta_T = \theta_s = \frac{c}{z} t_s , \quad (8.3.8b)$$

$$A_{IIc}(z,t) \sim -\mathrm{e}^{-z\alpha(\omega_c)} \sin(\beta(\omega_c)z - \omega_c t) ; \qquad \theta_T > \theta_s , \quad (8.3.8c)$$

as $z \to \infty$, where $t_{Ts} = t_s + T$.

8.3.3 The Delta-Function Pulse

For an input delta-function, the initial spectrum is given by (cf. (4.3.43)

$$\tilde{u}(\omega - \omega_c) = -\mathrm{i} ,$$

which is analytic throughout the entire complex ω-plane. Consequently, $\gamma = 0$ and there is no pole contribution to the delta-function field.

8.3.4 The Hyperbolic-Tangent Modulated Signal

The spectrum of the initial field envelope for the hyperbolic-tangent modulated signal is given by (cf. (4.3.46))

$$\tilde{u}(\omega-\omega_c)=\frac{\pi\mathrm{i}}{2\beta_T}\cdot\frac{1}{\sinh\left(\frac{\pi}{2\beta_T}(\omega-\omega_c)\right)}, \tag{8.3.9}$$

which possesses simple pole singularities located at

$$\omega_{pn}^{\pm}=\omega_c\pm 2n\beta_T i\;;\qquad n=0,1,2,3,\ldots, \tag{8.3.10}$$

where $\omega_c\geqslant 0$ is the real applied signal frequency. The parameter $\beta_T\geqslant 0$ is indicative of the rapidity of the turn-on time for the initial field; the value $\beta_T=\infty$ corresponds to the unit step-function modulated signal, and the value $\beta_T=0$ corresponds to a monochromatic plane wave, which is of no interest here. The values of the residues of the spectrum at the simple pole singularities given in (8.3.10) are

$$\gamma_{pn}^{\pm}=\lim_{\omega\to\omega_c\pm 2n\beta_T\mathrm{i}}\left[(\omega-\omega_c\mp 2n\beta_T\mathrm{i})\frac{\pi\mathrm{i}}{2\beta_T}\cdot\frac{1}{\sinh\left(\frac{\pi}{2\beta_T}(\omega-\omega_c)\right)}\right]=\mathrm{i}\;, \tag{8.3.11}$$

and hence, are independent of the particular pole singularity.

For an instantaneous turn-on time (i.e. for infinite β_T) the only pole singularity in the finite complex ω-plane is located along the positive real axis at $\omega_{p_0}=\omega_c$. In that case, the pole contribution is given entirely by the results of Sect. 8.3.1. As the parameter β_T is allowed to decrease, the other pole singularities $\omega_{pn}^{\pm}$, $n\neq 0$ will interact with the deformed contour of integration $P(\theta)$ which passes through the saddle points, and their contributions to the field $A_c(z,t)$ must be taken into account[7]. For that purpose, let $\theta=\theta_r$ be defined as the value of θ at which the real part of the interacting saddle point $\omega_{sp}(\theta)$ is equal to the real part of the simple pole singularity, viz

$$\mathrm{Re}[\omega_{sp}(\theta_r)]\equiv\omega_c\;. \tag{8.3.12}$$

The particular pole singularities $\omega_{pn}^{\pm}$ which interact with the relevant saddle point then depend upon the relative values of the quantity $\pm 2n\beta_T$ and the imaginary part of that saddle point at $\theta=\theta_r$. To illuminate the situation, the asymptotic approximation of $A_c(z,t)$ for applied signal frequencies in the range $\omega_c\geqslant\sqrt{\omega_1^2-\delta^2}$ is now obtained as the parameter β_T is allowed to decrease

[7] It is to be remembered that the original contour of integration, which extends from $ia-\infty$ to $ia+\infty$, is constrained to lie within the range $0<a<2\beta_T$ (see Sect. 4.3.4).

from infinity. In that case, ω_{sp} is the distant saddle point location $\omega_{sp_D^+}$, θ_r is given by (cf. (6.2.21 and 8.3.12)

$$\xi(\theta_r) \equiv \omega_c \geqslant \sqrt{\omega_1^2 - \delta^2}\ , \tag{8.3.13}$$

and $\operatorname{Im}[\omega_{sp_D^+}(\theta)] = -\delta(1 + \eta(\theta))$.

Consider first the case in which $\beta_T > \delta((1 + \eta(\theta_r))/2)$. The geometry of this situation is illustrated in Fig. 8.5(a). In this case the pole singularities at $\omega_{pn}^- = \omega_c - 2n\beta_T\mathrm{i}$, $n = 1, 2, 3, \ldots$, lie below the distant saddle point SP_D^+ for all θ. For small enough values of $\beta_T > \delta((1 + \eta(\theta_r))/2)$, the pole singularity at $\omega_{p1}^- = \omega_c - 2\beta\mathrm{i}$ will interact with the path $P(\theta)$ which passes through the saddle point SP_D^+. Now, according to (6.2.24), the approximate complex phase at the simple pole singularities in the lower-half of the complex ω-plane is given by

$$\begin{aligned}\phi(\omega_{pn}^-, \theta) &\cong \mathrm{i}(\omega_c - 2n\beta_T\mathrm{i})(1 - \theta) - \mathrm{i}\frac{b^2}{2(\omega_c - 2n\beta_T\mathrm{i} + 2\delta\mathrm{i})} \\ &= -2n\beta_T(\theta - 1) + b^2\frac{n\beta_T - \delta}{\omega_c^2 + 4(n\beta_T - \delta)^2} \\ &\quad - \mathrm{i}\left[\omega_c(\theta - 1) + \frac{b^2}{2}\cdot\frac{\omega_c}{\omega_c^2 + 4(n\beta_T - \delta)^2}\right],\end{aligned} \tag{8.3.14}$$

for $n = 1, 2, 3, \ldots$. The approximate complex phase at the distant saddle point SP_D^+ is given by (7.3.4), viz

$$\begin{aligned}\phi(\omega_{sp_D^+}, \theta) \cong &-\delta\left\{(1 + \eta(\theta))(\theta - 1) + \frac{b^2}{2}\cdot\frac{1 - \eta(\theta)}{\xi^2(\theta) + \delta^2(1 - \eta(\theta))^2}\right\} \\ &- \mathrm{i}\xi(\theta)\left\{\theta - 1 + \frac{b^2}{2}\cdot\frac{1}{\xi^2(\theta) + \delta^2(1 - \eta(\theta))^2}\right\}.\end{aligned} \tag{8.3.15}$$

From (8.3.14), it is seen that the other simple pole singularities located at $\omega_{pn}^- = \omega_c - 2n\beta_T\mathrm{i}$, $n = 2, 3, 4, \ldots$, have a larger exponential decay associated with their contribution (if any) than that at both $\omega_{p0} = \omega_c$ and $\omega_{p1}^- = \omega_c - 2\beta_T\mathrm{i}$, and hence, are asymptotically negligible in this case. Furthermore, for the simple pole singularities of the integrand located at $\omega_{pn}^+ = \omega_c + 2n\beta_T\mathrm{i}$, $n = 1, 2, 3, \ldots$, in the upper-half of the complex ω-plane, the deformation of the original contour of integration to the path $P(\theta)$ through the distant saddle point SP_D^+ does not cross any of these singularities for the situation depicted in Fig. 8.5(a). For smaller values of $\beta_T > \delta((1 + \eta(\theta_r))/2)$ than that depicted in Fig. 8.5(a), or for another choice of the Olver-type path $P(\theta)$ through the saddle point SP_D^+, the original contour of integration may be deformed across the simple pole singularity located at $\omega_{p1}^+ = \omega_c + 2\beta_T\mathrm{i}$. However, that pole contribution only occurs during those values of θ when the exponential attenuation associated with that simple pole singularity is less than that at the distant saddle point SP_D^+ (i.e.

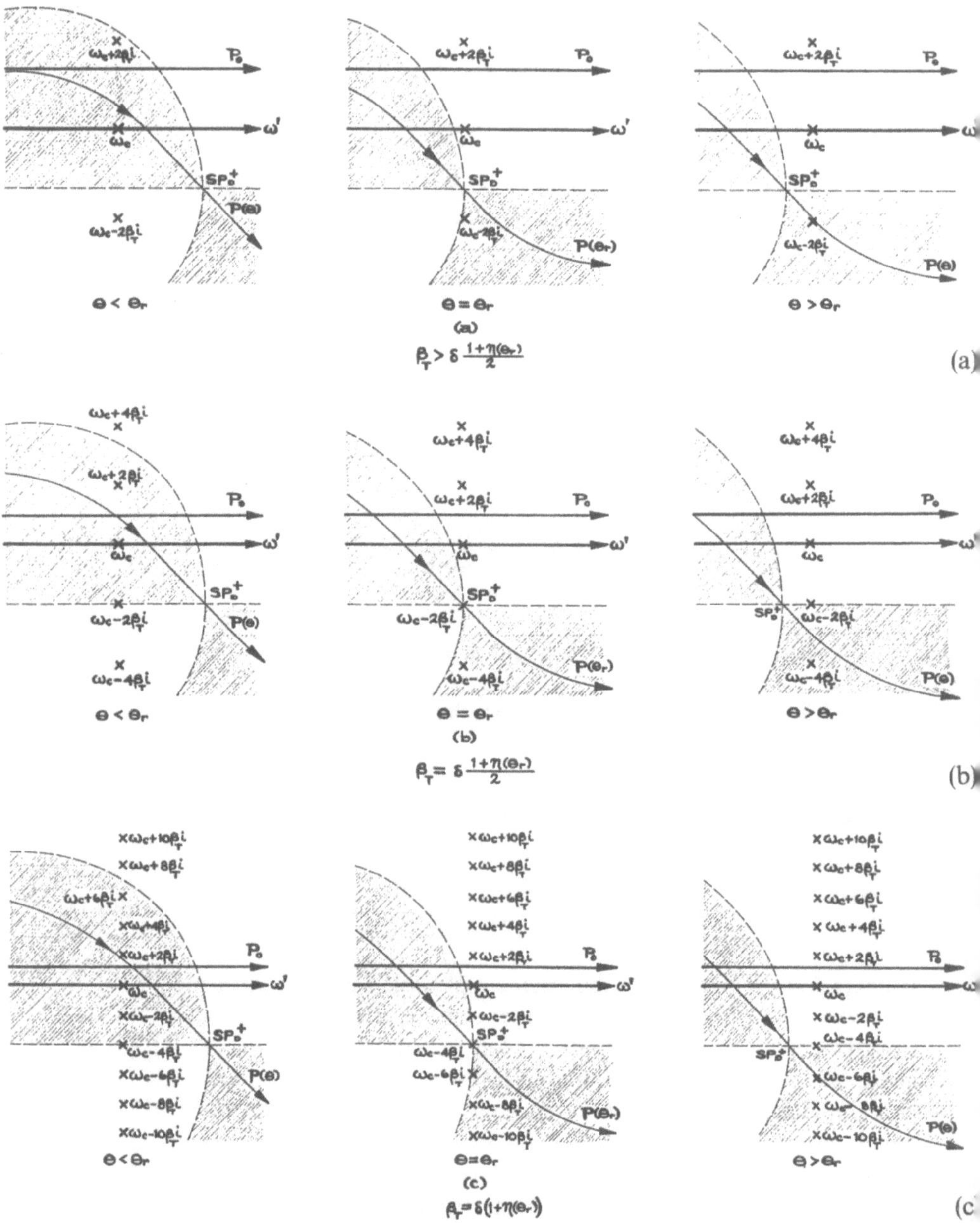

Fig. 8.5. The interaction of the deformed contour of integration through the distant saddle point SP_D^+ in the right-half of the complex ω-plane with the simple pole singularities of the initial envelope of the hyperbolic-targent modulated signal for $\omega_c > \sqrt{\omega_1^2 - \delta^2}$. The contour P_0 is the original contour of integration extending from $ia - \infty$ to $ia + \infty$ which is constrained to satisfy $0 < a < 2\beta_T$. The hatched area indicates the region of the complex ω-plane wherein the inequality $X(\omega, \theta) < X(\omega_{sp_D^+}, \theta)$ is satisfied.

$X(\omega_c + 2\beta_T i, \theta) < X(\omega_{sp_D^+}, \theta)$); for those values of θ when the value of $X(\omega, \theta)$ at that pole singularity is either comparable to or greater than that at the distant saddle point, the original contour of integration will not cross the pole singularity in being deformed to the path $P(\theta)$ through the distant saddle point SP_D^+ (i.e. P_0 and $P(\theta)$ lie on the same side of the pole in that case). Consequently, the contributions (if any) of the simple pole singularities $\omega_{pn}^+ = \omega_c + 2n\beta_T i$ located in the upper-half of the complex ω-plane to $A_c(z,t)$ are asymptotically negligible for $\beta_T > \delta((1 + \eta(\theta_r))/2)$ and are ignored here.

Therefore, for $\beta_T > \delta((1 + \eta(\theta_r))/2)$ and $\omega_c \geqslant \sqrt{\omega_1^2 - \delta^2}$, the distant saddle point SP_D^+ interacts with the simple pole singularities at $\omega_{p0} = \omega_c$ and $\omega_{p1}^- = \omega_c - 2\beta_T i$. In that case, $\Lambda(\theta)$ is given by the sum of the residues at these two pole singularities, viz

$$\Lambda(\theta) = -\frac{1}{2\pi}[e^{(z/c)\phi(\omega_c, \theta)}\mathscr{U}(\theta - \theta_s) + e^{(z/c)\phi(\omega_c - 2\beta_T i, \theta)}\mathscr{U}(\theta - \theta_{s1})] ,$$

where $\mathscr{U}(\theta) = 0$ for $\theta < 0$, $\mathscr{U}(0) = \frac{1}{2}$, and $\mathscr{U}(\theta) = 1$ for $\theta > 0$, and where θ_{s1} is defined below. The uniform asymptotic contribution of the simple pole singularity located at $\omega_{p0} = \omega_c$ to $A_c(z,t)$ is given by (8.3.3) with $\Delta(\theta)$ given by (8.2.14). The uniform asymptotic contribution of the simple pole singularity located at $\omega_{p1}^- = \omega_c - 2\beta_T i$ is given as follows. Let θ_{s1} denote that value of θ at which the path of steepest descent through the distant saddle point SP_D^+ crosses the simple pole singularity at ω_{p1}^-, viz

$$Y(\omega_{sp_D^+}, \theta_{s1}) = Y(\omega_{p1}^-, \theta_{s1}) . \tag{8.3.16}$$

As can easily be seen from Fig. 8.5(a),, $\theta_s < \theta_r < \theta_{s1}$. The particular form to be employed of the uniform asymptotic approximation of the contribution due to the simple pole singularity at $\omega_{p1}^- = \omega_c - 2\beta_T i$, as given by (8.2.2), depends upon the sign of the imaginary part of the quantity $\Delta_1(\theta)$, which is given by (cf. (8.2.3))

$$\Delta_1(\theta) \equiv [\phi(\omega_{sp_D^+}, \theta) - \phi(\omega_c - 2\beta_T i, \theta)]^{1/2} , \tag{8.3.17}$$

where $\phi(\omega_c - 2\beta_T i, \theta)$ is given by (8.3.14) with $n = 1$. The proper argument of this expression is determined by (8.2.4), where $\bar{\alpha}_c$ is the angle of slope of the vector from $\omega_{sp_D^+}$ to $\omega_{p1}^- = \omega_c - 2\beta_T i$ in the complex ω-plane, $\bar{\alpha}_{SD}$ is the angle of slope of the tangent vector to the path of steepest descent at the saddle point, and n is an integer which can be chosen so that the argument of $\Delta_1(\theta)$ lies within the principal range $(-\pi, \pi]$ for all $\theta \geqslant 1$. The angle of slope of the path of steepest descent leaving the distant saddle point SP_D^+ is $\bar{\alpha}_{SD} = -\pi/4$ for all $\theta > 1$[8]. The angle of slope $\bar{\alpha}_c$ of the vector from $\omega_{sp_D^+}$ to $\omega_c - 2\beta_T i$ increases

[8] For values of β_T approaching close to $\delta((1 + \eta(\theta_r))/2)$, a transition in the geometry of the saddle point interaction with the pole will occur in which the path of steepest descent approaching SP_D^+ will interact with the simple pole singularity at $\omega_{p1}^- = \omega_c - 2\beta_T i$ for $\theta < \theta_{s1}$, and the path of steepest descent leaving SP_D^+ will interact will that pole for $\theta > \theta_{s1}$. The analysis of this limiting situation is not treated here, but is similar to that treated for $\beta_T = \delta((1 + \eta(\theta_r))/2)$.

monotonically from $-\pi$ to $-\pi/4$ as θ increases from 1 to θ_{s1}, is equal to $-\pi/4$ at $\theta = \theta_{s1}$, and increases monotonically from $-\pi/4$ to approaching 0 as θ increases from θ_{s1} to infinity. Consequently, according to (8.2.4) with $n = 0$, $\arg(\Delta_1(\theta))$ increases monotonically from $-3\pi/4$ to 0 as θ increases from 1 to θ_{s1}, is equal to 0 at $\theta = \theta_{s1}$ (where $\Delta_1(\theta_{s1}) \neq 0$), and increases monotonically from 0 to approaching $\pi/4$ as θ increases from θ_{s1} to infinity.

The uniform asymptotic approximation of the contributions of the simple pole singularities at $\omega_{p0} = \omega_c$ and $\omega_{p1}^- = \omega_c - 2\beta i$ to $A_c(z,t)$, with $\beta_T > \delta((1+\eta(\theta_r))/2)$ and $\omega_c \geqslant \sqrt{\omega_1^2 - \delta^2}$, is then given as follows. For $1 \leqslant \theta < \theta_s$, $\mathrm{Im}(\Delta(\theta)) > 0$ and $\mathrm{Im}(\Delta_1(\theta)) < 0$, and (8.3.3a and 2.2a) yield

$$A_c(z,t) \sim \frac{1}{2\pi}\mathrm{Re}\left\{ -i\pi\,\mathrm{erfc}\left(-i\Delta(\theta)\sqrt{\frac{z}{c}}\right) e^{(z/c)\phi(\omega_c,\theta)} - \frac{1}{\Delta(\theta)}\sqrt{\frac{\pi c}{z}} e^{(z/c)\phi(\omega_{sp_D^+},\theta)} \right.$$
$$\left. + i\pi\,\mathrm{erfc}\left(i\Delta_1(\theta)\sqrt{\frac{z}{c}}\right) e^{(z/c)\phi(\omega_c - 2\beta_T i,\theta)} - \frac{1}{\Delta_1(\theta)}\sqrt{\frac{\pi c}{z}} e^{(z/c)\phi(\omega_{sp_D^+},\theta)} \right\} ;$$
$$1 \leqslant \theta \leqslant \theta_s \,, \qquad (8.3.18a)$$

as $z \to \infty$. At $\theta = \theta_s < \theta_{s1}$, $\mathrm{Im}(\Delta(\theta_s)) = 0$ with $\Delta(\theta_s) \neq 0$ and $\mathrm{Im}(\Delta_1(\theta_s)) < 0$, and (8.3.3b and 2.2a) yield

$$A_c(z,t_s) \sim \frac{1}{2\pi}\mathrm{Re}\left\{ 2\sqrt{\pi} e^{(z/c)\phi(\omega_{sp_D^+},\theta_s)} F\left(|\Delta(\theta_s)|\sqrt{\frac{z}{c}}\right) - \frac{1}{\Delta(\theta_s)}\sqrt{\frac{\pi c}{z}} e^{(z/c)\phi(\omega_{sp_D^+},\theta_s)} \right.$$
$$\left. + i\pi\,\mathrm{erfc}\left(i\Delta_1(\theta_s)\sqrt{\frac{z}{c}}\right) e^{(z/c)\phi(\omega_c - 2\beta_T i,\theta_s)} - \frac{1}{\Delta_1(\theta_s)}\sqrt{\frac{\pi c}{z}} e^{(z/c)\phi(\omega_{sp_D^+},\theta_s)} \right\}$$
$$- \frac{1}{2} e^{-z\alpha(\omega_c)} \sin(\beta(\omega_c)z - \omega_c t_s) \; ; \qquad \theta = \theta_s = \frac{ct_s}{z} \,, \qquad (8.3.18b)$$

as $z \to \infty$. For $\theta_s < \theta < \theta_{s1}$, $\mathrm{Im}(\Delta(\theta)) < 0$ and $\mathrm{Im}(\Delta_1(\theta)) < 0$, and (8.3.3c and 2a) yield

$$A_c(z,t) \sim \frac{1}{2\pi}\mathrm{Re}\left\{ i\pi\,\mathrm{erfc}\left(i\Delta(\theta)\sqrt{\frac{z}{c}}\right) e^{(z/c)\phi(\omega_c,\theta)} - \frac{1}{\Delta(\theta)}\sqrt{\frac{\pi c}{z}} e^{(z/c)\phi(\omega_{sp_D^+},\theta)} \right.$$
$$\left. + i\pi\,\mathrm{erfc}\left(i\Delta_1(\theta)\sqrt{\frac{z}{c}}\right) e^{(z/c)\phi(\omega_c - 2\beta_T i,\theta)} - \frac{1}{\Delta_1(\theta)}\sqrt{\frac{\pi c}{z}} e^{(z/c)\phi(\omega_{sp_D^+},\theta)} \right\}$$
$$- e^{-z\alpha(\omega_c)} \sin(\beta(\omega_c)z - \omega_c t) \; ; \qquad \theta_s < \theta < \theta_{s1} \,, \qquad (8.3.18c)$$

as $z \to \infty$. At $\theta = \theta_{s1}$, $\mathrm{Im}(\Delta(\theta_{s1})) < 0$ and $\mathrm{Im}(\Delta_1(\theta_{s1})) = 0$ with $\Delta_1(\theta_{s1}) \neq 0$, and (8.3.3c and 2.2b) yield

$$A_c(z, t_{s1}) \sim \frac{1}{2\pi}\mathbb{R}\mathrm{e}\left\{ i\pi \operatorname{erfc}\left(i\Delta(\theta_{s1})\sqrt{\frac{z}{c}}\right) e^{(z/c)\phi(\omega_c, \theta_{s1})} - \frac{1}{\Delta(\theta_{s1})}\sqrt{\frac{\pi c}{z}} e^{(z/c)\phi(\omega_{sp_D^+}, \theta_{s1})} \right.$$

$$+ 2\sqrt{\pi} e^{(z/c)\phi(\omega_{sp_D^+}, \theta_{s1})} F\left(|\Delta_1(\theta_{s1})| \sqrt{\frac{z}{c}}\right)$$

$$\left. - \frac{1}{\Delta_1(\theta_{s1})}\sqrt{\frac{\pi c}{z}} e^{(z/c)\phi(\omega_{sp_D^+}, \theta_{s1})} \right\}$$

$$- e^{-z\alpha(\omega_c)} \sin(\beta(\omega_c)z - \omega_c t_{s1})$$

$$+ \frac{1}{2}\mathbb{R}\mathrm{e}[i e^{(z/c)\phi(\omega_c - 2\beta_T i, \theta_{s1})}] \; ; \qquad \theta = \theta_{s1} = \frac{c t_{s1}}{z} \; , \tag{8.3.18d}$$

as $z \to \infty$. Finally, for $\theta_s > \theta_{s1}$, $\mathrm{Im}(\Delta(\theta)) < 0$ and $\mathrm{Im}(\Delta_1(\theta)) > 0$, and (8.3.3c and 2.2a) yield

$$A_c(z, t) \sim \frac{1}{2\pi}\mathbb{R}\mathrm{e}\left\{ i\pi \operatorname{erfc}\left(i\Delta(\theta)\sqrt{\frac{z}{c}}\right) e^{(z/c)\phi(\omega_c, \theta)} - \frac{1}{\Delta(\theta)}\sqrt{\frac{\pi c}{z}} e^{(z/c)\phi(\omega_{sp_D^+}, \theta)} \right.$$

$$\left. - i\pi \operatorname{erfc}\left(-i\Delta_1(\theta)\sqrt{\frac{z}{c}}\right) e^{(z/c)\phi(\omega_c - 2\beta_T i, \theta)} - \frac{1}{\Delta_1(\theta)}\sqrt{\frac{\pi c}{z}} e^{(z/c)\phi(\omega_{sp_D^+}, \theta)} \right\}$$

$$- e^{-z\alpha(\omega_c)} \sin(\beta(\omega_c)z - \omega_c t)$$

$$+ \mathbb{R}\mathrm{e}[i e^{(z/c)\phi(\omega_c - 2\beta_T i, \theta)}] \; ; \qquad \theta > \theta_{s1} \; , \tag{8.3.18e}$$

as $z \to \infty$. Taken together, (8.3.18a–e) represents the uniform asymptotic approximation of the pole contribution $A_c(z, t)$ for $\omega_c \geqslant \sqrt{\omega_1^2 - \delta^2}$ and $\beta_T > \delta((1 + \eta(\theta_r))/2)$. The residue contribution of the simple pole singularity located at $\omega_{p1}^- = \omega_c - 2\beta_T i$ appearing in (8.3.18d and 18e) is given by (cf. (8.3.14))

$$\mathbb{R}\mathrm{e}[i e^{(z/c)\phi(\omega_c - 2\beta_T i, \theta)} \cong \exp\left\{ -\frac{z}{c}\left[2\beta_T(\theta - 1) + b^2 \cdot \frac{\delta - \beta_T}{\omega_c^2 + 4(\beta_T - \delta)^2}\right]\right\}$$

$$\cdot \sin\left[\frac{z}{c}\omega_c\left(\theta - 1 + \frac{b^2}{2}\cdot\frac{1}{\omega_c^2 + 4(\beta_T - \delta)^2}\right)\right] .$$

For large values of $\beta_T \gg \delta((1 + \eta(\theta_r))/2)$ such that $|\Delta_1(\theta)|$ is bounded away from zero for all $\theta \geqslant 1$, and for sufficiently large values of the observation distance z such that the quantity $|\Delta_1(\theta)|\sqrt{z/c}$ is large for all $\theta \geqslant 1$, the large argument asymptotic expansions of the complementary error function $\operatorname{erfc}(\pm i\Delta_1(\theta)\sqrt{z/c})$ and *Dawson*'s integral $F(|\Delta_1(\theta_{s1})|\sqrt{z/c})$ may be employed in (8.3.18), yielding the following set of equations

$$A_c(z, t) \sim \frac{1}{2\pi}\mathbb{R}\mathrm{e}\left\{ -i\pi \operatorname{erfc}\left(-i\Delta(\theta)\sqrt{\frac{z}{c}}\right) e^{(z/c)\phi(\omega_c, \theta)} \right.$$

$$\left. - \frac{1}{\Delta(\theta)}\sqrt{\frac{\pi c}{z}} e^{(z/c)\phi(\omega_{sp_D^+}, \theta)} \right\} \; ; \qquad 1 \leqslant \theta \leqslant \theta_s \; , \tag{8.3.19a}$$

$$A_c(z,t_s) \sim \frac{1}{2\pi}\,\mathbb{R}\mathrm{e}\left\{2\sqrt{\pi}\,\mathrm{e}^{(z/c)\phi(\omega_{sp_D^+},\theta_s)}F\left(|\Delta(\theta_s)|\sqrt{\frac{z}{c}}\right)\right.$$
$$\left.-\frac{1}{\Delta(\theta_s)}\sqrt{\frac{\pi c}{z}}\,\mathrm{e}^{(z/c)\phi(\omega_{sp_D^+},\theta_s)}\right\}$$
$$-\frac{1}{2}\mathrm{e}^{-z\alpha(\omega_c)}\sin(\beta(\omega_c)z-\omega_c t_s)\ ; \qquad \theta=\theta_s=\frac{ct_s}{z}\ , \tag{8.3.19b}$$

$$A_c(z,t) \sim \frac{1}{2\pi}\,\mathbb{R}\mathrm{e}\left\{\mathrm{i}\pi\,\mathrm{erfc}\left(\mathrm{i}\Delta(\theta)\sqrt{\frac{z}{c}}\right)\mathrm{e}^{(z/c)\phi(\omega_c,\theta)}-\frac{1}{\Delta(\theta)}\sqrt{\frac{\pi c}{z}}\,\mathrm{e}^{(z/c)\phi(\omega_{sp_D^+},\theta)}\right\}$$
$$-\mathrm{e}^{-z\alpha(\omega_c)}\sin(\beta(\omega_c)z-\omega_c t)\ ; \qquad \theta_s<\theta<\theta_{s1}\ , \tag{8.3.19c}$$

$$A_c(z,t_{s1}) \sim \frac{1}{2\pi}\,\mathbb{R}\mathrm{e}\left\{\mathrm{i}\pi\,\mathrm{erfc}\left(\mathrm{i}\Delta(\theta_{s1})\sqrt{\frac{z}{c}}\right)\mathrm{e}^{(z/c)\phi(\omega_c,\theta_{s1})}\right.$$
$$\left.-\frac{1}{\Delta(\theta_{s1})}\sqrt{\frac{\pi c}{z}}\,\mathrm{e}^{(z/c)\phi(\omega_{sp_D^+},\theta_{s1})}\right\}$$
$$-\mathrm{e}^{-z\alpha(\omega_c)}\sin(\beta(\omega_c)z-\omega_c t_{s1})$$
$$+\frac{1}{2}\mathbb{R}\mathrm{e}\left[\mathrm{i}\mathrm{e}^{(z/c)\phi(\omega_c-2\beta_T\mathrm{i},\theta_{s1})}\right]\ ; \qquad \theta=\theta_{s1}=\frac{ct_{s1}}{z}\ , \tag{8.3.19d}$$

$$A_c(z,t) \sim \frac{1}{2\pi}\,\mathbb{R}\mathrm{e}\left\{\mathrm{i}\pi\,\mathrm{erfc}\left(\mathrm{i}\Delta(\theta)\sqrt{\frac{z}{c}}\right)\mathrm{e}^{(z/c)\phi(\omega_c,\theta)}-\frac{1}{\Delta(\theta)}\sqrt{\frac{\pi c}{z}}\,\mathrm{e}^{(z/c)\phi(\omega_{sp_D^+},\theta)}\right\}$$
$$-\mathrm{e}^{-z\alpha(\omega_c)}\sin(\beta(\omega_c)z-\omega_c t)$$
$$+\mathbb{R}\mathrm{e}\left[\mathrm{i}\mathrm{e}^{(z/c)\phi(\omega_c-2\beta_T\mathrm{i},\theta)}\right]\ ; \qquad \theta>\theta_{s1}\ , \tag{8.3.19e}$$

as $z\to\infty$. For values of $\beta_T \gg \delta((1+\eta(\theta_r))/2)$, the exponential decay associated with the simple pole singularity at $\omega_{p1}^- = \omega_c - 2\beta_T\mathrm{i}$ is much greater than that at $\omega_{p0} = \omega_c$. Consequently, for increasingly large values of $\beta_T \gg \delta((1+\eta(\theta_r))/2)$, the pole contribution due to the singularity located at $\omega_{p1}^- = \omega_c - 2\beta_T\mathrm{i}$ is exponentially negligible in comparison to that at $\omega_{p0} = \omega_c$. In that case, which corresponds to a very rapid initial signal turn-on time, (8.3.19) reduce to (8.3.3) for the unit step-function modulated signal, as is to be expected. However, as the parameter β_T approaches close to the value $\delta((1+\eta(\theta_r))/2)$, the contribution of the pole singularity at ω_{p1}^- is not insignificant and the asymptotic approximation of $A_c(z,t)$ is given by (8.3.18).

Consider now the case for which $\beta_T = \delta((1+\eta(\theta_r))/2)$. The geometry of this situation is illustrated in Fig. 8.5(b). In this case the pole singularity at $\omega_{p1}^- = \omega_c - 2\beta_T\mathrm{i}$ coalesces with the distant saddle point SP_D^+ at $\theta = \theta_{s1} = \theta_r$. In addition, the path of steepest descent leaving the saddle point SP_D^+ crosses the simple pole singularity at $\omega_{p2}^- = \omega_c - 4\beta_T i$ at $\theta = \theta_{s2} > \theta_{s1}$, where

$$Y(\omega_{sp_D^+}, \theta_{s2}) = Y(\omega_{p2}^-, \theta_{s2}) \ . \tag{8.3.20}$$

Therefore, for $\beta_T = \delta((1 + \eta(\theta_r))/2)$ and $\omega_c \geqslant \sqrt{\omega_1^2 - \delta^2}$, the distant saddle point SP_D^+ interacts with the simple pole singularities located at $\omega_{p0} = \omega_c$, $\omega_{p1}^- = \omega_c - 2\beta_T i$, and $\omega_{p2}^- = \omega_c - 4\beta_T i$. For this case $\Lambda(\theta)$ is given by the sum of the residues at these three pole singularities, viz

$$\Lambda(\theta) = -\frac{1}{2\pi}\left[e^{(z/c)\phi(\omega_c,\theta)}\mathscr{U}(\theta - \theta_s) + e^{(z/c)\phi(\omega_c - 2\beta_T i,\theta)}\mathscr{U}(\theta - \theta_{s1}) \right.$$
$$\left. + e^{(z/c)\phi(\omega_c - 4\beta_T i,\theta)}\mathscr{U}(\theta - \theta_{s2})\right] \ .$$

The other pole singularities located at $\omega_{pn}^- = \omega_c - 2n\beta_T i$, $n = 3, 4, 5, \ldots$, have a larger exponential decay associated with their contributions (if any) to the asymptotic approximation of $A_c(z,t)$ than that at $\omega_{p2}^- = \omega_c - 4\beta_T i$, and are consequently negligible in this case. The uniform asymptotic contribution of the simple pole singularity located at $\omega_{p0} = \omega_c$ to $A_c(z,t)$ is given by (8.3.3) with $\Delta(\theta)$ given by (8.2.14). The particular form to be employed of the uniform asymptotic contribution of the simple pole singularity at $\omega_{p1}^- = \omega_c - 2\beta_T i$, as given by (8.2.2), depends upon the sign of the imaginary part of the quantity $\Delta_1(\theta)$, which is given by

$$\Delta_1(\theta) \equiv [\phi(\omega_{sp_D^+}, \theta) - \phi(\omega_c - 2\beta_T i, \theta)]^{1/2} \ , \tag{8.3.21}$$

and that at $\omega_{p2}^- = \omega_c - 4\beta_T i$ depends upon the sign of the imaginary part of the quantity $\Delta_2(\theta)$, which is given by

$$\Delta_2(\theta) \equiv [\phi(\omega_{sp_D^+}, \theta) - \phi(\omega_c - 4\beta_T i, \theta)]^{1/2} \ . \tag{8.3.22}$$

The proper arguments of these two expressions are determined by (8.2.4) as follows. The angle of slope of the path of steepest descent approaching the distant saddle point SP_D^+ is $\bar{\alpha}_{SD} = 3\pi/4$ for all $\theta \geqslant 1$, and the angle of slope of the path of steepest descent leaving the distant saddle point is $\bar{\alpha}_{SD} = -\pi/4$ for all $\theta > 1$. The path of steepest descent approaching SP_D^+ interacts with the simple pole singularity at $\omega_{p1}^- = \omega_c - 2\beta_T i$ for values of θ in the range $1 \leqslant \theta < \theta_{s1}$, and the path of steepest descent leaving SP_D^+ interacts with that pole for values of θ in the range $\theta > \theta_{s1}$. The angle of slope $\bar{\alpha}_c$ of the vector from $\omega_{sp_D^+}$ to $\omega_c - 2\beta_T i$ is approximately equal to π as θ increases from 1 to θ_{s1}, and is approximately equal to zero as θ increases from θ_{s1} to infinity. Consequently, according to (8.2.4) with $n = 0$, $\arg(\Delta_1(\theta))$ is approximately equal to $\pi/4$ for all $\theta \geqslant 1$. The path of steepest descent leaving the distant saddle point SP_D^+ interacts with the simple pole singularity at $\omega_{p2}^- = \omega_c - 4\beta_T i$ for all $\theta > 1$. The angle of slope $\bar{\alpha}_c$ of the vector from $\omega_{sp_D^+}$ to $\omega_{p2}^- = \omega_c - 4\beta_T i$ increases monotonically from $-\pi$ to $-\pi/4$ as θ increases from 1 to θ_{s2}, and increases monotonically from $-\pi/4$ to approaching 0 as θ increases from θ_{s2} to infinity. Consequently, according to

(8.2.4) with $n = 0$, $\arg(\Delta_2(\theta))$ increases monotonically from $-3\pi/4$ to 0 as θ increases from 1 to θ_{s2}, and increases monotonically from 0 to approaching $\pi/4$ as θ increases from θ_{s2} to infinity.

The uniform asymptotic contributions of the simple pole singularities at $\omega_{p0} = \omega_c$, $\omega_{p1}^- = \omega_c - 2\beta_T\mathrm{i}$, and $\omega_{p2}^- = \omega_c - 4\beta_T\mathrm{i}$ to $A_c(z,t)$, with $\beta_T = \delta((1+\eta(\theta_r))/2)$ and $\omega_c \geqslant \sqrt{\omega_1^2 - \delta^2}$, is then given as follows. For $1 \leqslant \theta < \theta_s$, $\mathrm{Im}(\Delta(\theta)) > 0$, $\mathrm{Im}(\Delta_1(\theta)) > 0$, and $\mathrm{Im}(\Delta_2(\theta)) < 0$, and (8.3.3a and 2.2a) yield

$$
\begin{aligned}
A_c(z,t) \sim \frac{1}{2\pi}\mathrm{Re}\Bigg\{ & -\mathrm{i}\pi\,\mathrm{erfc}\left(-\mathrm{i}\Delta(\theta)\sqrt{\frac{z}{c}}\right)\mathrm{e}^{(z/c)\phi(\omega_c,\theta)} - \frac{1}{\Delta(\theta)}\sqrt{\frac{\pi c}{z}}\mathrm{e}^{(z/c)\phi(\omega_{sp_D^+},\theta)} \\
& - \mathrm{i}\pi\,\mathrm{erfc}\left(-\mathrm{i}\Delta_1(\theta)\sqrt{\frac{z}{c}}\right)\mathrm{e}^{(z/c)\phi(\omega_c-2\beta_T\mathrm{i},\theta)} - \frac{1}{\Delta_1(\theta)}\sqrt{\frac{\pi c}{z}}\mathrm{e}^{(z/c)\phi(\omega_{sp_D^+},\theta)} \\
& + \mathrm{i}\pi\,\mathrm{erfc}\left(\mathrm{i}\Delta_2(\theta)\sqrt{\frac{z}{c}}\right)\mathrm{e}^{(z/c)\phi(\omega_c-4\beta_T\mathrm{i},\theta)} - \frac{1}{\Delta_2(\theta)}\sqrt{\frac{\pi c}{z}}\mathrm{e}^{(z/c)\phi(\omega_{sp_D^+},\theta)}\Bigg\} ; \\
& \qquad 1 \leqslant \theta < \theta_s \, , \qquad (8.3.23\mathrm{a})
\end{aligned}
$$

as $z \to \infty$. At $\theta = \theta_s < \theta_{s1} < \theta_{s2}$, $\mathrm{Im}(\Delta(\theta_s)) = 0$ with $\Delta(\theta_s) \neq 0$, $\mathrm{Im}(\Delta_1(\theta_s)) > 0$, and $\mathrm{Im}(\Delta_2(\theta_s)) < 0$, and (8.3.3b and 2.2a) yield

$$
\begin{aligned}
A_c(z,t_s) \sim \frac{1}{2\pi}\mathrm{Re}\Bigg\{ & 2\sqrt{\pi}\mathrm{e}^{(z/c)\phi(\omega_{sp_D^+},\theta_s)}F\left(|\Delta(\theta_s)|\sqrt{\frac{z}{c}}\right) - \frac{1}{\Delta(\theta_s)}\sqrt{\frac{\pi c}{z}}\mathrm{e}^{(z/c)\phi(\omega_{sp_D^+},\theta_s)} \\
& - \mathrm{i}\pi\,\mathrm{erfc}\left(-\mathrm{i}\Delta_1(\theta_s)\sqrt{\frac{z}{c}}\right)\mathrm{e}^{(z/c)\phi(\omega_c-2\beta_T\mathrm{i},\theta_s)} \\
& - \frac{1}{\Delta_1(\theta_s)}\sqrt{\frac{\pi c}{z}}\mathrm{e}^{(z/c)\phi(\omega_{sp_D^+},\theta_s)} \\
& + \mathrm{i}\pi\,\mathrm{erfc}\left(\mathrm{i}\Delta_2(\theta_s)\sqrt{\frac{z}{c}}\right)\mathrm{e}^{(z/c)\phi(\omega_c-4\beta_T\mathrm{i},\theta_s)} \\
& - \frac{1}{\Delta_2(\theta_s)}\sqrt{\frac{\pi c}{z}}\mathrm{e}^{(z/c)\phi(\omega_{sp_D^+},\theta_s)}\Bigg\} \\
& - \frac{1}{2}\mathrm{e}^{-z\alpha(\omega_c)}\sin(\beta(\omega_c)z - \omega_c t_s) \; ; \qquad \theta = \theta_s = \frac{ct_s}{z} \, , \qquad (8.3.23\mathrm{b})
\end{aligned}
$$

as $z \to \infty$. For $\theta_s < \theta < \theta_{s1}$, $\mathrm{Im}(\Delta(\theta)) < 0$, $\mathrm{Im}(\Delta_1(\theta)) > 0$, and $\mathrm{Im}(\Delta_2(\theta)) < 0$, and (8.3.3c and 2.2a) yield

$$A_c(z,t) \sim \frac{1}{2\pi}\mathbb{R}\mathrm{e}\left\{ i\pi\,\mathrm{erfc}\left(i\Delta(\theta)\sqrt{\frac{z}{c}}\right)e^{(z/c)\phi(\omega_c,\theta)} - \frac{1}{\Delta(\theta)}\sqrt{\frac{\pi c}{z}}e^{(z/c)\phi(\omega_{sp_D^+},\theta)}\right.$$

$$- i\pi\,\mathrm{erfc}\left(-i\Delta_1(\theta)\sqrt{\frac{z}{c}}\right)e^{(z/c)\phi(\omega_c-2\beta_T i,\theta)} - \frac{1}{\Delta_1(\theta)}\sqrt{\frac{\pi c}{z}}e^{(z/c)\phi(\omega_{sp_D^+},\theta)}$$

$$\left. + i\pi\,\mathrm{erfc}\left(i\Delta_2(\theta)\sqrt{\frac{z}{c}}\right)e^{(z/c)\phi(\omega_c-4\beta_T i,\theta)} - \frac{1}{\Delta_2(\theta)}\sqrt{\frac{\pi c}{z}}e^{(z/c)\phi(\omega_{sp_D^+},\theta)}\right\}$$

$$- e^{-z\alpha(\omega_c)}\sin(\beta(\omega_c)z - \omega_c t)\ ; \qquad \theta_s < \theta < \theta_{s1}\ , \tag{8.3.23c}$$

as $z \to \infty$. At $\theta = \theta_{s1} = \theta_r$, $\mathrm{Im}(\Delta(\theta_r)) < 0$, $\Delta_1(\theta_r) = 0$, and $\mathrm{Im}(\Delta_2(\theta_r)) < 0$, and (8.3.3c, 2.2c, and 2.2a) yield

$$A_c(z,t_r) \sim \frac{1}{2\pi}\mathbb{R}\mathrm{e}\left\{ i\pi\,\mathrm{erfc}\left(i\Delta(\theta_r)\sqrt{\frac{z}{c}}\right)e^{(z/c)\phi(\omega_c,\theta_r)} - \frac{1}{\Delta(\theta_r)}\sqrt{\frac{\pi c}{z}}e^{(z/c)\phi(\omega_{sp_D^+},\theta_r)}\right.$$

$$+ \left[-\frac{2\pi c}{z\phi^{(2)}(\omega_{sp_D^+},\theta_r)}\right]^{1/2}\left[\frac{1}{\omega_{sp_D^+}(\theta_r) - (\omega_c - 2\beta_T i)} + \frac{\phi^{(3)}(\omega_{sp_D^+},\theta_r)}{6\phi^{(2)}(\omega_{sp_D^+},\theta_r)}\right]$$

$$\cdot e^{(z/c)\phi(\omega_{sp_D^+},\theta_r)} + i\pi\,\mathrm{erfc}\left(i\Delta_2(\theta_r)\sqrt{\frac{z}{c}}\right)e^{(z/c)\phi(\omega_c-4\beta_T i,\theta_r)}$$

$$\left. - \frac{1}{\Delta_2(\theta_r)}\sqrt{\frac{\pi c}{z}}e^{(z/c)\phi(\omega_{sp_D^+},\theta_r)}\right\} - e^{-z\alpha(\omega_c)}\sin(\beta(\omega_c)z - \omega_c t)$$

$$+ \frac{1}{2}\mathbb{R}\mathrm{e}\left[ie^{(z/c)\phi(\omega_c-2\beta_T i,\theta_r)}\right]\ ,$$

as $z \to \infty$. Upon employing the approximate expression (6.2.4) for the complex phase behavior in the region of the complex ω-plane removed from the origin, it is found that

$$\frac{\phi^{(3)}(\omega_{sp_D^+},\theta_r)}{\phi^{(2)}(\omega_{sp_D^+},\theta_r)} \cong -3\frac{\omega_c + \delta i(\eta(\theta_r) - 1)}{\omega_c^2 + \delta^2(\eta(\theta_r) - 1)^2}\ .$$

Furthermore, according to equation (7.2.5) and the appropriate branch specified by Olver's convention, one has that

$$\left[-\frac{1}{\phi^{(2)}(\omega_{sp_D^+},\theta_r)}\right]^{1/2} \cong \frac{\sqrt{2}}{b}e^{-i(\pi/4)}\omega_c^{1/2}\left[\omega_c + \frac{3}{2}\delta i(1 - \eta(\theta_r))\right]\ .$$

Upon substitution of these results into the above asymptotic approximation of $A_c(z,t_r)$, there results

$$A_c(z,t_r) \sim \frac{1}{2\pi}\mathrm{Re}\left\{ i\pi\,\mathrm{erfc}\left(i\Delta(\theta_r)\sqrt{\frac{z}{c}} \right) e^{(z/c)\phi(\omega_c,\theta_r)} - \frac{1}{\Delta(\theta_r)}\sqrt{\frac{\pi c}{z}} e^{(z/c)\phi(\omega_{sp_D^+},\theta_r)} \right.$$

$$+ \frac{2}{b}e^{-i(\pi/4)}\sqrt{\frac{\pi c\omega_c}{z}}\left[\omega_c + \frac{3}{2}\delta i(1-\eta(\theta_r))\right]\left[\frac{1}{\omega_{sp_D^+}(\theta_r) - (\omega_c - 2\beta_T i)}\right.$$

$$\left. - \frac{1}{2}\cdot\frac{\omega_c + \delta i(\eta(\theta_r)-1)}{\omega_c^2 + \delta^2(\eta(\theta_r)-1)^2}\right] e^{(z/c)\phi(\omega_{sp_D^+},\theta_r)}$$

$$\left. + i\pi\,\mathrm{erfc}\left(i\Delta_2(\theta_r)\sqrt{\frac{z}{c}} \right) e^{(z/c)\phi(\omega_c - 4\beta_T i,\theta_r)} - \frac{1}{\Delta_2(\theta_r)}\sqrt{\frac{\pi c}{z}} e^{(z/c)\phi(\omega_{sp_D^+},\theta_r)} \right\}$$

$$- e^{-z\alpha(\omega_c)}\sin(\beta(\omega_c)z - \omega_c t_r)$$

$$+ \frac{1}{2}\mathrm{Re}\left[ie^{(z/c)\phi(\omega_c - \beta_T i,\theta_r)}\right]; \qquad \theta = \theta_{s1} = \theta_r = \frac{ct_r}{z}, \tag{8.3.23d}$$

as $z \to \infty$. For $\theta_r < \theta < \theta_{s2}$, $\mathrm{Im}(\Delta(\theta)) < 0$, $\mathrm{Im}(\Delta_1(\theta)) > 0$, and $\mathrm{Im}(\Delta_2(\theta)) < 0$, and (8.3.3c and 2.2a) yield

$$A_c(z,t) \sim \frac{1}{2\pi}\mathrm{Re}\left\{ i\pi\,\mathrm{erfc}\left(i\Delta(\theta)\sqrt{\frac{z}{c}} \right) e^{(z/c)\phi(\omega_c,\theta)} - \frac{1}{\Delta(\theta)}\sqrt{\frac{\pi c}{z}} e^{(z/c)\phi(\omega_{sp_D^+},\theta)} \right.$$

$$- i\pi\,\mathrm{erfc}\left(-i\Delta_1(\theta)\sqrt{\frac{z}{c}} \right) e^{(z/c)\phi(\omega_c - 2\beta_T i,\theta)} - \frac{1}{\Delta_1(\theta)}\sqrt{\frac{\pi c}{z}} e^{(z/c)\phi(\omega_{sp_D^+},\theta)}$$

$$\left. + i\pi\,\mathrm{erfc}\left(i\Delta_2(\theta)\sqrt{\frac{z}{c}} \right) e^{(z/c)\phi(\omega_c - 4\beta_T i,\theta)} - \frac{1}{\Delta_2(\theta)}\sqrt{\frac{\pi c}{z}} e^{(z/c)\phi(\omega_{sp_D^+},\theta)} \right\}$$

$$- e^{-z\alpha(\omega_c)}\sin(\beta(\omega_c)z - \omega_c t)$$

$$+ \mathrm{Re}\left[ie^{(z/c)\phi(\omega_c - 2\beta_T i,\theta)}\right]; \qquad \theta_r < \theta < \theta_{s2}, \tag{8.3.23e}$$

as $z \to \infty$. At $\theta = \theta_{s2}$, $\mathrm{Im}(\Delta(\theta_{s2})) < 0$, $\mathrm{Im}(\Delta_1(\theta_{s2})) > 0$, and $\mathrm{Im}(\Delta_2(\theta_{s2})) = 0$ with $\Delta_2(\theta_{s2}) \neq 0$, so that (8.3.3c, 2.2a, and 2.2b) yield

$$A_c(z,t_{s2}) \sim \frac{1}{2\pi}\mathrm{Re}\left\{ i\pi\,\mathrm{erfc}\left(i\Delta(\theta_{s2})\sqrt{\frac{z}{c}} \right) e^{(z/c)\phi(\omega_c,\theta_{s2})} \right.$$

$$- \frac{1}{\Delta(\theta_{s2})}\sqrt{\frac{\pi c}{z}} e^{(z/c)\phi(\omega_{sp_D^+},\theta_{s2})}$$

$$- i\pi\,\mathrm{erfc}\left(-i\Delta_1(\theta_{s2})\sqrt{\frac{z}{c}} \right) e^{(z/c)\phi(\omega_c - 2\beta_T i,\theta_{s2})}$$

$$- \frac{1}{\Delta_1(\theta_{s2})}\sqrt{\frac{\pi c}{z}} e^{(z/c)\phi(\omega_{sp_D^+},\theta_{s2})}$$

$$+ 2\sqrt{\pi}e^{(z/c)\phi(\omega_{sp_D^+},\theta_{s2})}F\left(|\Delta_2(\theta_{s2})|\sqrt{\frac{z}{c}}\right)$$

$$\left.- \frac{1}{\Delta_2(\theta_{s2})}\sqrt{\frac{\pi c}{z}}e^{(z/c)\phi(\omega_{sp_D^+},\theta_{s2})}\right\}$$

$$- e^{-z\alpha(\omega_c)}\sin(\beta(\omega_c)z - \omega_c t_{s2})$$

$$+ \mathbb{R}\mathrm{e}\left[\mathrm{i}e^{(z/c)\phi(\omega_c - 2\beta_T\mathrm{i},\theta_{s2})} + \frac{1}{2}\mathrm{i}e^{(z/c)\phi(\omega_c - 4\beta_T\mathrm{i},\theta_{s2})}\right];$$

$$\theta = \theta_{s2} = \frac{ct_{s2}}{z}, \quad (8.3.23\text{f})$$

as $z \to \infty$. Finally, for $\theta > \theta_{s2}$, $\mathrm{Im}(\Delta(\theta)) < 0$, $\mathrm{Im}(\Delta_1(\theta)) > 0$, and $\mathrm{Im}(\Delta_2(\theta)) > 0$, so that (8.3.3c and 2.2a) yield

$$A_c(z,t) \sim \frac{1}{2\pi}\mathbb{R}\mathrm{e}\left\{\mathrm{i}\pi\,\mathrm{erfc}\left(\mathrm{i}\Delta(\theta)\sqrt{\frac{z}{c}}\right)e^{(z/c)\phi(\omega_c,\theta)} - \frac{1}{\Delta(\theta)}\sqrt{\frac{\pi c}{z}}e^{(z/c)\phi(\omega_{sp_D^+},\theta)}\right.$$

$$- \mathrm{i}\pi\,\mathrm{erfc}\left(-\mathrm{i}\Delta_1(\theta)\sqrt{\frac{z}{c}}\right)e^{(z/c)\phi(\omega_c - 2\beta_T\mathrm{i},\theta)} - \frac{1}{\Delta_1(\theta)}\sqrt{\frac{\pi c}{z}}e^{(z/c)\phi(\omega_{sp_D^+},\theta)}$$

$$\left.- \mathrm{i}\pi\,\mathrm{erfc}\left(-\mathrm{i}\Delta_2(\theta)\sqrt{\frac{z}{c}}\right)e^{(z/c)\phi(\omega_c - 4\beta_T\mathrm{i},\theta)} - \frac{1}{\Delta_2(\theta)}\sqrt{\frac{\pi c}{z}}e^{(z/c)\phi(\omega_{sp_D^+},\theta)}\right\}$$

$$- e^{-z\alpha(\omega_c)}\sin(\beta(\omega_c)z - \omega_c t)$$

$$+ \mathbb{R}\mathrm{e}[\mathrm{i}e^{(z/c)\phi(\omega_c - 2\beta_T\mathrm{i},\theta)} + \mathrm{i}e^{(z/c)\phi(\omega_c - 4\beta_T\mathrm{i},\theta)}];$$

$$\theta > \theta_{s2}, \quad (8.3.23\text{g})$$

as $z \to \infty$. Taken together, (8.3.23a–g) represents the uniform asymptotic approximation of the pole contribution $A_c(z,t)$ for $\omega_c \geqslant \sqrt{\omega_1^2 - \delta^2}$ and $\beta_T = \delta((1 + \eta(\theta_r))/2)$.

This analysis can be continued indefinitely as the parameter β_T is allowed to get successively smaller and increasing numbers of pole singularity contributions interact with the path $P(\theta)$ and must be taken into account in the asymptotic approximation of the component field $A_c(z,t)$. For example, for the situation depicted in Fig. 8.5(c) for $\beta_T = \delta(1 + \eta(\theta_r))$, the contributions due to the simple pole singularities at $\omega_{p0} = \omega_c$, $\omega_{p1}^- = \omega_c - 2\beta_T\mathrm{i}$, $\omega_{p2}^- = \omega_c - 4\beta_T\mathrm{i}$, $\omega_{p3}^- = \omega_c - 6\beta_T\mathrm{i}$, and possibly $\omega_{p4}^- = \omega_c - 8\beta_T\mathrm{i}$ must be taken into account in constructing $A_c(z,t)$. This analysis is not given here, but may be directly obtained by following the approach used in the previous two cases.

Consider now the contributions to $A_c(z,t)$ of the simple pole singularities of the spectrum located at $\omega_{pn}^+ = \omega_c + 2n\beta_T\mathrm{i}$, $n = 1, 2, 3, \ldots$; in the upper-half of the complex ω-plane. It is important to remember that the original contour of integration, which extends from $\mathrm{i}a - \infty$ to $\mathrm{i}a + \infty$ is constrained to lie between the real axis and the first pole singularity in the upper-half of the complex ω-plane (viz $0 < a < 2\beta_T$). As was discussed previously, for the situation depicted in Fig. 8.5(a) for $\beta_T > \delta((1 + \eta(\theta_r))/2)$, the deformation of the original contour of integration to the path $P(\theta)$ through SP_D^+ does not cross any of the simple pole singularities located at $\omega_{pn}^+ = \omega_c + 2n\beta_T i$, and hence, they make no contribution to the asymptotic approximation of $A_c(z,t)$. For smaller values of β_T, such as, for example, that depicted in Fig. 8.5(b) and (c), the original contour of integration will be deformed across at least one of the simple pole singularities in the upper half-plane. However, such a pole contribution occurs only during those values of θ when the exponential attenuation associated with the simple pole singularity is much greater than that associated with the distant saddle point SP_D^+ (e.g., $X(\omega_c + 2\beta_T i, \theta) \ll X(\omega_{SP_D^+}, \theta)$); for those values of θ when the value of $X(\omega_{pn}^+, \theta)$ at the pole singularity is either comparable to or greater than that at the distant saddle point SP_D^+, the original contour of integration will not cross that pole singularity in being deformed to the path $P(\theta)$ through SP_D^+. Consequently, for values of β_T on the order of δ or greater, the simple pole singularities in the upper-half of the complex ω-plane do not significantly contribute to the asymptotic behavior of the component field $A_c(z,t)$ for $\omega_c \geqslant \sqrt{\omega_1^2 - \delta^2}$ (for $\delta = 0.28 \times 10^{16}$/s, this corresponds to a turn-on time of $1/\beta$ on the order of a thousandth of a picosecond).

For small values of $\beta_T (1/\beta_T \gg \delta)$, however, the pole singularities located at $\omega_{pn}^+ = \omega_c + 2n\beta_T i$, $n = 1, 2, 3, \ldots$, approach close to the real axis and the interaction of the deformed contour of integration with them becomes more important. In that case, their contributions to the field $A_c(z,t)$ are no longer negligible in comparison to the other contributions at $\omega_{p0} = \omega_c$ and $\omega_{pn}^- = \omega_c - 2n\beta_T i$. Since the parameter β_T is small in that case, which corresponds to a very long turn-on time for the signal, the field is quasimonochromatic. That limit is not of central interest to this research and so is not treated here.

For applied signal frequencies in the range $0 < \omega_c \leqslant \sqrt{\omega_0^2 - \delta^2}$, the near saddle point SP_N^+ interacts with the simple pole singularities of the spectrum located at $\omega_{pn}^\pm = \omega_c \pm 2n\beta_T i$. In that case, θ_r is given by the relation (cf. (6.2.53 and 8.3.12)]

$$\psi(\theta_r) \equiv \omega_c \,, \tag{8.3.24}$$

and $\mathrm{Im}[\omega_{SP_N^+}(\theta)] = -\frac{2}{3}\delta\zeta(\theta)$. The approximate phase behavior in the complex ω-plane near the origin is given by (6.2.45), so that at $\omega_{pn}^- = \omega_c - 2n\beta_T i$, $n = 1, 2, 3, \ldots$,

$$\begin{aligned}\phi(\omega_{pn}^-, \theta) \cong &\left\{2n\beta_T(\theta_0 - \theta) + \frac{b^2}{\theta_0\omega_0^4}[(n\beta_T\alpha - \delta)(\omega_c^2 - 4n^2\beta_T^2) + 2n\beta_T\alpha\omega_c^2]\right\} \\ &+ i\omega_c\left\{\theta_0 - \theta + \frac{b^2}{2\theta_0\omega_0^4}[\alpha(\omega_c^2 - 4n^2\beta_T^2) - 8n\beta_T(n\beta_T\alpha - \delta)]\right\} .\end{aligned} \tag{8.3.25}$$

The analysis of the pole contributions to the uniform asymptotic approximation of $A_c(z,t)$ in this case follows the same procedure as that employed for applied frequencies in the range $\omega_c \geqslant \sqrt{\omega_1^2 - \delta^2}$. Only a qualitative discussion is given here. For $\beta_T > \delta(\zeta(\theta_r)/3)$, the path $P(\theta)$ through the near saddle point SP_N^+ interacts with the simple pole singularities at $\omega_{p0} = \omega_c$ and $\omega_{p1}^- = \omega_c - 2\beta_T i$. In that case, $\Lambda(\theta)$ is given by the sum of the residues at these two pole singularities, viz

$$\Lambda(\theta) = -\frac{1}{2\pi}[e^{(z/c)\phi(\omega_c,\theta)}\mathscr{U}(\theta - \theta_s) + e^{(z/c)\phi(\omega_c - 2\beta_T i,\theta)}\mathscr{U}(\theta - \theta_{s1})] ,$$

where the relations $Y(\omega_{sp_N^+}, \theta_s) = Y(\omega_c, \theta_s)$ and $Y(\omega_{sp_N^+}, \theta_{s1}) = Y(\omega_c - 2\beta_T i, \theta_{s1})$ define the quantities θ_s and θ_{s1}, respectively. The other pole singularities either do not contribute to the asymptotic behavior of $A_c(z,t)$ or are asymptotically negligible in comparison to these two contributions. For increasingly large values of $\beta_T \gg \delta(\zeta(\theta_r)/3)$, the exponential decay associated with the pole singularity at $\omega_{p1}^- = \omega_c - 2\beta_T i$ is much greater than that at $\omega_{p0} = \omega_c$, so that the pole contribution due to the singularity at ω_{p1}^- is exponentially negligible in comparison to that at $\omega_{p0} = \omega_c$. In that case, which corresponds to a very rapid turn-on time for the initial signal, the uniform asymptotic approximation of $A_c(z,t)$ is given by (8.3.2) for the unit step-function modulated signal, as is to be expected. However, as the parameter β_T approaches close to the quantity $\delta(\zeta(\theta_r)/3)$, the contribution of the simple pole singularity at $\omega_{p1}^- = \omega_c - 2\beta_T i$ to $A_c(z,t)$ must be taken into account. For $\beta_T = \delta(\zeta(\theta_r)/3)$, the pole singularity at $\omega_{p1}^- = \omega_c - 2\beta_T i$ coalesces with the near saddle point SP_N^+ at $\theta = \theta_{s1} = \theta_r$. In addition, the path of steepest descent approaching SP_N^+ crosses the simple pole singularity at $\omega_{p2}^- = \omega_c - 4\beta_T i$ at $\theta = \theta_{s2} > \theta_{s1}$, where θ_{s2} satisfies the relation $Y(\omega_{sp_N^+}, \theta_{s2}) = Y(\omega_{p2}^-, \theta_{s2})$. In that case, $\Lambda(\theta)$ is given by the sum of the residues at these three pole singularities, viz

$$\Lambda(\theta) = -\frac{1}{2\pi}[e^{(z/c)\phi(\omega_c,\theta)}\mathscr{U}(\theta - \theta_s) + e^{(z/c)\phi(\omega_c - 2\beta_T i,\theta)}\mathscr{U}(\theta - \theta_{s1})$$
$$+ e^{(z/c)\phi(\omega_c - 4\beta_T i,\theta)}\mathscr{U}(\theta - \theta_{s2})] .$$

For values of β_T in this range, the original contour of integration will be deformed across at least one of the simple pole singularities located at $\omega_{pn}^+ = \omega_c + 2n\beta_T i$, $n = 1, 2, 3, \ldots$, in the upper-half of the complex ω-plane. However, such a pole contribution occurs only during those values of θ when the exponential attenuation associated with the simple pole singularity is much greater than that associated with the near saddle point, and for those values of θ when the value of $X(\omega_{pn}^+, \theta)$ at the pole singularity is either comparable to or greater than that at the near saddle point SP_N^+, the original contour of integration will not cross that pole singularity in being deformed to the path $P(\theta)$ along the path of steepest descent through the saddle point SP_N^+. Consequently, for values of β_T on the order of δ or greater, the pole singularities in the upper-half of the complex ω-plane do not significantly contribute to the asymptotic approximation of $A_c(z,t)$ for values of the applied signal frequency ω_c in the range $0 < \omega_c \leqslant \sqrt{\omega_0^2 - \delta^2}$. In the quasimonochromatic limit of small $\beta_T \ll \delta$, however, the pole singularities located at $\omega_{pn}^+ = \omega_c + 2n\beta_T i$ approach close to the real axis and the interaction of the deformed contour of integration with them becomes more important. In that case, their contributions to the asymptotic approximation of the field $A_c(z,t)$ are no longer negligible in comparison to the other contributions at $\omega_{p0} = \omega_c$ and $\omega_{pn}^- = \omega_c - 2n\beta_T i$.

Finally, for applied signal frequencies ω_c in the range $\sqrt{\omega_0^2 - \delta^2} < \omega_c < \sqrt{\omega_1^2 - \delta^2}$, which is within the absorption band of the medium, neither the near nor distant saddle points comes within close proximity of any of the simple pole singularities at $\omega_{pn}^\pm = \omega_c \pm 2n\beta_T i$, $n = 0, 1, 2, 3, \ldots$. The results of (5.3.28) then apply in this case for each pole singularity crossed by the path $P(\theta)$. In particular, the uniform asymptotic approximation of the contribution of $\omega_{p0} = \omega_c$ is given by (8.3.4). Due to the presence of the branch cut $\omega_+\omega_+'$ in the lower-half of the complex ω-plane (cf. (6.1.7)), the pole singularities at $\omega_{pn}^- = \omega_c - 2n\beta_T i$, $n = 1, 2, 3, \ldots$, do not contribute unless $\beta_T < \delta/2n$. The value of θ_{sn} at which each of these contributions occurs is determined by that value of θ at which $P(\theta)$ crosses the pole singularity at ω_{pn}^-. As in the previous two cases, the contributions due to the simple pole singularities located at $\omega_{pn}^+ = \omega_c + 2n\beta_T i$, $n = 1, 2, 3, \ldots$, in the upper-half of the complex ω-plane are negligible unless $\beta_T \ll \delta$, in which case the quasimonochromatic limit is attained.

9 The Continuous Evolution of the Total Field

This chapter combines the results of Chap. 7 and 8 to obtain the uniform asymptotic evolution of the propagated field $A(z,t)$. In accordance with the discussion in Sect. 6.5 (cf. (6.5.8)), the propagated field may be expressed in the form

$$A(z,t) = A_s(z,t) + A_B(z,t) + A_c(z,t)$$

for $\theta = ct/z \geqslant 1$, where $A_B(z,t)$ and $A_c(z,t)$ are negligible when the first precursor field $A_s(z,t)$ is predominant, $A_s(z,t)$ and $A_c(z,t)$ are negligible when the second precursor field $A_B(z,t)$ is predominant, and $A_s(z,t)$ and $A_B(z,t)$ are negligible when the pole contribution $A_c(z,t)$ is predominant. Two or three terms become important at the same time during periods of transition giving a continuous asymptotic description of the time evolution of the total field $A(z,t)$ for large z for all $\theta = ct/z \geqslant 1$[1].

The chapter begins with the analysis of the total precursor field $A_p(z,t)$. The results are based upon the analysis of the individual precursor fields presented in Chap. 7. The possible occurrence of a resonance peak in one of the precursor fields and its complete cancellation by a similar resonance peak in the pole contribution when the two fields are combined is considered next in Sect. 9.2. Since this resonance phenomenon does not occur in the propagated field $A(z,t)$, it is simply an artifact of the asymptotic analysis and is totally nonphysical. The uniform transition from the total precursor field to the main signal and the interaction of the main signal with the *Sommerfeld* and *Brillouin* precursor fields are then considered in Sect. 9.3. Based upon these results, refined expressions for the signal velocity replacing those described by *Brillouin* [9.1, 2] are derived. The signal velocity for pulse propagation in *Lorentz* media is then compared to the phase velocity, group velocity, and velocity of energy propagation in such dispersive media. The chapter is concluded with a discussion of the total uniform asymptotic evolution of the unit step-function modulated signal, the rectangular-pulse modulated signal, the delta-function pulse, and the hyperbolic-tangent modulated signal.

[1] Recall that it was shown in Sect. 7.1 that the field behavior may be evaluated exactly by application of *Cauchy*'s Residue Theorem for $\theta < 1$.

9.1 The Total Precursor Field

The total precursor field $A_p(z,t)$ is the combined contribution to $A(z,t)$ due to both the near and distant saddle points for $\theta \geqslant 1$, i.e., the sum of the *Sommerfeld* and *Brillouin* precursor fields

$$A_p(z,t) = A_s(z,t) + A_B(z,t) \ . \tag{9.1.1}$$

For values of $\theta \geqslant 1$ bounded away from θ_{SB}, it follows from the results of Sect. 6.3 that for large enough z, the second precursor $A_B(z,t)$ is asymptotically negligible in comparison to the first precursor field $A_s(z,t)$ when $1 \leqslant \theta < \theta_{SB}$, and the first precursor $A_s(z,t)$ is asymptotically negligible in comparison to the second precursor field $A_B(z,t)$ when $\theta > \theta_{SB}$. Both $A_s(z,t)$ and $A_B(z,t)$ are important in the transition region between the first and second precursor fields in a neighborhood of $\theta = \theta_{SB}$. When asymptotic approximations of $A_s(z,t)$ and $A_B(z,t)$, uniformly valid for θ in a specific range, are applied in (9.1.1), it follows from the results of Sect. 5.2.1 that the result is an asymptotic approximation of $A_p(z,t)$ that is uniform over the same range of values of θ. Hence, application in (9.1.1) of the uniform asymptotic approximations of $A_s(z,t)$ and $A_B(z,t)$ obtained in Sect. 7.3 and 7.4, respectively, provides an asymptotic approximation of the total precursor field $A_p(z,t)$ that is valid uniformly for all $\theta \geqslant 1$.

Based upon these results and the results of Chap. 7, the general dynamic behavior of the total precursor field is as follows. At $\theta = 1$ the front of the first precursor arrives, and the first precursor is dominant for θ in the range $1 \leqslant \theta < \theta_{SB}$[2]. For values of θ soon after 1 the peak amplitude in the first precursor occurs[3], and for all later values of θ the amplitude of the field decays exponentially. Furthermore, the instantaneous angular frequency of the first precursor, which is initially infinite at $\theta = 1$, monotonically decreases with increasing θ and approaches the limiting value $+\sqrt{\omega_1^2 - \delta^2}$ as an asymptote as θ approaches infinity. For values of θ close to θ_{SB}, the transition from the first to the second precursor field takes place. Since $1 < \theta_{SB} < \theta_1$ and since the second precursor field is nonoscillatory for values of θ in the range $1 \leqslant \theta \leqslant \theta_1$, the total precursor field behavior for values of θ about θ_{SB} is that of the superposition of an exponentially decaying oscillatory field with an increasing nonoscillatory field. Finally, for values of $\theta > \theta_{SB}$, the second precursor is dominant. The magnitude of the second precursor increases as θ approaches θ_0 from below. At $\theta = \theta_0$, the second precursor has no exponential decay and decays with distance only as $z^{-1/2}$. Finally, for $\theta > \theta_0$, the exponential decay increases with θ, and for $\theta > \theta_1$ the second precursor is oscillatory. The instantaneous angular frequency of the

[2] Recall that for all values of θ in the range $1 \leqslant \theta < \theta_{SB}$, the exponential decay associated with the second precursor field is greater than that associated with the first precursor field, whereas the opposite is true for all $\theta > \theta_{SB}$.

[3] This is neglecting the nonphysical phenomenon of a so-called resonance peak which is discussed in Sect. 9.2.

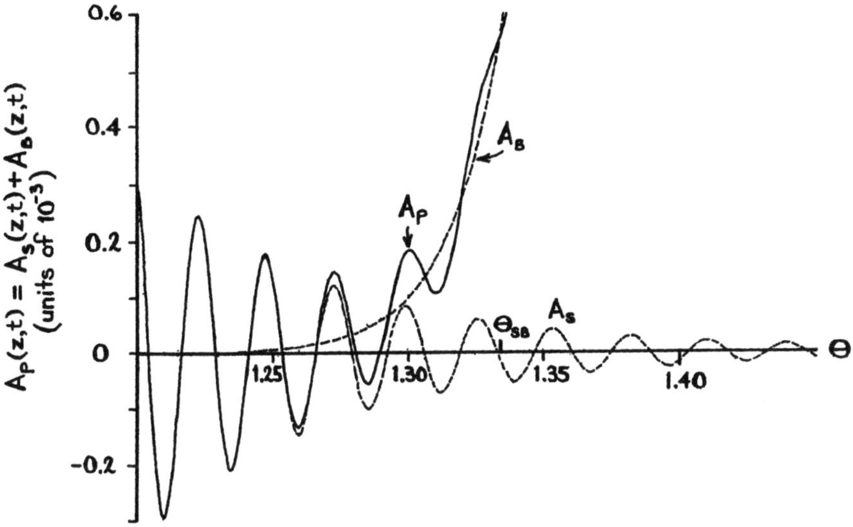

Fig. 9.1. Dynamic behavior of the transition from the first precursor field to the second precursor field in the total precursor evolution for the unit step-function modulated signal with applied signal frequency $\omega_c = 1.0 \times 10^{16}$/s at an observation distance of $z = 1 \times 10^{-4}$ cm, where *Brillouin*'s choice of the medium parameters are employed.

second precursor monotonically increases with increasing θ from zero at $\theta = \theta_1$ and approaches the limiting value $+\sqrt{\omega_0^2 - \delta^2}$ as an asymptote as θ approaches infinity.

The uniform transition of the total precursor field from the first to the second precursor for the unit step-function modulated signal is illustrated in Fig. 9.1. The oscillating dashed line indicates the first precursor field $A_s(z, t)$, the exponentially increasing nonoscillatory dashed line indicates the second precursor field $A_B(z, t)$, and the total precursor field $A_p(z, t) = A_s(z, t) + A_B(z, t)$ is shown by the solid line. These curves are calculated for Brillouin's choice of the medium parameters at an observation distance $z = 1 \times 10^{-4}$ cm and an applied signal frequency $\omega_c = 1 \times 10^{16}$/s. The critical value of θ about which the transition between the first and second precursor fields takes place is $\theta_{SB} \cong 1.334$. As can be seen from the figure, the transition takes place in three periods of osillation of the first precursor field. For larger values of the observation distance z this transition interval will decrease and become more centered about $\theta = \theta_{SB}$.

9.2 Resonance Peaks of the Precursors and the Main Signal

Examination of the *Sommerfeld* and *Brillouin* precursor fields $A_s(z, t)$ and $A_B(z, t)$, respectively, shows that each of them exhibits a resonance peak as θ varies if the relevant saddle point passes near a first-order pole of the spectral function.

Indeed, it is apparent from the general expression for the asymptotic contribution of a first-order saddle point as given in (5.1.14), viz

$$I(z,t) \sim q(\omega_{sp})\left[-\frac{2\pi}{zp^{(2)}(\omega_{sp},\theta)}\right]^{1/2} e^{zp(\omega_{sp},\theta)} , \tag{9.2.1}$$

that $I(z,t)$ becomes large if $\omega_{sp}(\theta)$ approaches a pole ω_p of $q(\omega)$ as θ varies. It is the purpose of this section to show that such a resonance peak is not exhibited by the total field $A(z,t)$, however. That resonance peak is cancelled by an identical resonance peak with opposite sign in the term $C(z,t)$ appearing in the uniform asymptotic expression for $A_c(z,t)$. That is, these resonance peaks are an artifact of the separation of the asymptotic behavior of the integral into the various component contributions.

Under the conditions that lead to the resonance peak in the saddle point contribution, the saddle point passes near a first-order pole. As a result, the uniform asymptotic expression for the pole contribution $A_c(z,t)$ must be included in the asymptotic approximation of the integral $A(z,t)$. The result can be written

$$A(z,t) \sim q(\omega_{sp})\left[-\frac{2\pi}{zp^{(2)}(\omega_{sp},\theta)}\right]^{1/2} e^{zp(\omega_{sp},\theta)} + A_c(z,t) . \tag{9.2.2}$$

According to (8.2.2), if the simple pole and saddle point do not coalesce, the uniform asymptotic approximation of $A_c(z,t)$ can be written

$$A_c(z,t) \sim \frac{\gamma}{\Delta(\theta)}\sqrt{\frac{\pi}{z}}\, e^{zp(\omega_{sp},\theta)} + f_0(\omega_p) , \tag{9.2.3}$$

where $f_0(\omega_p)$ is an analytic function of complex ω_p and γ is the residue of the pole of the spectral amplitude function $q(\omega)$. Since ω_{sp} is a first-order saddle point of $p(\omega,\theta)$, the phase function evaluated at $\omega = \omega_p$ can be expanded in a Taylor series about ω_{sp} in the form

$$p(\omega_p,\theta) = p(\omega_{sp},\theta) + \tfrac{1}{2}p^{(2)}(\omega_{sp},\theta)(\omega_p - \omega_{sp})^2 + \cdots \tag{9.2.4}$$

for ω_p close enough to ω_{sp}. Hence, according to the definition of $\Delta(\theta)$ given in (8.2.3), viz

$$\Delta(\theta) = [p(\omega_{sp},\theta) - p(\omega,\theta)]^{1/2} , \tag{9.2.5}$$

it can be written

$$\Delta(\theta) = [-\tfrac{1}{2}p^{(2)}(\omega_{sp},\theta)]^{1/2}(\omega_p - \omega_{sp}) + f_1(\omega_p) , \tag{9.2.6}$$

where $f_1(\omega_p)$ is an analytic function of ω_p that goes to zero as ω_p approaches ω_{sp}. As a result, (9.2.3) can be written

$$A_c(z,t) \sim \frac{\gamma}{[-\frac{1}{2}p^{(2)}(\omega_{sp},\theta)]^{1/2}(\omega_p-\omega_{sp})}\sqrt{\frac{\pi}{z}}\mathrm{e}^{zp(\omega_{sp},\theta)} + f_2(\omega_p) \ , \tag{9.2.7}$$

where $f_2(\omega_p)$ is an analytic function of ω_p.

Similarly, since $q(\omega)$ has a first-order pole at $\omega = \omega_p$ with residue γ, the first term in (9.2.2) can be written

$$q(\omega_{sp})\left[-\frac{2\pi}{zp^{(2)}(\omega_{sp},\theta)}\right]^{1/2}\mathrm{e}^{zp(\omega_{sp},\theta)} = \frac{\gamma}{\omega_{sp}-\omega_p}\left[-\frac{2\pi}{zp^{(2)}(\omega_{sp},\theta)}\right]^{1/2}\mathrm{e}^{zp(\omega_{sp},\theta)} + f_3(\omega_p) \ , \tag{9.2.8}$$

where $f_3(\omega_p)$ is an analytic function of ω_p. Substitution of (9.2.7 and 8) into (9.2.2) then yields

$$A(z,t) \sim f_2(\omega_p) + f_3(\omega_p) \ . \tag{9.2.9}$$

Hence, $A(z,t)$ is an analytic function of ω_p in a neighborhood of the saddle point ω_{sp}, and therefore cannot have a singularity at $\omega_p = \omega_{sp}$. Consequently, the resonance peak of the precursor field is exactly cancelled by an identical resonance peak of the pole contribution. The propagated field $A(z,t)$ does not exhibit the resonance peaks exhibited by its component sub-fields $A_p(z,t)$ and $A_c(z,t)$.

9.3 The Signal Arrival and the Signal Velocity

This section is devoted to describing the arrival of the main signal due to the contribution of a simple pole singularity (if any) appearing in the spectral function $\tilde{u}(\omega-\omega_c)$ in the integrand of the propagated scalar wave field (4.3.31), viz

$$A(z,t) = \frac{1}{2\pi}\mathbb{R}\mathrm{e}\,\mathrm{i}\int_{ia-\infty}^{ia+\infty} \tilde{u}(\omega-\omega_c)\mathrm{e}^{(z/c)\phi(\omega,\theta)}\,d\omega \ .$$

The entire uniform asymptotic approximation of the pole contribution $A_c(z,t)$ to the field $A(z,t)$ is given in Chap. 8. That pole contribution constitutes the main body of the propagated field $A(z,t)$. In this section, a detailed analysis is given of the transition of the total field from the precursor field to the main signal. The vaue of θ at which the main signal arrives is given and the resulting signal velocity is obtained. That velocity is then compared to the phase velocity, group velocity, and energy velocity in a *Lorentz* medium.

As in Chap. 8, the pole ω_p is taken to lie along the positive real axis of the complex ω-plane. Since the near saddle point SP_1 for $1 \leqslant \theta \leqslant \theta_1$ or SP_N^+ for $\theta > \theta_1$ and the distant saddle point SP_D^+ lie in (or on the boundary of) the right-half of the complex ω-plane, only those two saddle points may interact with the simple pole singularity. The other two saddle points are of no consequence to the pole contribution and are ignored in this analysis.

9.3.1 Transition from the Precursor Field to the Main Signal

According to the results of Chap. 8, the contribution of the simple pole singularity at ω_p occurs when the original contour of integration, which extends along the straight line from $ia - \infty$ to $ia + \infty$ in the upper-half of the complex ω-plane, lies on the opposite side of the pole singularity than does the Olver-type path $P(\theta)$. That is, $P(\theta)$ and the original contour of integration lie on the same side of the pole for $\theta < \theta_s$ and on opposite sides for $\theta > \theta_s$ (cf. (8.1.4)). Consequently, for $\theta < \theta_s$ the pole is not crossed when the original contour is deformed to $P(\theta)$ and there is no residue contribution, while for $\theta > \theta_s$ the pole is crossed in deforming the original contour of integraton to $P(\theta)$ and there is a residue contribution to the asymptotic behavior of the field. The value of θ_s depends upon which Olver-type path is chosen for $P(\theta)$. If that path is taken to lie along the path of steepest descent through the saddle point nearest the pole, then, since $Y(\omega, \theta)$ is constant along the path of steepest descent, it follows that the value of θ_s is given by the expression

$$Y(\omega_{sp}, \theta_s) = Y(\omega_p, \theta_s) \tag{9.3.1}$$

where $\omega_{sp}(\theta)$ denotes the saddle point which interacts with the pole singularity. At $\theta = \theta_s$, however, the pole contribution is negligible in comparison to the saddle point contribution since $P(\theta)$ is an Olver-type path. Consequently, the particular value of θ_s at which the pole crossing occurs is of little or no importance to the asymptotic behavior of the propagated field. An example of such an Olver-type path when the two near saddle points $SP_N^{\pm}$ are dominant over the two distant saddle points $SP_D^{\pm}$ is illustrated in Fig. 9.2. The path $P(\theta)$ through the near saddle points SP_N^- and SP_N^+ can lie anywhere within the shaded area of the figure, which is drawn for one fixed value of θ. With the path $P(\theta)$ shown, $\theta < \theta_s$ if the pole ω_p lies within the range $\omega_{c2} < \omega_p < \omega_{c4}$, $\theta > \theta_s$ if ω_p lies within either of the ranges $0 \leqslant \omega_p < \omega_{c2}$ or $\omega_p > \omega_{c4}$, and $\theta = \theta_s$ if $\omega_p = \omega_{c2}$ or $\omega_p = \omega_{c4}$. If the path $P(\theta)$ was completely in the lower-half of the complex ω-plane (such an Olver-type path is possible for the situation illustrated in Fig. 9.2), then $\theta > \theta_s$ for all positive ω_p.

The pole contribution at $\omega = \omega_p$ is the dominant contribution to the asymptotic behavior of the field for $\theta > \theta_c > \theta_s$, where θ_c is that value of θ which satisfies the relation

$$X(\omega_{sp}, \theta_c) = X(\omega_p) , \tag{9.3.2}$$

where ω_{sp} denotes the dominant saddle point at that value of θ. For values of $\theta < \theta_c$ such that the inequality $X(\omega_{sp}, \theta) > X(\omega_p)$ is satisfied, the saddle point is

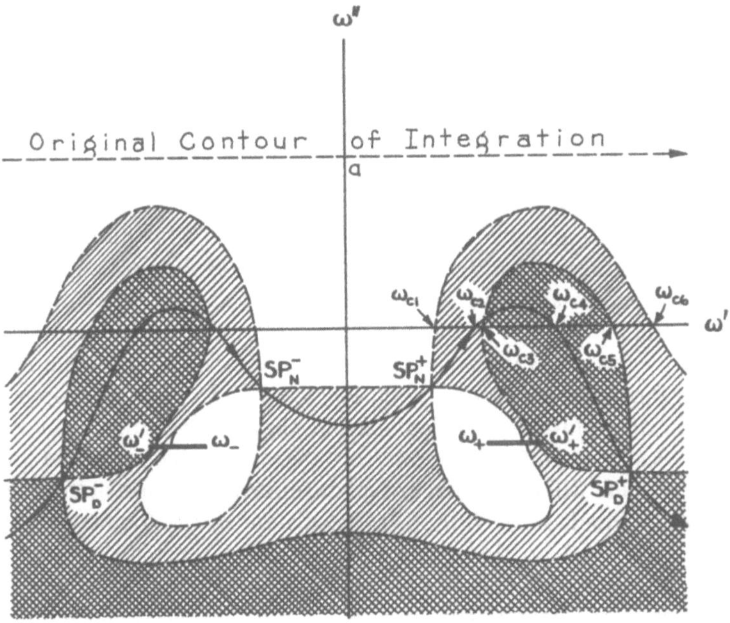

Fig. 9.2. A deformed contour of integration $P(\theta)$ through both the near and distant saddle points for a fixed value of θ. This contour is an Olver-type path with respect to the near saddle point SP_N^+ in the right-half of the complex ω-plane, and is an Olver-type path with respect to the near saddle point SP_N^- in the left-half of the complex ω-plane.

The hatched area ////// indicates the region of the complex ω-plane wherein the inequality $X(\omega < X(\omega_{sp_N^\pm})$ is satisfied.

The hatched area \\\\\\ indicates the region of the complex ω-plane wherein the inequality $X(\omega) < X(\omega_{sp_D^\pm})$ is satisfied.

the dominant contribution to the asymptotic behavior of the field and the pole contribution is asymptotically negligible by comparison. For values of $\theta > \theta_c$ such that the inequality $X(\omega_{sp}, \theta) < X(\omega_p)$ is satisfied, however, the pole contribution is the dominant contribution to the asymptotic behavior of the field and the saddle point is asymptotically negligible by comparison. For example, for the situation illustrated in Fig. 9.2, (9.3.2) is satisfied if $\omega_p = \omega_{c1}$ or $\omega_p = \omega_{c6}$ and the value of θ illustrated is θ_c for either of these two pole locations. Furthermore, for the value of θ depicted in that figure, $\theta < \theta_c$ and the inequality $X(\omega_{sp_N^+}, \theta) > X(\omega_p)$ is satisfied if ω_p lies within the range $\omega_{c1} < \omega_p < \omega_{c6}$, and $\theta > \theta_c$ and the inequality $X(\omega_{sp_N^+}, \theta) < X(\omega_p)$ is satisfied if ω_p lies within either of the ranges $0 \leqslant \omega_p < \omega_{c1}$ or $\omega_p > \omega_{c6}$. Consequently, the pole contribution is asymptotically negligible in comparison to the saddle point contribution to the field for the value of θ depicted in Fig. 9.2 if ω_p lies within the range $\omega_{c1} < \omega_p < \omega_{c6}$, whereas the pole contribution is the dominant contribution to the asymptotic behavior of the field and the saddle point contribution is asymptotically negligible in comparison to it if ω_p lies within either of the ranges $0 \leqslant \omega_p < \omega_{c1}$ or $\omega_p > \omega_{c6}$.

Based upon these results, it is seen that the arrival of the main signal for a fixed value of ω_p occurs at the value of $\theta = \theta_c$ satisfying (9.3.2). Notice that this definition of the signal arrival yields a signal velocity which depends only upon the properties of the medium and the value of ω_p. It is independent of the initial envelope of the field and the distance of propagation z.

A general overview of the arrival of the main signal and its interaction with the *Sommerfeld* and *Brillouin* precursor fields is now given. This description is based entirely upon (9.3.2) and the topography of $X(\omega, \theta)$ as a function of θ in the complex ω-plane presented in Fig. 6.4 through 6.9 for Brillouin's choice of the medium parameters. In addition, this description serves as a review of the results of Chap. 7 and 8.

Consider first the range of values of θ during which the distant saddle points are dominant over the upper near saddle point SP_1 (i.e. $1 \leqslant \theta < \theta_{SB}$). The topography of $X(\omega, \theta)$ in the right-half of the complex ω-plane in the range $1 \leqslant \theta \leqslant \theta_{SB}$ is given in Fig. 6.4 through 6.6. It is seen from these three diagrams that for high frequencies ω_p, there will be a signal arrival during the evolution of the first precursor field. At θ equals unity (Fig. 6.4), the distant saddle points are at $\pm\infty - 2\delta i$ (cf. (6.2.21)) and no signal can have arrived (except for the non-physical signal with infinite angular frequency ω_p). At $\theta = 1.25$ (Fig. 6.5), it is seen that for values of ω_p greater than $\omega_p \simeq 9.4 \times 10^{16}$/sec the signal has already arrived and is the dominant contribution to the field $A(z, t)$. Finally, at $\theta = \theta_{SB} \cong 1.33425$ (Fig. 6.6), the distant saddle points and the upper near saddle point are of equal exponential importance in their individual contributions to the asymptotic behavior of the total field $A(z, t)$. In that case, for values of ω_p greater than $\omega_{SB} \cong 8.6 \times 10^{16}/s$[4]. The signal has already arrived and is the dominant contribution to the field $A(z, t)$. For values of ω_p less than ω_{SB}, the signal has yet to arrive at $\theta = \theta_{SB}$ and the precursor field $A_p(z, t)$ is the dominant contribution to the total field $A(z, t)$.

For all values of θ larger than θ_{SB}, the upper near saddle point SP_1 for $\theta_{SB} < \theta \leqslant \theta_1$ and the two near saddle points $SP_N^{\pm}$ for $\theta > \theta_1$ are dominant over the distant saddle points. The asymptotic contribution of these near saddle points to the field $A(z, t)$ yields the second or *Brillouin* precursor. For values of θ in the range $\theta_{SB} < \theta \leqslant \theta_0$[5], during which the upper near saddle point SP_1 is the dominant saddle point, a consideration of the contour defined by $X(\omega) = X(\omega_{sp_1})$ (which is inclined at a positive angle of $\pi/4$ to the positive real axis at the saddle point SP_1 which lies along the imaginary axis in the upper-half of the complex ω-plane) reveals that the signal due to a pole singularity at $\omega_p > \omega_{SB}$ is no longer the dominant contribution to the asymptotic behavior of the field since it now possesses a larger exponential decay than does the evolving second precursor field. That is, as θ increases over the range $\theta_{SB} < \theta \leqslant \theta_0$, the contour $X(\omega) = X(\omega_{sp_1})$ through the dominant upper near saddle point recrosses any

[4] The approximate expression (6.3.15) for the value of ω_{SB} yield the value $\omega_{SB} \cong 7.23 \times 10^{16}$/s, which is in fair agreement with the actual value.

[5] In the present example for *Brillouin*'s choice of the medium parameters, $\theta_0 = 1.50$

pole singularity at $\omega_p > \omega_{SB}$ which had previously been crossed by the contour $X(\omega) = X(\omega_{sp_D^+})$ through the distant right saddle point when θ varied over the range $1 \leqslant \theta < \theta_{SB}$. Note, however, that the pole contribution at $\omega_p > \omega_{SB}$ has not been cancelled or negated by this phenomenon, but rather has only become less dominant than the evolving second precursor field. At $\theta = \theta_0 = 1.50$, $X(\omega_{sp_1})$ is identically zero at the dominant near saddle point SP_1, which is exactly located at the origin. The contour $X(\omega) = X(\omega_{sp_1}, \theta_0)$ intersects the real axis only at $\omega' = 0$ and infinity, and remains above the real axis for all other positive values of ω'. Consequently, at $\theta = \theta_0$ the second precursor field is exponentially dominant over all other contributions to the asymptotic behavior of the propagated field $A(z, t)$.

Consider finally the remaining three graphs of the topography of $X(\omega, \theta)$ in the complex ω-plane for values of θ greater than θ_0. At $\theta = 1.501$ (Fig. 6.7), it is seen that for values of ω_p in either the range $0 \leqslant \omega_p < 0.05 \times 10^{16}$/s or $\omega_p > 20 \times 10^{16}$/s, the pole contribution is the dominant contributon to the asymptotic behavior of the field $A(z, t)$, while for values of ω_p in the range $0.05 \times 10^{16}/\text{s} < \omega_p < 20 \times 10^{16}$/s, the second precursor is the dominant contribution to the asymptotic behavior of $A(z, t)$. At $\theta = 1.65$ (Fig. 6.8), the dominant near saddle points have already moved off of the imaginary axis and into the complex ω-plane, and it is seen that for values of ω_p in either the range $0 \leqslant \omega_p < 1.27 \times 10^{16}$/s or $\omega_p > 15.6 \times 10^{16}$/s, the pole contribution at $\omega = \omega_p$ is the dominant contribution to the asymptotic behavior of the field, while for values of ω_p in the range $1.27 \times 10^{16}/\text{s} < \omega_p < 15.6 \times 10^{16}$/s, the second precursor field is dominant. At $\theta = 5.0$ (Fig. 6.9), it is seen that for values of ω_p in either the range $0 \leqslant \omega_p < 3.2 \times 10^{16}$/s or $\omega_p > 6.35 \times 10^{16}$/s, the signal is the dominant contribution to the propagated field $A(z, t)$, while for values of ω_p in the range $3.2 \times 10^{16}/\text{s} < \omega_p < 6.35 \times 10^{16}$/s, the second precursor field is dominant. Since the near saddle points are dominant over the distant saddle points for all $\theta > \theta_{SB}$, and since $X(\omega_{sp})$ at the near saddle points monotonically decreases with increasing $\theta > \theta_0$, a value of $\theta = \theta_m$ will finally be attained at which the relation

$$X(\omega_{sp_N^\pm}, \theta_m) = X(\omega_{\min}) \tag{9.3.3}$$

is satisfied. Here $\omega_{\min}$ is that value of ω along the positive real axis at which $X(\omega)$ attains its minimum value [cf. (6.1.20)]. At this value of θ, the contour $X(\omega) = X(\omega_{sp_N^\pm}, \theta_m)$ lies entirely in the lower-half of the complex ω-plane and touches the real axis only at $\omega = \omega_{\min}$. This situation is illustrated in Fig. 9.3. As can be seen, for the frequency $\omega_p = \omega_{\min}$, the signal arrives at $\theta_c = \theta_m$ which is larger than any other value of θ_c for $\omega_p \neq \omega_{\min}$. That is, an input signal with $\omega_p = \omega_{\min}$ has a main signal whose front propagates with a minimum velocity in the medium. For all $\theta > \theta_m$, the main signal at any frequency value ω_p has already arrived and is the dominant contribution to the propagated field $A(z, t)$.

In summary, the signal arrival separates naturally into two distinct cases which depend upon the value of the real frequency ω_p. For values of ω_p in the range $0 \leqslant \omega_p \leqslant \omega_{SB}$, the signal arrival is due to the crossing of the contour

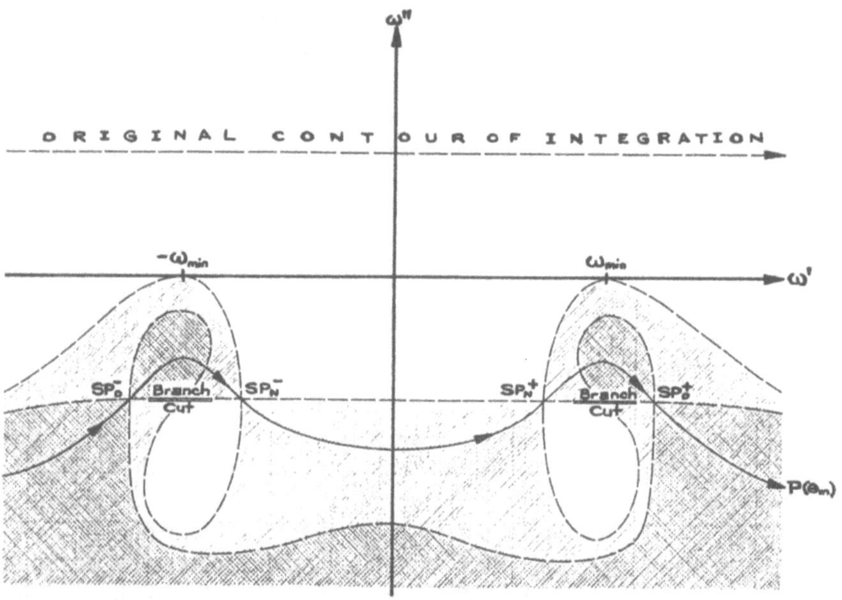

Fig. 9.3. The behavior of the topography of $X(\omega)$ in the complex ω-plane at $\theta = \theta_m$.

The hatched area ▨ indicates the region of the complex ω-plane wherein the inequality $X(\omega) < X(\omega_{sp_N^\pm})$ is satisfied.

The hatched area ▧ indicates the region of the complex ω-plane wherein the inequality $X(\omega) < X(\omega_{sp_D^\pm})$ is satisfied.

$X(\omega) = X(\omega_{sp_N})$ with the simple pole singularity at $\omega = \omega_p$, where ω_{sp_N} denotes the location of the upper near saddle point SP_1 for $\theta < \theta_1$, the second order near saddle point at $\theta = \theta_1$, and the near saddle point SP_N^+ for $\theta > \theta_1$. For such values of ω_p, the main signal due to the pole contribution at $\omega = \omega_p$ is preceded by the first and second precursor fields, and the main signal evolves essentially undisturbed. For values of the frequency ω_p in the range $\omega_p > \omega_{SB}$, the signal arrival first occurs due to the crossing of the contour $X(\omega) = X(\omega_{sp_D})$ with the simple pole singularity at $\omega = \omega_p$. This arrival occurs for a value of θ within the range $1 < \theta < \theta_{SB}$ for finite ω_p. However, for some value of θ within the range $\theta_{SB} < \theta < \theta_0$, the pole is again crossed in the opposite direction by the contour $X(\omega) = X(\omega_{sp_1})$ so that the pole contribution becomes less dominant than the second precursor field. Then, for some value of $\theta > \theta_0$, the pole is again crossed by the contour $X(\omega) = X(\omega_{sp_N})$ so that it finally becomes dominant over all other contributions to the asymptotic behavior of $A(z, t)$ for all remaining values of θ. Consequently, for sources with $\omega_p > \omega_{SB}$ there is the existence of a "prepulse" due to the interuption of the signal evolution by the second precursor field which becomes dominant over the pole contribution for some finite interval of θ. This "prepulse" formation is seen to be an integral part of the dynamic evolution of the second precursor field superimposed upon the evolution of the main signal. The signal evolution for such frequencies $\omega_p > \omega_{SB}$ may then be considered to be separated into three parts: the so-called "prepulse", which is followed by the dominant second precursor field superimposed upon the signal, which is then finally followed by the main signal which remains dominant for all later times. It is important to keep in mind that the prepulse is not independent of the main signal evolution. Indeed, the prepulse formation is simply a consequence of the superposition of the signal with the second precursor field which becomes dominant over the signal for a finite interval of θ.

As a final aside in this subsection, it is of interest to point out that the uniform asymptotic approximation of the pole contribution at $\omega = \omega_p$ takes on a particularly useful form for numerical calculations for $\theta = \theta_c$. For example, for values of ω_p in the range $0 < \omega_p \leqslant \sqrt{\omega_0^2 - \delta^2}$, the uniform asymptotic approximation of the pole contribution at $\theta = \theta_c > \theta_s$ is given by (8.2.11), where ω_{sp} denotes $\omega_{sp_N^+}$ and $\Delta(\theta)$ is given by (8.2.7c). At $\theta = \theta_c$, $\bar{\alpha}_{SD} = \pi/4$, $\arg(-i\Delta(\theta_c)\sqrt{z/c}) = -\pi/4$, and the complementary error function appearing in (8.2.11) may be replaced by the right-hand side of (5.3.17), so that

$$A_c(z,t_c) \sim \frac{1}{2\pi}\operatorname{Re} i\gamma\left\{i\sqrt{2}\pi e^{-i(\pi/4)}\left[\mathscr{C}\left(|\Delta(\theta_c)|\sqrt{\frac{2z}{\pi c}}\right) + i\mathscr{S}\left(|\Delta(\theta_c)|\sqrt{\frac{2z}{\pi c}}\right)\right]e^{(z/c)\phi(\omega_p,\theta_c)} + \frac{1}{\Delta(\theta_c)}\sqrt{\frac{\pi c}{z}}e^{(z/c)\phi(\omega_{sp_N^+},\theta_c)}\right\} + \frac{3}{2}\operatorname{Re}[\gamma e^{(z/c)\phi(\omega_p,\theta_c)}]\ ; \qquad \theta = \theta_c = \frac{ct_c}{z}\ , \tag{9.3.4}$$

as $z \to \infty$. Here $\mathscr{C}(x)$ and $\mathscr{S}(x)$ are the Cosine and Sine Fresnel integrals of a real-valued argument, respectively (see (5.3.18)). For the relevant examples treated in this text, $\gamma = i$ and $\omega_p = \omega_c$, the applied signal frequency. With these substitutions (9.3.4) becomes

$$A_c(z,t_c) \sim -\frac{1}{2}\left\{\mathscr{C}\left(|\Delta(\theta_c)|\sqrt{\frac{2z}{\pi c}}\right)[\cos(\beta(\omega_c)z - \omega_c t) - \sin(\beta(\omega_c)z - \omega_c t)] - \mathscr{S}\left(|\Delta(\theta_c)|\sqrt{\frac{2z}{\pi c}}\right)[\cos(\beta(\omega_c)z - \omega_c t) + \sin(\beta(\omega_c)z - \omega_c t)] + 3\sin(\beta(\omega_c)z - \omega_c t)\right\}e^{-z\alpha(\omega_c)} - \frac{1}{2\pi}\operatorname{Re}\left\{\frac{1}{\Delta(\theta_c)}\sqrt{\frac{\pi c}{z}}\,3^{(z/c)\phi(\omega_{sp_N^+},\theta_c)}\right\}\ ; \qquad \theta = \theta_c = \frac{ct_c}{z}\ , \tag{9.3.5}$$

as $z \to \infty$.

9.3.2 The Signal Velocity

The analysis presented in Chap. 7 and 8 furnishes a complete uniform asymptotic approximation of the propagated field in a dispersive *Lorentz* medium that is characterized by a single resonance frequency. According to that analysis, the following succession of events in the time evolution of the propagated field $A(z,t)$ is obtained. At θ equals unity, the first precursor field $A_s(z,t)$ arrives at the vacuum speed of light c. This field component rapidly builds up to a peak value immediately after its arrival, after which it is increasingly exponentially attenuated for increasing values of θ. At $\theta = \theta_{SB} > 1$, the second precursor field $A_B(z,t)$ becomes dominant over the first precursor field for $\theta > \theta_{SB}$. This field component rapidly builds up to a peak value around $\theta = \theta_0 > \theta_{SB}$, after which it is increasingly exponentially attenuated for increasing values of θ. If the spectral amplitude function has a pole singularity at positive ω_p, the final contribution to the propagated field arises from the contribution due to that pole. According to the results of Chap. 8, this pole contribution, when it is the dominant contribution to the asymptotic behavior of the field $A(z,t)$, is given by

$$A_c(z,t) \sim -e^{-z\alpha(\omega_p)} \sin[\beta(\omega_p)z - \omega_p t] \ , \tag{9.3.6}$$

as $z \to \infty$, where $\alpha(\omega_p)$ is the amplitude attenuation coefficient of the medium at the frequency ω_p (cf. (8.1.7)) and $\beta(\omega_p)$ is the propagation constant at the frequency ω_p (cf. (8.1.9)). The value of $\theta = \theta_c$ at which the pole contribution (9.3.6) is of equal dominance with the second precursor field is given by

$$X(\omega_{sp_N}, \theta_c) = X(\omega_p) \ ; \qquad \theta_c \geqslant \theta_0 \ , \tag{9.3.7}$$

where ω_{sp_N} denotes the location of the upper near saddle point SP_1 for $\theta_0 \leqslant \theta_c < \theta_1$, the second-order near saddle point SP_N at $\theta_0 = \theta_1$, and the near saddle point SP_N^+ for all $\theta_c > \theta_1$. For all values of $\theta > \theta_c$, the pole contribution (9.3.6) is the dominant contribution to the asymptotic behavior of the total field $A(z,t)$ for all values of the frequency $\omega_p \geqslant 0$. For frequencies ω_p in the range $\omega_p > \omega_{SB}$, the pole contribution (9.3.6) is the dominant contribution to the asymptotic behavior of the total field $A(z,t)$ for values of θ in the range $\theta_{c1} < \theta < \theta_{c2}$, where $\theta_{c2} < \theta_c$ for finite ω_p. At $\theta = \theta_{c1} < \theta_{SB}$, which is given by

$$X(\omega_{sp_D^+}, \theta_{c1}) = X(\omega_p) \ ; \qquad 1 < \theta_{c1} < \theta_{SB}, \omega_p > \omega_{SB} \ , \tag{9.3.8}$$

the pole contribution (9.3.6) is of equal dominance with the first precursor field, and for all $\theta > \theta_{c1}$, the pole contribution is dominant over the first precursor. However, at $\theta = \theta_{c2}$, given by

$$X(\omega_{sp_1}, \theta_{c2}) = X(\omega_p) \ ; \qquad \theta_{SB} < \theta_{c2} < \theta_0, \omega_p > \omega_{SB} \ , \tag{9.3.9}$$

the second precursor field is of equal dominance with the pole contribution, and for values of θ in the range $\theta_{c2} < \theta < \theta_c$, the second precursor field is dominant over the pole contribution. At $\theta = \theta_c$, given by (9.3.7), these two contributions are again of equal dominance, and for all $\theta > \theta_c$, the pole contribution is the dominant contribution to the asymptotic behavior of the field $A(z, t)$.

Physically, the first precursor field is due to the high frequency components which are present in the frequency spectrum of the initial pulse envelope, whereas the second precursor field is due to the low frequency components which are present in the initial envelope spectrum. The pole contribution (9.3.6) is physically due to the frequency component at the frequency ω_p, as can readily be seen from that equation. For the relevant examples treated in this thesis, the pole occurs at $\omega_p = \omega_c$, the applied signal frequency of the source. As a consequence, a well-defined signal velocity may be defined in the following manner.

The main signal arrival is defined to occur at that value of $\theta = \theta_c$ satisfying (9.3.7) at which the pole contribution (9.3.6) becomes the dominant contribution to the asymptotic behavior of the total field $A(z, t)$. The velocity at which this point in the field propagates through the dispersive medium is the *main signal velocity*, given by

$$v_c = \frac{c}{\theta_c} , \tag{9.3.10}$$

where c is the vacuum speed of light. Furthermore, for frequencies $\omega_p > \omega_{SB}$, there is the existence of a "prepulse", the front of which arrives at that value of $\theta = \theta_{c1}$ satisfying (9.3.8) at which the pole contribution (9.3.6) becomes the dominant contribution to the asymptotic behavior of the total field, and the back of which arrives at that value of $\theta = \theta_{c2}$ satisfying (9.3.9) at which the second precursor field becomes the dominant contribution to the asymptotic behavior of the total field. The velocity at which the front of the "prepulse" propagates through the dispersive medium is called in this research the *anterior pre-signal velocity*, given by

$$v_{c1} = \frac{c}{\theta_{c1}} ; \qquad \omega_p > \omega_{SB} , \tag{9.3.11}$$

and the velocity at which the back of the "prepulse" propagates through the dispersive medium is called in this research the *posterior pre-signal velocity*, given by

$$v_{c2} = \frac{c}{\theta_{c2}} ; \qquad \omega_p > \omega_{SB} . \tag{9.3.12}$$

According to (9.3.7) through (9.3.9), these three velocities satisfy the inequality

$$c > v_{c1} > \frac{c}{\theta_{SB}} > v_{c2} > \frac{c}{\theta_0} \geqslant v_c . \tag{9.3.13}$$

Note that the main signal, anterior pre-signal, and posterior pre-signal velocities depend only upon the value of ω_p and the medium parameters.

An explicit analytic expression for the main signal velocity defined by (9.3.10) is now obtained. This entails solving (9.3.7) algebraically for θ_c, which may, to a good approximation, be accomplished in the following manner. According to (6.2.48 and 55), the real phase behavior at the near saddle point for $\theta \geqslant \theta_0$ is given by

$$X(\omega_{sp_1}, \theta) \cong \left[|\psi(\theta)| - \frac{2}{3}\delta\zeta(\theta) \right](\theta - \theta_0) + \frac{b^2}{2\theta_0\omega_0^4}\left[|\psi(\theta)| - \frac{2}{3}\delta\zeta(\theta) \right]^2 \left[\alpha|\psi(\theta)| - \frac{2}{3}\alpha\delta\zeta(\theta) + 2\delta \right]; \quad \theta_0 \leqslant \theta \leqslant \theta_1 ,$$

$$X(\omega_{sp_N^+}, \theta) \cong -\frac{2}{3}\delta\zeta(\theta)(\theta - \theta_0) + \frac{\delta b^2}{2\theta_0\omega_0^4}\left\{ \frac{8}{9}\delta^2\zeta^2(\theta)\left(1 - \frac{1}{3}\alpha\zeta(\theta)\right) + 2\psi^2(\theta)(\alpha\zeta(\theta) - 1) \right\}; \quad \theta \geqslant \theta_1 .$$

Since θ_1 is very close to θ_0 and since it is known that $\theta_c = \theta_0$ at $\omega_p = 0$ at which $X(\omega_p) = 0$, then the first of the above two equations need not be considered any further. Substitution of (6.2.39 and 40) into the second of the above two equations with $\alpha \cong 1$ gives

$$X(\omega_{sp_N}, \theta) \cong -\delta(\theta - \theta_0)\cdot \frac{\theta^2 - \theta_0^2 + \dfrac{2b^2}{\omega_0^2}}{\theta^2 - \theta_0^2 + \dfrac{3b^2}{\omega_0^2}} \cdot \left\{ 1 - \frac{b^2}{2\theta_0\omega_0^2} \cdot \frac{\theta + \theta_0}{\theta^2 - \theta_0^2 + \dfrac{3b^2}{\omega_0^2}} \right\}, \tag{9.3.14}$$

where the terms multiplied by the factor δ^3 have been neglected in obtaining this expression. Note that this approximate expression describes the proper real phase behavior at $\theta = \theta_0$ so that it may be employed for all $\theta \geqslant \theta_0$. Substitution of this expression into (9.3.7) results in

$$X(\omega_p) \cong -\delta(\theta_c - \theta_0)\cdot \frac{\theta_c^2 - \theta_0^2 + \dfrac{2b^2}{\omega_0^2}}{\theta_c^2 - \theta_0^2 + \dfrac{3b^2}{\omega_0^2}} \cdot \left\{ 1 - \frac{b^2}{2\theta_0\omega_0^2} \cdot \frac{\theta_c + \theta_0}{\theta_c^2 - \theta_0^2 + \dfrac{3b^2}{\omega_0^2}} \right\}; \quad \theta_c \geqslant \theta_0 . \tag{9.3.15}$$

If the quantity within the brackets { } is approximated by unity, this expression becomes

$$X(\omega_p) + \delta(\theta_c - \theta_0)\cdot\frac{\theta_c^2 - \theta_0^2 + \dfrac{2b^2}{\omega_0^2}}{\theta_c^2 - \theta_0^2 + \dfrac{3b^2}{\omega_0^2}} \cong 0\ ; \qquad \theta_c \geqslant \theta_0\ . \tag{9.3.16}$$

For values of θ_c close to θ_0 (i.e., for $\omega_p \ll \omega_0$ or $\omega_p \gg \omega_1$), this equation yields

$$\theta_c \cong \theta_0 - \frac{3X(\omega_p)}{2\delta}\ ; \qquad 0 \leqslant \omega_p \ll \omega_0 \text{ or } \omega_p \gg \omega_1\ , \tag{9.3.17}$$

whereas for large values of θ_c (i.e., for $\omega_0 \leqslant \omega_p \leqslant \omega_1$), (9.3.16) yields

$$\theta_c \cong \theta_0 - \frac{X(\omega_p)}{\delta}\ ; \qquad \omega_0 \leqslant \omega_p \leqslant \omega_1\ . \tag{9.3.18}$$

Substitution of these results into (9.3.10) then yields the desired analytic approximations for the main signal velocity.

An explicit analytic expression for the anterior pre-signal velocity v_{c1} defined by (9.3.11) is now obtained. This entails solving (9.3.8) for θ_{c1}, which may be accomplished as follows. According to (6.2.26), the real part of the complex phase behavior at the distant saddle point SP_D^+ is, to a good approximation, given by

$$\begin{aligned} X(\omega_{sp_D^+}, \theta) &\cong \delta(1 + \eta(\theta))(1 - \theta) - \frac{b^2}{2}\cdot\frac{\delta(1 - \eta(\theta))}{\xi^2(\theta) + \delta^2(1 - \eta(\theta))^2} \\ &\cong 2\delta(1 - \theta)\ , \end{aligned} \tag{9.3.19}$$

Substitution of this result in (9.3.8) results in

$$\theta_{c1} \cong 1 - \frac{X(\omega_p)}{2\delta}\ ; \qquad 1 < \theta_{c1} < \theta_{SB}\ , \tag{9.3.20}$$

from which the anterior pre-signal velocity v_{c1} may be directly determined.

Finally, an explicit analytic expression for the posterior pre-signal velocity v_{c2} defined by (9.3.12) is now obtained. This entails solving (9.3.9) for θ_{c2}, which may be accomplished as follows. Since $\theta_{SB} < \theta_{c2} < \theta_0$, the first approximate expression (6.2.33) for the upper near saddle point location may be substituted into (6.2.48) with the result

$$\begin{aligned} X(\omega_{sp_1}, \theta) &\cong \left(\frac{2}{3}\delta - \chi(\theta)\right)\left[\theta_0 - \theta + \frac{b^2}{2\theta_0\omega_0^4}\left(\frac{2}{3}\delta - \chi(\theta)\right)\left(\frac{4}{3}\delta + \chi(\theta)\right)\right] \\ &\cong -\frac{\theta_0\omega_0^4}{4\delta b^2}(\theta_0 - \theta)^2\left[1 - \frac{\theta_0\omega_0^4}{4\delta^2 b^2}(\theta_0 - \theta)\right]\ , \end{aligned} \tag{9.3.21}$$

for $\theta_{SB} < \theta < \theta_0$. Substitution of this result into (9.3.9) results in

$$X(\omega_p) + \frac{\theta_0\omega_0^4}{4\delta b^2}(\theta_0 - \theta_{c2})^2 - \frac{\theta_0^2\omega_0^8}{16\delta^3 b^4}(\theta_0 - \theta_{c2})^3 \cong 0 \; ; \qquad \theta_{SB} < \theta_{c2} < \theta_0 \; . \tag{9.3.22}$$

The proper solution to this cubic equation corresponds to the one which yields the limiting values $\theta_{c2} = \theta_0$ as $\omega_p \to \infty$ and $\theta_{c2} = \theta_{SB}$ at $\omega_p = \omega_{SB}$.

Based upon (9.3.7) through (9.3.9), accurate numerical values of θ_c, θ_{c1}, and θ_{c2}, and hence, of the main signal, anterior pre-signal, and posterior pre-signal velocities, may be easily obtained by direct comparison of the numerical behavior of $X(\omega, \theta)$ at either the near or distant saddle points with the value of $X(\omega)$ at the real signal frequency $\omega_p = \omega_c$, as has been done in Table 9.1 for the case of Brillouin's choice of the medium parameters. The frequency dependence of these velocities is illustrated graphically in Fig. 9.4. In that figure, the solid lines represent the numerically determined behavior given in Table 9.1, and the dashed lines represent the approximate behavior given by (9.3.16, 20, and 22). As can be seen from this graph, the main signal velocity attains a minimum value near the resonance frequency of the medium. The actual minimum occurs at $\omega_c = \omega_{\min} \cong \omega_0$ (cf. (6.1.20)), at which value $X(\omega_c)$ attains its minimum value along the positive real axis. Consequently, the signal velocity does not peak to the vacuum speed of light c near resonance, as was indicated by Brillouin[6], but rather attains a minimum value near resonance [9.3].

9.3.3 Comparison of the Signal Velocity with the Other Velocities of Light

In continuation of the results of the preceeding subsection, the velocity of propagation of the applied or input signal in a *Lorentz* medium characterized by a single resonance frequency is now compared to the phase, group, and energy velocities in the medium. For a monochromatic plane wave of frequency ω_c, the propagation factor $\beta(\omega_c)$ of the plane wave is related to the real part of the complex index of refraction of the medium by

$$\beta(\omega_c) = \frac{\omega_c}{c} n_r(\omega_c) \; . \tag{9.3.23}$$

where c denotes the vacuum speed of light. Accordingly, the phase velocity $v_p(\omega_c)$ of the monochromatic plane wave is given by

$$v_p(\omega_c) = \frac{\omega_c}{\beta(\omega_c)} = \frac{c}{n_r(\omega_c)} \; . \tag{9.3.24}$$

[6] *Brillouin* [9.1, 2] incorrectly interpreted the signal arrival to occur when the simple pole singularity at the applied signal frequency was crossed in deforming the original contour of integration to the path of steepest descent through the relevant saddle points.

Table 9.1. Frequency dependence of the main signal, anterior pre-signal, posterior pre-signal, and energy transport velocities in a *Lorentz* medium, where *Brillouin*'s choice of the medium parameters are employed.

Applied Signal Frequency ω_c in units of 10^{16}/s	Main Signal Velocity		Anterior Pre-Signal Velocity		Posterior Pre-Signal Velocity		Energy Transport Velocity
	θ_c	$\frac{v_c}{c}=\frac{1}{\theta_c}$	θ_{c1}	$\frac{v_{c1}}{c}=\frac{1}{\theta_{c1}}$	θ_{c2}	$\frac{v_{c2}}{c}=\frac{1}{\theta_{c2}}$	$\frac{v_E}{c}$
0.0	1.500	0.6667					0.6667
0.5	1.522	0.6570					0.6580
1.0	1.5875	0.6299					0.6309
1.5	1.730	0.5780					0.5825
2.0	1.970	0.5076					0.5083
2.5	2.500	0.4000					0.4034
3.0	3.790	0.2639					0.2681
3.1	4.250	0.2353					0.2385
3.2	4.880	0.2049					0.2086
3.3	5.650	0.1770					0.1788
3.4	6.750	0.1481					0.1497
3.5	8.30	0.1205					0.1218
3.6	10.60	0.0943					0.09594
3.7	13.95	0.0717					0.07301
3.8	18.92	0.0529					0.05414
3.9	25.55	0.0391					0.04045
4.0	32.20	0.0311					0.03249
4.1	35.95	0.0278					0.02945
4.2	36.35	0.0275					0.02949
4.3	34.90	0.0287					0.03107
4.4	32.85	0.0304					0.03337
4.5	30.70	0.0326					0.03601
4.6	28.70	0.0348					0.03887
4.7	26.85	0.0372					0.04189
4.8	25.15	0.0398					0.04509
4.9	23.57	0.0424					0.04851
5.0	22.10	0.0452					0.05220
5.1	20.65	0.0484					0.05627
5.2	19.30	0.0518					0.06082
5.3	17.95	0.0557					0.06605
5.4	16.60	0.0602					0.07218
5.5	15.23	0.0657					0.07960
5.6	13.85	0.0722					0.08888
5.7	12.40	0.0806					0.1009
5.8	10.92	0.0916					0.1170
5.9	9.45	0.1058					0.1391
6.0	8.05	0.1242					0.1682
6.2	5.96	0.1678					0.2412
6.4	4.79	0.2088					0.3153
6.6	4.07	0.2457					0.3820
6.8	3.62	0.2762					0.4408
7.0	3.28	0.3049					0.4925

Table 9.1 (*Contd.*)

Applied Signal Frequency ω_c in units of 10^{16}/s	Main Signal Velocity		Anterior Pre-Signal Velocity		Posterior Pre-Signal Velocity		Energy Transport Velocity
	θ_c	$\frac{v_c}{c} = \frac{1}{\theta_c}$	θ_{c1}	$\frac{v_{c1}}{c} = \frac{1}{\theta_{c1}}$	θ_{c2}	$\frac{v_{c2}}{c} = \frac{1}{\theta_{c2}}$	$\frac{v_E}{c}$
7.5	2.76	0.3623					0.5966
8.0	2.45	0.4082					0.6734
8.5	2.25	0.4444					0.7309
9.0	2.13	0.4695	1.290	0.7752	1.350	0.7407	0.7748
9.5	2.03	0.4926	1.233	0.8110	1.360	0.7353	0.8088
10.0	1.95	0.5128	1.195	0.8368	1.375	0.7273	0.8357
10.5	1.90	0.5263	1.165	0.8584	1.390	0.7194	0.8573
11.0	1.85	0.5405	1.140	0.8772	1.397	0.7158	0.8748
11.5	1.815	0.5510	1.125	0.8889	1.404	0.7123	0.8892
12.0	1.78	0.5618	1.105	0.9050	1.410	0.7092	0.9012
12.5	1.75	0.5714	1.095	0.9132	1.414	0.7072	0.9114
13.0	1.73	0.5780	1.085	0.9217	1.421	0.7037	0.9200
13.5	1.71	0.5848	1.080	0.9259	1.423	0.7027	0.9273
14.0	1.695	0.5900	1.070	0.9346	1.429	0.6998	0.9337
14.5	1.68	0.5952	1.065	0.9390	1.433	0.6978	0.9393
15.0	1.66	0.6024	1.060	0.9434	1.435	0.6969	0.9441

A graph illustrating the frequency behavior of the phase velocity for the case of *Brillouin*'s choice of the medium parameters (viz $\omega_0 = 4.0 \times 10^{16}$/s, $b^2 = 20.0 \times 10^{32}/\text{s}^2$, $\delta = 0.28 \times 10^{16}$/s) is given in Fig. 9.5. The phase velocity is not a physically observable quantity, but rather is a mathematical construct from which the phase of a monochromatic plane wave at some point in space and time may be determined if the phase is known at some other point in space and time. Since a monochromatic plane wave exists for all time t, there is no violation of the relativistic principle of causality by the fact that the phase velocity may exceed the vacuum speed of light c since no real physical information is associated with the phase velocity [9.4]. Nevertheless, the concept of the phase velocity is kept as a useful mathematical construct since any physical wave field may be constructed by a suitable unique Fourier representation in terms of monochromatic plane waves. Indeed, such a superposition of monochromatic plane waves, each traveling at a different phase velocity and being attenuated at a different rate with increasing distance of propagation, forms the physical basis for understanding the underlying physics of pulse propagation in temporally dispersive, absorbing media. Furthermore, according to (9.3.24), the phase velocity serves in the definition of the real index of refraction of the medium, defined for a monochromatic plane wave.

Consider next the concept of the group velocity, as given in (4.4.7) for an arbitrary wave packet of mean frequency ω_c, viz

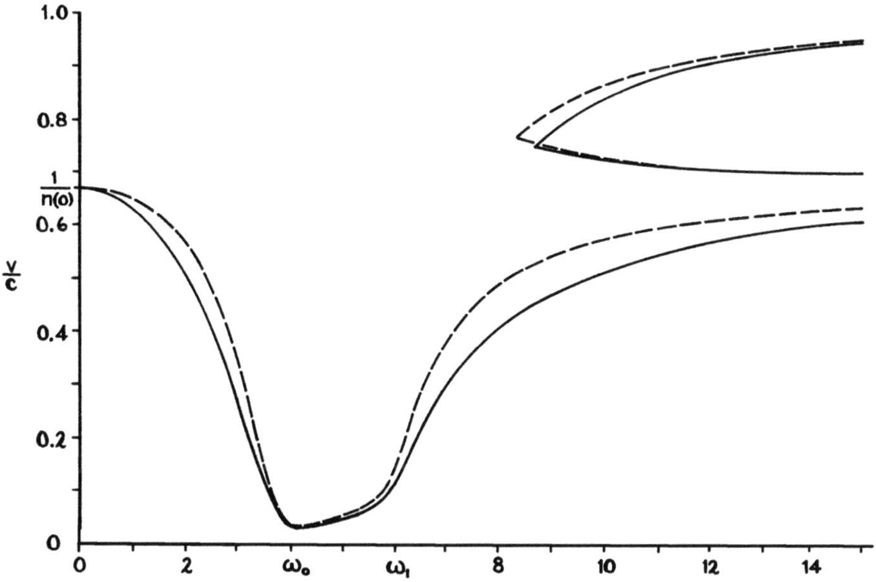

Fig. 9.4. Frequency dependence of the main signal, anterior pre-signal, and posterior pre-signal velocities in a *Lorentz* medium characterized by *Brillouin*'s medium parameters ($\omega_0 = 4.0 \times 10^{16}$/s, $b^2 = 20 \times 10^{32}/\text{s}^2$, and $\delta = 0.28 \times 10^{16}$/s). The solid curves represent the numerically determined behavior and the dashed line curves represent the approximate solutions.

$$v_g(\omega_c) = \left.\frac{1}{\dfrac{\partial\beta(\omega)}{\partial\omega}}\right|_{\omega=\omega_c} = \frac{c}{n_r(\omega_c) + \omega_c n_r'(\omega_c)} \,. \tag{9.3.25}$$

A graph illustrating the frequency behavior of the group velocity for the case of *Brillouin*'s choice of the medium parameters is given in Fig. 9.6. The group velocity is a mathematical construct intended to describe the propagation of the maximum of the pulse envelope under the ad hoc assumption of the quasi-monochromatic approximation. However, that assumption is not valid in a region of anomalous dispersion [9.5] wherein the group velocity may become negative, zero, or infinite, as is illustrated in Fig. 9.6. Consequently, the above standard expression for the group velocity fails to describe the motion of the peak of the pulse envelope in a region wherein there is either significant absorption or amplification, and hence, no real physical significance may be associated with it in such regions. Furthermore, it can be shown [9.6, 7] that the group velocity in resonant pulse propagation is a functon of the attenuation or amplifi-

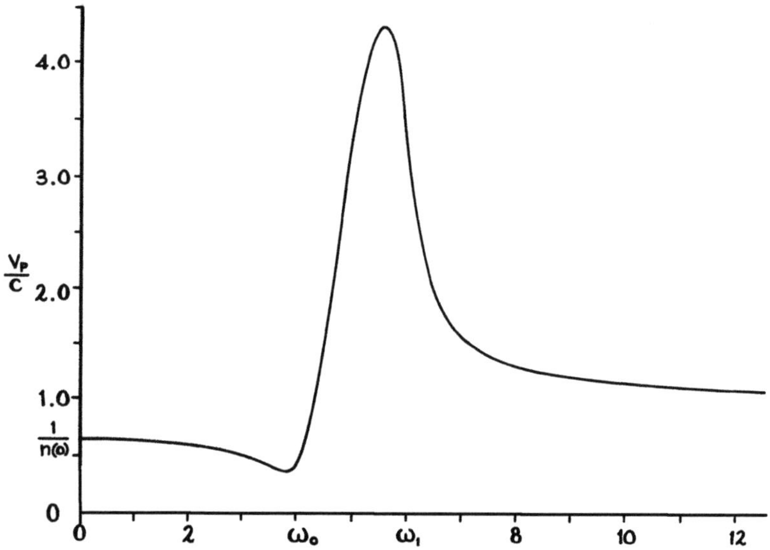

Applied Signal Frequency ω_c in units of 10^{16}/s

Fig. 9.5. Frequency dependence of the phase velocity in a *Lorentz* medium characterized by *Brillouin*'s medium parameters ($\omega_0 = 4.0 \times 10^{16}$/s, $b^2 = 20.0 \times 10^{32}/s^2$, and $\delta = 0.28 \times 10^{16}$/s).

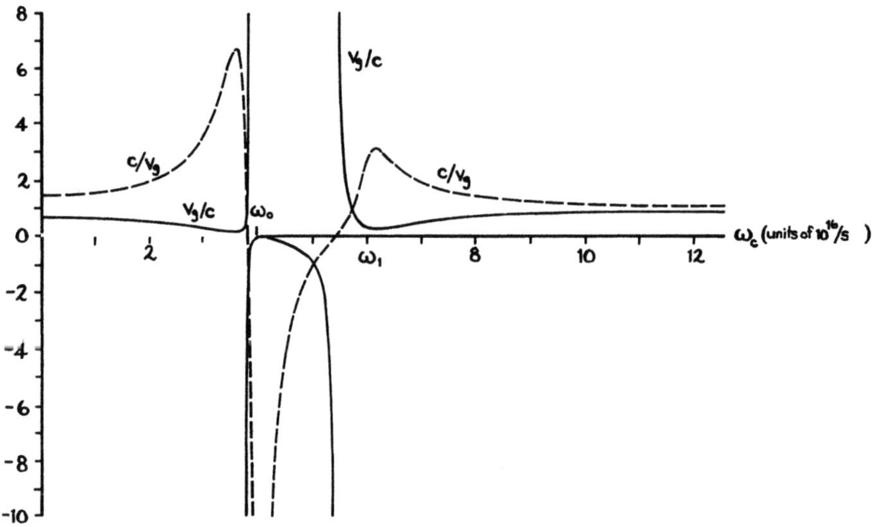

Fig. 9.6. Frequency dependence of the group velocity (solid line) and inverse group velocity (dashed line) in a *Lorentz* medium characterized by *Brillouin*'s medium parameters' ($\omega_0 = 4.0 \times 10^{16}$/s, $b^2 = 20.0 \times 10^{32}/s^2$, and $\delta = 0.28 \times 10^{16}$/s).

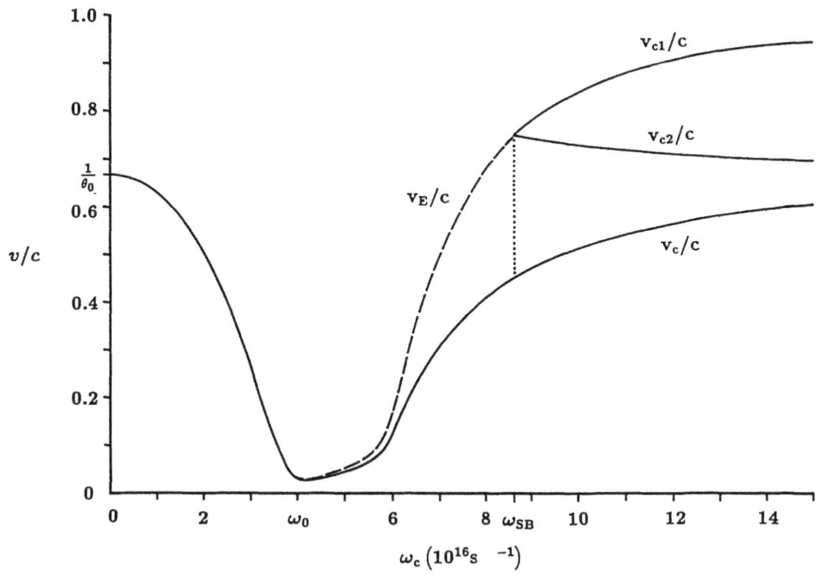

Fig. 9.7. Frequency dependence of the energy transport velocity (dashed curve) for a strictly monochromatic field in a single resonance *Lorentz* medium with $\omega_0 = 4 \times 10^{16}$/s, $b^2 = 20 \times 10^{32}/\mathrm{s}^2$, and $\delta = 0.28 \times 10^{16}$/s. The solid curves depict the frequency dependences of the main signal, anterior presignal, and posterior presignal velocities for a unit step-function modulated signal in the medium.

cation of the medium, the distance propagated in the medium, and the pulse shape. As a consequence, the definition of the group velocity has been reformulated [9.5] as the velocity at which the temporal center of gravity of the amplitude of the pulse envelope moves. This definition reduces to the standard definition when the quasimonochromatic approximation is valid. This revised definition of the group velocity [9.8, 9] is not pursued any further here, but it is noted here that such a reformulation of the concept of group velocity becomes untenable in both the case of a step-function modulated signal and the hyperbolic-tangent modulated signal.

Consider now the concept of the velocity of energy transport [9.8–12], which is defined to be the rate of energy flow as determined by the Poynting vector divided by the stored energy density of the wave. For a monochromatic plane wave of angular frequency ω_c in a *Lorentz* medium characterized by a single resonance frequency, the velocity of energy propagation is given by [cf. Sect. 2.4.2]

$$v_E(\omega_c) = \frac{\langle|\mathcal{S}|\rangle}{\langle\mathcal{U}\rangle} = \frac{c}{n_r(\omega_c) + \dfrac{\omega_c}{\delta} n_i(\omega_c)}\,, \tag{9.3.26}$$

where the magnitude of the Poynting vector $\mathcal{S}$ and the total energy density $\mathcal{U}$ have been averaged over one complete cycle of oscillation of the wave. The frequency dependence of the energy transport velocity is illustrated in Fig. 9.7 for the case of *Brillouin*'s choice of the medium parameters. In the limit of a nonabsorbing medium (i.e., in the limit as the phenomenological damping constant δ approaches zero), it can be shown that $n_i(\omega_c)/\delta \to n_r'(\omega_c)$ as $\delta \to 0$. Hence, in that limit the velocity of energy transport becomes identical with the standard definition of the group velocity. It is important to note that in this limit of vanishing absorptivity outside the absorption band, the medium remains causal [9.4]. Furthermore, it is seen that for applied signal frequencies ω_c sufficiently off resonance, the quantity δ^2 will become negligible in comparison to the factor $(\omega_c^2 - \omega_0^2)$ appearing in the expression for the complex index of refraction of the medium, so that in the regions of normal dispersion removed from the region of anomalous dispersion, the energy transport velocity and the group velocity will become approximately equal. Finally, it is important to note that the velocity of energy propagation is less than the vacuum speed of light c for all finite applied signal frequencies ω_c, and attains a minimum value near the resonance frequency of the medium.

The signal velocity as given in Sect. 9.3.2 is compared to the energy transport velocity in Table 9.1 and Fig. 9.7 for *Brillouin*'s choice of the medium parameters. As can be seen for this example, the main signal velocity is numerically identical with the energy velocity for all signal frequencies in the range $0 \leqslant \omega_c \leqslant \omega_0$, and the anterior pre-signal velocity is identical with the energy velocity for all signal frequencies in the range $\omega_c \geqslant \omega_{SB}$. This result can be proved in general by considering the approximate analytic form the quantity $\theta_E \equiv c/v_E(\omega_c)$ takes on in these ranges of ω_c. To that end, note that according to (9.3.26)

$$\theta_E = \frac{c}{v_E(\omega_c)} = n_r(\omega_c) + \frac{\omega_c}{\delta} n_i(\omega_c) \ . \tag{9.3.27}$$

Now, for values of ω_c in the range $0 \leqslant \omega_c \leqslant \omega_0$, the complex index of refraction is given approximately by (6.2.29), so that

$$n(\omega_c) \cong \theta_0 + \frac{b^2}{2\theta_0 \omega_0^4}\omega_c^2 + \mathrm{i}\frac{\delta b^2}{\theta_0 \omega_0^4}\omega_c$$

for $0 \leqslant \omega_c \ll \omega_0$. Hence

$$n_r(\omega_c) \cong \theta_0 + \frac{b^2}{2\theta_0 \omega_0^4}\omega_c^2 \ ,$$

$$n_i(\omega_c) \cong \frac{\delta b^2}{\theta_0 \omega_0^4}\omega_c \ ,$$

$$X(\omega_c) \cong -\frac{\delta b^2}{\theta_0 \omega_0^4}\omega_c^2 \ ,$$

for $0 \leqslant \omega_c \ll \omega_0$. Substitution of these approximate expressions into (9.3.27) results in

$$\theta_E \cong \theta_0 + \frac{3b^2}{2\theta_0\omega_0^4}\omega_c^2$$

$$\cong \theta_0 - \frac{3}{2}\frac{X(\omega_c)}{\delta}\ ; \qquad 0 \leqslant \omega_c \ll \omega_0\ , \tag{9.3.28}$$

which is precisely the result given in (9.3.17) for the signal velocity in this frequency range. For values of ω_c in the region $\omega_0 \leqslant \omega_c \leqslant \omega_1$ (i.e., within the region of anomalous dispersion of the medium), the complex index of refraction (6.1.5) may be approximated as follows

$$n(\omega_c) = \left[\frac{1 + \dfrac{\omega_c^2 - \omega_1^2}{2\delta i\omega_c}}{1 + \dfrac{\omega_c^2 - \omega_0^2}{2\delta i\omega_c}}\right]^{1/2}$$

$$\cong \left[1 + \frac{\omega_c^2 - \omega_1^2}{4\delta i\omega_c}\right]\left[1 - \frac{\omega_c^2 - \omega_0^2}{4\delta i\omega_c}\right]$$

$$\cong 1 + i\frac{b^2}{4\delta\omega_c}\ .$$

Hence

$$n_r(\omega_c) \cong 1\ ,$$

$$n_i(\omega_c) \cong \frac{b^2}{4\delta\omega_c}\ ,$$

$$X(\omega_c) \cong -\frac{b^2}{4\delta}\ ,$$

for $\omega_0 \leqslant \omega_c \leqslant \omega_1$. Substitution of these expressions into (9.3.27) results in

$$\theta_E \cong 1 + \frac{b^2}{4\delta^2} \cong 1 - \frac{X(\omega_c)}{\delta}\ ; \qquad \omega_0 \leqslant \omega_c \leqslant \omega_1\ , \tag{9.3.29}$$

which is agreement with the result given in (9.3.18) except for the absence of the term $b^2/\omega_0^2 \ll b^2/4\delta^2$. Finally, for values of $\omega_c > \omega_{SB}$, the complex index of refraction is given approximately by (6.2.7), so that

$$n(\omega_c) \cong 1 - \frac{b^2}{2}\cdot\frac{\omega_c - 2\delta i}{\omega_c^3}\ ,$$

and hence

$$n_r(\omega_c) \cong 1 - \frac{b^2}{2\omega_c^2} ,$$

$$n_i(\omega_c) \cong \frac{\delta b^2}{\omega_c^3} ,$$

$$X(\omega_c) \cong -\frac{\delta b^2}{\omega_c^2} .$$

Substitution of these results into (9.3.27) results in

$$\theta_E \cong 1 + \frac{b^2}{2\omega_c^2} \cong 1 - \frac{X(\omega_c)}{2\delta} ; \qquad \omega_c > \omega_{SB} , \tag{9.3.30}$$

which is precisely the result given in (9.3.20) for the anterior pre-signal velocity.

It has therefore been shown that, to a first approximation, the energy transport velocity and the main signal velocity are equal for applied signal frequencies in the range $0 \leqslant \omega_c \leqslant \omega_1$ (note, however, that the approximate equality between $v_E(\omega_c)$ and $v_c(\omega_c)$ begins to fail for values of ω_c greater than ω_0 in such a manner that $v_E(\omega_c) > v_c(\omega_c)$), and that the energy transport velocity and anterior pre-signal velocity are equal for applied signal frequencies in the range $\omega_c > \omega_{SB}$. The energy transport velocity does not, however, describe the signal velocity behavior for applied signal frequencies in the range $\omega_1 < \omega_c < \omega_{SB}$, or the break-up of the field into a prepulse and main signal for applied signal frequencies $\omega_c > \omega_{SB}$. This is due simply to the fact that the energy transport velocity (9.3.26) was derived strictly for the case of a monochromatic signal of frequency ω_c, and hence, does not take into account any of the precursor phenomena associated with a temporal pulse of light in a dispersive medium. The signal velocity, on the other hand, takes fully into account the precursor fields associated with a temporal pulse of light as it propagates through a dispersive, lossy medium. This transient phenomena, composed of the first and second precursor fields, is intimately related to the signal arrival and the prepulse formation for a unit step-function modulated signal. Indeed, the prepulse formation is due entirely to the superposition of the second precursor field with the pole contribution.

9.4 Special Pulses

The evolution of the total propagated field $A(z, t)$ for the special pulses considered in this research is given in this final section of the chapter. The results presented here are based upon the uniform asymptotic analysis of the precursor fields and pole contribution given in Chap. 7 and 8, respectively, as well as upon the analysis of Sec. 9.1 through 9.3 of this chapter. The remarkable accuracy of this asymptotic description is illustrated throughout with numerical examples.

9.4.1 The Unit Step-Function Modulated Signal

A discussion of the complete dynamic behavior of the propagated unit step-function modulated signal at sufficiently large observation distances z is now given [9.3, 13, 14]. This discourse is based upon the uniform asymptotic analysis of the various contributions to the behavior of the propagated signal in a dispersive *Lorentz* medium characterized by a single resonance frequency. Furthermore, it has been shown in Sect. 7.1 that the field $A(z, t)$ vanishes identically for all $\theta < 1$. The remaining evolution of the field for $\theta \geqslant 1$ is described below. To begin, the dynamics of the first and second precursor fields associated with the propagated unit step-function modulated signal are reviewed.

Consider first the evolution of the first precursor field (*Sommerfeld*'s precursor), which contribution is due to the distant saddle points. At θ equals unity, the amplitude of the first precursor identically vanishes, and as θ increases away from unity, the amplitude at first grows very rapidly, and then is exponentially damped out with increasing values of θ. However, due to the presence of the resonance term $(\omega_{sp_D^+} - \omega_c)^{-1}$ in the asymptotic approximation of the first precursor field, a resonance peak in the field may occur when the distant saddle point in the right-half of the complex ω-plane approaches close to the applied signal frequency ω_c, as discussed in Sect. 9.2. The amplitude of the first precursor field for values of $\theta \cong \theta_r$ is approximately lorentzian in shape with a resonance peak at the value of θ at which the equation $\xi(\theta_r) = \omega_c$ is satisfied. Since $\xi(\theta) > \sqrt{\omega_1^2 - \delta^2}$ for finite values of $\theta \geqslant 1$, such a resonance peak may only occur for applied signal frequencies in the range $\omega_c > \sqrt{\omega_1^2 - \delta^2}$. However, even for such a high applied signal frequency ω_c above the absorption band of the medium, such a resonance peak may not appear in the case of a highly absorptive medium (i.e., large δ) due to the distance between the saddle point and ω_c at $\theta = \theta_r$.

The dynamic behavior of the first precursor field is illustrated in the two diagrams of Fig. 9.8. The field illustrated in Fig. 9.8 (a) is typical of either a highly absorptive *Lorentz* medium or an arbitrary medium for an applied signal frequency in the ragne $\omega_c \leqslant \sqrt{\omega_1^2 - \delta^2}$, in either of which case the resonance peak will not be present. On the other hand, the field illustrated in Fig. 9.8 (b) is typical of a weakly absorbing medium for an applied signal frequency ω_c in the range $\omega_c > \sqrt{\omega_1^2 - \delta^2}$, in which case the resonance peak is present. According to the results of Sect. 9.2, a resonance peak of similar form but opposite sign will occur at $\theta = \theta_r$ in the pole contribution to the propagated field such that the resonance phenomena will not appear in the total field evolution. Of final importance to a consideration of the dynamics of the first precursor field is the

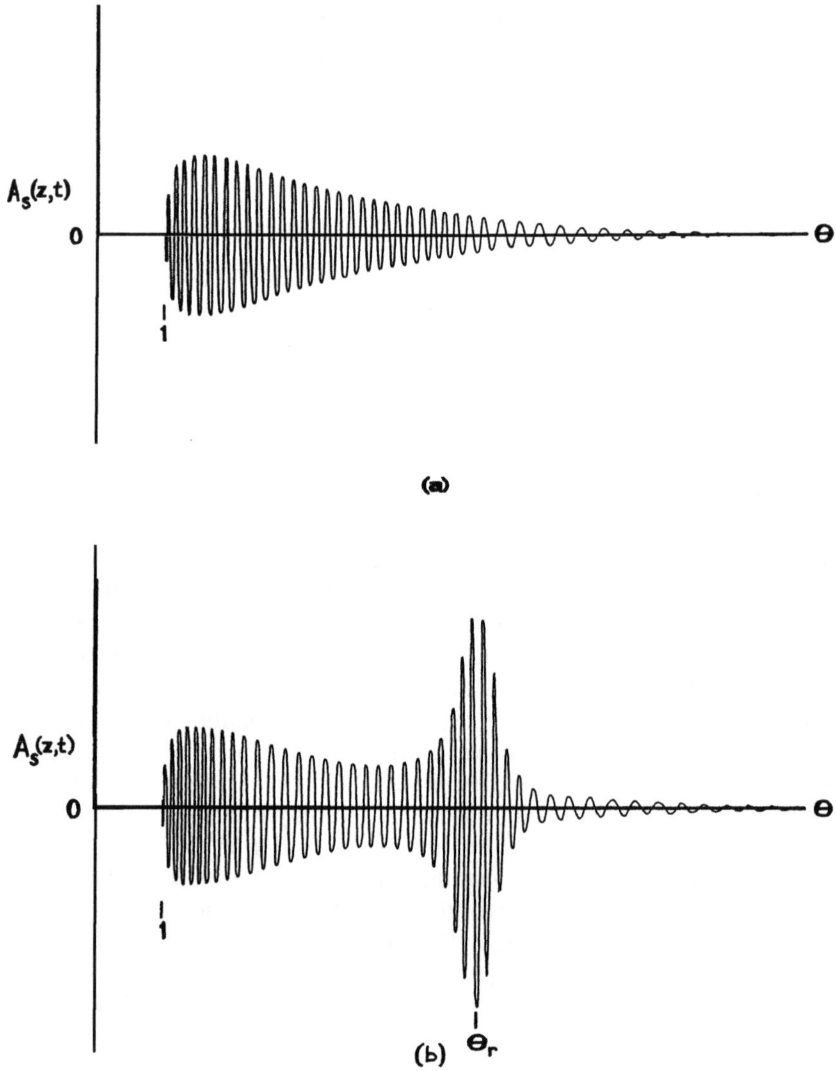

Fig. 9.8. **(a)** Dynamic behavior of the first precursor field for either a highly absorptive medium or for applied signal frequencies $\omega_c \leqslant \sqrt{\omega_1^2 - \delta^2}$. **(b)** Dynamic behavior of the first precursor field for applied signal frequencies $\omega_c > \sqrt{\omega_1^2 - \delta^2}$ in a weakly absorbing medium.

evolution of the angular frequency of oscillation of the field. The instantaneous angular frequency of this first forerunner, which is approximately equal to the real part of the distant saddle point location in the right-half of the complex ω-plane, is at first infinite (when $\theta = 1$ and the field vanishes), and then rapidly decreases as θ initially increases away from unity, this decrease becoming less and less rapid as θ continues to increase and the distant saddle point SP_D^+ approaches ever closer to the branch point ω'_+.

Consider next the evolution of the second precursor field (*Brillouin*'s precursor), which contribution is due to the near saddle points. As θ increases away from unity and up to θ_0, the amplitude of the second precursor steadily increases, the increase becoming more rapid as θ approaches closer to θ_0, until at $\theta = \theta_0$, a maximum in amplitude is reached at which point there is no exponential decay in the second precursor field. As θ increases away from θ_0, the amplitude steadily decreases due to the monotonically increasing exponential decay with θ, but remains larger than the corresponding amplitude of the first precursor field, which, for all $\theta > \theta_{SB}$, possesses a larger exponential decay than does the second precursor field. However, due to the presence of the resonance term $(\omega_{sp_N^+} - \omega_c)^{-1}$ in the asymptotic description of the second precursor field, a resonance peak in the field may now occur when the near saddle point in the right-half of the complex ω-plane approaches close to the applied signal frequency ω_c, as discussed in Sect. 9.2. As in the first precursor field, the amplitude of the second precursor for $\theta \cong \theta_r$ is approximately lorentzian in shape with a resonance peak at the value of θ at which the equation $\psi(\theta_r) = \omega_c$ is satisfied. Since $|\psi(\theta)| < \sqrt{\omega_0^2 - \delta^2}$ for finite values of $\theta \geqslant 1$, such a resonance peak may only occur for applied signal frequencies ω_c in the range $0 \leqslant \omega_c < \sqrt{\omega_0^2 - \delta^2}$. However, even for such applied signal frequencies in the spectral region below the absorption band of the medium, such a resonance peak may not appear in the case of a highly absorptive medium (i.e., large δ) due to the distance between the saddle point and ω_c at $\theta = \theta_r$.

The dynamic behavior of the second precursor field is illustrated in the two diagrams of Fig. 9.9. The behavior depicted in Fig. 9.9 (a) is typical of either a highly absorptive *Lorentz* medium or for an arbitrary *Lorentz* medium for an applied signal frequency in the range $\omega_c \geqslant \sqrt{\omega_0^2 - \delta^2}$, in either of which case the resonance peak will not be present. The behavior illustrated in Fig. 9.9 (b) is typical of a weakly absorbing medium for an applied signal frequency ω_c in the range $0 < \omega_c < \sqrt{\omega_0^2 - \delta^2}$, in which case the resonance peak is present. According to the results of Sect. 9.2, a resonance peak of similar form but opposite sign will occur at $\theta = \theta_r$ in the pole contribution to the propagated field such that the resonance phenomena will not appear in the total field evolution. Of final importance to a consideration of the dynamics of the second precursor field is the evolution of the angular frequency of oscillation of that field. The instantaneous angular frequency of this second forerunner, which is approximately equal to the real part of the near saddle point location in the right-half of the complex ω-plane, is at first zero and remains so up to $\theta = \theta_1$, after which it rapidly increases as θ initially increases away from θ_1, the rate of increase becoming less and less rapid as θ continues to increase and the near saddle point SP_N^+ approaches ever closer to the branch point ω_+.

The complete time evolution of the propagated field $A(z,t)$ for an input unit step-function modulated signal after propagating through a *Lorentz* medium to a fixed observation point a distance z from the source is now considered. According as to whether the applied signal frequency ω_c is in either of the ranges $0 < \omega_c \leqslant \sqrt{\omega_0^2 - \delta^2}$, $\sqrt{\omega_0^2 - \delta^2} < \omega_c < \sqrt{\omega_1^2 - \delta^2}$, or $\sqrt{\omega_1^2 - \delta^2} \leqslant \omega_c$, the to-

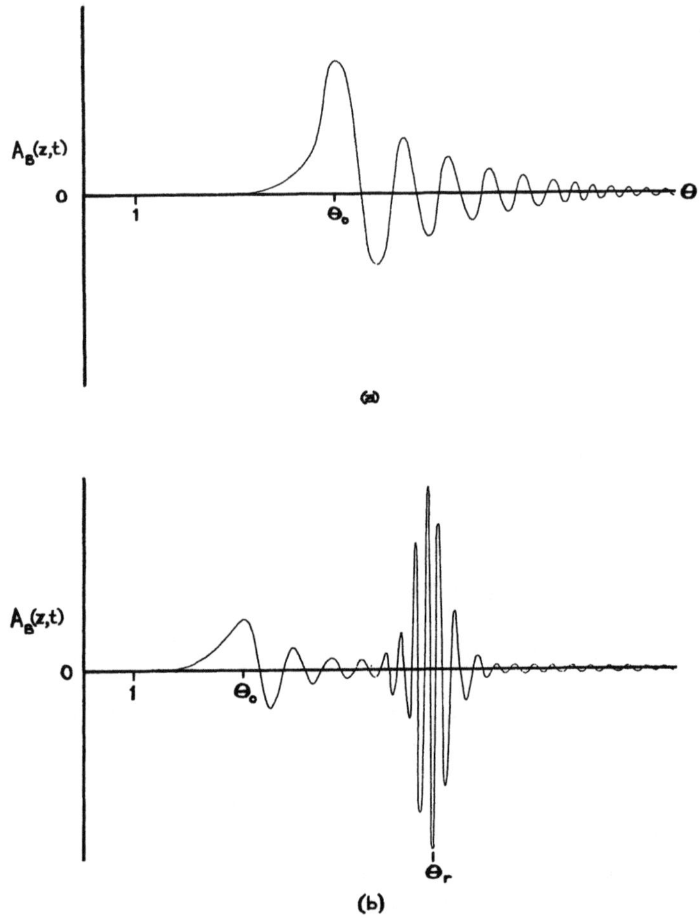

Fig. 9.9. (a) Dynamic behavior of the second precursor field for either a highly absorptive medium or for applied signal frequencies $\omega_c \geqslant \sqrt{\omega_0^2 - \delta^2}$. **(b)** Dynamic behavior of the second precursor field for applied signal frequencies $\omega_c < \sqrt{\omega_0^2 - \delta^2}$ in a weakly absorbing medium.

tal field behavior will differ considerably. Since the resonance phenomena which may occur in the precursor field is identically cancelled by the associated resonance phenomena which will then occur in the pole contribution, the total field behavior in a weakly dispersive medium is the same as that in a strongly dispersive medium. As a consequence, only the total field behavior in a strongly dispersive medium is considered in the following discousre.

Consider first the behavior of $A(z, t)$ for applied signal frequencies ω_c within the range $0 < \omega_c \leqslant \sqrt{\omega_0^2 - \delta^2}$ for a highly absorptive medium, as is illustrated graphically in Fig. 9.10*. The first diagram in Fig. 9.10 illustrates the dynamic

*The field behavior illustrated in Fig. 9.10 through 9.12 have only been drawn for illustrative purposes in order to depict general properties of the propagated field behavior. They do not represent an actual calculation for a specific set of medium and signal parameters.

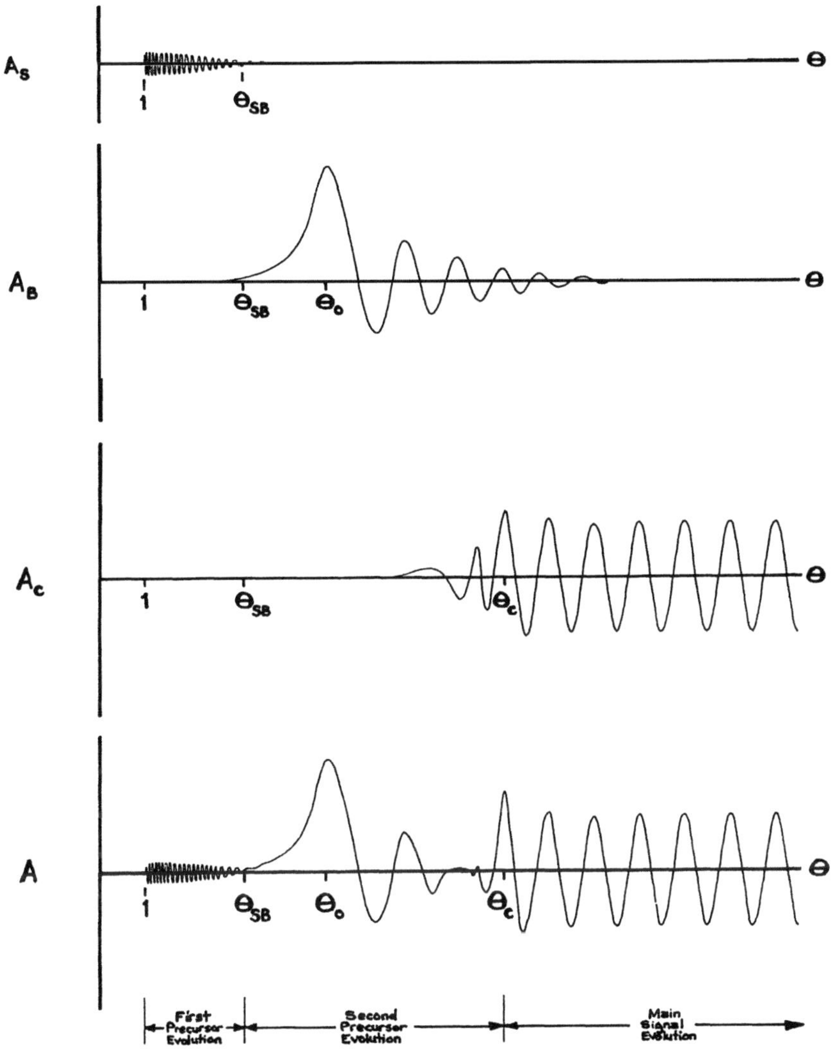

Fig. 9.10. Dynamic behavior of the propagated input unit step-modulated signal in a highly absorptive medium for applied signal frequencies $\omega_c \leqslant \sqrt{\omega_0^2 - \delta^2}$.

evolution of the first precursor field $A_s(z,t)$, the second diagram illustrates the dynamic evolution of the second precursor field $A_B(z,t)$, and the third diagram illustrates the dynamic evolution of the pole contribution $A_c(z,t)$. The fourth and most important diagram illustrates the dynamic evolution of the total field $A(z,t) = A_s(z,t) + A_B(z,t) + A_c(z,t)$, which is obtained from the superposition of the previous three diagrams. As can be seen, the first precursor evolves essentially undisturbed for values of θ in the range $1 \leqslant \theta < \theta_{SB}$. For values of θ about θ_{SB} there is a transition between the first and second precursor fields. The second precursor then evolves essentially undisturbed for values of θ in the range $\theta_{SB} < \theta < \theta_c$. For values of θ about θ_c there is a transition between the second precursor field and the main signal due to the pole contribution at $\omega = \omega_c$. Finally, the main signal then evolves essentially undisturbed for all $\theta \geqslant \theta_c$. Notice the interference effects between the second precursor field and the transient phenomena associated with the main signal arrival. These transient interference effects serve to degrade the main signal behavior about $\theta = \theta_c$. As θ increases away from θ_c, these transient effects rapidly attenuate and the field settles down to a steady-state behavior given by the simple pole contribution alone (cf. equation (9.3.6)). Note finally that as θ approaches θ_c from below, the instantaneous angular frequency $\omega_B \cong \psi(\theta)$ of the second precursor field approaches the applied signal frequency ω_c from below and is approximately equal to ω_c for values of θ about θ_c (i.e., during the transition between the second precursor and the main signal).

For the applied signal frequency $\omega_c = 0$ (i.e., for an input unit step-function modulated static field), the main signal arrival occurs at $\theta_c = \theta_0$ when the upper near saddle point SP_1 coalesces with the simple pole singularity at $\omega_c = 0$. As in the previous case, the *Sommerfeld* precursor evolves essentially undisturbed for values of θ within the range $1 \leqslant \theta < \theta_{SB}$ with a transition about $\theta = \theta_{SB}$ to the *Brillouin* precursor field, which evolves essentially undisturbed for values of θ in the range $\theta_{SB} < \theta < \theta_0$, during which this component field grows exponentially with increasing θ up towards a peak value at $\theta = \theta_0$, at which point the exponential attenuation of the second precursor field identically vanishes. The main signal arrival occurs at $\theta = \theta_0$. For values of θ about θ_0 there is a transition between the second precursor field and the zero frequency main signal which experiences no attenuation. For all values of θ larger than θ_0, this static main signal then evolves essentially undisturbed. Since all other frequency components present in the propagated field experience some attenuation, by proceeding to a large enough observation distance they can be made exponentially negligible in comparison to the main signal evolution and the second precursor field about $\theta = \theta_0$.

Consider now the behavior of the propagated field $A(z, t)$ for applied signal frequencies in the range $\omega_c \geqslant \sqrt{\omega_1^2 - \delta^2}$ for a highly absorptive *Lorentz* medium. In this case there are two separate subcases to be considered, depending upon the value of the applied signal frequency ω_c in comparison to ω_{SB}. For values of ω_c in the range $\sqrt{\omega_1^2 - \delta^2} \leqslant \omega_c \leqslant \omega_{SB}$, the main signal arrival occurs at $\theta_c > \theta_0$ after the second precursor field evolution. However, for applied signal frequencies in the range $\omega_c > \omega_{SB}$, which is the case illustrated in Fig. 9.11, the signal arrival first occurs prior to the second precursor evolution. In that case, the *Sommerfeld* precursor evolves essentially undisturbed within the range $1 \leqslant \theta < \theta_{c1}$. At $\theta = \theta_{c1} < \theta_{SB}$, the prepulse signal arrives and evolves along with the decaying first precursor field up to $\theta = \theta_{SB}$, at which point there is a transition to the second precursor field which also evolves along with the prepulse. At $\theta = \theta_{c2}$, the second precursor field becomes dominant over the pole contribution. For values of θ in the range $\theta_{c2} < \theta < \theta_c$, the second precursor field evolves along with the pole contribution, with the second precursor being the dominant

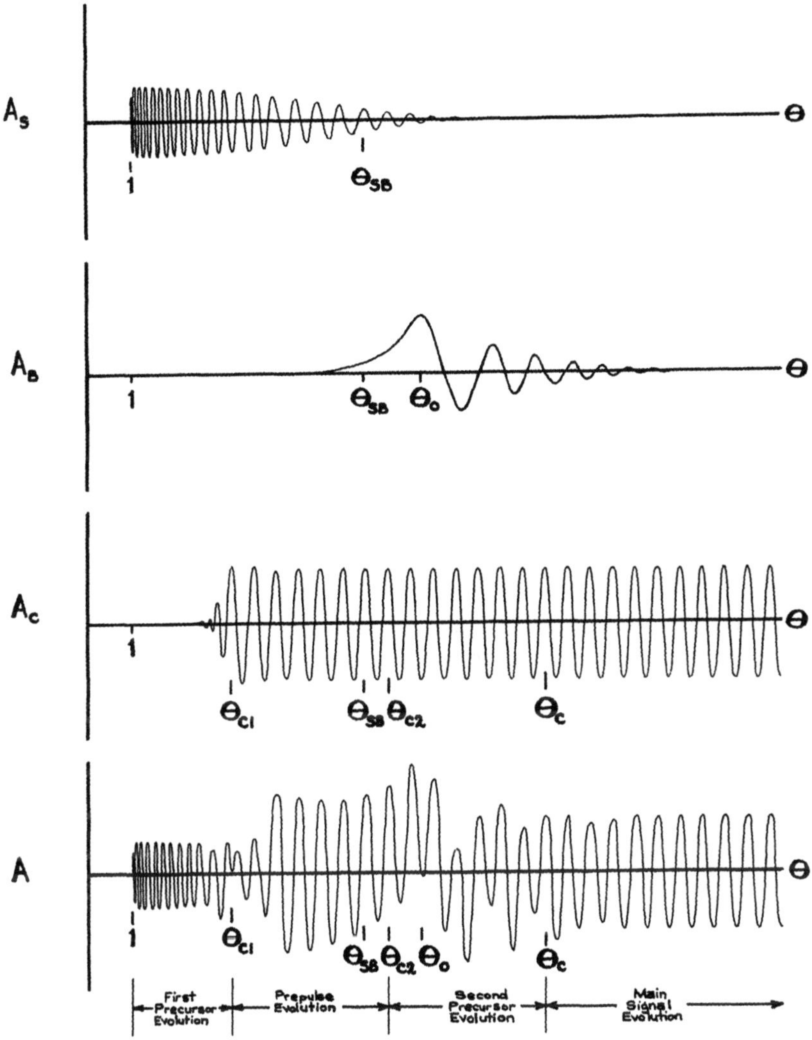

Fig. 9.11. Dynamic behavior of the propagated input unit step-modulated signal in a highly absorptive medium for applied signal frequencies $\omega_c > \omega_{SB} > \sqrt{\omega_1^2 - \delta^2}$.

contribution of the two. The second precursor field becomes negligible in comparison to the pole contribution at $\theta = \theta_c$, so that for all later values of θ the main signal evolves essentially undisturbed, as shown in Fig. 9.11. Consequently, for values of θ in the range $\theta_{c1} \leqslant \theta \leqslant \theta_{SB}$, interference exists between the pole contribution and the first precursor field, which primarily serves to degrade the front of the prepulse, while for values of θ in the range $\theta_{SB} \leqslant \theta \leqslant \theta_c$, interference exists between the pole contribution and the second precursor field. Since the second precursor field experiences virtually zero attenuation for values of θ about θ_0 (at $\theta = \theta_0$ the exponential attenuation of the second precursor field identically vanishes), for sufficiently large observation distances z, the total field behavior about $\theta = \theta_0$ will be essentially a high frequency ripple due to the pole contribution superimposed upon the slowly-varying, large amplitude second precursor field. As can easily be seen, this phenomenon becomes more pronounced the higher the absorptivity of the host medium is made. For such an extremely high absorptive medium, the break up of the pole contribution into the prepulse and the main signal is more pronounced than that depicted in Fig. 9.11. Furthermore, this same result is true as the observation distance z is allowed to become very large. Note finally that as θ approaches θ_{c1} from below, the instantaneous angular frequency $\omega_s \cong \xi(\theta)$ of the first precursor field approaches the applied signal frequency ω_c from above and is approximately equal to ω_c for values of θ about θ_{c1} (i.e., during the transition between the first precursor field and the prepulse).

The final case left to be considered for the behavior of the propagated field $A(z,t)$ is that for applied signal frequencies ω_c which lie in the range $\sqrt{\omega_0^2 - \delta^2} < \omega_c < \sqrt{\omega_1^2 - \delta^2}$. That case is depicted in Fig. 9.12. Such applied signal frequencies are within the absorption band of the medium. The total propagated field behavior at some (sufficiently large) fixed observation distance z then evolves simply as follows. The first precursor field evolves essentially undisturbed for values of θ in the range $1 \leqslant \theta < \theta_{SB}$. For values of θ about $\theta = \theta_{SB}$, there is a transition between the first and second precursor fields. The second precursor field then evolves essentially undisturbed for values of θ in the range $\theta_{SB} < \theta < \theta_0$. For values of θ about $\theta = \theta_c$, there is a transition between the second precursor field and the main signal. The main signal then evolves essentially undisturbed for all $\theta > \theta_c$. Unlike the previous two cases, as θ approaches θ_c from below, the instantaneous angular frequency of either of the precursor fields cannot attain the applied signal frequency ω_c of the main signal. The instantaneous angular frequency of the first precursor field approaches the value $\sqrt{\omega_1^2 - \delta^2}$ from above, while the instantaneous angular frequency of the second precursor field approaches the value $\sqrt{\omega_0^2 - \delta^2}$ from below, where these limiting values are only attained in the limit as θ approaches infinity. Consequently, there is a discontinuous change in the instantaneous angular frequency of the total field in the region about the main signal arrival, as depicted in Fig. 9.12.

The actual computed behavior of the total field evolution is illustrated in Fig. 9.13 for the case of *Brillouin*'s choice of the *Lorentz* medium parameters (viz, $\omega_0 = 4.0 \times 10^{16}$/s, $b^2 = 20.0 \times 10^{32}/\text{s}^2$, $\delta = 0.28 \times 10^{16}$/s, in which case $\theta_0 = 1.50$), which is representative of a highly absorptive medium. The observation distance z at which the field $A(z,t)$ has been calculated here is $z = 1.0 \times 10^{-3}$ cm, and the applied signal frequency of the source is $\omega_c = 1.0 \times 10^{16}$/s. In that case, which corresponds to the depiction of the field construction given in Fig. 9.10, the main signal arrival occurs at $\theta_c \cong 1.5875$, as indicated in Fig. 9.13. Note the continuous transition of the second precursor field as the dominant contribution to the total field over to the main signal as the dominant contribution to $A(z,t)$, which occurs for values of θ about $\theta = \theta_c$. Each individual contribution to the total field evolution illustrated here is presented in Fig. 7.4, 7.9, and 8.4

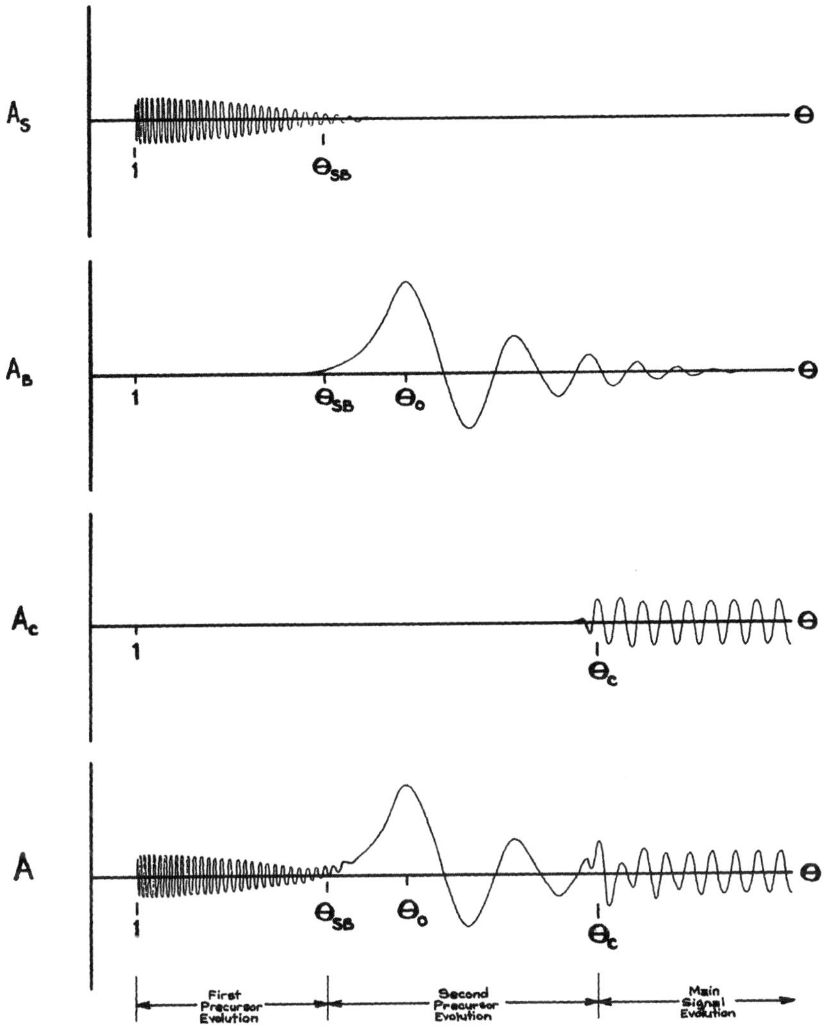

Fig. 9.12. Dynamic behavior of the propagated input unit step modulated signal for applied signal frequencies in the range $\sqrt{\omega_0^2 - \delta^2} < \omega_c < \sqrt{\omega_1^2 - \delta^2}$.

for the same set of medium and field parameters. For the range of values of θ depicted in Fig. 9.13, the amplitude of the first precursor field is less than 10^{-78}.

A calculated example of the resonance peak phenomena occuring in the component fields for the unit step-function modulated signal in a weakly dispersive medium ($\omega_0 = 2.0 \times 10^{16}/\text{s}, b^2 = 0.4 \times 10^{32}/\text{s}^2, \delta = 0.00017 \times 10^{16}/\text{s}$) is illustrated in Fig. 9.14. The applied signal frequency is $\omega_c = 0.1 \times 10^{16}/\text{s}$, which is below the absorption band of the medium, and the observation distance is $z = 100.0$ cm. In that case, $\theta_0 = 1.0488$ and the resonance peak present in the second precursor field, as well as in the pole contribution or main signal, occurs

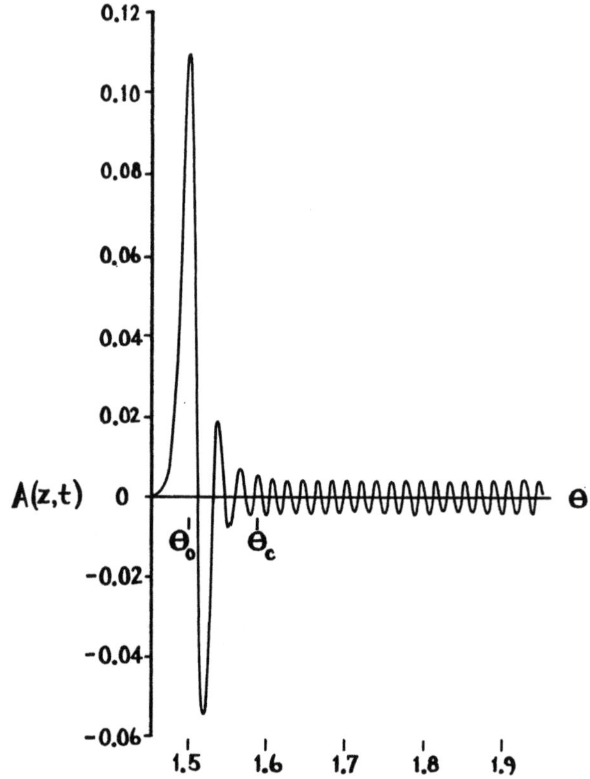

Fig. 9.13. Dynamic behavior of the total propagated field $A(z, t)$ for an input unit step-function modulated signal with applied signal frequency $\omega_c = 1.0 \times 10^{16}$/s at an observation distance of $z = 1.0 \times 10^{-3}$ cm, where *Brillouin*'s choice of the medium parameters are employed.

at $\theta_r \cong 1.04917$. The evolution of the second precursor field is given in part (a) of the figure. The evolution of the pole contribution (as given by (8.3.2)) is given in part (b) of the figure, and the total field evolution, which is given by the sum of these two fields (the amplitude of the first precursor field in this range of values of θ is less than 10^{-72}), is given in part (c) of the figure. Note that there is no resonance phenomena present in the total field.

The remarkable accuracy of this asymptotic description is readily evident in Figs. 9.15–9.27 which presents the numerically determined [9.15] dynamical field evolution due to an input unit step-function modulated signal in a single resonance *Lorentz* medium with parameters $\omega_0 = 4.0 \times 10^{16}$/s, $b^2 = 20.0 \times 10^{32}/\text{s}^2$, $\delta = 0.14 \times 10^{16}$/s. Each figure depicting the field evolution is for a particular carrier frequency ω_c at several increasing values of the propagation distance z and has been calculated using a numerical procedure for inverting Laplace transform type integrals [9.16–17]. The results are displayed most conveniently as a function of the space-time parameter $\theta = ct/z$ with z fixed, since critical aspects of the field evolution are then independent of the propagation

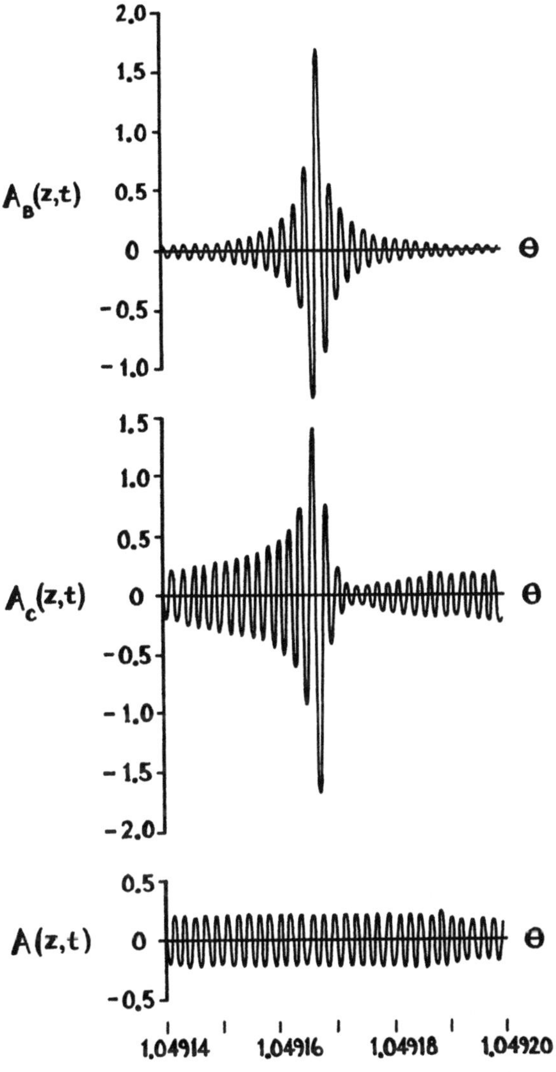

Fig. 9.14. Dynamic behavior of (a) the second precursor field $A_B(z, t)$, (b) the main signal $A_c(z, t)$, and (c) the total propagated field $A(z, t) = A_s(z, t) + A_B(z, t) + A_c(z, t)$ for an input unit step-function modulated signal with applied signal frequency $\omega_c = 0.1 \times 10^{16}$/s at an observation distance of $z = 100$ cm in a weakly dispersive medium ($\omega_0 = 2.0 \times 10^{16}$/s, $b^2 = 0.4 \times 10^{32}/\text{s}^2$, $\delta = 0.00017 \times 10^{16}$/s).

distance. Figures 9.17, 9.24, and 9.26 depict the evolution of the instantaneous angular frequency of oscillation for the field evolution in the corresponding preceeding figure. Each numerically determined value of the instantaneous oscillation frequency is determined by $\omega_j = \pi c/z\Delta\theta_j$, where $\Delta\theta_j = \theta_{j+1} - \theta_j$ represents the absolute difference between the θ values at successive zero crossings in

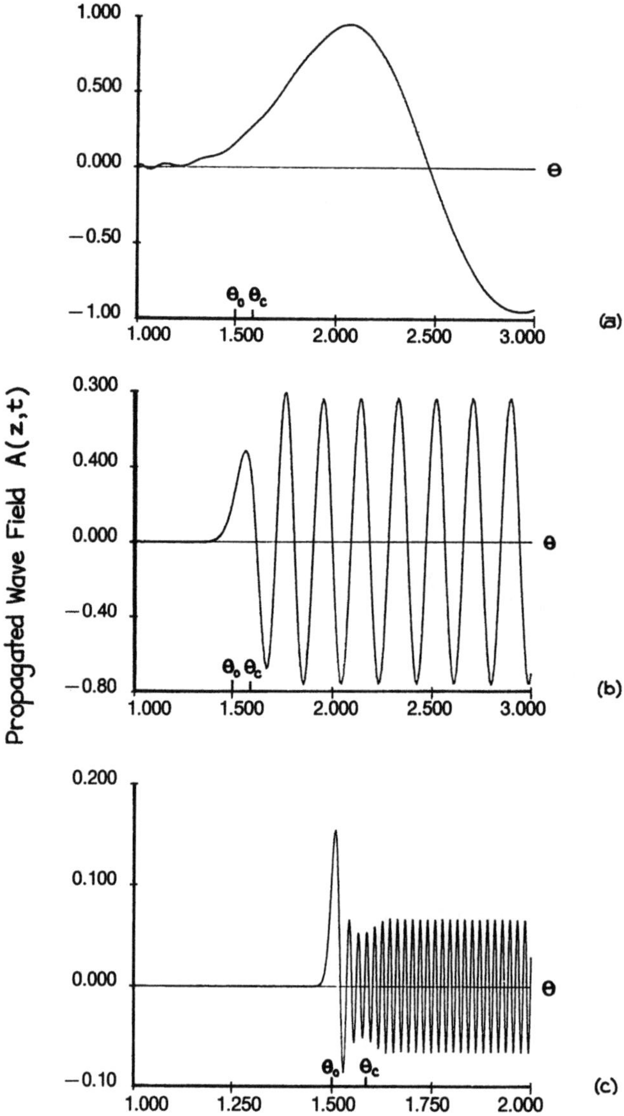

Fig. 9.15. Dynamical field evolution of an input unit step-function modulated signal with below resonance carrier frequency $\omega_c = 1.0 \times 10^{16}$/s at the propagation distances (a) $z = 1 \times 10^{-5}$ cm, (b) $z = 1 \times 10^{-4}$ cm, and (c) $z = 1 \times 10^{-3}$ cm.

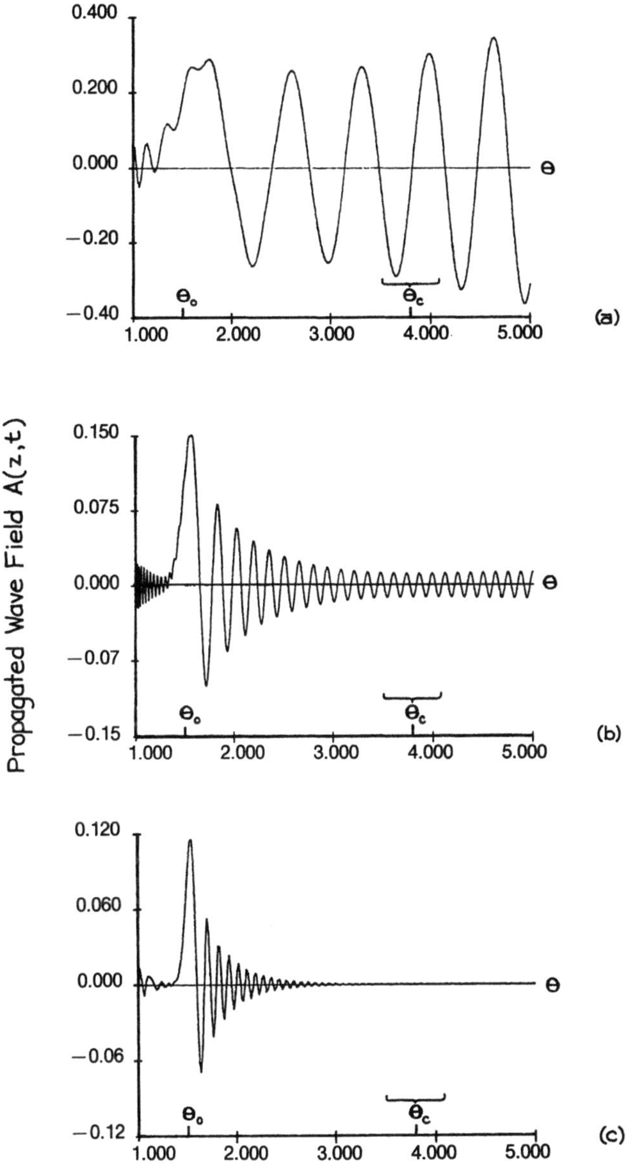

Fig. 9.16. Dynamical field evolution of an input unit step-function modulated signal with below resonance carrier frequency $\omega_c = 3.0 \times 10^{16}$/s at the propagation distances (a) $z = 1 \times 10^{-5}$ cm, (b) $z = 5 \times 10^{-5}$ cm, and (c) $z = 1 \times 10^{-4}$ cm.

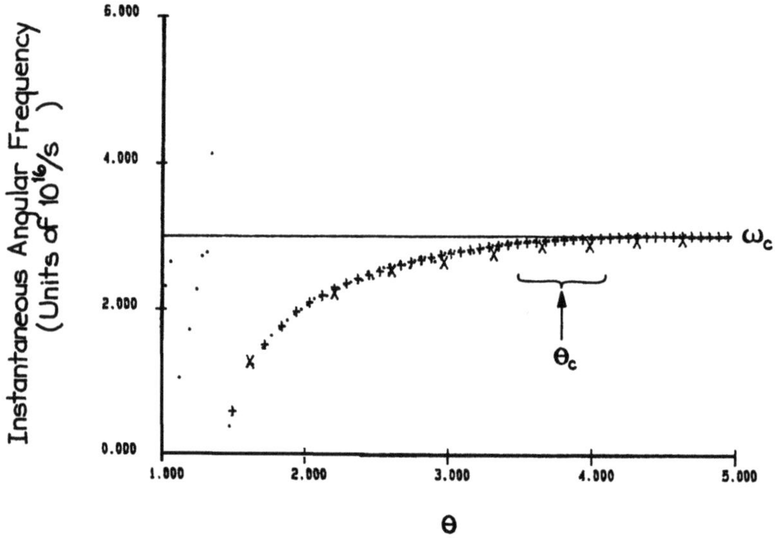

Fig. 9.17. Evolution of the instantaneous angular frequency of oscillation for the dynamical field evolution of Fig. 9.16.

the calculated field evolution at a fixed value of the propagation distance. The calculated frequency value ω_j is then assigned to the midpoint of the interval (θ_j, θ_{j+1}).

In Figs. 9.15 and 9.16 the applied signal frequency ω_c is below the undamped resonance frequency ω_0 of the medium and hence is below the absorption band of the medium. The dynamical field evolution begins with the *Sommerfeld* precursor evolution over $1 \leqslant \theta < \theta_{SB}$, followed by the *Brillouin* precursor, which evolves over $\theta_{SB} < \theta < \theta_c$, and then by the main signal evolution for $\theta > \theta_c$. Because of the low carrier frequency used in Fig. 9.15, the *Sommerfeld* precursor amplitude is nearly negligible in comparison with both the *Brillouin* precursor and main signal amplitudes and is barely visible in Fig. 9.15 (a). Because of this, these calculations were not focused on ensuring that the *Sommerfeld* precursor evolution was always sampled properly; for example, it is sampled adequately in Fig. 9.16 (b) but not in Fig. 9.16 (c). Notice that the propagation distance used in part (a) of both figures is sufficiently small that the propagated field evolution depicted is still in the immature dispersion region. The propagation distances in the latter two diagrams of each figure is sufficiently large for the propagated field evolution to be in the mature dispersion limit for which the asymptotic theory applies.

The evolution of the instantaneous oscillation frequency for the field evolution sequence of Fig. 9.16 is given in Fig. 9.17. The × symbols indicate the numerical values for part (a) of Fig. 9.16, the + symbols indicate the numerical values for part (b) of that figure, and the · symbols for part (c) of that figure. The instantaneous oscillation frequency of the *Brillouin* precsursor is clearly seen to

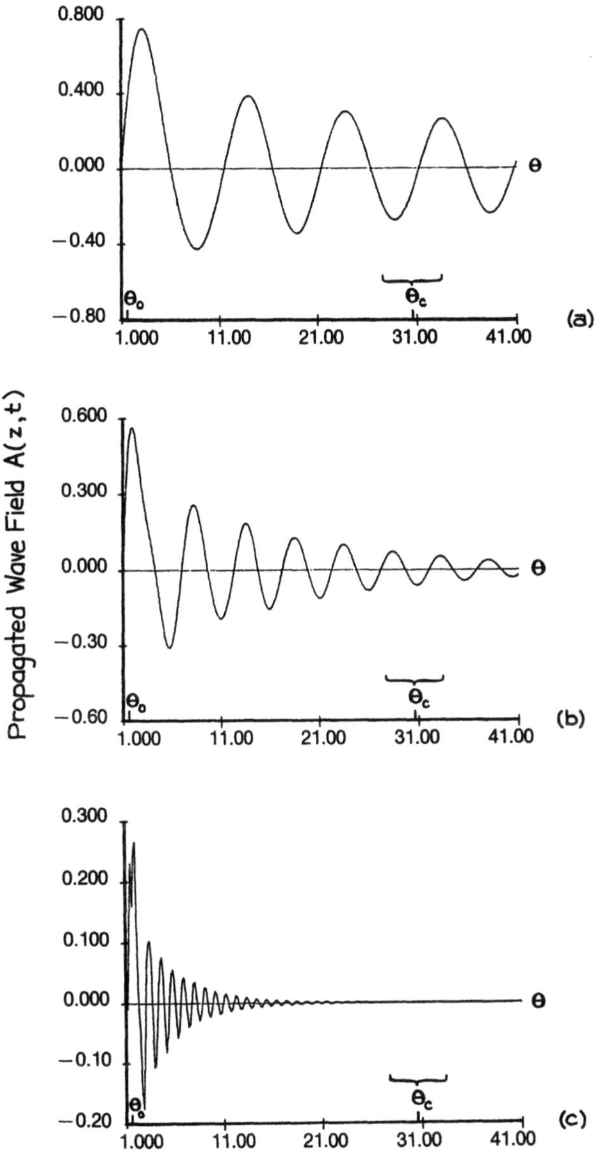

Fig. 9.18. Dynamical field evolution of an input unit step-function modulated signal with resonant carrier frequency $\omega_c = \omega_0 = 4.0 \times 10^{16}$/s at the propagation distances (a) $z = 5 \times 10^{-7}$ cm, (b) $z = 1 \times 10^{-6}$ cm, and (c) $z = 5 \times 10^{-6}$ cm.

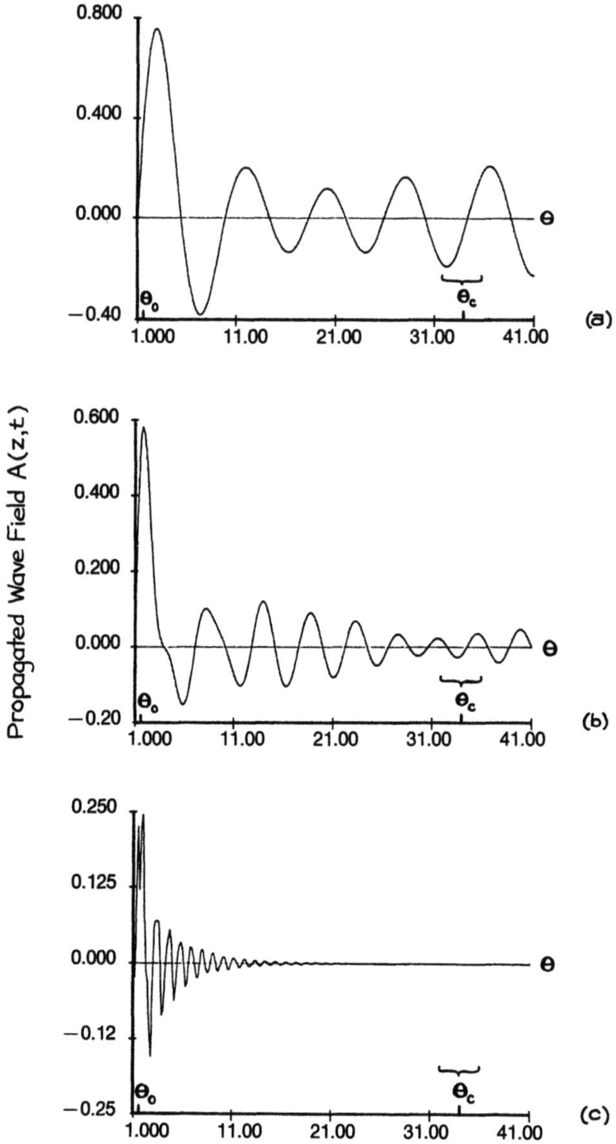

Fig. 9.19. Dynamical field evolution of an input unit step-function modulated signal with intra-absorption band carrier frequency $\omega_c = 4.4 \times 10^{16}$/s at the propagation distances (a) $z = 5 \times 10^{-7}$ cm, (b) $z = 1 \times 10^{-6}$ cm, and (c) $z = 5 \times 10^{-6}$ cm.

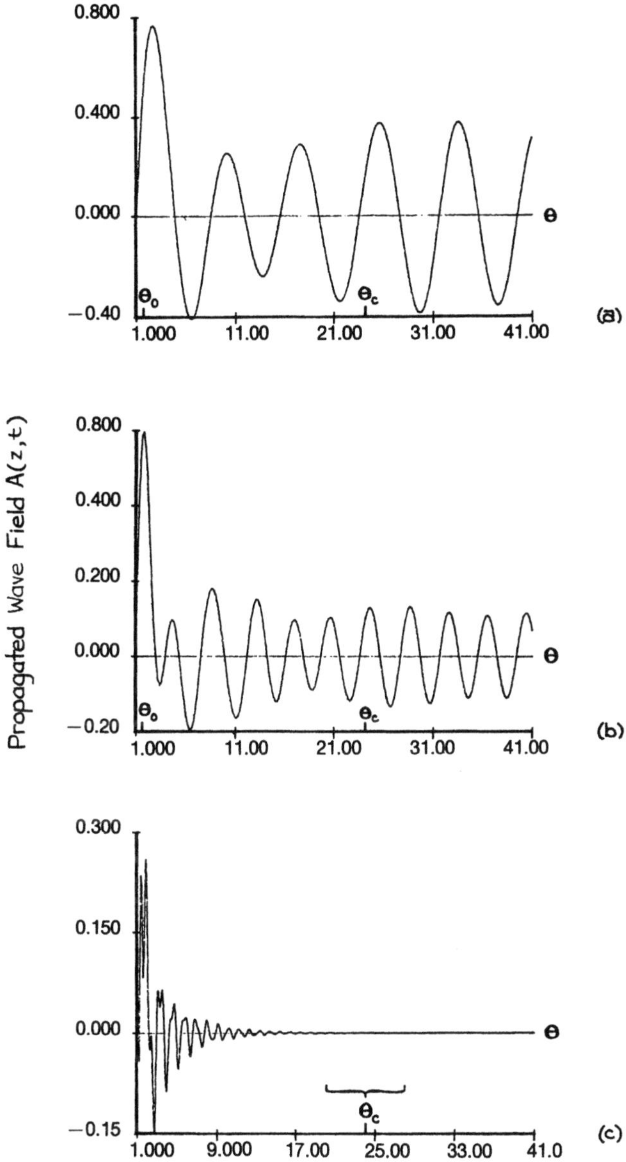

Fig. 9.20. Dynamical field evolution of an input unit step-function modulated signal with intra-absorption band carrier frequency $\omega_c = 4.8 \times 10^{16}$/s at the propagation distances (a) $z = 5 \times 10^{-7}$ cm, (b) $z = 1 \times 10^{-6}$ cm, and (c) $z = 5 \times 10^{-6}$ cm.

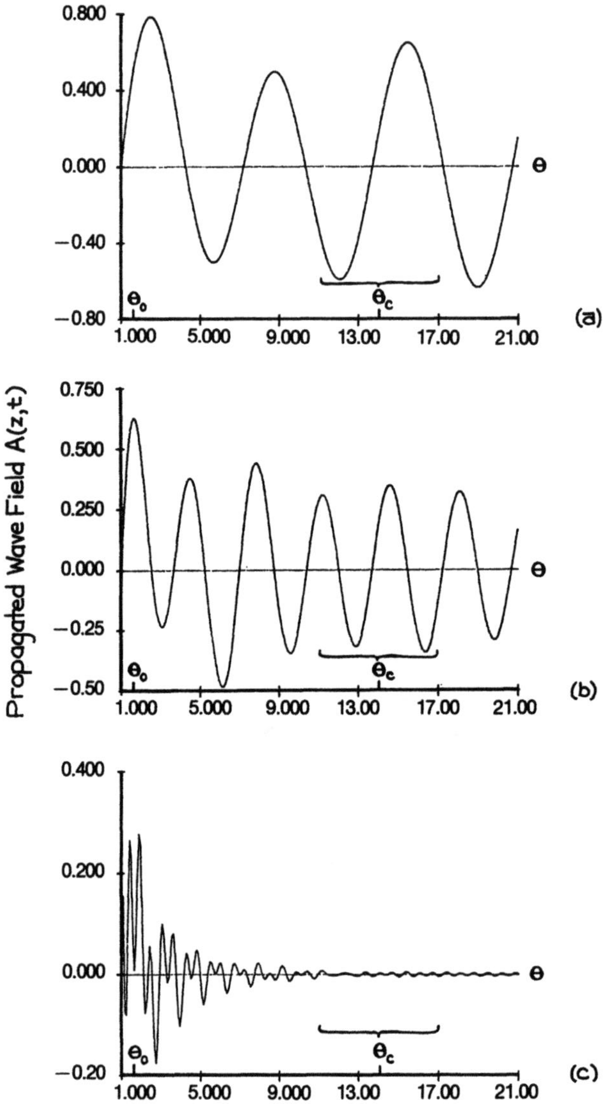

Fig. 9.21. Dynamical field evolution of an input unit step-function modulated signal with intra-absorption band carrier frequency $\omega_c = 5.5 \times 10^{16}$/s at the propagation distances (a) $z = 5 \times 10^{-7}$ cm, (b) $z = 1 \times 10^{-6}$ cm, and (c) $z = 5 \times 10^{-6}$ cm.

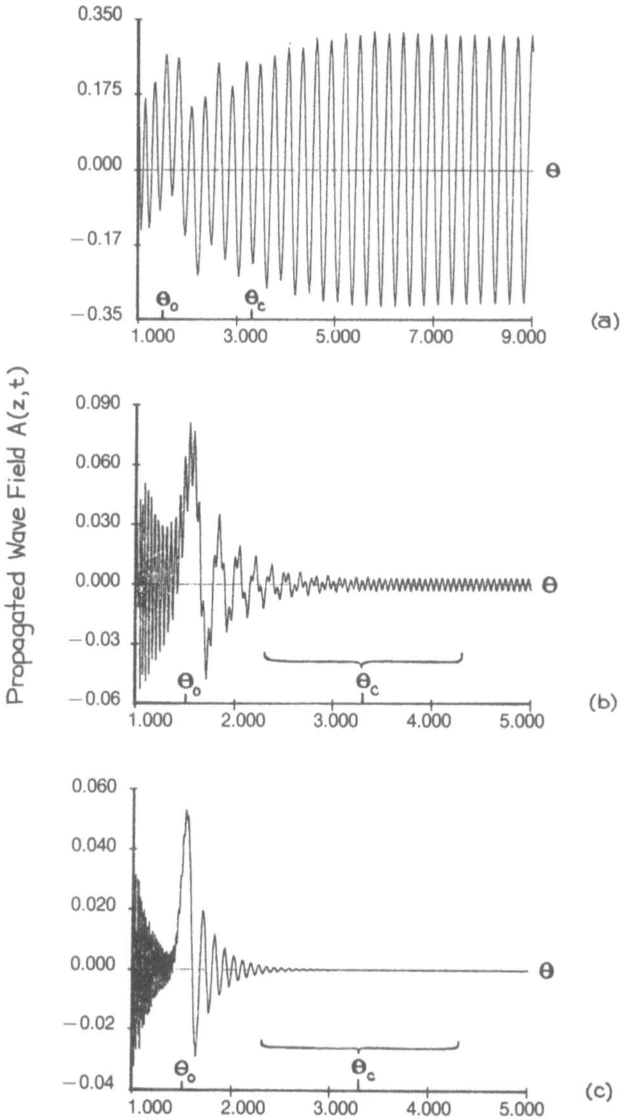

Fig. 9.22. Dynamical field evolution of an input unit step-function modulated signal with above absorption band carrier frequency $\omega_c = 6.5 \times 10^{16}$/s $< \omega_{SB}$ at the propagation distances (a) $z = 1 \times 10^{-5}$ cm, (b) $z = 5 \times 10^{-5}$ cm, and (c) $z = 1 \times 10^{-4}$ cm.

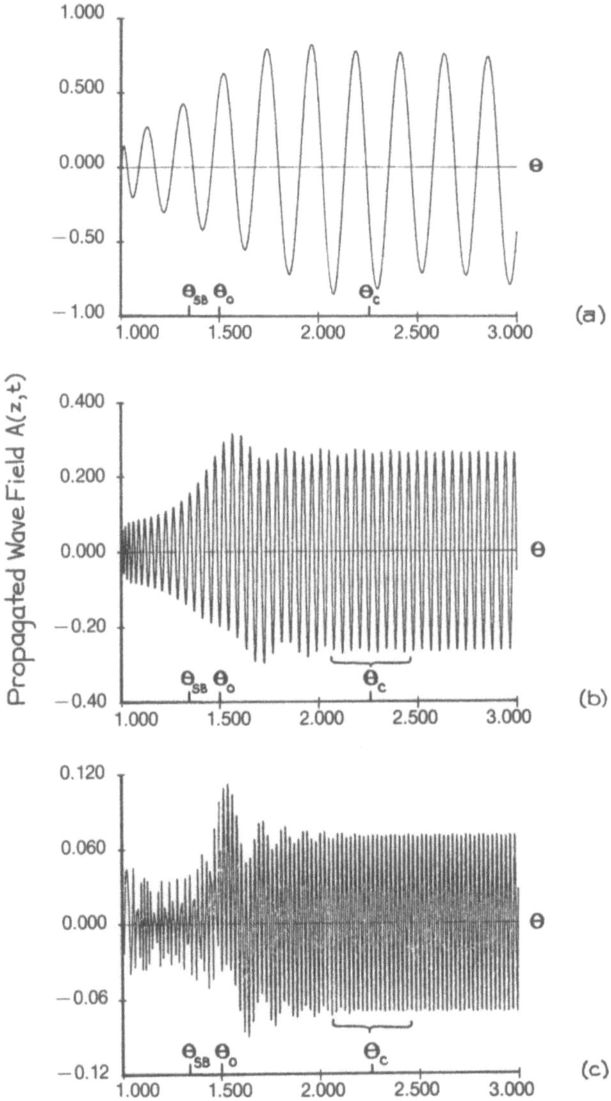

Fig. 9.23. Dynamical field evolution of an input unit step-function modulated signal with above absorption band carrier frequency $\omega_c = 8.5 \times 10^{16}$/s $< \omega_{SB}$ at the propagation distances (a) $z = 1 \times 10^{-5}$ cm, (b) $z = 5 \times 10^{-5}$ cm, and (c) $z = 1 \times 10^{-4}$ cm.

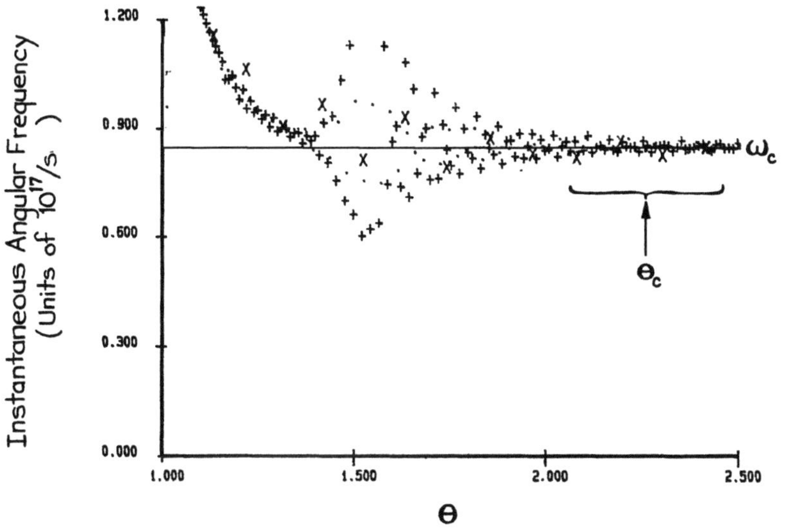

Fig. 9.24. Evolution of the instantaneous angular frequency of oscillation for the dynamical field evolution of Fig. 9.23.

increase monotonically from zero such that the oscillation frequency of the total field reaches ω_c from below and remains fixed at ω_c for all $\theta > \theta_c$. The numerically determined value of θ_c (obtained when the frequency first stabilizes at the input carrier frequency ω_c) is indicated in the figure along with its associated error. This value (as well as its error) is also indicated in Fig. 9.16.

The calculations illustrated in Figs. 9.18 through 9.21 are for input carrier frequencies that increase up through the absorption band of the medium. Again, the signal arrival is preceded by the *Brillouin* precursor, and the instantaneous angular frequency of oscillation of the field is found [9.16] to approach ω_c from below and remains fixed at ω_c for all $\theta > \theta_c$. The numerically determined values of θ_c as well as their associated uncertainties are indicated in each figure. Notice that the field evolution depicted in this set of figures is dominated by the precursor evolution, particularly at the larger propagation distances, since the associated attenuation is much less than that at the input carrier frequency. For applied signal frequencies at the lower end of the absorption band ($\omega_c \approx \omega_0$) the dominant field structure at large propagation distances is dominated by the *Brillouin* precursor, as seen in Fig. 9.18. As the input signal frequency increases up through the absorption band to ω_1 the trailing edge of the *Sommerfeld* precursor becomes more noticeable as it interferes increasingly with the *Brillouin* precursor evolution. Because the propagation distances are so small, the precursor evolution depicted here is primarily in the immature dispersion regime. It is clear that an experimental measurement of this propagated field structure would detect only this precursor field structure (unless the propagation distance was exceedingly small) and consequently would measure a field velocity that is asso-

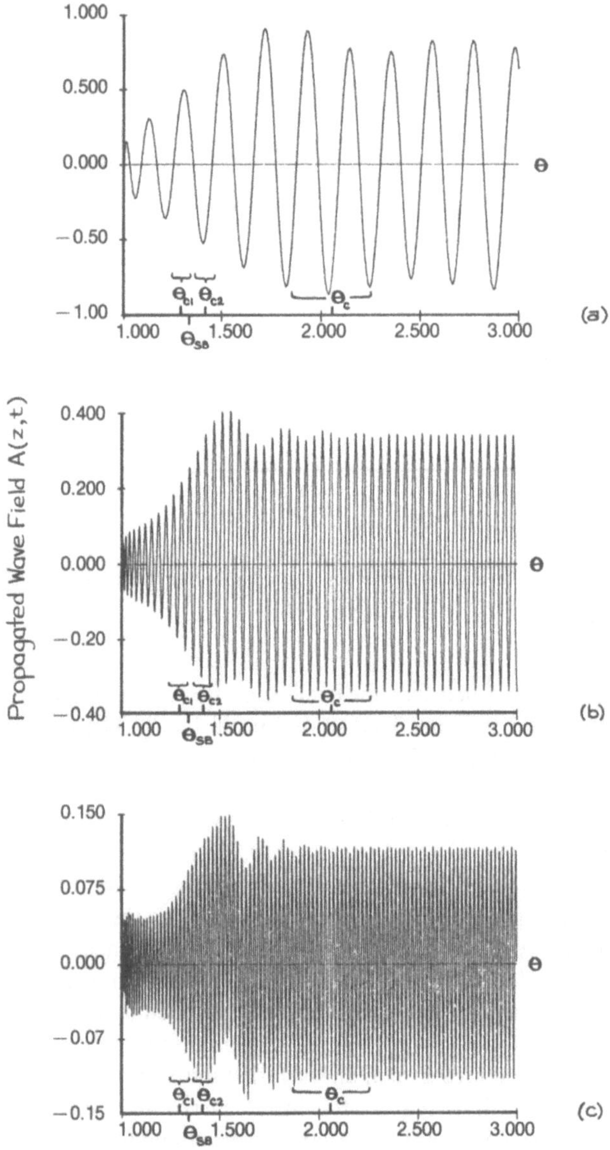

Fig. 9.25. Dynamical field evolution of an input unit step-function modulated signal with above absorption band carrier frequency $\omega_c = 9.0 \times 10^{16}/\text{s} > \omega_{SB}$ at the propagation distances (a) $z = 1 \times 10^{-5}$ cm, (b) $z = 5 \times 10^{-5}$ cm, and (c) $z = 1 \times 10^{-4}$ cm.

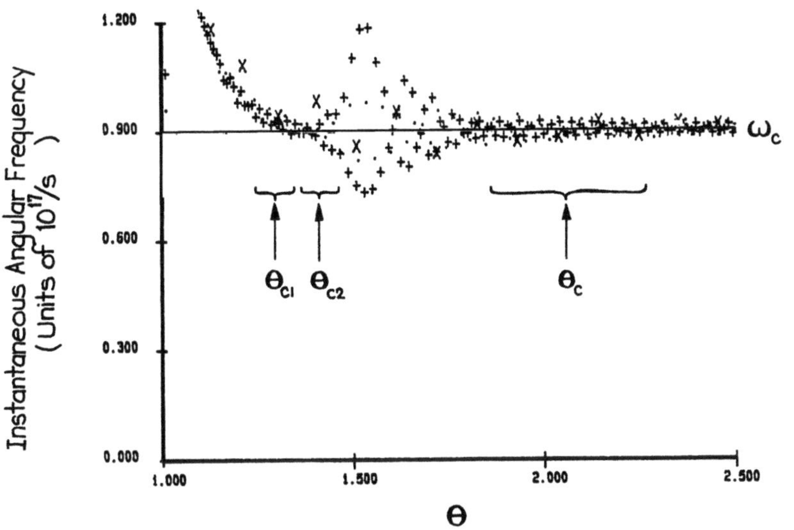

Fig. 9.26. Evolution of the instantaneous angular frequency of oscillation for the dynamical field evolution of Fig. 9.25.

ciated with the pronounced peak in the precursor field. This velocity would be close to the vacuum speed of light, which is much greater than the proper signal velocity.

As the applied signal frequency ω_c is increased to a value above the medium absorption band, the *Sommerfeld* precursor field becomes more pronounced in the total field evolution, as is readily evident in Fig. 9.22. In part (a) of the figure the propagation distance is sufficiently small to yield a propagated field structure that is in the immature dispersion regime wherein the precursor field structure is not yet well defined, whereas in part (c) it is large enough to be in the mature dispersion regime wherein the precursor field structure is clearly well defined. The propagated field structure is between these two regimes in part (b) of the figure, in which the interference between the trailing edge of the *Sommerfeld* precursor and entire *Brillouin* precursor evolution is clearly evident. Notice that the medium attenuation at ω_c is still much larger then that over much of the precursor field evolution, so that for the largest propagation distance considered, the amplitude of the main signal evolution is nearly negligibe in comparison to that of either precursor field.

For the dynamical field evolution depicted in Fig. 9.23 the input signal frequency $\omega_c = 8.5 \times 10^{16}$/s is slightly less than the value $\omega_{SB} = 8.7 \times 10^{16}$/s. Figure 9.24 shows that the instantaneous angular frequency of oscillation of the *Sommerfeld* precursor approaches ω_c from above but does not reach this value because of the subsequent evolution of the *Brillouin* precursor field (here the × symbols indicate the numerical values of the oscillation frequency for part (a) of Fig. 8.23, the · symbols indicate the numerical values for part (b) of that figure,

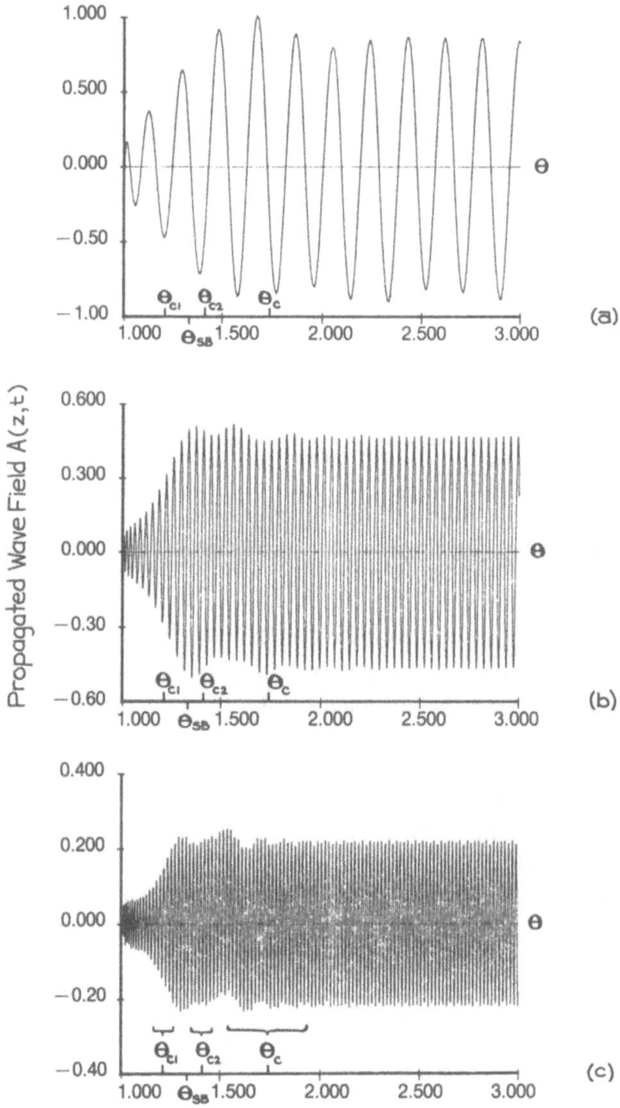

Fig. 9.27. Dynamical field evolution of an input unit step-function modulated signal with above absorption band carrier frequency $\omega_c = 10.0 \times 10^{16}/\mathrm{s} > \omega_{SB}$ at the propagation distances (a) $z = 1 \times 10^{-5}$ cm, (b) $z = 5 \times 10^{-5}$ cm, and (c) $z = 1 \times 10^{-4}$ cm.

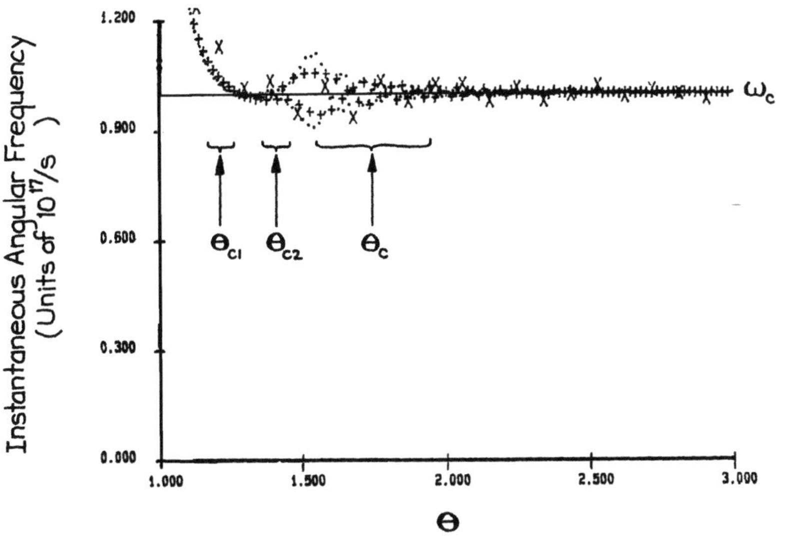

Fig. 9.28. Evolution of the instantaneous angular frequency of oscillation for the dynamical field evolution of Fig. 9.27.

and the + symbols for part (c) of that figure). Thus no prepulse evolution is observed when $\omega_c < \omega_{SB}$, in agreement with the modern asymptotic theory. Because of the interference between the trailing edge of the *Sommerfeld* precursor and the *Brillouin* precursor field, the instantaneous oscillation frequency takes values within a range above and below ω_c and then finally stabilizes at $\omega = \omega_c$ for $\theta = \theta_c$, as indicated in Fig. 9.24.

The dynamical field and associated instantaneous angular frequency evolution of the propagated field for several values of the input carrier frequency $\omega_c > \omega_{SB}$ are illustrated in Figs. 9.25–9.28. Figures 9.26 and 9.28 show that the instantaneous oscillation frequency first reaches ω_c from above at $\theta = \theta_{c1}$, is equal to ω_c for $\theta_{c1} < \theta < \theta_{c2}$, then takes values greater than or less than ω_c when θ increases above θ_{c2} (because of interference with the *Brillouin* precursor evolution), and finally stabilizes at ω_c at $\theta = \theta_c$ and remains at that value thereafter (the symbols here are the same as those in Fig. 9.24). The prepulse formation predicted by the asymptotic theory is therefore clearly obtained in these two cases.

The numerically determined values of θ_{c1}, θ_{c2}, and θ_c at each value of ω_c for both these and additional field calculations may then be used to calculate the relative signal velocity values $v_{c1}/c = 1/\theta_{c1}$, $v_{c2}/c = 1/\theta_{c2}$, and $v_c/c = 1/\theta_c$ and compared with that predicted by the asymptotic theory for this particular medium. The result is illustrated in Fig. 9.29. The solid curves in the figure depict the frequency dependence of the corresponding signal velocity branches as given by the asymptotic theory, and the dashed curve describes the frequency dependence of the energy transport velocity for a monochromatic plane wave field in the *Lorentz* medium. As is readily evident, excellent agreement between the

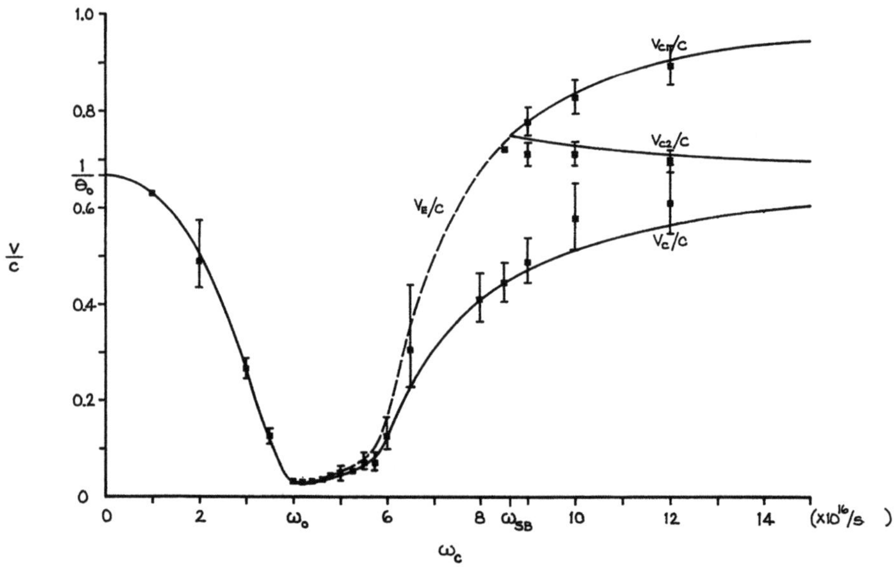

Fig. 9.29. Frequency dependence of the signal velocity for pulse propagation in a single resonance *Lorentz* medium as described by the asymptotic theory (solid curves) and the numerically obtained results of Ref. [9.15]. The dashed curve indicates the frequency dependence of the energy transport velocity for a monochromatic plane wave field of angular frequency ω_c in the *Lorentz* medium with parameters $\omega_0 = 4 \times 10^{16}/s$, $b^2 = 20 \times 10^{32}/s^2$, and $\delta = 0.14 \times 10^{16}/s$.

purely numerical results and the description afforded by the modern asymptotic theory is maintained over the entire frequency domain. In addition to demonstrating the accuracy of the modern asymptotic theory in describing the complete evolution of the unit step-function modulated signal in a single resonance *Lorentz* medium, these results also provide a physical measure of the signal velocity that is based solely on the instantaneous angular frequency of oscillation of the propagated field in the mature dispersion regime.

9.4.2 The Rectangular-Pulse Modulated Signal

According to the results of Sect. 4.3.2, the propagated field due to an initial rectangular-modulated pulse of initial time duration T may be represented as the difference between two unit step function modulated signals separated in time by the amount T (cf. (4.3.37)), viz

$$\begin{aligned} A(z,t) &= A_I(z,t) - A_{II}(z,t) = U(z,t,0) - U(z,t,T) \\ &= -\frac{1}{2\pi}\mathrm{Re}\left\{\int_{ia-\infty}^{ia+\infty} \frac{1}{\omega-\omega_c} e^{(z/c)\phi_0(\omega,\theta)}\, d\omega \right. \\ &\quad \left. - e^{-i\omega_c T}\int_{ia-\infty}^{ia+\infty} \frac{1}{\omega-\omega_c} e^{(z/c)\phi_T(\omega,\theta_T)}\, d\omega \right\}, \end{aligned} \tag{9.4.1}$$

where

$$\phi_0(\omega, \theta) = \phi(\omega, \theta) = i\omega[n(\omega) - \theta] \ , \tag{9.4.2a}$$

$$\theta = \frac{ct}{z} \ , \tag{9.4.2b}$$

and

$$\phi_T(\omega, \theta_T) = i\omega[n(\omega) - \theta_T] \ , \tag{9.4.3a}$$

$$\theta_T = \frac{c}{z}(t - T) \ . \tag{9.4.3b}$$

Consequently, the results of the previous subsection may be directly applied to obtain an understanding of the asymptotic behavior for large z of a given initial rectangular modulated pulse. The uniform asymptotic behavior of the front of the pulse, denoted by $A_I(z, t) = U(z, t, 0)$, is completely described by the results of Sect. 9.4.1. The uniform asymptotic behavior of the back of the pulse, denoted by $A_{II}(z, t) = U(z, t, T)$, is also described by the results of Sect. 9.4.1 provided that the retarded parameter θ_T is substituted for θ and the constant phase factor $e^{-i\omega_c T}$ is incorporated into those results. The total uniform asymptotic behavior of the propagated initial rectangular modulated pulse is then given by the difference between the front of the pulse $A_I(z, t)$ and the back of the pulse $A_{II}(z, t)$.

Consider first the asymptotic behavior of the field (9.4.1) for values of $\theta \gg \theta_{CT}$, where θ_{CT} is that value of θ_T at which the pole contribution to the back of the pulse $A_{II}(z, t)$ is the dominant contribution to that field component. Since the value of θ at which the second pole contribution occurs is necessarily greater than θ_c for all nonzero T and finite positive z, so that the first pole contribution is the dominant contribution to the front of the pulse $A_I(z, t)$, the asymptotic behavior of the total field behavior is then given by

$$\begin{aligned} A(z, t) \sim & -e^{-(z/c)n_i(\omega_c)\omega_c} \sin\left[\frac{z}{c}\omega_c(n_r(\omega_c) - \theta)\right] \\ & + e^{-(z/c)n_i(\omega_c)\omega_c} \sin\left[\frac{z}{c}\omega_c(n_r(\omega_c) - \theta_T) - \omega_c T\right] \\ = & \ 0 \ ; \qquad \theta \gg \theta_{CT} \ , \end{aligned} \tag{9.4.4}$$

as $z \to \infty$. That is, the asymptotic contribution of the back of the pulse $A_{II}(z, t)$ identically cancels the asymptotic contribution of the front of the pulse $A_I(z, t)$ for all $\theta \gg \theta_{CT}$. Since

$$\theta_T = \frac{c}{z}(t - T) = \theta - \frac{c}{z}T \ , \tag{9.4.5}$$

it is then seen that the evolution in θ of the back of the pulse is retarded in θ by the amount $(c/z)T$ from the corresponding θ evolution of the front of the pulse. The value of θ at which the pole contribution to the back of the pulse arrives is given by $\theta_{CT} = \theta_c + (c/z)T$. Hence, the θ width of the propagated pulse between the front and back pole contributions is

$$\Delta\theta_c = \frac{c}{z}T , \tag{9.4.6}$$

and the corresponding time width of the pulse between these two points is

$$\Delta t_c = \frac{z}{c}\Delta\theta_c = T . \tag{9.4.7}$$

Consequently, any pulse broadening and envelope degradation of the initial pulse envelope in propagating through the dispersive medium is due solely to the precursor fields associated with the front and back of the pulse. If the precursor fields are neglected and only the pole contributions are considered, then the propagated field may be written (cf. (8.3.4 and 8))

$$A(z,t) \sim -e^{-z\alpha(\omega_c)}\sin[\beta(\omega_c)z - \omega_c t] \tag{9.4.8}$$

for all values of θ in the range $\theta_c \leqslant \theta \leqslant \theta_{CT}$, and is zero for all other values of θ. Consequently, by neglecting the precursor phenomena associated with the front and back of the pulse, the propagated pulse retains its initial envelope function and temporal width, and is simply attenuated with the propagation distance z by the amplitude factor $e^{-z\alpha(\omega_c)}$.

In order to obtain a complete understanding of the propagation of an initial rectangular modulated pulse through a temporally dispersive *Lorentz* medium, the interaction between the precursor phenomena associated with the front and back edges of the pulse with both the pole contributions and each other must be fully taken into account. That understanding is afforded by the uniform asymptotic description given in this monograph. As is described in detail in Sect. 9.4.1, the propagation characteristics of the front and back edges of the pulse depend upon whether the applied signal frequency ω_c lies in the spectral region below the absorption band of the medium $(0 \leqslant \omega_c \leqslant \sqrt{\omega_0^2 - \delta^2})$, above the absorption band of the medium $(\omega_c \geqslant \sqrt{\omega_1^2 - \delta^2})$, or within the absorption band of the medium $(\sqrt{\omega_0^2 - \delta^2} < \omega_c < \sqrt{\omega_1^2 - \delta^2})$. To complicate matters further in the present case, the total field behavior depends strongly upon both the initial time width T of the input pulse and the propagation distance z at which the dynamic behavior of the field is observed. For a sufficiently short propagation distance z and/or a sufficiently long initial pulse width T such that $\theta_c < (c/z)T$, the first and second precursor fields associated with the leading edge of the pulse will evolve undisturbed by the precursor field contributions associated with the trailing edge of the pulse, and the first and second precursor fields associated with the back of the pulse will interfere with the pole contribution associated with the front of the pulse. For larger propagation distances z or shorter initial pulse widths T such that $\theta_{SB} < (c/z)T < \theta_c$, the first precursor field associated with the leading edge of the pulse will still evolve undisturbed, but the second precursor field associated with the leading edge of the pulse will overlap and interfere with the first precursor field associated with the trailing edge of the pulse, and the second precursor field associated with the trailing edge of the pulse will interfere with the pole contribution associated with the leading edge of the pulse. Finally, for larger propagation distances z or even shorter initial pulse widths T such that $1 < (c/z)T < \theta_{SB}$, the first and second precursor fields associated with the leading edge of the pulse and the trailing edge of the pulse will overlap and interfere with each other. Consequently, for any given nonzero finite initial pulse width T, its θ-value $(c/z)T$ progressively falls into each one of these θ domains as the propagation distance z increases, so that for sufficiently large z the inequality $1 < (c/z)T < \theta_{SB}$ is satisfied for a given finite value of T.

For values of z and T in the first range $(c/z)T \geqslant \theta_c$, the degradation and temporal spread of the pulse is minimal. As z increases, the degradation and width of the pulse increase as the precursor fields associated with the leading and trailing edges of the pulse interfere more with each other and the two pole contributions come closer together in θ-space. For values of z and T in the final range $1 < (c/z)T < \theta_{SB}$, the degradation and temporal spread of the pulse is a maximum. For short initial pulse widths T, this final range is obtained for small values of z, whereas for long initial pulse widths T, the pulse must propagate farther to obtain this final range. Consequently, pulses with short initial pulse widths T degrade much faster than those with long initial pulse widths T. That is, the shorter the initial pulse width, the shorter is the propagation distance over which a given amount of pulse degradation occurs. For two initial rectangular-modulated pulses with widths related by $T_1 = mT_2, m > 0$, their propagation distances at which their field behavior in the θ domain is identical are related by $z_1 = mz_2$. Consequently, increasing (decreasing) the initial width of the pulse by the positive factor m increases (decreases) the propagation distance through the dispersive medium by the same factor m for an equal level of resulting pulse degradation in the two pulses.

Before any full consideration is given to a detailed description [9.18] of the dynamical evolution of an input rectangular modulated pulse with carrier frequency ω_c, the signal arrival and subsequent signal velocity must first be carefully described. For an input rectangular modulated signal $A(z,t) = U(z,t,0) - U(z,t,T)$ of initial time duration T and carrier frequency $\omega_c \in [0, \omega_{SB})$, the signal arrival occurs at $\theta = \theta_c$ and the propagated field ceases to oscillate at ω_c when $\theta_T = \theta - cT/z = \theta_c$. Both of these transition points propagate with the signal velocity $v_c = c/\theta_c$. The main body of the propagated pulse that is oscillating at $\omega = \omega_c$ then evolves over the space-time interval from $\theta = \theta_c$ to $\theta = \theta_c + cT/z$. The θ-width of the propagated pulse between the front and back pole contributions is then $\Delta\theta_c = cT/z$ so that the corresponding temporal width of the pulse between these two points at any fixed propagation distance z is

$$\Delta t_c = \frac{z}{c}\Delta\theta_c = T \tag{9.4.9}$$

Consequently, any pulse broadening and envelope degradation of the input rectangular modulated pulse with $\omega_c \in [0, \omega_{SB}]$ is due primarily to the precursor field structure of the propagated field that arises from the leading and trailing edges of the input field. This interference of the pole contribution with the precursor field structure will shorten the θ-domain over which the propagated field oscillates predominantly at the input carrier frequency ω_c. *Strictly speaking, the temporal width of the signal is then found to decrease with increasing propagation distance.* The commonly observed phenomenon [9.9, 12, 19–22] of pulse spreading is obtained only when the propagated signal is redefined so as to include a range of frequencies about ω_c, and this in turn implies the incorporation of some portion of the precursor fields in the definition of the main body of the pulse.

For $\omega_c > \omega_{SB}$ the signal arrival first occurs at $\theta = \theta_{c1}$ when the simple pole at $\omega = \omega_c$ is crossed, and the propagated field finally ceases to oscillate predominantly at ω_c when $\theta_T = \theta - cT/z = \theta_{c1}$ when the pole contribution is subtracted out. The overall θ-width of the propagated pulse between the front and back contributions is then $\Delta\theta_c = cT/z$ so that the corresponding temporal width of the pulse between these two points at any fixed propagation distance z is

$$\Delta t_c = \frac{z}{c}\Delta\theta_c = T\ . \tag{9.4.10}$$

Between these two space-time points there are, at most, two other distinct transition points at $\theta = \theta_{c2}$ and $\theta = \theta_c$ at which the propagated field either ceases to oscillate predominantly at ω_c or begins again to oscillate predominantly at $\omega = \omega_c$ due to the asymptotic dominance of the leading edge *Brillouin* precursor between these two points. Because of this, the propagated field due to the input rectangular pulse separates into, at most, two subpulses (provided that $cT/z > \theta_c - \theta_{c1}$) which reduce to a single pulse at sufficiently large propagation distances. Apart from this pulse breakup, the only other envelope degradation and any pulse broadening of the input rectangular modulated pulse with $\omega_c > \omega_{SB}$ is due to the precursor field structure of the propagated field that arises from the front and back of the input field.

The resultant frequency dependence of the signal velocity for a finite duration rectangular pulse with $cT/z < \theta_c - \theta_{c1}$ is illustrated in Fig. 9.30. As in Fig. 9.7, the dashed curve represents the energy transport velocity of a strictly monochromatic field in the medium. The discontinuous jump in the signal velocity at $\omega_c = \omega_{SB}$ is fundamentally due to the change in dominance of the precursor fields at $\theta = \theta_{SB}$. Below ω_{SB} the signal arrival occurs following the evolution of the *Sommerfeld* and *Brillouin* precursors, while above ω_{SB} the signal arrival occurs during the evolution of the *Sommerfeld* precursor field. When the opposite inequality is satisfied (i.e. when $cT/z > \theta_c - \theta_{c1}$) by choosing a sufficiently

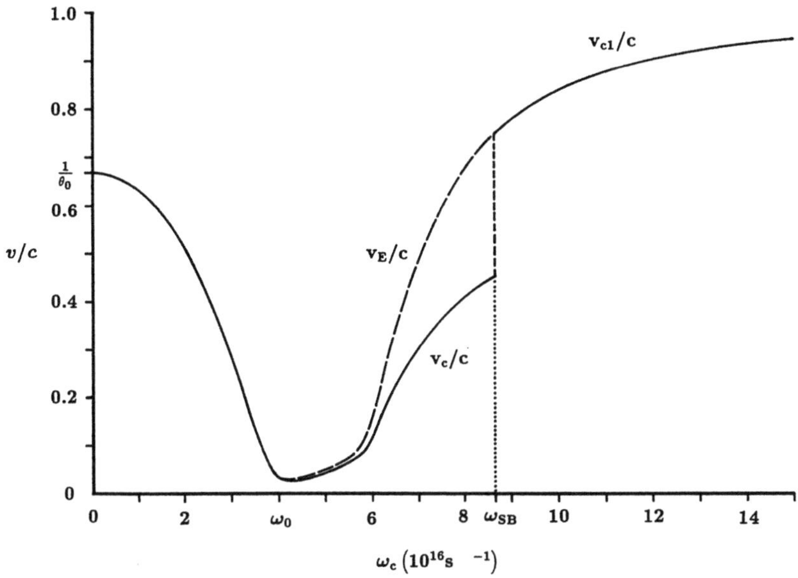

Fig. 9.30. Frequency dependence of the signal velocity for a rectangular pulse of finite duration with $cT/z < \theta_c - \theta_{c1}$ in a *Lorentz* medium with $\omega_0 = 4 \times 10^{16}/\text{s}$, $b^2 = 20 \times 10^{32}/\text{s}^2$, and $\delta = 0.28 \times 10^{16}/\text{s}$. The behavior of the energy transport velocity for a strictly monochromatic field in the medium is indicated by the long dashed curve.

long initial pulse width T or a sufficiently short propagation distance z (but which is still large enough to be in the mature dispersion regime), then the propagated field is found to be separated into two pulses whose velocities are described in Fig. 9.7. The first is a prepulse with front velocity v_{c1} and back velocity v_{c2}, and the second subpulse has front velocity v_c and back velocity v_{c1}.

From (9.4.1) and the uniform asymptotic representation of the unit step-function modulated signal given in (6.5.8), the uniform asymptotic description of the propagated field due to an input rectangular-modulated signal of initial time duration T is given by the representation

$$A(z,t) \sim U_S(z,t,0) + U_B(z,t,0) + U_c(z,t,0) - U_S(z,t,T) - U_B(z,t,T) - U_c(z,t,T) \,, \tag{9.4.11}$$

which is simply the difference between the propagated fields due to an input unit step-function-modulated signal $U(z,t,0) \sim U_S(z,t,0) + U_B(z,t,0) + U_c(z,t,0)$ that begins to oscillate at time $t = 0$ in the $z = 0$ plane and the unit step-function-modulated signal $U(z,t,T) \sim U_S(z,t,T) + U_B(z,t,T) + U_c(z,t,T)$ that begins to oscillate at time $t = T$ in the $z = 0$ plane. It is now shown that this representation of the propagated field in the matrue dispersion limit provides a complete, accurate representation of the dynamical field evolution over the entire range of signal frequencies $\omega_c \in [0, \infty)$.

Consider first the below resonance range $\omega_c \in [0, \omega_c)$. In this case the peak *Sommerfeld* precursor field amplitude is typically several orders of magnitude less than the peak amplitude of the *Brillouin* precursor so that the entire propagated field structure in the mature dispersion limit is dominated by the *Brillouin* precursor and the pole contribution at $\omega = \omega_c$. For a sufficiently long initial pulse width T and/or a sufficiently short propagation distance z such that $(c/z)T > \theta_c - 1$ the precursor fields associated with the leading edge $U(z, t, 0)$ of the pulse will completely evolve prior to the arrival of the precursor fields associated with the trailing edge $U(z, t, T)$ of the pulse. Indeed, the trailing edge precursors will arrive only after the main signal of the leading edge has arrived (at $\theta = \theta_c$) and is evolving. Hence, when this condition prevails the interference between the precursor fields of the leading and trailing edges of the pulse is minimal and the pulse distortion is also minimal, as is evident in Fig. 9.31. For shorter initial pulse widths T or large propagation distances z such that $\theta_c - 1 > (c/z)T > \theta_{SB} - 1$ the first precursor field associated with the leading edge of the pulse will still evolve undisturbed, but during the evolution of the second precursor field $U_B(z, t, 0)$ the arrival and evolution of the precursor fields associated with the trailing edge of the pulse will occur. Hence, when this condition prevails there will be interference between the *Brillouin* precursor of the leading edge and the *Sommerfeld* precursor of the trailing edge of the pulse, as illustrated in Fig. 9.32, and the *Brillouin* precursor of the trailing edge of the pulse will occur soon after the signal arrival at $\theta = \theta_c$ so that the resultant pulse distortion is found to be moderate. Finally, for even shorter initial pulse widths T or greater propagation distances z such that $0 < (c/z)T \leqslant \theta_{SB} - 1$ there will be a nearly complete overlap of these two sets of precursor fields and the resultant pulse distortion is severe, as illustrated in Fig. 9.33.

In each case the pole contribution to the total field evolution occurs at $\theta = \theta_c$ and is subtracted out at $\theta = \theta_c + cT/z$ so that the overall temporal width of the propagated signal is T, as given by (9.4.9). However, because of the asymptotic dominance of the *Brillouin* precursor $U_B(z, t, T)$ from the back edge of the input pulse, the propagated signal is the dominant contribution to the total field evolution only over the θ-domain from $\theta = \theta_c$ to $\theta \cong \theta_0 + cT/z$. The corresponding temporal width of the propagated signal is then

$$\Delta t_s \cong T - \frac{z}{c}(\theta_c - \theta_0) \tag{9.4.12}$$

provided that $\theta_0 + cT/z > \theta_c$ which is satisfied up through most of the moderate distortion range. When the opposite inequality (i.e. $\theta_0 + ct/z \leqslant \theta_c$) is satisfied the pulse distortion is severe and the total propagated field is dominated by the precursor fields over the entire space-time domain.

Similar results hold in the signal frequency domain $\omega_c \in [\omega_0, \omega_{SB}]$ which contains the medium absorption band $[\omega_0, \omega_1]$ and proceeds up to the signal frequency ω_{SB} at which the signal velocity of the unit-step-function-modulated signal bifurcates. The only difference from the preceeding case is that the *Som-*

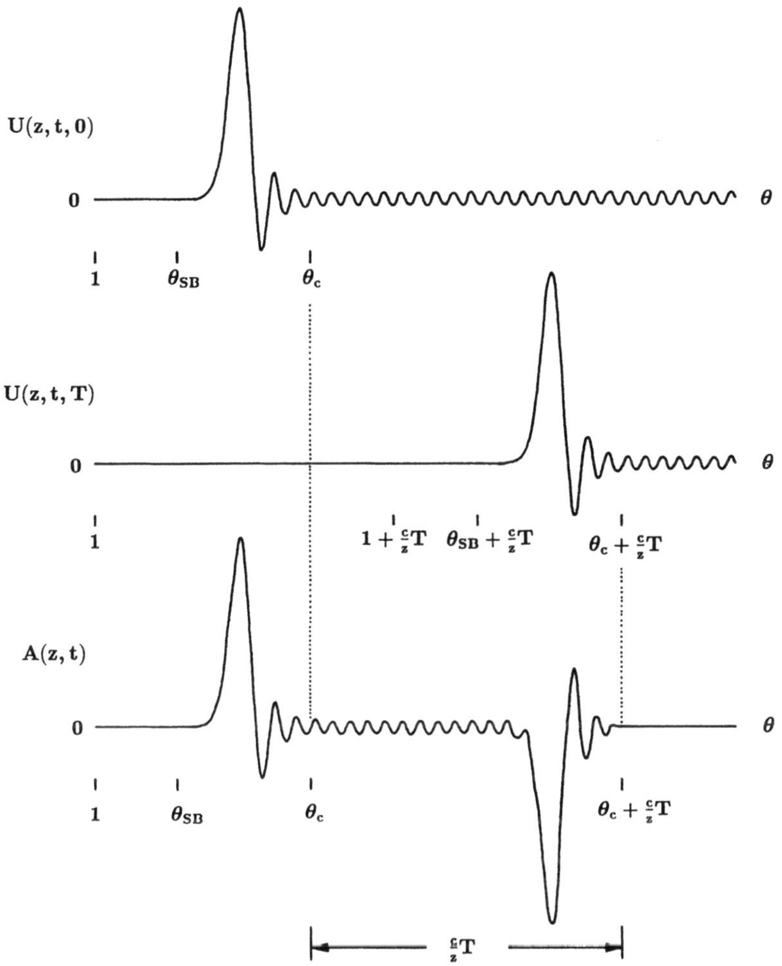

Fig. 9.31. Construction of the dynamical structure of the propagated field $A(z,t) = U(z,t,0) - U(z,t,T)$ in the below resonance signal frequency domain $0 < \omega_c < \omega_0$ when $(c/z)T > \theta_c - 1$. When this situation prevails the interference between the precursor fields of the leading and trailing edges of the pulse is minimal and the resultant pulse distortion is also minimal.

merfeld precursor field structure now becomes more pronounced in the total field evolution as the spectral amplitude of each unit-step-function-modulated signal increases in the above resonance frequency domain. The construction of the propagated field structure in the severe distortion limit $0 < (c/z)T \leqslant \theta_{SB} - 1$ is illustrated in Fig. 9.34 for a signal frequency ω_c that is near resonance. For initial pulse widths T and propagation distances z satisfying this inequality there is a nearly complete overlap of the two sets of precursor fields so that the propagated field structure $A(z,t)$ is dominanted by a pair of interfering *Sommerfeld* precursors followed by a pair of interfering *Brillouin* precursors which

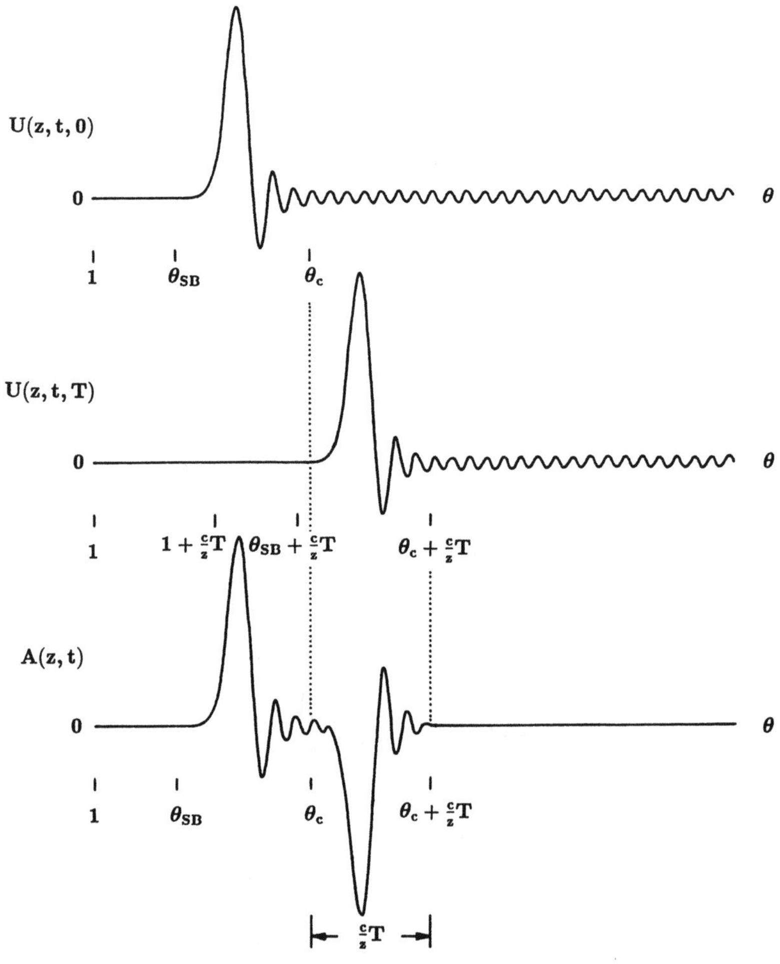

Fig. 9.32. Construction of the dynamical structure of the propagated field $A(z,t) = U(z,t,0) - U(z,t,T)$ in the below resonance signal frequency domain $0 < \omega_c < \omega_0$ when $\theta_c - 1 > (c/z)T > \theta_{SB} - 1$. When this situation prevails the interference between the precursor fields of the leading and trailing edges of the pulse is moderate and the resultant pulse distortion is also moderate.

is then followed by the signal oscillating at $\omega = \omega_c$ that evolves over the space-time interval from $\theta = \theta_c$ to $\theta = \theta_c + cT/z$, as illustrated. Just prior to the signal arrival at $\theta = \theta_c$ the field is dominated by the interfering pair of *Brillouin* precursor fields whose instantaneous oscillation frequency is less than ω_c.

In the low frequency region of the signal frequency domain $[\omega_0, \omega_{SB}]$ the *Sommerfeld* precursor field is relatively insignificant in comparison to both the *Brillouin* precursor field and the pole contribution so that (9.4.12) applies for the temporal width of the propagated signal. On the other hand, in the high frequency region of this signal frequency domain the *Sommerfeld* precursor is a dominant feature in the total propagated field over the space-time domains

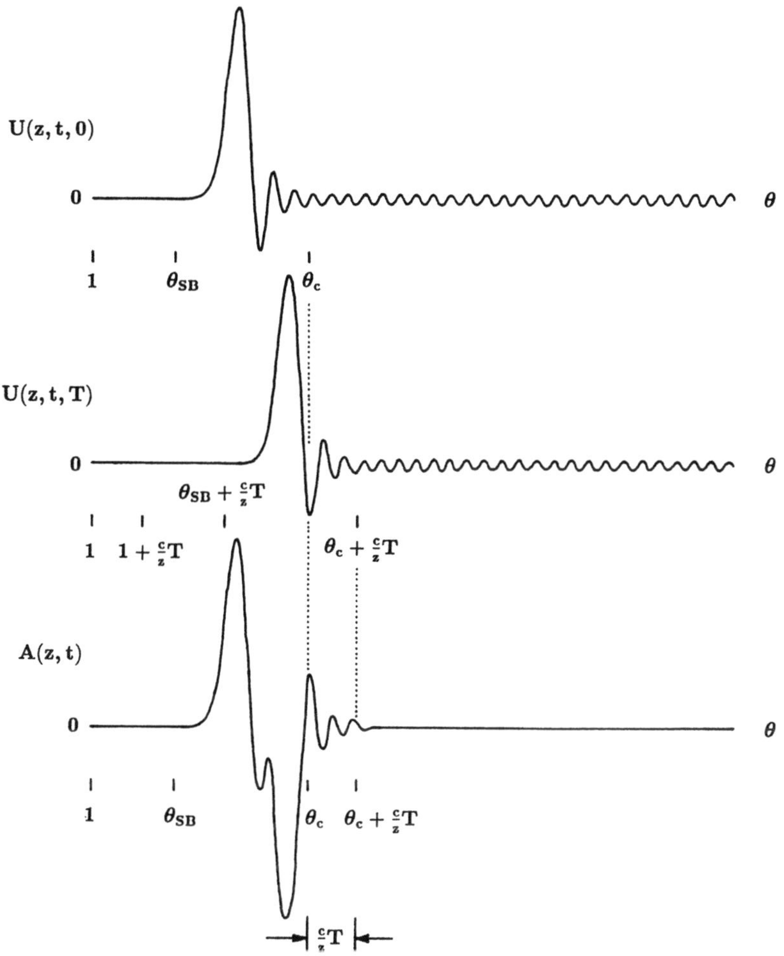

Fig. 9.33. Construction of the dynamical structure of the propagated field $A(z,t) = U(z,t,0) - U(z,t,T)$ in the below resonance signal frequency domain $0 < \omega_c < \omega_0$ when $(c/z)T \leqslant \theta_{SB} - 1$. When this situation prevails the interference between the precursor fields of the leading and trailing edges of the pulse is nearly complete and the resultant pulse distortion is severe.

$\theta \in [1, \theta_{SB}]$ and $\theta_T \in [1, \theta_{SB})$. The propagated signal, which arises from the pole contribution that evolves over the space-time domain from $\theta = \theta_c$ to $\theta = \theta_c + cT/z$, is then seen to be the dominant contribution to the total field over the θ-domain from $\theta = \theta_c$ to $\theta = 1 + cT/z$ when $cT/z > \theta_c - 1$, with temporal width

$$\Delta t_s = T - \frac{z}{c}(\theta_c - 1) , \qquad (9.4.13)$$

and is then again the dominant contribution over a small θ-interval about the

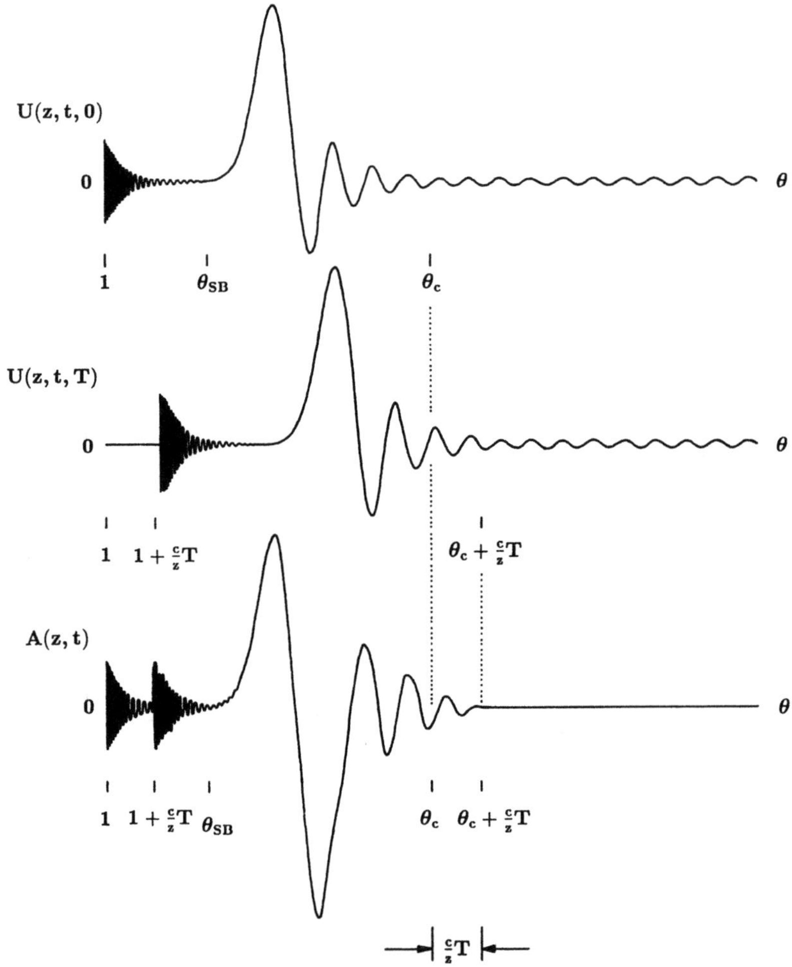

Fig. 9.34. Construction of the dynamical structure of the propagated field $A(z,t) = U(z,t,0) - U(z,t,T)$ for a near resonance signal frequency $\omega_c \in [\omega_0, \omega_1]$ when $(c/z)T \leqslant \theta_{SB} - 1$. When this situation prevails the interference between the precursor fields of the leading and trailing edges of the pulse is nearly complete and the resultant pulse distortion is severe.

space-time point $\theta = \theta_{SB} + cT/z$ provided that $cT/z \gtrsim \theta_c - \theta_{SB}$. When the first inequality (i.e. $cT/z > \theta_c - 1$) is satisfied the propagated signal is seen to be separated into two pulses oscillating at $\omega = \omega_c$. These reduce to a single pulse oscillating at $\omega = \omega_c$ when $\theta_c - 1 \geqslant cT/z \gtrsim \theta_c - \theta_{SB}$. Finally, when the inequality $cT/z \lesssim \theta_c - \theta_{SB}$ is satisfied the pulse distortion is severe and the total propagated field is dominated by the precursor fields over the entire space-time domain.

Consider finally the high frequency domain $\omega_c > \omega_{SB}$ in which the propagated unit step-function-modulated signal separates into a prepulse that evolves over the space-time domain $\theta \in [\theta_{c1}, \theta_{c2}]$ and a main signal that evolves over the space-time domain $\theta > \theta_c$, these two signal components being separated by the *Brillouin* precursor field which is the asymptotically dominant field contribution over the domain $\theta \in (\theta_{c2}, \theta_c)$, as described in Sect. 9.4.1. In this high frequency domain the *Sommerfeld* precursor field is a dominant feature in the total field evolution of the propagated unit step-function-modulated signal and is the asymptotically dominant field contribution over the space-time domain $\theta \in [1, \theta_{c1})$. For a sufficiently long initial pulse width T and/or a sufficiently small propagation distance z such that $(c/z)T > \theta_c - 1$ the precursor fields and prepulse associated with the leading edge $U(z,t,0)$ of the pulse will completely evolve prior to the arrival and evolution of the precursors and prepulse associated with the trailing edge $U(z,t,T)$ of the pulse so that their interference is minimal. When this condition prevails the total propagated field evolves in the following sequential manner [the leading term in each asymptotic expression given here and in the following expressions of this section indicates that it is asymptotically dominant over the remaining terms (if any)]:

$1 \leqslant \theta < \theta_{c1}$	$: A(z,t) \sim U_S(z,t,0)$	: *Sommerfeld* precursor
$\theta_{c1} \leqslant \theta \leqslant \theta_{c2}$	$: A(z,t) \sim U_c(z,t,0)$	: Signal
$\theta_{c2} < \theta < \theta_c$	$: A(z,t) \sim U_B(z,t,0) + U_c(z,t,0)$	: *Brillouin* precursor plus signal
$\theta_c \leqslant \theta < 1 + \frac{c}{z}T$	$: A(z,t) \sim U_c(z,t,0)$	: Signal
$1 + \frac{c}{z}T \leqslant \theta < \frac{c}{z}T + \theta_{c1}$	$: A(z,t) \sim -U_S(z,t,T) + U_c(z,t,0)$	: *Sommerfeld* precursor plus signal
$\theta_{c1} + \frac{c}{z}T \leqslant \theta < \theta_{SB} + \frac{c}{z}T$	$: A(z,t) \sim -U_S(z,t,T)$	: *Sommerfeld* precursor
$\theta_{SB} + \frac{c}{z}T < \theta$	$: A(z,t) \sim -U_B(z,t,T)$	: *Brillouin* precursor.

The propagated signal is thus separated into a pre-pulse that evolves over the space-time domain $\theta \in [\theta_{c1}, \theta_{c2}]$ with temporal width

$$\Delta t_p = \frac{z}{c}(\theta_{c2} - \theta_{c1}) \tag{9.4.14}$$

that increases linearly with the propagation distance, and a main pulse that evolves over the space-time domain $\theta \in [\theta_c, 1 + (c/z)T]$ with temporal width

$$\Delta t_s = T - \frac{z}{c}(\theta_c - 1) \tag{9.4.15}$$

that decreases to zero linearly with the propagation distance z over the propagation domain $(c/z)T > \theta_c - 1$. The front and back of the prepulse then propagate with the pre-signal velocities $v_{c1} = c/\theta_{c1}$ and $v_{c2} = c/\theta_{c2}$, respectively, while the front of the main pulse propagates with the signal velocity $v_c = c/\theta_c$, just as for the unit step-function-modulated signal. A moments reflection on the limiting behavior of this pulse structure as the propagation z becomes small (ignoring momentarily that the above results are derived from asymptotic theory) shows that the prepulse width (9.4.14) approaches zero while the main pulse width (9.4.15) approaches the initial pulse width T. The main pulse is then clearly associated with the initial rectangular pulse. As the propagation distance z increases so that the inequality $(c/z)T < \theta_c - 1$ is satisfied, the main pulse vanishes from the propagated field structure and all that remains is the prepulse and the precursor fields. The prepulse remains intact, evolving undisturbed over the space-time domain $[\theta_{c1}, \theta_{c2}]$ until the inequality $(c/z)T < \theta_{c2} - 1$ is satisfied. When this latter condition prevails the prepulse becomes distorted as the *Sommerfeld* precursor associated with the back edge of the pulse evolves over this space-time domain. This situation is illustrated in Fig. 9.35 when $\theta_{c1} - 1 < (c/z)T < \theta_{c2} - 1$. When this condition prevails the total propagated field evolves in the following sequential manner:

$1 \leqslant \theta < \theta_{c1}$	$: A(z,t) \sim U_S(z,t,0)$	: *Sommerfeld* precursor
$\theta_{c1} \leqslant \theta < 1 + \frac{c}{z}T$	$: A(z,t) \sim U_c(z,t,0)$	: Signal
$1 + \frac{c}{z}T \leqslant \theta \leqslant \theta_{c2}$	$: A(z,t) \sim -U_S(z,t,T) + U_c(z,t,0)$	: *Sommerfeld* precursor plus signal
$\theta_{c2} < \theta < \theta_{c1} + \frac{c}{z}T$	$: A(z,t) \sim A_B(z,t,0) - U_S(z,t,T) + U_c(z,t,0)$	: *Brillouin* and *Sommerfeld* precursors plus signal
$\theta_{c1} + \frac{c}{z}T < \theta < \theta_{c2} + \frac{c}{z}T$	$: A(z,t) \sim A_B(z,t,0)$	: *Brillouin* precursor
$\theta_{c2} + \frac{c}{z}T < \theta$	$: A(z,t) \sim -A_B(z,t,T) + A_B(z,t,0)$	: *Brillouin* precursors.

The propagated field structure is then seen to be dominated by the precursor fields associated with the front and back of the pulse over all but the domain $\theta \in [\theta_{c1}, 1 + (c/z)T]$. The temproal width of the prepulse is now given by

$$\Delta t_p = T - \frac{z}{c}(\theta_{c1} - 1) \tag{9.4.16}$$

which decreases from its maxium value of $(z/c)(\theta_{c2} - \theta_{c1})$ when $(c/z)T = \theta_{c2} - 1$ to zero when $(c/z)T = \theta_{c1} - 1$ as the propagation distance z increases.

The validity of this uniform asymptotic description of rectangular pulse propagation in a single resonance *Lorentz* medium is borne out by comparison with detailed numerical calculations of the dynamical pulse evolution. The cal-

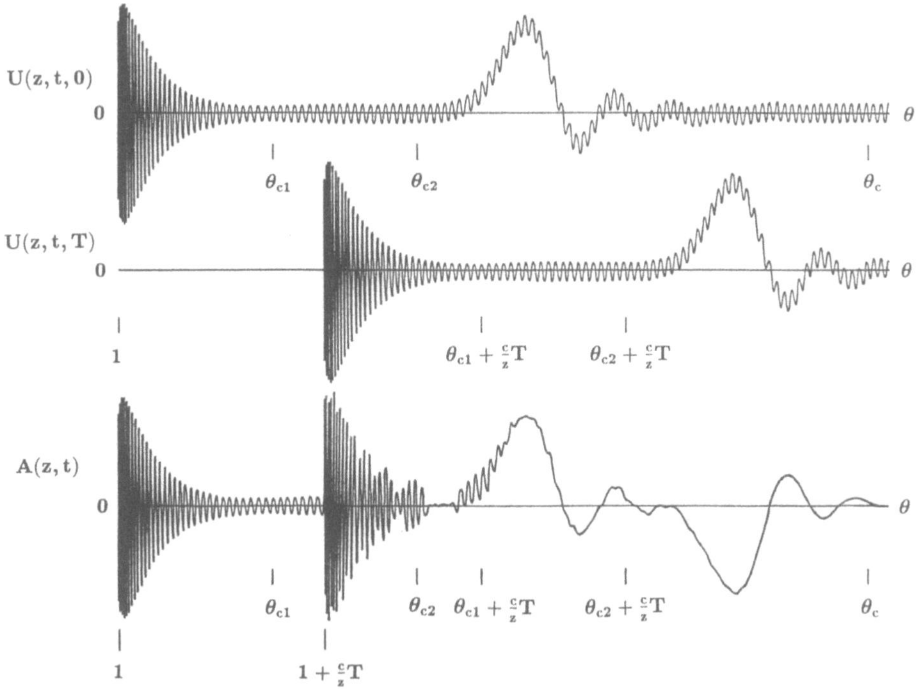

Fig. 9.35. Construction of the dynamical structure of the propagated field $A(z,t) = U(z,t,0) - U(z,t,T)$ in the high signal frequency domain $\omega_c > \omega_{SB} > \omega_0$ when $\theta_{c1} - 1 \leqslant (c/z)T < \theta_{c2} - 1$. When this situation prevails the interference between the precursor fields of the leading and trailing edges of the pulse is moderate to severe and the resultant pulse distortion is becoming severe.

culations presented here are for a strongly absorptive medium with parameters

$$\omega_0 = 4.0 \times 10^{16}/\mathrm{s} \ ,$$

$$b^2 = 20.0 \times 10^{32}/\mathrm{s} \ ,$$

$$\delta = 0.28 \times 10^{16}/\mathrm{s} \ ,$$

which are the same as those chosen by *Brillouin* [9.2]. The dynamical evolution of the propagated field at several increasing values of the propagation distance z is illustrated in Figs. 9.36–39 for the below resonance signal frequency $\omega_c = 1.0 \times 10^{16}$/s. The e^{-1} penetration depth at this signal frequency is $d = 1.82 \times 10^{-4}$ cm. In each sequence of propagated waveforms the time origin has been shifted by the amount $\theta_c z/c$ so that the signal arrival at $\theta_c z/c$ and signal departure at $t = \theta_c z/c + T$ are aligned at each propagation distance; these time instances are indicated by the dotted lines in each figure. The initial pulse width in Fig. 9.36 is $T = 6.283 \times 10^{-16}$ s and corresponds to a single period of oscillation of the signal. In this case the pulse distortion becomes severe ($cT/z <$

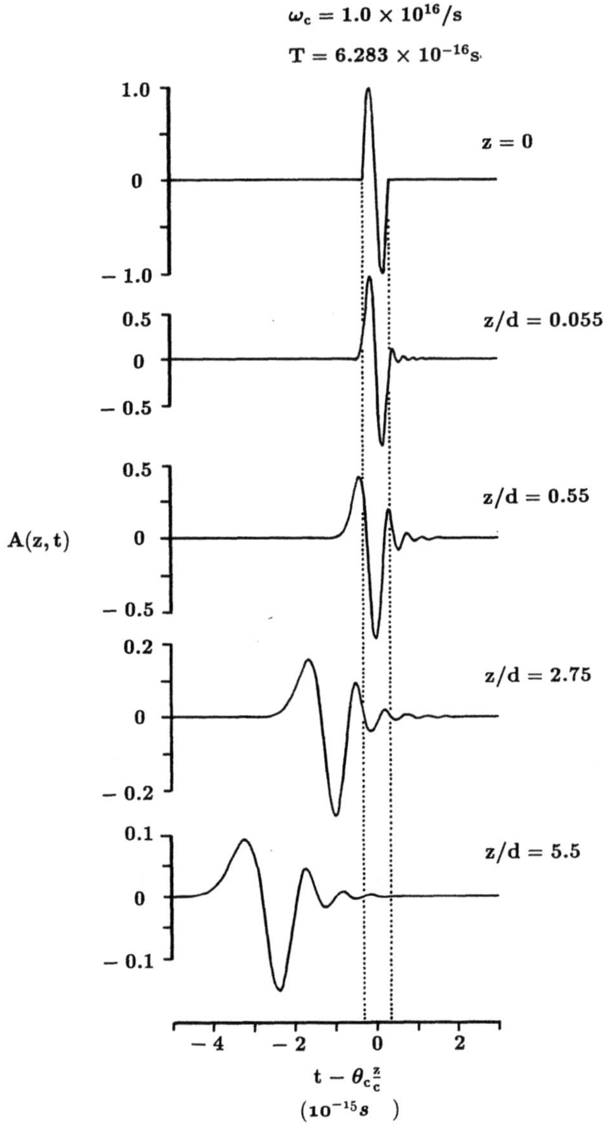

Fig. 9.36. Dynamical evolution of the propagated field due to an input rectangular modulated signal with below resonance carrier frequency $\omega_c = 1.0 \times 10^{16}$/s and initial pulse width $T = 6.283 \times 10^{-16}$ s in a strongly dispersive and absorptive medium. The e^{-1} penetration depth at this signal frequency is $d = 1.82 \times 10^{-4}$ cm.

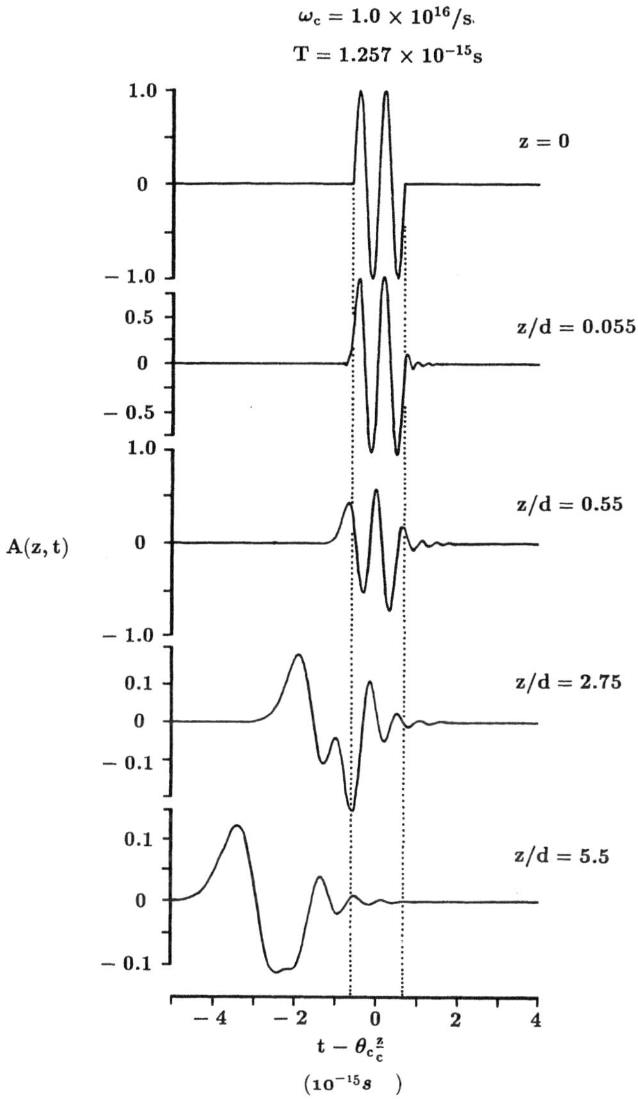

Fig. 9.37. Dynamical evolution of the propagated field due to an input rectangular modulated signal with below resonance carrier frequency $\omega_c = 1.0 \times 10^{16}$/s and initial pulse width $T = 1.257 \times 10^{-15}$ s in a strongly dispersive and absorptive medium.

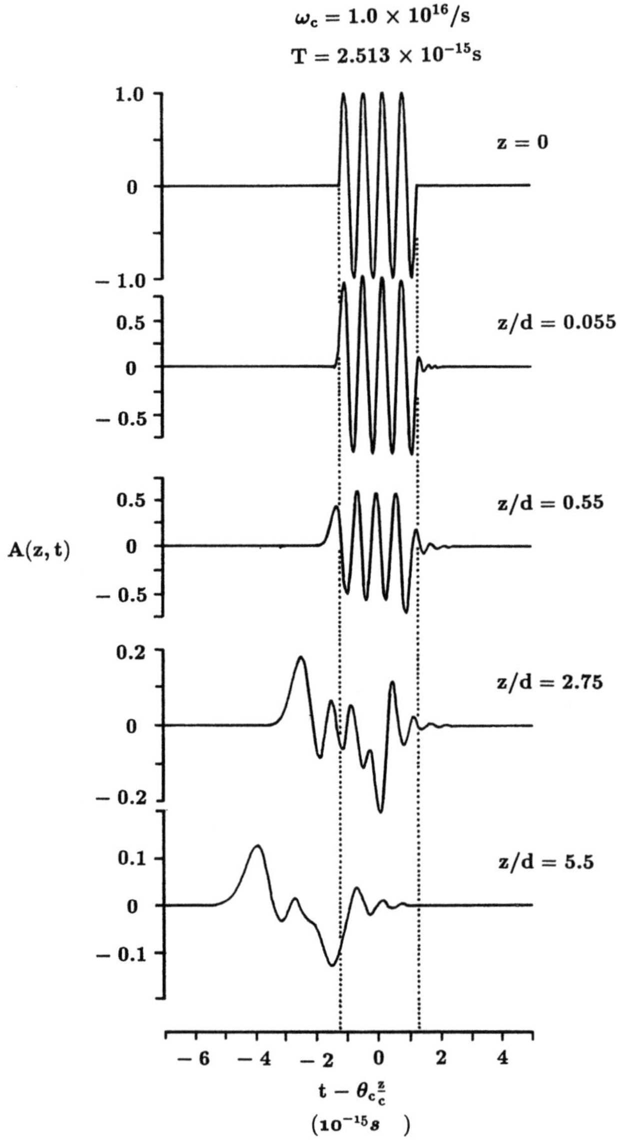

Fig. 9.38. Dynamical evolution of the propagated field due to an input rectangular modulated signal with below resonance carrier frequency $\omega_c = 1.0 \times 10^{16}$/s and initial pulse width $T = 2.513 \times 10^{-15}$ s in a strongly dispersive and absorptive medium.

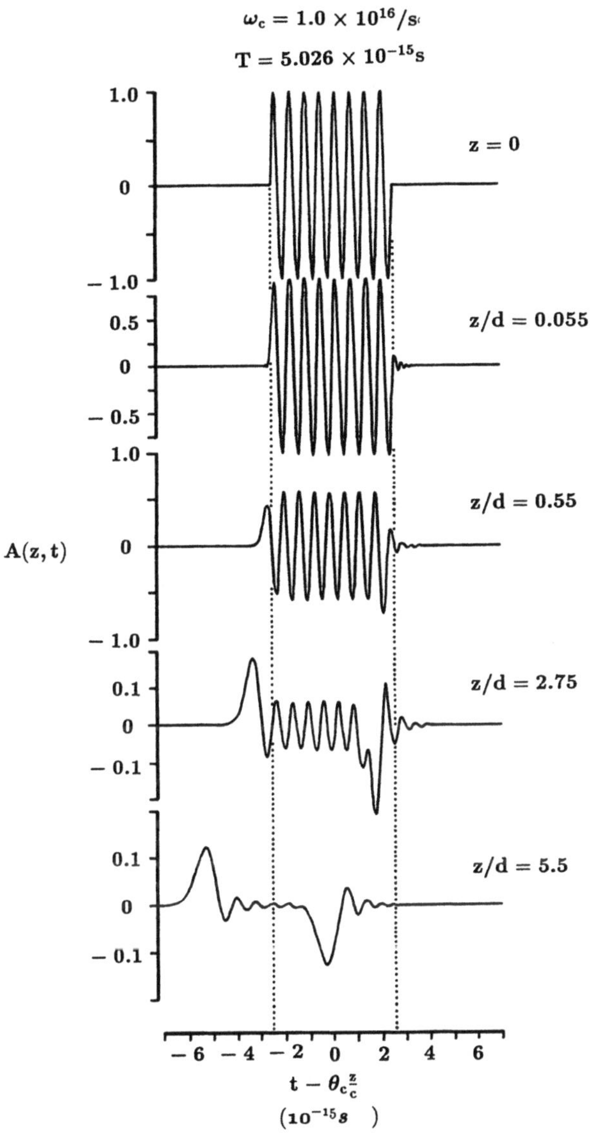

Fig. 9.39. Dynamical evolution of the propagated field due to an input rectangular modulated signal with below resonance carrier frequency $\omega_c = 1.0 \times 10^{16}$/s and initial pulse width $T = 5.026 \times 10^{-15}$ s in a strongly dispersive and absorptive medium.

$\theta_{SB} - 1$) after only $\sim 1/3$ of an absorption depth into the medium, after which the propagated waveform is dominated by the interfering *Brillouin* precursors from the leading and trailing edges of the pulse. In Fig. 9.37 the initial pulse width is doubled to $T = 1.257 \times 10^{-15}$ sec and corresponds to two periods of oscillation of the input signal. In this case the pulse distortion is minimal when $z/d = 0.055$, moderate when $z/d = 0.55$, and severe when $z/d = 2.75$ and all larger propagation distances. Each of these cases corresponds qualitatively to the constructions depicted in Figs. 9.31–33, respectively. The initial pulse width is again doubled to $T = 2.513 \times 10^{-15}$ s in Fig. 9.38. In this case the pulse distortion is minimal when $z/d \leqslant 0.7d$ and becomes severe when $z/d = 1.24d$, after which the propagated waveform is dominated by the interfering *Brillouin* precursors. Finally, the initial pulse width is doubled once more to $T = 5.026 \times 10^{-15}$ s in Fig. 9.39, which corresponds to eight periods of oscillation of the input signal. In this case the transition from minimal to moderate pulse distortion occurs when $z = 1.41d$ and the transition to severe pulse distortion occurs when $z = 2.48d$. By comparison, the transition to the severe pulse distortion regime for a picosecond pulse occurs when $z \sim 500d$. Again, in the severe pulse distortion regime the propagated waveform is dominated by the interfering *Brillouin* precursors. Similar results have been obtained numerically by *Barakat* [9.23] for a *Lorentz* medium and by *Albanese* et al. [9.24] for a Debye model medium.

Careful inspection of Figs. 9.36–39 shows that the propagated pulse-signal width given by (9.4.12) correctly describes the time duration over which the propagated waveform is dominated by the signal oscillating at the input signal frequency ω_c. In particular, this pulse-signal width is seen to decrease with increasing propagation distance z from its input value T to zero at the transition point to the severe distortion limit. Nevertheless, the overall temporal width of the entire propagated waveform is seen to increase with the propagation distance z. Up into the severe distortion regime the propagated pulse waveform is seen to be defined between $\theta = \theta_0$ and $\theta = \theta_c + cT/z$, with corresponding temporal width

$$\Delta t = T + \frac{z}{c}(\theta_c - \theta_0) \ . \tag{9.4.17}$$

Once into the severe distortion regime the propagated field structure becomes completely dominated by the front and back *Brillouin* precursors whose peak values occur at $\theta = \theta_0$ and $\theta = \theta_0 + cT/z$ and are thus separated in time by the initial pulse width T. Since these two points in the field evolution experience no exponential decay, but rather decrease with the propagation distance only as $z^{-1/2}$, they will remain the prominent feature in the propagated field structure long after the signal contribution has attenuated away. This behavior applies throughout the below resonance frequency domain $\omega_c \in [0, \omega_0)$ and remains applicable up through most of the absorption band $\omega_c \in [\omega_0, \omega_1]$. In the upper region of the absorption band and for signal frequencies $\omega_c > \omega_1$ the *Sommerfeld* precursor becomes a dominant feature in the propagated waveform and must be included in any description of its overall temporal width.

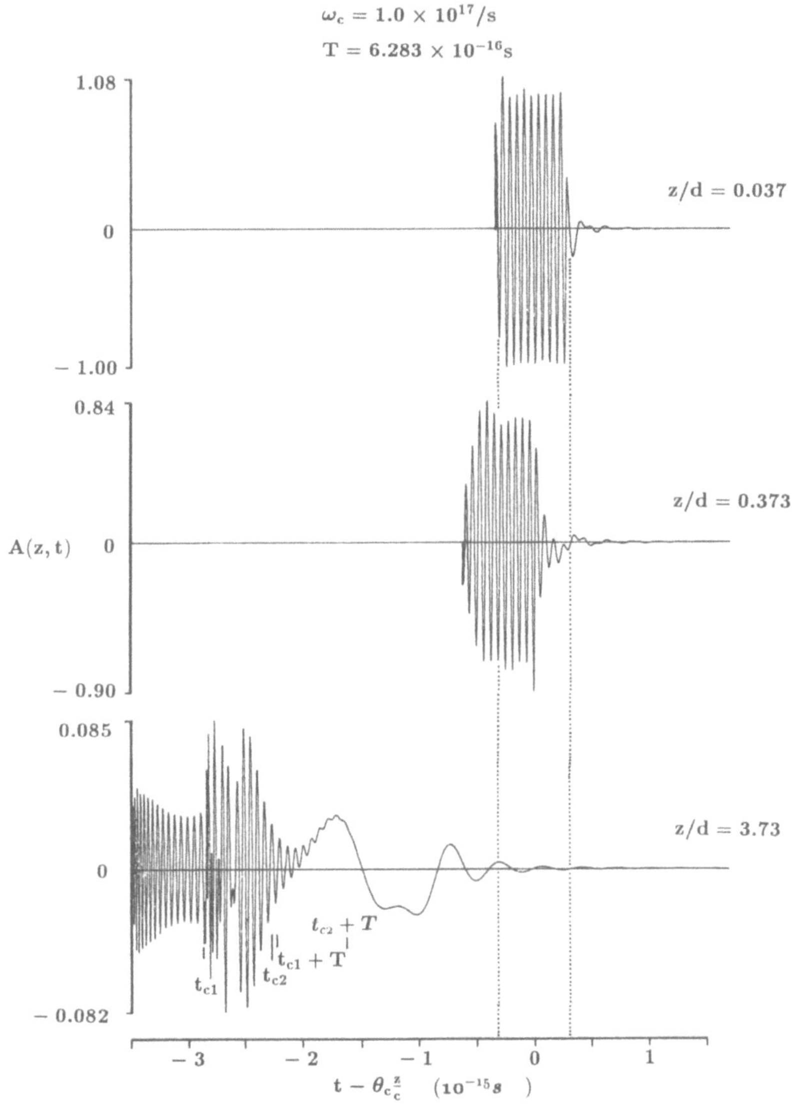

Fig. 9.40. Dynamical evolution of the propagated field due to an input rectangular modulated signal with above resonance carrier frequency $\omega_c = 1.0 \times 10^{17}$/s and initial pulse width $T = 6.283 \times 10^{-16}$ s in a strongly dispersive and absorptive medium.

The dynamical evolution of the propagated field at several increasing values of the propagation distance z is illustrated in Fig. 9.40 for the above resonance signal frequency $\omega_c = 1.0 \times 10^{17}$/s, where $\omega_c > \omega_{SB}$. The e^{-1} penetration depth at this signal frequency is $d = 2.68 \times 10^{-5}$ cm. The initial pulse width here is $T = 6.283 \times 10^{-16}$ s which corresponds to 10 oscillation periods of the signal. At the smallest propagation distance illustrated in the figure, $z/d = 0.037$ while at the intermediate propagation distance illustrated, $z/d = 0.37$, so that these two propagated waveforms are in the immature dispersion regime. In both cases $cT/z > \theta_c - 1$ so that the pulse distortion is minimal. At the largest propagation distance illustrated in the figure, $z/d = 3.73$ so that the propagated waveform is in the mature dispersion regime. In this last case $cT/z \approx \theta_{c1} - 1$ so that the prepulse is almost fully distorted (due to interference with the trailing edge *Sommerfeld* precursor) and the main pulse has disappeared, being replaced by the interfering *Brillouin* precursors. The time origin for each propagated wave form illustrated has been shifted by the amount $\theta_c z/c$, and the dotted lines in the figure depict the locations of the front and back of the initial, undistorted pulse which propagate at the main signal velocity $v_c = c/\theta_c$.

The temporal width Δt_s of the main pulse is then seen to decrease from T to zero as the propagation distance z increases from zero, as described by Eq. (9.4.15). In addition, the temporal width Δt_p of the pre-pulse is seen to first increase with increasing propagation distance z, as described by (9.4.14), and then decrease with increasing propagation distance z as the pulse distortion becomes severe, as described by (9.4.16). Nevertheless, the overall temporal width of the entire propagated waveform is seen to increase with the propagation distance z. If one includes just the high frequency structure in the mature dispersion regime, which evolves over the θ-domain from $\theta = 1$ to $\theta \approx \theta_0$, the overall temporal width is given by

$$\Delta t' \cong \frac{z}{c}(\theta_0 - 1) \ ; \qquad \frac{c}{z}T < \theta_c - 1 \ , \tag{9.4.18}$$

while if one includes both the low and high frequency structure, which evolves over the θ-domain from $\theta = 1$ to $\theta \simeq \theta_0 + (c/z)T$, the overall temporal width is found to be

$$\Delta t \cong T + \frac{z}{c}(\theta_0 - 1) \ , \tag{9.4.19}$$

which is valid for all $(c/z)T < \theta_c - 1$. The relation in (9.4.18) is the appropriate measure of the overall propagated pulse width in the mature dispersion regime if one only detects the high frequency content of the field, while (9.4.19) is the appropriate measure if one includes all significant frequency components.

9.4.3 The Delta-Function Pulse

The impulse response of a dispersive medium due to an input delta-function pulse that is applied at the plane $z = 0$ at the time $t = 0$ is given by the contour integral (cf. (4.3.44))

$$A(z,t) = \frac{1}{2\pi} \mathrm{Re} \int_{ia-\infty}^{ia+\infty} e^{(z/c)\phi(\omega,\theta)} \, d\omega \ . \tag{9.4.20}$$

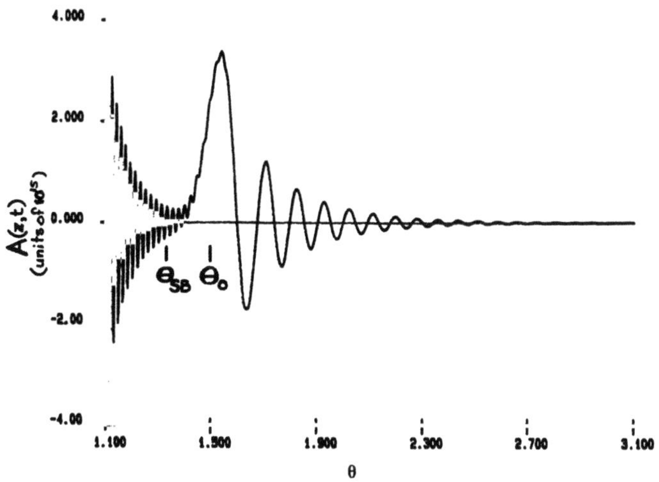

Fig. 9.41. Numerically determined impulse response of a single resonance *Lorentz* medium with parameters $\omega_0 = 4 \times 10^{16}/\mathrm{s}$, $b^2 = 20 \times 10^{32}/\mathrm{s}^2$, $\delta = 0.28 \times 10^{16}/\mathrm{s}$ at a fixed propagation distance $z = 1 \times 10^{-4}$ cm.

Since the spectral amplitude function $\tilde{f}(\omega)$ is unity in this case there is then no pole contribution so that, for a single resonance *Lorentz* medium, the uniform asymptotic description of the propagated field (9.4.20) is given by

$$A(z,t) = A_s(z,t) + A_B(z,t) \ , \tag{9.4.21}$$

which is the superposition of the *Sommerfeld* and *Brillouin* precursor fields for the delta-function pulse. Here $A_B(z,t)$ is negligible when the first precursor is predominant, $A_s(z,t)$ is negligible when the second precursor is predominant, and both $A_s(z,t)$ and $A_B(z,t)$ are important at the same time during the transition from the first to the second precursor field for values of θ about $\theta = \theta_{SB}$.

The numerically determined [9.17] impulse response of a single resonance *Lorentz* medium with parameters $\omega_0 = 4 \times 10^{16}/\mathrm{s}$, $b^2 = 20 \times 10^{32}/\mathrm{s}^2$, $\delta = 0.28 \times 10^{16}/\mathrm{s}$ is illustrated in Fig. 9.41 as a function of the space-time parameter $\theta = ct/z$ at a fixed propagation distance $z = 1 \times 10^{-4}$ cm. The propagated field vanishes for all $\theta < 1$. At $\theta = 1$ the first precursor arrives with infinite amplitude and oscillation frequency. The amplitude of the first precursor is finite for all $\theta > 1$ and monotonically decreases as θ increases. The first precursor evolves essentially undisturbed up until $\theta = \theta_{SB}$, at which point there is a transition in asymptotic dominance over to the second precursor field. The second precursor then evolves essentially undisturbed for all $\theta > \theta_{SB}$. A maximum in amplitude of the second precursor is attained near θ_0 since there is no exponential decay in the field at $\theta = \theta_0$. As θ increases away from θ_0, the amplitude of the second precursor steadily decreases due to the monotonically increasing exponential decay with θ increasing above θ_0.

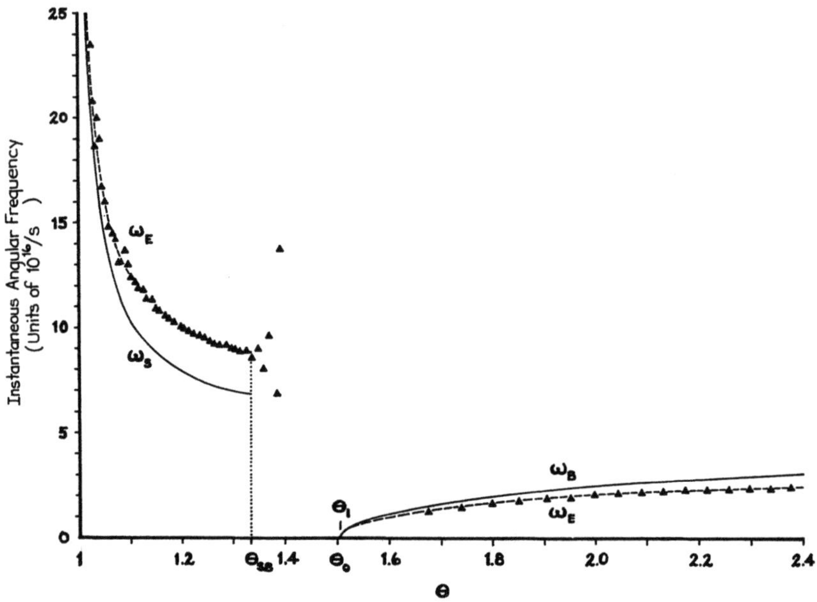

Fig. 9.42. Evolution of the instantaneous angular frequency of oscillation of the propagated field for an input delta-function pulse in a single resonance *Lorentz* medium with $\omega_0 = 4 \times 10^{16}$/s, $b^2 = 20 \times 10^{32}/\text{s}^2$, $\delta = 0.28 \times 10^{16}$/s. The solid curves depict the angular frequencies ω_S and ω_B for the *Sommerfeld* and *Brillouin* precursor fields, respectively, as predicted by the asymptotic theory. The dashed curves depict the angular frequency of oscillation ω_E obtained from the energy-transport velocity in the dispersive medium. The triangular data points are obtained from the numerical calculation of the field evolution.

The numerically determined [9.17] oscillation frequency of the impulse response is illustrated in Fig. 9.42 as a function of θ at a fixed propagation distance and is compared with the theoretical curves for the instantaneous angular frequency ω_s for the *Sommerfeld* precursor [cf. (7.2.18)] and ω_B for the *Brillouin* precursor [cf. (7.3.61)], as obtained from the asymptotic theory. The dashed curve in the figure depicts the frequency ω_E that is obtained from the energy transport velocity [cf. (9.3.26)]. The observed difference between the numerically determined oscillation frequency and the asymptotic description of this quantity is due to the approximations made in determining the saddle point locations and the complex phase behavior at them. On the other hand, the agreement between the numerically determined behavior and the frequency ω_E is seen to be exceptionally good. This is due to the excellent agreement between the appropriate signal velocity branch and the energy transport velocity (see Fig. 8.7).

9.4.4 The Hyperbolic-Tangent Modulated Signal

In order to illustrate the effects of a finite turn-on time on the precursor fields associated with dispersive pulse propagation, the general behavior of the hyperbolic-tangent modulated signal is now briefly considered. The initial envelope of this field is given by (4.3.45), viz

$$u(t) = \frac{1}{2}[1 + \tanh(\beta_T t)] = \frac{1}{1 + e^{-2\beta_T t}} , \tag{9.4.22}$$

where the parameter $\beta_T > 0$ is a measure of the rapidity of turn-on of the initial signal (see Fig. 4.9). In the limit as β_T approaches infinity (i.e. in the limit of an instantaneous initial turn-on time), the hyperbolic-tangent modulated signal reduces to the unit step-function modulated signal. In the opposite limit as β_T is allowed to approach zero from above, the input hyperbolic-tangent modulated signal simplifies to a monochromatic wave of amplitude 1/2 for all time t at the input plane. The spectrum of this initial pulse envelope is given by [cf. (4.3.46)]

$$\tilde{u}(\omega) = \frac{\pi i}{2\beta_T} \cdot \frac{1}{\sinh\left(\dfrac{\pi}{2\beta_T}\omega\right)} \tag{9.4.23}$$

with $0 < \mathrm{Im}(\omega) < 2\beta_T$, which possesses simple pole singularities at

$$\omega = \pm 2n\beta_T i \ ; \qquad n = 0, 1, 2, 3, \ldots , \tag{9.4.24}$$

so that $\tilde{u}(\omega - \omega_c)$ has simple pole singularities at $\omega_{pn}^{\pm} = \omega_c \pm i2n\beta_T$.

As was shown in Sect. 8.3.4, for large values of $\beta_T \gg \delta$, where δ is the phenomenological damping constant of the single resonance *Lorentz* medium, only the simple pole singularity at $\omega = \omega_c$ contributes to $A_c(z, t)$. In that case the dynamical evolution of the propagated field due to an input hyperbolic-tangent modulated signal is very similar to that for the unit step-function modulated signal. As the initial rise-time parameter β_T is allowed to approach δ from above, the pole singularities at $\omega_{pn}^{-} = \omega_c - 2n\beta_T i$ begin to contribute to the asymptotic behavior of the propagated field for values of $\theta > \theta_c$. For small values of initial rise-time parameter $\beta_T < \delta$, however, a large number of the simple pole singularities located at $\omega_{pn}^{+} = \omega_c + 2n\beta_T i$ approach close to the real axis, and the interaction of the deformed contour of integration with them becomes increasingly important so that their contributions to the asymptotic behvaior of the propagated field are no longer negligible. In the limit as β_T becomes very small ($\beta_T \ll \delta$), the original contour of integration is constrained to lie very near to the real axis. In deforming this contour to the Olver-type path $P(\theta)$, a large number of pole contributions due to the singularities in the upper-half of the complex ω-plane will be picked up for values of θ prior to the value θ_c when the pole contribution at ω_c becomes dominant, and a large number of pole contributions due to the singularities in the lower-half of the complex ω-plane will be picked up for values of θ after the value θ_c. These pole contributions, in addition to the pole contribution at $\omega = \omega_c$, will then constitute a major portion of the propagated field for all $\theta \geqslant 1$ and the precursor fields due to the saddle point contributions will be negligible by comparison.

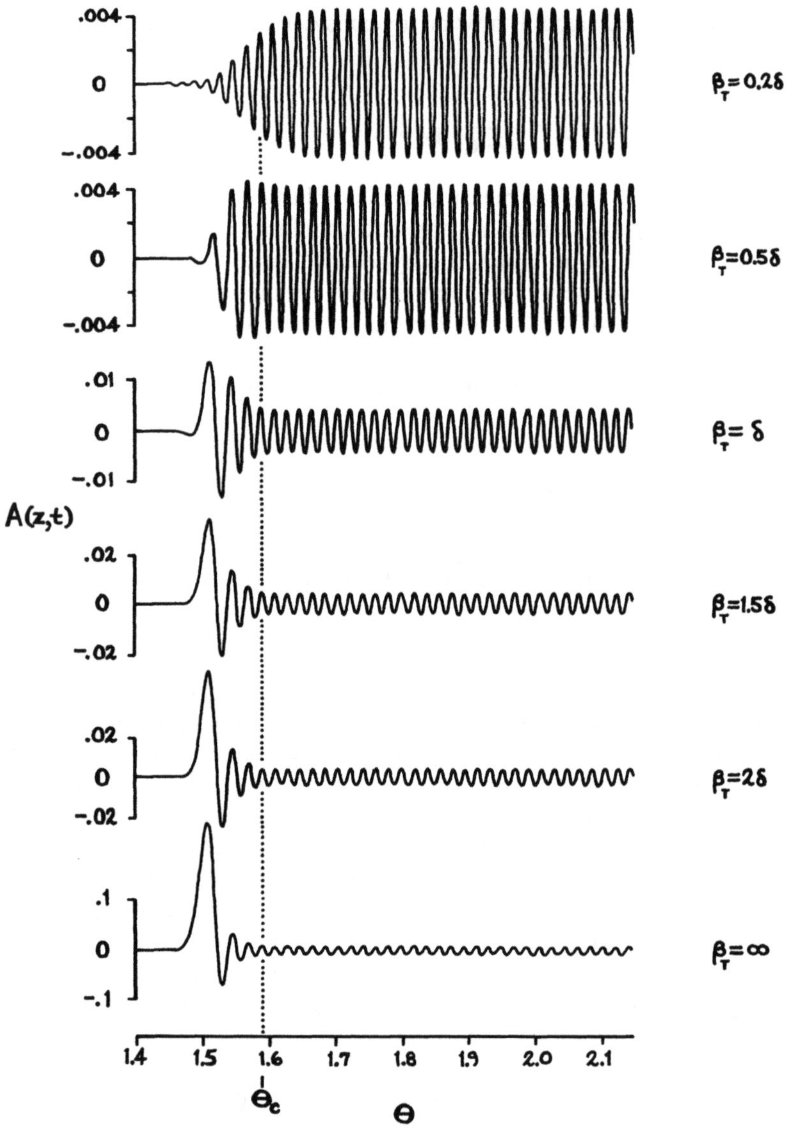

Fig. 9.43. Propagated field evolution at the fixed propagation distance $z/d = 5.495$ due to an input hyperbolic-tangent modulated signal at a below resonance carrier frequency $\omega_c = 1 \times 10^{16}$/s for several values of the initial rise-time parameter β_T.

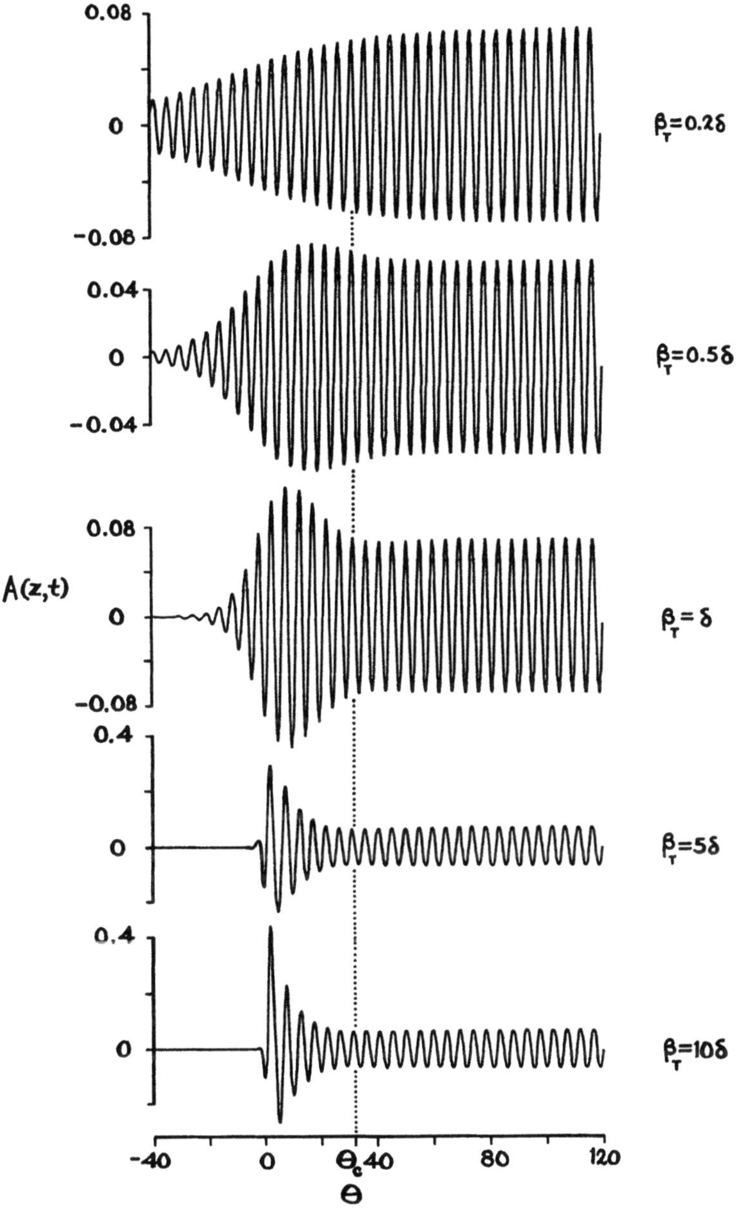

Fig. 9.44. Propagated field evolution at the fixed propagation distance $z/d = 2.67$ due to an input hyperbolic-tangent modulated signal with resonant carrier frequency $\omega_c = \omega_0 = 4 \times 10^{16}$/s for several values of the initial rise-time parameter.

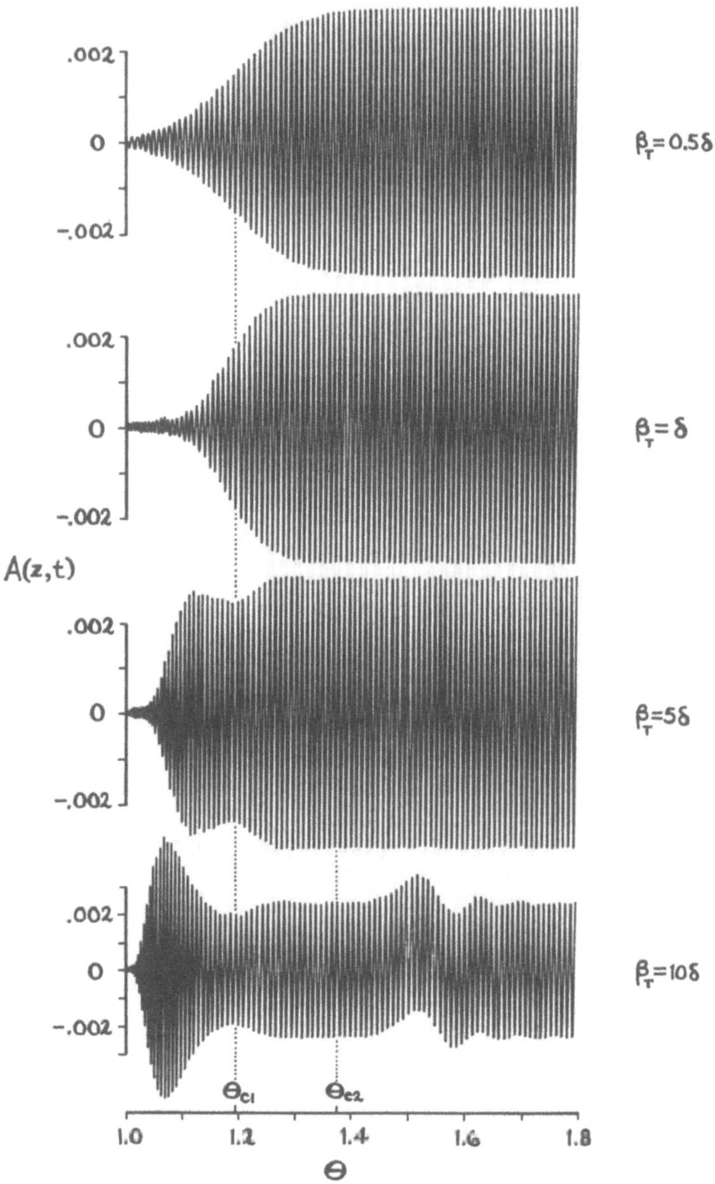

Fig. 9.45. Propagated field evolution at the fixed propagation distance $z/d = 7.46$ due to an input hyperbolic-tangent modulated signal at an above resonance carrier frequency $\omega_c = 1 \times 10^{17}$/s for several values of the initial rise-time parameter β_T.

The accuracy of this asymptotic description is completely borne out by numerical calculations of the dynamical field evolution, illustrated in Figs. 9.43–9.45. Each of these figures depicts the propagated field evolution (at a fixed propagation distance z) due to an input hyperbolic-tangent modulated signal at a fixed value of the carrier frequency ω_c for several values of the initial rise-time parameter β_T. The medium parameters used here are (once again) $\omega_0 = 4 \times 10^{16}/\text{s}$, $b^2 = 20 \times 10^{32}/\text{s}^2$, and $\delta = 0.28 \times 10^{16}/\text{s}$. The below resonance field behavior is illustrated in Fig. 9.43 with $\omega_c = 1 \times 10^{16}/\text{s}$ and $z = 1 \times 10^{-3}$ cm ($z/d = 5.495$, where d is the e^{-1} penetration depth of the dispersive medium at the carrier frequency). Notice the change in scale for the field amplitude for different values of β_T; in each case the steady-state amplitude of the field is the same. The *Brillouin* precursor is clearly well-defined for all $\beta_T \gtrsim \delta$, in agreement with the asymptotic description. The field behavior when the carrier frequency is equal to the undamped resonance frequency ω_0 of the *Lorentz* medium is illustrated in Fig. 9.44 with $z = 2.67d = 1 \times 10^{-6}$ cm. As in the below resonance case, the *Brillouin* precursor is seen to be well-defined for all $\beta_T \gtrsim \delta$; however, its definition for values of β_T near δ is not as pronounced as that in the below resonance case illustrated in Fig. 9.43. This is due to the smaller spectral amplitude around $\omega = 0$ for larger values of ω_c. Finally, the field behavior when the carrier frequency $\omega_c > \omega_{SB}$ is well above the medium absorption band is illustrated in Fig. 9.45 with $z = 7.46d = 2 \times 10^{-4}$ cm and $\omega_c = 1 \times 10^{17}/\text{s}$. Because of the high carrier frequency, it is now the *Sommerfeld* precursor that provides any significant distortion to the leading edge of the propagated field. The *Sommerfeld* precursor is found to be well-defined for $\beta_T > \delta$; its definition becomes more pronounced as ω_c increases in value.

10 Physical Interpretation of the Pulse Dynamics

The effects of dispersion and absorption on the evolution of an electromagnetic pulse as it propagates through a homogeneous linear dielectric, developed originally by *Sommerfeld* [10.1] and *Brillouin* [10.2, 3] have been described in detail by the mathematical description presented in the preceeding chapters of this monograph. The results show that after the pulse has propagated sufficiently far in the material, its dynamics settle into a relatively simple regime (known as "mature dispersion") for the rest of the propagation. In this regime, the field becomes locally quasi-monochromatic with fixed local frequency and wave number in small regions of space which move with their own characteristic constant velocity. The theory provides approximate analytic expressions for the local wave properties to be expected at a given space-time point of observation. The expressions are very complicated, however, and neither the results nor their derivations have provided insight into the physical reasons for the field having the particular local properties it does in the various subregions of space moving with specific velocities.

The mature dispersion regime is well known in the theory of propagation of rather general linear waves in homogeneous dispersive media in which there is no absorption or gain. It is exhibited by all waves whose monochromatic components are described by the *Helmholtz* equation with real propagation constant such as electromagnetic, acoustic, elastic, or gravity waves in lossless, gainless linear systems. Furthermore, a physical explanation is available for the local properties of all of these waves based on the concept of the group velocity of time-harmonic waves [10.4–6]. When absorption or gain is present in the medium, however, the group-velocity description breaks down [10.7]. This is a severe limitation of a basic kind because a lossless, gainless dispersive system is noncausal [10.8]. Hence, the previously accepted physical understanding of the details of dispersive pulse propagation in terms of the well-known group velocity description is confined to the case of no absorption or gain which is itself fundamentally unphysical.

In 1981, *Sherman* and *Oughstun* [10.9] published a physical explanation of the local wave properties of electromagnetic pulses propagating in absorbing dielectrics in the mature-dispersion regime. The explanation is similar to the group-velocity description valid for lossless, gainless energy in time-harmonic waves and the attenuation of the waves is included. This is the only such physical description of the details of pulse dynamics that is known for any dispersive system that includes absorption (or gain) or that is causal.

In addition to explaining the details of the local behavior of the pulse in physical terms, the energy-velocity description provides a simple mathematical algorithm for predicting those details quantitatively.

10.1 Review of the Physical Problem and Its Asymptotic Description

For purpose of review, the physical problem of interest is briefly restated here and the exact integral solution and its asymptotic approximation summarized. For convenience, we apply the notation used in the preceeding chapters with the single minor modification that only the electric field of a linearly polarized plane wave is considered.

Consider then a plane electromagnetic wave with real electric field $\mathscr{E}(z,t)$ linearly polarized along the x axis and traveling in the positive z direction in the dielectric which occupies the half-space $z > 0$. The field is taken to be zero for $t > 0$ and therefore can be expressed in a Laplace representation as

$$\mathscr{E}(z,t) = \int_{ia-\infty}^{ia+\infty} E(z,\omega)e^{-i\omega t}\,d\omega \ , \tag{10.1.1}$$

where a is a positive constant. The contour of integration for the integral in (10.1.1) is a straight horizontal line in the complex ω plane a distance a above the real axis. The spectral amplitude $E(z,\omega)$ satisfies the scalar *Helmholtz* equation

$$\nabla^2 E(z,\omega) + \tilde{k}^2(\omega)E(z,\omega) = 0 \tag{10.1.2}$$

in the half-space $z > 0$.

The complex propagation constant $\tilde{k}(\omega)$ is given in terms of the complex index of refraction $n(\omega)$ by $\tilde{k}(\omega) = \omega n(\omega)/c$ where c is the speed of light in vacuo, and where the complex index of refraction due to a single resonance *Lorentz* medium is given by

$$n(\omega) = \left[1 - \frac{b^2}{\omega^2 - \omega_0^2 + 2\delta\omega i}\right]^{1/2} , \tag{10.1.3}$$

where ω_0 is the undamped resonance frequency, δ is the phenomenological damping constant, and b is the plasma frequency of the medium. For most of the numerical computations presented in this monograph, we apply the same numerical values for the medium parameters ω_0, δ, and b as used by *Brillouin* [10.2, 3];

$$b^2 = 2 \times 10^{33}\ \mathrm{s}^{-2} , \tag{10.1.4a}$$

$$\delta = 2.8 \times 10^{15}\ \mathrm{s}^{-1} , \tag{10.1.4b}$$

$$\omega_0 = 4 \times 10^{16}\ \mathrm{s}^{-1} . \tag{10.1.4c}$$

These values correspond to a highly absorbing medium. For comparison, we also include in this chapter some computations with the same parameter values except that δ is taken to be nearly zero, corresponding to a highly transparent medium.

Let the field satisfy the boundary value

$$\mathscr{E}(0,t) = f(t) \ , \tag{10.1.5}$$

where $f(t)$ is a real function that satisfies $f(t) = 0$ for $t < 0$. The case treated by *Sommerfeld* and *Brillouin* was the special case $f(t) = \sin(\omega_c t)$ for $t > 0$. We are concerned here with $f(t)$ that is zero after some finite positive time. The case of primary interest is the delta-function pulse having $f(t) = A\delta(t)$ where A is a constant and $\delta(t)$ is the Dirac delta function (not to be confused with the damping constant δ). In that case, $\mathscr{E}(z,t)/A$ is the impulse response of the medium.

The exact integral solution to this boundary-value problem can be expressed in the form (see Sect. 4.3)

$$\mathscr{E}(z,t) = \frac{1}{2\pi} \int_{ai-\infty}^{ai+\infty} \tilde{f}(\omega) \exp\left[\frac{z}{c}\phi(\omega)\right] d\omega \ , \tag{10.1.6}$$

where

$$\tilde{f}(\omega) = \int_{-\infty}^{\infty} f(t) e^{i\omega t} \, dt \ , \tag{10.1.7}$$

$$\phi(\omega) = i\omega[n(\omega) - \theta] \ , \tag{10.1.8}$$

$$\theta = \frac{ct}{z} \ . \tag{10.1.9}$$

For the delta-function pulse, the pulse spectrum $\tilde{f}(\omega)$ is simply A.

Sommerfeld [10.1] showed that it follows from (10.1.6) that $\mathscr{E}(z,t) = 0$ for $\theta < 1$, thereby proving that the field is relativistically causal (i.e. it does not travel faster than c). *Brillouin* [10.2, 3] showed that as z tends to infinity with fixed $\theta > 1$, the asymptotic approximation of the field may be expressed as (see Sect. 8.3.3)

$$\mathscr{E}(z,t) \approx \mathscr{E}_s(z,t) + \mathscr{E}_b(z,t) \ , \tag{10.1.10}$$

where

$$\mathscr{E}_s(z,t) = \frac{\sqrt{2\pi c}\tilde{f}(\omega_s)}{\sqrt{-z\phi^{(2)}(\omega_s)}} \exp\left[\frac{z}{c}\phi(\omega_s)\right] + \frac{\sqrt{2\pi c}\tilde{f}(-\omega_s^*)}{\sqrt{-z\phi^{(2)}(-\omega_s^*)}} \exp\left[\frac{z}{c}\phi(-\omega_s^*)\right] , \tag{10.1.11}$$

$$\mathscr{E}_b(z,t) = \frac{\sqrt{2\pi c}\tilde{f}(\omega_b)}{\sqrt{z\phi^{(2)}(\omega_b)}} \exp\left[\frac{z}{c}\phi(\omega_b)\right] \quad \text{for } 1 < \theta < \theta_1 \ , \tag{10.1.12a}$$

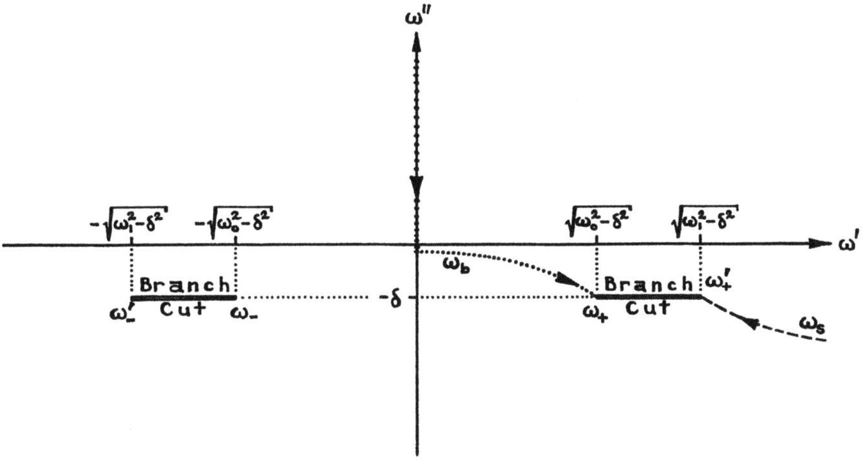

Fig. 10.1. Schematic of the paths traced out by the near saddle point ω_b and the distant saddle point ω_s as θ increases from 1 to infinity.

and

$$\mathscr{E}_b(z,t) = \frac{\sqrt{2\pi c \tilde{f}(\omega_b)}}{\sqrt{-z\phi^{(2)}(\omega_b)}} \exp\left[\frac{z}{c}\phi(\omega_b)\right] + \frac{\sqrt{2\pi c \tilde{f}(-\omega_b^*)}}{\sqrt{-z\phi^{(2)}(-\omega_b^*)}} \exp\left[\frac{z}{c}\phi(-\omega_b^*)\right]$$

$$\text{for } \theta > \theta_1 \;. \qquad (10.1.12\text{b})$$

Here and throughout this chapter, we apply the notation $g^{(n)}(\omega_\alpha)$ to indicate the nth derivative of an arbitrary function $g(\omega)$ with respect to ω evaluated at ω_α. The quantities ω_s ω_b are first-order saddle points which are complex solutions of the saddle point equation

$$\phi^{(1)}(\omega) = 0. \qquad (10.1.13)$$

These saddle points are functions of the dimensionless space-time parameter θ. It is shown in Chap. 6 that they trace out the paths in the complex ω plane that are indicated in Fig. 10.1 as θ increases from 1. The saddle point ω_b moves down the positive imaginary axis until θ reaches a value θ_1 which depends on the medium parameters. For θ increasing beyond θ_1, ω_b moves into the fourth quadrant as shown in the figure.

The above results are referred to as nonuniform asymptotic results because they break down at certain critical values of the parameter θ. In particular, the right-hand sides of (10.1.12) become infinite at $\theta = \theta_1$ and give discontinuous asymptotic behaviors on opposite sides of that value because $\phi^{(2)}(\omega_b) = 0$ there. Uniform asymptotic results which do not have this behavior have been derived in Chap. 7. We employ the nonuniform results for our initial analysis here because they are much simpler to understand. We employ the uniform results in the final analysis at the end of the chapter, however, to obtain an energy-velocity description that is continuous at $\theta = \theta_1$.

It is shown in Chap. 7 that the second terms on the right-hand sides of (10.1.11 and 12b) are the complex conjugates of the corresponding first terms. Also, it is readily seen that the quantity on the right-hand side of (10.1.12a) is real since $\phi(\omega)$ is real for imaginary ω. Hence, it follows from (10.1.10)–(10.1.12) that $\mathscr{E}(z,t)$ can be expressed as the real part of a complex function $E(z,t)$

$$\mathscr{E}(z,t) = \mathrm{Re}[E(z,t)] \tag{10.1.14}$$

with asymptotic approximation given by

$$E(z,t) = E_s(z,t) + E_b(z,t) \ , \tag{10.1.15}$$

with

$$E_s(z,t) = \frac{2\sqrt{2\pi c}\tilde{f}(\omega_s)}{\sqrt{-z\phi^{(2)}(\omega_s)}} \exp\left[\frac{z}{c}\phi(\omega_s)\right] \ , \tag{10.1.16}$$

$$E_b(z,t) = \frac{\sqrt{2\pi c}\tilde{f}(\omega_b)}{\sqrt{-z\phi^{(2)}(\omega_b)}} \exp\left[\frac{z}{c}\phi(\omega_b)\right] \quad \text{for } 1 < \theta < \theta_1 \ , \tag{10.1.17a}$$

and

$$E_b(z,t) = \frac{2\sqrt{2\pi c}\tilde{f}(\omega_b)}{\sqrt{-z\phi^{(2)}(\omega_b)}} \exp\left[\frac{z}{c}\phi(\omega_b)\right] \quad \text{for } \theta > \theta_1 \ . \tag{10.1.17b}$$

Equations (10.1.14)–(10.1.17) predict the behavior of the pulse after it has travelled far enough to enter into the mature dispersion regime. The details of the dynamics are determined by the motion of the saddle points in the complex ω plane with increasing θ. Crude analytic approximations for the saddle point locations as functions of θ are derived in Refs. [10.2, 3] with substantial improvements derived in Ref. [10.10] and described in Chap. 6. Unfortunately, the formulas are very complicated and neither the results nor their derivations provide insight into the physical reasons behind the pulse behavior.

10.2 Approximations Having Physical Interpretations

In this section, approximations of (10.1.16–17) are obtained which do have physical interpretations and which provide a simple physical model of pulse dynamics in the mature dispersion regime. The approximations are valid for δ much smaller than ω_0 and b. This requires that the medium not be too highly absorbing. We show in the next section that the requirement is not overly restrictive, however, by demonstrating numerically that the approximations are sufficiently accurate for a medium (the *Lorentz* medium with the parameter values studied by *Brillouin*) which is so absorbing that the material would be considered to be opaque at nearly all nonzero, finite frequency values.

To obtain the desired simplifications, we replace the saddle points in the asymptotic expressions by other frequencies which yield approximately the same results but which have clearer physical interpretations. In particular, we replace the saddle points in (10.15 and 16b) with specific real frequencies leading to quasi time-harmonic (quasimonochromatic) waves with local frequency, phase, and amplitude which are easily understood in physical terms. Similarly, we replace the saddle point in (10.1.16a) with a specific purely imaginary frequency leading to a non-oscillatory field with local amplitude and growth rate which is easily understood in physical terms.

10.2.1 The Quasimonochromatic Contribution

In order to identify the real frequencies of interest, our attention is focused on the attenuation of the field with increasing z. It is important that the resultant approximation have the correct attenuation since we are interested in the properties of an exponentially decaying field after it has propagated a large distance in the dispersive medium. Hence, it is desired to determine those time-harmonic waves (with real frequencies) which are attenuated in the medium at the same rate as the fields given in (10.1.16 and 17b). To begin, define the notation that g^r and g^i represent respectively the real and imaginary parts of the arbitrary complex quantity g. Then, for fixed θ, the attenuation with increasing z of a wave of the form $\exp\{z\phi(\omega)/c\}$ for complex ω is determined by $\phi^r(\omega)$. For given θ, therefore, we define ω_{Ej} to be the real frequencies nearest the saddle points ω_j that satisfy

$$\phi^r(\omega_{Ej}) = \phi^r(\omega_j) \tag{10.2.1}$$

with $j = s, b$. Then, a time-harmonic plane wave with real frequency ω_{Ej} has the same attenuation as the pulse in the mature-dispersion regime given in (10.1.16 and 17b). The locations of these frequencies in the complex ω plane are indicated in Fig. 10.2 (for some value of θ greater than θ_1) as the intersections with the real axis of the contours of constant $\phi^r(\omega)$ that pass through the saddle points.

The frequencies ω_{Ej} so defined are intimately connected to the physics of the propagation of time-harmonic waves in the medium. It is shown in Sect. 9.3 that to a good approximation, they satisfy

$$v_E(\omega_{Ej}) = \frac{z}{t} \tag{10.2.2}$$

for δ much smaller than ω and b, where $v_E(\omega)$ is the velocity of energy in time-harmonic waves (with real frequencies) defined by

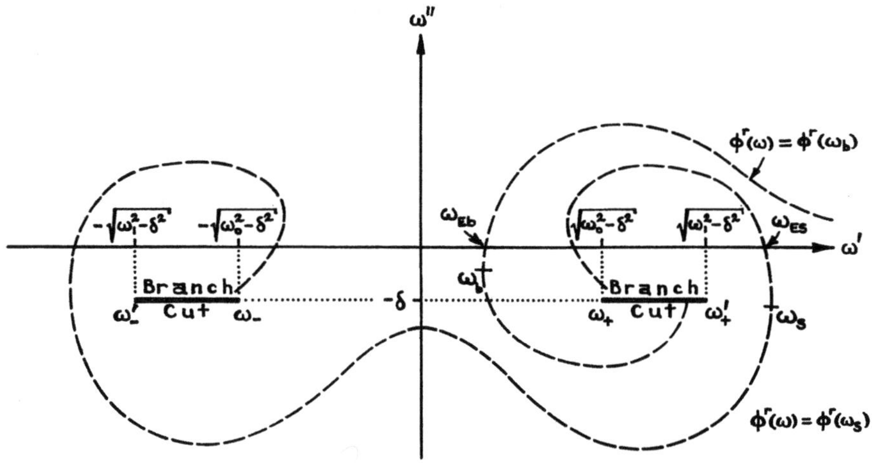

Fig. 10.2. Location in the complex ω-plane of the real frequencies ω_{Eb} and ω_{ES} relative to the locations of the saddle points ω_b and ω_S, respectively, for a fixed value of $\theta > \theta_1$. The dashed curves are the isotimic contours of constant $\phi^r(\omega)$ that pass through the saddle points and cross the real axis.

$$v_E(\omega) = \frac{\langle S(\omega)\rangle}{\langle u(\omega)\rangle} \,. \tag{10.2.3}$$

Here, $\langle S(\omega)\rangle$ is the time-averaged magnitude of the Poynting vector and $\langle u(\omega)\rangle$ is the time-averaged energy density. *Loudon* [10.11] has shown that for a single resonance *Lorentz* medium the energy transport velocity is given by (cf. Sect. 2.4.2)

$$v_E(\omega) = \frac{c}{n^r(\omega) + n^i(\omega)\omega/\delta} \,. \tag{10.2.4}$$

This then establishes a fundamental connection between the attenuation of the pulse and the physics of the problem. In particular, it is seen that in the mature dispersion regime, the attenuation of $E_s(z,t)$ as the point of observation moves with fixed velocity (i.e. for fixed θ) is approximately the same as the attenuation of a time-harmonic wave with real frequency that has energy velocity equal to the velocity of the point of observation. The same is true of $E_b(z,t)$ for $\theta > \theta_1$. Although this result does not tell us anything about the phase of the field or the amplitude of the exponential term, it does give a description of the main dynamics of the energy of the pulse in physical terms. Moreover, it provides a simpler mathematical algorithm for calculating those dynamics than has been available previously.

This result is now extended to include the rest of the dynamics of the pulse. First, consider the amplitude of the exponential term. If the damping constant δ is not too large, the slowly varying functions of ω appearing in (10.1.16 and

17b) can be approximated by replacing ω_j by ω_{Ej}. This can be seen by applying as follows some general properties of the saddle point locations (see Sect. 6.2). The saddle point ω_s is less than 2δ below the real axis and its real part is greater than $\sqrt{\omega_0^2 + b^2 - \delta^2}$. Even for the highly absorbing medium treated by *Brillouin*, the former value is much smaller than the latter. Hence, the imaginary part of the saddle point can be neglected compared to its real part. The saddle point ω_b is less than δ below the real axis and its real part starts at zero for $\theta = \theta_1$ and increases rapidly towards $\sqrt{\omega_0^2 - \delta^2}$ for increasing θ. Again the numbers are such that even for *Brillouin*'s parameter values, the imaginary part of ω_b can be neglected compared to the real part in slowly varying functions for θ sufficiently large.

Next, notice that the relevant contour of constant $\phi^r(\omega)$ through each saddle point is vertical in the vicinity of the saddle point. This means that the real part of the saddle point ω_j can be approximated by ω_{Ej} for values of δ that are not too large. Again, this approximation is found to be valid for the highly absorbing case treated by *Brillouin*. Combination of these two approximations for each saddle point, it is concluded that the slowly varying functions of ω appearing in (10.1.16 and 17b) can be approximated by replacing ω_j by ω_{Ej} (provided that θ is not too close to θ_1 in the case of (10.1.17b)). Hence, the expressions given in (10.1.16 and 17b) may be respectively approximated as

$$E_s(z,t) \cong \frac{2\sqrt{2\pi c}\tilde{f}(\omega_{Es})}{\sqrt{-z\phi^{(2)}(\omega_{Es})}} \exp\left[\frac{z}{c}\phi(\omega_s)\right] , \tag{10.2.5}$$

and

$$E_b(z,t) \cong \frac{2\sqrt{2\pi c}\tilde{f}(\omega_{Eb})}{\sqrt{-z\phi^{(2)}(\omega_{Eb})}} \exp\left[\frac{z}{c}\phi(\omega_b)\right] , \tag{10.2.6}$$

where (10.2.6) is valid for $\theta \gg \theta_1$.

Greater care must be taken in approximating the exponential function in the above equations because it is a more rapidly varying function of ω. The attenuation has already been expressed in terms of ω_{Ej} in (10.2.1). That expression can be expressed in terms of the index of refraction by noting that for real frequencies, we have

$$\phi^r(\omega) = -\omega n^i(\omega) . \tag{10.2.7}$$

Equation (10.1.1) therefore can be written

$$\phi^r(\omega_j) = -\omega_{Ej} n^i(\omega_{Ej}) = -c\tilde{k}^i(\omega_{Ej}) . \tag{10.2.8}$$

The oscillatory portion is determined by

$$iz\phi^i(\omega)/c = i[-\omega^r t + zn^r(\omega)\omega^r/c - zn^i(\omega)\omega^i/c] \tag{10.2.9}$$

evaluated at the relevant saddle point $\omega = \omega_j$. If we replace ω_j by ω_{Ej} in the index of refraction terms and set $\omega_j^r = \omega_{Ej}$ elsewhere in this expression, it becomes

$$iz\phi^i(\omega_j)/c \cong i[-\omega_{Ej}t + z\tilde{k}^r(\omega_{Ej}) - zn^i(\omega_{Ej})\omega_j^i/c] \ . \tag{10.2.10}$$

With the combination of (10.2.8 and 10), the quantity occurring in the exponential in (10.2.5 and 6) can be written

$$z\phi(\omega_j)/c \cong i[-\omega_{Ej}t + z\tilde{k}(\omega_{Ej})] - izn^i(\omega_{Ej})\omega_j^i/c \ . \tag{10.2.11}$$

In the case $j = b$, (10.2.11) applies when $\theta > \theta_1$ and is accurate only when θ is not too close to θ_1.

The last term in (10.2.11) includes the location of the saddle point ω_j. That term is not very important, however, because it is negligible except when ω_{Ej} is in the absorption band and it contributes only a small phase shift to the field even then. Hence, that term is ignored in our physical model of pulse dynamics with the understanding that the phase of the field that is obtained using the model may be a little off when ω_{Ej} is in the absorption band. In the same spirit, we apply the approximations given in (10.2.6 and 11) with $j = b$ for all $\theta > \theta_1$ with the understanding that they become inaccurate for θ too close to θ_1. The region where this problem exists becomes smaller as δ decreases compared to ω_0. The numerical computations of the field using the physical model that are presented later in this chapter display the effects of these simplifications for highly absorbing and highly transparent cases.

10.2.2 The Non-Oscillatory Contribution

We now turn our attention to the contribution to the propagated field given in (10.1.17a). This contribution is non-oscillatory and is important only for points of observation in the region about the space-time point given by

$$\frac{z}{t} = v_E(0) \ . \tag{10.2.12}$$

The group velocity or phase velocity for time-harmonic waves with zero frequency could be applied in (10.2.12) equally well since all three velocities are equal for zero frequency. Hence, this field is essentially quasi-static. Consider then non-oscillatory electromagnetic fields of the form

$$w(z, t, \tilde{\omega}) = \exp[\tilde{\omega}t - \tilde{\tilde{k}}(\tilde{\omega})z] \tag{10.2.13}$$

where the growth rate $\tilde{\omega}$ is a real constant. Then, it is easy to show that the field is a solution to *Maxwell*'s equations in a *Lorentz* medium if $\tilde{\tilde{k}}(\tilde{\omega})$ is given by

$$\tilde{\tilde{k}}(\tilde{\omega}) = -\mathrm{i}\tilde{k}(\mathrm{i}\tilde{\omega}) = \frac{\tilde{\omega}}{c}\sqrt{1 + \frac{b^2}{\tilde{\omega}^2 + \omega_0^2 + 2\delta\tilde{\omega}}} \tag{10.2.14}$$

which is real.

Since ω_b is purely imaginary for $1 < \theta < \theta_1$, the exponential term appearing in (10.1.17a) is a field of the form given in (10.2.13–14) except that the growth rate is a function of position and time. The equation for the saddle point ω_b (10.1.13)) can be written in the form

$$\omega_b = \mathrm{i}\tilde{\omega}_b \ , \tag{10.2.15}$$

where $\tilde{\omega}_b$ is the real solution to the equation

$$\tilde{v}_G(\tilde{\omega}_b) = \frac{z}{t} \ , \tag{10.2.16}$$

with $\tilde{v}_G(\tilde{\omega})$ defined by

$$\tilde{v}_G(\tilde{\omega}) = \frac{1}{\dfrac{d\tilde{\tilde{k}}(\tilde{\omega})}{d\tilde{\omega}}} \ . \tag{10.2.17}$$

It is then seen from the definition of $\tilde{v}_G(\tilde{\omega})$ that it can be taken to be the group velocity of the non-oscillatory waves given in (10.2.13). This identification does not provide much physical insight, however, because the group velocity is primarily a mathematical object rather than a physical one. In order to connect it with a more physical quantity, consideration is now given to the velocity of energy in the non-oscillatory waves.

Take the velocity of energy in fields of the form given in (10.2.13) to be given by (10.2.3) with the change that the Poynting vector and energy density are not time averaged (since the field is nonoscillatory). With this definition, it is shown in Ref. [10.12] that an electromagnetic field of the form given in (10.2.13–14) has energy velocity $\tilde{v}_E(\tilde{\omega})$ given by

$$\tilde{v}_E(\tilde{\omega}) = \frac{cn(\mathrm{i}\tilde{\omega})Q^2(\tilde{\omega})}{n^2(\mathrm{i}\tilde{\omega})Q^2(\tilde{\omega}) - b^2\tilde{\omega}\delta} \ , \tag{10.2.18}$$

where

$$Q(\tilde{\omega}) = \tilde{\omega}^2 + \omega_0^2 + 2\delta\tilde{\omega} \ , \tag{10.2.19}$$

$$n(\mathrm{i}\tilde{\omega}) = \sqrt{1 + \frac{b^2}{Q(\tilde{\omega})}} \ . \tag{10.2.20}$$

It is shown also in Ref. [10.12] that the group velocity of the non-oscillatory waves is given by

$$\tilde{v}_G(\tilde{\omega}) = \frac{cn(\mathrm{i}\tilde{\omega})Q^2(\tilde{\omega})}{n^2(\mathrm{i}\tilde{\omega})Q^2(\tilde{\omega}) - b^2\tilde{\omega}\delta - b^2\tilde{\omega}^2} \ . \tag{10.2.21}$$

Comparison of (10.2.18 and 21) shows that the two velocities differ only by an extra term $-b^2\tilde{\omega}^2$ in the denominator of (10.2.21). This extra term is negligible for all growth rates that satisfy the inequality

$$\tilde{\omega} \ll \omega_0 \ ; \tag{10.2.22}$$

notice that this is a sufficient condition but that it is far from necessary. Since the attenuation with z of the non-oscillatory waves increases with increasing growth rate, it is clear that for z sufficiently large, the non-oscillatory waves which do not satisfy (10.2.22) will be negligible compared to those that do. In particular, it is shown in Ref. [10.12] that if z satisfies the inequality

$$z > \frac{K}{r^2}\sqrt{1 + \frac{b^2}{\omega_0^2}\frac{4c(\omega_0 + \delta)^2}{b^2\delta}} \ , \tag{10.2.23}$$

then $w(z, t, \tilde{\omega})/w(z, t, 0) < \exp(-K)$ for $\tilde{\omega} \leqslant r\omega_0$, where K and r are arbitrary positive constants which can be chosen to give as good an approximation as desired when neglecting the non-oscillatory waves that do not satisfy (10.2.22) well enough to approximate $\tilde{v}_G(\tilde{\omega})$ by $\tilde{v}_E(\tilde{\omega})$. For *Brillouin*'s parameter values given in (10.1.4) and with $K = 2$ and $r = 0.1$, (10.2.23) gives $z > 0.012$ cm. It is shown by a numerical example in the next section, that the approximation is useful even for z much smaller than this.

Hence, for sufficiently large z, a good approximation of the saddle point location can be obtained by taking $\tilde{\omega}_b$ to be the solution to the equation

$$\tilde{v}_E(\tilde{\omega}) = \frac{z}{t} \tag{10.2.24}$$

that is closest to the saddle point.

One more approximation is useful for the formulation of the physical model. It follows from (6.2.34) that θ_1 cana be approximated by θ_0 defined by

$$\theta_0 = \frac{c}{v_E(0)} = \frac{c}{\tilde{v}_E(0)} = n(0) \ , \tag{10.2.25}$$

which is simply the index of refraction for static fields. The validity of the approximation decreases with increasing δ but is still very good even for the parameter values used by *Brillouin*. (For those parameter values, one obtains $\theta_0 = 1.500$ and $\theta_1 = 1.503$.) The result implies that the second precursor field $E_b(z, t)$ changes from the non-oscillatory form with growth rate equal to zero to the time-harmonic form with frequency equal to zero at the point of observation which is moving with velocity $v_E(0) = \tilde{v}_E(0)$.

10.3 Physical Model of Pulse Dynamics

10.3.1 The Nonuniform Physical Model

We now present the physical model of pulse dynamics based on the above results. As z tends to infinity with fixed θ, the electric field $\mathscr{E}(z,t)$ can be expressed as the real part of a complex quantity $E(z,t)$ which, for δ much smaller than ω_0 and b, can be approximated by

$$E(z,t) \approx E_{TH}(z,t) + E_{QS}(z,t) \ , \tag{10.3.1}$$

where

$$E_{TH}(z,t) = \sum_j \frac{2\sqrt{2\pi c}\tilde{f}(\omega_{Ej})}{\sqrt{-z\phi^{(2)}(\omega_{Ej})}} \exp\{\mathrm{i}[z\tilde{k}(\omega_{Ej}) - \omega_{Ej}t]\} \ , \tag{10.3.2}$$

and

$$E_{QS}(z,t) = \frac{\sqrt{2\pi c}\tilde{f}(\mathrm{i}\tilde{\omega}_E)}{\sqrt{-z\phi^{(2)}(\mathrm{i}\tilde{\omega}_E)}} \exp[\tilde{\omega}_E t - \tilde{\tilde{k}}(\tilde{\omega}_E)z] \ . \tag{10.3.3}$$

Here, the frequency values ω_{Ej} are defined to be the nonnegative, real solutions to

$$v_E(\omega_{Ej}) = \frac{z}{t} \equiv \frac{c}{\theta} \ , \tag{10.3.4}$$

and $\tilde{\omega}_E$ is defined to be the positive solution to

$$\tilde{v}_E(\tilde{\omega}_E) = \frac{z}{t} \equiv \frac{c}{\theta} \ . \tag{10.3.5}$$

The quantity $E_{TH}(z,t)$ given in (10.3.2) is the *time-harmonic* component discussed Sect. 10.2.1. It is shown in the following, that the sum in (10.3.2) includes only one term, the *Sommerfeld* precursor $E_s(z,t)$, for $1 < \theta < \theta_0$ and includes two terms, the *Sommerfeld* precursor $E_s(z,t)$ and the *Brillouin* precursor $E_b(z,t)$, for $\theta \geqslant \theta_0$. The quantity $E_{QS}(z,t)$ is the *non-oscillatory* component discussed in Sect. 10.2.2.

We refer to the above results as a physical model because they can be used to describe the local dynamics of the pulse in physical terms. They are similar to the mathematical results that lead to the group-velocity description valid for lossless, gainless media but are different in three respects:

a. the non-oscillatory contribution is included in addition to the time-harmonic contribution,
b. the energy velocity is used to determine the pulse dynamics in place of the group velocity,

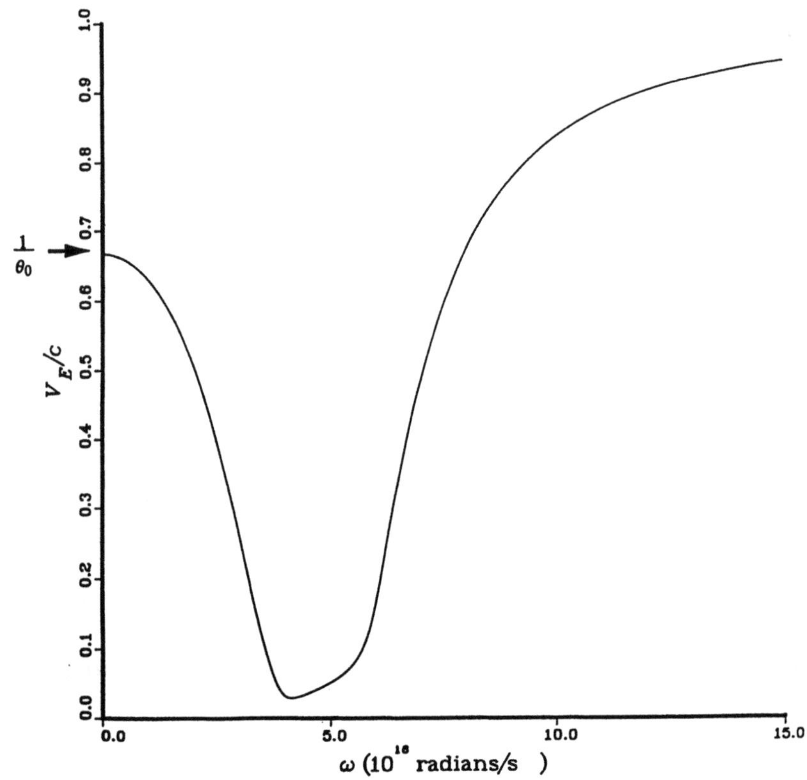

Fig. 10.3. Frequency dispersion of the normalized energy velocity of a monochromatic wave of carrier frequency ω in a single resonance *Lorentz* medium.

c. the pulse dynamics are strongly affected by the relative attenuation of the various time-harmonic and non-oscillatory contributions.

It follows from (10.3.1–5) that the primary quantities that determine the local dynamics of the pulse are the energy velocity and attenuation coefficient α of time-harmonic waves as functions of frequency and the energy velocity and attenuation coefficient $\tilde{\alpha}$ of non-oscillatory waves as a function of growth rate. Plots of these functions for a medium with *Brillouin*'s parameter values (specified in (10.1.4)) are given respectively in Figs. 10.3–10.6. The energy velocities were computed by applying (10.2.4 and 18). The attenuation coefficients of the waves are the rates of exponential decay of the waves as z increases with constant θ. For time-harmonic waves with real frequencies, α is equal to the imaginary part of the propagation constant $\tilde{k} = \beta + i\alpha = \omega n(\omega)/c$, with the index of refraction given in (10.1.3). For the non-oscillatory waves, $\tilde{\alpha}$ is given by

$$\tilde{\alpha} = \tilde{\tilde{k}}(\tilde{\omega}_E) - \frac{\tilde{\omega}_E \theta}{c} . \qquad (10.3.6)$$

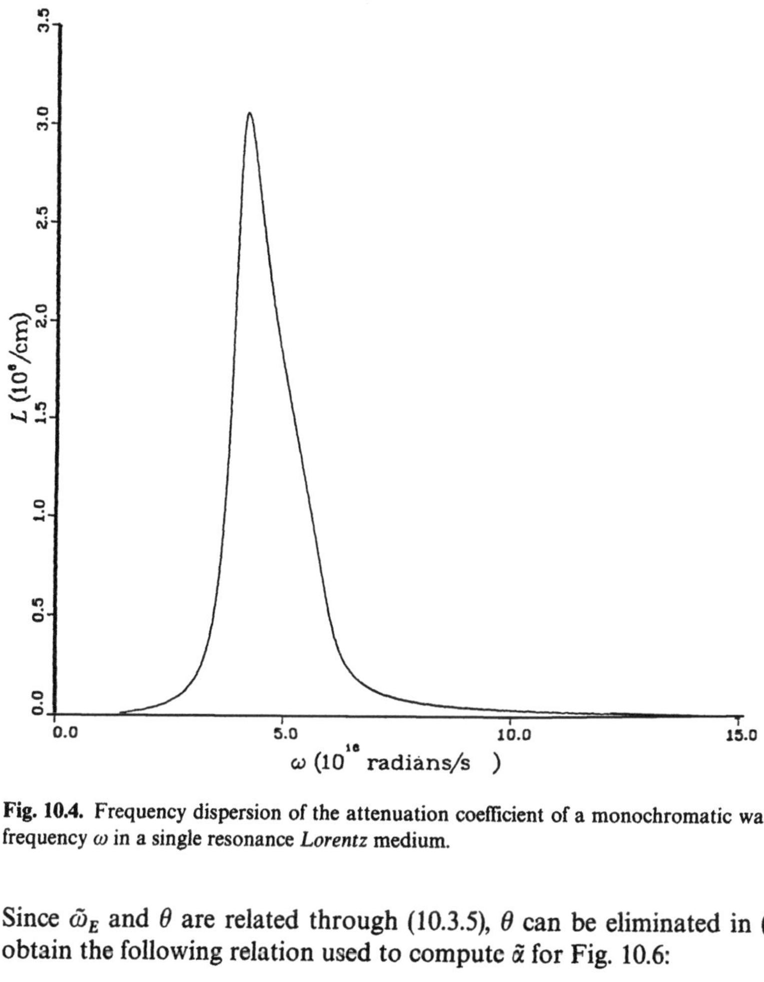

Fig. 10.4. Frequency dispersion of the attenuation coefficient of a monochromatic wave of carrier frequency ω in a single resonance *Lorentz* medium.

Since $\tilde{\omega}_E$ and θ are related through (10.3.5), θ can be eliminated in (10.3.6) to obtain the following relation used to compute $\tilde{\alpha}$ for Fig. 10.6:

$$\tilde{\alpha} = \tilde{\tilde{k}}(\tilde{\omega}_E) - \frac{\tilde{\omega}_E}{\tilde{v}_E(\tilde{\omega}_E)}, \tag{10.3.7}$$

where $\tilde{\tilde{k}}(\tilde{\omega})$ is given by (10.2.14) and $\tilde{v}_E(\tilde{\omega})$ is given by (10.2.18).

A qualitative description of all of the main features of pulse dynamics can be obtained by studying these plots. We first notice from Fig. 10.3 (by recalling that $\theta_0 = 1.5$ in a *Lorentz* medium with *Brillouin*'s parameter values) that for $1 < \theta < \theta_0$, there is only one solution to (10.3.4) so that the sum appearing in (10.3.2) includes only one term. The frequency ω_E of this term is large for θ near 1 and decreases monotonically with increasing θ. From Fig. 10.4, it is seen that as the frequency decreases from a large value, the attenuation increases. This means that this high-frequency, quasi-time-harmonic term decreases in frequency and amplitude as θ increases, in agreement with the asymptotic description of the *Sommerfeld* precursor (see Sect. 7.2.6).

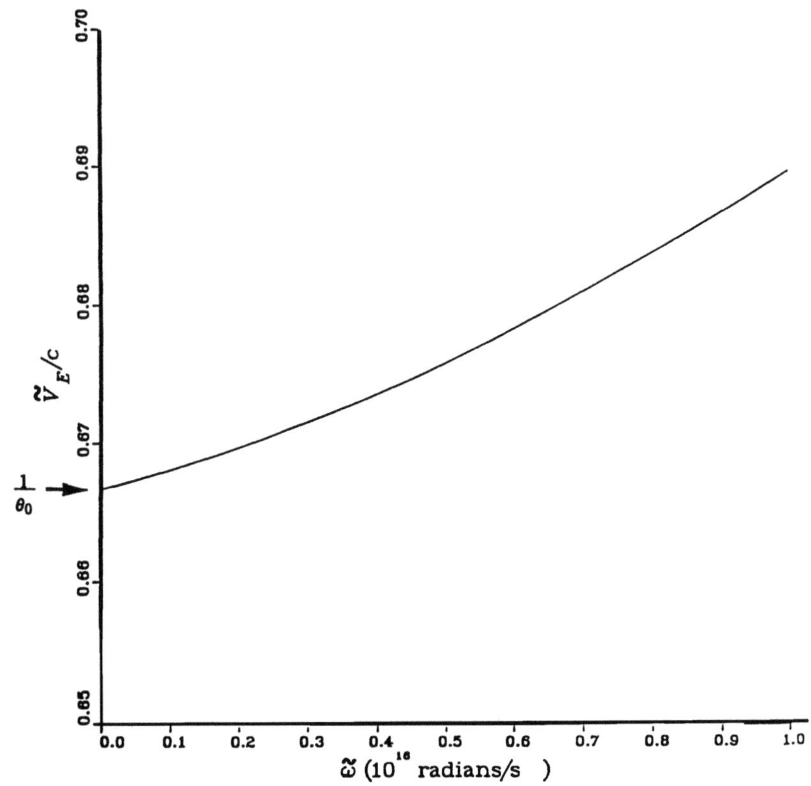

Fig. 10.5. Normalized energy velocity of non-oscillatory wave components.

Continuing along with the physical description, it is noticed in Fig. 10.5 that there is one positive solution $\tilde{\omega}_E$ to (10.3.5) with $1 < \theta < \theta_0$. The growth rate $\tilde{\omega}_E$ decreases with increasing θ, tending towards 0 as θ approaches θ_0. From Fig. 10.6, it is seen that the attenuation with z of the non-oscillatory contribution is large for large growth rate but gradually decreases to zero as $\tilde{\omega}_E$ approaches 0. Hence, the non-oscillatory contribution is negligible compared to the *Sommerfeld* precursor for small $\theta > 1$ but gradually increases until it dominates the *Sommerfeld* precursor field as θ approaches θ_0 from below. This marks the arrival of the *Brillouin* precursor. Finally, notice from Fig. 10.5 that there is no positive solution to (10.3.5) for $\theta \geqslant \theta_0$. Hence, the non-oscillatory contribution no longer contributes. Notice that a non-oscillatory contribution with $\tilde{\omega}_E = 0$ at $\theta = \theta_0$ has been disallowed by including only positive solutions to (10.3.5). This has been done so as to avoid the inclusion of the zero frequency solution twice. It is included in the time-harmonic contribution since all nonnegative solutions of (10.3.4) have been included.

Returning to Fig. 10.3, notice that for $\theta \geqslant \theta_0$, there are now two nonnegative solutions to (10.3.4). The first is a high-frequency solution which is the continuation of the *Sommerfeld* precursor. The new solution is a low-

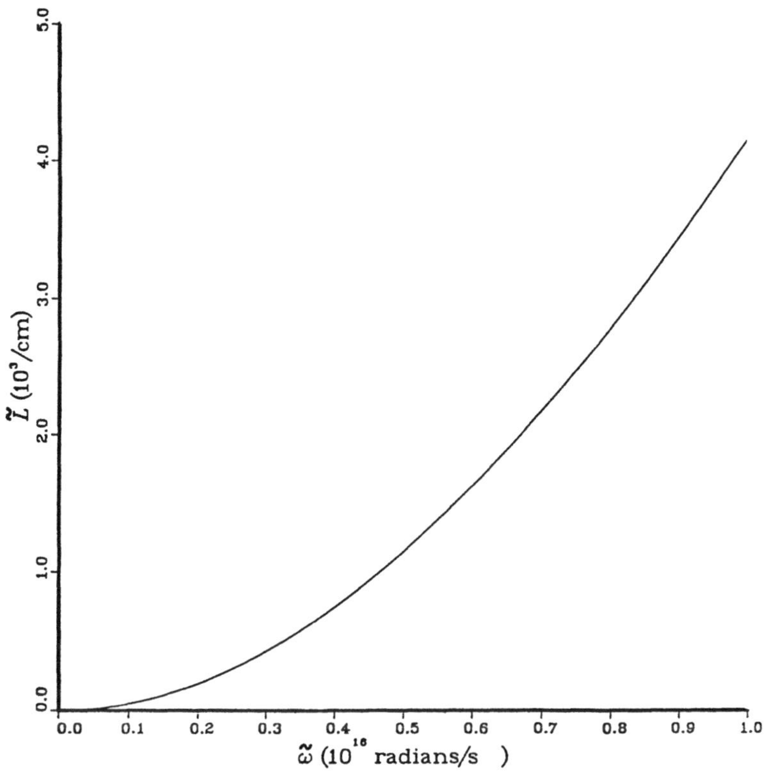

Fig. 10.6. Attenuation coefficient of non-oscillatory wave components.

frequency solution with frequency which starts at zero for $\theta = \theta_0$ and increases with increasing θ. Consideration of Fig. 10.4 shows that the attenuation of this new wave is much less than that of the high-frequency contribution. Hence, this wave dominates the *Sommerfeld* precursor. Fig. 10.4 shows also that the attenuation of this wave increases with increasing ω causing the wave to decrease in amplitude with increasing θ. Hence, this contribution has increasing frequency and decreasing amplitude with increasing θ, in agreement with the asymptotic description of the *Brillouin* precursor (see Sect. 7.3.6).

In addition to providing a description of the qualitative pulse behavior in physical terms, the physical model gives approximate analytical formulas which predict the propagated pulse dynamics quantitatively without requiring the evaluation of the saddle-point locations in the complex plane. Of course, the model does require the solution of (10.3.4–5), which are transcendental equations, but these equations are simpler to deal with than the saddle-point equations because they involve only real quantities.

In order to investigate the accuracy of these approximate expressions, they have been evaluated numerically for the case of the delta-function pulse. Recall that the delta-function pulse is the field that results when the boundary value

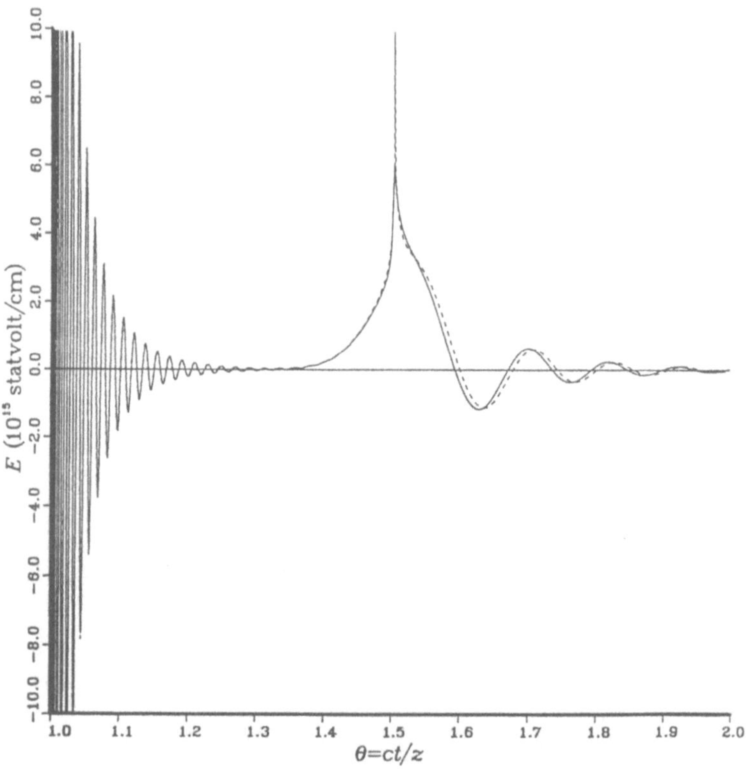

Fig. 10.7. Nonuniform results for the propagated electric field due to an input delta-function pulse. The solid curve is the result of the physical model while the dashed curve is the result of the nonuniform asymptotic theory.

$f(t)$ occurring in (10.1.5) is given to be the Dirac delta function $A\delta(t)$ with $A = 1$ statvolt-s/cm (in cgs units). In this case, $\tilde{f}(\omega) = 1$ statvolt-s/cm. For the numerical computations, *Brillouin*'s parameter values for the medium as given in (10.1.4) were used and the propagation distance $z = 10^{-4}$ cm was chosen. Equations (10.3.4–5) were than solved numerically using *Mueller*'s method [10.12].

The numerical results of the physical model are plotted as a solid curve in Fig. 10.7, superimposed on a dotted curve, which is a plot of the nonuniform asymptotic result given by (10.1.13–16) for the same parameter values. In order to evaluate the asymptotic result, *Mueller*'s method was used to solve (10.1.13) in order to determine the precise saddle point locations numerically. It is apparent from the figure that the accuracy of the physical model is very good. The main discrepancy is a minor shift in the phase of the *Brillouin* precursor as discussed at the end of Sect. 10.2.1.

The scale in Fig. 10.7 was chosen so that the transition between the two precursors can be seen clearly. Since the curves are off scale for small θ in that figure, the same two quantities are replotted for small θ with an appropriate

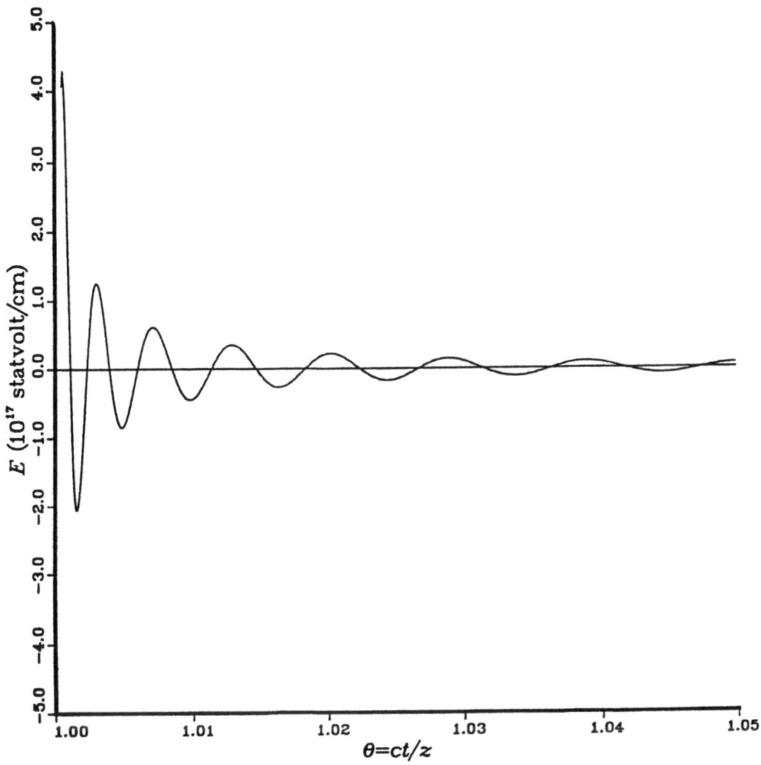

Fig. 10.8. Nonuniform results for the propagated electric field due to an input delta-function pulse for values of θ near 1. The solid curve is the result of the physical model while the dashed curve is the result of the nonuniform asymptotic theory. The dashed curve is barely visible because it falls almost exactly on the solid curve.

scale in Fig. 10.8. The results of the physical model agree so well with the asymptotic results that the two curves are indistinguishable making it impossible to see the dotted curve.

The large peak that occurs in both the physical-model results and asymptotic results in Fig. 10.7 for θ near θ_0 is a consequence of the nonuniform nature of the results as explained in Sect. 10.1. This is an artifact of the asymptotic analysis which makes the results invalid in that region. In order to obtain results that are useful there, it is necessary to employ the uniform asymptotic approximation of the field. This is done in the following section.

10.3.2 The Uniform Physical Model

The asymptotic results that have been used so far are nonuniform in the vicinity of two values of θ:

a. $\theta = 1$, which corresponds to the arrival of the *Sommerfeld* precursor field, and
b. $\theta = \theta_1$, which occurs during the arrival of the *Brillouin* precursor field.

This means that in order for the results to be useful approximations for large z, the propagation distance must be taken larger and larger as θ approaches one of these critical values. Furthermore, the functional form of the results are different for θ on opposite sides of a critical value. These difficulties may be removed from the physical model by using the appropriate uniform asymptotic approximations described in Chap. 5 and applied in Chap. 7. The field is then expressed in terms of special functions (e.g., the Bessel and Airy functions) which are more complicated than the exponential functions occurring in the nonuniform expressions. The arguments of these functions involve the same saddle points as applied in the nonuniform analysis. As the value of θ tends away from the critical values, the uniform results tend asymptotically to the same formulas as given in the nonuniform results.

In order to utilize the uniform asymptotic results of Chap. 7 in the physical model, we simply replace the saddle points occurring in the uniform asymptotic expressions with the approximations used in the nonuniform physical model, i.e. the corresponding real solutions to (10.3.4–5). Because the results involve Bessel functions and Airy functions instead of quasi-time-harmonic waves and non-oscillatory exponentially growing waves, the physical interpretation of the results is not as obvious as with the nonuniform results. Nevertheless, such functions frequently arise in the theory of wave propagation and can be considered to be physical waves. Moreover, in the uniform physical model, the arguments of these functions are real, involving real frequencies and growth rates which are clearly connected with the physics of time-harmonic waves and non-oscillatory waves in the medium through (10.3.4–5).

For the case of the delta function pulse (the only canonical problem considered here), we do not take the trouble to make the physical model uniform in the vicinity of the point $\theta = 1$ because the nonuniform model is good for θ quite close to 1. This can be seen as follows. Figure 10.9 presents as a solid curve, the results of the numerical evaluation of the exact integral representation of the wave field as given in (10.1.6) for the same parameter values as used for Fig. 10.8. The algorithm used to evaluate the integral numerically for this and subsequent plots is described in Ref. [10.13]. As discussed in that reference, the algorithm begins to produce artifacts in the results as θ approaches 1 from above for the delta-function pulse. This is a consequence of the fact that the integral itself is divergent at $\theta = 1$. In Fig. 10.9, it is seen that the curve has the same form as the curve in Fig. 10.8 but with a high-frequency ripple superimposed. It has been verified [10.12] that the ripple is an artifact of the numerical algorithm by showing that its frequency changed when the numerical value of the parameter k used in the algorithm was changed whereas the rest of the curve remained unchanged. In order to compare the results of the physical model with the numerical-integration results, the results of the nonuniform physical model are

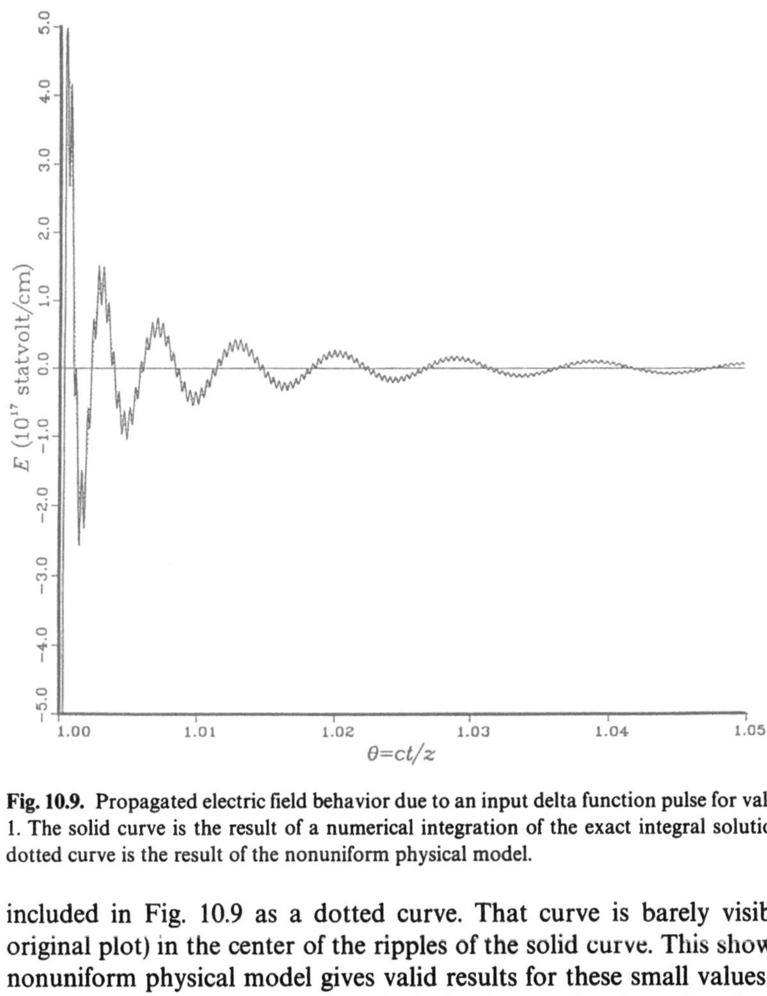

Fig. 10.9. Propagated electric field behavior due to an input delta function pulse for values of θ near 1. The solid curve is the result of a numerical integration of the exact integral solution, while the dotted curve is the result of the nonuniform physical model.

included in Fig. 10.9 as a dotted curve. That curve is barely visible (in the original plot) in the center of the ripples of the solid curve. This shows that the nonuniform physical model gives valid results for these small values of θ for θ greater than 1.00055 where the calculation began. For θ smaller than this, one would need to make the physical model uniform in the vicinity of the point $\theta = 1$.

We now proceed to make the physical model uniform in the vicinity of the point $\theta = \theta_1$. We begin with (10.1.14–17) with the exception that the expressions in (10.1.17) are replaced by the uniform asymptotic approximation obtained in Sect. 7.3. The asymptotic approximation is made uniform in the vicinity of $\theta = \theta_1$ by replacing the expression given in (10.1.17) for $A_b(z,t)$ with

$$\begin{aligned} A_b(z,t) = \frac{\mathrm{i}}{2}\exp(z\alpha_0/c)\{&(c/z)^{1/3}\exp(-\mathrm{i}2\pi/3)[\tilde{f}(\omega_+)h_+ + \tilde{f}(\omega_-)h_-] \\ &\cdot Ai[|\alpha_1|(z/c)^{2/3}] + (c/z)^{2/3}\exp(-\mathrm{i}4\pi/3)[\tilde{f}(\omega_+)h_+ - \tilde{f}(\omega_-)h_-] \\ &\cdot Ai^{(1)}[|\alpha_1|(z/c)^{2/3}]/\alpha_1^{1/2}\} \ . \end{aligned} \tag{10.3.8}$$

In Equation (10.3.8), $Ai(x)$ represents the Airy function of argument x. For $1 \leqslant \theta \leqslant \theta_1, \omega_+$ represents the upper near saddle point ω_b which moves down the positive real axis as θ increases whereas ω_- represents the lower near saddle point which moves up the negative imaginary axis as θ increases. These two saddle points coalesce when θ reaches θ_1. For $\theta \geqslant \theta_1, \omega_+$ represents the saddle point ω_b as it moves into the fourth quadrant of the ω plane with increasing θ and ω_- represents $-\omega_b^*$. The other new quantities occurring in (10.3.8) are defined by (cf. Sect. 7.3)

$$\alpha_0 = \frac{1}{2}[\phi(\omega_+) + \phi(\omega_-)] \ , \tag{10.3.9}$$

$$\alpha_1^{1/2} = \{3/4[\phi(\omega_+) - \phi(\omega_-)]\}^{1/3} \ , \tag{10.3.10}$$

$$h_\pm = \left[\mp \frac{2\alpha_1^{1/2}}{\phi^{(2)}(\omega_\pm)}\right]^{1/2} \tag{10.3.11}$$

for $\theta \neq \theta_1$. At the critical value θ_1 of θ at which the two saddle points coalesce, these coefficients take on the limiting values

$$\lim_{\theta \to \theta_1} h_\pm = \left[-\frac{2}{\phi^{(3)}(\omega_1)}\right]^{1/3} \equiv h_1 \ , \tag{10.3.12}$$

$$\lim_{\theta \to \theta_1} [\tilde{f}(\omega_+)h(\omega_+) + \tilde{f}(\omega_-)h(\omega_-)] = 2\tilde{f}(\omega_1)h_1 \ , \tag{10.3.13}$$

$$\lim_{\theta \to \theta_1} [\tilde{f}(\omega_+)h(\omega_+) - \tilde{f}(\omega_-)h(\omega_-)]/\alpha_1^{1/2} = 2\tilde{f}^{(1)}(\omega_1)h_1^2 \ , \tag{10.3.14}$$

and α_0 at θ_1 is given by (10.3.9). The specification of the branches used to make the multivalued functions occurring in the above equations single-valued is given in Sect. 5.2.2.

The uniform physical model is obtained from the above uniform asymptotic approximation by making the same approximations as were used in the non-uniform asymptotic approximation to obtain the nonuniform physical model. The critical value θ_1 is approximated by θ_0 and the saddle points are approximated by using the appropriate solutions to the energy velocity equations (10.3.4 and 5). In particular, for $1 \leqslant \theta \leqslant \theta_0, \omega_+$ is approximated by $i\tilde{\omega}_+$ and ω_- is approximated by $i\tilde{\omega}_-$, where $\tilde{\omega}_\pm$ are the real solutions to (10.3.5) such that $\tilde{\omega}_+ \geqslant \tilde{\omega}_-$. Similarly, for $\theta \geqslant \theta_0, \omega_+$ is approximated by the low-frequency real solution to (10.3.4) and ω_- is approximated by minus that value. As in the nonuniform physical model, ω_s is approximated in (10.1.16) by the high-frequency real solution to (10.3.4).

As θ moves away from θ_0 in either direction, the uniform physical model approaches the nonuniform physical model asymptotically. Hence, we can apply the nonuniform model for all θ except in the vicinity of θ_0 where we apply the uniform model. To examine the validity of the uniform physical model, the

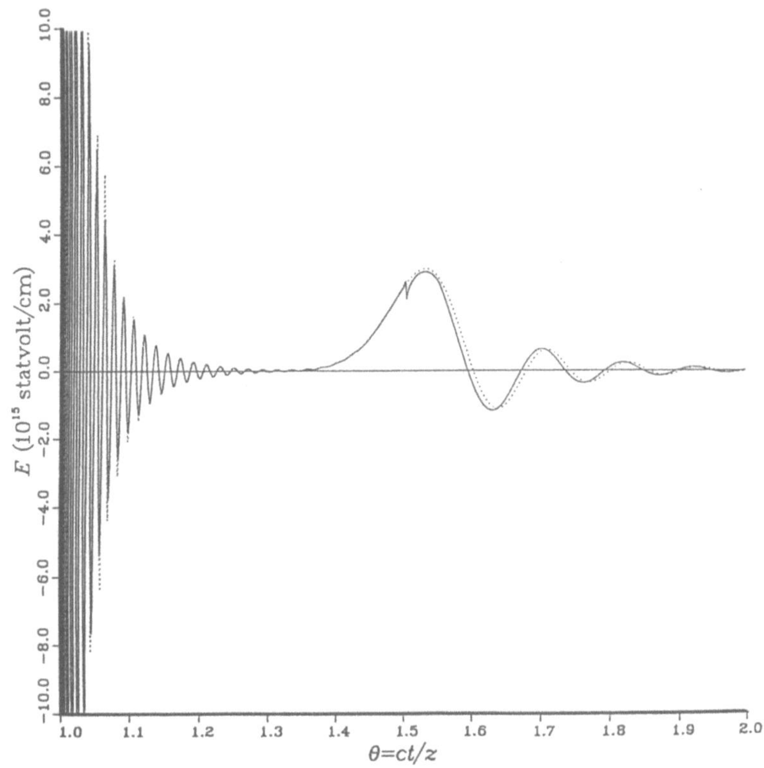

Fig. 10.10. Comparison of the uniform physical model (solid curve) with the numerical integration of the exact integral solution (dotted curve) for the propagated electric field due to an input delta function pulse in a highly absorbing *Lorentz* medium.

computations that were used to obtain Fig. 10.7 were repeated with the change that the uniform physical model was applied in the region $1.43 < \theta \leqslant 1.55$. Since the expressions for some of the coefficients in (10.3.8) become indeterminate for $\theta = \theta_0$, the limiting expressions given in (10.3.12–14) were used for $1.45 \leqslant \theta \leqslant 1.503$. The results are presented as the solid curve superimposed on the results of numerical integration of the exact integral solution (10.1.6) shown as a dotted curve in Fig. 10.10. The discontinuity in the solid curve at $\theta = \theta_0$ and its departure from the dotted curve for θ larger than but near θ_0 is a result of the fact that the approximation of the saddle point location by a real solution to the energy velocity equation (10.3.4) is not very good for θ in this region, as discussed in Sect. 10.2.1. The small shift in phase between the two curves for θ larger than θ_0 has been discussed already in Sect. 10.3.1. Apart from these rather minor discrepancies, the results of the uniform physical model are in excellent agreement with the exact integral solution for the field for θ varying over the entire range shown.

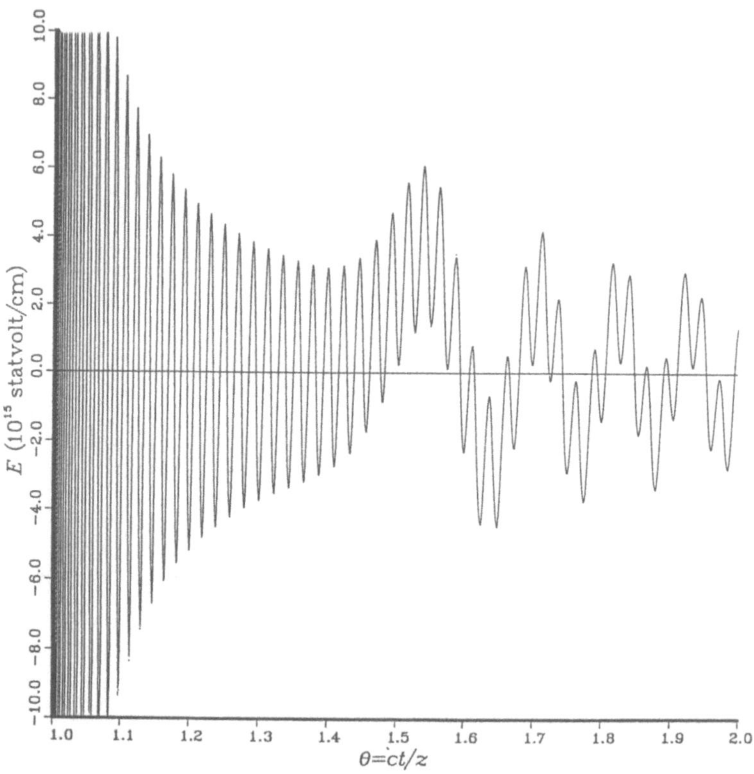

Fig. 10.11. Comparison of the uniform physical model (solid curve) with the numerical integration of the exact integral solution (dotted curve) for the propagated electric field due to an input delta function pulse in a highly transparent *Lorentz* medium.

As has already been mentioned, the medium parameter values chosen for our numerical computations correspond to a medium with very high absorption, much too high for the material to be considered to be transparent. Since the approximations become better as the overall material absorption decreases, these numerical results can be considered to be a worst case test of the validity of the physical model for describing pulse propagation in absorbing dielectrics of practical interest. In order to verify the utility of the physical model in the opposite extreme of a highly transparent medium, the same computations have been repeated with the same parameter values except that δ is now taken to be $1.0\ s^{-1}$ for both the uniform physical model computations and the numerical evaluation of the exact integral solution. The results are shown in Fig. 10.11. The physical model agrees so well with the exact solution that the two curves can be distinguished from each other only at a few isolated points.

One might expect the group-velocity description to be applicable for obtaining the results plotted in Fig. 10.11 since that case is nearly lossless. Indeed, it can be shown that the energy velocity given in (10.2.4) approaches the group velocity for all frequencies for which the medium is lossless as δ approaches zero. The medium is not lossless for all frequencies, however, even when δ is identically zero because the index of refraction given by (10.1.3) is purely imaginary for ω^2 slightly larger than ω_0^2. As a result, the group-velocity description is not strictly valid for the *Lorentz* medium even when δ is identically zero. Nevertheless, the group velocity description may still be applied to that case in order to see what it would yield. The results were found [10.12] to be identical to those shown in Fig. 10.11 except in the region $1.35 \leqslant \theta \leqslant 1.5$ where the field failed to rise as it does in Fig. 10.11. It is apparent that all that is missing in the group-velocity description in this case is the non-oscillatory contribution.

10.4 Summary and Conclusions

In this concluding chapter a simple model has been presented which accurately describes in physical terms all of the dynamics of an electromagnetic pulse as it propagates through a single resonance *Lorentz* medium in the mature dispersion regime. Its accuracy decreases as the absorption increases but the accuracy is still excellent even for media too absorbing to be of practical interest as absorbing transparent materials. According to the model, once the pulse propagates far enough to be in the mature-dispersion regime, it separates into two distinct components at any given space-time point. Each component is either a quasi-monochromatic wave of the form

$$\exp\{i[\tilde{k}(\omega)z - \omega t)]\}$$

with real frequency ω which is a slowly varying function of position and time or a non-oscillatory wave of the form

$$\exp[\tilde{\omega}t - \tilde{\tilde{k}}(\tilde{\omega})z]$$

with real growth rate $\tilde{\omega}$ which is a slowly varying function of position and time. The propagation constant in the quasi-monochromatic wave is $\tilde{k}(\omega) = \underset{\sim}{\omega} n(\omega)/c$ where the index of refraction $n(\omega)$ is given in (10.1.3) and the quantity $\tilde{\tilde{k}}(\tilde{\omega})$ in the non-oscillatory waves is given in (10.2.14). The frequencies of the quasi-monochromatic components satisfy

$$v_E(\omega) = \frac{z}{t}, \tag{10.4.1}$$

where $v_E(\omega)$ is the velocity of energy in a monochromatic wave with real frequency ω. The growth rates of the non-oscillatory components satisfy

$$\tilde{v}_E(\tilde{\omega}) = \frac{z}{t}, \tag{10.4.2}$$

where $\tilde{v}_E(\tilde{\omega})$ is the velocity of energy in a non-oscillatory wave with growth rate $\tilde{\omega}$.

The easiest way to describe the components is to consider the point of observation to be moving with a fixed velocity $v = z/t$. Then according to (10.4.1–2), the components that contribute at a specific point of observation are the ones with energy velocities equal to the velocity of that point. If v satisfies the inequality

$$v < \frac{c}{n(0)}, \tag{10.4.3}$$

where $n(0)$ is the index of refraction at zero frequency, then there is only one real solution to each of the above two equations. Hence, in the region of space-time where (10.4.3) is satisfied, the pulse has one quasi-monochromatic component (the *Sommerfeld* precursor) and one non-oscillatory component (the initial rise of the *Brillouin* precursor). When v is larger than $c/n(0)$, then there is no real solution to (10.4.2) and there are two real solutions to (10.4.1). Hence, in that region of space-time, the pulse has two quasi-monochromatic components (the *Sommerfeld* and *Brillouin* precursors) and no non-oscillatory component. In the transition region where v is approximately equal to $c/n(0)$, the functional form of the field is more complicated but its dynamics are still controlled by the solutions to (10.4.1–2).

The variations of the precursor fields with changes in v are determined by the variation of the energy velocities and attenuation rates of monochromatic waves and non-oscillatory waves as functions of frequency and growth rate respectively. These latter variations are depicted by the simple smooth curves plotted in Figs. 10.3–6. They are determined by the physics of energy flow and absorption in the medium through the interaction of the electromagnetic field with the molecules according to the microscopic model of the material.

After having described the field behavior as above, one can now describe it from a different viewpoint. If one observes the pulse in a small region of space at fixed time in the mature-dispersion regime, it will be found to be made up of the superposition of a quasi-monochromatic component of some real frequency and another component which is either a quasi-monochromatic wave with a lower real frequency or a non-oscillatory wave with real growth rate. These two components will have the same energy velocity. Moreover, if these components are then followed through space as time progresses, they move together with that velocity and each one will decay exponentially as it propagates with the attenuation coefficient corresponding to that component. This will be true throughout the pulse with the components with higher energy velocities being up ahead of those with lower velocities and the separation between the various components will be increasing with time. As a consequence of this spreading of the components, the energy in any given frequency (or growth rate) interval will be spreading out over an increasing region of space causing the component amplitudes to decrease (in addition to the exponential attenuation) by a factor of one over the square root which occurs in the denominator of (10.3.2–3).

The accuracy of the physical model is illustrated in Fig. 10.10 for a highly absorbing medium and Fig. 10.11 for a nearly nonabsorbing medium by superimposing the result of the model on the results of numerical evaluation of the exact integral solution. In both cases, the agreement between the model and exact results is excellent. Since these computations were made for a propagation distance of only 10^{-4} cm, it follows that the pulse reaches the mature-dispersion regime after a very short propagation distance. Hence, the physical model is applicable nearly everywhere in the medium (i.e. at least for all z greater than or equal to 10^{-4} cm).

The agreement between the physical model and the exact solution is especially striking in Fig. 10.11 where the plots describe a very complicated and strange looking curve. If one had only the results of the numerical integration, the strange appearance of the curve might lead one to question the validity of either the numerical algorithm used or the computer code itself. But since both the physical model and the numerical integration give precisely the same curve even though they use very different mathematical algorithms and computer codes, there can be no doubt about the validity of both approaches. Moreover, without the physical model, one would be unable to explain the peculiar features of the curve. The physical model, however, reveals the source of this behavior. Because the attenuation of the monochromatic waves is so low in this case, the high-frequency component does not decay rapidly like it does for the case depicted in Fig. 10.10. Hence, it interferes, first with the non-oscillatory component and later with the low-frequency quasi-monochromatic component, to give the observed complicated behavior.

The physical model that has been presented is intuitively satisfying from a physical point of view. It seems like the natural extension of the group-velocity description known to be valid for lossless, gainless dispersive media. In fact, the group-velocity description can be considered to be an energy-velocity description since, under very general conditions, it has been shown that the energy and the group velocities are identical in lossless, gainless media [10.14, 15]. Hence, an energy-velocity description of pulse dynamics is now available for all lossless, gainless dispersive media and for one dispersive medium with loss, the single resonance *Lorentz* medium. At this time, we are not aware of a physical description of pulse dynamics in any other dispersive system that includes loss or gain. But it is tempting to speculate that the energy-velocity model is valid for very general dispersive media with small (or zero) loss or gain because the model appears to account for the effects of small loss or gain in terms of physical principles that transcend the details of the particular dispersion relation of the medium. With this in mind, it may be possible to avoid a lengthy and complicated asymptotic analysis (such as that given in Chap. 6–9) when confronted with the need to determine the functional behavior of the pulse in the mature-dispersion regime for a new dispersive system. If the velocity of energy in monochromatic and non-oscillatory exponentially-growing waves in that system can be determined, then one could try the energy-velocity model given here and compare it with numerical integration of the exact integral solution (10.1.6–19) with the appropriate dispersion relation.

References

Chapter 1

1.1 H.M. Nussenzveig: *Causality and Dispersion Relations* (Academic, New York 1972) Chap. 1

1.2 A. Sommerfeld: Ann. Phys. (Leipzig) *44*, 177–202 (1914)

1.3 L. Brillouin: Ann. Phys. (Leipzig) *44*, 203–240 (1914)

1.4 L. Brillouin: *Wave Propagation and Group Velocity* (Academic, New York 1960)

1.5 K.E. Oughstun: Propagation of optical pulses in dispersive media. Ph.D. Thesis, University of Rochester (1978)

1.6 K.E. Oughstun, G.C. Sherman: J. Opt. Soc. Am. B *5*, 817–849 (1988)

1.7 K.E. Oughstun, G.C. Sherman: J. Opt. Soc. Am. A *6*, 1394–1420 (1989)

1.8 K.E. Oughstun, G.C. Sherman: Phys. Rev. A *41*, 6090–6113 (1990)

1.9 S. Shen, K.E. Oughstun: J. Opt. Soc. Am. B *6*, 948–963 (1989)

1.10 K.E. Oughstun, J.E.K. Laurens: Radio Sci. *26*, 245–258 (1991)

1.11 K.E. Oughstun: Proc. IEEE *79*, 1379–1390 (1991)

1.12 J.A. Stratton: *Electromagnetic Theory* (McGraw-Hill, New York 1941) Sect. 5.18

1.13 J.D. Jackson: *Classical Electrodynamics* 2nd edn. (Wiley, New York 1975) Chap. 7

1.14 F.W.J. Olver: SIAM Rev. *12*, 228–247 (1970)

1.15 C. Chester, B. Friedman, F. Ursell: Proc. Cambridge Phil. Soc. *53*, 599–611 (1957)

1.16 N. Bleistein: Commun. Pure Appl. Math. *19*, 353–370 (1966)

1.17 N. Bleisten: J. Math. Mech. *17*, 533–559 (1967)

1.18 R.A. Handelsman, N. Bleistein: Arch. Rational Mech. Analysis *35*, 267–283 (1969)

1.19 L.B. Felsen, N. Marcuvitz: *Radiation and Scattering of Waves* (Prentice-Hall, New Jersey 1973) Chap. IV

1.20 K.E. Oughstun, P. Wyns, D. Foty: J. Opt. Soc. Am. A *6*, 1430–1440 (1989)

1.21 C.M. Balictsis, K.E. Oughstun: Phys. Rev. E *47*, 3645–3669 (1993)

1.22 R.L. Fork, C.H. Brito Cruz, P.C. Becker, C.V. Shank: Opt. Lett. *12*, 483–485 (1987)

1.23 W.R. Hamilton: Proc. Royal Irish Academy *1*, 341–349 (1839)

1.24 Lord Rayleigh: Proc. London Math. Society *IX*, 21–26 (1877)

1.25 Lord Rayleigh: Nature *XXV*, 52 (1881)

1.26 H.A. Lorentz: *Theory of Electrons*, 2nd edn. (Dover, New York 1952)

1.27 L. Rosenfeld: *Theory of Electrons* (Dover, New York 1965)

1.28 A. Sommerfeld: *Optics*, Lectures on Theoretical Physics, vol. IV (Academic, New York 1972) Chap. 3

1.29 M. Born, E. Wolf: *Principles of Optics*, 5th edn. (Pergamon, Oxford 1975) Sect. 2.3

1.30 P. Drude: *The Theory of Optics* (Dover, New York 1959) p. 382

1.31 C.J.F. Böttcher, P. Bordewijk: *Theory of Electric Polarization*, (Elsevier, Amsterdam 1978) Vol. 2

1.32 P. Debye: Math. Ann. *67*, 535–558 (1909)

1.33 R. Mosseri: Léon Brillouin: une vie à la croisée des ondes, Sciences et Vie, numero special "200 ans de sciences 1789–1989," 256–261 (1989)

1.34 H. Baerwald: Ann. Phys. *7*, 731–760 (1930)

1.35 D.B. Trizna, T.A. Weber: Radio Sci. *17*, 1169–1180 (1982)

1.36 C. Eckart: Rev. Modern Physics *20*, 399–417 (1948)

1.37 G.B. Whitham: Commun. Pure Appl. Math. *14*, 675–691 (1961)
1.38 M.J. Lighthill: J. Inst. Maths. Applics. *1*, 1–28 (1964)
1.39 R.M. Lewis: Archive Rational Mech. and Analysis *20*, 191–250 (1965)
1.40 R.M. Lewis: In *Electromagnetic Wave Theory*, ed. by J. Brown (Pergamon, New York 1967) pp. 845–869
1.41 L.B. Felsen: SIAM Rev. *12*, 424–448 (1970)
1.42 K.A. Connor, L.B. Felsen: Proc. IEEE *62*, 1586–1598 (1974)
1.43 Yu.A. Kravtsov, L.A. Ostrovsky, N.S. Stepanov: Proc. IEEE *62*, 1492–1510 (1974)
1.44 L.B. Felsen: IEEE Trans. AP-*17*, 191–200 (1969)
1.45 G.M. Whitman, L.B. Felsen: IEEE Trans. AP-*17*, 200–208 (1969)
1.46 L.B. Felsen: IEEE Trans. AP-*19*, 424–432 (1971)
1.47 C.M. Knop: IEEE Trans. AP-*12*, 494–496 (1964)
1.48 C.T. Case, R.E. Haskell: IEEE Trans. AP-*14*, 401 (1966)
1.49 R.S. Elliott: IRE Trans. MTT-*5*, 254–257 (1958)
1.50 M.P. Forrer: Proc. IRE *46*, 1830–1835 (1958)
1.51 R.D. Wanselow: J. Franklin Inst. *274*, 178–184 (1962)
1.52 C.M. Knop, G.I. Cohn: IEEE Trans. MTT-*11*, 445–447 (1963)
1.53 L.E. Vogler: Radio Sci. *5*, 1469–1474 (1970)
1.54 J.R. Wait: Radio Sci. *69D*, 1387–1401 (1965)
1.55 R.E. Haskell, C.T. Case: IEEE Trans. AP-*15*, 458–464 (1967)
1.56 D. Ludwig: SIAM Review *12*, 325–331 (1970)
1.57 J.R. Wait: Elect. Lett. 7, 285–286 (1971)
1.58 J.R. Wait: Radio Sci. *5*, 733–735 (1970)
1.59 J. Jones: Am. J. Physics *42*, 43–46 (1974)
1.60 R.E. Eaves: J. Math. Physics *14*, 432–433 (1973)
1.61 M.A. Biot: Phys. Rev. *105*, 1129–1137 (1957)
1.62 E.O. Schulz-DuBois: Proc. IEEE *57*, 1748–1757 (1969)
1.63 R.L. Smith: Am. J. Physics *38*, 978–984 (1970)
1.64 R. Loudon: J. Phys. A *3*, 233–245 (1970)
1.65 J. Askne and B. Lind: Phys. Rev. A *2*, 2335–2340 (1970)
1.66 D. Anderson, J. Askne: Proc. IEEE (Lett.) *60*, 901–902 (1972)
1.67 D. Anderson, J. Askne, M. Lisak: Proc. IEEE *63*, 715–717 (1975)
1.68 M. Lisak: J. Phys. A *9*, 1145–1158 (1976)
1.69 F.R. Faxvog, C.N.Y. Chow, T. Bieber, J.A. Carruthers: Appl. Phys. Lett. *17*, 192–193 (1970)
1.70 N.G. Basov, R.V. Ambartsumyan, V.S. Zuev, P.G. Kryukov, V.S. Letokhov: Doklady Akademii Nauk SSSR, *165*, 58–60 (1965). [English trans: Sov. Phys.-Doklady *10*, 1039–1040 (1966)]
1.71 S. Chu, S. Wong: Phys. Rev. Lett. *48*, 738–741 (1982)
1.72 A. Katz, R.R. Alfano: Phys. Rev. Lett. *49*, 1292 (1982)
1.73 D.G. Anderson, J.I.H. Askne: Proc. IEEE *62*, 1518–1523 (1974)
1.74 D. Anderson, J. Askne, M. Lisak: Phys. Rev. A *12*, 1546–1552 (1975)
1.75 L.A. Vainshtein: Usp. Fiz. Nauk *118*, 339–367 (1976). [English transl.: Sov. Phys.-Usp. *19*, 189–205 (1976)]
1.76 S.A. Akhmanov, V.A. Vysloukh, A.S. Chirkin: Usp. Fiz. Nauk *149*, 449–509 (1986). [English transl.: Sov. Phys.-Usp. *29*, 642–677 (1986)]
1.77 G.P. Agrawal: *Nonlinear Fiber Optics* (Academic, Boston 1989) Chap. 3
1.78 Yu.A. Kravtsov, L.A. Ostrovsky, N.S. Stepanov: Proc. IEEE *62*, 1492–1510 (1974)
1.79 D. Censor: J. Phys. A *10*, 1781–1790 (1977)
1.80 D. Censor: Phys. Rev. A *24*, 1452–1459 (1981)
1.81 J. Arnaud: Opt. Quantum Electron. *16*, 125–130 (1984)
1.82 E.G. Skrotskaya, A.N. Makhlin, V.A. Kashin, and G.V. Skrotskii: Zh. Eksp. Teor. Fiz. *56*, 220–226 (1969). [English transl.: Sov. Phys. JETP *29*, 123–125 (1969)]
1.83 J.L. Birman, M.J. Frankel: Opt. Comm. *13*, 303–306 (1975)
1.84 E. Gitterman, M. Gitterman: Phys. Rev. A *13*, 763–776 (1976)

1.85 M.J. Frankel, J.L. Birman: Phys. Rev. A *15*, 2000–2008 (1977)
1.86 D.B. Trizna, T.A. Weber: Radio Sci. *17*, 1169–1180 (1982)
1.87 O. Avenel, M. Rouff, E. Varoquaux, G.A. Williams: Phys. Rev. Lett. *50*, 1591–1594 (1983)
1.88 R.J. Vidmar, F.W. Crawford, K.J. Harker: Radio Sci. *18*, 493–502 (1983)
1.89 R.J. Vidmar, F.W. Crawford, K.J. Harker: Radio Sci. *18*, 503–511 (1983)
1.90 N.D. Hoc, I.M. Besieris, M.E. Sockell: IEEE Trans. AP-*33*, 1237–1248 (1985)
1.91 R. Barakat: J. Opt. Soc. Am. B *3*, 1602–1604 (1986)
1.92 V.A. Vasilev, M.Ya. Kelbert, I.A. Sazonov, I.A. Chaban: Opt. Spektrosk. *64*, 862–868 (1988). [English transl.: Opt. Spectrosc. *64*, 513–517 (1988)]
1.93 R. Albanese, J. Penn, R. Medina: J. Opt. Soc. Am. A *6*, 1441–1446 (1989)
1.94 L. Gagnon: Phys. Lett. A *148*, 452–456 (1990)
1.95 P. Pleshko and I. Palócz: Phys. Rev. Lett. *22*, 1201–1204 (1969)
1.96 J. Aaviksoo, J. Lippmaa, J. Kuhl: J. Opt. Soc. Am. B *5*, 1631–1635 (1988)
1.97 J. Aaviksoo, J. Kuhl, K. Ploog: Phys. Rev. A. *44*, 5353–5356 (1991)

Chapter 2

2.1 R.N. Bracewell: *The Fourier Transform and Its Applications*, 2nd edn. (McGraw-Hill, New York 1986)
2.2 M.J. Lighthill: *Introduction to Fourier Analysis and Generalised Functions* (Cambridge Univ. Press, London 1970)
2.3 G.C. Sherman, K.E. Oughstun: Phys. Rev. Lett. *47*, 1451–1454 (1981)
2.4 J.A. Stratton: *Electromagnetic Theory* (McGraw-Hill, New York 1941) p. 135
2.5 F.N.H. Robinson: *Macroscopic Electromagnetism* (Pergamon, Oxford 1973)
2.6 J.D. Jackson: *Classical Electrodynamics*, 2nd edn. (Wiley, New York 1975) Sect. 6.7
2.7 Yu.S. Barash, V.L. Ginzburg: Usp. Fiz. Nauk *118*, 523–530 (1976). [English trans.: Sov. Phys.-Usp. *19*, 263–270 (1976)]
2.8 R. Loudon: J. Phys. A *3*, 233–245 (1970)
2.9 K.E. Oughstun, S. Shen: J. Opt. Soc. Am. B *5*, 2395–2398 (1988)
2.10 H.A. Lorentz: *Theory of Electrons* (Dover, New York 1952)
2.11 L. Rosenfeld: *Theory of Electrons* (Dover, New York 1965)
2.12 P. Drude: *The Theory of Optics* (Dover, New York 1959) Sect. II, Chap V
2.13 H.M. Nussenzveig: *Causality and Dispersion Relations* (Academic, New York 1972) Chap. 1
2.14 R. Loudon: *The Quantum Theory of Light* (Oxford Univ. Press, London 1973)
2.15 K.E. Oughstun: Propagation of optical pulses in dispersive media. Ph.D. Thesis, University of Rochester (1978)
2.16 L. Brillouin: *Wave Propagation and Group Velocity* (Academic, New York 1960) Chap. V

Chapter 3

3.1 J.A. Stratton: *Electromagnetic Theory* (McGraw-Hill, New York 1941) Sect. 5.12
3.2 J.D. Jackson: *Classical Electrodynamics*, 2nd edn. (Wiley, New York 1975) Sect. 6.5
3.3 E.T. Copson: *An Introduction to the Theory of Functions of a Complex Variable* (Oxford Univ. Press, London 1972) Sect. 5.5
3.4 E.T. Whittaker, G.N. Watson: *A Course of Modern Analysis*, 4th edn. (Cambridge Univ. Press, London 1963) Sect. 6.222
3.5 [Ref. 3.3, Sect. 6.1]
3.6 H.B. Phillips: *Vector Analysis* (Wiley, New York 1933) Sect. 55
3.7 H. Weyl: Ann. Phys. (Leipzig) *60*, 481–500 (1919)
3.8 E.T. Whittaker: Math. Ann. *57*, 333–355 (1902)

3.9 G.C. Sherman, A.J. Devaney, L. Mandel: Opt. Commun. *6*, 115–118 (1972)
3.10 G.C. Sherman, J.J. Stamnes, A.J. Devaney, E. Lalor: Opt. Commun. *8*, 271–274 (1973)
3.11 A.J. Devaney, G.C. Sherman: SIAM Rev. *15*, 765–786 (1973)
3.12 P.C. Clemmow: *The Plane Wave Spectrum Representation of Electromagnetic Fields* (Pergamon, Oxford 1966)
3.13 G.N. Watson: *A Treatise on the Theory of Bessel Functions*, 2nd edn. (Cambridge Univ. Press, Cambridge 1958) Sect. 3.62, Eq. (5)
3.14 A Sommerfeld: Ann. Phys. (Leipzig) *28*, 665–737 (1909)
3.15 A. Baños: *Dipole Radiation in the Presence of a Conducting Half-Space* (Pergamon, Oxford 1966) Sect. 2.12
3.16 H. Ott: Ann. Phys. (Leipzig) *41*, 443–467 (1942)

Chapter 4

4.1 R.F. Harrington: *Time-Harmonic Electromagnetic Fields* (McGraw-Hill, New York 1961) Sect. 4.2
4.2 J.W. Goodman: *Introduction to Fourier Optics* (McGraw-Hill, New York 1968)
4.3 C.J. Bouwkamp: Rept. Progr. Phys. *17*, 35–100 (1954)
4.4 W.H. Carter: Optica Acta *21*, 871–892 (1974)
4.5 J.J. Stamnes: *Waves in Focal Regions* (Hilger, Bristol 1986)
4.6 M. Born and E. Wolf: *Principles of Optics*, 5th edn. (Pergamon, Oxford 1975) Sects. 8.2, 3
4.7 G.C. Sherman: J. Opt. Soc. Am. *57*, 1490–1498 (1967)
4.8 G.C. Sherman: J. Opt. Soc. Am. *57*, 546–547 (1967)
4.9 E. Lalor: J. Opt. Soc. Am. *58*, 1235–1237 (1968)
4.10 H.L. Royden: *Real Analysis*, 2nd edn. (Macmillan, New York 1968) p. 269
4.11 A. Sommerfeld: *Optics*, Lectures on Theoretical Physics, vol. IV (Academic, New York 1964) pp. 199–201
4.12 M. Born and E. Wolf: *Principles of Optics*, 5th edn. (Pergamon, Oxford 1970) Sect. 1.4.3
4.13 A. Nisbet, E. Wolf: Proc. Camb. Phil. Soc. *50*, 614–622 (1954)
4.14 K.E. Oughstun: J. Opt. Soc. Am. A *9*, 578–584 (1992)
4.15 A. Sommerfeld: Ann. Phys. (Leipzig) *44*, 177–202 (1914)
4.16 L. Brillouin: Ann. Phys. (Leipzig) *44*, 203–240 (1914)
4.17 L. Brillouin: *Wave Propagation and Group Velocity* (Academic, New York 1960)
4.18 H. Baerwald: Ann. Phys. *7*, 731–760 (1930)
4.19 K. E. Oughstun: Propagation of optical pulses in dispersive media. Ph.D. Thesis, University of Rochester (1978)
4.20 K.E. Oughstun, G.C. Sherman: J. Opt. Soc. Am. B *5*, 817–849 (1988)
4.21 K.E. Oughstun, G.C. Sherman: J. Opt. Soc. Am. A *6*, 1394–1420 (1989)
4.22 J. Jones: Am. J. Phys. *42*, 43–46 (1974)
4.23 M.J. Lighthill: J. Inst. Math. Applics. *1*, 1–28 (1965)
4.24 D. Anderson, J. Askne: Proc. IEEE *60*, 901–902 (1972)
4.25 D. Anderson, J. Askne: Proc. IEEE *62*, 1518–1523 (1974)
4.26 D. Anderson, J. Askne, M. Lisak: Proc. IEEE *63*, 715–717 (1975)
4.27 D. Anderson, J. Askne, M. Lisak: Phys. Rev. A *12*, 1546–1552 (1975)
4.28 D. Censor: J. Phys. A *10*, 1781–1790 (1977)
4.29 D. Marcuse: Appl. Opt. *19*, 1653–1660 (1980)
4.30 N.D. Hoc, I.M. Besieris, M.E. Sockell: IEEE Trans. AP-*33*, 1237–1248 (1985)
4.31 S.A. Akhmanov, V.A. Vysloukh, A.S. Chirkin: Usp. Fiz. Nauk, *149*, 449–509 (1986) [English transl.: Sov. Phys.-Usp. *29*, 642–677 (1986)]
4.32 D. Anderson, M. Lisak: Phys. Rev. A *35*, 184–187 (1987)

Chapter 5

5.1 H.M. Nussenzveig: *Causality and Dispersion Relations* (Academic, New York 1972) Chap 1
5.2 E.T. Copson: *Asymptotic Expansions* (Cambridge Univ. Press, London 1971)
5.3 F.W.J. Olver: SIAM Rev. *12*, 228–247 (1970)
5.4 L.B. Felsen and N. Marcuvitz: *Radiation and Scattering of Waves* (Prentice-Hall, New Jersey 1973) Chap. IV
5.5 C. Chester, B. Friedman, F. Ursell: Proc. Cambridge Phil. Soc. *53*, 599–611 (1957)
5.6 N. Bleistein: Commun. Pure Appl. Math. *19*, 353–370 (1966)
5.7 N. Bleistein: J. Math. Mech. *17*, 533–559 (1967)
5.8 R.A. Handelsman, N. Bleistein: Arch. Rational Mech. Analysis *35*, 267–283 (1969)
5.9 L. Brillouin: Ann. Phys. (Leipzig) *44*, 203–240 (1914)
5.10 L. Brillouin: *Wave Propagation and Group Velocity* (Academic, New York 1960)
5.11 P.M. Morse, H. Feshbach: *Methods of Theoretical Physics* (McGraw-Hill, New York 1953) Vol. I
5.12 A. Baños, Jr.: *Dipole Radiation in the Presence of a Conducting Space* (Pergamon, Oxford 1966) Sect. 3.3

Chapter 6

6.1 L. Brillouin: *Wave Propagation and Group Velocity* (Academic, New York 1960)
6.2 L. Brillouin: Ann. Phys. (Leipzig) *44*, 203–240 (1914)
6.3 S.M. Selby, ed: Standard Mathematical Tables, 19th edn. (CRC, Cleveland 1971) p. 103
6.4 K.E. Oughstun, G.C. Sherman: J. Opt. Soc. Am. B *5*, 817–849 (1988)
6.5 E.T. Copson: *An Introduction to the Theory of Functions of a Complex Variable* (Oxford Univ. Press, London 1963) Sect. 6.1

Chapter 7

7.1 A. Sommerfeld: Ann. Phys. (Leipzig) *44*, 177–202 (1914)
7.2 K.E. Oughstun, G.C. Sherman: J. Opt. Soc. Am. B *5*, 817–849 (1988)
7.3 H.M. Nussenzveig: *Causality and Dispersion Relations* (Academic, New York 1972) Chap. 1
7.4 L. Brillouin: Ann. Phys. (Leipzig) *44*, 203–240 (1914)
7.5 L. Brillouin: *Wave Propagation and Group Velocity*, (Academic, New York 1960)
7.6 E.T. Whittaker, G.N. Watson: *A Course of Modern Analysis*, (Cambridge Univ. Press, London 1963)
7.7 E.T. Copson: *An Introduction to the Theory of Functions of a Complex Variable*, (Oxford Univ. Press, London 1972)
7.8 R.A. Handelsman, N. Bleistein: Arch. Rational Mech. Analysis *35*, 267–283 (1969)
7.9 L. Mandel: Am. J. Phys. *42*, 840–846 (1974)

Chapter 8

8.1 L. Brillouin: Ann. Phys. (Leipzig) *44*, 203–240 (1914)
8.2 L. Brillouin: *Wave Propagation and Group Velocity*, (Academic, New York 1960)

Chapter 9

9.1 L. Brillouin: Ann. Phys. (Leipzig) *44*, 203–240 (1914)
9.2 L. Brillouin: *Wave Propagation and Group Velocity*, (Academic, New York 1960)
9.3 K.E. Oughstun, G.C. Sherman: J. Opt. Soc. Am. B *5*, 817–849 (1988)
9.4 H.M. Nussenzveig: *Causality and Dispersion Relations* (Academic, New York 1972) Chap. 1
9.5 R.L. Smith: Am. J. Phys. *38*, 978–984 (1970)
9.6 N.G. Basov, R.V. Ambartsumyan, V.S. Zuev, P.G. Kryukov, V.S. Letokhov: Dokl. Akad. Nauk SSSR *165*, 58–60. [English transl.: Sov. Phys.-Dokl. *10*, 1039–1040 (1966)]
9.7 M.D. Crisp: Phys. Rev. A *4*, 2104–2108 (1971)
9.8 D. Anderson, J. Askne, M. Lisak: Proc. IEEE *63*, 715–717 (1975)
9.9 D. Anderson, J. Askne, M. Lisak: Phys. Rev. A *12*, 1546–1552 (1975)
9.10 R. Loudon: J. Phys. A *3*, 233–245 (1970)
9.11 J. Askne, B. Lind: Phys. Rev. A *2*, 2335–2340 (1970)
9.12 D. Anderson, J. Askne: Proc. IEEE *60*, 901–902 (1972)
9.13 K.E. Oughstun: Propagation of optical pulses in dispersive media. Ph.D. Thesis, University of Rochester (1978)
9.14 K. E. Oughstun, G.C. Sherman: J. Opt. Soc. Am. A *6*, 1394–1420 (1989)
9.15 K.E. Oughstun, P. Wyns, D. Foty: J. Opt. Soc. Am. A *6*, 1430–1440 (1989)
9.16 T. Hosono: Proceedings of the 1980 International Union of Radio Science International Symposium on Electromagnetic Waves (International Union of Radio Science, Munich F.R.G., 1980), pp. C1–C4
9.17 P. Wyns, D.P. Foty, K.E. Oughstun: J. Opt. Soc. Am. A *6*, 1421–1429 (1989)
9.18 K.E. Oughstun, G.C. Sherman: Phys. Rev. A *41*, 6090–6113 (1990)
9.19 C.G.B. Garret, D.E. McCumber: Phys. Rev. A *1*, 305–313 (1970)
9.20 J. Jones: Am. J. Phys. *42*, 43–46 (1974)
9.21 K.A. Connor, L.B. Felsen: Proc. IEEE *62*, 1586–1598 (1974)
9.22 D. Anderson, M. Lisak: Phys. Rev. A *35*, 184–187 (1987)
9.23 R. Barakat: J. Opt. Soc. Am. B *3*, 1602–1604 (1986)
9.24 R. Albanese, J. Penn, R. Medina: J. Opt. Soc. Am. A *6*, 1441–1446 (1989)

Chapter 10

10.1 A. Sommerfeld: Ann. Phys. (Leipzig) *44*, 177–202 (1914)
10.2 L. Brillouin: Ann. Phys. (Leipzig) *44*, 203–240 (1914)
10.3 L. Brillouin: *Wave Propagation and Group Velocity*, (Academic, New York, 1960)
10.4 B.R. Baldock, T. Bridgeman: *Mathematical Theory of Wave Motion*, (Halsted, New York 1981) Chap. 5
10.5 L.A. Segel, G.H. Handelman: *Mathematics Applied to Continuum Mechanics*, (Macmillan, New York 1977) Chap. 9
10.6 I. Tolstoy: *Wave Propagation*, (McGraw-Hill, New York 1973) Chap. 1–2
10.7 L.B. Felsen: *Transient Electromagnetic Fields*, ed. by L.B. Felsen (Springer-Verlag, New York 1976), Chap. 1, p. 65
10.8 H.M. Nussenzveig: *Causality and Dispersion Relations*, (Academic, New York 1972) Chap. 1
10.9 G.C. Sherman, K.E. Oughstun: Phys. Rev. Lett. *47*, 1451–1454 (1981)
10.10 K.E. Oughstun, G.C. Sherman: J. Opt. Soc. Am. B *5*, 817–849 (1988)
10.11 R. Loudon: J. Phys. A *3*, 233–245 (1970)
10.12 G.C. Sherman, K.E. Oughstun: J. Opt. Soc. Am. (submitted)
10.13 P. Wyns, D.P. Foty, K.E. Oughstun: J. Opt. Soc. Am. A *6*, 1421–1429 (1989)
10.14 M.A. Biot: Phys. Rev. *105*, 1129–1137 (1957)
10.15 M.J. Lighthill: J. Inst. Math. Appl. *1*, 1–28 (1965)

Subject Index

Springer Series on

Wave Phenomena

Editors: L. M. Brekhovskikh L. B. Felsen H. A. Haus
Managing Editor: H. K. V. Lotsch

Volume 1
Mechanics of Continua and Wave Dynamics 2nd Edition
By L. M. Brekhovskikh, V. Goncharov

Volume 2
Rayleigh-Wave Theory and Application
Editors: E. A. Ash, E. G. S. Paige

Volume 3
Electromagnetic Surface Excitations
Editors: R. F. Wallis, G. I. Stegeman

Volume 4
Short-Wavelength Diffraction Theory
Asymptotic Methods
By V. M. Babič, V. S. Buldyrev

Volume 5
Acoustics of Layered Media I
Plane and Quasi-Plane Waves
By L. M. Brekhovskikh, O. A. Godin

Volume 6
Geometrical Optics of Inhomogeneous Media
By Yu. A. Kravtsov, Yu. I. Orlov

Volume 7
Recent Developments in Surface Acoustic Waves
Editors: D. F. Parker, G. A. Maugin

Volume 8
Fundamentals of Ocean Acoustics
2nd Edition
By L. M. Brekhovskikh, Yu. P. Lysanov

Volume 9
Nonlinear Optics in Solids
Editor: O. Keller

Volume 10
Acoustics of Layered Media II
Point Sources and Bounded Beams
By L. M. Brekhovskikh, O. A. Godin

Volume 11
Resonance Acoustic Spectroscopy
By N. Veksler

Volume 12
Scalar Wave Theory
Green's Functions and Applications
By J. A. De Santo

Volume 13
Radar Target Imaging
Editors: W.-M. Boerner, H. Überall

Volume 14
Random Media and Boundaries
Unified Theory, Two-Scale Method, and Applications
By K. Furutsu

Volume 15
Caustics, Catastrophes, and Wave Fields
By Yu. A. Kravtsov, Yu. I. Orlov

Volume 16
Electromagnetic Pulse Propagation in Causal Dielectrics
By K. E. Oughstun, G. C. Sherman

Volume 17
Wave Scattering from Rough Surfaces
By A. S. Voronovich

Volume 18
Electromagnetic Wave Propagation in Turbulence
Evaluation and Application of Mellin Transforms
By R. J. Sasiela

MIX
Papier aus verantwortungsvollen Quellen
Paper from responsible sources
FSC® C105338

If you have any concerns about our products,
you can contact us on
ProductSafety@springernature.com

In case Publisher is established outside the EU,
the EU authorized representative is:
Springer Nature Customer Service Center GmbH
Europaplatz 3, 69115 Heidelberg, Germany

Printed by Libri Plureos GmbH
in Hamburg, Germany